Addition and Multiplication Facts

ADDITION FACTS

+	0	1	2	3	4	5	6	7	8	9
0	0	1	2	3	4	5	6	7	8	9
1	1	2	3	4	5	6	7	8	9	10
2	2	3	4	5	6	7	8	9	10	11
3	3	4	5	6	7	8	9	10	11	12
4	4	5	6	7	8	9	10	11	12	13
5	5	6	7	8	9	10	11	12	13	14
6	6	7	8	9	10	11	12	13	14	15
7	7	8	9	10	11	12	13	14	15	16
8	8	9	10	11	12	13	14	15	16	17
9	9	10	11	12	13	14	15	16	17	18

MULTIPLICATION FACTS

·	0	1	2	3	4	5	6	7	8	9
0	0	0	0	0	0	0	0	0	0	0
1	0	1	2	3	4	5	6	7	8	9
2	0	2	4	6	8	10	12	14	16	18
3	0	3	6	9	12	15	18	21	24	27
4	0	4	8	12	16	20	24	28	32	36
5	0	5	10	15	20	25	30	35	40	45
6	0	6	12	18	24	30	36	42	48	54
7	0	7	14	21	28	35	42	49	56	63
8	0	8	16	24	32	40	48	56	64	72
9	0	9	18	27	36	45	54	63	72	81

Jack Barker James Rogers James Van Dyke

Portland Community College Portland, Oregon

FUNDAMENTALS OF
MATHEMATICS

FIFTH EDITION

SAUNDERS COLLEGE PUBLISHING
Philadelphia • Fort Worth • Chicago • San Francisco
Montreal • Toronto • London • Sydney • Tokyo

Text Typeface • *Century Schoolbook*
Compositor • *York Graphic Services, Inc.*
Acquisitions Editor • *Robert B. Stern*
Developmental Editor • *Ellen Newman*
Managing Editor • *Carol Field*
Copy Editor • *York Production Services*
Manager of Art and Design • *Carol Bleistine*
Art Director • *Carol Bleistine*
Art and Design Coordinator • *Doris Bruey*
Text and Cover Designer • *York Production Services*
Director of EDP • *Tim Frelick*
Production Manager • *Bob Butler*

Cover Credit • *York Production Services*

Printed in the United States of America

FUNDAMENTALS OF MATHEMATICS, fifth edition

ISBN 0-03-032222-7

Library of Congress Catalog Card Number: 90-48952

1234 061 987654321

To our wives:

Mary Barker

Elinore Rogers

Carol Van Dyke

Preface

The fifth edition of *Fundamentals of Mathematics* provides a write-in text for students at the college level who need a review of the basic skills of arithmetic in order to fulfill competency requirements, prepare for business mathematics, achieve adequate scores on placement exams, or complete the prerequisites for elementary algebra.

Special Features The fifth edition contains the following new features:

- All of the exercises (including Pre-Tests, True-False Concept Reviews, Post-Tests, and end-of-section exercises) can be torn out and handed in for grading without disturbing text.
- A True-False Concept Review has been added to the end of each chapter as a check on student understanding of the language, properties, and concepts of arithmetic.
- Pre-Tests and Post-Tests are now keyed to sections of the text as well as to objectives.
- Four-color pedagogical system is used to highlight the various pedagogical features throughout the text. Multiple colors are useful, for example, to distinguish between different portions of the figure being discussed. The complete color system is described in more detail on pages xxii and xxiii.
- The exercises have been revised to include 25 to 30% new problems.
- Answers to Warm Ups in the Model Problem-Solving section will now appear at the bottom of the page so that students can check their answers more readily.
- Expanded and clarified *Strategy* comments.
- More fully worked examples in each section.
- Caution comments, highlighted in red, alert students to common errors and help them avoid pitfalls.

The fifth edition has also added coverage of the following topics:

- Section 6.10 (Draw and Interpret Graphs) gives the student an understanding of how "pictorial" displays of data are constructed and how data is retrieved from these displays. Emphasis is on data commonly found in business and everyday life.
- Section 6.11 (Read and Interpret Tables) takes displayed data and has the student apply newly acquired arithmetic skills to interpret and extend the data.
- Section 7.9 (Reasoning) has been added to expose the student to the types of reasoning used on a daily basis and to prepare for a formal study of geometry.

- Appendix V (Plane Geometry Supplement) has been added to provide a concise coverage of some topics of Euclidean geometry. These topics include angles, triangles, congruent triangles, parallel lines, similar triangles, square roots, and the Pythagorean Theorem. This supplement provides the student with a good foundation for later mathematics classes.
- Particular attention has been paid to the testing requirements for various states (e.g., ELM, TASP, CLAST, etc.). Please see pages xiv–xvi for additional information on how this text meets requirements for these schools.

Format The write-in text gives the student space to practice arithmetic skills with ready reference to step-by-step directions and worked examples. Beginning with section-by-section objectives, each section is a complete treatment of the particular topic and contains the following features: Objectives, Application, Vocabulary, How and Why, Model Problem-Solving, and Exercises. The chapter begins with a Pre-Test to determine which objectives the student needs to study and ends with a Post-Test that may serve as either a chapter review or a chapter evaluation; both the Pre-Test and the Post-Test are referenced to the section number and objective so the student can easily refer to a specific section for assistance. A True-False Concept Review in each chapter helps to reinforce student understanding. This format makes it possible for students to work their way through the material in a Math Lab or Learning Center.

In **Chapter 1** we begin with the numeration system including the concepts of place value, word name, expanded form, inequality, and rounding, so the student will have a thorough understanding of the concept of "number" prior to performing operations. The basic operations of addition, subtraction, multiplication, and division of whole numbers are reviewed. An introduction to the solution of equations is presented in Getting Ready for Algebra exercises at the end of Section 1.4 an others throughout the text to prepare the student for the study of algebra. Exponents and powers of ten are examined in presentation for order of operations involving whole numbers and for writing prime factors using exponential notation in Chapter 2. At the end of the chapter, order of operations is presented to expose the student to problems involving multiple operations. This is followed by a section on the idea of average which is applied to real-life situations.

In **Chapter 2** to prepare the student for operations on fractions, the whole number classifications of multiple, divisor, factor, prime, composite, and least common multiple are covered. The section also serves to reinforce the student's skills with whole numbers with practice in multiplication and division.

In **Chapter 3** the meaning of a fraction is modeled using shaded rectangles. This approach provides a visual presentation to back the mathematical concept. We study operations on both fractions and mixed numbers. Building and reducing fractions are based on the skills acquired in Chapter 2. The least common multiple of Chapter 2 leads here to lowest common denominator for adding fractions. Again the operations are reinforced with a review of order of operations and average incorporating fractions and mixed numbers.

In **Chapter 4** decimals are brought in, employing an approach parallel to that for whole numbers. Coverage includes place value, word names, expanded form, inequalities, and rounding for decimal numbers. The basic operations on decimals comes next. Multiplication and division, using powers of ten, give the student an opportunity to utilize the exponent skills learned in Chapter 1. Both fraction to decimal and decimal to fraction conversions demonstrate the relationship between the two ways of writing a rational number. The fact that an exact decimal representation is not always possible supports the need for fractions. The chapter ends with the review of order of operations and average using decimals.

In **Chapter 5** ratio and proportion leads to useful applications of the fundamentals of mathematics to practical situations. Here the student is introduced to the formal process of translating from a written statement of facts to a mathematical statement that can be solved. This skill is re-enforced in the next two chapters.

In **Chapter 6** percent is developed as a useful way of describing a numerical comparison. Students practice is changing from percent to decimal to fraction so that they can see the relationship of percent to the different numbers and can become skilled at expressing a number in any one of the forms. Solutions of percent problems are covered using either ratios or the formula, $R\%$ of $A = B$. Practical applications are presented showing general applications with a separate section dedicated to business applications. This approach is designed to prepare a student for a business mathematics course. More applications show the use of bar, circle, broken line, and pictorial graphs. Students are given the opportunity to draw conclusions from a visual display. In the last section, tables and charts are used to display data where students extract information and use their mathematical skills to draw conclusions.

In **Chapter 7** both English and metric measurements are formally introduced. Conversion within the systems and between the systems is investigated. The metric system gives a student a skill that can be used in science courses and an appreciation of the measurement system used by the majority of the world's population. Measurement is applied to geometric figures covering perimeter, area of common geometric figures, area of compound figures, and volume. The chapter ends with a discussion of reasoning in preparation for later studies in mathematics or formal logic.

In **Chapter 8** the number system is expanded to include signed numbers. Operations on signed numbers include absolute value, opposites, addition, subtraction, multiplication, and division. This chapter, together with sections in prior chapters on Getting Ready for Algebra, serves to bridge the gap to the student's first study of algebra.

PRE-TEST

Each chapter begins with a Pre-Test to help the student determine which areas require the most attention. Questions are keyed to both the section number and objective to allow the student to refer to that portion of the text for assistance. If the instructor wishes, the Pre-Tests can be torn out and handed in as a classroom assignment.

OBJECTIVES

In each chapter, objectives are identified in the beginning of each section; the student may take the chapter Pre-Test to determine which of the objectives require the most study. In the classroom the instructor can use the Pre-Test to determine which objectives should be emphasized for that particular group of students. The individual student or class can then work on problem objectives and take follow-up exams when ready.

APPLICATION

This section contains a posed problem that is a practical application of the section objectives; the student will be able to solve the problem after reading the How and Why and Model Problem-Solving sections. The solution to the application, an integral part of the worked-out Examples, follows the Examples and serves as additional reinforcement of the concepts presented; the application is also reinforced in the Exercises with similar problems. Including these applications serves as a bridge between the mathematics classroom and the fields of business, shop mathematics, health science, consumer mathematics, and other fields.

VOCABULARY

As in previous editions, this section provides the student with definitions or examples of words that have not been previously used in the text.

HOW AND WHY

This section explains the concepts presented in the objective and application while developing the necessary methods of solution. Throughout the text the explanations are primarily intuitive. Procedures and rules (methods of solution) are highlighted by bold type and color screens for quick reference and easy review. The explanation in this section, paired with the Model Problem-Solving section that follows, provides an immediate linking of theory and practice.

MODEL PROBLEM-SOLVING

Examples This section uses examples to illustrate the concepts explained in the How and Why section. Each example has a strategy column; as each example is worked out, there is a step-by-step explanation that expands on the procedures and the thinking necessary to work the example. The examples also illustrate common shortcuts. Where applicable, the strategy includes caution comments about errors and pitfalls, highlighted in red for the student.

Warm Ups Each example is paired with a Warm Up problem of the same type and level of difficulty, which reinforces the procedures used to solve the example. These Warm Ups are useful to check the student's understanding of the material before advancing to a more difficult example.

Some examples contain problems solved by using a calculator. These examples, set off with a calculator symbol, demonstrate how to use a calculator and signal to the student that these problems are suited for calculator practice. However, the use of the calculator is left to the discretion of the instructor and/or the student. Nowhere is the use of a calculator required, and all sections of the text can be studied without a calculator.

Here is an example of the presentation used in the text:

MODEL PROBLEM-SOLVING

Examples | **Strategy**

a. Find the quotient correct to the nearest thousandth: $0.47891 \div 0.072$

$0.072\overline{)0.47891}$

Move both decimals 3 places to the right; that is, multiply both by 1000.

```
        6.6515
  72)478.9100
     432
      46 9
      43 2
       3 71
       3 60
         110
          72
         380
         360
          20
```

Since we will round to the thousandth, we carry the division to one place past the thousandths, that is, to four places.

Hence, $0.47891 \div 0.072 \approx 6.652$.

Warm Up a. Find the quotient correct to the nearest thousandth: $0.75593 \div 0.043$

b. Calculator Example

78.1936 ÷ 8.705 = ? Round to the nearest thousandth.

ENTER	DISPLAY	
78.1936	78.1936	**Enter 78.1936.**
÷	78.1936	**Press the "÷" key.**
8.705	8.705	**Enter 8.705.**
=	8.9826077	**Press the "=" key.**

The quotient is 8.983 to the nearest thousandth.

Warm Up b. 103.843 ÷ 4.088 = ? Round to the nearest thousandth.

Answers to the Warm Ups appear at the bottom of each page.

EXERCISES

The fifth edition of *Fundamentals of Mathematics* continues to organize the exercises in order of difficulty. As before, "A" indicates that the problems are relatively easy; the "A" problems have been increased in number and can be used as class or "oral" exercises. The "B" problems are more difficult and involve computation with larger numbers. The "C" problems in most instances offer a challenge for the more advanced students and may, at the discretion of the instructor, provide calculator drill for the students. "D" problems follow up the featured application of the section, provide practical application of the skill, and call upon the student to set up the computation.

Maintain Your Skills This section immediately follows the "D" problems in the exercises. The function of this portion of the exercises is to review material previously covered. The exercises in each Maintain Your Skills section have been referenced to the sections reviewed. All of the exercises can be torn out and handed in as classroom assignments without disturbing the text.

▲▲ Getting Ready for Algebra The purpose of this section is to show students that they can use basic skills in an elementary algebra setting; this section was favorably received in the fourth edition. In those sections of Chapters 1, 3, 4, and 8 where operations lend themselves to solving simple equations, this topic is included in the exercises immediately following the *Maintain Your Skills* section. Preceded by a logo of two solid triangles so the student always knows it is the algebra section, Getting Ready for Algebra is completely separate and contains an explanation section, a model problem-solving section, and its own exercises. If the instructor deems this section inappropriate, it can easily be omitted at the instructor's discretion without loss of continuity.

In all of the exercises, the problems are paired so that the set of odd-numbered problems is equivalent in type and difficulty to the set of even-numbered problems. Answers to the odd-numbered exercises, as well as answers to all the Pre-Tests, Post-Tests, and True-False Concept Reviews are provided in the back of the book.

TRUE-FALSE CONCEPT REVIEW

Appearing toward the end of each chapter, this test helps remind the student of the new concepts before the student proceeds to the Post-Test. Answers to all of the questions in the True-False Concept Review appear in the back of the book.

POST-TEST

Each chapter concludes with a Post-Test as a final review of student comprehension. Questions on the Post-Test are referenced to a section number and objective so the student can easily refer back to the text for assistance. Answers to all of the questions in the Post-Test appear in the back of the book.

TIMETABLE

The text can be used in a variety of classroom situations, depending upon the needs of the students. Two such possibilities are:

1. A one-quarter or one-semester course given as a review of basic arithmetic. Such a course would cover Chapters 1 through 6, with Chapter 7 optional.
2. A one-quarter or one-semester course given as a review before beginning basic algebra. This course would cover Chapters 1 through 6 and Chapter 8, with Chapter 7 optional.

Since each chapter is relatively independent of the others, an instructor may select chapters that meet the student's career goals.

ANCILLARY ITEMS

The following supplements are available to the students to accompany this text:*

MathCue Interactive Software This program disk contains practice problems for every section of the worktext. Using an interactive approach, the software provides students with an alternate way to learn the material and, at the same time, gives the student individualized attention. The program will automatically advance to the next level of difficulty once the student has successfully solved a few problems; the student may also ask to see the solution to check the process used. The software is keyed to the worktext and will refer the student to the appropriate section of the text if an incorrect answer is input. A useful tool to check skills and to identify and correct any difficulties in finding solutions, this software is available for the Apple II and IBM PC microcomputers.

MathCue Solution Finder Software Available for IBM, this software allows students to input their own questions through the use of an expert system, a branch of artificial intelligence. Students may check their answers or receive help as if they were working with a tutor. The software will refer the student to the appropriate section of the text and will record the number of problems entered and correct answers given.

Videotapes A complete set of videotapes (17 hours) is available to give added assistance or to serve as a quick review of the book. The tapes use a newscaster format to review problem-solving methods and guide students through practice problems; students can stop the tape to work the problems and begin it again to check their solutions. Keyed to the text and providing coverage of every section of the book, these tapes are provided as another approach to mastery of the given topic.

Student Solutions Manual This guide contains worked-out solutions to one quarter of the problems in the exercise sets (every other odd problem) to help the student learn and practice the techniques used in solving problems.

*For the instructor we have an Instructor's Manual with complete solutions to all of the exercises, in the Student Solutions Manual; Prepared Tests with six tests for each chapter, as well as two final examinations and a Diagnostic test; a Computerized Test Bank for the Apple II and IBM computers; and a Printed Test Bank containing tests generated from the Computerized Test Bank.

ACKNOWLEDGMENTS

The authors appreciate the unfailing patience and continuing support of their wives, Mary Barker, Elinore Rogers, and Carol Van Dyke, who made the completion of this work possible. Thanks go to Sue O'Rielly of Portland Community College for her help revising the exercises. We also thank our colleagues for their help and suggestions for the improved fifth edition.

We are grateful to Bob Stern and Ellen Newman of Saunders College Publishing for their suggestions during the preparation of the text and to Kirsten Kauffman of York Production Services. We would also like to express our gratitude to the following reviewers for their many excellent contributions to the development of the text in this edition and the prior edition:

James Bennett, Eastfield College
Tim Cavanaugh, University of Northern Colorado
Laurence Chernoff, Miami-Dade Community College
Linda Desue, Edison State Community College
Louise Ettline, Trident Technical College
Roy D. Frysinger, Harrisburg Area Community College
Virginia Hamilton, Shawnee State University
Calvin Holt, Paul D. Camp Community College
Jean L. Holton, Tidewater Community College
Elizabeth Koball, Tidewater Community College
Robert Langston, Tarrant County Junior College
Shelba Mormon, North Lake College
Frederic Norwood, William Paterson College
Sue O'Rielly, Portland Community College
Julienne K. Pendleton, Brookhaven College
Roy Pearson, St. Louis Community College at Florrissant Valley
David Price, Tarrant County Junior College
Greg St. George, University of Montana
Michele A. Sassone, Bergen Community College
Ara B. Sullenberger, Tarrant County Junior College
Mary Jane Smith, Tidewater Community College
Phyllis Steinmann, Scottsdale Community College
Lenore Vest, Lower Columbia College
Priscilla Wake, San Jacinto College
Juanita Woods, Chattanooga State Technical College

Special thanks to John R. Martin of Tarrant County Junior College and Forest Simmons of Portland Community College for their accuracy reviews of all the problems and exercises in the text.

We would also like to thank the following people for their excellent work on the various ancillary items that accompany *Fundamentals of Mathematics:*

George W. Bergeman, Northern Virginia Community College (MathCue Interactive Software and MathCue Solution Finder Software)

Robert Finnell and Hollis Adams, Portland Community College (Video Tapes)

Grace Malaney, Donnelly College (Student Solutions Manual)

Jack Barker
Jim Rogers
Jim Van Dyke

ELM Mathematical Skills

The following table lists the California **ELM MATHEMATICAL SKILLS** and where coverage of these skills can be found in the text. Skills not covered in this text can be found in Basic Algebra, 3rd Edition, or Intermediate Algebra, 3rd Edition. Location of skills are indicated by chapter section or chapter.

Skill	Location in Text
Whole numbers and their operations	Ch. 1
Fractions and their operations	Ch. 3
Decimals and their operations	Ch. 4
Exponentiation and square roots	1.7, 2.2, 2.5, Appendix* V
Fraction-decimal conversion	4.3, 4.11
Applications (averages, percents, word problems)	1.9, 3.14, 4.12, 6.8 6.9, In applications throughout the text
Ratio, proportion and variance	Ch. 5
Reading data from graphs and charts	6.10, 6.11
Perimeter and area of triangles, squares, rectangles and parallelograms	7,5, 7.6, 7.7
Circumference and area of circles	7.5, 7.6, 7.7
Volumes of cubes, cylinders, rectangular solids, and spheres	7.8
Sum of interior angles of a triangle	Appendix V
Properties of isosceles and equilateral triangles	Appendix V
Properties of similar and congruent triangles	Appendix V
Pythagorean theorem and special triangles	Appendix V
Parallel and perpendicular lines	Appendix V

TASP Mathematics Skills

The following table lists the Texas **TASP MATHEMATICS SKILLS** and where coverage of these skills can be found in the text. Skills not covered in this text can be found in Basic Algebra, 3rd Edition, or Intermediate Algebra, 3rd Edition. Location of skills are indicated by chapter section or chapter.

Skill	Location in Text
Use number concepts and computation skills.	Ch. 1, Ch. 2, Ch. 3, Ch. 4, Ch. 6, Ch. 8
Solve word problems involving integers, fractions, or decimals (including percents, ratios, and proportions.	Ch. 3, Ch. 4, Ch. 5, Ch. 6
Solve one- and two-variable equations.	1.4, 1.6, 3.6, 3.13, 4.7, 4.10, 4.12, 8.6
Solve problems involving geometric figures.	Ch. 7, Appendix V
Apply reasoning skills.	7.9, Appendix V

CLAST Mathematical Skills

The following table lists the Florida **CLAST MATHEMATICAL SKILLS** and where coverage of these skills can be found in the text. Skills not covered in this text can be found in Basic Algebra, 3rd Edition, or Intermediate Algebra, 3rd Edition. Location of skills are indicated by chapter section or chapter.

Skill	Location in Text
1A1a—Adds and subtracts rational numbers	3.9, 3.10, 3.11, 3.12, 3.13
1A1b—Multiplies and divides rational numbers	3.4, 3.5, 3.6
1A2a—Adds and subtracts rational numbers in decimal form	4.6, 4.7
1A2b—Multiplies and divides rational numbers in decimal form	4.8, 4.9, 4.10, 4.11
1A3 —Calculates percent increase and percent decrease	6.8, 6.9
2A1 —Recognizes the meaning of exponents	1.7, 2.2, 2.5
2A2 —Recognizes the role of the base number in determining place value in the base-ten numeration system and in systems that are patterned after it	1.1, 1.2, 4.1, 4.2
2A3 —Identifies equivalent forms of positive rational numbers involving decimals, percents, and fractions	6.2, 6.3, 6.4, 6.5 6.6
2A4 —Determines the order-relation between magnitudes	1.2, 3.8, 4.4
4A1 —Solves real-world problems which do not require the use of variables and which do not require the use of percent	Ch. 1, Ch. 2, Ch. 3, Ch. 4, Ch. 5
4A2 —Solves real-world problems which do not require the use of variables and which do require the use of percent	6.8, 6.9
4A3 —Solves problems that involve the structure and logic of arithmetic	In applications throughout text
1B1 —Rounds measurements to the nearest given unit of the measuring device	4.5
1B2a—Calculates distances	7.5
1B2b—Calculates areas	7.6, 7.7
1B2c—Calculates volumes	7.8
2B1 —Identifies relationships between angle measures	Appendix V
2B2 —Classifies simple plane figures by recognizing their properties	Ch. 7, Appendix II
2B3 —Recognizes similar triangles and their properties	Appendix V
3B1 —Infers formulas for measuring geometric figures	Ch. 7
3B2 —Identifies applicable formulas for computing measures of geometric figures	Ch. 7
4B1 —Solves real-world problems involving perimeters, areas, and volumes of geometric figures	7.5, 7.6, 7.7, 7.8
4B2 —Solves real-world problems involving the Pythagorean property	Appendix V
1C1a—Adds and subtracts real numbers	8.2, 8.3
1C1b—Multiplies and divides real numbers	8.4, 8.5
1C2 —Applies the order-of-operations agreement to computations involving numbers and variables	8.6
1C4 —Solves linear equations and inequalities	1.4, 1.6, 3.6, 3.13, 4.7, 4.10, 4.12
1C5 —Uses given formulas to compute results when geometric measurements are not involved	In applications throughout text
2C3 —Recognizes statements and conditions of proportionality and variation	Ch. 5
1D1 —Identifies information contained in bar, line, and circle graphs	6.10
4D1 —Interprets real-world data from tables and charts	6.11
4E1 —Draws logical conclusions when facts warrant them	7.9

Student Preface

"It looks so easy when you do it, but when I get home . . ." is a popular lament of many students studying mathematics.

Taking our cue from a saying that is current in mathematical circles today—"Mathematics is not a spectator sport"—we believe that the only way to learn mathematics is to get involved with the subject. It is useful for you to read the text and go to class or lab, but you really learn math when you work problems (or better yet, help explain it to one of your classmates outside of class).

Here are some steps that will help you to get involved:

1. Work the Pre-Test in the beginning of the chapter. The answers to the Pre-Test are in the back of the book.
2. Read the section objectives. Note that the application problem that follows is a real-life illustration of how the objectives are used.
3. Read the Vocabulary and How and Why sections once or twice to get a general idea of the topic. Then read the How and Why section more slowly, looking for definitions and rules. Definitions, procedures, and rules are highlighted by various color screens for quick reference and easy review. See chart on pages xxii and xxiii for more detail.
4. Now read the Model Problem-Solving section. A good procedure is to cover up the worked-out solution, work the problem by yourself, and then uncover the solution. Compare your solution to the one in the text, noting the step-by-step explanation in the strategy column. Then work the Warm Up problem that follows to reinforce what you have learned. You can check your answer to the Warm Up problem by looking at the bottom of the page.
5. Read the solution to the Application that appeared earlier in the section. You should now understand how the problem was solved and be ready to go on to the Exercises. If you are not sure how the Application Solution was worked out, reread the How and Why and Model Problem-Solving sections.
6. Now work the Exercises at the end of the section. The Group A problems can usually be done mentally. Group B problems will probably require paper and pencil. Group C problems are more difficult; you may find that a calculator will save some time here. Group D problems follow up the application in each section. The application solution immediately preceding the Exercise section is a useful guide for these problems. Group E problems should not be skipped. They are for review and will help you practice earlier procedures so you do not become "rusty."

Some sections have problems immediately following Group E called "▲▲ Getting Ready for Algebra." This section is included where the skills you have learned can be applied in an elementary algebra setting. This section is preceded by a logo of two solid triangles and contains its own explanation section, model problem-solving section, and exercises. Whenever you see these triangles, you will know you're in the algebra section.

The answers to the odd-numbered exercises are provided in the back of the book so you can check your progress. If you do not get the correct answers after 45 minutes of study, stop and get help before you continue.

7. Work the True-False Concept Review. You should understand the basic concepts before proceeding to the Post-Test.
8. Work the Post-Test at the end of the chapter to review all of the concepts you have learned. The answers to each Post-Test are included in the back of the book. If you miss one of the Post-Test problems, go back to that section (the section number and objective are shown in parentheses at the beginning of each question) and practice a little more.

The chart on the facing page illustrates these steps. If you follow these steps and ask questions when you don't understand, you will have an excellent chance for success in this course. But remember, you must get involved because math is not a spectator sport.

Jack Barker
Jim Rogers
Jim Van Dyke

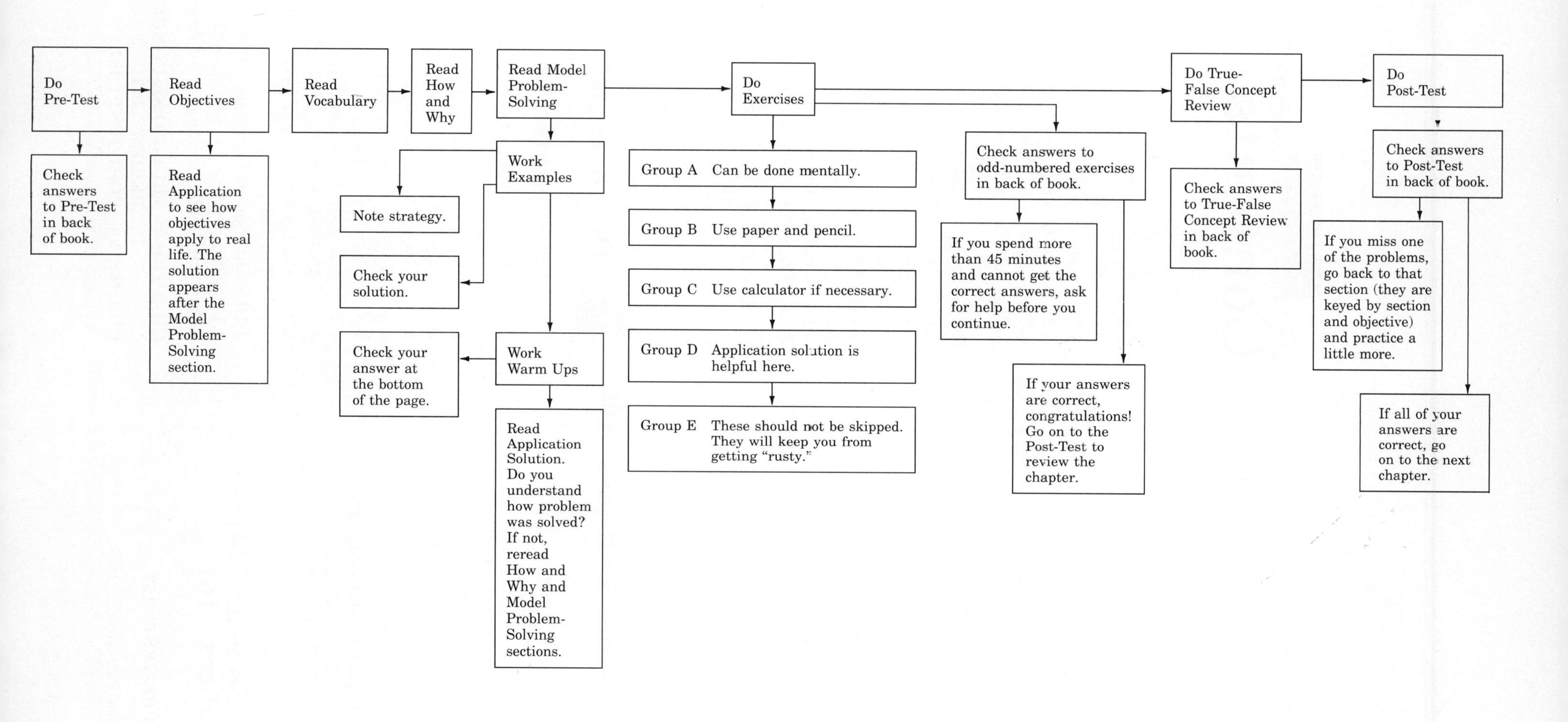

Do Pre-Test
Check answers to Pre-Test in back of book.
Read Objectives
Read Application to see how objectives apply to real life. The solution appears after the Model Problem-Solving section.
Read Vocabulary
Read How and Why
Read Model Problem-Solving
Work Examples
Note strategy.
Check your solution.
Work Warm Ups
Check your answer at the bottom of the page.
Read Application Solution. Do you understand how problem was solved? If not, reread How and Why and Model Problem-Solving sections.
Do Exercises
Group A Can be done mentally.
Group B Use paper and pencil.
Group C Use calculator if necessary.
Group D Application solution is helpful here.
Group E These should not be skipped. They will keep you from getting "rusty."
Check answers to odd-numbered exercises in back of book.
If you spend more than 45 minutes and cannot get the correct answers, ask for help before you continue.
If your answers are correct, congratulations! Go on to the Post-Test to review the chapter.
Do True-False Concept Review
Check answers to True-False Concept Review in back of book.
Do Post-Test
Check answers to Post-Test in back of book.
If you miss one of the problems, go back to that section (they are keyed by section and objective) and practice a little more.
If all of your answers are correct, go on to the next chapter.

Contents

Ratio and Proportion 469

Percent 513

Measurement 649

Algebra Preview: Signed Numbers 759

Pedagogical Use of Color

The various colors in the text figures are used to improve clarity and understanding. Any figures with three-dimensional representations are shown in various colors to make them as realistic as possible. Color is used in those figures where different portions are being highlighted and discussed.

In addition to the use of color in the figures, the pedagogical system in the text has been enhanced with color as well. We have used the following colors to distinguish the various pedagogical features:

PROPERTY

PROCEDURE

CAUTION

RULE

FORMULA

Table Title

Table Head

Introducing the Book

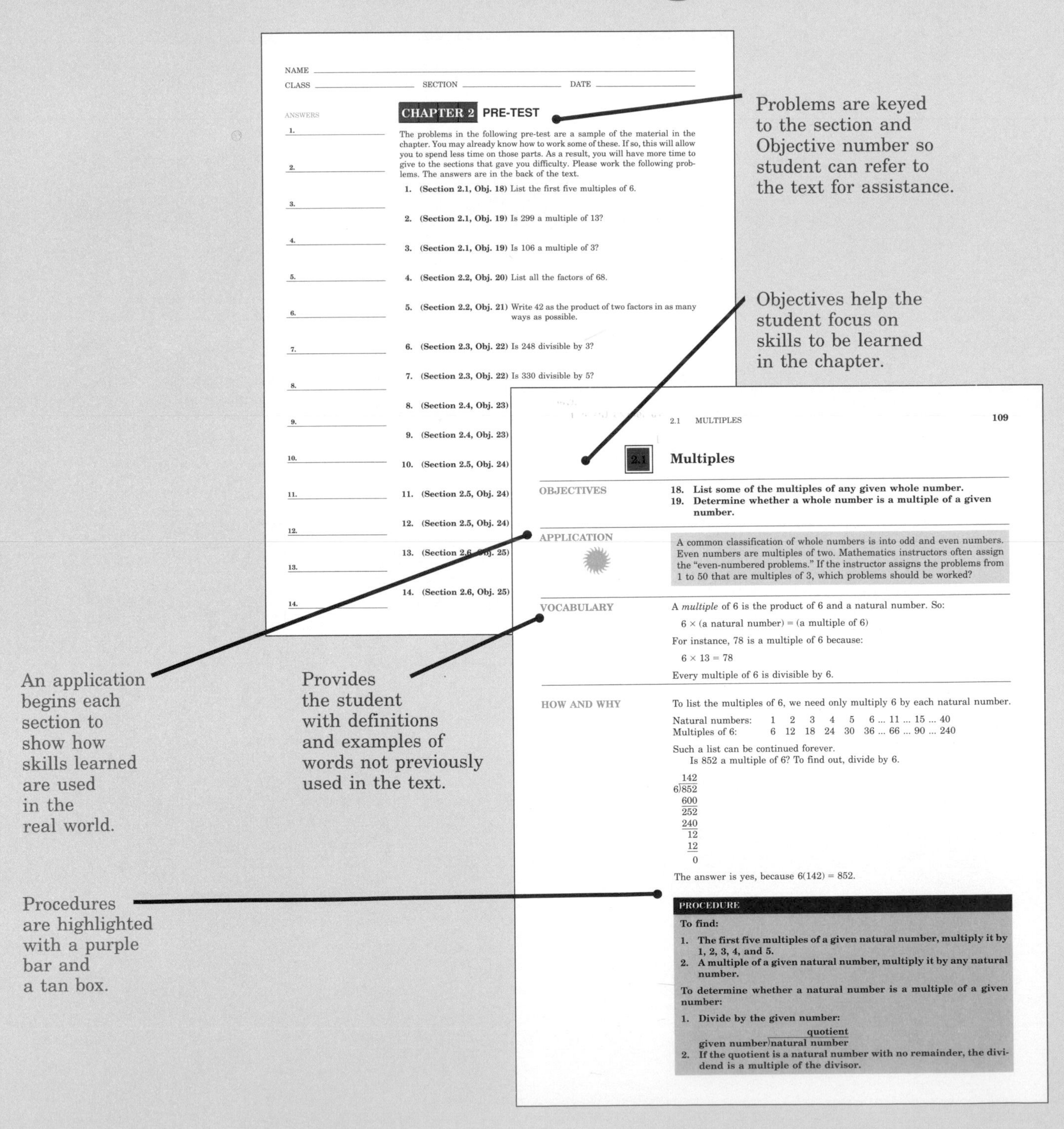

NAME
CLASS
SECTION
DATE
ANSWERS
CHAPTER 2 PRE-TEST
The problems in the following pre-test are a sample of the material in the chapter. You may already know how to work some of these. If so, this will allow you to spend less time on those parts. As a result, you will have more time to give to the sections that gave you difficulty. Please work the following problems. The answers are in the back of the text.
1. (Section 2.1, Obj. 18) List the first five multiples of 6.
2. (Section 2.1, Obj. 19) Is 299 a multiple of 13?
3. (Section 2.1, Obj. 19) Is 106 a multiple of 3?
4. (Section 2.2, Obj. 20) List all the factors of 68.
5. (Section 2.2, Obj. 21) Write 42 as the product of two factors in as many ways as possible.
6. (Section 2.3, Obj. 22) Is 248 divisible by 3?
7. (Section 2.3, Obj. 22) Is 330 divisible by 5?
8. (Section 2.4, Obj. 23)
9. (Section 2.4, Obj. 23)
10. (Section 2.5, Obj. 24)
11. (Section 2.5, Obj. 24)
12. (Section 2.5, Obj. 24)
13. (Section 2.6, Obj. 25)
14. (Section 2.6, Obj. 25)
Problems are keyed to the section and Objective number so student can refer to the text for assistance.
Objectives help the student focus on skills to be learned in the chapter.
2.1 MULTIPLES
109
2.1 Multiples
OBJECTIVES
18. List some of the multiples of any given whole number.
19. Determine whether a whole number is a multiple of a given number.
APPLICATION
A common classification of whole numbers is into odd and even numbers. Even numbers are multiples of two. Mathematics instructors often assign the "even-numbered problems." If the instructor assigns the problems from 1 to 50 that are multiples of 3, which problems should be worked?
VOCABULARY
A multiple of 6 is the product of 6 and a natural number. So:
6 × (a natural number) = (a multiple of 6)
For instance, 78 is a multiple of 6 because:
6 × 13 = 78
Every multiple of 6 is divisible by 6.
HOW AND WHY
To list the multiples of 6, we need only multiply 6 by each natural number.
Natural numbers: 1 2 3 4 5 6 ... 11 ... 15 ... 40
Multiples of 6: 6 12 18 24 30 36 ... 66 ... 90 ... 240
Such a list can be continued forever.
Is 852 a multiple of 6? To find out, divide by 6.
142
6)852
600
252
240
12
12
0
The answer is yes, because 6(142) = 852.
PROCEDURE
To find:
1. The first five multiples of a given natural number, multiply it by 1, 2, 3, 4, and 5.
2. A multiple of a given natural number, multiply it by any natural number.
To determine whether a natural number is a multiple of a given number:
1. Divide by the given number:
quotient
given number)natural number
2. If the quotient is a natural number with no remainder, the dividend is a multiple of the divisor.
An application begins each section to show how skills learned are used in the real world.
Provides the student with definitions and examples of words not previously used in the text.
Procedures are highlighted with a purple bar and a tan box.

3.5 Division of Fractions

OBJECTIVES

33. Find the reciprocal of a number.
34. Divide two fractions.

APPLICATION

If the distance a nut moves on a bolt with one turn is $\frac{3}{16}$ inch, how many turns will it take to move the nut $\frac{3}{4}$ inch?

VOCABULARY

If two fractions have a product of 1, either fraction is called the *reciprocal* of the other. For example, $\frac{2}{3}$ is the reciprocal of $\frac{3}{2}$.

HOW AND WHY

Writing a reciprocal is often called "inverting" the fraction. For instance, the reciprocal of $\frac{3}{7}$ is $\frac{7}{3}$. We check by showing that the product is 1:

$$\frac{3}{7} \cdot \frac{7}{3} = \frac{21}{21} = 1$$

PROCEDURE

The reciprocal of a fraction is found by interchanging the numerator and the denominator.

The reciprocal of a whole number is found by first writing the whole number as a fraction: the reciprocal of 21 $\left(21 = \frac{21}{1}\right)$ is $\frac{1}{21}$.

CAUTION

The number zero, 0, does not have a reciprocal.

It is pointed out in Chapter 1 that division is the inverse of multiplication. That is, the answer to a division problem is the number that is multiplied times the divisor (second number), which will give the first number as an answer. Another way of thinking of division is to ask, "How many groups of a certain size are contained in a number?"

	THINK	ANSWER
$6 \div 2$	How many twos in six?	3
$\frac{4}{5} \div \frac{1}{10}$	How many one-tenths in four-fifths?	See Figure 3.8.

Caution boxes are highlighted with a red bar and a tan box.

Fractions

fraction by finding the missing numerator.

ions of building are in comparing, adding, and subtracting frac-

actions are fractions that are different names for the same num- tance, $\frac{3}{6}$ and $\frac{4}{8}$ are equivalent, since both represent one-half $\left(\frac{1}{2}\right)$

of renaming fractions (see the description that follows) is some- d to as "building" fractions. It is the opposite of the process called actions, which we studied in Section 3.3.

$$3 \cdot \frac{2}{2} = \frac{6}{10}$$

l" fractions by multiplying by a fraction that is equal to one.

Tables are highlighted with a lavender bar.

	MULTIPLY BY: $\frac{2}{2}$	$\frac{3}{3}$	$\frac{4}{4}$	$\frac{5}{5}$	$\frac{10}{10}$	$\frac{15}{15}$
$\frac{3}{5}$ =	$\frac{6}{10}$ =	$\frac{9}{15}$ =	$\frac{12}{20}$ =	$\frac{15}{25}$ =	$\frac{30}{50}$ =	$\frac{45}{75}$
$\frac{1}{2}$ =	$\frac{2}{4}$ =	$\frac{3}{6}$ =	$\frac{4}{8}$ =	$\frac{5}{10}$ =	$\frac{10}{20}$ =	$\frac{15}{30}$
$\frac{3}{8}$ =	$\frac{6}{16}$ =	$\frac{9}{24}$ =	$\frac{12}{32}$ =	$\frac{15}{40}$ =	$\frac{30}{80}$ =	$\frac{45}{120}$
$\frac{2}{9}$ =	$\frac{4}{18}$ =	$\frac{6}{27}$ =	$\frac{8}{36}$ =	$\frac{10}{45}$ =	$\frac{20}{90}$ =	$\frac{30}{135}$
$\frac{7}{15}$ =	$\frac{14}{30}$ =	$\frac{21}{45}$ =	$\frac{28}{60}$ =	$\frac{35}{75}$ =	$\frac{70}{150}$ =	$\frac{105}{225}$

To find a missing numerator such as:

$$\frac{3}{5} = \frac{?}{60}$$

divide 60 by 5 to find out what fraction to multiply by:

$60 \div 5 = 12$

The correct multiplier is $\frac{12}{12}$:

$$\frac{3}{5} = \frac{3}{5} \cdot \frac{12}{12} = \frac{36}{60}$$

For example, find the perimeter of Figure 7.2. P (the perimeter) is the sum of the lengths of the four sides; that is:

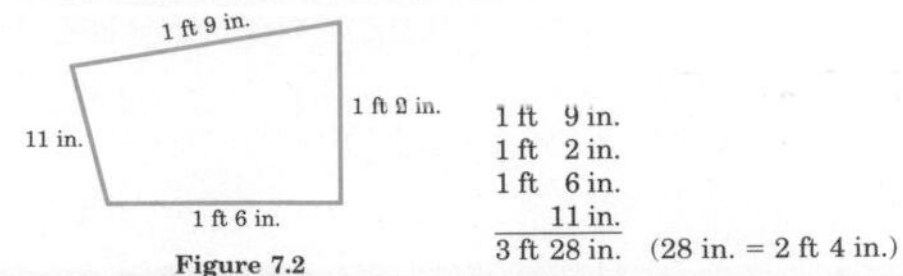

Figure 7.2

1 ft 9 in.
1 ft 2 in.
1 ft 6 in.
11 in.
3 ft 28 in. (28 in. = 2 ft 4 in.)

So, P = 5 ft 4 in.

There are special formulas for the perimeters of squares, rectangles, and circles and for the length of a semicircle.

FORMULAS

Square

If P is the perimeter and s is the length of one side, then the formula is:

$$P = 4 \cdot s$$

Rectangle

If P is the perimeter, ℓ is the length, and w is the width, then the formula is:

$$P = 2 \cdot \ell + 2 \cdot w$$

Circle

If C is the circumference and d is the diameter, then the formula is:

$$C = \pi$$

(π is read "pi," and its value is

Semicircle

If L is the length and r is the

$$L = \pi \cdot$$

PROCEDURE

To find the perimeter of a geo above, add the lengths of the s

Formulas are highlighted with a green bar and a tan box.

division are inverse operations. Division by 100 will move the decimal point two places to the left, and so on. Thus:

347.1 ÷ 100 = 3.47.1 = 3.471, and

0.763 ÷ 1000 = 0.000763

Three zeros are placed on the left so that the decimal point may be moved three places to the left.

PROCEDURE

To multiply a number by a power of ten, move the decimal point to the right. The number of places to move is shown by the number of zeros in the power of ten.

To divide a number by a power of ten, move the decimal point to the left. The number of places to move is shown by the number of zeros in the power of ten.

Scientific notation is widely used in science, technology, and industry to write large and small numbers. Every "scientific calculator" has a key for entering numbers in scientific notation. This notation makes it possible for a calculator or computer to deal with much larger or smaller numbers than those that take up eight, nine, or ten spaces on the display of a calculator.

Properties are highlighted with an orange bar and a tan box.

PROPERTY

A number in scientific notation is written as the product of two numbers. The first number is between 1 and 10 (including 1 but not 10) and the second number is a power of ten.

For example:

WORD FORM	PLACE VALUE (NUMERAL FORM)	SCIENTIFIC NOTATION
one million	1,000,000	1×10^6
five billion	5,000,000,000	5×10^9
one trillion, three billion	1,003,000,000,000	1.003×10^{12}

Small numbers are shown by writing the power of ten using a negative exponent. (You will learn more about this when you take a course in algebra.) For now, remember that multiplying by a negative power of ten is the same as *dividing* by a power of ten, which means you will be moving the decimal point to the left:

WORD FORM	PLACE VALUE (NUMERAL FORM)	SCIENTIFIC NOTATION
seven thousandths	0.007	7×10^{-3}
six ten-millionths	0.0000006	6×10^{-7}
fourteen hundred-billionths	0.00000000014	1.4×10^{-10}

The shortcut for multiplying by a power of ten is to move the decimal to the right, and the shortcut for dividing by a power of ten is to move the decimal to the left.

2.3 DIVISIBILITY TESTS 127

2.3 Divisibility Tests

OBJECTIVE 22. **Determine whether a natural number is divisible by 2, 3, and/or 5.**

APPLICATION

The following tests show how to tell whether a number is divisible by 2, 3, or 5 even more quickly than by using a calculator.

VOCABULARY No new words.

HOW AND WHY

Look at the following table.

SOME NATURAL NUMBERS	MULTIPLES OF 2	MULTIPLES OF 3	MULTIPLES OF 5
1	2	3	5
2	4	6	10
3	6	9	15
4	8	12	20
5	10	15	25
6	12	18	30
21	42	63	105
65	130	195	325
133	266	399	665

In the column of multiples of 2, eac
In the column of multiples of 3, the su
by 3. For example, the digits of 18 (1
the digits of 195 (1 + 9 + 5 = 15).
In the column of multiples of 5, eac
In the column of multiples of 2, each
is also in the column of multiples of 5
and hence by 10.

From the table the following co

RULE

Divisibility by 2: If the digit ber is 0, 2, 4 divisible by

Divisibility by 3: If the sum divisible by ble by 3.

Divisibility by 5: If the digit ber is 0 or 5 ble by 5.

Divisibility by 10: If the digit ber is 0, the 10.

Examples illustrate concepts explained in the How and Why section.

110 2 PRIMES AND MULTIPLES

For example, 104 is a multiple of 8, for when we divide by 8, as shown:

$$8\overline{)104}\quad 13$$

the quotient is 13, which is a natural number. We can state this in different ways:

1. 104 is a multiple of 8
2. 8 is a factor of 104
3. 104 divided by 8 is a natural number
4. 8 divides 104
5. 104 is divisible by 8

MODEL PROBLEM-SOLVING

Examples	Strategy
a. List the first five multiples of 3.	
The first five multiples of 3 are $1 \cdot 3$, $2 \cdot 3$, $3 \cdot 3$, $4 \cdot 3$, $5 \cdot 3$; that is, 3, 6, 9, 12, 15.	**Multiply 3 by 1, 2, 3, 4, and 5.**
Warm Up a. List the first five multiples of 4.	
b. List the first five multiples of 18.	
The first five multiples of 18 are: $1 \cdot 18$, $2 \cdot 18$, $3 \cdot 18$, $4 \cdot 18$, $5 \cdot 18$; that is, 18, 36, 54, 72, 90.	**Multiply 18 by 1, 2, 3, 4, and 5.** **Some calculators will find these multiples by repeated addition.** **Try pressing these keys: [18] followed by the [+] key *twice*. Then, if you press the [=] key repeatedly, the multiples of 18 will appear if your calculator has this feature. See example e also.**
Warm Up b. List the first five multiples of 11.	
c. Is 117 a multiple of 9?	
Since $117 \div 9 = 13$, we can say that:	**To find whether 117 is a multiple of 9, divide 117 by 9.**
117 is a multiple of 9.	**Because $117 = 9 \cdot 13$**
Warm Up c. Is 39 a multiple of 3?	

ANSWERS TO WARM UPS (2.1) **a.** 4, 8, 12, 16, 20 **b.** 11, 22, 33, 44, 55 **c.** Yes

Rules are highlighted with a blue bar and a tan box.

Warm Up Exercises of the same type and level of difficulty reinforce student's understanding.

Answers to Warm Ups are provided at the bottom of the page so student can double-check the answer.

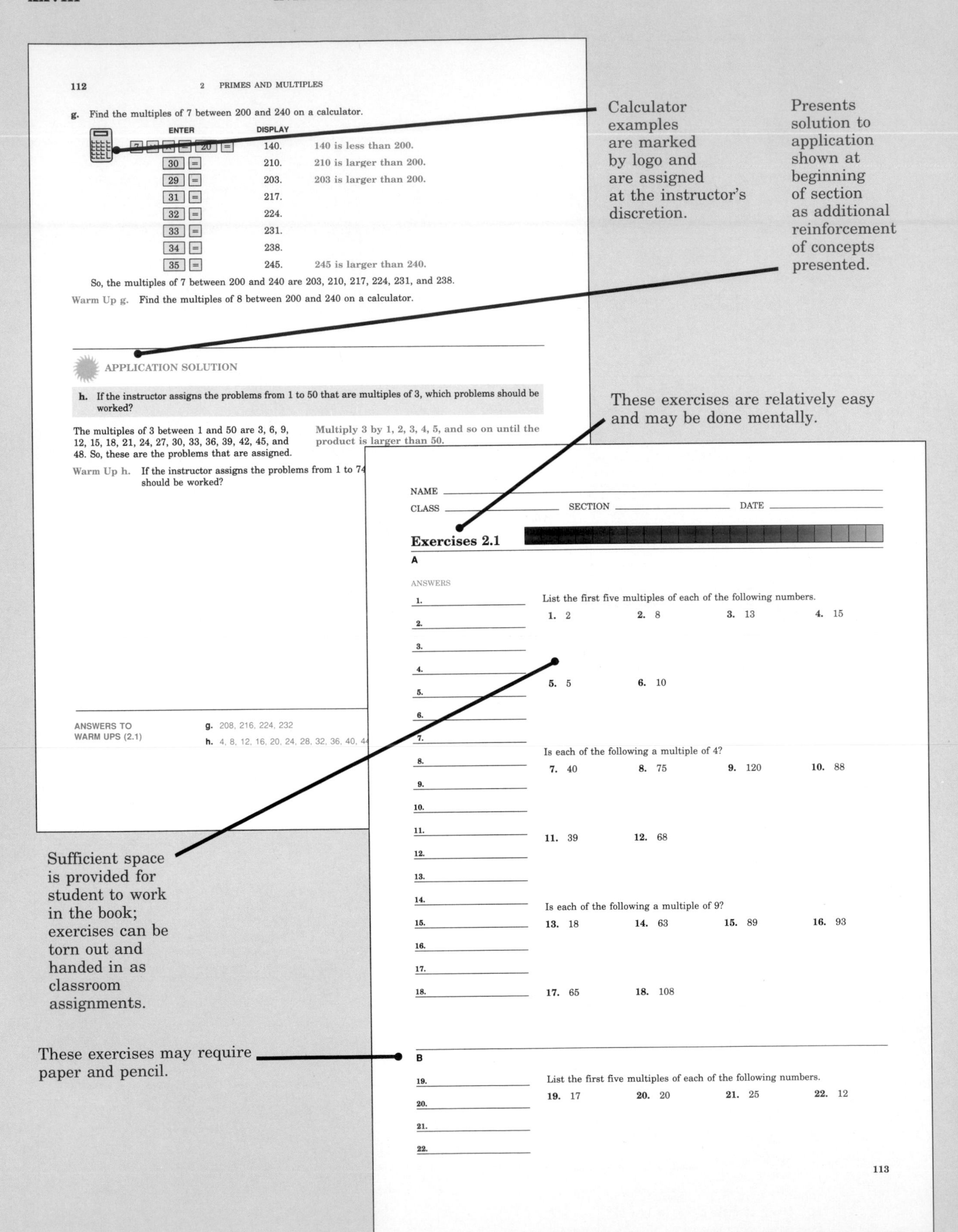

Calculator examples are marked by logo and are assigned at the instructor's discretion.
Presents solution to application shown at beginning of section as additional reinforcement of concepts presented.
These exercises are relatively easy and may be done mentally.
Sufficient space is provided for student to work in the book; exercises can be torn out and handed in as classroom assignments.
These exercises may require paper and pencil.

112 2 PRIMES AND MULTIPLES

g. Find the multiples of 7 between 200 and 240 on a calculator.

ENTER	DISPLAY	
7 × × = 20 =	140.	140 is less than 200.
30 =	210.	210 is larger than 200.
29 =	203.	203 is larger than 200.
31 =	217.	
32 =	224.	
33 =	231.	
34 =	238.	
35 =	245.	245 is larger than 240.

So, the multiples of 7 between 200 and 240 are 203, 210, 217, 224, 231, and 238.

Warm Up g. Find the multiples of 8 between 200 and 240 on a calculator.

APPLICATION SOLUTION

h. If the instructor assigns the problems from 1 to 50 that are multiples of 3, which problems should be worked?

The multiples of 3 between 1 and 50 are 3, 6, 9, 12, 15, 18, 21, 24, 27, 30, 33, 36, 39, 42, 45, and 48. So, these are the problems that are assigned.

Multiply 3 by 1, 2, 3, 4, 5, and so on until the product is larger than 50.

Warm Up h. If the instructor assigns the problems from 1 to 74 ... should be worked?

ANSWERS TO WARM UPS (2.1)
g. 208, 216, 224, 232
h. 4, 8, 12, 16, 20, 24, 28, 32, 36, 40, 4...

NAME
CLASS SECTION DATE

Exercises 2.1

A

ANSWERS
1. 2. 3. 4. 5. 6. 7. 8. 9. 10. 11. 12. 13. 14. 15. 16. 17. 18.

List the first five multiples of each of the following numbers.

1. 2 **2.** 8 **3.** 13 **4.** 15

5. 5 **6.** 10

Is each of the following a multiple of 4?

7. 40 **8.** 75 **9.** 120 **10.** 88

11. 39 **12.** 68

Is each of the following a multiple of 9?

13. 18 **14.** 63 **15.** 89 **16.** 93

17. 65 **18.** 108

B

19. 20. 21. 22.

List the first five multiples of each of the following numbers.

19. 17 **20.** 20 **21.** 25 **22.** 12

113

ANSWERS

23. ______
24. ______
25. ______
26. ______
27. ______
28. ______
29. ______
30. ______
31. ______
32. ______
33. ______
34. ______
35. ______
36. ______
37. ______
38. ______

23. 31 **24.** 14 **25.** 40 **26.** 100

Is each of the following a multiple of 7?

27. 28 **28.** 42 **29.** 47 **30.** 74

31. 728 **32.** 364

Is each of the following a multiple of 6?

33. 42 **34.** 99 **35.** 129 **36.** 954

37. 150 **38.** 1008

C

39. ______
40. ______
41. ______
42. ______
43. ______
44. ______
45. ______
46. ______
47. ______
48. ______
49. ______
50. ______

List the first five multiples of ea

39. 35 **40.** 42

43. 96 **44.** 88

Is each of the following numbers

45. 96 **46.** 480

49. 120 **50.** 900

114

These problems are more challenging. At the discretion of the instructor, "C" problems may sometimes be solved with a calculator.

"D" problems are similar to the application problem posed at the beginning of the section.

NAME ______

CLASS ______ SECTION ______ DATE ______

D

51. ______

51. If a mathematics instructor assigns problems from 1 to 60 that are multiples of 4, which problems should be done?

52. ______

52. If a statistics teacher assigns problems from 18 to 117 that are multiples of 9, which problems should her students do?

53. ______

53. List all of the multiples of 8 from 72 to 112.

54. ______

54. List all of the multiples of 7 from 231 to 266.

55. ______

55. List all of the multiples of 9 between 305 and 330.

56. ______

56. List all of the multiples of 13 between 400 and 460.

E *Maintain your skills* (Sections 1.3, 1.4, 1.5, 1.6)

57. ______

57. Find the sum of 53, 1689, 308, 9, 417, and 92.

115

These exercises review material previously covered and are referenced to sections of the text.

NAME ______________________

CLASS ____________ SECTION ____________ DATE ____________

CHAPTER 2 TRUE–FALSE CONCEPT REVIEW

ANSWERS

1. ______
2. ______
3. ______
4. ______
5. ______
6. ______
7. ______
8. ______
9. ______
10. ______
11. ______
12. ______
13. ______
14. ______
15. ______
16. ______
17. ______
18. ______
19. ______
20. ______
21. ______
22. ______
23. ______
24. ______
25. ______

Check your understanding of the language of basic mathematics. Tell whether each of the following statements is True (always true) or False (not always true.)

1. Every multiple of 6 ends with the digit 6.
2. Every multiple of 10 ends with the digit 0.
3. Every multiple of 13 is divisible by 13.
4. Every multiple of 7 is the product of 7 and some natural number.
5. Every whole number, except the number 1, has at least two different factors.
6. Every factor of 200 is also a divisor of 200.
7. Every multiple of 200 is also a factor of 200.
8. The square of 200 is 100.
9. Every natural number ending in 4 is divisible by 4.
10. Every natural number ending in 6 is divisible by 2.
11. Every natural number ending in 9 is divisible by 3.
12. The number 123,321,231 is divisible by 3
13. The number 123,321,234 is di
14. The number 123,321,235 is di
15. All prime numbers are odd.
16. Every composite number ends
17. Every composite number has
18. Every prime number has exac
19. It is possible for a composite
20. All of the prime factors of a nat
21. The least common multiple (LC product of the three numbers.
22. Some natural numbers have e
23. The largest divisor of the least is the largest of the three nun
24. It is possible for some groups
25. If any one number in a group least common multiple (LCM)

158

Helps pinpoint any remaining conceptual errors before student proceeds to Post-Test. Answers to all questions are in the back of the book.

NAME ______________________

CLASS ____________ SECTION ____________ DATE ____________

CHAPTER 2 POST-TEST

ANSWERS

1. ______
2. ______
3. ______
4. ______
5. ______
6. ______
7. ______
8. ______
9. ______
10. ______

1. **(Section 2.3, Obj. 22)** Is 552 divisible by 5?
2. **(Section 2.2, Obj. 20)** List all the factors of 110.
3. **(Section 2.1, Obj. 19)** Is 623 a multiple of 7?
4. **(Section 2.3, Obj. 22)** Is 411 divisible by 3?
5. **(Section 2.6, Obj. 25)** What is the LCM of 20 and 16?
6. **(Section 2.2, Obj. 21)** Write 75 as the product of two factors in as many ways as possible.
7. **(Section 2.5, Obj. 24)** Write the prime factorization of 260.
8. **(Section 2.6, Obj. 25)** Find the least common multiple (LCM) of 9, 42, and 84.
9. **(Section 2.5, Obj. 24)** Write the prime factorization of 846.
10. **(Section 2.1, Obj. 18)** List the first five multiples of 11.

159

All of the Post-Test answers are in the back of the book for the student.

This logo identifies the material as an algebra section.

This section shows students that they can use basic skills in an elementary algebra setting; it is included in sections where operations lend themselves to solving simple equations.

"Getting Ready for Algebra" sections have their own Model Problem-Solving sections.

▲▲ *Getting Ready for Algebra*

An *equation* is a statement about numbers that says two expressions are equal. Examples of equations are:

$8 = 8$ $12 = 12$ $100 = 100$ $20 + 5 = 25$ $49 - 9 = 40$

In algebra we often use a letter to represent a number. These letters are called *variables* or *unknowns*. When we use variables, the equations then can look like:

$x = 2$ $x = 5$ $y = 12$ $x + 3 = 10$ $y - 7 = 13$

This kind of equation will only be true when the letter is replaced by a specific number. For example:

$x = 2$ is true only when x is replaced by 2.
$x = 5$ is true only when x is replaced by 5.
$y = 12$ is true only when y is replaced by 12.
$x + 3 = 10$ is true only when x is replaced by 7, so that $7 + 3 = 10$.

$y - 7 = 13$ is true only when y is replaced by 20, so that $20 - 7 = 13$.

The numbers that make equations true are called *solutions*. Solutions of equations such as $x - 4 = 7$ can be found by trial and error, but let's develop a more practical way.

Addition and subtraction are inverse or opposite operations. For example, if 12 is added to a number and then 12 is subtracted from that sum, the difference is the original number.

As a specific example, add 11 to 15: $15 + 11 = 26$; the sum is 26. If 11 is subtracted from that sum, $26 - 11$, the result is 15, which was the original number.

We will use this idea to solve th

$x + 15 = 19$ — 15 is add
$x + 15 - 15 = 19 - 15$ — To remov the left s 15. To ke subtract
$x = 4$ — This equ replaced

To check, replace x in the original eq statement:

$x + 15 = 19$
$4 + 15 = 19$
$19 = 19$ — The state

We can also use the idea of inverses subtracted from a variable (letter):

$x - 4 = 7$ — Since 4 i
$x - 4 + 4 = 7 + 4$ — eliminate both side addition
$x = 11$ — This equ replaced

To check, replace x in the original equation with 11 and see if the result is a true statement:

$x - 4 = 7$
$11 - 4 = 7$
$7 = 7$ — The statement is true, so the solution is 11.

PROCEDURE

If a number is subtracted from the variable, in order to eliminate the subtraction, do the inverse; that is, add that number to both sides of the equation. The result is an equation in which the solution can be determined by inspection. If a number is added to the variable, in order to eliminate the addition, subtract that number from both sides of the equation. The result is an equation in which the solution can be determined by inspection.

▲▲ MODEL PROBLEM-SOLVING

Examples	Strategy
a. Solve:	
$x + 4 = 9$ $x + 4 - 4 = 9 - 4$	Since 4 is added to the variable, we eliminate that addition by subtracting 4 from both sides of the equation.
$x = 5$	Simplify.
Check: $x + 4 = 9$ $5 + 4 = 9$ $9 = 9$	We check by substituting 5 for x in the original equation. We get a true statement, so 5 is the solution.
Warm Up a. Solve: $x + 7 = 11$	
b. Solve:	
$x - 9 = 7$ $x - 9 + 9 = 7 + 9$	Since 9 is subtracted from the variable, we eliminate the subtraction by adding 9 to both sides of the equation.
$x = 16$	Simplify.
Check: $x - 9 = 7$ $16 - 9 = 7$ $7 = 7$	We substitute 16 for x in the original equation. We get a true statement, so 16 is the solution.

GETTING READY FOR ALGEBRA **a.** $x = 4$

Exercises are included in "Getting Ready for Algebra" sections.

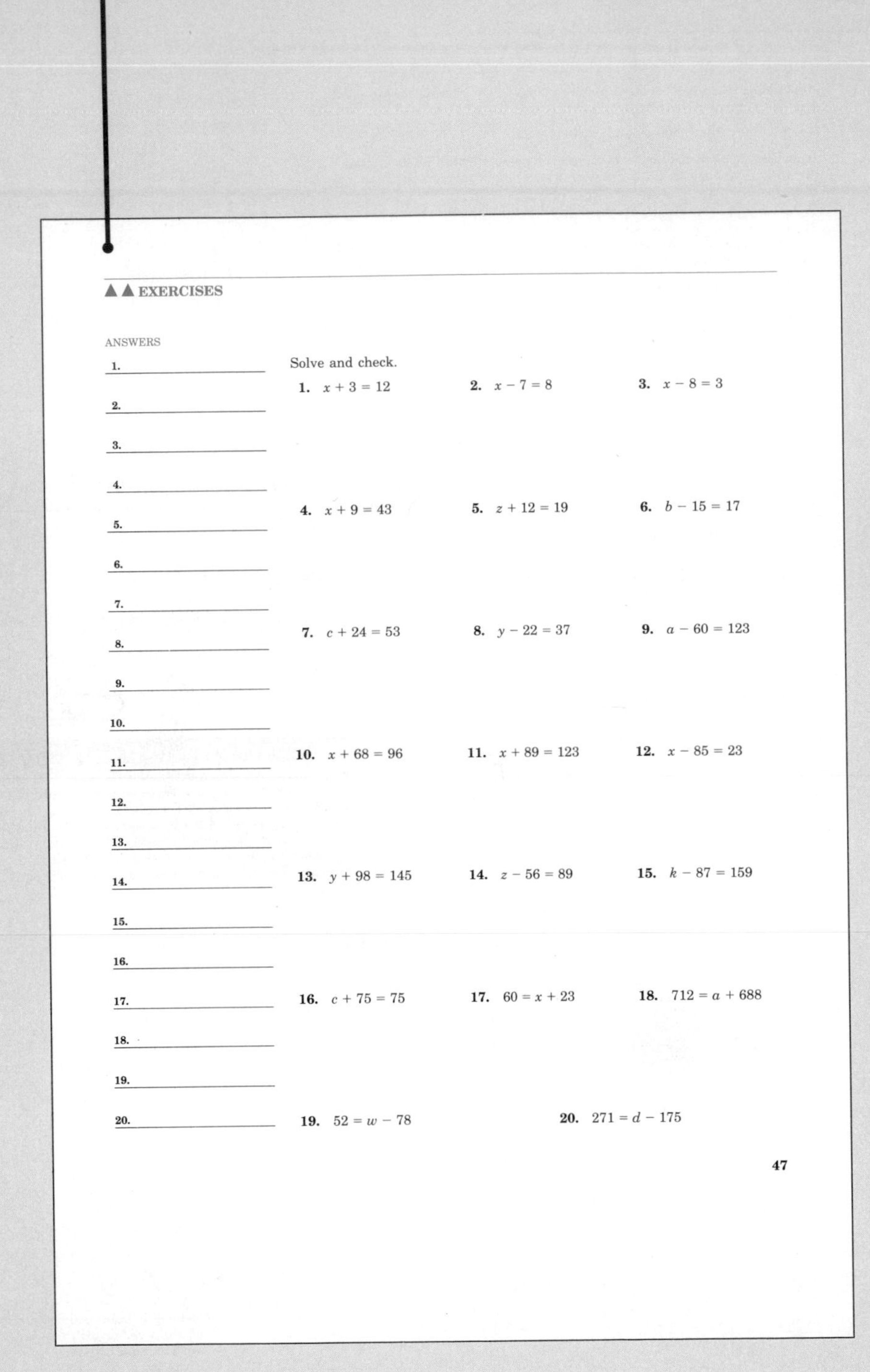

▲▲ EXERCISES

ANSWERS

1. ____
2. ____
3. ____
4. ____
5. ____
6. ____
7. ____
8. ____
9. ____
10. ____
11. ____
12. ____
13. ____
14. ____
15. ____
16. ____
17. ____
18. ____
19. ____
20. ____

Solve and check.

1. $x + 3 = 12$
2. $x - 7 = 8$
3. $x - 8 = 3$
4. $x + 9 = 43$
5. $z + 12 = 19$
6. $b - 15 = 17$
7. $c + 24 = 53$
8. $y - 22 = 37$
9. $a - 60 = 123$
10. $x + 68 = 96$
11. $x + 89 = 123$
12. $x - 85 = 23$
13. $y + 98 = 145$
14. $z - 56 = 89$
15. $k - 87 = 159$
16. $c + 75 = 75$
17. $60 = x + 23$
18. $712 = a + 688$
19. $52 = w - 78$
20. $271 = d - 175$

47

CHAPTER 1

Whole Numbers

NAME ____________________

CLASS ____________ SECTION ____________ DATE ____________

ANSWERS

1. ____________

2. ____________

3. ____________

4. ____________

5. ____________

6. ____________

7. ____________

8. ____________

9. ____________

10. ____________

11. ____________

12. ____________

13. ____________

CHAPTER 1 PRE-TEST

The problems in the following pre-test are a sample of the material in the chapter. You may already know how to work some of these. If so, this will allow you to spend less time on those parts. As a result, you will have more time to give to the sections that gave you difficulty. Please work the following problems. The answers are in the back of the text.

1. **(Section 1.1, Obj. 1)** Write the place value of the digit 7 in 73,558.

2. **(Section 1.1, Obj. 2)** Write the place value name for fourteen thousand, four hundred two.

3. **(Section 1.1, Obj. 3)** Write the word name for 3,821.

4. **(Section 1.1, Obj. 3)** Write the word name for 70,006.

5. **(Section 1.2, Obj. 4)** Write 9,874 in expanded form.

6. **(Section 1.2, Obj. 5)** Write the place value name for $7000 + 800 + 20 + 3$.

7. **(Section 1.2, Obj. 6)** Round to the nearest thousand: 373,489

8. **(Section 1.2, Obj. 6)** Round to the nearest hundred thousand: 1,666,666

9. **(Section 1.2, Obj. 7)** State whether the following is true or false: $67 < 59$

10. **(Section 1.3, Obj. 8)** Add: $74 + 683 + 7 + 18{,}432$

11. **(Section 1.3, Obj. 8)** Add:

$$\begin{array}{r} 68 \\ 321 \\ 4814 \\ 17 \\ \underline{973} \end{array}$$

12. **(Section 1.4, Obj. 9)** Subtract:

$$\begin{array}{r} 9617 \\ \underline{7204} \end{array}$$

13. **(Section 1.4, Obj. 9)** Subtract: $6803 - 1471$

ANSWERS

14. ______

15. ______

16. ______

17. ______

18. ______

19. ______

20. ______

21. ______

22. ______

23. ______

24. ______

25. ______

26. ______

27. ______

28. ______

29. ______

30. ______

14. **(Section 1.4, Obj. 9)** Subtract: $3004 - 298$

15. **(Section 1.5, Obj. 10)** Multiply: $(672)(87)$

16. **(Section 1.5, Obj. 10)** Multiply:

$$\begin{array}{r} 876 \\ \underline{403} \end{array}$$

17. **(Section 1.6, Obj. 11)** Divide: $48\overline{)5904}$

18. **(Section 1.6, Obj. 11)** Divide: $26{,}934 \div 67$

19. **(Section 1.6, Obj. 12)** Divide: $430\overline{)68388}$

20. **(Section 1.6, Obj. 12)** Divide: $29\overline{)19985}$

21. **(Section 1.7, Obj. 13)** Find the value of 6^3.

22. **(Section 1.7, Obj. 14)** Multiply: 326×10^5

23. **(Section 1.7, Obj. 15)** Divide: $8{,}600{,}000 \div 10^4$

24. **(Section 1.8, Obj. 16)** Perform the indicated operations:
$6 + 3 \cdot 8 - 6 \div 2$

25. **(Section 1.8, Obj. 16)** Perform the indicated operations:
$25 - 5^2 + 10 \div 2$

26. **(Section 1.9, Obj. 17)** Find the average of 144, 260, 196, and 364.

27. **(Section 1.9, Obj. 17)** Find the average of 361, 385, and 727.

28. The Wilsons have the following monthly payments: rent, $475; car, $210; insurance, $85; furniture, $112. What is the total of the payments they make each month?

29. A secretary can type 85 words per minute. At this rate how many words can she type in 145 minutes?

30. Sixteen people are to share in the Lotto Jackpot in the state lottery. If the jackpot is worth $1,818,480, how much will each person receive? (Round to the nearest hundred dollars.)

1.1 Whole Numbers: Place Value and Word Names

OBJECTIVES

1. **Write the place value of any digit, given the place value name for a number.**
2. **Write place value names from word names.**
3. **Write word names from place value names.**

APPLICATION

Mr. Joe Bates is writing a check to the Brian Company for payment of his monthly rent. If the rent is $325, what will be the word name written out on the check?

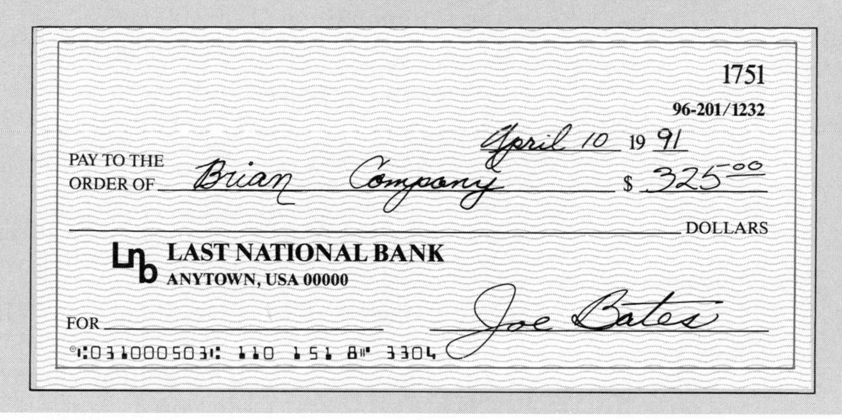

VOCABULARY

The *digits* are 0, 1, 2, 3, 4, 5, 6, 7, 8, and 9.
The *natural numbers (counting numbers)* are 1, 2, 3, 4, 5, and so on.
The *whole numbers* are 0, 1, 2, 3, 4, 5, and so on. Numbers larger than 9, such as 125, are written with place value names by putting the digits in positions having standard *place values.* Words, spoken or written, that name numbers are called *word names,* such as one hundred twenty-five.

HOW AND WHY

In our number system (called the Hindu–Arabic system), digits and commas are the only symbols used to write whole numbers. This system is also called a base ten (decimal) system. From right to left, the first three place value names are one, ten, and hundred.

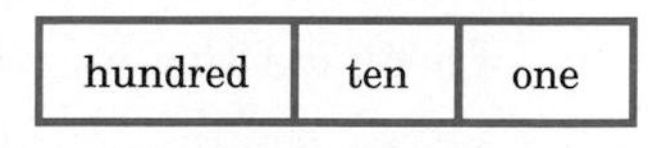

Figure 1.1

In 573, the digit 3 has place value one (1)
the digit 7 has place value ten (10)
the digit 5 has place value hundred (100)
so 573 is 5 hundreds + 7 tens + 3 ones.

Continuing to the left, the digits are grouped in threes. (See Fig. 1.2.) The first five groups are (from right to left) unit, thousand, million, billion, and trillion. The group on the far left may have one, two, or three digits. All other groups *must* have three digits. In each group the names are the same.

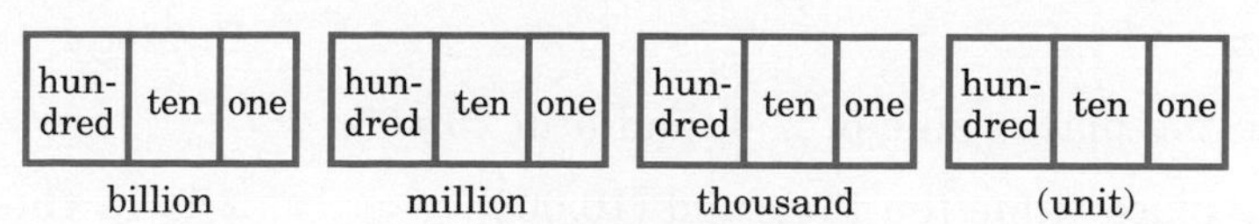

Figure 1.2

The place value of any position is one, ten, or hundred, followed by the group name.

Figure 1.3

In Figure 1.3:

DIGIT	PLACE VALUE
2	ten (10)
8	one thousand (1,000)
7	hundred thousand (100,000)
3	ten million (10,000,000)
5	one billion (1,000,000,000)
6	hundred billion (100,000,000,000)

Consider the number 1,321,208.

Place value name:	1,	321,	208
Word name of group:	one	three hundred twenty-one	two hundred eight
Group name:	million	thousand	(unit)

Word name: One million, three hundred twenty-one thousand, two hundred eight.

CAUTION

Do not insert the word "and" when reading or writing a whole number.

Although most people would know you meant 573 when you say "five hundred and seventy-three" it is not correct to read or write 573 with "and."

Work backward to write the place value name.

PROCEDURE

To write the word name from the place value name:

1. **Write the word name for each group of 3 digits.**
2. **Write the group name.**

To write the place value name, reverse these steps.

Commas are often used to separate groups, but they are not absolutely necessary. We will use commas except when writing four-digit numbers.

MODEL PROBLEM-SOLVING

Examples

a. Write the place value of 2, 4, and 5 in 23,465.

2 has place value ten thousand (10,000).

Strategy

2 is in the tens position in the thousands group.

4 has place value hundred (100). **4 is in the hundreds position in the units group.**

5 has place value one (1). **5 is in the ones position in the units group.**

Warm Up a. Write the place value of 3, 6, and 7 in 175,306.

b. Write the place value of 4, 6, and 3 in 546,213.

4 has place value ten thousand. **4 is in the tens position in the thousands group.**

6 has place value thousand. **6 is in the ones position in the thousands group.**

3 has place value one. **3 is in the ones position in the units group.**

Warm Up b. Write the place value of 4, 5, and 9 in 1,934,056.

c. What digit is in the ten thousands place in 835,601?

The digit 3. **The fifth position from the right is the ten thousands place.**

Warm Up c. What digit is in the hundreds place in 835,601?

d. Write the word name of the number whose place value name is 17,698,453.

17,	698,	453
seventeen	six hundred ninety-eight	four hundred fifty-three
million	thousand	(unit)

Place value name

Write the word name of each group.

Write the group name.

CAUTION

The group name, units, is not written.

Seventeen million, six hundred ninety-eight thousand, four hundred fifty-three. **Write the word name by writing the name of each group followed by its group name, going from left to right.**

Warm Up d. Write the word name for 6,455,091.

ANSWERS TO WARM UPS (1.1)

a. 3 has place value hundred (100).
6 has place value one (1).
7 has place value ten thousand (10,000).

b. 4 has place value thousand (1000).
9 has place value hundred thousand (100,000).
5 has place value ten (10).

c. 6

d. Six million, four hundred fifty-five thousand, ninety-one

e. Write the place value name for two million, thirty-seven thousand, five hundred sixty-four.

Two million, thirty-seven thousand,
(group name) (group name)

Identify the group names.

five hundred sixty-four

The group name "units" is understood but not written.

2,037,564

Write the place value name of each group. Use a comma in place of the group name. Notice the relationship between the commas in the word name and in the place value name.

Warm Up e. Write the place value name for thirty-two million, twenty-seven thousand, nine hundred ten.

APPLICATION SOLUTION

f. Mr. Joe Bates is writing a check to the Brian Company for payment of his monthly rent. If the rent is $325, what will be the word name written out on the check?

1751
96-201/1232
April 10 19 91
PAY TO THE ORDER OF Brian Company $ 325.00
Three hundred twenty-five DOLLARS
LAST NATIONAL BANK
ANYTOWN, USA 00000
FOR
Joe Bates
⑆0310005031⑆ 110 151 8⑈ 3204

To write the word name for 325, write: Three hundred twenty-five. (The group name "units" is not written.)

Warm Up f. Joe Bates also makes his car payment of $178 by check. What will be the word name written on the check?

ANSWERS TO WARM UPS (1.1)

e. 32,027,910

f. One hundred seventy-eight

NAME ______________________

CLASS ______________ SECTION ______________ DATE ______________

Exercises 1.1

A

ANSWERS

1. ______
2. ______
3. ______
4. ______
5. ______
6. ______
7. ______
8. ______
9. ______
10. ______
11. ______
12. ______
13. ______
14. ______
15. ______
16. ______

Write the place value of the digit 7 in each of the following.

1. 17 **2.** 673 **3.** 67,823 **4.** 51,725 **5.** 70,002

Write the word names of these numbers.

6. 542 **7.** 540 **8.** 504 **9.** 5,042 **10.** 5,402

Write the place value names for these numbers.

11. Seven thousand five hundred

12. Seven hundred fifty

13. Two hundred forty-three

14. Sixty-five

15. Six hundred five

16. Six thousand five

B

17. ______
18. ______
19. ______
20. ______
21. ______
22. ______

Write the place value of the digit 8 in each of the following.

17. 227,844 **18.** 138,472 **19.** 513,085

20. 813,000 **21.** 86,599 **22.** 721,846

ANSWERS

23. ____________

24. ____________

25. ____________

26. ____________

27. ____________

28. ____________

29. ____________

30. ____________

31. ____________

32. ____________

33. ____________

34. ____________

35. ____________

Write the word names of these numbers.

23. 25,310 **24.** 25,031 **25.** 2,531

26. 25,301 **27.** 205,310 **28.** 250,031

29. What digit is in the hundreds place in 4,168?

30. What digit is in the tens place in 17,284?

31. What digit is in the ones place in 65,024?

32. What digit is in the thousands place in 19,236?

Write the place value names for these numbers.

33. Two hundred forty-three thousand, seven

34. Two hundred three thousand, forty-seven

35. Twenty-three thousand, four hundred seven

NAME ______________________

CLASS ______________ SECTION ______________ DATE ______________

ANSWERS

36. ______________

37. ______________

38. ______________

36. Two hundred four thousand, seventy

37. Two thousand, four hundred thirty-seven

38. Two hundred forty thousand, four hundred seventy

C

39. ______________

40. ______________

41. ______________

42. ______________

43. ______________

44. ______________

45. ______________

46. ______________

47. ______________

48. ______________

Write the place value of the digit 7 in each of the following.

39. 67,211,384

40. 171,055,032

41. 7,055,011,200

42. 53,711,491

43. 273,560,681

44. 1,725,006,114

Write the word names of these numbers.

45. 502,520,052

46. 5,713,522,117

47. 50,050,500

48. 10,101,010

ANSWERS

49. ____________

50. ____________

51. ____________

52. ____________

53. ____________

54. ____________

55. ____________

56. ____________

57. ____________

58. ____________

49. 3,756,489

50. 17,892,054

51. What digit is in the hundred thousands place in 315,726?

52. What digit is in the ten thousands place in 724,590?

53. What digit is in the one thousands place in 362,718?

54. What digit is in the tens place in 503,274?

Write the place value names for these numbers.

55. Four hundred six million, two hundred forty-two thousand, seven hundred thirteen

56. Three hundred fifteen million, five hundred seventy-two

57. Six million, six hundred six

58. Fifty-four billion, fifty-five million, fifty-six thousand, fifty-seven

NAME ______________________________

CLASS ______________ SECTION ______________ DATE ______________

D

ANSWERS

59. ______________

60. ______________

61. ______________

62. ______________

63. ______________

64. ______________

59. Sue decided to have a "math costume" party. She sent everyone invitations with this number on it: 356,789,024. She asked Mel to dress as the digit in the hundred thousands place. What costume does Mel need?

60. Elena wanted to go to Sue's party (see Exercise 59) as a six. What place did she represent?

61. Sally bought a used motorcycle for $832. She wrote a check to pay for it. What word name must be written on the check?

62. Jim bought a used Ford Mustang for $1565 and wrote a check to pay for it. What word name must be written on the check?

63. Tom's company earned $34,079,021 in 1989. He gets $1000 for every million dollars earned as a bonus. What was Tom's bonus in 1989?

64. In 1980, Tom's company (see Exercise 63) earned $1,089,234. What was Tom's bonus that year?

ANSWERS

65. ______________________

65. Lynn Wang is paid 23 cents per mile when she uses her car to go from one plant to another. If she travels a total of 684 miles during the month of July, what word name should she write for her mileage?

66. ______________________

66. Michelle is paid 25 cents per mile when she uses her car for company business. If she drives 785 miles during the month of June, what word name should she write for the mileage?

67. ______________________

67. The world population during 1976 exceeded four billion. Write the place value name for four billion.

68. ______________________

68. The purchasing agent for the Upright Corporation received a telephone bid of twenty-five thousand sixty-three dollars as the price of a new printing press. What is the place value name for the bid?

Whole Numbers: Expanded Form, Rounding, and Inequalities

OBJECTIVES

4. **Write the expanded form from the place value name.**
5. **Write the place value name from the expanded form.**
6. **Round off a given whole number.**
7. **Tell whether an inequality statement involving two whole numbers is true or false.**

APPLICATION

Mary Kay bought a new Honda Accord for $13,568. To the nearest hundred dollars, how much did she pay for the car?

VOCABULARY

Expanded form is the sum of the values of each of the digits.

To *round off* a whole number means to give an approximate value. The approximate value is found by rounding to an indicated place. The number 987 rounded to the nearest ten is 990, so $987 \approx 990$ ($\approx$ means "approximately equal to").

The symbols *less than*, "<", and *greater than*, ">", are used to relate two whole numbers that are not equal to each other.

HOW AND WHY

PLACE VALUE NAME	EXPANDED FORM
2,584	2000 + 500 + 80 + 4 or 2 thousands + 5 hundreds + 8 tens + 4 ones
190	100 + 90 + 0 or 1 hundred + 9 tens + 0 ones
42,037	40,000 + 2,000 + 0 + 30 + 7 or 4 ten thousands + 2 thousands + 0 hundreds + 3 tens + 7 ones

PROCEDURE

To change from place value name to expanded form, multiply each digit by its place value and insert plus signs.

To change from the expanded form to the place value name, add.

If two whole numbers are not equal, then one is either *less than* or *greater than* the other.

Look at this number line (or ruler):

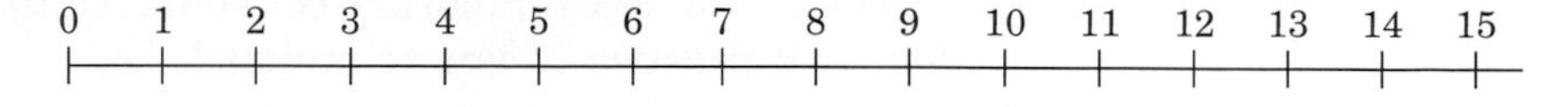

Numbers to the right of a given number are greater than the given number (the measure is larger), so:

$8 > 5$	8 is greater than 5.
$7 > 1$	7 is greater than 1.
$14 > 12$	14 is greater than 12.
$15 > 0$	15 is greater than 0.

Numbers to the left of any given number are less than the given number (the measure is smaller), so:

$2 < 6$	2 is less than 6.
$7 < 10$	7 is less than 10.
$5 < 9$	5 is less than 9.
$11 < 13$	11 is less than 13.

For larger numbers imagine the number line to be extended as far as you like. Note that the symbols $<$ and $>$ always point to the smaller number. So:

$109 < 405$
$34 > 25$
$1009 > 1007$

We can also use the number line to see how whole numbers are rounded. Suppose we wish to "round off" 27 to the nearest ten. Look at the following number line:

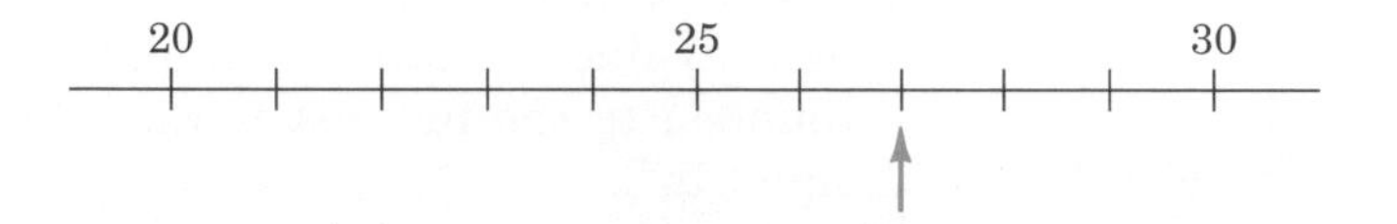

The arrow (under 27) is closer to 30 than to 20. We say, "27 rounds off to 30."

To round 34,568 to the nearest thousand, without a number line, draw an arrow under the digit in the thousands place to identify the round-off position.

34,568
↑

We must choose between 34,000 and 35,000. Since the digit to the right of the arrow is 5 (in the hundreds place) and since 500 is half of 1000, 34,568 is at least halfway to 35,000. Whenever the number is halfway or closer to the larger number, we choose the larger number. So:

$34{,}568 \approx 35{,}000$ (34,568 is closer to 35,000 than to 34,000.)

PROCEDURE

To round off a whole number to a given place value:

1. **Draw an arrow under the given place value.**
2. **If the digit to the right of the arrow is 5, 6, 7, 8, or 9, add one to the digit above the arrow. (Round to the larger number.)**
3. **If the digit to the right of the arrow is 0, 1, 2, 3, or 4, keep the digit above the arrow. (Round to the smaller number.)**
4. **Replace all digits to the right of the arrow with zeros.**

The arrow is a temporary convenience to help identify the round-off position; with practice it can be omitted.

MODEL PROBLEM-SOLVING

Examples | **Strategy**

Write in expanded form.

a. 625 = 600 + 20 + 5

Each value is the product of the digit and its place value.

Warm Up a. Write 782 in expanded form.

b. 1,675 = 1000 + 600 + 70 + 5

Warm Up b. Write 23,612 in expanded form.

c. 3,002 = 3000 + 0 + 0 + 2

Warm Up c. Write 60,070 in expanded form.

Write the place value name.

d. 60,000 + 3,000 + 0 + 40 + 7 = 63,047

Add the numbers.

Warm Up d. Write the place value name for 100,000 + 70,000 + 0 + 900 + 0 + 8

e. Is the following statement true or false?

395 > 401

False

395 is to the left of 401 on a number line, so 395 < 401.

Warm Up e. Is the following statement true or false?

199 > 311

f. Is the following statement true or false?

675 < 678

True

675 is to the left of 678 on a number line.

Warm Up f. Is the following statement true or false?

3499 < 3611

ANSWERS TO WARM UPS (1.2)

a. 700 + 80 + 2
b. 20,000 + 3000 + 600 + 10 + 2
c. 60,000 + 0 + 0 + 70 + 0
d. 170,908
e. False
f. True

g. Round 25,349 to the nearest hundred.

25,349
 ↑ — **Draw an arrow under the hundreds place.**

25,300 — **Since the digit to the right of the arrow is 4, keep the 3 and replace the digits on the right with zeros.**

Warm Up g. Round 37,489 to the nearest thousand.

h. Round 127,341 to the nearest ten thousand.

127,341
↑ — **Draw an arrow under the ten thousands place.**

130,000 — **Since the digit to the right of the arrow is 7, add one to the digit above the arrow (2 + 1).**

CAUTION

Don't forget to replace all digits to the right of the arrow with zeros.

Warm Up h. Round 99,858 to the nearest ten.

APPLICATION SOLUTION

i. Mary Kay bought a new Honda Accord for $13,568. How much did she pay for the car, to the nearest hundred dollars?

13568
 ↑ — **Draw an arrow under the hundreds place.**

13600 — **Since the digit to the right of the arrow is 6, add one to the digit above the arrow (5 + 1). Replace all digits on the right with zeros.**

Mary Kay paid approximately $13,600, to the nearest hundred dollars, for the Honda.

Warm Up i. The budget of the local community college was adopted at $36,419,850. Round the amount of the budget to the nearest ten thousand dollars.

ANSWERS TO WARM UPS (1.2) **g.** 37,000 **h.** 99,860 **i.** $36,420,000

NAME ______________________

CLASS ______________ SECTION ______________ DATE ______________

Exercises 1.2

A

ANSWERS

1. ______
2. ______
3. ______
4. ______
5. ______
6. ______
7. ______
8. ______
9. ______
10. ______
11. ______
12. ______
13. ______
14. ______
15. ______
16. ______
17. ______
18. ______
19. ______
20. ______
21. ______
22. ______

Write each of the following in expanded form.

1. 401 **2.** 712 **3.** 3005

4. 6080 **5.** 126 **6.** 4157

Write the place value name for each of the following.

7. $700 + 0 + 1$ **8.** $300 + 50 + 0$

9. $30{,}000 + 3000 + 400 + 40 + 5$ **10.** $70{,}000 + 0 + 500 + 20 + 3$

State whether the following statements are true or false.

11. $14 < 21$ **12.** $3 > 12$ **13.** $26 > 15$

14. $24 < 19$ **15.** $28 < 20$ **16.** $37 > 35$

Round each of the following to the indicated place value.

17. 311 (ten) **18.** 562 (hundred) **19.** 1,658 (ten)

20. 3,450 (hundred) **21.** 789 (hundred) **22.** 2164 (ten)

B

ANSWERS

23. ______

24. ______

25. ______

26. ______

27. ______

28. ______

29. ______

30. ______

31. ______

32. ______

33. ______

34. ______

35. ______

36. ______

37. ______

38. ______

39. ______

40. ______

41. ______

42. ______

Write each of the following in expanded form.

23. 60,000 **24.** 30,284 **25.** 50,204

26. 32,875 **27.** 302,084 **28.** 470,302

Write the place value name for each of the following.

29. 60,000 + 2000 + 800 + 30 + 2 **30.** 30,000 + 7000 + 400 + 0 + 6

31. 50,000 + 8000 + 800 + 70 + 3 **32.** 70,000 + 2000 + 0 + 0 + 8

33. 7 ten thousands + 2 thousands + 6 hundreds + 3 tens + 4

34. 8 ten thousands + 9 thousands + 8 hundreds + 2 tens + 3

State whether the following statements are true or false.

35. $33 > 44$ **36.** $62 < 26$ **37.** $57 > 39$ **38.** $100 > 98$

Round each of the following to the indicated place value.

39. 134,999 (ten thousand) **40.** 134,999 (thousand)

41. 134,999 (ten) **42.** 139,494 (hundred)

NAME ______________________

CLASS ____________ SECTION ____________ DATE ____________

ANSWERS

43. ______________

44. ______________

45. ______________

46. ______________

43. 134,999 (hundred)

44. 134,999 (hundred thousand)

45. 214,982 (thousand)

46. 406,127 (ten thousand)

C

47. ______________

48. ______________

49. ______________

50. ______________

51. ______________

52. ______________

53. ______________

54. ______________

Write each of the following in expanded form.

47. 654,377 **48.** 184,230 **49.** 9,443,205 **50.** 4,578,927

Write the place value name for each of the following.

51. 500,000 + 0 + 7,000 + 400 + 30 + 1

52. 800,000 + 30,000 + 7,000 + 0 + 60 + 0

53. 9,000,000 + 800,000 + 70,000 + 4,000 + 300 + 20 + 1

54. 4,000,000 + 0 + 50,000 + 2,000 + 400 + 0 + 8

ANSWERS

55. ____________

56. ____________

57. ____________

58. ____________

59. ____________

60. ____________

61. ____________

62. ____________

63. ____________

64. ____________

55. 3 million + 6 ten thousands + 9 thousands

56. 5 hundred millions + 8 hundred thousands + 7 hundreds

State whether each of the following is true or false.

57. $256 < 345$

58. $5634 > 5637$

59. $10{,}276 > 10{,}199$

60. $189{,}544 < 188{,}544$

Round each of the following to the indicated place value.

61. 672,456,895 (thousand)

62. 672,456,895 (ten million)

63. 34,870,675 (ten thousand)

64. 34,870,675 (thousand)

NAME ____________________

CLASS ____________ SECTION ____________ DATE ____________

ANSWERS

65. ____________

65. 75,347,216 (hundred)

66. ____________

66. 75,347,216 (hundred thousand)

D

67. ____________

67. On the first day of the Oregon State Lottery, $267,975 worth of tickets were sold. What was the value of the tickets sold, to the nearest thousand dollars?

68. ____________

68. An office building in downtown Memphis sold for $1,657,340. Give the price of the building to the nearest hundred thousand dollars.

69. ____________

69. Ten thousand shares of Greyhound Corp. sold for $278,750. What was the value of the sale, to the nearest thousand dollars?

ANSWERS

70. ____________________

70. Hawthorne Farms harvested 15,912 bushels of wheat from 234 acres. Express the bushels of wheat to the nearest hundred bushels.

71. ____________________

71. China placed an order for oscilloscopes with an electronics firm in Texas. The value of the order was $2,341,895. What was this value to the nearest hundred thousand dollars?

72. ____________________

72. The total land area of the earth is approximately 52,425,000 square miles. What is the land area to the nearest million square miles?

Addition of Whole Numbers

OBJECTIVE

8. Add whole numbers.

APPLICATION

The OK Floral Shop had sales of \$610 on Monday, \$845 on Tuesday, \$1425 on Wednesday, \$305 on Thursday, \$1705 on Friday, and \$2816 on Saturday. Find the total sales for the week and round to the nearest hundred dollars.

VOCABULARY

The answer to an addition problem is called the *sum* or *total.* "+" is the symbol used to show addition.

HOW AND WHY

In order to be able to add, you must memorize certain facts. The facts are listed in the "addition table." See inside front cover.

$$\begin{aligned} 32 &= 3 \text{ tens} + 2 \text{ ones} \\ \underline{27} &= \underline{2 \text{ tens} + 7 \text{ ones}} \\ 59 &= 5 \text{ tens} + 9 \text{ ones} \end{aligned}$$

In the column form there is a natural grouping of the ones and the tens. To find the sum of 48 and 34, we use the same idea:

$$\begin{aligned} 48 &= 4 \text{ tens} + 8 \text{ ones} = 40 + 8 \\ \underline{34} &= \underline{3 \text{ tens} + 4 \text{ ones}} = \underline{30 + 4} \\ & \quad 7 \text{ tens} + 12 \text{ ones} = 70 + 12 \end{aligned}$$

Since 8 + 4 is 12, which is a two-digit number, we must rename:

$$\begin{aligned} 70 &= 70 + 0 \\ 12 &= 10 + 2 \\ & \quad 80 + 2 = 82 \end{aligned}$$

The common shortcut is:

$$\begin{array}{r} {\scriptstyle 1} \\ 48 \\ \underline{34} \\ 82 \end{array}$$

8 + 4 = 12 = 1 *ten* + 2 *ones*
The "2" is placed in the ones column below, and the "1" is carried to be added in the tens column.

PROCEDURE

To add whole numbers:

1. **Write the numbers in columns so that place values are lined up.**
2. **Add each column, starting with the ones.**
3. **If the sum in any column is more than nine, write the ones digit and carry the tens.**

MODEL PROBLEM-SOLVING

Examples

a. Add:

```
  1
 274
 382
 ---
 656
```

Strategy

Add the numbers in the ones column: 4 + 2 = 6. Since "6" is a single-digit number, there is no carry.
Add the numbers in the tens column: 7 + 8 = 15. Since "15" is a two-digit number, write the "5" in the tens column and carry the "1" to the hundreds column.
Add the numbers in the hundreds column, 1 + 2 + 3 = 6. There is no carry, so this completes the addition.

Warm Up a. Add:

```
 362
 204
 ---
```

b. Add:

```
  121
  1682
  4491
  7629
 -----
 13802
```

Add the numbers in the ones column: 2 + 1 + 9 = 12. Carry the "1" to the tens column.
Add the numbers in the tens column: 1 + 8 + 9 + 2 = 20. Carry the "2" to the hundreds column.
Add the numbers in the hundreds column: 2 + 6 + 4 + 6 = 18. Carry the "1" to the thousands column.
Add the numbers in the thousands column: 1 + 1 + 4 + 7 = 13. Since there are no digits in the ten thousands column, no carrying is necessary. Write the "3" in the thousands column and the "1" in the ten thousands column.

Warm Up b. Add:

```
 2342
 1568
 7842
 ----
```

ANSWERS TO WARM UPS (1.3) **a.** 566 **b.** 11,752

c. Add: 48 + 232 + 4 + 2834

111 48 232 4 2834 3118	**Write the problem in columns, making sure to line up the place values correctly.** **Add.**

Warm Up c. Add: 72 + 7963 + 11 + 495

d. Calculator example*

Add: 378 + 4091 + 46

ENTER	DISPLAY	
378	378.	**Enter 378.**
+	378.	**Press the "+" key.**
4091	4091.	**Enter 4091.**
+	4469.	**Press the "+" key.**
46	46.	**Enter 46.**
=	4515.	**Press the "=" key.**

The sum is 4515.

Warm Up d. Add: 4695 + 7887 + 309 + 1105

APPLICATION SOLUTION

e. The OK Floral Shop had sales of $610 on Monday, $845 on Tuesday, $1425 on Wednesday, $305 on Thursday, $1705 on Friday, and $2816 on Saturday. Find the total sales for the week and round to the nearest hundred dollars.

610 845 1425 305 1705 2816 7706	**Add the daily sales. Write the amounts in columns. Make sure the place values are lined up correctly.**
$7706 \approx 7700$	**Round to the nearest hundred.**

The total sales were approximately $7700.

Warm Up e. The Friendly Puppy Pet Shop had sales of $1113, $945, $1612, $2349, and $3415 in five consecutive days. Find the total sales in the five days to the nearest hundred dollars.

*The exercises in sections with calculator examples lend themselves to calculator use. Calculators may be used at the discretion of the student and/or instructor.

ANSWERS TO WARM UPS (1.3) **c.** 8541 **d.** 13,996 **e.** $9400

NAME ______________________________

CLASS ______________ SECTION ______________ DATE ______________

Exercises 1.3

A

ANSWERS

1. ________
2. ________
3. ________
4. ________
5. ________
6. ________
7. ________
8. ________
9. ________
10. ________
11. ________
12. ________
13. ________
14. ________
15. ________
16. ________
17. ________
18. ________
19. ________
20. ________

Add:

1. $\begin{array}{r}12\\ \underline{37}\end{array}$ **2.** $\begin{array}{r}47\\ \underline{28}\end{array}$ **3.** $\begin{array}{r}48\\ \underline{51}\end{array}$ **4.** $\begin{array}{r}56\\ \underline{39}\end{array}$

5. $\begin{array}{r}14\\ \underline{19}\end{array}$ **6.** $\begin{array}{r}12\\ \underline{49}\end{array}$ **7.** $\begin{array}{r}121\\ \underline{384}\end{array}$ **8.** $\begin{array}{r}146\\ \underline{214}\end{array}$

9. $\begin{array}{r}214\\ \underline{194}\end{array}$ **10.** $\begin{array}{r}178\\ \underline{259}\end{array}$ **11.** $\begin{array}{r}490\\ \underline{186}\end{array}$ **12.** $\begin{array}{r}365\\ \underline{507}\end{array}$

13. $\begin{array}{r}1145\\ \underline{2036}\end{array}$ **14.** $\begin{array}{r}7856\\ \underline{3333}\end{array}$ **15.** $\begin{array}{r}9078\\ \underline{1309}\end{array}$ **16.** $\begin{array}{r}10360\\ \underline{9315}\end{array}$

17. $7 + 18 + 35 + 4$

18. $6 + 19 + 22 + 7$

19. $823 + 716$

20. $127 + 849$

B

ANSWERS

21. ____________

22. ____________

23. ____________

24. ____________

25. ____________

26. ____________

27. ____________

28. ____________

29. ____________

30. ____________

31. ____________

32. ____________

33. ____________

34. ____________

35. ____________

36. ____________

37. ____________

38. ____________

39. ____________

40. ____________

21.
78
59
27

22.
93
42
71

23.
62
78
43

24.
92
76
12

25.
414
165
258

26.
349
237
184

27.
17
82
23
46

28.
148
261
382
146

29.
207
626
315
177

30.
147
236
874
921

31.
491
309
380
976
621

32.
654
382
197
846
321

33.
1684
2873
4091
5004

34.
5693
7562
3043
4537

35.
2145
1086
3290
8762
1314
1222

36.
1438
2174
6832
1456
7890
1249

37. 76 + 98 + 23

38. 25 + 78 + 92

39. 216 + 459 + 244

40. 345 + 592 + 327

NAME ______________________

CLASS ____________ SECTION ____________ DATE ____________

C

ANSWERS

41. ____________

42. ____________

43. ____________

44. ____________

45. ____________

46. ____________

47. ____________

48. ____________

49. ____________

50. ____________

51. ____________

52. ____________

53. ____________

54. ____________

41.
```
  48
 276
   5
4176
  29
----
```

42.
```
 965
  42
3012
   8
 603
----
```

43.
```
47021
  762
 1904
   18
  217
-----
```

44.
```
96311
 5016
   92
 7009
-----
```

45.
```
20000
60904
 9130
   72
-----
```

46.
```
48416
18729
 9762
   34
-----
```

47.
```
19036
 4023
  201
    5
43129
-----
```

48.
```
29762
    4
 9999
  388
  777
-----
```

49. 21 + 46 + 87 + 91

50. 48 + 271 + 4 + 195

51. 8004 + 78 + 987 + 5

52. 402 + 38 + 9003

53. 13 + 8271 + 4082 + 16,752

54. 27 + 1849 + 5927 + 32,154

ANSWERS

55. ____________

56. ____________

55. 16,003 + 49 + 831 + 5391

56. 708 + 47 + 9 + 10,005

D

57. ____________

57. On Monday, Mr. John drove his car 182 miles, on Tuesday 364 miles, on Wednesday 235 miles, on Thursday 317 miles, and on Friday 293 miles. What was the total mileage driven in the five days?

58. ____________

58. A new car costs $1,598 more than last year's model. If the price of last year's model is $13,937, find the price of the new car.

59. ____________

59. The attendance at three basketball games was 17,842, 16,021, and 18,045. What was the total attendance for the three games?

60. ____________

60. It is known that in a city in the Northwest there are 21,274 people under 23 years of age, 20,274 between 23 and 65 years of age, and 5,864 over 65 years of age. What is the population of the city?

NAME ______

CLASS ______ SECTION ______ DATE ______

ANSWERS

61. ______

61. The Mendosa family has the following payments to make each month; rent, $425; car, $175; TV, $34; washer and dryer, $58; bedroom furniture, $29. What is the total of the payments that they make each month?

62. ______

62. Five rooms of a house contain lamps, which have the following wattages: kitchen, 100 W; dining room, 125 W; bathroom, 60 W; living room, 150 W; and bedroom, 100 W. What is the total wattage when all lamps are lit?

63. ______

63. In estimating the number of concrete blocks needed for a building, a contractor lists them as follows: east wall, 288 blocks; west wall, 288 blocks; north wall, 144 blocks; and south wall, 144 blocks. How many concrete blocks will be needed?

64. ______

64. The attendance at four professional football games was 65,890, 72,684, 48,975, and 53,672. Find the total attendance for the four games, to the nearest thousand.

ANSWERS

65. ____________________

66. ____________________

67. ____________________

65. The mileage log for the company car showed the following numbers of miles for the last ten days: 457, 346, 89, 125, 297, 430, 610, 35, 57, 385. Find the total mileage for the ten days, to the nearest ten miles.

66. John Cagney works for $250 a week plus commissions. If his commissions for the week were $110, $65, $76, $150, and $240, what were his total earnings for the week? Find to the nearest ten dollars.

67. Susan Summers is paid $325 a week plus commissions. If her commissions for the week were $325, $240, $32, $157, and $75, what were her total earnings for the week? Find to the nearest hundred dollars.

Subtraction of Whole Numbers

OBJECTIVE

9. Subtract whole numbers.

APPLICATION

Nina Carretta received an electricity bill, which showed that 1572 kilowatt hours (KWH) of energy had been used. If it is known that 682 KWH were used for lighting and appliances and the rest for hot water, how many kilowatt hours were used for heating water?

VOCABULARY

"−" is the symbol showing subtraction and is read "minus."
The answer to a subtraction problem is called the *difference.*

HOW AND WHY

To subtract 9 − 5 = ? asks 5 + ? = 9. Since 5 + 4 = 9, we can tell that 9 − 5 = 4.

47 − 15 = ? Since 15 + 32 = 47, this tells us that 47 − 15 = 32.
274 − 162 = ? Write in column form and subtract columns:

		Check by adding:
274 =	2 hundreds + 7 tens + 4 ones	162
162 =	1 hundred + 6 tens + 2 ones	112
112 =	1 hundred + 1 ten + 2 ones	274

453 − 238 − ?

453 = 4 hundreds + 5 tens + 3 ones

238 = 2 hundreds + 3 tens + 8 ones

We cannot subtract 8 ones from 3 ones, so we must rename by borrowing one of the tens from the 5 tens (1 ten = 10 ones) and adding this to the 3 ones.

		Check by adding:
453 =	4 hundreds + 4 tens + 13 ones	238
238 =	2 hundreds + 3 tens + 8 ones	215
215 =	2 hundreds + 1 ten + 5 ones	453

The examples show a shortcut for calculations like this one:

PROCEDURE

To subtract whole numbers:

1. **Write the numbers in columns so that place values are lined up.**
2. **Subtract each column, starting with the ones.**
3. **When the column cannot be subtracted, rename by borrowing.**

MODEL PROBLEM-SOLVING

Examples / **Strategy**

a. Subtract: 75 − 32

```
 75    Check:
 32     32
 --     --
 43     43
        --
        75
```

Write in columns with place values lined up.
Subtract the ones column: 5 − 2 = 3
Subtract the tens column: 7 − 3 = 4
Check by adding.

Warm Up a. Subtract: 96 − 53

b. Subtract:

```
 42
 29
 --
```

```
 3  12
 4̸  2̸    Check:
 2  9     29
 ----     13
 1  3     --
          42
```

Since 9 ones cannot be subtracted from 2 ones, we borrow 1 ten from the 4 tens.
We add it to the 2 ones (10 + 2 = 12), for a total of 12 ones.
Now subtract in each column.

Warm Up b. Subtract:

```
 64
 48
 --
```

c. Subtract:

```
 752
 295
 ---
```

```
 6  14
    4̸  12
 7̸  5̸  2̸    Check:
 2  9  5     295
 -------     457
 4  5  7     ---
             752
```

We must rename twice.
We cannot subtract 5 ones from 2 ones or 9 tens from 5 tens.
We borrow 1 ten (10 ones) from the tens to rename the ones (10 + 2 = 12).
Borrow 1 hundred (10 tens) from the hundreds to rename the tens (10 + 4 = 14).
Subtract in each column.

Warm Up c. Subtract:

```
 567
 398
 ---
```

ANSWERS TO WARM UPS (1.4) **a.** 43 **b.** 16 **c.** 169

d. Subtract: 2500
  689

We cannot subtract 9 from 0, and since there are 0 tens, we cannot borrow from the tens place.

$$\begin{array}{cccc} & ^{4} & ^{\overset{9}{\cancel{10}}} & ^{10} \\ 2 & \cancel{5} & \cancel{0} & \cancel{0} \\ & 6 & 8 & 9 \\ \hline \end{array}$$

We go to the hundreds place and borrow 1 hundred (10 tens).
Now borrow a ten to make 10 ones.
We can now subtract in the ones and the tens columns but not in the hundreds column.
So, we must borrow to subtract in the hundreds column.

$$\begin{array}{cccc} ^{1} & ^{\overset{14}{\cancel{4}}} & ^{\overset{9}{\cancel{10}}} & ^{10} \\ \cancel{2} & \cancel{5} & \cancel{0} & \cancel{0} \\ & 6 & 8 & 9 \\ \hline 1 & 8 & 1 & 1 \end{array}$$

Check:
$$\begin{array}{r} 689 \\ 1811 \\ \hline 2500 \end{array}$$

Warm Up d. Subtract: 4600
  1343

e. Calculator example

Subtract: 3582 − 2785

ENTER	DISPLAY	
3582	3582.	Enter 3582.
−	3582.	Press the "−" key.
2785	2785.	Enter 2785.
=	797.	Press the "=" key.

The difference is 797.

Warm Up e. Subtract: 5677
  −3529

ANSWERS TO WARM UPS (1.4) **d.** 3257 **e.** 2148

APPLICATION SOLUTION

f. Nina Carretta received an electricity bill, which showed that 1572 kilowatt hours (KWH) of energy had been used. If it is known that 682 KWH were used for lighting and appliances and the rest for hot water, how many kilowatt hours were used for heating water?

$$\begin{array}{r} 1572 \\ \underline{682} \\ 890 \end{array}$$

To find the number of kilowatt hours used for heating water, subtract the number used for lighting and appliances from the total used.

Nina used 890 KWH for heating water.

Warm Up f. Nina also paid the water and garbage bills for the next three months. She paid a total of $125. If the water bill was $48, how much was the garbage bill?

ANSWERS TO WARM UPS (1.4)

f. $77

NAME ______________________________

CLASS ____________ SECTION ____________ DATE ____________

Exercises 1.4

A

ANSWERS

1. ____
2. ____
3. ____
4. ____
5. ____
6. ____
7. ____
8. ____
9. ____
10. ____
11. ____
12. ____
13. ____
14. ____
15. ____
16. ____
17. ____
18. ____
19. ____
20. ____

Subtract:

1. $\begin{array}{r} 9 \\ \underline{2} \end{array}$ **2.** $\begin{array}{r} 7 \\ \underline{3} \end{array}$ **3.** $\begin{array}{r} 17 \\ \underline{14} \end{array}$ **4.** $\begin{array}{r} 18 \\ \underline{12} \end{array}$

5. $\begin{array}{r} 24 \\ \underline{13} \end{array}$ **6.** $\begin{array}{r} 15 \\ \underline{11} \end{array}$ **7.** $\begin{array}{r} 36 \\ \underline{23} \end{array}$ **8.** $\begin{array}{r} 88 \\ \underline{35} \end{array}$

9. $\begin{array}{r} 47 \\ \underline{25} \end{array}$ **10.** $\begin{array}{r} 91 \\ \underline{15} \end{array}$ **11.** $\begin{array}{r} 56 \\ \underline{8} \end{array}$ **12.** $\begin{array}{r} 62 \\ \underline{8} \end{array}$

13. $\begin{array}{r} 92 \\ \underline{64} \end{array}$ **14.** $\begin{array}{r} 72 \\ \underline{49} \end{array}$ **15.** $\begin{array}{r} 86 \\ \underline{59} \end{array}$ **16.** $\begin{array}{r} 70 \\ \underline{43} \end{array}$

17. $27 - 8$ **18.** $34 - 19$ **19.** $103 - 78$ **20.** $110 - 82$

B

ANSWERS

21. ____
22. ____
23. ____
24. ____
25. ____
26. ____
27. ____
28. ____
29. ____
30. ____
31. ____
32. ____
33. ____
34. ____
35. ____
36. ____
37. ____
38. ____
39. ____
40. ____

21. 274 181	**22.** 514 373	**23.** 867 495	**24.** 218 149
25. 408 47	**26.** 509 73	**27.** 205 66	**28.** 523 60
29. 608 59	**30.** 504 306	**31.** 899 500	**32.** 731 427
33. 627 314	**34.** 384 138	**35.** 721 439	**36.** 3591 2185

37. 9000
3937

38. 800
338

39. 727 − 582

40. 4150 − 3983

C

41. ____
42. ____
43. ____
44. ____

41. 2184 998	**42.** 6802 4765	**43.** 7002 3654	**44.** 3005 709

NAME ______________________________

CLASS ____________ SECTION ____________ DATE ____________

ANSWERS

45. ____________

46. ____________

47. ____________

48. ____________

49. ____________

50. ____________

51. ____________

52. ____________

53. ____________

54. ____________

55. ____________

56. ____________

57. ____________

58. ____________

45.
$$\begin{array}{r} 40000 \\ \underline{18214} \end{array}$$

46.
$$\begin{array}{r} 8000 \\ \underline{2016} \end{array}$$

47.
$$\begin{array}{r} 46709 \\ \underline{30510} \end{array}$$

48.
$$\begin{array}{r} 27082 \\ \underline{10910} \end{array}$$

49. 7982 − 5697

50. 6000 − 4818

51. 4125 − 687

52. 2008 − 193

53. 5678 − 4987

54. 3491 − 2973

55. 62,004 − 45,782

56. 75,060 − 52,673

57. 102,456 − 97,567

58. 215,678 − 105,821

D

59. ____________

59. If there are 58,275 tickets available for a certain baseball game, and all but 2,863 are sold, how many tickets have been sold?

60. ____________

60. A carload of fertilizer contained 3,684 one-hundred-pound bags. If 1,793 bags are sold, how many bags are left?

ANSWERS

61. ______________________

62. ______________________

63. ______________________

64. ______________________

65. ______________________

66. ______________________

67. ______________________

68. ______________________

61. A casting of aluminum weighs 75 lb. The same casting of iron weighs 210 lb. How much more does the iron weigh?

62. The bids to remodel the gym at the Cascade Center of Portland Community College were $236,500 and $219,300. How much was saved by taking the smaller bid?

63. Larry owes Apex Construction $3,086. He makes payments of $212, $359, and $610. How much does he still owe, rounded to the nearest hundred dollars?

64. The highest possible score on a test was 120 points. If Mary missed 28 points, what was the number of points she had correct?

65. The Cleer Glass Company had $12,862 in their checking account. If the treasurer of the company wrote checks totaling $8,275, how much, to the nearest hundred dollars, was left in the account?

66. The population of Toonerville was 82,645 in 1970. In 1980 the population had increased to 102,764. What was the amount of increase?

67. A mason orders 448 cubic yards of concrete and uses 435 cubic yards. How many cubic yards of concrete are not used?

68. Rich has $5,182 in his checking account. If he writes a check for one thousand eight hundred twenty-seven dollars, how much money is left in his account?

▲▲ *Getting Ready for Algebra*

An *equation* is a statement about numbers that says two expressions are equal. Examples of equations are:

$8 = 8$ $\quad 12 = 12$ $\quad 100 = 100$ $\quad 20 + 5 = 25$ $\quad 49 - 9 = 40$

In algebra we often use a letter to represent a number. These letters are called *variables* or *unknowns.* When we use variables, the equations then can look like:

$x = 2$ $\quad x = 5$ $\quad y = 12$ $\quad x + 3 = 10$ $\quad y - 7 = 13$

This kind of equation will only be true when the letter is replaced by a specific number. For example:

$x = 2$ is true only when x is replaced by 2.
$x = 5$ is true only when x is replaced by 5.
$y = 12$ is true only when y is replaced by 12.
$x + 3 = 10$ is true only when x is replaced by 7, so that $7 + 3 = 10$.

$y - 7 = 13$ is true only when y is replaced by 20, so that $20 - 7 = 13$.

The numbers that make equations true are called *solutions.* Solutions of equations such as $x - 4 = 7$ can be found by trial and error, but let's develop a more practical way.

Addition and subtraction are inverse or opposite operations. For example, if 12 is added to a number and then 12 is subtracted from that sum, the difference is the original number.

As a specific example, add 11 to 15: $15 + 11 = 26$; the sum is 26. If 11 is subtracted from that sum, $26 - 11$, the result is 15, which was the original number.

We will use this idea to solve the following equation:

$x + 15 = 19$	**15 is added to the number represented by x.**
$x + 15 - 15 = 19 - 15$	**To remove the addition and have only x on the left side of the equal sign, we subtract 15. To keep a true equation, we must subtract 15 from both sides.**
$x = 4$	**This equation will be true when x is replaced by 4.**

To check, replace x in the original equation with 4 and see if the result is a true statement:

$x + 15 = 19$	
$4 + 15 = 19$	
$19 = 19$	**The statement is true, so the solution is 4.**

We can also use the idea of inverses to solve an equation in which a number is subtracted from a variable (letter):

$x - 4 = 7$	**Since 4 is subtracted from the variable, we eliminate the subtraction by adding 4 to both sides of the equation. Recall that addition is the inverse of subtraction.**
$x - 4 + 4 = 7 + 4$	
$x = 11$	**This equation will be true when x is replaced by 11.**

To check, replace x in the original equation with 11 and see if the result is a true statement:

$$x - 4 = 7$$
$$11 - 4 = 7$$
$$7 = 7$$

The statement is true, so the solution is 11.

PROCEDURE

If a number is subtracted from the variable, in order to eliminate the subtraction, do the inverse; that is, add that number to both sides of the equation. The result is an equation in which the solution can be determined by inspection. If a number is added to the variable, in order to eliminate the addition, subtract that number from both sides of the equation. The result is an equation in which the solution can be determined by inspection.

▲▲ MODEL PROBLEM-SOLVING

Examples	Strategy
a. Solve:	
$x + 4 = 9$ $x + 4 - 4 = 9 - 4$	**Since 4 is added to the variable, we eliminate that addition by subtracting 4 from both sides of the equation.**
$x = 5$	**Simplify.**
Check: $x + 4 = 9$ $5 + 4 = 9$ $9 = 9$	**We check by substituting 5 for x in the original equation. We get a true statement, so 5 is the solution.**

Warm Up a. Solve: $x + 7 = 11$

b. Solve:	
$x - 9 = 7$ $x - 9 + 9 = 7 + 9$	**Since 9 is subtracted from the variable, we eliminate the subtraction by adding 9 to both sides of the equation.**
$x = 16$	**Simplify.**
Check: $x - 9 = 7$ $16 - 9 = 7$ $7 = 7$	**We substitute 16 for x in the original equation. We get a true statement, so 16 is the solution.**

a. $x = 4$

Warm Up b. Solve: $x - 11 = 8$

c. Solve:

$$22 = y + 15$$
$$22 - 15 = y + 15 - 15$$

Since 15 is added to the variable, we eliminate that addition by subtracting 15 from both sides of the equation.

$$7 = y$$

Simplify.

Check:
$22 = 7 + 15$
$22 = 22$

We substitute 7 for *y* in the original equation. We get a true statement, so 7 is the solution.

Warm Up c. Solve: $z + 18 = 37$

d. Solve:

$$z - 27 = 35$$
$$z - 27 + 27 = 35 + 27$$

Since 27 is subtracted from the variable, we eliminate the subtraction by adding 27 to both sides of the equation.

$$z = 62$$

Simplify.

Check:
$62 - 27 = 35$
$35 = 35$

We substitute 62 for *z* in the original equation. We get a true statement, so 62 is the solution.

Warm Up d. Solve: $b - 47 = 45$

GETTING READY FOR ALGEBRA

b. $x = 19$ **c.** $z = 19$ **d.** $b = 92$

NAME ______________________

CLASS ____________ SECTION ____________ DATE ____________

▲▲ EXERCISES

ANSWERS

1. ____________
2. ____________
3. ____________
4. ____________
5. ____________
6. ____________
7. ____________
8. ____________
9. ____________
10. ____________
11. ____________
12. ____________
13. ____________
14. ____________
15. ____________
16. ____________
17. ____________
18. ____________
19. ____________
20. ____________

Solve and check.

1. $x + 3 = 12$

2. $x - 7 = 8$

3. $x - 8 = 3$

4. $x + 9 = 43$

5. $z + 12 = 19$

6. $b - 15 = 17$

7. $c + 24 = 53$

8. $y - 22 = 37$

9. $a - 60 = 123$

10. $x + 68 = 96$

11. $x + 89 = 123$

12. $x - 85 = 23$

13. $y + 98 = 145$

14. $z - 56 = 89$

15. $k - 87 = 159$

16. $c + 75 = 75$

17. $60 = x + 23$

18. $712 = a + 688$

19. $52 = w - 78$

20. $271 = d - 175$

Multiplication of Whole Numbers

OBJECTIVE

10. Multiply whole numbers.

APPLICATION

The Sweet & Sour Company shipped 62 cartons of packaged candy to the Chewum Candy Store. Each carton contained 48 packages of candy. What was the total number of packages of candy shipped?

VOCABULARY

When two or more whole numbers are multiplied, each is called a *factor*. The answer is called the *product*.

$$(7) \quad (8) \quad = \quad 56$$

$$(\text{factor})\ (\text{factor}) = \text{product}$$

Five ways to write a multiplication problem are:

$7 \times 8 \qquad 7 \cdot 8 \qquad 7(8) \qquad (7)8 \qquad (7)(8)$

Any number times zero is zero. (See Appendix I.)

HOW AND WHY

Multiplying whole numbers is a shortcut for repeated addition:

$$\underbrace{8 + 8 + 8 + 8 + 8 + 8}_{\text{6 EIGHTS}} = 48 \quad \text{or} \quad 6 \cdot 8 = 48$$

To do multiplication without repeated addition, you must memorize the basic multiplication facts. If you have not yet memorized these, or if you have forgotten any, see inside the front cover and memorize them.

As the numbers get larger, the shortcut saves time. Imagine adding 156 eights:

$$\underbrace{8 + 8 + 8 + 8 + \cdots + 8}_{\text{156 EIGHTS}} = ?$$

We can use the expanded form to show the multiplication:

$$\begin{array}{rcl} 156 & = & 100 + \ \ 50 + \ \ 6 \\ \underline{8} & = & \underline{\qquad\qquad\qquad 8} \\ 1248 & = & 800 + 400 + 48 \end{array}$$

In the example just given, 8 is multiplied by the 6, then by the 50, and finally by the 100. The answer can be written in a vertical form without expanding the number:

$$\begin{array}{rlcr} 156 & & & \\ \underline{8} & & & \\ 48 & \text{8 times 6} & & 156 \\ 400 & \text{8 times 50} & \text{or} & \underline{8} \\ \underline{800} & \text{8 times 100} & & 1248 \\ 1248 & & & \end{array}$$

The second form shows the usual shortcut. The addition is done mentally as you go. Study this example:

$$\begin{array}{r} 629 \\ \underline{46} \end{array}$$

First multiply 629 by 6.

```
 15
 629
  46
3774
```

6 · 9 = 54. Carry the five to the tens column.
6 · 2 tens = 12 tens. Add the 5 tens that were carried: (12 + 5) tens = 17 tens. Carry the 1 to the hundreds column.
6 · 6 hundreds = 36 hundreds. Add the 1 hundred that was carried: (36 + 1) hundreds.

```
  13
  1̸5̸
  629
   46
 3774
25160
```

Now multiply 629 by 40.
40 · 9 = 360, or 36 tens. Carry the 3 to the hundreds column.
40 · 20 = 800, or 8 hundreds. Add the 3 hundreds that were carried: (8 + 3) hundreds = 11 hundreds.
Carry the 1 to the thousands column.
40 · 600 = 24,000, or 24 thousands. Add the 1 thousand that was carried: (24 + 1) thousands = 25 thousands.
Since the multiplication is complete, write the 5 in the thousands column and the 2 in the ten thousands column.

```
  13
  1̸5̸
  629
   46
 3774
25160
28934
```

Add the products.

When you work a problem this way, the entire process should resemble the last step. Think through the procedure in Example c.

An important property of multiplication is called the **zero-product property:**

PROPERTY

Any number times zero is zero.

So $0 \cdot 8 = 0$; $0 \cdot 0 = 0$; and $340 \cdot 0 = 0$.

MODEL PROBLEM-SOLVING

Examples	Strategy
a. $0 \cdot 37 = 0$	**Zero-product property.**
Warm Up a. Multiply: (1369)(0)	
b. 351 (carries) / 4592 × 6 = 27552	**Multiply 6 times each digit, carry when necessary, then add the number carried to the next product.**

ANSWERS TO WARM UPS (1.5)

a. 0

Warm Up b. Multiply:
```
3456
   8
----
```

c. 38 · 74

```
 1
 3
74
38
---
 592
2220
----
2812
```

Write in vertical form.
When multiplying by the three in the tens place, write a 0 in the ones column to keep the place values lined up.

Warm Up c. Multiply: 59 · 84

d. The answer in Example c can be checked by reversing the numbers and multiplying again.

```
 5
 3
38
74
---
 152
2660
----
2812
```

Warm Up d. Check Warm Up c by reversing the numbers and multiplying again.

e. (513)(205)

```
   513
   205
------
  2565
102600
------
105165
```

CAUTION

When multiplying by the zero in the tens place, rather than showing a row of zeros, just put a zero in the tens column and then multiply by the 2 in the hundreds column.

ANSWERS TO WARM UPS (1.5) **b.** 27,648 **c.** 4956 **d.** 4956

Warm Up e. Multiply: (302)(707)

f. (162)(300)

$$\begin{array}{r} 162 \\ \underline{300} \\ 48600 \end{array}$$

When multiplying by the zeros in the ones and tens places, put zeros in the ones and tens columns and then multiply by three in the hundreds place. Write the product starting with the hundreds column. This avoids having two rows of zeros.

Warm Up f. Multiply: (245)(400)

g. Calculator Example

Multiply: 346 · 76

ENTER	DISPLAY	
346	346.	**Enter 346.**
×	346.	**Press the "×" key.**
76	76.	**Enter 76.**
=	26296.	**Press the "=" key.**

The product is 26,296.

Warm Up g. Multiply: 398 · 148

APPLICATION SOLUTION

h. The Sweet & Sour Company shipped 62 cartons of packaged candy to the Chewum Candy Store. Each carton contained 48 packages of candy. What was the total number of packages of candy shipped?

$$\begin{array}{r} 62 \\ \underline{48} \\ 496 \\ \underline{2480} \\ 2976 \end{array}$$

To find the number of packages of candy shipped, multiply the number of cartons by the number of packages per carton.

The company shipped 2976 packages of candy.

Warm Up h. The Sweet & Sour Company shipped 68 cartons of chewing gum to the Sweet Shop. If each carton contained 72 packages of gum, how many packages of gum were shipped to the Sweet Shop?

ANSWERS TO WARM UPS (1.5)

e. 213,514 **f.** 98,000 **g.** 58,904 **h.** 4896 packages

NAME ______________________________

CLASS ______________ SECTION ______________ DATE ______________

Exercises 1.5

A

ANSWERS

1. ____
2. ____
3. ____
4. ____
5. ____
6. ____
7. ____
8. ____
9. ____
10. ____
11. ____
12. ____
13. ____
14. ____
15. ____
16. ____
17. ____
18. ____
19. ____
20. ____
21. ____
22. ____

Multiply:

1. $\begin{array}{r} 42 \\ \underline{3} \end{array}$ **2.** $\begin{array}{r} 63 \\ \underline{3} \end{array}$ **3.** $\begin{array}{r} 59 \\ \underline{4} \end{array}$ **4.** $7 \cdot 21$

5. 4×92 **6.** $8(66)$ **7.** $(64)(0)$ **8.** $0 \cdot 25$

9. 6×410 **10.** $\begin{array}{r} 140 \\ \underline{8} \end{array}$ **11.** $\begin{array}{r} 60 \\ \underline{40} \end{array}$ **12.** $\begin{array}{r} 52 \\ \underline{40} \end{array}$

13. $\begin{array}{r} 114 \\ \underline{6} \end{array}$ **14.** $\begin{array}{r} 108 \\ \underline{7} \end{array}$ **15.** $\begin{array}{r} 1050 \\ \underline{9} \end{array}$ **16.** $\begin{array}{r} 2156 \\ \underline{20} \end{array}$

17. $(7)(82)$ **18.** $(76)(8)$ **19.** $(23)(15)$ **20.** $(25)(12)$

21. $(8)(10)(0)(21)$ **22.** $(7)(0)(20)(15)$

B

23. ____
24. ____
25. ____
26. ____

23. $\begin{array}{r} 147 \\ \underline{6} \end{array}$ **24.** $\begin{array}{r} 509 \\ \underline{3} \end{array}$ **25.** $\begin{array}{r} 4060 \\ \underline{6} \end{array}$ **26.** $3020(4)$

ANSWERS

27. ____________

28. ____________

29. ____________

30. ____________

31. ____________

32. ____________

33. ____________

34. ____________

35. ____________

36. ____________

37. ____________

38. ____________

39. ____________

40. ____________

27. 47(39)

28. (97)(37)

29. $\begin{array}{r}64\\ \underline{57}\end{array}$

30. $\begin{array}{r}85\\ \underline{67}\end{array}$

31. $\begin{array}{r}469\\ \underline{15}\end{array}$

32. $\begin{array}{r}317\\ \underline{19}\end{array}$

33. $\begin{array}{r}400\\ \underline{93}\end{array}$

34. $\begin{array}{r}700\\ \underline{26}\end{array}$

35. $\begin{array}{r}321\\ \underline{20}\end{array}$

36. $\begin{array}{r}167\\ \underline{40}\end{array}$

37. $\begin{array}{r}215\\ \underline{300}\end{array}$

38. $\begin{array}{r}195\\ \underline{400}\end{array}$

39. $\begin{array}{r}507\\ \underline{84}\end{array}$

40. $\begin{array}{r}260\\ \underline{308}\end{array}$

C

41. ____________

42. ____________

43. ____________

44. ____________

45. ____________

46. ____________

47. ____________

48. ____________

49. ____________

41. $\begin{array}{r}837\\ \underline{49}\end{array}$

42. $\begin{array}{r}684\\ \underline{94}\end{array}$

43. $\begin{array}{r}797\\ \underline{83}\end{array}$

44. $\begin{array}{r}772\\ \underline{235}\end{array}$

45. $\begin{array}{r}585\\ \underline{709}\end{array}$

46. $\begin{array}{r}399\\ \underline{687}\end{array}$

47. 91(2006)

48. 87(4007)

49. (38)(4656)

NAME ____________________

CLASS ______________ SECTION ______________ DATE ______________

ANSWERS

50. ____________

51. ____________

52. ____________

53. ____________

54. ____________

55. ____________

56. ____________

57. ____________

58. ____________

59. ____________

50. $46 \cdot 23008$

51. (309)(7056)

52. 908(4009)

53. (3148)(798)

54. (8512)(651)

55. (407)(5048)

56. (305)(7093)

57. (50,009)(364)

58. (72)(17)(25)

59. (35)(9)(210)

D

60. ____________

60. The Good Food Grocery Store ordered 315 cases of pears. If each case cost 20 dollars, what was the total cost of the pears?

61. ____________

61. Ellie ordered three hundred twenty-two 8-packs of Hot'n Cola for sale at her store. How many bottles will there be?

ANSWERS

62. ______________________

63. ______________________

64. ______________________

65. ______________________

66. ______________________

67. ______________________

68. ______________________

69. ______________________

70. ______________________

62. A family car averages 26 miles to the gallon. The gas tank has a capacity of 16 gallons. How far can the family go on a full tank of gas?

63. What is the total number of watts, to the nearest hundred, used by 32 fluorescent lamps that use 40 watts each?

64. A brick layer lays 53 concrete blocks per hour. How many blocks can he lay in 5 hours?

65. An automobile coil has 53 layers of wire wound on it, with 42 turns per layer. How many turns of wire are on the coil?

66. Each assembly worker at Exact Electronics earns $240 per week. Find the total amount paid 86 assembly workers in one week.

67. One hundred twenty-six taxpayers each paid $536 in income taxes. To the nearest thousand dollars, find the total tax paid.

68. Janet ordered two hundred eighty-seven radios for her store. If she sells them for $85 each, how much money will they earn for her?

69. If Janet paid $58 for each radio (see Exercise 68), how much did the radios cost her?

70. How much profit did Janet make on these radios? (See Exercises 68 and 69)

Division of Whole Numbers

OBJECTIVES

11. Divide whole numbers.
12. Divide whole numbers when the quotient is not a whole number.

APPLICATION

The Belgium Bulb Company packs 6 hyacinth bulbs per package for shipment to the United States. How many packages can be made from 8,724 bulbs?

VOCABULARY

In $8 \div 2 = 4$, 8 is called the *dividend,* 2 the *divisor,* and 4 the *quotient.* A division problem is written:

$8 \div 2$ or $2\overline{)8}$ or $\frac{8}{2}$

The expression $75 \div 8$ does not have a whole-number quotient. It can be written $8\overline{)75} = 9$ remainder 3. In this case, 9 is commonly called the *quotient* (actually, 9 is only a partial quotient), and 3 is called the *remainder.*

HOW AND WHY

To divide $96 \div 24 = ?$ (read 96 divided by 24) is to ask what number times the second number equals the first:

$\underbrace{96 \div 24 = ?}_{\text{QUOTIENT}}$ asks $\underbrace{24 \times ? = 96}_{\text{MISSING FACTOR}}$

In $96 \div 24 = ?$, we can find the missing factor by repeatedly subtracting 24 from 96:

$$\begin{array}{r} 96 \\ \underline{24} \\ 72 \\ \underline{24} \\ 48 \\ \underline{24} \\ 24 \\ \underline{24} \\ 0 \end{array}$$

Four subtractions, so $96 \div 24 = 4$.

More quickly, we can guess the number of 24's and subtract from 96:

$$\begin{array}{r|l} 24\overline{)96} & \\ \underline{48} & 2 \text{ twenty-fours} \\ 48 & \\ \underline{48} & \underline{2} \text{ twenty-fours} \\ 0 & 4 \end{array} \quad \text{or} \quad \begin{array}{r|l} 24\overline{)96} & \\ \underline{72} & 3 \text{ twenty-fours} \\ 24 & \\ \underline{24} & \underline{1} \text{ twenty-four} \\ 0 & 4 \end{array}$$

$$\text{or} \quad \begin{array}{r|l} 24\overline{)96} & \\ \underline{96} & \underline{4} \text{ twenty-fours} \\ 0 & 4 \end{array}$$

In each case, $96 \div 24 = 4$.

This process works regardless of the size of the numbers. If the divisor is considerably smaller than the dividend, you will want to guess a rather large number.

36)7308			
3600	100		
3708			
3600	100	or	7308 ÷ 36 = 203
108			
108	3		
0	203		

All division problems can be done by this method. However, the process can be shortened by finding the number of groups, starting with the largest place value on the left, in the dividend, and then working toward the right. Study the following example. Note that the answer is written above the problem for convenience.

23)17135

We will work from left to right. Notice that 23 does not divide 1, and 23 does not divide 17.

745
23)17135
161
103
92
115
115
0

However, 23 will divide 171 seven times (remember that 171 is really 17,100, so 23 divides it 700 times). Place the 7 in the hundreds column above the 1.
Now multiply 7 times 23: (7)(23) = 161. Remember that 161 is really 16,100. We do not show the zeros because the placement of the 161 under the 171 keeps the place values lined up. Subtract. Bring down the next digit (3).

Now, 23 divides 103 four times (remember that 103 is really 1030, so 23 divides it 40 times), so a 4 goes in the tens column in the answer.
Now multiply 4 times 23, (4)(23) = 92, and subtract from 103. Bring down the next digit (5).

Finally, 23 divides 115 five times, so a 5 goes in the ones column in the answer. Now multiply 5 times 23, (5)(23) = 115, and subtract. Since the difference is zero, the division is complete.

Check:

745
23
2235
1490
17135

Check by multiplying the answer times the divisor.

The answer (quotient) is 745.

Not all division problems "come out even" (have a zero remainder). In

3
12)41
36
5

we see that 41 contains 3 twelves and 5 toward the next group of twelve. The answer is written as 3 remainder 5. The word "remainder" is abbreviated "R," and then the result is:

3 R 5

A check can be made by taking (12)(3) and adding the remainder:

$$\begin{array}{r} 12 \\ \underline{3} \\ 36 \\ \underline{5} \\ 41 \end{array}$$

Division by Zero

Recall that $45 \div 0 = ?$ asks what number times 0 is 45: $0 \times ? = 45$. By the zero-product property we know that $0 \times ? = 0$, so it cannot equal 45.

CAUTION

Division by zero is not defined. **It is an operation that cannot be performed.**

MODEL PROBLEM-SOLVING

Examples

Divide:

$$\begin{array}{r} 601 \\ 7\overline{)4207} \\ \underline{42} \\ 0 \\ \underline{0} \\ 7 \\ \underline{7} \\ 0 \end{array}$$

a. (division shown above)

Strategy

7 divides 42 six times.
7 divides 0 zero times.
7 divides 7 one time.
The answers are written above the dividend and in the correct column to give the right place value.

CAUTION

A zero must be placed above the tens digit if the place values of the 6 and 1 are to be correct.

Check:
(7)(601) = 4207

Check by multiplication.

Warm Up a. Divide: $9\overline{)5238}$

ANSWERS TO WARM UPS (1.6)

a. 582

b.
```
       806
   25)20150
      200
        15
         0
       150
       150
         0
```

25 does not divide 2 or 20 but divides 201 eight times: (8)(25) = 200. Write the 8 above the one (hundreds column) in the answer. Subtract and bring down the next digit, 5.

25 does not divide 15, so place a 0 in the tens column of the answer (above the 5). Multiply and subtract. Bring down the next digit, 0.

25 divides 150 six times: (6)(25) = 150. Place the six above the 0 (ones column) in the answer. Multiply and subtract.

Warm Up b. Divide: 36)18252

c.
```
          56
   365)20440
       1825
        2190
        2190
           0
```

365 does not divide 2.
365 does not divide 20.
365 does not divide 204.
365 divides 2044 five times: (5)(365) = 1825.
365 divides 2190 six times: (6)(365) = 2190.

Warm Up c. Divide: 453)35334

d.
```
        4 R 12
   27)120
      108
       12
```

Check:

```
    27
     4
   108
    12
   120
```

12 is a remainder because it does not contain a single group of 27.

To check, multiply the divisor by the partial quotient and add the remainder.

Warm Up d. Divide: 39)3049

ANSWERS TO WARM UPS (1.6) **b.** 507 **c.** 78 **d.** 78 R 7

e.

```
      30 R 47
  55)1697
     1650
       47
```

Check:

```
   55
   30
   00
 165
 1650
   47
 1697
```

Warm Up e. Divide: 17)45931

f. Calculator example

56)47432

ENTER	DISPLAY	
47432	47432.	**Enter 47432.**
÷	47432.	**Press the "÷" key.**
56	56.	**Enter 56.**
=	847.	**Press the "=" key.**

The quotient is 847.

Warm Up f. Divide: 78)7488

g. Calculator example

37)6483

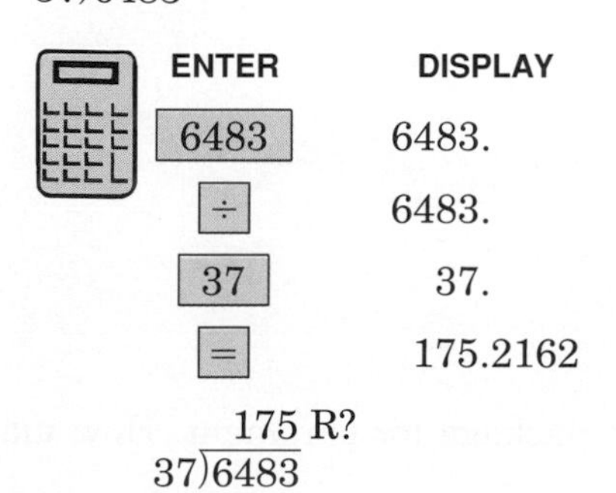


```
  175 R?
37)6483
```

The partial quotient is the whole number 175.

ANSWERS TO WARM UPS (1.6) **e.** 2701 R 14 **f.** 96

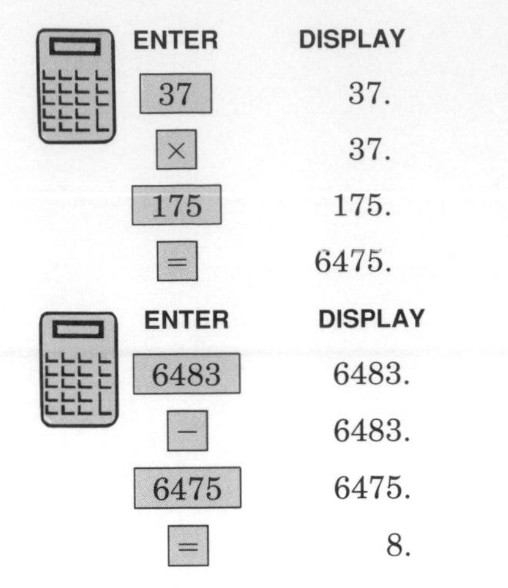

ENTER	DISPLAY
37	37.
×	37.
175	175.
=	6475.

To find the remainder, first find the product (175)(37).

ENTER	DISPLAY
6483	6483.
−	6483.
6475	6475.
=	8.

Now subtract this product from the original dividend to get the remainder.

The answer is 175 R 8.

Warm Up g. Divide: $306\overline{)43947}$

APPLICATION SOLUTION

h. The Belgium Bulb Company packs 6 hyacinth bulbs per package for shipment to the United States. How many packages can be made from 8724 bulbs?

```
  1454
6)8724
  6
  27
  24
   32
   30
    24
    24
     0
```

To find the number of packages of 6 bulbs each that can be made from 8724 bulbs, divide 8724 by 6.

Check:

```
1454
   6
----
8724
```

The Belgium Bulb Company can make 1454 packages.

Warm Up h. The Belgium Bulb Company packs 18 daffodil bulbs per package for shipment. How many packages can be made from 51,750 bulbs?

ANSWERS TO WARM UPS (1.6) **g.** 143 R 189 **h.** 2875 packages

NAME ______________________________

CLASS ______________ SECTION ______________ DATE ______________

Exercises 1.6

A

ANSWERS

1. ______
2. ______
3. ______
4. ______
5. ______
6. ______
7. ______
8. ______
9. ______
10. ______
11. ______
12. ______
13. ______
14. ______
15. ______
16. ______
17. ______
18. ______
19. ______
20. ______

Divide:

1. $7\overline{)63}$ **2.** $8\overline{)56}$ **3.** $9\overline{)54}$ **4.** $4\overline{)32}$

5. $15\overline{)15}$ **6.** $25\overline{)75}$ **7.** $5\overline{)255}$ **8.** $6\overline{)126}$

9. $6\overline{)426}$ **10.** $5\overline{)105}$ **11.** $15\overline{)600}$ **12.** $25\overline{)4000}$

13. $21\overline{)54}$ **14.** $17\overline{)40}$ **15.** $20\overline{)308}$ **16.** $30\overline{)519}$

17. $128 \div 8$ **18.** $222 \div 6$ **19.** $560 \div 16$ **20.** $486 \div 18$

B

21. ______
22. ______
23. ______

21. $9\overline{)1008}$ **22.** $8\overline{)736}$ **23.** $9\overline{)2349}$

ANSWERS

24. ______

25. ______

26. ______

27. ______

28. ______

29. ______

30. ______

31. ______

32. ______

33. ______

34. ______

35. ______

36. ______

37. ______

38. ______

39. ______

40. ______

41. ______

42. ______

43. ______

44. ______

45. ______

46. ______

24. $7\overline{)2205}$

25. $5\overline{)15095}$

26. $7\overline{)14035}$

27. $1288 \div 8$

28. $936 \div 24$

29. $768 \div 32$

30. $632 \div 79$

31. $(58)(?) = 4292$

32. $(?)(47) = 1551$

33. $7015 \div 23$

34. $9856 \div 32$

35. $25\overline{)319}$

36. $21\overline{)666}$

37. $25\overline{)6978}$

38. $33\overline{)99165}$

39. $14\overline{)29302}$

40. $68\overline{)2614}$

41. $76\overline{)4142}$

42. $83\overline{)14721}$

43. $51\overline{)30957}$

44. $79\overline{)32311}$

45. $52\overline{)4108}$

46. $48\overline{)54816}$

NAME ____________________

CLASS ____________ SECTION ____________ DATE ____________

C

ANSWERS

47. ____________

48. ____________

49. ____________

50. ____________

51. ____________

52. ____________

53. ____________

54. ____________

55. ____________

56. ____________

57. ____________

58. ____________

59. ____________

60. ____________

47. $146\overline{)31244}$ **48.** $126\overline{)30870}$ **49.** $235\overline{)21385}$ **50.** $325\overline{)28275}$

51. $101\overline{)69003}$ **52.** $123\overline{)139005}$ **53.** $257\overline{)49786}$ **54.** $108\overline{)46215}$

55. $214\overline{)46812}$ **56.** $408\overline{)126075}$ **57.** $507\overline{)104949}$ **58.** $903\overline{)93912}$

59. $241{,}110 \div 342$

60. $150{,}176 \div 247$

D

61. ____________

61. The Belgium Bulb Company has 124,380 daffodil bulbs. If they pack them for sale in packages of 12, how many packages can they make?

62. ____________

62. The Belgium Bulb Company has 607,704 tulip bulbs. If they are packed in packages of 8 bulbs, how many packages will they have?

ANSWERS

63. ______

64. ______

65. ______

66. ______

67. ______

68. ______

69. ______

70. ______

63. If a contractor can build a house in 27 days, how many houses can be completed in 297 days?

64. The Little Red Shed Company can build an 8 ft by 10 ft garden shed in 14 hours. How many sheds can the company make in 210 hours?

65. The estate of Ken Barker totaled $27,835. It is to be shared equally by his five nephews. How much money will each nephew receive?

66. Eight people are to share equally in the sale of a business. If the selling price is $16,280, how much did each person receive?

67. The Nippon Electronics firm assembles radios for export to the United States. Each radio is constructed using 12 resistors. How many radios can be assembled using the 23,469 resistors that they have in stock? How many resistors will not be used?

68. The Nippon Electronics firm in Exercise 67 assembles a second model of radio that contains 23 resistors. How many of these radios could be assembled if the company has 23,469 resistors in stock? How many resistors would not be used?

69. If it takes 33 bales of alfalfa hay to make a ton, how many tons and how many extra bales are there in a field that yields 3,691 bales?

70. If it takes 42 bales of grass hay to make a ton, how many tons and how many extra bales are there in a field that yields 3,691 bales?

▲▲ *Getting Ready for Algebra*

In Section 1.4 the equations involved the inverse operations addition and subtraction. Multiplication and division are also inverse operations. We can use this idea to solve equations containing those operations.

For example, if 4 is multiplied by 2, $4 \cdot 2 = 8$, the product is 8. If the product, 8, is divided by 2, $8 \div 2$, the result is 4, the original number. In the same manner, if 12 is divided by 3, $12 \div 3 = 4$, the quotient is 4. If the quotient, 4, is multiplied by 3, $4 \cdot 3$, the result is 12, the original number. We use this idea to solve equations in which the variable is either multiplied or divided by a number.

When a variable is multiplied or divided by a number, the multiplication symbols ($\cdot$ or $\times$) and the division symbol ($\div$) normally are not written. We write $3x$ for "three times x" and $\frac{x}{3}$ for x divided by 3.

Consider the following:

$$3x = 9$$

$$\frac{3x}{3} = \frac{9}{3}$$ **Division will eliminate multiplication.**

or $x = 3$

If x in the original equation is replaced by 3, we have:

$$3x = 9$$
$$3 \cdot 3 = 9$$
$9 = 9$, which is a true statement.

Therefore, the solution is 3.

If the variable is divided by a number:

$$\frac{x}{5} = 20$$

$$5 \cdot \frac{x}{5} = 5 \cdot 20$$ **Multiplication will eliminate division.**

Thus, $x = 100$

If x in the original equation is replaced by 100, we have:

$$\frac{100}{5} = 20$$

$20 = 20$, which is a true statement.

Therefore, the solution is 100.

PROCEDURE

If a number is multiplied by a variable, in order to eliminate the multiplication, do the inverse; that is, divide both sides of the equation by that number. The result is an equation that can be solved by inspection. If a variable is divided by a number, in order to eliminate the division, do the inverse; multiply both sides of the equation by that number. The result is an equation that can be solved by inspection.

▲▲ MODEL PROBLEM-SOLVING

Examples

a. Solve:

$2x = 12$

$\frac{2x}{2} = \frac{12}{2}$

$x = 6$

Check:

$2 \cdot 6 = 12$
$12 = 12$

Strategy

Since the variable is multiplied by 2, eliminate the multiplication by dividing both sides of the equation by 2.

Simplify both sides.

Substitute 6 for x in the original equation. We get a true statement, so 6 is the solution.

Warm Up a. Solve: $3y = 15$

b. Solve:

$\frac{x}{5} = 4$

$5 \cdot \frac{x}{5} = 4 \cdot 5$

$x = 20$

Check:

$\frac{20}{5} = 4$

$4 = 4$

The variable is divided by 5, so eliminate the division by multiplying both sides of the equation by 5.

Simplify both sides.

Substitute 20 for x in the original equation. We get a true statement, so 20 is the solution.

Warm Up b. Solve: $\frac{a}{6} = 7$

a. $y = 5$ **b.** $a = 42$

c. Solve:

$\frac{b}{2} = 9$ — **Eliminate the division by multiplying both sides of the equation by 2.**

$\frac{b}{2} \cdot 2 = 9 \cdot 2$

$b = 18$ — **Simplify both sides.**

Check:

$\frac{18}{2} = 9$ — **Substitute 18 for *b* in the original equation. We get a true statement, so 18 is the solution.**

$9 = 9$

Warm Up c. Solve: $\frac{c}{3} = 12$

d. Solve:

$3y = 12$ — **Eliminate the multiplication by dividing both sides of the equation by 3.**

$\frac{3y}{3} = \frac{12}{3}$

$y = 4$ — **Simplify both sides.**

Check:

$3 \cdot 4 = 12$ — **Substitute 4 for *y* in the original equation. We get a true statement, so 4 is the solution.**

$12 = 12$

Warm Up d. $5z = 35$

GETTING READY FOR ALGEBRA

c. $c = 36$ **d.** $z = 7$

NAME ______________________

CLASS ______________ SECTION ______________ DATE ______________

▲▲ EXERCISES

ANSWERS

1. ______
2. ______
3. ______
4. ______
5. ______
6. ______
7. ______
8. ______
9. ______
10. ______
11. ______
12. ______
13. ______
14. ______
15. ______
16. ______
17. ______
18. ______
19. ______
20. ______

Solve:

1. $3x = 15$ **2.** $\frac{z}{4} = 5$ **3.** $\frac{c}{3} = 6$

4. $5x = 30$ **5.** $12x = 48$ **6.** $\frac{y}{8} = 12$

7. $\frac{b}{8} = 15$ **8.** $15a = 135$ **9.** $12x = 144$

10. $\frac{x}{14} = 12$ **11.** $\frac{y}{13} = 24$ **12.** $23c = 184$

13. $27x = 648$ **14.** $\frac{a}{32} = 1536$ **15.** $\frac{b}{29} = 1566$

16. $63z = 2457$ **17.** $80 = 16x$ **18.** $288 = 9y$

19. $71 = \frac{w}{18}$ **20.** $57 = \frac{c}{23}$

1.7 Exponents and Powers of Ten

OBJECTIVES

13. Find the value of an expression written with an exponent.
14. Multiply a whole number by a power of ten.
15. Divide a whole number by a power of ten.

APPLICATION

A recent fund-raising campaign raised an average of \$123 per donor. How much was raised if there were 10,000 (10^4) donors?

VOCABULARY

Whole-number *exponents* greater than one show how many times a number is to be used as a factor:

EXPONENTS

$8 = 2 \cdot 2 \cdot 2 = 2^3$ $\qquad$ $(10)(10)(10)(10)(10) = 10^5$

A *power of ten* is the value obtained when ten is written with an exponent. The expression 10^5 is read "ten to the fifth power." Exponents of 2 and 3 are often read "squared" and "cubed," respectively. The expression 7^2 is read "seven squared," and 5^3 is read "five cubed."

HOW AND WHY

Whole-number exponents greater than one are used to write repeated multiplications in shorter form. For example, $3^4 = 3 \cdot 3 \cdot 3 \cdot 3$. The value of 3^4 is 81 since $3 \cdot 3 \cdot 3 \cdot 3 = 81$, and 81 is sometimes referred to as the fourth power of three:

EXPONENT ↓

BASE → $3^4 = 81$ ← VALUE

If one is used as the exponent, the number named is equal to the base. That is, $7^1 = 7$. If zero is used as the exponent, the number named is one (unless the base is zero). That is, $7^0 = 1$.

PROCEDURE

To find the value of an expression with an exponent:

1. **If the exponent is 0 and the base number is not zero, the value is 1.**
2. **If the exponent is 1, the value is the base number.**
3. **If the exponent is larger than 1, use the base as a factor as many times as shown by the exponent.**

It is particularly easy to multiply a whole number by 10 or a power of 10. When a number represented by a single digit is multiplied by 10, the product is that single digit with a zero written on the right:

$5 \times 10 = 50 \qquad 7 \times 10 = 70 \qquad 3 \times 10 = 30$

If a whole number larger than 9 is multiplied by 10, the place value of every digit becomes ten times larger:

$$
\begin{array}{rll}
24 \times 10 = & 2 \text{ tens} + 4 \text{ ones} & \\
 & \underline{\qquad\qquad\qquad 10} & \\
 & 2 \text{ hundreds} + 4 \text{ tens} & \text{since ten} \times \text{ten} = \text{hundred} \\
 & & \text{and one} \times \text{ten} = \text{ten} \\
 = & 240 &
\end{array}
$$

So, to multiply by 10, we need merely write a zero on the right of the whole number. If a whole number is multiplied by 10 more than once, a zero is written on the right for each 10. So,

$24 \times 10^4 = 240{,}000$ (Four zeros were written on the right, one for each 10.)

Since division is the inverse of multiplication, dividing by ten will eliminate the last zero on the right of a whole number. So:

$240{,}000 \div 10 = 24{,}000$ (Eliminate the final zero on the right.)

If we divide by ten more than once, one zero is eliminated for each 10. So:

$240{,}000 \div 10^3 = 240$ (Eliminate three zeros.)

PROCEDURE

To multiply a whole number by a power of 10, write zeros to the right of the whole number. Write the same number of zeros as the exponent of 10.

To divide using a power of 10, eliminate zeros on the right of the whole number. Eliminate the same number of zeros as the exponent of 10.

We now have another method of writing a number in expanded form.

$$
\begin{aligned}
2{,}345 &= 2000 + 300 + 40 + 5 \\
&= 2 \text{ thousands} + 3 \text{ hundreds} + 4 \text{ tens} + 5 \text{ ones} \\
&= 2 \cdot 10^3 + 3 \cdot 10^2 + 4 \cdot 10^1 + 5 \cdot 10^0
\end{aligned}
$$

MODEL PROBLEM-SOLVING

Examples | **Strategy**

a. Find the value of 2^3.

$$
\begin{aligned}
2^3 &= 2 \cdot 2 \cdot 2 \\
&= 8
\end{aligned}
$$

The exponent tells us to use 2 as a factor three times.

Warm Up a. Find the value of 4^2.

ANSWERS TO WARM UPS (1.7)

a. 16

b. Find the value of 17^1.

$17^1 = 17$

If the exponent is 1, the value is the base number.

Warm Up b. Find the value of 9^0.

c. Find the value of 10^7.

$10^7 = (10)(10)(10)(10)(10)(10)(10)$
$= 10{,}000{,}000$

Ten is used as a factor seven times.

Warm Up c. Find the value of 3^5.

d. Find the product: $12{,}784 \times 10^5$.

$12{,}784 \times 10^5 = 1{,}278{,}400{,}000$

To multiply by a power of 10, place as many zeros on the right as the exponent of 10.

Warm Up d. Find the product: $1{,}699 \times 10^8$

e. Find the product: 346×10^2

$346 \times 10^2 = 34{,}600$

Warm Up e. Find the product: 57×10^4

ANSWERS TO WARM UPS (1.7)

b. 1 **c.** 243 **d.** 169,900,000,000
e. 570,000

f. Find the quotient: $\frac{957{,}000}{10^2}$

$\frac{957{,}000}{10^2} = 9570$

To divide by 10^2, eliminate two zeros on the right.

Warm Up f. Find the quotient: $\frac{1{,}860{,}000}{10^4}$

g. Find the quotient: $496{,}230{,}000 \div 10^4$

$496{,}230{,}000 \div 10^4 = 49{,}623$

To divide by 10^4, eliminate four zeros on the right.

Warm Up g. Find the quotient: $281{,}000 \div 10^2$.

h. Write 56,875 in expanded form using powers of ten:

$56{,}875 = 5 \cdot 10^4 + 6 \cdot 10^3 + 8 \cdot 10^2 + 7 \cdot 10^1 + 5 \cdot 10^0$

Multiply each digit by its place value and write the sum. Recall that $10^1 = 10$ and $10^0 = 1$.

Warm Up h. Write 169,536 in expanded form using powers of ten.

APPLICATION SOLUTION

i. A recent fund-raising campaign averaged $123 per donor. How much was raised if there were 10,000 (10^4) donors?

123×10^4
$123 \times 10^4 = 1{,}230{,}000$

Multiply the average donation by the number of donors. Place four zeros on the right.

The campaign raised $1,230,000.

Warm Up i. A survey of 100,000 (10^5) people indicated that they paid an average of $3,785 in federal taxes. What was the total paid in taxes? (Write in place value notation.)

ANSWERS TO WARM UPS (1.7)

f. 186 **g.** 2810

h. $1 \times 10^5 + 6 \times 10^4 + 9 \times 10^3 + 5 \times 10^2 + 3 \times 10^1 + 6 \times 10^0$ **i.** $378,500,000

NAME ______________________________

CLASS ______________ SECTION ______________ DATE ______________

Exercises 1.7

A

ANSWERS

1. ______
2. ______
3. ______
4. ______
5. ______
6. ______
7. ______
8. ______
9. ______
10. ______
11. ______
12. ______
13. ______
14. ______
15. ______
16. ______

Find the value of each expression.

1. 11^1 **2.** 19^1 **3.** 8^2 **4.** 7^2

5. 1^5 **6.** 1^{10} **7.** 9^0 **8.** 14^2

Multiply or divide as indicated.

9. 36×10^2 **10.** 420×10^1 **11.** 7×10^3 **12.** 13×10^0

13. $150 \div 10^1$ **14.** $25{,}000 \div 10^3$ **15.** $9700 \div 10^2$ **16.** $121 \div 10^0$

B

17. ______
18. ______
19. ______
20. ______
21. ______
22. ______
23. ______
24. ______

Find the value of each expression.

17. 2^4 **18.** 4^4 **19.** 5^3 **20.** 10^3

21. 18^2 **22.** 11^2 **23.** 24^0 **24.** 27^1

ANSWERS

25. ____________

26. ____________

27. ____________

28. ____________

29. ____________

30. ____________

31. ____________

32. ____________

Multiply or divide as indicated.

25. 853×10^4 **26.** 35×10^5 **27.** $100{,}000 \div 10^4$

28. $1{,}000{,}000 \div 10^3$ **29.** 1783×10^3 **30.** 2501×10^2

31. $\dfrac{200{,}100}{100}$ **32.** $\dfrac{800{,}400}{10^1}$

C

33. ____________

34. ____________

35. ____________

36. ____________

37. ____________

38. ____________

39. ____________

40. ____________

41. ____________

42. ____________

43. ____________

44. ____________

45. ____________

46. ____________

Find the value of each expression.

33. 3^7 **34.** 5^6 **35.** 2^9 **36.** 6^5

37. 11^4 **38.** 9^6 **39.** 7^4 **40.** 8^3

Multiply or divide as indicated.

41. 506×10^7 **42.** 1300×10^6 **43.** $\dfrac{800{,}000{,}000}{10^7}$

44. $1{,}505{,}000{,}000 \div 10^5$ **45.** 2123×10^0 **46.** 2123×10^5

NAME ______________________________

CLASS ______________ SECTION ______________ DATE ______________

ANSWERS

47. ______________

48. ______________

49. ______________

50. ______________

51. ______________

52. ______________

53. ______________

54. ______________

55. ______________

56. ______________

47. $200{,}100 \div 10^2$

48. $\frac{750{,}000{,}000{,}000}{10^8}$

Write the following in expanded form, using powers of ten.

49. 782

50. 180,000

51. 6835

52. 16,932

53. 70,961

54. 35,836

55. 16,003,930

56. 1,400,603

D

57. ______________

57. The budget of the local community college is approximately 44×10^6 dollars. Write this amount in place value form.

58. ______________

58. A congressional committee proposes to increase the national debt by 13×10^9 dollars. Write this amount in place value form.

59. ______________

59. A western state had a fiscal surplus of 16×10^7 dollars for a given year. Write this number of dollars in both place value and word form.

ANSWERS

60. ______________

61. ______________

62. ______________

60. The distance from Earth to the nearest star outside our solar system (Alpha Centauri) is approximately 255×10^{11} miles. Write this distance in place value form.

61. The Joneses deposited $2 on their son's first birthday, and they deposited double the amount of the prior deposit each birthday thereafter (first, 2; second, 2^2, third, 2^3, and so on). How much must they deposit on his fifteenth birthday?

62. A high roller in Atlantic City placed 9 consecutive bets at the "Twenty-one" table. The first bet was $3 and each succeeding bet tripled the one before. How much did she wager on the ninth bet?

1.8 Order of Operations (Problems with Two or More Operations)

OBJECTIVE

16. Perform any combination of operations on whole numbers.

APPLICATION

The Lend A Helping Hand Association is preparing two types of food baskets for distribution to the needy. The family pack is to contain 9 cans of vegetables and the elderly pack is to contain 4 cans of vegetables. How many cans of vegetables are needed for 125 family packs and 50 elderly packs?

VOCABULARY

No new words.

HOW AND WHY

Without a rule, it is possible to interpret $2 \cdot 3 + 4$ two ways:

$2 \cdot 3 + 4 = 6 + 4$ **Multiply first.**
$= 10$ **Then add.**

$2 \cdot 3 + 4 = 2 \cdot 7$ **Add first.**
$= 14$ **Then multiply.**

In order to decide which answer to use, it is agreed to multiply first. So:

$2 \cdot 3 + 4 = 10$

Other agreements are shown as follows:

$8 - 10 \div 2 = 8 - 10 \div 2$ **Divide first.**
$= 8 - 5$ **Then subtract.**
$= 3$

$(6 - 4) \cdot 5 = (6 - 4) \cdot 5$ **Subtract in parentheses first.**
$= 2 \cdot 5$ **Then multiply.**
$= 10$

$16 - 2^3 = 16 - 2^3$ **Do exponents first.**
$= 16 - 8$ **Then subtract.**
$= 8$

ORDER OF OPERATIONS

PROCEDURE

In an expression with more than one operation:

Step 1. PARENTHESES—Do the operations within grouping symbols first (parentheses, fraction bar, etc.), in the order given in steps 2, 3, and 4.
Step 2. EXPONENTS—Do the operations indicated by exponents.
Step 3. MULTIPLY and DIVIDE—Do only multiplication and division as they appear from left to right.
Step 4. ADD and SUBTRACT—Do addition and subtraction as they appear from left to right.

MODEL PROBLEM-SOLVING

Examples / **Strategy**

a. $7 \cdot 9 + 6 \cdot 2 = 7 \cdot 9 + 6 \cdot 2$
$= 63 + 12$ — **Multiply.**
$= 75$ — **Add.**

Warm Up a. Perform the indicated operations: $4 \cdot 3 + 6 \cdot 5$

b. $25 - 6 \div 3 + 8 \cdot 4 = 25 - 6 \div 3 + 8 \cdot 4$
$= 25 - 2 + 32$ — **Divide and multiply.**
$= 55$ — **Subtract and add.**

Warm Up b. Perform the indicated operations: $4 \cdot 18 - 9 - 3 + 6 \cdot 2$

c. $17 - 5 + 3^2 = 17 - 5 + 3^2$
$= 17 - 5 + 9$ — **Do exponents first.**
$= 21$ — **Subtract and add.**

Warm Up c. Perform the indicated operations: $4 \cdot 22 - 5^2 + 3 \cdot 4$

d. $5 \cdot 9 + 9 - 6(7 + 1) = 5 \cdot 9 + 9 - 6(7 + 1)$
$= 5 \cdot 9 + 9 - 6 \cdot 8$ — **Add in parentheses first.**
$= 45 + 9 - 48$ — **Multiply.**
$= 6$ — **Add and then subtract.**

Warm Up d. Perform the indicated operations: $24 - 6 + 6 - 3(5 - 3)$

e. $3 \cdot 4^3 - 8 \cdot 3^2 + 11 = 3 \cdot 4^3 - 8 \cdot 3^2 + 11$
$= 3 \cdot 64 - 8 \cdot 9 + 11$ — **Do exponents first.**
$= 192 - 72 + 11$ — **Multiply.**
$= 131$ — **Subtract and then add.**

Warm Up e. Perform the indicated operations: $5 \cdot 2^3 - 2 \cdot 4^2 + 25 - 7 \cdot 3$

ANSWERS TO WARM UPS (1.8)

a. 42 **b.** 72 **c.** 75
d. 18 **e.** 12

f. $(2^2 + 2 \cdot 3)^2 = (2^2 + 2 \cdot 3)^2$

$= (4 + 2 \cdot 3)^2$ — **Do what is in parentheses first, starting with the exponent.**

$= (4 + 6)^2$ — **Multiply in parentheses.**

$= (10)^2$ — **Add in parentheses.**

$= 100$ — **Do the exponent.**

Warm Up f. Perform the indicated operations: $(3^3 - 4 \cdot 2)^2 + 5^2$

g. $45 \div 9 - 2 + 28 \div 14 + 10 = 45 \div 9 - 2 + 28 \div 14 + 10$

$= 5 - 2 + 2 + 10$

$= 15$

Warm Up g. Perform the indicated operations: $64 \div 4 + 7 - 35 \div 5 + 2$

h. $8(3 \cdot 4 - 6) \div 6 - 3 = 8(3 \cdot 4 - 6) \div 6 - 3$

$= 8(12 - 6) \div 6 - 3$

$= 8(6) \div 6 - 3$

$= 48 \div 6 - 3$

$= 8 - 3$

$= 5$

Warm Up h. Perform the indicated operations: $6(5 \cdot 4 - 8 \cdot 2) \div 8 + 4$

i. Calculator example

$8 + 6 \cdot 5 \div 3 = ?$

If your calculator is a four-function, nonalgebraic calculator, do the addition last:

ENTER	DISPLAY
6	6.
×	6.
5	5.
÷	30.
3	3.
+	10.
8	8.
=	18.

The answer is 18.

ANSWERS TO WARM UPS (1.8) **f.** 386 **g.** 18 **h.** 7

If your calculator has algebraic logic (see Appendix IV):

ENTER	DISPLAY
8	8.
+	8.
6	6.
×	6.
5	5.
÷	30.
3	3.
=	18.

The answer is 18.

Warm Up i. Perform the indicated operations: $13 + 9 \cdot 20 \div 12$

APPLICATION SOLUTION

j. The Lend A Helping Hand Association is preparing two types of food baskets for distribution to the needy. The family pack is to contain 9 cans of vegetables and the elderly pack is to contain 4 cans of vegetables. How many cans of vegetables are needed for 125 family packs and 50 elderly packs?

$(125)(9) + (50)(4)$
$1125 + 200 = 1325$

To find the number of cans of vegetables needed for the packs, multiply the number of packs by the number of cans per pack. Then add the two amounts.

The Lend A Helping Hand Association needs 1325 cans of vegetables.

Warm Up j. The Fruit of the Month Club is preparing two types of boxes for shipment. Box A contains 6 apples and Box B contains 10 apples. How many apples are needed for 96 orders of Box A and 82 orders of Box B?

ANSWERS TO WARM UPS (1.8)

i. 28

j. 1396 apples

NAME ______________________

CLASS ____________ SECTION ____________ DATE ____________

Exercises 1.8

A

ANSWERS

1. ____________
2. ____________
3. ____________
4. ____________
5. ____________
6. ____________
7. ____________
8. ____________
9. ____________
10. ____________
11. ____________
12. ____________
13. ____________
14. ____________
15. ____________
16. ____________
17. ____________
18. ____________
19. ____________
20. ____________

Perform the indicated operations.

1. $4 \cdot 7 + 12$

2. $15 + 9 \cdot 2$

3. $12 \div 3 \cdot 2$

4. $4^2 + 7 \cdot 5$

5. $18 + 3 \div 3$

6. $16 \div 8 + 8$

7. $16 \div (8 + 8)$

8. $(14 - 2) \div 6$

9. $20 \div 5 \times 2$

10. $30 \div 3 \times 5$

11. $9 + 54 \div 6$

12. $(9 + 54) \div 7$

13. $20 - 8 \cdot 2$

14. $36 \div 6 + 3$

15. $28 + 4 \cdot 5 - 3$

16. $10 \div 5 \cdot 2$

17. $30 \cdot 6 \div 3$

18. $40 - 8 + 10 \cdot 2$

19. $3^2 + 4^3$

20. $5^3 - 2^5 + 6^2$

B

ANSWERS

21. ____________________

22. ____________________

23. ____________________

24. ____________________

25. ____________________

26. ____________________

27. ____________________

28. ____________________

29. ____________________

30. ____________________

31. ____________________

32. ____________________

33. ____________________

34. ____________________

21. $14 \cdot 2 + 5 \cdot 3$

22. $12 \cdot 3 - 5 \cdot 4$

23. $4^3 - 9 \cdot 2 + 6 \cdot 3$

24. $6^2 + 44 \div 2 + 11$

25. $28 \div 4 + 3 - 1$

26. $(3 + 25) \div (7 - 3)$

27. $(3 + 25) \div 7 - 3$

28. $4(8) \div 2 + 5$

29. $68 \div 4 \cdot 2^3 + 5 \cdot 2$

30. $5^2 \times 15 \div 3 + 4(11)$

31. $60 \div 3 \cdot 4$

32. $44 \div 2 \cdot 11$

33. $56 \div 7(2) + 8 - 10$

34. $(5)(10) \div (2) - 6 + 8$

C

35. ____________________

36. ____________________

37. ____________________

38. ____________________

39. ____________________

40. ____________________

35. $12(4 + 2) - 24 \div 12$

36. $(5 \cdot 2 + 3) \cdot 2 - 10$

37. $8 \div 2 - 2 + 9 \cdot 2 - 8$

38. $120 \div (5 \cdot 5 - 5)$

39. $40 - 9 \div 3 - 15 \div 5 + 2$

40. $5(2 + 4) \div 3 \cdot 2 - 4 \cdot 2$

NAME ______________________________

CLASS ______________ SECTION ______________ DATE ______________

ANSWERS

41. ______________

42. ______________

43. ______________

44. ______________

45. ______________

46. ______________

47. ______________

48. ______________

49. ______________

50. ______________

41. $6(2^3 \cdot 3 - 4) \div 4 - 2$

42. $5(15 \div 3) + (4^2 \cdot 11 - 8)$

43. $5(7 - 2)^3 + 3^2$

44. $3(2 + 3)^2 - 8(3 - 1)^3$

45. What is the quotient of the sum of 15 and 51 and the difference of 57 and 46?

46. What is the difference of the product of 6 and 8 and the quotient of 143 and 11?

47. Add the product of 7 and 288 to the difference of 882 and 77.

48. Subtract the sum of 321 and 548 from the product of 54 and 26.

49. What is the product of the sum of 9 and 14 and the difference of 78 and 29?

50. Divide the product of 28 and 15 by the quotient of 126 and 3.

D

51. ______________

51. The Lend A Helping Hand Association is preparing two types of food baskets for the needy. The family pack is to contain 5 cans of fruit and the elderly pack is to contain 3 cans of fruit. How many total cans of fruit are needed for 125 family packs and 50 elderly packs?

ANSWERS

52. ____________

53. ____________

54. ____________

55. ____________

56. ____________

57. ____________

58. ____________

52. As an added feature for the Easter holiday, the Lend A Helping Hand Association put a dozen decorated eggs in the family pack and four decorated eggs in the elderly pack. How many decorated eggs were needed if the association distributed 148 family packs and 74 elderly packs?

53. The Neat-n-Trim clothing store last week sold 24 suits at $178 each. They also sold 35 sport coats at $80 each. What was the total sale from the two items?

54. During a year-end sale the Neat-n-Trim clothing store hired two extra clerks. One was paid $8 per hour and the other one $6 per hour. What was the total additional payroll if each of the clerks worked 36 hours?

55. The OK Delivery Service agreed to deliver Christmas catalogs for a base fee of $500 plus $1 for every catalog delivered. The catalogs were shipped in boxes containing 15 catalogs. What was the delivery charge if 315 boxes of catalogs were delivered?

56. The OK Floral Company has 984 red roses to sell on an advertised special. The special is a bouquet of 8 roses for $11 plus a $2 delivery charge. If all of the roses were sold and delivered, what was the income from the sale?

57. Marie's Hair Salon ordered 62 bottles of shampoo at $2 each, 45 bottles of conditioner at $3 each, and 53 perms at $15 each. During this month they used 52 bottles of shampoo, 23 bottles of conditioner, and 27 perms. How much money in supplies is remaining at the end of this month.

58. Knight's Camera Shop ordered 35 cameras that cost $215 each and 21 lenses at $132 each. Then they sold the cameras for $325 and lenses for $185. If they sold 14 cameras and 8 lenses, how much more money will they need to make in order to pay for their initial purchases?

▲▲ Getting Ready for Algebra

Recall that we have solved equations involving only one operation. Let's look at some equations that involve two operations.

To solve $x - 4 = 2$, we added 4 to both sides of the equation. To solve $3x = 12$, we divided both sides of the equation by 3. The following equation requires both steps.

Solve: $3x - 4 = 2$

$3x - 4 + 4 = 2 + 4$	**First eliminate the subtraction by adding 4 to both sides.**
$3x = 6$	**Simplify both sides.**
$\frac{3x}{3} = \frac{6}{3}$	**Eliminate the multiplication. Divide both sides by 3.**
$x = 2$	**Simplify.**

Check:

$3x - 4 = 2$	
$3 \cdot 2 - 4 = 2$	**Replace x by 2.**
$6 - 4 = 2$	**Multiply.**
$2 = 2$	**Subtract.**

Thus, if in the original equation x is replaced by 2, the statement is true. So the solution is 2.

Solve: $2x + 3 = 9$

$2x + 3 - 3 = 9 - 3$	**First, eliminate the addition by subtracting 3 from both sides.**
$2x = 6$	**Simplify.**
$\frac{2x}{2} = \frac{6}{2}$	**Eliminate the multiplication by dividing both sides by 2.**
$x = 3$	**Simplify.**

Check:

$2 \cdot 3 + 3 = 9$	**Replace x by 3.**
$6 + 3 = 9$	**Multiply.**
$9 = 9$	**Add.**

Thus, if in the original equation x is replaced by 3, the statement is true. So the solution is 3.

PROCEDURE

To solve an equation in which more than one operation is involved:

1. **Eliminate either the addition or subtraction.**
2. **Eliminate either the multiplication or division.**
3. **Check.**

Note that the operations are eliminated in the opposite order in which they are performed. Eliminate addition and subtraction first, then eliminate multiplication and division.

▲▲ MODEL PROBLEM-SOLVING

Examples — **Strategy**

a. Solve:

$2x - 7 = 9$

$2x - 7 + 7 = 9 + 7$ — **Eliminate the subtraction first. Add 7 to both sides.**

$2x = 16$ — **Simplify both sides.**

$\frac{2x}{2} = \frac{16}{2}$ — **Eliminate the multiplication. Divide both sides by 2.**

$x = 8$ — **Simplify.**

Check:

$2 \cdot 8 - 7 = 9$ — **Substitute 8 for x in the original equation.**

$16 - 7 = 9$ — **The multiplication is done first.**

$9 = 9$ — **Do the subtraction.**

The solution is 8. — **The statement is true.**

Warm Up a. $5x - 8 = 17$

b. Solve:

$\frac{y}{5} + 4 = 5$ — **Eliminate the addition first.**

$\frac{y}{5} + 4 - 4 = 5 - 4$ — **Subtract 4 from both sides.**

$\frac{y}{5} = 1$ — **Simplify both sides.**

$\frac{y}{5} \cdot 5 = 1 \cdot 5$ — **Eliminate the division. Multiply both sides by 5.**

$y = 5$ — **Simplify.**

Check:

$5 \div 5 + 4 = 5$ — **Substitute 5 for y in the original equation.**

$1 + 4 = 5$ — **Do the division first.**

$5 = 5$ — **Do the addition.**

The solution is 5. — **The statement is true.**

a. $x = 5$

Warm Up b. $\frac{a}{3} + 8 = 12$

c. Solve:

$\frac{z}{2} - 6 = 4$	**Eliminate the subtraction first.**
$\frac{z}{2} - 6 + 6 = 4 + 6$	**Add 6 to both sides.**
$\frac{z}{2} = 10$	**Simplify.**
$\frac{z}{2} \cdot 2 = 10 \cdot 2$	**Eliminate the division. Multiply both sides by 2.**
$z = 20$	**Simplify.**

Check:

$20 \div 2 - 6 = 4$	**Substitute 20 for *z* in the original equation.**
$10 - 6 = 4$	**Do the division first.**
$4 = 4$	**Do the subtraction.**
The solution is 20.	**The statement is true.**

Warm Up c. $\frac{x}{5} - 8 = 7$

d. Solve:

$3b + 4 = 22$	**Eliminate the addition first.**
$3b + 4 - 4 = 22 - 4$	**Subtract 4 from both sides.**
$3b = 18$	**Simplify both sides.**
$\frac{3b}{3} = \frac{18}{3}$	**Eliminate the multiplication. Divide both sides by 3.**
$b = 6$	**Simplify both sides.**

Check:

$3 \cdot 6 + 4 = 22$	**Substitute 6 for *b* in the original equation.**
$18 + 4 = 22$	**Do the multiplication.**
$22 = 22$	**Do the addition.**
The solution is 6.	**The statement is true.**

Warm Up d. $4c + 9 = 25$

GETTING READY FOR ALGEBRA **b.** $a = 12$ **c.** $x = 75$ **d.** $c = 4$

NAME ______________________

CLASS ____________ SECTION ____________ DATE ____________

▲▲ EXERCISES

ANSWERS

1. ______
2. ______
3. ______
4. ______
5. ______
6. ______
7. ______
8. ______
9. ______
10. ______
11. ______
12. ______
13. ______
14. ______
15. ______
16. ______

Solve:

1. $4x - 12 = 8$

2. $\frac{a}{3} + 7 = 12$

3. $\frac{y}{2} - 9 = 3$

4. $31 = 5x + 6$

5. $25 = 4x + 9$

6. $\frac{a}{6} + 7 = 12$

7. $\frac{c}{8} + 14 = 25$

8. $12x - 8 = 28$

9. $9x + 24 = 78$

10. $5y + 36 = 151$

11. $12c - 56 = 88$

12. $2 = \frac{w}{15} - 45$

13. $54 = \frac{a}{32} + 29$

14. $\frac{x}{41} + 79 = 187$

15. $429 = 23b - 77$

16. $556 = 36c + 124$

1.9 Average

OBJECTIVE

17. Find the average of a group of whole numbers.

APPLICATION

In order to help Pete lose weight, the dietician had him record his caloric intake for a week. He recorded the following:

Monday	3165
Tuesday	1795
Wednesday	1500
Thursday	2615
Friday	1407
Saturday	1850
Sunday	1913

What was Pete's average caloric intake per day?

VOCABULARY

Given a group of whole numbers, find the sum of the numbers in the group. Then divide the sum by the total number of members in the group. The result will be the *average (arithmetic mean)* of all the numbers in the group. If we replace each member with the average, the sum remains unchanged. For instance, 5 is the average of 3, 7, and 5 because:

$5 + 5 + 5 = 3 + 7 + 5 = 15$

HOW AND WHY

The average or mean of a group of numbers is used in statistics. It is one of the ways used to measure central tendency. The average of 7, 8, and 9 is 8 since:

$(7 + 8 + 9) \div 3 = 24 \div 3 = 8$

The "central" number or average may not be one of the numbers in the group. The average of 7, 9, and 14 is:

$(7 + 9 + 14) \div 3 = 30 \div 3 = 10$

PROCEDURE

To find the average of a group of whole numbers:

1. **Add the numbers.**
2. **Divide the sum by the number of numbers.**

MODEL PROBLEM-SOLVING

Examples	Strategy

a. Find the average of 39 and 47.

$39 + 47$	**Add the numbers.**
$(39 + 47) \div 2$ $86 \div 2 = 43$	**Since there are two numbers, divide the sum by 2 to find the average.**

So the average is 43.

Warm Up a. Find the average of 49 and 83.

b. Find the average of 103, 98, and 123.

$103 + 98 + 123$	**Add the numbers.**
$(103 + 98 + 123) \div 3$ $324 \div 3 = 108$	**Divide by 3 (there are three numbers).**

So the average is 108.

Warm Up b. Find the average of 313, 129, and 500.

c. Find the average of 7, 40, 122, and 211.

$7 + 40 + 122 + 211$	**Add the numbers.**
$(7 + 40 + 122 + 211) \div 4$ $380 \div 4 = 95$	**Divide by 4.**

So the average is 95.

Warm Up c. Find the average of 9, 27, 46, 58, and 65.

d. Calculator example

Find the average of 345, 567, 824, and 960.

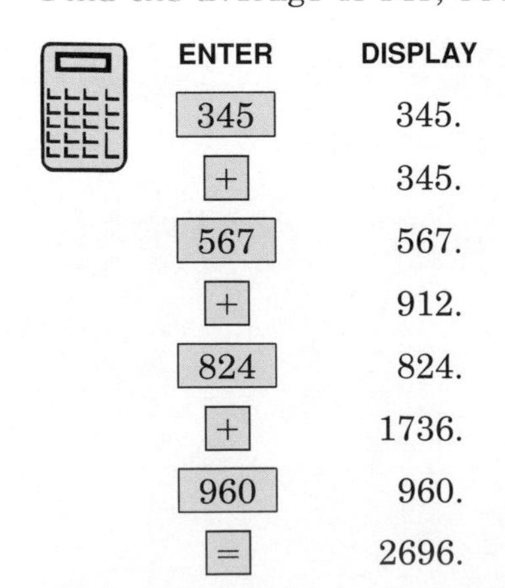

ENTER	DISPLAY
345	345.
+	345.
567	567.
+	912.
824	824.
+	1736.
960	960.
=	2696.

ANSWERS TO WARM UPS (1.9) **a.** 66 **b.** 314 **c.** 41

÷	2696.
4	4.
=	674.

The average is 674.

Warm Up d. Find the average of 917, 855, 1014, and 622.

APPLICATION SOLUTION

e. In order to help Pete lose weight, the dietician had him record his caloric intake for a week. He recorded the following:

Monday	3165
Tuesday	1795
Wednesday	1500
Thursday	2615
Friday	1407
Saturday	1850
Sunday	1913

What was Pete's average caloric intake per day?

To find the average caloric intake, add the calories and divide by 7.

```
 3165        2035
 1795     7)14245
 1500       14
 2615        2
 1407        0
 1850        24
 1913        21
14245         35
              35
               0
```

Pete's average intake per day was 2035 calories.

Warm Up e. The local dog food company shipped the following cases of dog food:

Monday	3059
Tuesday	2175
Wednesday	3755
Thursday	1851
Friday	2875

What was the average number of cases shipped per day?

ANSWERS TO WARM UPS (1.9) **d.** 852 **e.** 2743 cases

f. During the annual Salmon Fishing Derby 35 fish were entered. The weights of the fish were recorded as shown below.

NUMBER OF FISH	WEIGHT PER FISH
2	6 pounds
4	7 pounds
8	10 pounds
10	12 pounds
5	15 pounds
4	21 pounds
1	22 pounds
1	34 pounds

What was the average weight of a fish entered in the Derby?

$$\begin{aligned} 2(6) &= 12 \\ 4(7) &= 28 \\ 8(10) &= 80 \\ 10(12) &= 120 \\ 5(15) &= 75 \\ 4(21) &= 84 \\ 1(22) &= 22 \\ 1(34) &= \underline{34} \\ & 455 \end{aligned}$$

To find the average weight per fish, we must first find the total weight of all 35 fish. Since there are 2 fish that weigh 6 pounds, the total weight of the two fish is 2(6) = 12 pounds.

$$\begin{array}{r} 13 \\ 35\overline{)455} \\ \underline{35} \\ 105 \\ \underline{105} \end{array}$$

Once the total weight is found, divide by 35 to find the average.

So the average weight per fish was 13 pounds.

Warm Up f. In a class of 30 seniors the following weights were recorded on Health Day:

NUMBER OF STUDENTS	WEIGHT PER STUDENT
1	120
3	128
7	153
4	175
5	182
5	195
3	200
2	215

What is the average weight of a student in the class?

ANSWERS TO WARM UPS (1.9)

f. 173 pounds

NAME ______________________

CLASS ______________ SECTION ______________ DATE ______________

Exercises 1.9

A

ANSWERS

1. ______
2. ______
3. ______
4. ______
5. ______
6. ______
7. ______
8. ______
9. ______
10. ______
11. ______
12. ______
13. ______
14. ______
15. ______
16. ______
17. ______
18. ______

Find the average of these groups of numbers.

1. 2, 6 **2.** 6, 10 **3.** 5, 7 **4.** 5, 9

5. 2, 10 **6.** 14, 18 **7.** 9, 15, 18 **8.** 10, 12, 14

9. 8, 12, 16 **10.** 6, 9, 15 **11.** 6, 10, 14 **12.** 20, 25, 27

13. 2, 4, 6, 8 **14.** 4, 4, 8, 8 **15.** 1, 3, 4, 4 **16.** 1, 2, 3, 6

17. 8, 12, 24, 32 **18.** 9, 21, 17, 18, 45

B

19. ______
20. ______
21. ______
22. ______

19. 2, 3, 4, 7 **20.** 1, 3, 4, 6, 11

21. 12, 27, 36 **22.** 15, 45, 54

ANSWERS

23. ______
24. ______
25. ______
26. ______
27. ______
28. ______
29. ______
30. ______
31. ______
32. ______
33. ______
34. ______
35. ______
36. ______

23. 12, 16, 28, 40

24. 12, 35, 42, 63

25. 20, 30, 40, 50

26. 7, 8, 9, 10, 11

27. 2, 9, 11, 18, 25

28. 12, 5, 24, 48, 36

29. 19, 37, 42, 26

30. 16, 24, 36, 48

31. 15, 35, 65, 45, 25

32. 6, 9, 17, 26, 42, 74

33. 25, 36, 44, 82, 98

34. 23, 24, 24, 25, 29

35. 27, 126, 234, 273

36. 85, 136, 213, 130

C

37. ______
38. ______
39. ______
40. ______
41. ______
42. ______

37. 97, 101, 103, 119

38. 444, 460, 470

39. 278, 778, 510, 410

40. 223, 415, 111, 303

41. 264, 384, 196, 736, 685

42. 183, 526, 682, 589, 720

NAME ______________________________

CLASS ______________ SECTION ______________ DATE ______________

ANSWERS

43. ______________

44. ______________

43. 1214, 387, 463, 80, 1476

44. 3, 482, 7964, 62, 4683, 2184

D

45. ______________

45. Ms. Obese counted her caloric intake for one week prior to starting a diet. She reported the following intake:

Monday	4310
Tuesday	3800
Wednesday	3215
Thursday	4000
Friday	3800
Saturday	4700
Sunday	4903

What was her average caloric intake per day?

46. ______________

46. A service station sold the following numbers of gallons of gasoline:

Monday	2680
Tuesday	1976
Wednesday	1821
Thursday	1974
Friday	4214
Saturday	3862
Sunday	2940

What was the average number of gallons of gasoline sold each day?

47. ______________

47. A basketball player scored 22, 25, 16, 12, 18, 22, 36, and 9 points in eight games. What was her average score per game?

ANSWERS

48. ________

49. ________

50. ________

51. ________

52. ________

48. A football team had scores of 28, 13, 17, 26, 28, 21, 6, 20, 31, and 20 in ten games. What was their average score per game?

49. A consumer magazine had 15 makes of cars tested for gas mileage. The results are shown below:

NUMBER OF MAKES	GAS MILEAGE BASED ON 200 MILES
1	12 miles per gallon
2	18 miles per gallon
2	28 miles per gallon
4	31 miles per gallon
3	35 miles per gallon
2	38 miles per gallon
1	41 miles per gallon

What was the average gas mileage of the cars?

50. A home economist listed the costs of 12 brands of canned fruit drinks. The results are shown below:

NUMBER OF BRANDS	PRICE PER CAN
2	81 cents
1	80 cents
3	76 cents
3	68 cents
1	60 cents
2	53 cents

What is the average price per can?

51. A student received the following grades on math tests: 67, 92, 88, 93, and 75. What is the average grade?

52. A professional bowler bowled the following games: 282, 256, 272. What was the average score?

NAME ______

CLASS ______ SECTION ______ DATE ______

ANSWERS

1. ______

2. ______

3. ______

4. ______

5. ______

6. ______

7. ______

8. ______

9. ______

10. ______

11. ______

12. ______

13. ______

14. ______

CHAPTER 1 TRUE–FALSE CONCEPT REVIEW

Check your understanding of the language of basic mathematics. Tell whether each of the following statements is True (always true) or False (not always true).

1. All whole numbers can be written using nine digits.

2. In the number 6731, the digit "7" represents 700.

3. The word "and" is not used when writing the word names of whole numbers.

4. The symbols, $5 < 18$, can be read "five is greater than eighteen."

5. $1345 < 1344$

6. To the nearest thousand, 8498 rounds to 9000.

7. It is possible for the rounded value of a number to be equal to the original number.

8. The expanded form of a whole number shows the plus signs that are usually not written.

9. The sum of 60 and 4 is 604.

10. The process of "carrying" when doing an addition problem with pencil and paper is based on the place values of the numbers.

11. The product of 7 and 4 is 11.

12. It is possible to subtract 37 from 55 without "borrowing."

13. The number 4 is a factor of 56.

14. The multiplication sign is sometimes omitted when writing a multiplication problem.

ANSWERS

15. ____________

16. ____________

17. ____________

18. ____________

19. ____________

20. ____________

21. ____________

22. ____________

23. ____________

24. ____________

25. ____________

26. ____________

27. ____________

28. ____________

29. ____________

30. ____________

15. Whenever a number is multiplied by zero, the value remains unchanged.

16. There is more than one method for doing division problems.

17. In $92 \div 4 = 23$, the quotient is 92.

18. If a division exercise has a remainder then we know that there is no whole number quotient.

19. When zero is divided by any whole number from 33 to 78, the result is 0.

20. The result of zero divided by zero can either be 1 or 0.

21. The value of 8^2 is 16.

22. The value of 111^0 is 1.

23. The number 1 is a power of 10.

24. The product 340×10^2 is equal to 3400.

25. The quotient of 7000 and 100 is 70.

26. In the order of operations, exponents always take precedence over addition.

27. In the order of operations, multiplication always takes precedence over subtraction.

28. The value of $2^3 + 2^3$ is the same as the value of 2^4.

29. The average of three different numbers is smaller than the largest of the three numbers.

30. The word "mean" sometimes has the same meaning as "average."

NAME ________________________________

CLASS ____________ SECTION ____________ DATE ____________

CHAPTER 1 POST-TEST

ANSWERS

1. ____________
2. ____________
3. ____________
4. ____________
5. ____________
6. ____________
7. ____________
8. ____________
9. ____________
10. ____________
11. ____________
12. ____________
13. ____________
14. ____________
15. ____________

1. **(Section 1.3, Obj. 8)** Add: $206 + 3982 + 16 + 145 + 9 + 2731$

2. **(Section 1.2, Obj. 6)** Round to the nearest ten thousand: 693,856,639

3. **(Section 1.5, Obj. 10)** Multiply: (370)(48)

4. **(Section 1.9, Obj. 17)** Find the average of 2344, 461, and 1944.

5. **(Section 1.6, Obj. 11)** Divide: $85\overline{)6715}$

6. **(Section 1.6, Obj. 11)** Divide: $20{,}196 \div 66$

7. **(Section 1.1, Obj. 3)** Write the word name for 4,205.

8. **(Section 1.2, Obj. 6)** Round to the nearest hundred: 5642

9. **(Section 1.1, Obj. 2)** Write the place value name for three hundred nine thousand, nine hundred sixty-three.

10. **(Section 1.1, Obj. 3)** Write the word name for 120,355.

11. **(Section 1.7, Obj. 14)** Multiply: 13×10^2

12. **(Section 1.2, Obj. 5)** Write the place value name for $40{,}000 + 3000 + 600 + 80 + 1$.

13. **(Section 1.3, Obj. 8)** Add:
$$\begin{array}{r} 337 \\ 8 \\ 25 \\ 874 \\ \underline{2283} \end{array}$$

14. **(Section 1.6, Obj. 12)** Divide: $176\overline{)62{,}197}$

15. **(Section 1.4, Obj. 9)** Subtract:
$$\begin{array}{r} 8277 \\ \underline{3047} \end{array}$$

ANSWERS

16. ____________

17. ____________

18. ____________

19. ____________

20. ____________

21. ____________

22. ____________

23. ____________

24. ____________

25. ____________

26. ____________

27. ____________

28. ____________

29. ____________

30. ____________

16. **(Section 1.7, Obj. 13)** Find the value of 4^4.

17. **(Section 1.3, Obj. 8)** Add: $382 + 77 + 5280 + 9$

18. **(Section 1.2, Obj. 4)** Write 937 in expanded form.

19. **(Section 1.5, Obj. 10)** Multiply: (498)(976)

20. **(Section 1.7, Obj. 15)** Divide: $3{,}006{,}000{,}000 \div 10^3$

21. **(Section 1.6, Obj. 11)** Divide: $98\overline{)20{,}482}$

22. **(Section 1.4, Obj. 9)** Subtract: $19{,}864 - 2876$

23. **(Section 1.9, Obj. 17)** Find the average of 274, 682, 1924, 361, and 1294.

24. **(Section 1.8, Obj. 16)** Perform the indicated operations: $18 \div 3 + 3 \cdot 5 - 1$

25. **(Section 1.2, Obj. 7)** State whether the following is true or false: $299 > 312$

26. **(Section 1.1, Obj. 1)** Write the place value of the digit 4 in 34,388.

27. **(Section 1.6, Obj. 12)** Divide: $234\overline{)20{,}845}$

28. Pete owes $4015 on his beach lot. After making payments of $212, $125, and $305, how much does he still owe on the lot?

29. A farmer delivers 87 tote boxes of cucumbers. If each box contains 874 pounds of cucumbers, how many pounds of cucumbers did the farmer deliver? (Round to the nearest thousand pounds.)

30. A walnut grower sold 11,580 50-pound sacks of nuts. Of these, 5,450 sacks were sold at $15 per sack and the rest were sold at $13 per sack. What was the total income from the two sales?

CHAPTER 2

Primes and Multiples

NAME ______________________

CLASS ____________ SECTION ____________ DATE ____________

ANSWERS

1. ____________

2. ____________

3. ____________

4. ____________

5. ____________

6. ____________

7. ____________

8. ____________

9. ____________

10. ____________

11. ____________

12. ____________

13. ____________

14. ____________

15. ____________

CHAPTER 2 PRE-TEST

The problems in the following pre-test are a sample of the material in the chapter. You may already know how to work some of these. If so, this will allow you to spend less time on those parts. As a result, you will have more time to give to the sections that gave you difficulty. Please work the following problems. The answers are in the back of the text.

1. **(Section 2.1, Obj. 18)** List the first five multiples of 6.

2. **(Section 2.1, Obj. 19)** Is 299 a multiple of 13?

3. **(Section 2.1, Obj. 19)** Is 106 a multiple of 3?

4. **(Section 2.2, Obj. 20)** List all the factors of 68.

5. **(Section 2.2, Obj. 21)** Write 42 as the product of two factors in as many ways as possible.

6. **(Section 2.3, Obj. 22)** Is 248 divisible by 3?

7. **(Section 2.3, Obj. 22)** Is 330 divisible by 5?

8. **(Section 2.4, Obj. 23)** Is 59 a prime number or a composite number?

9. **(Section 2.4, Obj. 23)** Is 91 a prime number or a composite number?

10. **(Section 2.5, Obj. 24)** Write the prime factorization of 126.

11. **(Section 2.5, Obj. 24)** Write the prime factorization of 407.

12. **(Section 2.5, Obj. 24)** Write the prime factorization of 490.

13. **(Section 2.6, Obj. 25)** Write the least common multiple (LCM) of 18 and 24.

14. **(Section 2.6, Obj. 25)** Write the least common multiple (LCM) of 25, 30, and 40.

15. What is the smallest natural number that 14, 21, and 30 will divide evenly?

Multiples

OBJECTIVES

18. List some of the multiples of any given whole number.
19. Determine whether a whole number is a multiple of a given number.

APPLICATION

A common classification of whole numbers is into odd and even numbers. Even numbers are multiples of two. Mathematics instructors often assign the "even-numbered problems." If the instructor assigns the problems from 1 to 50 that are multiples of 3, which problems should be worked?

VOCABULARY

A *multiple* of 6 is the product of 6 and a natural number. So:

$6 \times$ (a natural number) $=$ (a multiple of 6)

For instance, 78 is a multiple of 6 because:

$6 \times 13 = 78$

Every multiple of 6 is divisible by 6.

HOW AND WHY

To list the multiples of 6, we need only multiply 6 by each natural number.

Natural numbers:	1	2	3	4	5	6 ... 11 ... 15 ... 40
Multiples of 6:	6	12	18	24	30	36 ... 66 ... 90 ... 240

Such a list can be continued forever.

Is 852 a multiple of 6? To find out, divide by 6.

$$\begin{array}{r} 142 \\ 6\overline{)852} \\ \underline{600} \\ 252 \\ \underline{240} \\ 12 \\ \underline{12} \\ 0 \end{array}$$

The answer is yes, because 6(142) = 852.

PROCEDURE

To find:

1. **The first five multiples of a given natural number, multiply it by 1, 2, 3, 4, and 5.**
2. **A multiple of a given natural number, multiply it by any natural number.**

To determine whether a natural number is a multiple of a given number:

1. **Divide by the given number:**

   $$\text{given number}\overline{)\text{natural number}}^{\text{quotient}}$$

2. **If the quotient is a natural number with no remainder, the dividend is a multiple of the divisor.**

For example, 104 is a multiple of 8, for when we divide by 8, as shown:

$$8\overline{)104}\quad 13$$

the quotient is 13, which is a natural number. We can state this in different ways:

1. 104 is a multiple of 8
2. 8 is a factor of 104
3. 104 divided by 8 is a natural number
4. 8 divides 104
5. 104 is divisible by 8

MODEL PROBLEM-SOLVING

Examples | **Strategy**

a. List the first five multiples of 3.

The first five multiples of 3 are $1 \cdot 3$, $2 \cdot 3$, $3 \cdot 3$, $4 \cdot 3$, $5 \cdot 3$; that is, 3, 6, 9, 12, 15.

Multiply 3 by 1, 2, 3, 4, and 5.

Warm Up a. List the first five multiples of 4.

b. List the first five multiples of 18.

The first five multiples of 18 are: $1 \cdot 18$, $2 \cdot 18$, $3 \cdot 18$, $4 \cdot 18$, $5 \cdot 18$; that is, 18, 36, 54, 72, 90.

Multiply 18 by 1, 2, 3, 4, and 5.

Some calculators will find these multiples by repeated addition.

Try pressing these keys: [18] followed by the [+] key *twice*. Then, if you press the [=] key repeatedly, the multiples of 18 will appear if your calculator has this feature. See example e also.

Warm Up b. List the first five multiples of 11.

c. Is 117 a multiple of 9?

Since $117 \div 9 = 13$, we can say that:

To find whether 117 is a multiple of 9, divide 117 by 9.

117 is a multiple of 9.

Because $117 = 9 \cdot 13$

Warm Up c. Is 39 a multiple of 3?

ANSWERS TO WARM UPS (2.1) **a.** 4, 8, 12, 16, 20 **b.** 11, 22, 33, 44, 55 **c.** Yes

d. Is 236 a multiple of 21?

To find whether 236 is a multiple of 21, divide 236 by 21. Since $236 \div 21 = 11 \text{ R } 5$, we can say that 236 is *not* a multiple of 21.

$$\begin{array}{r} 11 \text{ R } 5 \\ 21\overline{)236} \\ \underline{21} \\ 26 \\ \underline{21} \\ 5 \end{array}$$

Warm Up d. Is 32 a multiple of 6?

e. Is 8 a multiple of 48?

No, 8 is *not* a multiple of 48.

Remember that the multiples of 48 are found by multiplying 48 by 1, 2, 3, 4, ... The smallest multiple of 48 is $1 \cdot 48 = 48$.

Warm Up e. Is 5 a multiple of 15?

f. Calculator example

Find the 24th, the 6th, the 29th, and the 468th multiples of 3 on a calculator.

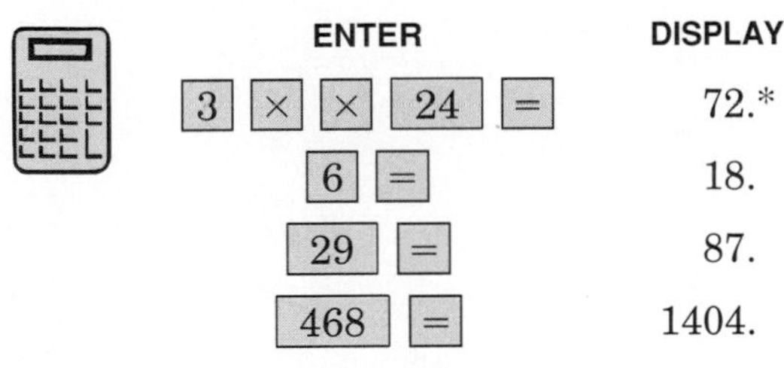

ENTER	DISPLAY
3 × × 24 =	72.*
6 =	18.
29 =	87.
468 =	1404.

If the display reads 72, you can now find other multiples of 3 just by entering any whole number and pressing the = key.

The requested multiples are 72, 18, 87, and 1404.

Warm Up f. Find the 24th, the 6th, the 29th, and the 468th multiples of 7.

*If your calculator does not give the above result, see the instruction manual for the calculator. (Some calculators use a "constant" key instead.)

ANSWERS TO WARM UPS (2.1) **d.** No **e.** No **f.** 168, 42, 203, 3276

g. Find the multiples of 7 between 200 and 240 on a calculator.

ENTER	DISPLAY	
7 × × = 20 =	140.	**140 is less than 200.**
30 =	210.	**210 is larger than 200.**
29 =	203.	**203 is larger than 200.**
31 =	217.	
32 =	224.	
33 =	231.	
34 =	238.	
35 =	245.	**245 is larger than 240.**

So, the multiples of 7 between 200 and 240 are 203, 210, 217, 224, 231, and 238.

Warm Up g. Find the multiples of 8 between 200 and 240 on a calculator.

APPLICATION SOLUTION

h. If the instructor assigns the problems from 1 to 50 that are multiples of 3, which problems should be worked?

The multiples of 3 between 1 and 50 are 3, 6, 9, 12, 15, 18, 21, 24, 27, 30, 33, 36, 39, 42, 45, and 48. So, these are the problems that are assigned.

Multiply 3 by 1, 2, 3, 4, 5, and so on until the product is larger than 50.

Warm Up h. If the instructor assigns the problems from 1 to 74 that are multiples of 4, which problems should be worked?

ANSWERS TO WARM UPS (2.1)

g. 208, 216, 224, 232

h. 4, 8, 12, 16, 20, 24, 28, 32, 36, 40, 44, 48, 52, 56, 60, 64, 68, 72

NAME ____________________

CLASS ____________ SECTION ____________ DATE ____________

Exercises 2.1

A

ANSWERS

1. ________
2. ________
3. ________
4. ________
5. ________
6. ________
7. ________
8. ________
9. ________
10. ________
11. ________
12. ________
13. ________
14. ________
15. ________
16. ________
17. ________
18. ________

List the first five multiples of each of the following numbers.

1. 2 **2.** 8 **3.** 13 **4.** 15

5. 5 **6.** 10

Is each of the following a multiple of 4?

7. 40 **8.** 75 **9.** 120 **10.** 88

11. 39 **12.** 68

Is each of the following a multiple of 9?

13. 18 **14.** 63 **15.** 89 **16.** 93

17. 65 **18.** 108

B

19. ________
20. ________
21. ________
22. ________

List the first five multiples of each of the following numbers.

19. 17 **20.** 20 **21.** 25 **22.** 12

ANSWERS

23. ______
24. ______
25. ______
26. ______
27. ______
28. ______
29. ______
30. ______
31. ______
32. ______
33. ______
34. ______
35. ______
36. ______
37. ______
38. ______

23. 31 **24.** 14 **25.** 40 **26.** 100

Is each of the following a multiple of 7?

27. 28 **28.** 42 **29.** 47 **30.** 74

31. 728 **32.** 364

Is each of the following a multiple of 6?

33. 42 **34.** 99 **35.** 129 **36.** 954

37. 150 **38.** 1008

C

39. ______
40. ______
41. ______
42. ______
43. ______
44. ______
45. ______
46. ______
47. ______
48. ______
49. ______
50. ______

List the first five multiples of each of the following numbers.

39. 35 **40.** 42 **41.** 110 **42.** 140

43. 96 **44.** 88

Is each of the following numbers a multiple of 4? of 6? of 15?

45. 96 **46.** 480 **47.** 1350 **48.** 210

49. 120 **50.** 900

NAME ______

CLASS ______ SECTION ______ DATE ______

D

ANSWERS

51. ______

51. If a mathematics instructor assigns problems from 1 to 60 that are multiples of 4, which problems should be done?

52. ______

52. If a statistics teacher assigns problems from 18 to 117 that are multiples of 9, which problems should her students do?

53. ______

53. List all of the multiples of 8 from 72 to 112.

54. ______

54. List all of the multiples of 7 from 231 to 266.

55. ______

55. List all of the multiples of 9 between 305 and 330.

56. ______

56. List all of the multiples of 13 between 400 and 460.

E *Maintain your skills* (Sections 1.3, 1.4, 1.5, 1.6)

57. ______

57. Find the sum of 53, 1689, 308, 9, 417, and 92.

ANSWERS

58. ____________________

59. ____________________

60. ____________________

61. ____________________

62. ____________________

63. ____________________

64. ____________________

65. ____________________

66. ____________________

58. Find the sum of 22, 981, 4589, 17, and 1088.

59. Find the difference of 9284 and 4738.

60. Find the difference of 9284 and 7483.

61. Find the difference of 5000 and 2983.

62. Find the product of 25, 32, and 8.

63. Find the product of 186 and 7.

64. Find the product of 309 and 38.

65. A storage tank holds 1850 gallons of gasoline. If 15 gallons are pumped into each of 30 cars per day (average) from the tank, how many full days will the tank last? How many gallons remain in the tank?

66. A convenience store has 8 cases of soup, each of which contains 36 cans of soup. If the store sells 19 cans per day (average), how many days will the cases last? How many cans of soup are left?

Divisors and Factors

OBJECTIVES

20. List all factors (divisors) of a given whole number.
21. Write a whole number as the product of two factors in all possible ways.

APPLICATION

A television station has 130 minutes of programming to fill. In what ways can the time be scheduled if all the programs are the same length and are a whole number of minutes? (For example, 10 programs each 13 minutes long: $10 \cdot 13 = 130$)

VOCABULARY

Remember:

6 is a *factor* of 24, since $6 \cdot 4 = 24$.
6 is a *divisor* of 24, since $24 \div 6 = 4$.

Another way to say this: 24 is *divisible* by 6.
Recall that $16^2 = 16 \cdot 16 = 256$, so the *square* of 16 is 256.

HOW AND WHY

We could list all factors (or divisors) of 250 by trial and error, but dividing 250 by all smaller numbers would take too long. The following steps take less time.

First: Write all the counting numbers from 1 to the first number whose square is larger than 250.

1	6	11	16
2	7	12	
3	8	13	
4	9	14	
5	10	15	

We can stop at 16 since 256 divided by any number larger than 16 gives a quotient that is less than 16. But all the possible factors less than 16 are already in the chart.

Second: Divide each of these into 250. If it divides evenly, write the factors. If not, cross out the number.

$1 \cdot 250$	~~6~~	~~11~~	~~16~~
$2 \cdot 125$	~~7~~	~~12~~	
~~3~~	~~8~~	~~13~~	
~~4~~	~~9~~	~~14~~	
$5 \cdot 50$	$10 \cdot 25$	~~15~~	

These steps give us a list of all factors (divisors) and products.

Factors of 250: 1, 2, 5, 10, 25, 50, 125, 250

250 as a product: $1 \cdot 250$, $2 \cdot 125$, $5 \cdot 50$, $10 \cdot 25$

A calculator can be used to do these divisions. If the quotient displayed is a whole number, the divisor is a factor.

MODEL PROBLEM-SOLVING

Examples

a. List all the factors of 68.

Make a chart of all possible factors that are less than 9:

1 · 68	4 · 17	~~7~~
2 · 34	~~5~~	~~8~~
~~3~~	~~6~~	~~9~~

The list of factors is:

1, 2, 4, 17, 34, 68

Strategy

We can stop with the factor 9 because the square of 9, $9^2 = 81$, is larger than 68. All the factors of 68 that are larger than 9 will appear when we divide 68 by the numbers in the chart.

Warm Up a. List all the factors of 10.

b. List the factors of 180, and write 180 as the product of two factors in all possible ways.

Make a chart of all possible factors that are less than the first number whose square is larger than 180:

1 · 180	6 · 30	~~11~~
2 · 90	~~7~~	12 · 15
3 · 60	~~8~~	~~13~~
4 · 45	9 · 20	~~14~~
5 · 36	10 · 18	

The number 14 is the smallest number whose square is larger than 180 ($13^2 = 169$ and $14^2 = 196$), so we can stop at 14. All the factors larger than 14 will appear when we divide by the numbers in the chart.

The list of factors is:

1, 2, 3, 4, 5, 6, 9, 10, 12, 15, 18, 20, 30, 36, 45, 60, 90, 180

The list of products is:

1 · 180, 2 · 90, 3 · 60, 4 · 45, 5 · 36, 6 · 30, 9 · 20, 10 · 18, 12 · 15

Warm Up b. List the factors of 28, and write 28 as the product of two factors in all possible ways.

ANSWERS TO WARM UPS (2.2)

a. 1, 2, 5, 10

b. Factors: 1, 2, 4, 7, 14, 28 Products: 1 · 28, 2 · 14, 4 · 7

c. List the factors of 29, and write 29 as the product of two factors in all possible ways.

Make a chart of all possible factors that are less than the first number whose square is larger than 29:

$1 \cdot 29$	~~3~~	~~5~~
~~2~~	~~4~~	~~6~~

The square of 6 is larger than 29, so we stop.

The list of factors is 1, 29.
The list of products is $1 \cdot 29$.

Warm Up c. List the factors of 73, and write 73 as the product of two factors in all possible ways.

APPLICATION SOLUTION

d. A television station has 130 minutes of programming to fill. In what ways can the time be scheduled if all the programs are the same length and are a whole number of minutes?

Write 130 as a product in all possible ways:

$1 \cdot 130$ — **One program that is 130 minutes long (maybe a movie) or 130 programs that are 1 minute long (probably too short to use).**

$2 \cdot 65$ — **Two programs that are 65 minutes long or 65 programs that are 2 minutes long.**

~~3~~
~~4~~
$5 \cdot 26$ — **Five programs that are 26 minutes long or 26 programs that are 5 minutes long.**

~~6~~
~~7~~
~~8~~
~~9~~
$10 \cdot 13$ — **Ten programs that are 13 minutes long or 13 programs that are 10 minutes long.**

~~11~~
~~12~~ — **Stop, since 12^2 is larger than 130.**

Warm Up d. A television station has 114 minutes of programming to fill. In what ways can the time be scheduled if all the programs are the same length and are a whole number of minutes?

ANSWERS TO WARM UPS (2.2)

c. Factors: 1, 73 Products: $1 \cdot 73$

d. One program that is 114 minutes long or 114 programs that are 1 minute long, or two programs that are 57 minutes long or 57 programs that are 2 minutes long, or three programs that are 38 minutes long or 38 programs that are 3 minutes long, or six programs that are 19 minutes long or 19 programs that are 6 minutes long.

NAME ______________________

CLASS ____________ SECTION ____________ DATE ____________

Exercises 2.2

A

ANSWERS

1. ____
2. ____
3. ____
4. ____
5. ____
6. ____
7. ____
8. ____
9. ____
10. ____
11. ____
12. ____
13. ____
14. ____
15. ____
16. ____
17. ____
18. ____
19. ____
20. ____

List all the factors (divisors) of each number.

1. 12 **2.** 14 **3.** 31 **4.** 25

5. 35 **6.** 40 **7.** 26 **8.** 50

9. 70 **10.** 72

Write each number as the product of two factors in all possible ways.

11. 12 **12.** 15 **13.** 16 **14.** 66

15. 49 **16.** 52 **17.** 50 **18.** 102

19. 45 **20.** 60

B

ANSWERS

21. ______

22. ______

23. ______

24. ______

25. ______

26. ______

27. ______

28. ______

29. ______

30. ______

31. ______

32. ______

33. ______

34. ______

35. ______

36. ______

37. ______

38. ______

39. ______

40. ______

List all the factors (divisors) of each number.

21. 42 **22.** 36 **23.** 99 **24.** 80

25. 100 **26.** 64 **27.** 32 **28.** 28

29. 88 **30.** 96

Write each number as the product of two factors in all possible ways.

31. 96 **32.** 101 **33.** 220 **34.** 111

35. 120 **36.** 128 **37.** 305 **38.** 405

39. 210 **40.** 320

NAME ____________

CLASS ____________ SECTION ____________ DATE ____________

C

ANSWERS

41. ____________
42. ____________
43. ____________
44. ____________
45. ____________
46. ____________
47. ____________
48. ____________
49. ____________
50. ____________
51. ____________
52. ____________

List all the factors (divisors) of each number.

41. 245 **42.** 847 **43.** 500 **44.** 555

45. 360 **46.** 480

Write each number as the product of two factors in all possible ways.

47. 245 **48.** 847 **49.** 720 **50.** 134

51. 510 **52.** 810

D

53. ____________

53. In what ways can a television station schedule 150 minutes of time if all the programs are the same length and each runs a whole number of minutes?

ANSWERS

54. ____________

55. ____________

56. ____________

57. ____________

58. ____________

54. In what ways can a television station schedule 3 hours of time if all the programs are the same length and each runs a whole number of minutes?

55. In what ways can a radio station schedule 90 minutes of news features if each program is the same length and each runs a whole number of minutes?

56. In what ways can a radio station schedule 100 minutes of news features if each program is the same length and each runs a whole number of minutes?

57. Elizabeth has 4 yards of ribbon (1 yard is 36 inches). In how many ways can she divide the ribbon into equal lengths (using lengths that are whole number inches)? The ribbon must be cut at least one time.

58. Matthew has 240 centimeters of fishing line. In how many ways can he divide it into equal lengths (using lengths that are whole number centimeters)? The ribbon must be cut at least one time.

NAME ______________________________

CLASS ______________ SECTION ______________ DATE ______________

E *Maintain your skills* (Sections 1.5, 1.6, 1.8)

ANSWERS

59. ______________

60. ______________

61. ______________

62. ______________

63. ______________

64. ______________

65. ______________

66. ______________

67. ______________

68. ______________

Multiply.

59. 59(41) **60.** 59(502) **61.** 77(245) **62.** 407(209)

Divide.

63. $59\overline{)1829}$ **64.** $59\overline{)17{,}759}$ **65.** $307\overline{)7370}$ **66.** $307\overline{)12{,}890}$

67. How many stereo speakers can be wired from a spool containing 880 feet of wire if each speaker requires 15 feet of wire? How much wire is left?

68. A consumer magazine had 15 brands of tires tested for wear. The results are in the table below.

NUMBER OF BRANDS TESTED	MILEAGE BEFORE TIRE WAS REPLACED (NEAREST FIVE THOUSAND MILES)
1	30,000
2	35,000
3	40,000
2	45,000
4	50,000
3	55,000

What was the average mileage of the 15 brands of tires?

Divisibility Tests

OBJECTIVE

22. Determine whether a natural number is divisible by 2, 3, and/or 5.

APPLICATION

The following tests show how to tell whether a number is divisible by 2, 3, or 5 even more quickly than by using a calculator.

VOCABULARY

No new words.

HOW AND WHY

Look at the following table.

SOME NATURAL NUMBERS	MULTIPLES OF 2	MULTIPLES OF 3	MULTIPLES OF 5
1	2	3	5
2	4	6	10
3	6	9	15
4	8	12	20
5	10	15	25
6	12	18	30
21	42	63	105
65	130	195	325
133	266	399	665

In the column of multiples of 2, each ones digit is either 0, 2, 4, 6, or 8.

In the column of multiples of 3, the sum of the digits of each number is divisible by 3. For example, the digits of 18 $(1 + 8 = 9)$ add up to a multiple of 3, as do the digits of 195 $(1 + 9 + 5 = 15)$.

In the column of multiples of 5, each ones digit is either 0 or 5.

In the column of multiples of 2, each multiple that has a zero in the ones place is also in the column of multiples of 5 and therefore is divisible by both 2 and 5 and hence by 10.

From the table the following conclusions seem plausible:

RULE

Divisibility by 2: **If the digit in the ones place of a natural number is 0, 2, 4, 6, or 8, then the natural number is divisible by 2.**

Divisibility by 3: **If the sum of the digits of a natural number is divisible by 3, then the natural number is divisible by 3.**

Divisibility by 5: **If the digit in the ones place of a natural number is 0 or 5, then the natural number is divisible by 5.**

Divisibility by 10: **If the digit in the ones place of a natural number is 0, then the natural number is divisible by 10.**

Although the table shows some multiples for only nine natural numbers, it can be shown, using algebra, that these divisibility tests work for all other natural numbers.

MODEL PROBLEM-SOLVING

Examples	Strategy
a. State whether 48 is divisible by 2, 3, or 5.	
48 is divisible by 2.	**The digit in the ones place is 8.**
48 is divisible by 3.	**4 + 8 = 12, which is divisible by 3.**
48 is *not* divisible by 5.	**The digit in the ones place is neither 0 or 5.**

Warm Up a. State whether 75 is divisible by 2, 3, or 5.

b. State whether 150 is divisible by 2, 3, or 5.	
150 is divisible by 2.	**The digit in the ones place is 0.**
150 is divisible by 3.	**1 + 5 + 0 = 6, which is divisible by 3.**
150 is divisible by 5.	**The digit in the ones place is 0.**

Warm Up b. State whether 218 is divisible by 2, 3, or 5.

c. State whether 689 is divisible by 2, 3, or 5.	
689 is *not* divisible by 2.	**The digit in the ones place is not 0, 2, 4, 6, or 8**
689 is *not* divisible by 3.	**6 + 8 + 9 = 23, which is *not* divisible by 3.**
689 is *not* divisible by 5.	**The digit in the ones place is neither 0 or 5.**

Warm Up c. State whether 487 is divisible by 2, 3, or 5.

ANSWERS TO WARM UPS (2.3)

a. 75 is *not* divisible by 2.
75 is divisible by 3.
75 is divisible by 5.

b. 218 is divisible by 2.
218 is *not* divisible by 3.
218 is *not* divisible by 5.

c. 487 is *not* divisible by 2.
487 is *not* divisible by 3.
487 is *not* divisible by 5.

NAME ______________________________

CLASS ______________ SECTION ______________ DATE ______________

Exercises 2.3

A

ANSWERS

1. ______
2. ______
3. ______
4. ______
5. ______
6. ______
7. ______
8. ______
9. ______
10. ______
11. ______
12. ______
13. ______
14. ______
15. ______
16. ______
17. ______
18. ______
19. ______
20. ______
21. ______
22. ______
23. ______
24. ______
25. ______
26. ______
27. ______
28. ______
29. ______
30. ______
31. ______
32. ______

State whether the following are divisible by 2.

1. 16 **2.** 18 **3.** 30 **4.** 45 **5.** 49

6. 109 **7.** 200 **8.** 201 **9.** 243 **10.** 480

11. 755 **12.** 2754 **13.** 4801 **14.** 45,890 **15.** 685,470

16. 6,238,452

State whether the following are divisible by 5.

17. 16 **18.** 18 **19.** 30 **20.** 45 **21.** 49

22. 109 **23.** 200 **24.** 201 **25.** 243 **26.** 480

27. 755 **28.** 2754 **29.** 4801 **30.** 45,890 **31.** 685,470

32. 6,238,452

B

ANSWERS

33. ______
34. ______
35. ______
36. ______
37. ______
38. ______
39. ______
40. ______
41. ______
42. ______
43. ______
44. ______
45. ______
46. ______
47. ______
48. ______

State whether the following are divisible by 3.

33. 16 **34.** 18 **35.** 30 **36.** 45

37. 49 **38.** 109 **39.** 200 **40.** 201

41. 243 **42.** 480 **43.** 755 **44.** 2754

45. 4801 **46.** 45,890 **47.** 6,238,452 **48.** 685,470

C

49. ______
50. ______
51. ______
52. ______
53. ______
54. ______
55. ______
56. ______
57. ______
58. ______

State whether the following are divisible by 2, 3, or 5:

49. 145 **50.** 210 **51.** 175 **52.** 315 **53.** 250

54. 350 **55.** 472 **56.** 596 **57.** 4180 **58.** 5280

NAME ______________________________

CLASS ______________ SECTION ______________ DATE ______________

D

ANSWERS

59. ______________

59. Pedro buys a package of gumdrops which contains 111 pieces. Can the gumdrops be divided evenly among Pedro and his two friends?

60. ______________

60. Janna and her four friends decide to run a distance of 120 miles in relays. Is it possible for each runner to run the same whole number of miles?

61. ______________

61. Ed and his five partners made a profit of $1550 in their stereo installation business. Can the profits be divided evenly among them?

62. ______________

62. Four merchandisers agree to build one new store in each of 310 cities. Is it possible for each merchandiser to build the same number of new stores?

E *Maintain your skills* (Sections 1.2, 1.8)

63. ______________

63. Round 67,853 to the nearest thousand.

64. ______________

64. Round 67,853 to the nearest ten.

65. ______________

65. Round 6,238,452 to the nearest hundred.

ANSWERS

66. ______

67. ______

68. ______

69. ______

70. ______

71. ______

72. ______

73. ______

74. ______

75. ______

66. Round 765,456,528 to the nearest ten million.

67. Round 765,456,528 to the nearest hundred thousand.

Perform the indicated operations.

68. $(28 + 3 \cdot 4) - 2 \cdot 3 + 4$

69. $28 + 3 \cdot 4 - 2(3 + 4)$

70. $800 - 8^2 \cdot 2^3$

71. $800 - 2^8 + 3^2$

72. $2(3 + 4)^2 - 5^2$

73. $2 \cdot 3^2 + 4^2 - 5^2$

74. On the first day of a canned goods sale, a grocery store had 34 boxes of macaroni and cheese on the shelf. During the week the shelf was restocked (several times) from 8 cases of 50 boxes each. At the end of the sale there were 37 boxes left. How many boxes of macaroni and cheese were sold during the sale?

75. Gina has filled the tank of her new car seven times. She paid

34 cents per liter the first time,
35 cents per liter the next time,
37 cents per liter the next two times,
36 cents per liter the next two times, and
37 cents per liter the last time.

What was the average price per liter that Gina paid?

Primes and Composites

OBJECTIVE

23. Tell whether a given whole number is prime or composite.

APPLICATION

Whole numbers can be classified in many different ways. One of these is the familiar odd and even classification. Primes and composites are another useful way to classify whole numbers.

VOCABULARY

A *prime number* is a whole number greater than one with exactly two different factors (divisors). These factors are 1 and the number itself. Whole numbers greater than 1 with more than two different factors are called *composite numbers.*

HOW AND WHY

The whole numbers zero (0) and one (1) are neither prime nor composite.

Two (2) is the first prime number ($2 = 2 \cdot 1$), since 2 and 1 are the only factors of 2.

Three (3) is a prime number, since 1 and 3 are the only factors of 3.

Four (4) is a composite number ($4 = 4 \cdot 1$ and $4 = 2 \cdot 2$), since it has more than two factors.

To tell whether a number is prime or composite, list its factors or divisors in a table like those in Section 2.2. For instance, the charts for 299 and 307 are:

$1 \cdot 299$	~~6~~	~~11~~	
~~2~~	~~7~~	~~12~~	
~~3~~	~~8~~	$13 \cdot 23$	**We can stop here since 299 has at least four factors.**
~~4~~	~~9~~		
~~5~~	~~10~~		

So, 299 is a composite number:

$1 \cdot 307$	~~6~~	~~11~~	~~16~~	
~~2~~	~~7~~	~~12~~	~~17~~	
~~3~~	~~8~~	~~13~~	~~18~~	**Stop here since $18 \cdot 18 = 324$.**
~~4~~	~~9~~	~~14~~		
~~5~~	~~10~~	~~15~~		

We see that 307 has exactly two factors (1, 307), so 307 is a prime number.

All the primes up to a given number may be found by a method called the Sieve of Eratosthenes. Eratosthenes (born ca. 230 B.C.) is remembered for both the prime sieve and his method of measuring the circumference of the earth. The accuracy of his measurement, compared with modern methods, is within 50 miles, or six-tenths of one percent. To use the famous sieve to find the primes up to 30, list the numbers from 2 to 30.

2	3	~~4~~	5	~~6~~	7
~~8~~	9	~~10~~	11	~~12~~	13
~~14~~	15	~~16~~	17	~~18~~	19
~~20~~	21	~~22~~	23	~~24~~	25
~~26~~	27	~~28~~	29	~~30~~	

All multiples of 2, except 2, are not prime, so they are crossed off.

2	3	~~4~~	5	~~6~~	7
~~8~~	~~9~~	~~10~~	11	~~12~~	13
~~14~~	~~15~~	~~16~~	17	~~18~~	19
~~20~~	~~21~~	~~22~~	23	~~24~~	25
~~26~~	~~27~~	~~28~~	29	~~30~~	

All remaining multiples of 3, except 3, are not prime, so they are crossed off.

2	3	~~4~~	5	~~6~~	7
~~8~~	~~9~~	~~10~~	11	~~12~~	13
~~14~~	~~15~~	~~16~~	17	~~18~~	19
~~20~~	~~21~~	~~22~~	23	~~24~~	~~25~~
~~26~~	~~27~~	~~28~~	29	~~30~~	

All remaining multiples of 5, except 5, are not prime, so they are crossed off.

The multiples of the other numbers, except themselves, have been crossed off. We need to test divisors only up to the first number whose square ($6 \cdot 6 = 36$) is larger than 30. So, the primes less than 30 are: 2, 3, 5, 7, 11, 13, 17, 19, 23, 29.

From the preceding Sieve we see that we can shorten the factor chart by omitting all the numbers except those that are prime. For instance, is 371 prime or composite?

$1 \cdot 371$	11	
~~2~~	13	
~~3~~	17	
~~5~~	19	
	23	$23^2 = 529$
$7 \cdot 53$	Stop here, since we do not need all factors.	

Since $7 \cdot 53 = 371$, 371 is composite. (It has *at least* four factors: 1, 7, 53, and 371). We know that a number is prime if no smaller prime divides it evenly (see example d).

PROCEDURE

To tell whether a whole number is prime or composite:

1. **The numbers 0 and 1 are neither prime nor composite.**
2. **List 1 and all the possible prime factors of the number in a chart.**
   a. **If the number has exactly two factors, it is prime.**
   b. **If the number has more than two factors, it is composite.**

MODEL PROBLEM-SOLVING

Examples **Strategy**

a. Tell whether 101 is a prime number or a composite number.

Make a chart to find the factors of 101:

$1 \cdot 101$	~~5~~	**The factor chart can be shortened by omitting all the numbers except those that are prime.**
~~2~~	~~7~~	
~~3~~	~~11~~	**Stop, since $11^2 > 101$.**

So, 101 is a prime number, since it has exactly two factors—itself and 1.

Warm Up a. Tell whether 71 is a prime number or a composite number.

b. Tell whether 91 is a prime number or a composite number.

Make a chart to find the factors of 91:

$1 \cdot 91$	~~5~~	
~~2~~	$7 \cdot 13$	
~~3~~	~~11~~	**Stop, since $11^2 > 91$.**

So, 91 is a composite number, since it has more than two factors.

Warm Up b. Tell whether 51 is a prime number or a composite number.

c. Tell whether 323 is a prime number or a composite number.

Make a chart to find the factors of 323:

$1 \cdot 323$	~~5~~	~~13~~	
~~2~~	~~7~~	$17 \cdot 19$	
~~3~~	~~11~~	19	**Stop, since $19^2 > 323$.**

So, 323 is a composite number, since it has more than two factors.

Warm Up c. Tell whether 221 is a prime number or a composite number.

ANSWERS TO WARM UPS (2.4) **a.** Prime **b.** Composite **c.** Composite

d. Tell whether 457 is a prime number or a composite number.

Make a chart to find the factors of 457:

$1 \cdot 457$ ~~5~~ ~~13~~

~~2~~ ~~7~~ ~~17~~

~~3~~ ~~11~~ ~~19~~

~~23~~ **Stop, since $23^2 > 457$.**

So, 457 is a prime number, since it has exactly two factors—itself and 1.

Warm Up d. Tell whether 547 is a prime number or a composite number.

ANSWERS TO WARM UPS (2.4)

d. Prime

NAME ______________________

CLASS ______________ SECTION ______________ DATE ______________

Exercises 2.4

A

ANSWERS

1. ______
2. ______
3. ______
4. ______
5. ______
6. ______
7. ______
8. ______
9. ______
10. ______
11. ______
12. ______
13. ______
14. ______
15. ______
16. ______
17. ______
18. ______
19. ______
20. ______
21. ______
22. ______

Tell whether each number is prime or composite.

1. 9 **2.** 6 **3.** 13 **4.** 17

5. 19 **6.** 15 **7.** 23 **8.** 33

9. 21 **10.** 41 **11.** 27 **12.** 37

13. 42 **14.** 52 **15.** 67 **16.** 77

17. 25 **18.** 65 **19.** 29 **20.** 33

21. 72 **22.** 51

B

ANSWERS

23. ______
24. ______
25. ______
26. ______
27. ______
28. ______
29. ______
30. ______
31. ______
32. ______
33. ______
34. ______
35. ______
36. ______
37. ______
38. ______
39. ______
40. ______
41. ______
42. ______
43. ______
44. ______
45. ______
46. ______

23. 39 **24.** 31 **25.** 53 **26.** 78

27. 97 **28.** 57 **29.** 90 **30.** 87

31. 109 **32.** 197 **33.** 183 **34.** 201

35. 211 **36.** 177 **37.** 270 **38.** 243

39. 151 **40.** 309 **41.** 323 **42.** 364

43. 427 **44.** 711 **45.** 297 **46.** 397

C

47. ______
48. ______
49. ______
50. ______

47. 497 **48.** 597 **49.** 697 **50.** 797

NAME ________________

CLASS ________ SECTION ________ DATE ________

ANSWERS

51. ________

52. ________

53. ________

54. ________

55. ________

56. ________

57. ________

58. ________

59. ________

51. 897 **52.** 997 **53.** 977 **54.** 1177

55. 979 **56.** 1097 **57.** 1197 **58.** 1297

59. Complete Eratosthenes' Sieve for the numbers from 2 to 150.

	2	3	4	5	6	7	8	9	10
11	12	13	14	15	16	17	18	19	20
21	22	23	24	25	26	27	28	29	30
31	32	33	34	35	36	37	38	39	40
41	42	43	44	45	46	47	48	49	50
51	52	53	54	55	56	57	58	59	60
61	62	63	64	65	66	67	68	69	70
71	72	73	74	75	76	77	78	79	80
81	82	83	84	85	86	87	88	89	90
91	92	93	94	95	96	97	98	99	100
101	102	103	104	105	106	107	108	109	110
111	112	113	114	115	116	117	118	119	120
121	122	123	124	125	126	127	128	129	130
131	132	133	134	135	136	137	138	139	140
141	142	143	144	145	146	147	148	149	150

D

60. ________

60. The year 1987 was the last year that was a prime number. What is the next year that is a prime number?

ANSWERS

61. ____________

62. ____________

63. ____________

61. What was the last year before 1987 that was a prime number?

62. What is the first year of the twenty-first century that will be a prime number?

63. Tell whether the year of your birth is a prime number or a composite number.

E *Maintain your skills* (Sections 1.2, 1.7, 1.8)

ANSWERS

64. ____________

65. ____________

66. ____________

67. ____________

68. ____________

69. ____________

70. ____________

71. ____________

72. ____________

73. ____________

74. ____________

64. True or false? $78 < 81$

65. True or false? $199 > 200$

66. Find the value of 13^3.

67. Find the value of 4^7.

68. Multiply: 45×10^7

69. Divide: $820{,}000 \div 10^4$.

70. Add the product of 8 and 19 to 753.

71. Subtract the quotient of 51 and 3 from 122.

72. Multiply the quotient of 323 and 19 by 18.

73. Divide the difference of 822 and 85 by 67.

74. Dan had 48 baseball cards from the American League and 82 cards from the National League. Sharon had three times as many in the American League and twice as many in the National League. How many baseball cards do they have together?

2.5 Prime Factorization

OBJECTIVES

24. Write the prime factorization of a given whole number.

APPLICATION

The prime factorization of whole numbers can be used to:

1. Reduce fractions.
2. Find the Least Common Multiple (LCM) of two or more whole numbers. The LCM is used to find common denominators for fractions (see Section 2.6) and has similar uses in algebra.

VOCABULARY

The *prime factorization* of a number is the number written as a product of prime factors:

$21 = 3 \cdot 7$ AND $30 = 2 \cdot 3 \cdot 5$ are prime factorizations.
$30 = 3 \cdot 10$ is *not* a prime factorization, since 10 is not a prime.

Recall that *exponents* show repeated factors. This can save space in writing:

$2 \cdot 2 \cdot 2 = 2^3$ $\quad 3 \cdot 3 \cdot 3 \cdot 3 \cdot 7 \cdot 7 \cdot 7 = 3^4 \cdot 7^3$

HOW AND WHY

To find all prime factors, repeatedly divide by primes until the quotient is 1. Then write the number as the product of these primes (the divisors). To find the prime factors of 48, divide 48 by 2 and then divide the quotient, 24, by 2, and so on. When the quotient can no longer be divided by 2, we divide by 3, 5, 7, 11, . . . to check for other prime factors. To save time we do not rewrite each division problem, but simply divide each quotient, starting at the top and dividing down until the quotient is 1:

```
2)48
2)24
2)12      48 = 2 · 2 · 2 · 2 · 3 = 2⁴ · 3
2) 6
3) 3
   1
```

$48 = 2 \cdot 2 \cdot 2 \cdot 2 \cdot 3 = 2^4 \cdot 3$

If there is a large prime factor, you will find it when you have tried each prime whose square is smaller than the number:

```
  3)1179
  3) 393
131) 131     131 is not divisible by 2, 3, 5, 7, or 11; also,
       1     13 · 13 = 169 > 131. Therefore, 131 is prime.
```

131 *is not divisible by* 2, 3, 5, 7, or 11; also, $13 \cdot 13 = 169 > 131$. Therefore, 131 is prime.

$1179 = 3 \cdot 3 \cdot 131 = 3^2 \cdot 131$

Examples c and d show the "tree method" for prime factorization.

MODEL PROBLEM-SOLVING

Examples / **Strategy**

a. Write the prime factorization of 42:

$$\begin{array}{r} 2\underline{)42} \\ 3\underline{)21} \\ 7\underline{)\ 7} \\ 1 \end{array}$$

Begin at the top by dividing 42 by 2. Since 21 is not divisible by 2, we divide by 3. The last quotient is 1, so we are done.

$42 = 2 \cdot 3 \cdot 7$

Check by multiplying the prime factors.

Warm Up a. Write the prime factorization of 28.

b. Write the prime factorization of 848:

$$\begin{array}{r} 2\underline{)848} \\ 2\underline{)424} \\ 2\underline{)212} \\ 2\underline{)106} \\ 53\underline{)\ 53} \\ 1 \end{array}$$

$848 = 2 \cdot 2 \cdot 2 \cdot 2 \cdot 53 = 2^4 \cdot 53$

Warm Up b. Write the prime factorization of 464.

c. Use the tree method to write the prime factorization of 68:

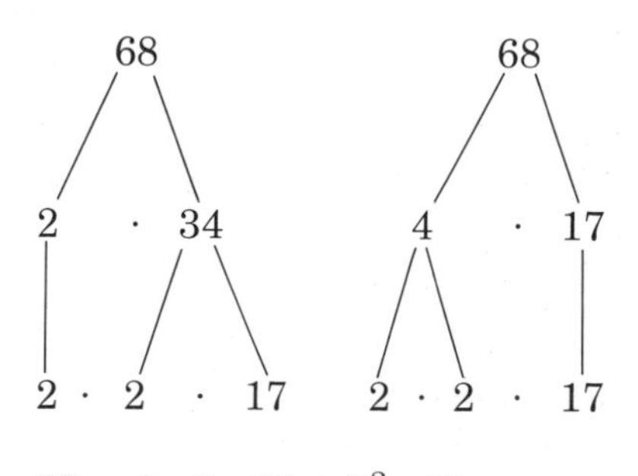

$68 = 2 \cdot 2 \cdot 17 = 2^2 \cdot 17$

To use the tree method, we draw "factor branches" using *any* two factors whose product is 68: either $2 \cdot 34$ or $4 \cdot 17$.

Now form additional branches from these new branches by using factors of the numbers at the end of each branch. A branch stops splitting when it ends in a prime number.

The prime factorization is the product of the primes at the ends of the branches.

Warm Up c. Write the prime factorization of 50.

ANSWERS TO WARM UPS (2.5) **a.** $2 \cdot 2 \cdot 7$ **b.** $2 \cdot 2 \cdot 2 \cdot 2 \cdot 29 = 2^4 \cdot 29$ **c.** $2 \cdot 5 \cdot 5 = 2 \cdot 5^2$

d. Use the tree method to write the prime factorization of 468:

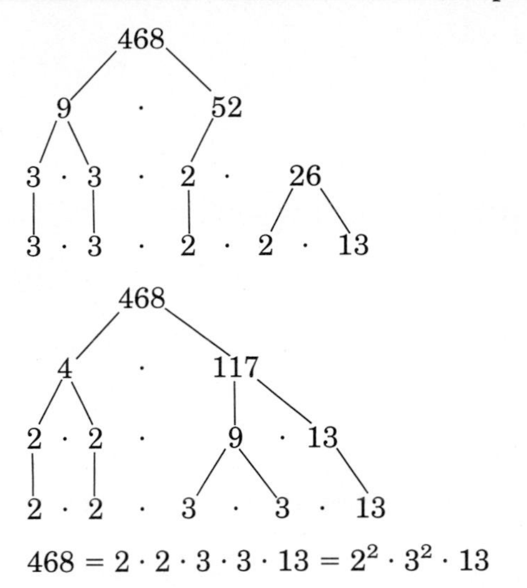

Although any two factors of 468 may be used to begin, the tree will be shorter if large factors are used first.

Here is another tree starting with different factors. Note that we still end up with the same prime factorization.

$468 = 2 \cdot 2 \cdot 3 \cdot 3 \cdot 13 = 2^2 \cdot 3^2 \cdot 13$

Warm Up d. Write the prime factorization of 612.

e. Calculator example

Write the prime factorization of 900.
Since 900 is divisible by 2, we begin by dividing by 2:

ENTER	DISPLAY	DISPLAY NUMBER DIVISIBLE BY
900	900.	2
÷	900.	
2	2.	
=	450.	2
÷ 2 =	225.	3
÷ 3 =	75.	3
÷ 3 =	25.	5
÷ 5 =	5.	5
÷ 5 =	1.	

So, 900 in prime-factored form is $2^2 \cdot 3^2 \cdot 5^2$.

Warm Up e. Write the prime factorization of 360.

ANSWERS TO WARM UPS (2.5)

d. $2 \cdot 2 \cdot 3 \cdot 3 \cdot 17 = 2^2 \cdot 3^2 \cdot 17$

e. $2 \cdot 2 \cdot 2 \cdot 3 \cdot 3 \cdot 5 = 2^3 \cdot 3^2 \cdot 5$

NAME ____________________

CLASS ____________ SECTION ____________ DATE ____________

Exercises 2.5

A

ANSWERS

1. ______
2. ______
3. ______
4. ______
5. ______
6. ______
7. ______
8. ______
9. ______
10. ______
11. ______
12. ______
13. ______
14. ______
15. ______
16. ______
17. ______
18. ______

Write the prime factorization of each number.

1. 15 **2.** 22 **3.** 35 **4.** 77

5. 34 **6.** 65 **7.** 45 **8.** 63

9. 66 **10.** 70 **11.** 48 **12.** 56

13. 80 **14.** 81 **15.** 82 **16.** 83

17. 51 **18.** 76

B

19. ______
20. ______
21. ______
22. ______

19. 36 **20.** 64 **21.** 71 **22.** 47

ANSWERS

23. ______

24. ______

25. ______

26. ______

27. ______

28. ______

29. ______

30. ______

31. ______

32. ______

33. ______

34. ______

35. ______

36. ______

23. 91 **24.** 96 **25.** 75 **26.** 120

27. 150 **28.** 122 **29.** 183 **30.** 119

31. 133 **32.** 180 **33.** 182 **34.** 184

35. 185 **36.** 270

C

37. ______

38. ______

39. ______

40. ______

41. ______

42. ______

43. ______

44. ______

37. 104 **38.** 136 **39.** 156 **40.** 234

41. 225 **42.** 310 **43.** 414 **44.** 563

NAME ______________________

CLASS ______________ SECTION ______________ DATE ______________

ANSWERS

45. ______________

46. ______________

47. ______________

48. ______________

49. ______________

50. ______________

51. ______________

52. ______________

53. ______________

54. ______________

55. ______________

56. ______________

45. 555 **46.** 485 **47.** 888 **48.** 657

49. 659 **50.** 1428 **51.** 900 **52.** 902

53. 903 **54.** 904 **55.** 1560 **56.** 1240

E *Maintain your skills* (Sections 2.1, 2.2)

57. ______________

57. List the first seven multiples of 31.

58. ______________

58. List the first seven multiples of 42.

59. ______________

59. Is 3003 a multiple of 7?

60. ______________

60. Is 4004 a multiple of 7?

ANSWERS

61. ____________

62. ____________

63. ____________

64. ____________

65. ____________

66. ____________

67. ____________

68. ____________

61. Is 1001 a multiple of 7?

62. Then, will 5005 be a multiple of 7?

63. Is 9 a divisor of 3003?

64. Is 7 a divisor of 3003?

65. Is 11 a factor of 3003?

66. Is 13 a factor of 3003?

67. List all the multiples of 17 between 500 and 600.

68. List all the multiples of 17 between 750 and 850.

Least Common Multiple

OBJECTIVE

25. Find the least common multiple of two or more numbers.

APPLICATION

Jane and Robin have each been saving coins. Jane saved dimes and Robin saved quarters. The girls went shopping together and each bought the same item, each spending all of her coins. What is the least amount the item could cost?

VOCABULARY

The *least common multiple* of two or more whole numbers is:

1. the smallest natural number that is a multiple of each, and
2. the smallest natural number that has each as a factor, and
3. the smallest natural number that has each as a divisor, and
4. the smallest natural number that each will divide "evenly."

These statements are four ways of expressing the same idea.
LCM is an abbreviation of Least Common Multiple.

HOW AND WHY

The factorizations of this chapter are most often used to simplify fractions and find LCMs. LCMs are used to compare, add, and subtract fractions. In algebra, LCMs are useful in equation solving.

Find the LCM of 21 and 35. This can be done by listing the multiples of each and finding the smallest one in both lists:

Multiples of 21: 21, 42, 63, 84, $\boxed{105}$, 126, 147, 168, 189, $\boxed{210}$, 231, . . .
Multiples of 35: 35, 70, $\boxed{105}$, 140, 175, $\boxed{210}$, 245, 280, 315, 350, . . .

The LCM of 21 and 35 is 105, since this is the smallest number in both lists. This fact can be stated in four (equivalent) ways:

1. 105 is the smallest natural number that is a multiple of 21 and 35.
2. 105 is the smallest natural number that has both 21 and 35 as factors.
3. 105 is the smallest natural number that has both 21 and 35 as divisors.
4. 105 is the smallest natural number that both 21 and 35 will divide "evenly."

Finding the LCM by this method has one drawback: You might have to list hundreds of multiples. For this reason, we look for a shortcut.

To find the LCM of 12 and 18, write the prime factorization of each. Write these prime factors in columns, so that the prime factors of 18 are under the prime factors of 12. Leave blank spaces for prime factors that do not match.

	PRIMES WITH LARGEST EXPONENTS
$12 = 2 \cdot 2 \cdot 3 \qquad = 2^2 \cdot 3^1$	
$18 = 2 \qquad \cdot 3 \cdot 3 = 2^1 \cdot 3^2$	2^2 and 3^2

LCM of 12 and 18 $= 2 \cdot 2 \cdot 3 \cdot 3 = 2^2 \cdot 3^2 = 36$

Since 12 must divide the LCM, the LCM must have $2 \cdot 2 \cdot 3$ as part of its factors; and since 18 must divide the LCM, the LCM must also have $2 \cdot 3 \cdot 3$ as part of its factors. Thus, the LCM is $2 \cdot 2 \cdot 3 \cdot 3$, or 36. Note that the LCM is the product of the highest power of each prime factor.

What is the LCM of 16, 10, and 24?

	PRIMES WITH LARGEST EXPONENTS
$16 = 2^4$	
$10 = 2^1 \cdot 5^1$	
$24 = 2^3 \cdot 3^1$	$2^4, 3^1, 5^1$
$\text{LCM} = 2^4 \cdot 3^1 \cdot 5^1 = 16 \cdot 3 \cdot 5 = 240$	

PROCEDURE

To find the Least Common Multiple (LCM) of two or more numbers:

1. **Write each number in prime-factored form using exponents.**
2. **Write the product of the highest power of each prime factor.**

MODEL PROBLEM-SOLVING

Examples | **Strategy**

a. Find the LCM of 12 and 20:

$12 = 2 \cdot 2 \cdot 3 = 2^2 \cdot 3^1$

First, write the prime factorization of 12 and 20.

$20 = 2 \cdot 2 \cdot 5 = 2^2 \cdot 5^1$

The prime factors needed for the LCM are 2, 3, and 5. The largest exponent of the factor 2 is 2; the largest exponent of the factor 3 is 1; and the largest exponent of the factor 5 is 1.

$\mathbf{2^2, 3^1, 5^1}$

$\text{LCM} = 2^2 \cdot 3 \cdot 5 = 60$

Here is another way to think of the LCM. The LCM here must contain $2 \cdot 2 \cdot 3$ for 12 to be a factor or divisor. For 20 to be a factor also, we need a factor of 5, so

$\mathbf{LCM = (2^2 \cdot 3) \cdot 5 = 60}$

Warm Up a. Find the LCM of 4 and 10.

b. Find the LCM of 18, 24, and 30:

$18 = 2 \cdot 3 \cdot 3 = 2^1 \cdot 3^2$
$24 = 2 \cdot 2 \cdot 2 \cdot 3 = 2^3 \cdot 3^1$
$30 = 2 \cdot 3 \cdot 5 = 2^1 \cdot 3^1 \cdot 5^1$

Write the prime factorizations of the three numbers.

ANSWERS TO WARM UPS (2.6) **a.** 20

$\text{LCM} = 2^3 \cdot 3^2 \cdot 5 = 360$

The prime factors needed for the LCM are 2, 3, and 5. The largest exponent of the factor 2 is 3; the largest exponent of the factor 3 is 2; and the largest exponent of the factor 5 is 1.

2^3, 3^2, 5^1

Warm Up b. Find the LCM of 10, 12, and 30.

c. Find the LCM of 48 and 40:

$$48 = 2 \cdot 2 \cdot 2 \cdot 2 \cdot 3 = 2^4 \cdot 3^1$$
$$40 = 2 \cdot 2 \cdot 2 \cdot 5 = 2^3 \cdot 5^1$$

$\text{LCM} = 2^4 \cdot 3 \cdot 5 = 240$

Warm Up c. Find the LCM of 30 and 35.

d. Find the LCM of 12, 16, 24, and 36:

$$12 = 2 \cdot 2 \cdot 3 = 2^2 \cdot 3^1$$
$$16 = 2 \cdot 2 \cdot 2 \cdot 2 = 2^4$$
$$24 = 2 \cdot 2 \cdot 2 \cdot 3 = 2^3 \cdot 3^1$$
$$36 = 2 \cdot 2 \cdot 3 \cdot 3 = 2^2 \cdot 3^2$$

$\text{LCM} = 2^4 \cdot 3^2 = 144$

Warm Up d. Find the LCM of 2, 5, 10, and 14.

e. Find the LCM of the denominators of these fractions:

$\frac{1}{6}, \frac{4}{9}, \frac{5}{12}$, and $\frac{7}{18}$

$$6 = 2 \cdot 3$$
$$9 = 3^2$$
$$12 = 2^2 \cdot 3$$
$$18 = 2 \cdot 3^2$$

The denominators are 6, 9, 12, and 18. We prime factor each of them.

$\text{LCM} = 2^2 \cdot 3^2 = 36$

The lowest common denominator is 36.

The lowest common denominator is used to add and subtract fractions in Chapter 3.

ANSWERS TO WARM UPS (2.6) **b.** 60 **c.** 210 **d.** 70

Warm Up e. Find the LCM of the denominators of these fractions:

$$\frac{1}{3}, \frac{3}{5}, \frac{7}{12}, \text{ and } \frac{9}{10}$$

APPLICATION SOLUTION

f. Jane and Robin have each been saving coins. Jane saved dimes and Robin saved quarters. The girls went shopping together and each bought the same item, each spending all of her coins. What is the least amount the item could cost?

$10 = 2 \cdot 5$
$25 = 5^2$

$\text{LCM} = 2 \cdot 5^2 = 50$

The least amount the item could cost is the smallest number that 10 and 25 both divide evenly. That number is the LCM of 10 and 25.

The least the item could cost is 50 cents.

Five dimes or two quarters.

Warm Up f. If Jane had saved nickels and Robin had saved quarters, what is the least each could pay for the same item?

ANSWERS TO WARM UPS (2.6)

e. 60

f. 25 cents

NAME ______________________________

CLASS ______________ SECTION ______________ DATE ______________

Exercises 2.6

A

ANSWERS

1. ______
2. ______
3. ______
4. ______
5. ______
6. ______
7. ______
8. ______
9. ______
10. ______
11. ______
12. ______
13. ______
14. ______
15. ______
16. ______
17. ______
18. ______
19. ______
20. ______

Find the LCM of each group of numbers.

1. 3, 12 **2.** 5, 15 **3.** 5, 8 **4.** 7, 8

5. 6, 7 **6.** 11, 3 **7.** 9, 10 **8.** 7, 28

9. 10, 20 **10.** 6, 8 **11.** 5, 50 **12.** 6, 18

13. 3, 5, 6 **14.** 2, 3, 6 **15.** 2, 5, 10 **16.** 3, 6, 9

17. 4, 6, 8 **18.** 3, 6, 8 **19.** 4, 5, 8 **20.** 6, 5, 10

B

21. ______
22. ______
23. ______
24. ______
25. ______
26. ______
27. ______
28. ______

21. 10, 15 **22.** 20, 18 **23.** 36, 60 **24.** 42, 28

25. 48, 60 **26.** 15, 36 **27.** 30, 18 **28.** 45, 60

ANSWERS

29. ______

30. ______

31. ______

32. ______

33. ______

34. ______

35. ______

36. ______

37. ______

38. ______

39. ______

40. ______

29. 6, 8, 10 **30.** 4, 8, 12 **31.** 24, 36, 40 **32.** 12, 16, 18

33. 10, 15, 25 **34.** 10, 12, 15 **35.** 2, 6, 12, 24

36. 4, 12, 10, 15 **37.** 25, 30, 40 **38.** 16, 24, 48

39. 8, 9, 10 **40.** 6, 8, 10

C

41. ______

42. ______

43. ______

44. ______

45. ______

46. ______

47. ______

48. ______

49. ______

41. 8, 14, 28, 32 **42.** 12, 18, 24, 36 **43.** 18, 24, 28

44. 50, 30, 20 **45.** 12, 42, 39 **46.** 15, 33, 55

47. 120, 96, 48 **48.** 80, 48, 72 **49.** 250, 75, 150

NAME ____________________

CLASS ____________ SECTION ____________ DATE ____________

ANSWERS

50. ____________

50. 32, 72, 60

51. 90, 70, 21

52. 36, 240, 90

51. ____________

52. ____________

53. 17, 51, 68, 12

54. 91, 14, 35, 49

55. 38, 57, 114, 171

53. ____________

54. ____________

56. 144, 128, 300, 180

55. ____________

56. ____________

57. 50, 70, 56

57. ____________

58. ____________

58. 15, 20, 25, 30

D

ANSWERS

59. ______

60. ______

61. ______

62. ______

63. ______

64. ______

65. ______

66. ______

Find the lowest common denominator for each set of fractions:

59. $\frac{3}{2}, \frac{2}{3}, \frac{5}{6}$

60. $\frac{4}{5}, \frac{3}{4}, \frac{7}{8}$

61. $\frac{5}{12}, \frac{9}{16}, \frac{7}{24}$

62. $\frac{5}{12}, \frac{3}{16}, \frac{9}{20}$

63. $\frac{11}{12}, \frac{5}{36}, \frac{7}{54}$

64. $\frac{3}{14}, \frac{8}{21}, \frac{7}{8}$

65. $\frac{1}{4}, \frac{7}{10}, \frac{2}{15}, \frac{5}{12}$

66. $\frac{3}{5}, \frac{11}{20}, \frac{1}{12}, \frac{17}{30}$

E *Maintain your skills* (Sections 1.6, 1.8, 2.2)

67. ______

67. List all the factors of 376.

68. ______

68. List all the factors of 378.

69. ______

69. List all the divisors of 388.

NAME ______________________

CLASS ______________ SECTION ______________ DATE ______________

ANSWERS

70. ______________

71. ______________

72. ______________

73. ______________

74. ______________

70. List all the divisors of 393.

71. Is 7007 divisible by 13?

72. Is 7007 divisible by 9?

73. A marketing researcher checked the weekly attendance at 14 theaters. The results are shown in the following table:

NUMBER OF THEATERS	ATTENDANCE (NEAREST HUNDRED)
1	900
2	1000
2	1200
3	1300
2	1400
2	1600
1	1800
1	1900

What was the average attendance at the 14 theaters?

74. The Sav-Mor Department Store has made a profit on appliances of $112 so far this week. If their profit on each appliance is $7, how many more appliances must they sell so that the profit for the entire week will be more than $170?

NAME ____________________

CLASS ____________ SECTION ____________ DATE ____________

ANSWERS

1. ______
2. ______
3. ______
4. ______
5. ______
6. ______
7. ______
8. ______
9. ______
10. ______
11. ______
12. ______
13. ______
14. ______
15. ______
16. ______
17. ______
18. ______
19. ______
20. ______
21. ______
22. ______
23. ______
24. ______
25. ______

CHAPTER 2 TRUE–FALSE CONCEPT REVIEW

Check your understanding of the language of basic mathematics. Tell whether each of the following statements is True (always true) or False (not always true.)

1. Every multiple of 6 ends with the digit 6.
2. Every multiple of 10 ends with the digit 0.
3. Every multiple of 13 is divisible by 13.
4. Every multiple of 7 is the product of 7 and some natural number.
5. Every whole number, except the number 1, has at least two different factors.
6. Every factor of 200 is also a divisor of 200.
7. Every multiple of 200 is also a factor of 200.
8. The square of 200 is 100.
9. Every natural number ending in 4 is divisible by 4.
10. Every natural number ending in 6 is divisible by 2.
11. Every natural number ending in 9 is divisible by 3.
12. The number 123,321,231 is divisible by 3.
13. The number 123,321,234 is divisible by 4.
14. The number 123,321,235 is divisible by 5.
15. All prime numbers are odd.
16. Every composite number ends in 1, 3, 7, or 9.
17. Every composite number has four or more factors.
18. Every prime number has exactly two multiples.
19. It is possible for a composite number to have exactly three divisors.
20. All of the prime factors of a natural number are smaller than the number.
21. The least common multiple (LCM) of three different prime numbers is the product of the three numbers.
22. Some natural numbers have exactly five different prime factors.
23. The largest divisor of the least common multiple (LCM) of three numbers is the largest of the three numbers.
24. It is possible for some groups of numbers to have two LCMs.
25. If any one number in a group of numbers is an even number, then the least common multiple (LCM) of the group is an even number.

NAME ______________________________

CLASS ______________ SECTION ______________ DATE ______________

ANSWERS

CHAPTER 2 POST-TEST

1. ______________

1. (Section 2.3, Obj. 22) Is 552 divisible by 5?

2. ______________

2. (Section 2.2, Obj. 20) List all the factors of 110.

3. ______________

3. (Section 2.1, Obj. 19) Is 623 a multiple of 7?

4. ______________

4. (Section 2.3, Obj. 22) Is 411 divisible by 3?

5. ______________

5. (Section 2.6, Obj. 25) What is the LCM of 20 and 16?

6. ______________

6. (Section 2.2, Obj. 21) Write 75 as the product of two factors in as many ways as possible.

7. ______________

7. (Section 2.5, Obj. 24) Write the prime factorization of 260.

8. ______________

8. (Section 2.6, Obj. 25) Find the least common multiple (LCM) of 9, 42, and 84.

9. ______________

9. (Section 2.5, Obj. 24) Write the prime factorization of 846.

10. ______________

10. (Section 2.1, Obj. 18) List the first five multiples of 11.

ANSWERS

11. ____________________

12. ____________________

13. ____________________

14. ____________________

15. ____________________

11. (Section 2.1, Obj. 19) Is 161 a multiple of 322?

12. (Section 2.4, Obj. 23) Is 107 a prime number or a composite number?

13. (Section 2.4, Obj. 23) Is 99 a prime number or a composite number?

14. (Section 2.5, Obj. 24) Write the prime factorization of 847.

15. What is the smallest natural number that 15, 21, and 70 will divide evenly?

CHAPTER 3

Fractions and Mixed Numbers

NAME ______________________________

CLASS ______________ SECTION ______________ DATE ______________

ANSWERS

1. ______________

2. ______________

3. ______________

4. ______________

5. ______________

6. ______________

7. ______________

8. ______________

9. ______________

10. ______________

11. ______________

12. ______________

13. ______________

CHAPTER 3 PRE-TEST

The problems in the following pre-test are a sample of the material in the chapter. You may already know how to work some of these. If so, this will allow you to spend less time on those parts. As a result, you will have more time to devote to the sections that gave you difficulty. Please work the following problems. The answers are in the back of the text.

1. **(Section 3.1, Obj. 26)** Write the fraction for the shaded part of the following figure:

2. **(Section 3.1, Obj. 27)** Which of these fractions are proper?

   $\frac{10}{11}, \frac{11}{12}, \frac{12}{10}, \frac{11}{10}, \frac{12}{11}, \frac{10}{12}, \frac{11}{11}$

3. **(Section 3.1, Obj. 28)** Which of these fractions represent the number one?

   $\frac{13}{14}, \frac{13}{13}, \frac{13}{12}, \frac{14}{13}, \frac{14}{14}, \frac{14}{15}$

4. **(Section 3.2, Obj. 29)** Change to a mixed number: $\frac{14}{5}$

5. **(Section 3.2, Obj. 30)** Change to an improper fraction: 17

6. **(Section 3.2, Obj. 30)** Change to an improper fraction: $4\frac{2}{3}$

7. **(Section 3.3, Obj. 31)** Reduce to lowest terms: $\frac{18}{24}$

8. **(Section 3.3, Obj. 31)** Reduce to lowest terms: $\frac{42}{70}$

9. **(Section 3.4, Obj. 32)** Multiply: $\frac{2}{3} \cdot \frac{4}{5}$

10. **(Section 3.4, Obj. 32)** Multiply: $\frac{2}{5} \cdot \frac{5}{8}$

11. **(Section 3.4, Obj. 32)** Multiply: $\frac{6}{35} \cdot \frac{7}{9} \cdot \frac{15}{21}$

12. **(Section 3.5, Obj. 33)** What is the reciprocal of $\frac{3}{5}$?

13. **(Section 3.5, Obj. 33)** What is the reciprocal of $1\frac{2}{3}$?

NAME ______________________________

CLASS ______________ SECTION ______________ DATE ______________

ANSWERS

14. ______________

15. ______________

16. ______________

17. ______________

18. ______________

19. ______________

20. ______________

21. ______________

22. ______________

23. ______________

24. ______________

25. ______________

26. ______________

14. **(Section 3.5, Obj. 34)** Divide: $\frac{4}{5} \div \frac{7}{8}$

15. **(Section 3.6, Obj. 35)** Multiply. Write the result as a mixed number:

$$\left(2\frac{1}{5}\right) \cdot \left(3\frac{3}{4}\right)$$

16. **(Section 3.6, Obj. 36)** Divide: $3\frac{3}{5} \div 1\frac{1}{5}$

17. **(Section 3.7, Obj. 37)** Find the missing numerator: $\frac{7}{9} = \frac{?}{36}$

18. **(Section 3.8, Obj. 38)** List these fractions from smallest to largest:

$$\frac{2}{5}, \frac{1}{4}, \frac{3}{8}$$

19. **(Section 3.8, Obj. 39)** Is the following statement true or false? $\frac{2}{3} < \frac{8}{9}$

20. **(Section 3.9, Obj. 40)** Add: $\frac{3}{7} + \frac{2}{7}$

21. **(Section 3.10, Obj. 41)** Add: $\frac{1}{6} + \frac{3}{4}$

22. **(Section 3.10, Obj. 41)** Add: $\frac{3}{10} + \frac{8}{15} + \frac{1}{6}$

23. **(Section 3.11, Obj. 42)** Add:

$$\begin{array}{r} 2\frac{3}{4} \\ 4\frac{7}{10} \\ \hline \end{array}$$

24. **(Section 3.12, Obj. 43)** Subtract: $\frac{7}{10} - \frac{1}{2}$

25. **(Section 3.13, Obj. 44)** Subtract:

$$\begin{array}{r} 12\frac{7}{9} \\ 4\frac{1}{6} \\ \hline \end{array}$$

26. **(Section 3.13, Obj. 44)** Subtract: $9\frac{5}{6} - 8$

ANSWERS

27. ____________________

28. ____________________

29. ____________________

30. ____________________

31. ____________________

32. ____________________

27. **(Section 3.13, Obj. 45)** Subtract: $10\frac{1}{3}$

$$\begin{array}{r} 10\frac{1}{3} \\ 3\frac{1}{2} \\ \hline \end{array}$$

28. **(Section 3.13, Obj. 45)** Subtract: 21

$$\begin{array}{r} 21 \\ 18\frac{2}{3} \\ \hline \end{array}$$

29. **(Section 3.14, Obj. 46)** Perform the indicated operations: $\frac{2}{3} + \frac{3}{4} \div \frac{1}{2}$

30. **(Section 3.14, Obj. 47)** Find the average: $1\frac{1}{4}, 2\frac{1}{3}, \frac{5}{12}$

31. Find the perimeter of (distance around) a triangle whose sides are $2\frac{1}{2}$ inches, $4\frac{3}{4}$ inches, and $3\frac{11}{12}$ inches.

32. Mr. Mack's auto has a gasoline tank that holds $18\frac{3}{4}$ gallons when full. He starts with a full tank and drives for 3 hours. He checks the gasoline gauge and sees that he has $\frac{1}{4}$ of a tank remaining. How many gallons still remain in the tank?

What Is a Fraction?

OBJECTIVES

26. **Write a fraction to describe parts of units.**
27. **Select proper or improper fractions from a list of fractions.**
28. **Write fractions that represent the number one (1).**

APPLICATION

At the beginning of the month the High Pressure Tire Company had 53 used tires in stock. During the month they sold 40 of those tires. What fraction of the total amount of used tires was sold?

VOCABULARY

A *fraction* $\left(\text{such as } \frac{6}{7}\right)$ is a name for a number. The upper numeral (6) is the *numerator.* The lower numeral (7) is the *denominator,* that is, $\frac{\text{numerator}}{\text{denominator}}$.

A fraction is another way of writing a division problem. The fraction $\frac{6}{7}$ also means $6 \div 7$.

If the numerator is smaller than the denominator $\left(\text{such as } \frac{4}{9}\right)$, the fraction is called a *proper fraction.* If the numerator is not smaller than the denominator $\left(\text{such as } \frac{8}{3} \text{ or } \frac{5}{5}\right)$, the fraction is called an *improper fraction.*

HOW AND WHY

A unit (here we use a rectangle) may be divided into smaller parts of equal size in order to picture a fraction. The rectangle in Figure 3.1 is divided into 7 parts, and 6 of the parts are shaded. The fraction $\frac{6}{7}$ represents the shaded part. The denominator (7) tells the number of parts in the unit. The numerator (6) tells the number of shaded parts.

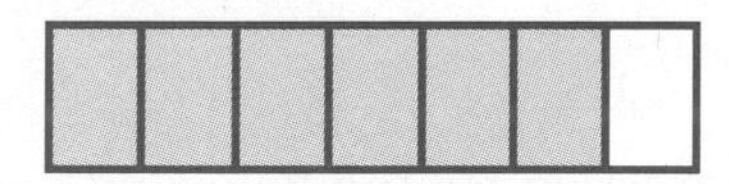

Figure 3.1

Since fractions are another way of writing division and since division by zero is not defined, the denominator can never be zero. There will be at least one part in a unit.

The unit may also be shown on a ruler.

The fraction $\frac{6}{10}$ represents the distance from 0 to the arrow in Figure 3.2.

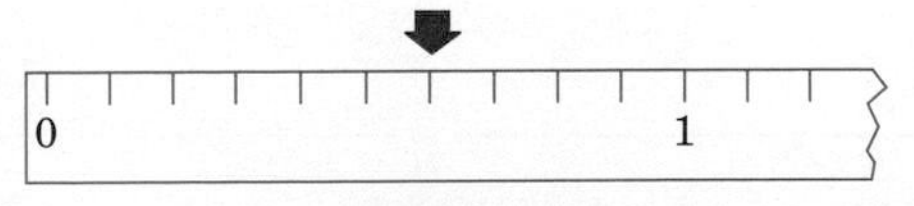

Figure 3.2

PROCEDURE

When using a shaded unit to picture a fraction:

$$\frac{\textbf{numerator}}{\textbf{denominator}} = \frac{\textbf{number of shaded parts}}{\textbf{total number of parts in one unit}}$$

When using a ruler to show a fraction:

$$\frac{\textbf{numerator}}{\textbf{denominator}} = \frac{\textbf{number of spaces between 0 and the arrow}}{\textbf{number of spaces between 0 and 1}}$$

MODEL PROBLEM-SOLVING

Examples

Strategy

a. Write the fraction for the following figure:

The unit is divided into five parts, and three of them are shaded.

The figure represents $\frac{3}{5}$.

Warm Up a. Write the fraction for the following figure:

b. Write the fraction for the following figure:

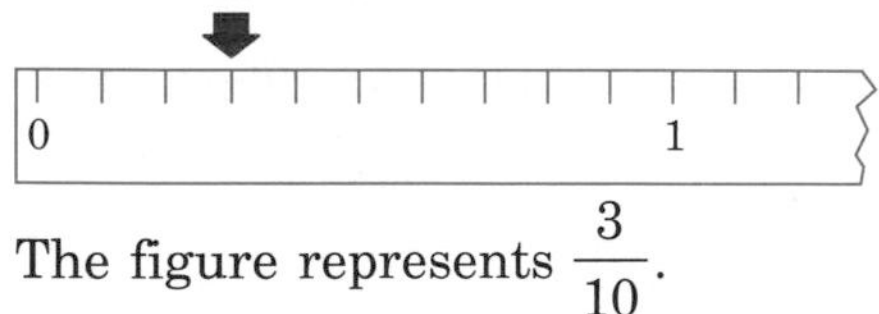

There are ten spaces between 0 and 1; therefore, the unit is divided into ten parts. There are three spaces between 0 and the arrow, indicating three out of the ten spaces.

The figure represents $\frac{3}{10}$.

Warm Up b. Write the fraction for the following figure:

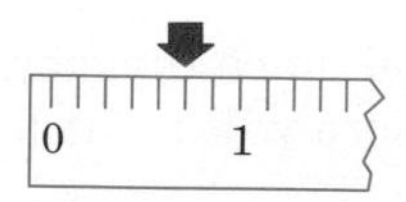

ANSWERS TO WARM UPS (3.1)

a. $\frac{2}{4}$ b. $\frac{5}{7}$

c. Write the fraction for the following figure:

The figure represents $\frac{6}{6}$, or 1.

The unit is divided into six parts, and all six of them are shaded. This is an example of an improper fraction.

Warm Up c. Write the fraction for the following figure:

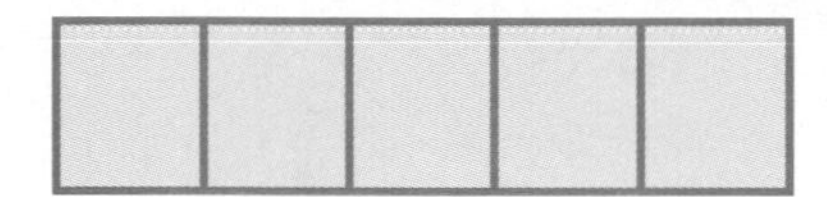

d. Write the fraction for the following figure:

The figure represents $\frac{4}{3}$.

There are two units, and each is divided into three parts. All three in the first unit are shaded, and one of the three parts is shaded in the second unit. Thus, a number greater than one is represented. A total of four of the three equal parts is shaded. This is an example of an improper fraction.

Warm Up d. Write the fraction for the following figure:

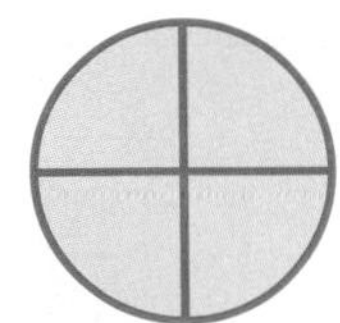

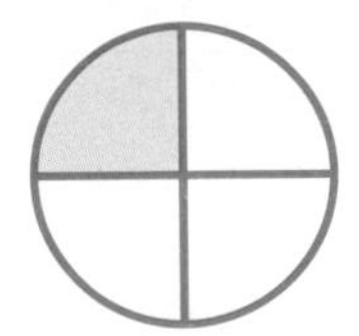

e. Write the fraction for the following figure:

(No shaded parts)

The figure represents $\frac{0}{3}$, or 0.

The unit is divided into three parts, and none of them is shaded.

Warm Up e. Write the fraction for the following figure:

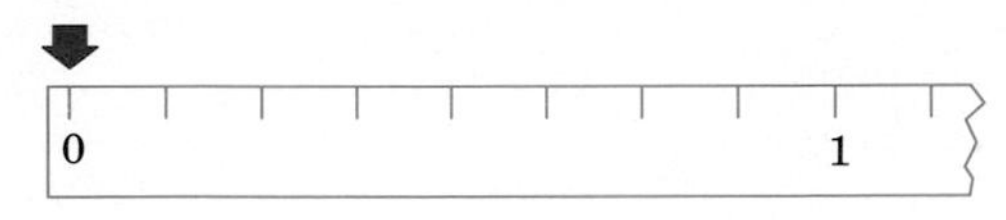

ANSWERS TO WARM UPS (3.1)

c. $\frac{5}{5}$ or 1 **d.** $\frac{5}{4}$ **e.** $\frac{0}{8}$ or 0

f. Which of these fractions are improper?

$\frac{4}{5}, \frac{2}{4}, \frac{5}{5}, \frac{7}{6}, \frac{22}{23}, \frac{23}{23}, \frac{24}{23}, \frac{20}{30}$

$\frac{5}{5}, \frac{7}{6}, \frac{23}{23}$, and $\frac{24}{23}$ are improper.

For a fraction to be improper, the numerator must be equal to or larger than the denominator.

Warm Up f. Which of these fractions are improper?

$\frac{9}{9}, \frac{13}{15}, \frac{6}{9}, \frac{11}{10}, \frac{5}{4}, \frac{14}{18}, \frac{7}{3}, \frac{18}{18}$

g. Which of the fractions in Example f are proper?

$\frac{4}{5}, \frac{2}{4}, \frac{22}{23}$, and $\frac{20}{30}$ are proper.

For a fraction to be proper, the numerator must be smaller than the denominator.

Warm Up g. Which of the fractions in Warm Up f are proper?

APPLICATION SOLUTION

h. At the beginning of the month the High Pressure Tire Company had 53 used tires in stock. During the month they sold 40 of those tires. What fraction of the total amount of used tires was sold?

The fraction representing the number of tires sold is 40 out of 53, or $\frac{40}{53}$. Therefore, the number of tires sold was $\frac{40}{53}$ of the total they had in stock.

The fraction is $\frac{\text{number sold}}{\text{number in stock}}$.

Warm Up h. In one month the Hardly Used Car agency sold 11 automobiles. Three of those automobiles were blue. What fraction of the cars sold was blue?

ANSWERS TO WARM UPS (3.1)

f. $\frac{9}{9}, \frac{11}{10}, \frac{5}{4}, \frac{7}{3}, \frac{18}{18}$ g. $\frac{13}{15}, \frac{6}{9}, \frac{14}{18}$ h. $\frac{3}{11}$

NAME ______________________

CLASS ____________ SECTION ____________ DATE ____________

Exercises 3.1

A

ANSWERS

1. ____________
2. ____________
3. ____________
4. ____________
5. ____________
6. ____________
7. ____________
8. ____________
9. ____________
10. ____________

Select the proper fractions from each list:

1. $\frac{4}{6}, \frac{5}{6}, \frac{6}{6}, \frac{7}{6}, \frac{8}{6}$

2. $\frac{3}{4}, \frac{6}{5}, \frac{12}{13}, \frac{14}{16}, \frac{17}{15}$

3. $\frac{6}{12}, \frac{7}{14}, \frac{9}{12}, \frac{10}{14}, \frac{11}{22}$

4. $\frac{8}{10}, \frac{8}{9}, \frac{8}{8}, \frac{8}{7}, \frac{8}{6}$

5. Select the fractions that represent the number one from the list in Exercise 1.

6. Select the improper fractions from this list: $\frac{5}{2}, \frac{8}{9}, \frac{11}{11}, \frac{18}{20}, \frac{1}{1}$

7. Select the fractions that represent the number one from the list in Exercise 6.

8. Select the fractions that represent the number one from the list in Exercise 4.

9. Select the improper fractions from this list:

$\frac{3}{8}, \frac{7}{5}, \frac{8}{3}, \frac{7}{9}, \frac{6}{6}$

10. Select the fraction that represents the number 1 in the previous exercise.

ANSWERS

11. __________

12. __________

13. __________

14. __________

15. __________

16. __________

17. __________

18. __________

Write the fraction for each figure:

11.

12.

13. 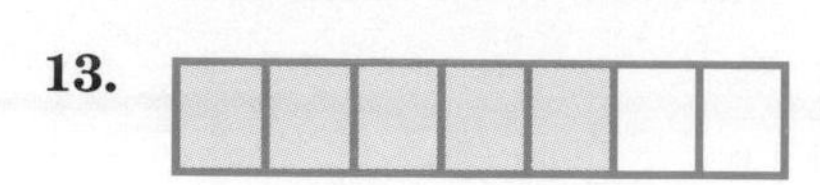

14. 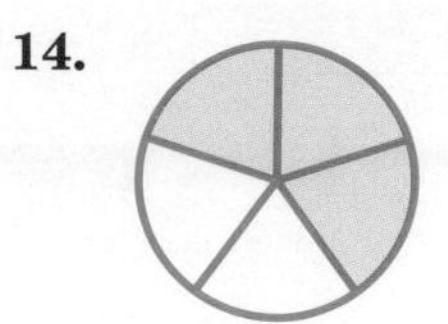

15. 

16. 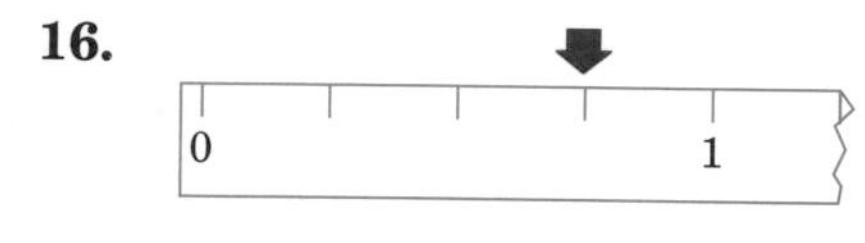

17. 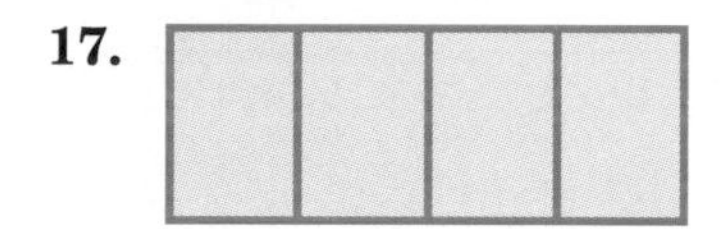

18. 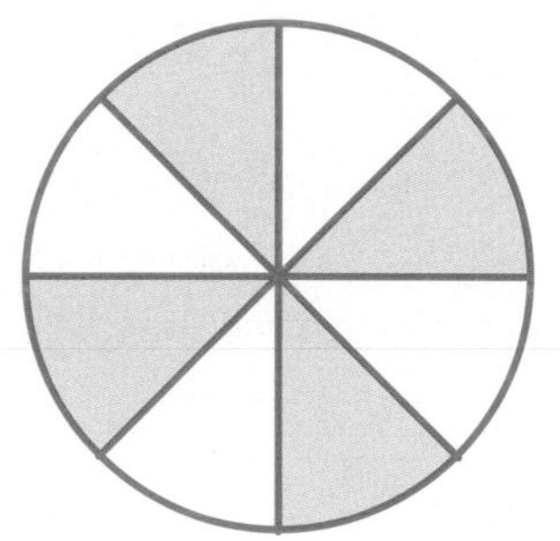

B

19. __________

20. __________

21. __________

22. __________

Which fractions in each list represent the number one?

19. $\frac{3}{3}, \frac{4}{5}, \frac{15}{15}, \frac{16}{17}, \frac{99}{100}, \frac{144}{144}$

20. $\frac{2}{3}, \frac{3}{4}, \frac{4}{5}, \frac{5}{6}, \frac{7}{7}, \frac{8}{9}, \frac{9}{10}, \frac{11}{10}, \frac{12}{12}$

21. Select the improper fractions from the list in Exercise 19.

22. Select the proper fractions from the list in Exercise 20.

NAME ____

CLASS ______________ SECTION ______________ DATE ______________

ANSWERS

23. ______________

24. ______________

25. ______________

26. ______________

27. ______________

28. ______________

29. ______________

30. ______________

Write the fraction for each figure:

23. One unit One unit

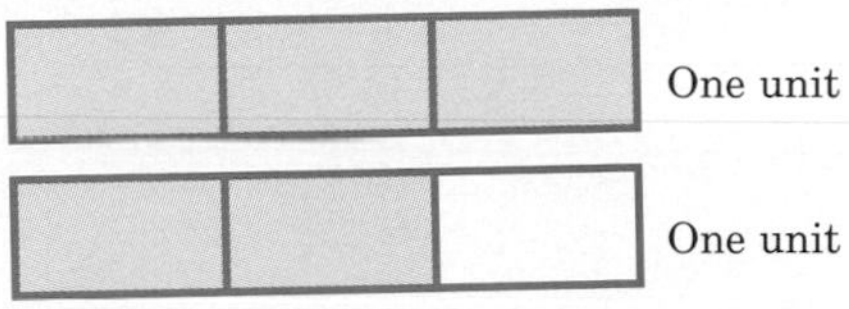

24. One unit One unit

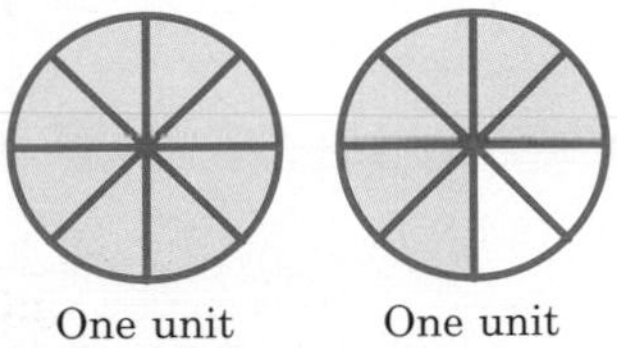

25. One unit One unit

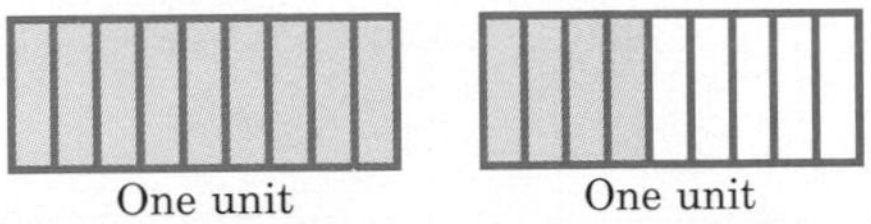

26. One unit One unit

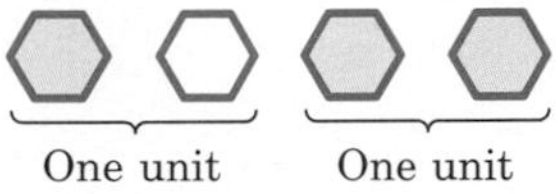

27. One unit

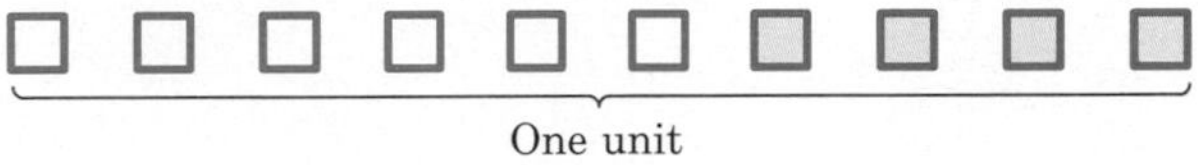

28. One unit

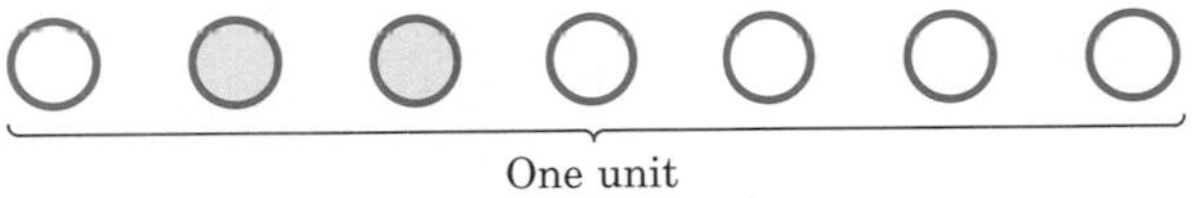

29. One unit

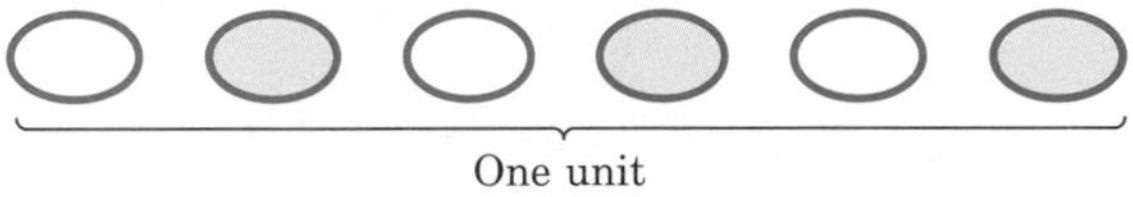

30. One unit

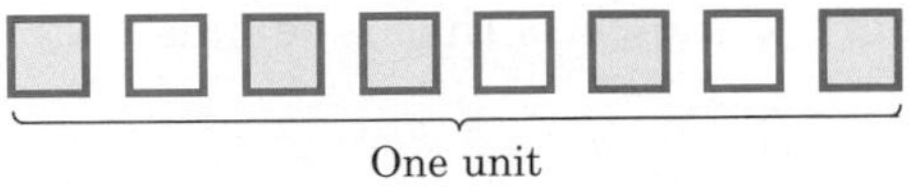

C

31. ______________

32. ______________

What fraction represents the distance from 0 to the arrow?

31. 0 1

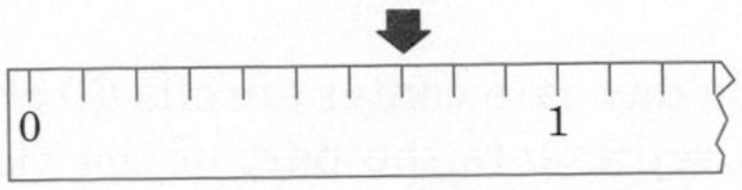

32. 0 1

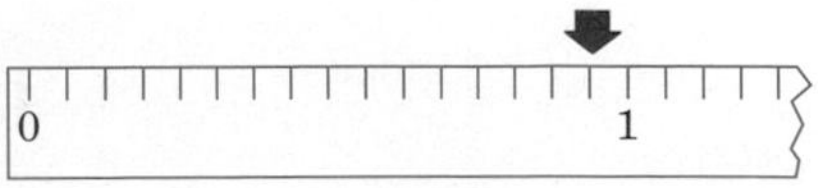

ANSWERS

33. ______________

34. ______________

35. ______________

33.

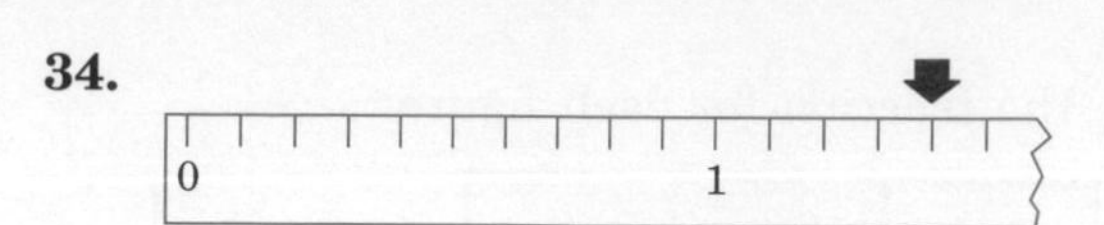

34.

35. Write a fraction that represents the number one, using only the digit 4.

36. Draw a unit divided into parts of equal size that shows $\frac{5}{5}$.

37. Draw a unit divided into parts of equal size that shows $\frac{5}{8}$.

38. Draw units divided into parts of equal size that show $\frac{7}{6}$.

D

39. ______________

39. In a class of 15 students there are 8 women. What fraction represents the part of the class that is female?

40. ______________

40. In a certain mathematics class there are 40 students; 19 of them are male. What fraction represents the part of the class that is male?

41. ______________

41. A certain day care center enrolls 60 children, 37 of whom are girls. What fraction represents the part of the class that is boys?

NAME ____________________

CLASS ____________ SECTION ____________ DATE ____________

ANSWERS

42. ____________

42. In the cooler of Hank's supermarket there are 7 bottles of cola and 3 bottles of noncola. What fraction of the bottles are cola?

43. ____________

43. In a sample of bricks taken from a certain job 25 were used and 32 were new. What fractional part was used?

44. ____________

44. If a six-cylinder motor has one cylinder that is not firing, what fractional part of the cylinders is not firing?

45. ____________

45. In a certain math class of 41 students 7 received a grade of B. What fraction represents the part of the class that received a B?

46. ____________

46. Twelve runners entered the mile race at the Last Chance track meet. Five of those runners were timed in less than four minutes. What fractional part does this represent?

47. ____________

47. Of the 12 women on a basketball team, 7 were less than 5 ft 9 in. tall. What fractional part of the team is *taller* than 5 ft 9 in.?

48. ____________

48. If 137 people applied for jobs and only 8 were hired, what fractional part of the applicants was hired?

E *Maintain your skills* (Sections 1.2, 2.3, 2.4, 2.5)

ANSWERS

49. ____________

50. ____________

51. ____________

52. ____________

53. ____________

54. ____________

55. ____________

56. ____________

57. ____________

58. ____________

49. Is 263 prime or composite?

50. Is 763 prime or composite?

51. Prime factor 231.

52. Prime factor 143.

53. Round 6007 to the nearest ten.

54. Round 6001 to the nearest thousand.

55. Is 550,033 divisible by 3?

56. Is 4,000,003 divisible by 5?

57. Bonnie's motorcycle averages 55 miles per gallon of gasoline, and the tank holds 3 gallons. How far can she travel on one tank of gasoline?

58. After a tune-up, Bonnie traveled 177 miles on a tank of gasoline. What was her gas mileage (number of miles per gallon)?

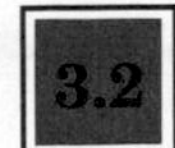

Improper Fractions and Mixed Numbers

OBJECTIVE

29. Change improper fractions to mixed numbers.
30. Change mixed numbers to improper fractions.

APPLICATION

Jill has 31 quarters. Write a mixed number to represent the number of dollars she has.

VOCABULARY

A *mixed number* is the sum of a whole number and a fraction $\left(3 + \frac{1}{2}\right)$ with the plus sign left out $\left(3\frac{1}{2}\right)$. The fraction part is usually a proper fraction.

HOW AND WHY

Some improper fractions can be changed to whole numbers:

$$\frac{6}{3} = 2$$

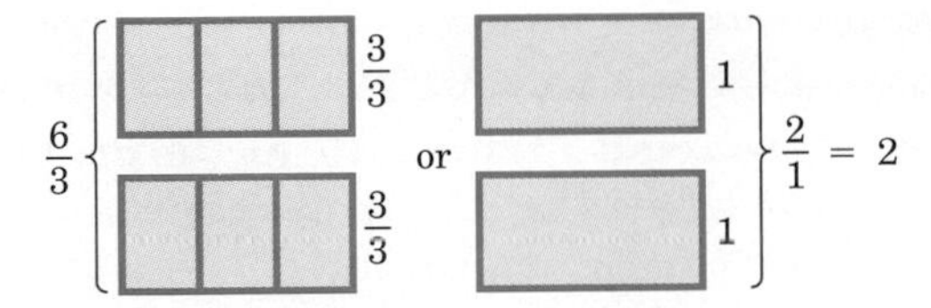

Some improper fractions can be changed to mixed numbers:

$$\frac{9}{7} = 1\frac{2}{7}$$

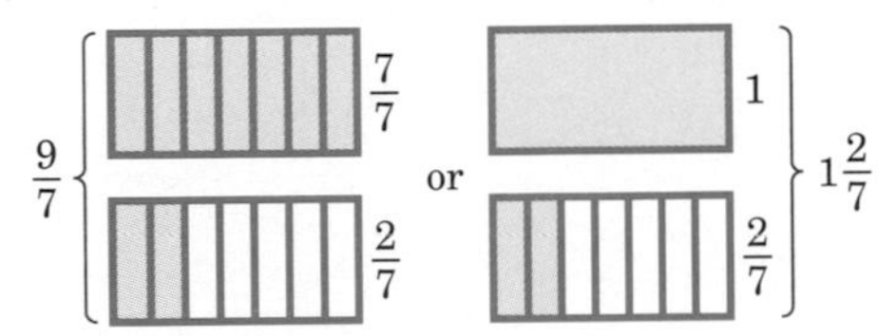

The shortcut for changing an improper fraction to a mixed number is to divide:

$$\frac{14}{7} = 14 \div 7 = 2$$

$$\frac{10}{7} = 10 \div 7 = 7\overline{)10}\ \ (\text{quotient } 1;\ 10 - 7 = 3) = 1\frac{3}{7}$$

PROCEDURE

To change an improper fraction to a mixed number, divide the numerator by the denominator. If there is a remainder, write the whole number and then write the fraction: $\frac{\text{remainder}}{\text{divisor}}$.

Despite the value judgment attached to the name "improper," in many cases improper fractions are a more convenient and useful form than mixed numbers. Thus, it is important to be able to convert from mixed numbers to improper fractions.

Every mixed number can be changed to an improper fraction:

$$1\frac{3}{7} = \frac{?}{7}$$

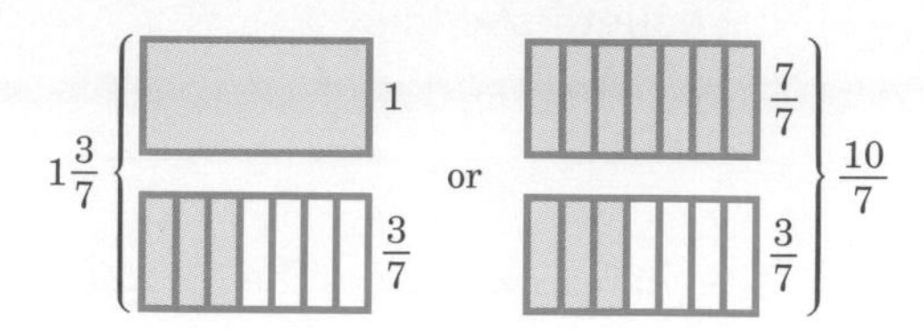

$$1\frac{3}{7} = \frac{10}{7}$$

The shortcut uses multiplication and addition:

$$1\frac{3}{7} = \frac{7 \cdot 1 + 3}{7} = \frac{7 + 3}{7} = \frac{10}{7}$$

PROCEDURE

To change a mixed number to an improper fraction multiply the denominator times the whole number, add the numerator, and put the sum over the denominator.

MODEL PROBLEM-SOLVING

Examples | **Strategy**

a. Change $\frac{8}{7}$ to a mixed number:

$$\frac{8}{7} = 7\overline{)8}\ (\text{quotient } 1,\ \text{remainder } 1) = 1\frac{1}{7}$$

Divide 8 by 7. The remainder is written as a fraction: $\frac{\text{remainder}}{\text{divisor}}$.

Warm Up a. Change $\frac{9}{8}$ to a mixed number.

ANSWERS TO WARM UPS (3.2)

a. $1\frac{1}{8}$

b. Change $\frac{12}{5}$ to a mixed number:

$$\frac{12}{5} = 5\overline{)12} = 2\frac{2}{5}$$
$$\begin{array}{r} 2 \\ 5\overline{)12} \\ \underline{10} \\ 2 \end{array}$$

Divide 12 by 5. The remainder is written as a fraction: $\frac{\text{remainder}}{\text{divisor}}$.

Warm Up b. Change $\frac{19}{7}$ to a mixed number.

c. Change $\frac{11}{3}$ to a mixed number:

$$\frac{11}{3} = 3\overline{)11} = 3\frac{2}{3}$$
$$\begin{array}{r} 3 \\ 3\overline{)11} \\ \underline{9} \\ 2 \end{array}$$

Divide 11 by 3.

Warm Up c. Change $\frac{19}{6}$ to a mixed number.

d. Change $\frac{114}{6}$ to a mixed number:

$$\frac{114}{6} = 6\overline{)114} = 19$$
$$\begin{array}{r} 19 \\ 6\overline{)114} \\ \underline{6} \\ 54 \\ \underline{54} \\ 0 \end{array}$$

Divide 114 by 6.

You could write 19 as $19\frac{0}{6}$; however, the fraction $\frac{0}{6}$ is seldom written. The exception is when we subtract mixed numbers.

Warm Up d. Change $\frac{115}{5}$ to a mixed number.

ANSWERS TO WARM UPS (3.2) **b.** $2\frac{5}{7}$ **c.** $3\frac{1}{6}$ **d.** 23

e. Change $\frac{211}{12}$ to a mixed number:

$$\frac{211}{12} = 12\overline{)211} = 17\frac{7}{12}$$

$$\begin{array}{r} 17 \\ 12\overline{)211} \\ \underline{12} \\ 91 \\ \underline{84} \\ 7 \end{array}$$

Divide 211 by 12.

Warm Up e. Change $\frac{319}{15}$ to a mixed number.

f. Change $2\frac{4}{5}$ to an improper fraction:

$$2\frac{4}{5} = \frac{5 \cdot 2 + 4}{5} = \frac{10 + 4}{5} = \frac{14}{5}$$

Multiply the denominator, 5, by the whole number, 2, and add the numerator, 4. Write this sum over the denominator, 5.

Warm Up f. Change $3\frac{5}{6}$ to an improper fraction.

g. Change $3\frac{1}{10}$ to an improper fraction:

$$3\frac{1}{10} = \frac{10 \cdot 3 + 1}{10} = \frac{30 + 1}{10} = \frac{31}{10}$$

Multiply the denominator, 10, by the whole number, 3, and add the numerator, 1. Write this sum over the denominator, 10.

Warm Up g. Change $4\frac{1}{12}$ to an improper fraction.

h. Change $4\frac{5}{9}$ to an improper fraction:

$$4\frac{5}{9} = \frac{9 \cdot 4 + 5}{9} = \frac{36 + 5}{9} = \frac{41}{9}$$

Find $9 \cdot 4 + 5$. Write this sum over the denominator, 9.

Warm Up h. Change $5\frac{5}{8}$ to an improper fraction.

ANSWERS TO WARM UPS (3.2) e. $21\frac{4}{15}$ f. $\frac{23}{6}$ g. $\frac{49}{12}$ h. $\frac{45}{8}$

i. Change 7 to an improper fraction:

$$7 = 7\frac{0}{1} = \frac{1 \cdot 7 + 0}{1} = \frac{7}{1}$$

Rewrite the whole number as a mixed number whose numerator is 0 and whose denominator is 1. Then follow the same procedure as in the previous examples.

Warm Up i. Change 8 to an improper fraction.

APPLICATION SOLUTION

j. Jill has 31 quarters. Write a mixed number to represent the number of dollars she has.

$$\frac{31}{4} = \begin{array}{r} 7 \\ 4\overline{)31} \\ \underline{28} \\ 3 \end{array} = 7\frac{3}{4}$$

Since there are four quarters in a dollar, we can write the 31 quarters as a mixed number by dividing 31 by 4. The remainder is written as a fraction: $\dfrac{\textbf{remainder}}{\textbf{divisor}}$**.**

So, Jill has $7\frac{3}{4}$ dollars.

Warm Up j. James has 49 dimes. Write a mixed number to represent the number of dollars he has.

ANSWERS TO WARM UPS (3.2)

i. $\frac{8}{1}$ or $\frac{16}{2}$ or $\frac{24}{3}$ etc

j. $4\frac{9}{10}$ dollars

NAME ______________________________

CLASS ______________ SECTION ______________ DATE ______________

Exercises 3.2

A

ANSWERS

1. ______________
2. ______________
3. ______________
4. ______________
5. ______________
6. ______________
7. ______________
8. ______________
9. ______________
10. ______________
11. ______________
12. ______________
13. ______________
14. ______________
15. ______________
16. ______________
17. ______________
18. ______________
19. ______________
20. ______________

Change each improper fraction to a mixed number:

1. $\frac{17}{4}$ **2.** $\frac{17}{5}$ **3.** $\frac{9}{2}$ **4.** $\frac{11}{5}$ **15.** $\frac{19}{4}$

6. $\frac{23}{6}$ **7.** $\frac{22}{9}$ **8.** $\frac{15}{7}$ **9.** $\frac{40}{8}$ **10.** $\frac{56}{7}$

Write these numbers as improper fractions:

11. $4\frac{3}{7}$ **12.** $3\frac{2}{7}$ **13.** 9 **14.** 11 **15.** $5\frac{3}{4}$

16. $4\frac{5}{6}$ **17.** $7\frac{1}{6}$ **18.** $10\frac{7}{8}$ **19.** $8\frac{2}{5}$ **20.** $8\frac{7}{9}$

B

ANSWERS

21. ________
22. ________
23. ________
24. ________
25. ________
26. ________
27. ________
28. ________
29. ________
30. ________
31. ________
32. ________
33. ________
34. ________
35. ________
36. ________
37. ________
38. ________
39. ________
40. ________

Change each improper fraction to a mixed number:

21. $\frac{195}{3}$ **22.** $\frac{195}{4}$ **23.** $\frac{195}{8}$ **24.** $\frac{195}{10}$ **25.** $\frac{48}{7}$

26. $\frac{73}{6}$ **27.** $\frac{78}{13}$ **28.** $\frac{108}{12}$ **29.** $\frac{148}{7}$ **30.** $\frac{145}{11}$

Write these numbers as improper fractions:

31. $27\frac{5}{8}$ **32.** $48\frac{1}{2}$ **33.** $15\frac{4}{5}$ **34.** $14\frac{2}{3}$ **35.** $16\frac{5}{6}$

36. $18\frac{7}{8}$ **37.** 15 **38.** 12 **39.** $32\frac{3}{7}$ **40.** $26\frac{2}{9}$

C

41. ________
42. ________
43. ________
44. ________
45. ________

Change each improper fraction to a mixed number:

41. $\frac{52}{13}$ **42.** $\frac{90}{25}$ **43.** $\frac{200}{31}$ **44.** $\frac{213}{41}$ **45.** $\frac{317}{11}$

NAME ______________________

CLASS ______________ SECTION ______________ DATE ______________

ANSWERS

46. ______________
47. ______________
48. ______________
49. ______________
50. ______________
51. ______________
52. ______________
53. ______________
54. ______________
55. ______________
56. ______________
57. ______________
58. ______________
59. ______________
60. ______________

46. $\frac{77}{20}$ **47.** $\frac{82}{25}$ **48.** $\frac{200}{33}$ **49.** $\frac{281}{19}$ **50.** $\frac{506}{25}$

Write these numbers as improper fractions:

51. $47\frac{7}{12}$ **52.** $30\frac{8}{15}$ **53.** $105\frac{3}{4}$ **54.** $146\frac{7}{10}$ **55.** $46\frac{7}{100}$

56. $46\frac{71}{100}$ **57.** $123\frac{23}{111}$ **58.** $210\frac{6}{123}$ **59.** 101 **60.** 1001

D

61. ______________

61. A scale is marked (as usual) with a whole number at each pound. What whole number mark will be closest to a measurement of $\frac{50}{16}$ lb?

62. ______________

62. A ruler is marked (as usual) with a whole number at each centimeter. What whole number mark will be closest to a measurement of $\frac{87}{10}$ cm?

ANSWERS

63. ______________

63. What is the total number of dollars in 31 dimes? Write as a mixed number.

64. ______________

64. Myra saved 76 cans, Judy saved 83 cans, and Ann saved 90 cans. Each can is worth a nickel. Write the total dollar amount as a mixed number.

65. ______________

65. Tad collected 75 bottles, Trevor collected 52, and Tyler collected 63. Each bottle is worth 10 cents. Write the total dollar amount collected as a mixed number.

66. ______________

66. Sue's construction company must place section barriers, each $\frac{1}{8}$ mile long, between the two sides of a freeway. How many such sections will be needed for $24\frac{3}{8}$ miles of freeway?

67. ______________

67. If $\frac{1}{2}$ inch represents 10 feet on a scale drawing, how many feet do $2\frac{1}{2}$ inches represent?

68. ______________

68. Big Burger advertises a quarter-pound hamburger for $2. How many of these burgers can be made from $23\frac{3}{4}$ lb of hamburger?

69. ______________

69. How much money can be made from the hamburgers in Exercise 68?

NAME ______________________

CLASS ______________ SECTION ______________ DATE ______________

ANSWERS

70. ______________

71. ______________

72. ______________

70. Jane purchased $12\frac{1}{2}$ yards of cement to pour slabs, each of which took $\frac{1}{2}$ yard. How many slabs did she pour?

71. Suppose the cement in Exercise 70 cost \$5 per $\frac{1}{2}$ yard. What was Jane's total cost?

72. James has $17\frac{3}{4}$ yards of rope to donate to the Daisy Day Care Center. It will be cut into $\frac{1}{4}$-yard pieces. How many pieces will there be?

E *Maintain your skills* (Sections 1.4, 1.5, 1.6, 1.7, 2.4, 2.5)

73. ______________

74. ______________

75. ______________

76. ______________

77. ______________

78. ______________

79. ______________

80. ______________

73. Divide: $49\overline{)245{,}098}$

74. Multiply: (702)(505)

75. Subtract: 8000 − 682

76. Divide: $8{,}030{,}000 \div 10^2$

77. Is 648 divisible by 3?

78. Prime factor 648

79. Find the value of 16^3

80. Multiply: $560 \times 10{,}000$

NAME ____

CLASS ______________ SECTION ______________ DATE ______________

ANSWERS

81. ______________

82. ______________

83. ______________

84. ______________

81. Divide: 12,400,000 ÷ 10,000

82. Find the value of 3^{11}

83. An electric motor turns 4500 revolutions per minute. How many revolutions does it turn if it runs 8 minutes?

84. The Lakehurst Gymnasium rents its handball court from 6 A.M. to 7 P.M. during the week. Each group of players is scheduled for 45 minutes. If maintenance on the courts is performed twice a day, morning and afternoon for a period of 30 minutes each, how many groups of players can use the court?

3.3 Reducing Fractions

OBJECTIVE

31. Reduce a fraction to lowest terms.

APPLICATION

Morris washes cars on Saturday to earn extra money. On a certain Saturday he has twelve cars to wash. After he has washed eight of them, what fraction of the total has he washed? Express your answer as a fraction reduced to lowest terms.

VOCABULARY

Reducing a fraction is the process of renaming it by using a smaller numerator and denominator. Reducing a fraction to *lowest terms* is writing an equivalent fraction such that the numerator and denominator have no common natural number factors. For instance, $\frac{10}{20} = \frac{1}{2}$. The process of dividing out the common factors from the numerator and denominator is called *canceling*.

HOW AND WHY

When we compare the two units in Figure 3.3, we note that each is divided into four parts. The shaded part on the left is named $\frac{2}{4}$, whereas the shaded part on the right is labeled $\frac{1}{2}$. It is clear that the two are the same size, and therefore we can say $\frac{2}{4} = \frac{1}{2}$.

The arithmetical way of showing that $\frac{2}{4} = \frac{1}{2}$ is to eliminate the common factors by canceling or dividing:

$$\frac{2}{4} = \frac{1 \cdot \overset{1}{\cancel{2}}}{2 \cdot \underset{1}{\cancel{2}}} = \frac{1}{2} \quad \text{or} \quad \frac{2}{4} = \frac{2 \div 2}{4 \div 2} = \frac{1}{2}$$

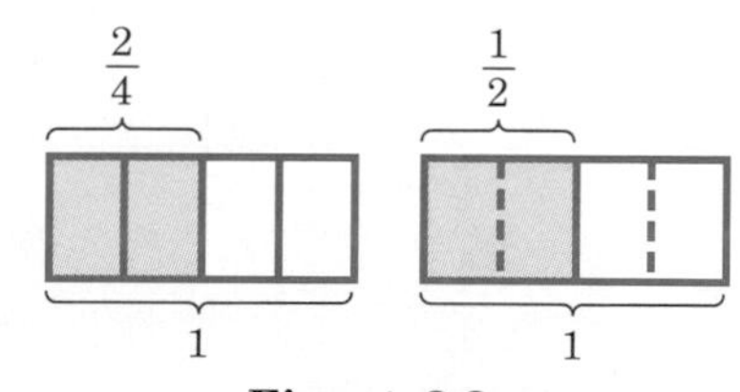

Figure 3.3

These methods work for reducing all fractions. Consider reducing $\frac{15}{18}$:

$$\frac{15}{18} = \frac{5 \cdot \overset{1}{\cancel{3}}}{6 \cdot \underset{1}{\cancel{3}}} = \frac{5}{6}$$ **Cancel out the common factors.**

$$\frac{15}{18} = \frac{15 \div 3}{18 \div 3} = \frac{5}{6}$$ **Divide out the common factors.**

When all common factors have been eliminated (canceled or divided out), the fraction is reduced to lowest terms:

$\frac{24}{40} = \frac{12}{20} = \frac{6}{10} = \frac{3}{5}$ **Reduced to lowest terms.**

If the common factors cannot be discovered easily, they can always be found by writing the numerator and denominator in prime-factored form. (See Example d.)

PROCEDURE

To reduce a fraction to lowest terms, eliminate all common factors (other than 1) in the numerator and denominator by canceling or dividing.

MODEL PROBLEM-SOLVING

Examples | **Strategy**

a. Reduce: $\frac{28}{21}$

$\frac{28}{21} = \frac{4 \cdot \overset{1}{\cancel{7}}}{3 \cdot \underset{1}{\cancel{7}}} = \frac{4}{3}$

There is a factor of 7 in both the numerator and denominator. Cancel the common factors by dividing out the 7s.

Warm Up a. Reduce: $\frac{36}{24}$

b. Reduce: $\frac{16}{24}$

$\frac{16}{24} = \frac{8 \cdot \overset{1}{\cancel{2}}}{12 \cdot \underset{1}{\cancel{2}}} = \frac{8}{12}$

The common factor of 2 is eliminated from both the numerator and denominator.

$= \frac{4 \cdot \overset{1}{\cancel{2}}}{6 \cdot \underset{1}{\cancel{2}}} = \frac{4}{6}$

There is still a factor of 2 in both the numerator and the denominator. The common factor is eliminated by dividing.

$= \frac{2 \cdot \overset{1}{\cancel{2}}}{3 \cdot \underset{1}{\cancel{2}}} = \frac{2}{3}$

Again, a common factor of 2 is divided out of both the numerator and the denominator. Since the numerator and the denominator no longer have a common factor, the fraction is reduced to lowest terms.

ANSWERS TO WARM UPS (3.3)

a. $\frac{3}{2}$

or

$$\frac{16}{24} = \frac{2 \cdot \overset{1}{\cancel{8}}}{3 \cdot \underset{1}{\cancel{8}}} = \frac{2}{3}$$

Rather than divide both the numerator and denominator by 2 three times, divide both by 8, and the fraction is reduced to lowest terms.

Warm Up b. Reduce: $\frac{18}{45}$

c. Reduce: $\frac{36}{54}$

$$\frac{36}{54} = \frac{36 \div 18}{54 \div 18} = \frac{2}{3}$$

Divide both the numerator and denominator by 18.

or

$$\frac{36}{54} = \frac{\overset{2}{\cancel{36}}}{\underset{3}{\cancel{54}}} = \frac{2}{3}$$

Warm Up c. Reduce: $\frac{48}{64}$

d. Reduce: $\frac{100}{600}$

$$\frac{100}{600} = \frac{100 \div 100}{600 \div 100} = \frac{1}{6}$$

Divide both numerator and denominator by 100.

or

$$\frac{1\cancel{0}\cancel{0}}{6\cancel{0}\cancel{0}} = \frac{1}{6}$$

Divide by 100 mentally using the shortcut division by powers of ten (section 1.7).

Warm Up d. Reduce: $\frac{4000}{14000}$

ANSWERS TO WARM UPS (3.3) **b.** $\frac{2}{5}$ **c.** $\frac{3}{4}$ **d.** $\frac{2}{7}$

e. Reduce: $\frac{126}{144}$

$$\frac{126}{144} = \frac{\overset{1}{\cancel{2}} \cdot \overset{1}{\cancel{3}} \cdot \overset{1}{\cancel{3}} \cdot 7}{\underset{1}{\cancel{2}} \cdot 2 \cdot 2 \cdot 2 \cdot \underset{1}{\cancel{3}} \cdot \underset{1}{\cancel{3}}}$$

Write in prime-factored form, since the numbers are large. Eliminate the common factors.

$$= \frac{7}{2 \cdot 2 \cdot 2}$$

$$= \frac{7}{8}$$

Multiply.

Warm Up e. Reduce: $\frac{160}{256}$

f. Reduce: $\frac{8}{9}$

$$\frac{8}{9} = \frac{2 \cdot 2 \cdot 2}{3 \cdot 3} = \frac{8}{9}$$

There are no common factors other than 1. The fraction cannot be reduced.

Warm Up f. Reduce: $\frac{16}{25}$

APPLICATION SOLUTION

g. Morris washes cars on Saturday to earn extra money. On a certain Saturday he has twelve cars to wash. After he has washed eight of them, what fraction of the total has he washed? Express your answer as a fraction reduced to lowest terms.

$$\frac{\text{Number of cars washed}}{\text{Total number of cars}} = \frac{8}{12}$$

To write the fraction that represents the part of the cars he has washed, we write the fraction as shown.

$$= \frac{2 \cdot \cancel{4}}{3 \cdot \cancel{4}}$$

$$= \frac{2}{3}$$

Eliminate the common factor of 4.

Therefore, Morris has washed $\frac{2}{3}$ of the cars.

Warm Up g. If in Example f Morris had washed only 4 cars, what fraction of the total had he washed?

ANSWERS TO WARM UPS (3.3) **e.** $\frac{5}{8}$ **f.** $\frac{16}{25}$ **g.** $\frac{1}{3}$

NAME ______________________

CLASS ______________ SECTION ______________ DATE ______________

Exercises 3.3

A

ANSWERS

1. ______
2. ______
3. ______
4. ______
5. ______
6. ______
7. ______
8. ______
9. ______
10. ______
11. ______
12. ______
13. ______
14. ______
15. ______
16. ______
17. ______
18. ______
19. ______
20. ______
21. ______
22. ______
23. ______
24. ______

Reduce to the lowest terms:

1. $\frac{5}{10}$ **2.** $\frac{4}{12}$ **3.** $\frac{3}{9}$ **4.** $\frac{8}{10}$

5. $\frac{10}{15}$ **6.** $\frac{14}{18}$ **7.** $\frac{10}{50}$ **8.** $\frac{30}{70}$

9. $\frac{12}{14}$ **10.** $\frac{12}{24}$ **11.** $\frac{18}{20}$ **12.** $\frac{14}{22}$

13. $\frac{28}{40}$ **14.** $\frac{20}{30}$ **15.** $\frac{48}{36}$ **16.** $\frac{33}{22}$

17. $\frac{12}{18}$ **18.** $\frac{20}{36}$ **19.** $\frac{28}{35}$ **20.** $\frac{45}{25}$

21. $\frac{16}{4}$ **22.** $\frac{40}{8}$ **23.** $\frac{45}{27}$ **24.** $\frac{55}{35}$

B

ANSWERS

25. ______
26. ______
27. ______
28. ______
29. ______
30. ______
31. ______
32. ______
33. ______
34. ______
35. ______
36. ______
37. ______
38. ______
39. ______
40. ______
41. ______
42. ______
43. ______
44. ______
45. ______
46. ______
47. ______
48. ______
49. ______
50. ______
51. ______
52. ______

25. $\frac{14}{24}$ **26.** $\frac{24}{36}$ **27.** $\frac{26}{36}$ **28.** $\frac{27}{36}$

29. $\frac{28}{36}$ **30.** $\frac{29}{36}$ **31.** $\frac{30}{36}$ **32.** $\frac{25}{45}$

33. $\frac{25}{75}$ **34.** $\frac{30}{75}$ **35.** $\frac{45}{75}$ **36.** $\frac{65}{75}$

37. $\frac{500}{800}$ **38.** $\frac{100}{400}$ **39.** $\frac{15}{40}$ **40.** $\frac{35}{40}$

41. $\frac{21}{36}$ **42.** $\frac{30}{48}$ **43.** $\frac{60}{12}$ **44.** $\frac{80}{16}$

45. $\frac{45}{32}$ **46.** $\frac{64}{72}$ **47.** $\frac{90}{126}$ **48.** $\frac{36}{100}$

49. $\frac{121}{132}$ **50.** $\frac{72}{144}$ **51.** $\frac{96}{144}$ **52.** $\frac{160}{144}$

NAME ______________________________

CLASS ______________ SECTION ______________ DATE ______________

ANSWERS

53. ______________

54. ______________

55. ______________

53. $\frac{200}{144}$ **54.** $\frac{255}{900}$ **55.** $\frac{279}{810}$

C

56. ______________

57. ______________

58. ______________

59. ______________

60. ______________

61. ______________

62. ______________

63. ______________

64. ______________

65. ______________

56. $\frac{98}{210}$ **57.** $\frac{268}{402}$ **58.** $\frac{97}{101}$ **59.** $\frac{98}{102}$

60. $\frac{153}{255}$ **61.** $\frac{200}{330}$ **62.** $\frac{546}{910}$ **63.** $\frac{630}{1050}$

64. $\frac{504}{1764}$ **65.** $\frac{294}{1617}$

D

66. ______________

66. Mary did a tune-up on her automobile. She found that two of the eight spark plugs were fouled. What fraction represents the number of fouled plugs? Reduce to lowest terms.

67. ______________

67. The float on a tank registers 8 feet. If the tank is full when it registers 12 feet, what fractional part of the tank is filled? Reduce to lowest terms.

ANSWERS

68. ____________________

69. ____________________

70. ____________________

71. ____________________

72. ____________________

73. ____________________

74. ____________________

68. If 12 of a total of 16 rolls of wire have been used, what fractional part of the wire is left? Reduce to lowest terms.

69. Gyrid has worked eight hours of a twelve-hour shift. Write the fraction that represents the part of her shift that is left.

70. An opening that is 3 inches wide is to be cut out of a piece of sheet metal that is 12 inches across. The opening is what fractional part of the width of the sheet metal? Reduce to lowest terms.

71. Jerry answered 18 of 20 problems correctly on a test. What fractional part did he answer correctly? Reduce your answer to lowest terms.

72. The local baseball team won 48 of the 56 games they played. What fractional part did they win? Reduce your answer to lowest terms.

73. Sharon earned $100 at her after school job in June. A new bike cost $75. What fractional part of her money will remain after she purchases the bike? Reduce to lowest terms.

74. A local basketball team won 25 games and lost 35. What fractional part of the games did the team win? Reduce to lowest terms.

NAME ______________________________

CLASS ____________ SECTION ____________ DATE ____________

ANSWERS

75. ______________

75. On a certain math test a student answers 24 problems correctly and 12 incorrectly. What fraction (in lowest terms) of the problems was correct?

76. ______________

76. Patty Mason solved 84 cases successfully while leaving 12 cases unsolved. What fractional part of the cases did she solve? Reduce to lowest terms.

E *Maintain your skills* (Sections 1.3, 1.4, 2.5, 2.6)

77. ______________

77. How much less than 5982 is 3456?

78. ______________

78. How much more is 6723 than 599?

79. ______________

79. Find the difference of 5111 and 3988.

80. ______________

80. Subtract 876 from 10,000.

81. ______________

81. Find the sum of 987, 789, 879, 87, 9, and 79.

ANSWERS

82. ____________

83. ____________

84. ____________

85. ____________

86. ____________

82. Find the total of 60234, 3078, 407, 20, 27011, and 78.

83. Prime factor 210.

84. Prime factor 187.

85. How many bricks, each weighing 4 pounds, will it take to weigh 6000 lb?

86. Trudy has to read all of her psychology text before the final week of the term. If the text is 980 pages long and Trudy does not want to read during weekends, how many pages must she read each day, on the average, to read the text before the start of the final week? Assume that the semester is 15 weeks long.

Multiplication of Fractions

OBJECTIVE

32. Multiply two or more fractions.

APPLICATION

During one year $\frac{7}{8}$ of all the cars sold by Trust-em Used Cars had automatic transmissions. Of the cars sold with automatic transmissions, $\frac{1}{70}$ had to be repaired before they were sold. What fractional part of the total cars sold had automatic transmissions and had to be repaired?

VOCABULARY

A *product* is the answer to a multiplication problem.

HOW AND WHY

What is $\frac{1}{2}$ of $\frac{1}{3}$ or $\frac{1}{2} \cdot \frac{1}{3} = ?$ See Figure 3.4. The rectangle is divided into three parts. One part, $\frac{1}{3}$, is shaded blue. To find $\frac{1}{2}$ of the shaded third, divide each of the thirds into two parts (halves). Figure 3.5 shows the rectangle divided into six parts. So, $\frac{1}{2}$ of the shaded third is $\frac{1}{6}$ of the rectangle which is shaded yellow:

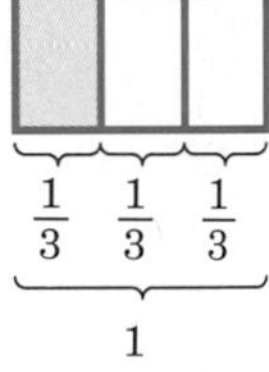

Figure 3.4

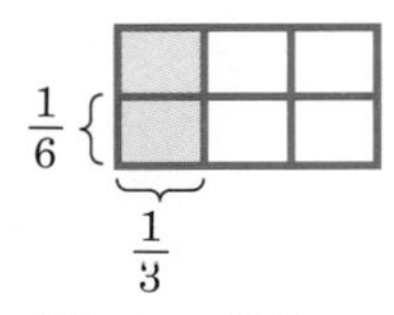

Figure 3.5

$$\frac{1}{2} \text{ of } \frac{1}{3} = \frac{1}{2} \cdot \frac{1}{3} = \frac{1}{6} = \frac{\text{number of parts double-shaded}}{\text{total number of parts}}$$

What is $\frac{1}{4}$ of $\frac{3}{4}$? $\left(\frac{1}{4} \cdot \frac{3}{4} = ?\right)$

In Figure 3.6 the rectangle has been divided into four parts, and $\frac{3}{4}$ is represented by the parts that are shaded blue. To find $\frac{1}{4}$ of the $\frac{3}{4}$, divide each of the fourths into four parts. The rectangle is now divided into 16 parts, so that $\frac{1}{4}$ of each of the three original fourths is shaded yellow and represents $\frac{3}{16}$. (See Fig. 3.7.)

$$\frac{1}{4} \text{ of } \frac{3}{4} = \frac{1}{4} \cdot \frac{3}{4} = \frac{3}{16}$$

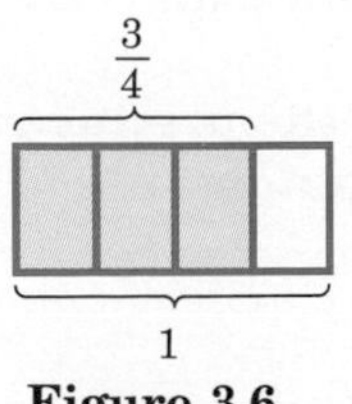

Figure 3.6

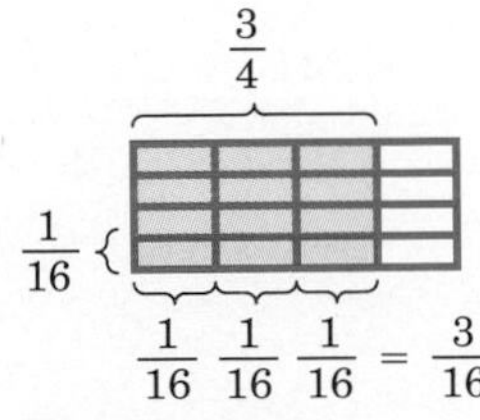

Figure 3.7

We have seen that $\frac{1}{2} \cdot \frac{1}{3} = \frac{1}{6}$ and $\frac{1}{4} \cdot \frac{3}{4} = \frac{3}{16}$. The shortcut is to multiply the numerators and multiply the denominators.

To multiply two or more fractions, write the product of the numerators over the product of the denominators. So:

$$\frac{12}{35} \cdot \frac{25}{18} = \frac{300}{630}$$

$$= \frac{300 \div 10}{630 \div 10}$$ **Divide both the numerator and denominator by 10.**

$$= \frac{30}{63}$$

$$= \frac{30 \div 3}{63 \div 3}$$ **Divide both the numerator and denominator by 3.**

$$= \frac{10}{21}$$ **The fraction is reduced to lowest terms.**

Multiplying two or more fractions, such as the preceding, can be done more quickly by reducing before multiplying:

$$\frac{12}{35} \cdot \frac{25}{18} = \frac{\overset{6}{\cancel{12}}}{35} \cdot \frac{25}{\underset{9}{\cancel{18}}}$$ **Divide 12 and 18 by 2.**

$$= \frac{\overset{2}{\cancel{6}}}{\underset{7}{\cancel{35}}} \cdot \frac{\overset{5}{\cancel{25}}}{\underset{3}{\cancel{9}}}$$ **Divide 25 and 35 by 5.**
Divide 6 and 9 by 3.

$$= \frac{10}{21}$$ **Multiply.**

The next example shows all of the reducing done in one step.

$$\frac{\overset{\overset{1}{\cancel{3}}}{\cancel{24}}}{\underset{\underset{1}{\cancel{4}}}{\cancel{32}}} \cdot \frac{\overset{2}{\cancel{8}}}{\underset{3}{\cancel{9}}} = \frac{2}{3}$$ **Divide 24 and 32 by 8, then divide 3 and 9 by 3, and finally divide 8 and 4 by 4. Then multiply.**

If the numbers are large, prime factor each numerator and denominator. (See Example g.)

PROCEDURE

To multiply two or more fractions:

1. **Eliminate all common factors in the numerators and denominators.**
2. **Write the product of the numerators over the product of the denominators.**

MODEL PROBLEM-SOLVING

Examples | **Strategy**

a. Multiply:

$$\frac{1}{2}\cdot\frac{3}{4}\cdot\frac{5}{8}=\frac{1\cdot 3\cdot 5}{2\cdot 4\cdot 8}=\frac{15}{64}$$

No common factors. Write the product of the numerators over the product of the denominators.

Warm Up a. Multiply: $\frac{1}{3}\cdot\frac{4}{7}\cdot\frac{8}{5}$

b. Multiply:

$$\frac{5}{2}\cdot 3=\frac{5}{2}\cdot\frac{3}{1}=\frac{15}{2}$$

Write the whole number as an improper fraction and multiply.

Warm Up b. Multiply: $\frac{8}{3}\cdot 4$

c. Multiply:

$$\frac{3}{4}\cdot\frac{6}{7}$$

Note that 4 and 6 have a common factor of 2.

$$\frac{3}{4}\cdot\frac{6}{7}=\frac{3}{\underset{2}{\cancel{4}}}\cdot\frac{\overset{3}{\cancel{6}}}{7}=\frac{9}{14}$$

Eliminate the common factors. Multiply.

Warm Up c. Multiply: $\frac{7}{8}\cdot\frac{4}{5}$

d. Multiply:

$$\frac{8}{15}\cdot\frac{5}{12}$$

Note that 8 and 12 have a common factor of 4 and that 5 and 15 have a common factor of 5.

$$\frac{8}{15}\cdot\frac{5}{12}=\frac{\overset{2}{\cancel{8}}}{\underset{3}{\cancel{15}}}\cdot\frac{\overset{1}{\cancel{5}}}{\underset{3}{\cancel{12}}}=\frac{2}{9}$$

Those common factors are divided out. Multiply.

Warm Up d. Multiply: $\frac{12}{15}\cdot\frac{9}{8}$

ANSWERS TO WARM UPS (3.4) **a.** $\frac{32}{105}$ **b.** $\frac{32}{3}$ **c.** $\frac{7}{10}$ **d.** $\frac{9}{10}$

e. Multiply:

$$\frac{7}{9}\cdot\frac{18}{5}\cdot\frac{10}{21}=\frac{\overset{1}{\cancel{7}}}{\underset{1}{\cancel{9}}}\cdot\frac{\overset{2}{\cancel{18}}}{\underset{1}{\cancel{5}}}\cdot\frac{\overset{2}{\cancel{10}}}{\underset{3}{\cancel{21}}}=\frac{4}{3}$$

The common factors of 9, 5, and 7 are eliminated.

Warm Up e. Multiply: $\frac{8}{9}\cdot\frac{12}{18}\cdot\frac{15}{16}$

f. Multiply:

$$\frac{20}{30}\cdot\frac{15}{88}=\frac{2\cdot 2\cdot 5}{2\cdot 3\cdot 5}\cdot\frac{3\cdot 5}{2\cdot 2\cdot 2\cdot 11}$$

$$=\frac{\cancel{2}\cdot\cancel{2}\cdot\cancel{5}\cdot\cancel{3}\cdot 5}{\cancel{2}\cdot\cancel{3}\cdot\cancel{5}\cdot\cancel{2}\cdot 2\cdot 2\cdot 11}$$

$$=\frac{5}{2\cdot 2\cdot 11}=\frac{5}{44}$$

Prime factorization is used because there are so many common factors that it is not easy to see all of them.

Warm Up f. Multiply: $\frac{32}{45}\cdot\frac{35}{24}$

APPLICATION SOLUTION

g. During one year $\frac{7}{8}$ of all the cars sold by Trust-em Used Cars had automatic transmissions. Of the cars sold with automatic transmissions, $\frac{1}{70}$ had to be repaired before they were sold. What fractional part of the total cars sold had automatic transmissions and had to be repaired?

$$\frac{7}{8}\cdot\frac{1}{70}=$$

To find what part of the total cars sold had automatic transmissions and had repairs, we multiply the fractional part that had automatic transmissions by the fractional part that had repairs.

$$\frac{\overset{1}{\cancel{7}}}{8}\cdot\frac{1}{\underset{10}{\cancel{70}}}=\frac{1}{80}$$

Divide out the factor 7.

Therefore, $\frac{1}{80}$ of the total cars sold had automatic transmissions and had to be repaired.

Warm Up g. During one year $\frac{2}{3}$ of all the tires sold by the Tire Factory were highway tread tires. If $\frac{1}{40}$ of the highway treads had to be repaired, what fractional part of the tires sold had to be repaired?

ANSWERS TO WARM UPS (3.4)

e. $\frac{5}{9}$ **f.** $\frac{28}{27}$ or $1\frac{1}{27}$ **g.** $\frac{1}{60}$

NAME ______________________

CLASS ____________ SECTION ____________ DATE ____________

Exercises 3.4

A

ANSWERS

1. ____________
2. ____________
3. ____________
4. ____________
5. ____________
6. ____________
7. ____________
8. ____________
9. ____________
10. ____________
11. ____________
12. ____________
13. ____________
14. ____________
15. ____________
16. ____________
17. ____________
18. ____________

Multiply. Reduce to the lowest terms:

1. $\frac{1}{2} \cdot \frac{3}{4}$

2. $\frac{1}{3} \cdot \frac{2}{3}$

3. $\frac{7}{8} \cdot \frac{5}{6}$

4. $\frac{7}{5} \cdot \frac{6}{11}$

5. $\frac{2}{3} \cdot \frac{2}{5}$

6. $\frac{1}{2} \cdot \frac{3}{14}$

7. $\frac{5}{4} \cdot \frac{7}{12}$

8. $\frac{1}{2} \cdot \frac{1}{2}$

9. $\frac{1}{2} \cdot \frac{2}{3}$

10. $\frac{2}{3} \cdot \frac{3}{8}$

11. $\frac{5}{8} \cdot \frac{4}{15}$

12. $\frac{15}{20} \cdot \frac{8}{12}$

13. $\frac{9}{12} \cdot \frac{10}{15}$

14. $\frac{5}{24} \cdot \frac{8}{10}$

15. $\frac{4}{9} \cdot \frac{3}{8}$

16. $\frac{7}{9} \cdot \frac{3}{5}$

17. $5 \cdot \frac{25}{40}$

18. $\frac{25}{36} \cdot 12$

B

ANSWERS

19. ______
20. ______
21. ______
22. ______
23. ______
24. ______
25. ______
26. ______
27. ______
28. ______
29. ______
30. ______
31. ______
32. ______
33. ______
34. ______
35. ______
36. ______

19. $\frac{4}{5} \cdot \frac{7}{8}$

20. $\frac{7}{8} \cdot \frac{8}{21}$

21. $\frac{5}{11} \cdot 6$

22. $7 \cdot \frac{2}{3}$

23. $\frac{6}{11} \cdot \frac{11}{6}$

24. $\frac{7}{8} \cdot \frac{4}{21}$

25. $\frac{4}{6} \cdot \frac{9}{30} \cdot \frac{5}{3}$

26. $\frac{2}{3} \cdot \frac{4}{5} \cdot \frac{6}{7}$

27. $5 \cdot \frac{1}{4} \cdot \frac{8}{15}$

28. $\frac{15}{2} \cdot 8 \cdot \frac{1}{9}$

29. $\frac{3}{5} \cdot \frac{5}{2} \cdot \frac{2}{3}$

30. $\frac{8}{10} \cdot \frac{3}{4} \cdot \frac{2}{9}$

31. $\frac{1}{2} \cdot \frac{2}{3} \cdot \frac{3}{4} \cdot \frac{4}{5}$

32. $4 \cdot \frac{1}{2} \cdot \frac{3}{4} \cdot \frac{2}{5}$

33. $\frac{5}{3} \cdot \frac{4}{7} \cdot 5 \cdot \frac{2}{3}$

34. $\frac{7}{9} \cdot \frac{5}{3} \cdot \frac{11}{13} \cdot 4$

35. $\frac{5}{8} \cdot \frac{12}{15} \cdot \frac{18}{30}$

36. $\frac{20}{25} \cdot \frac{15}{16} \cdot \frac{36}{24}$

C

37. ______
38. ______
39. ______

37. $\frac{32}{40} \cdot \frac{25}{28}$

38. $\frac{7}{2} \cdot \frac{4}{3} \cdot \frac{6}{35}$

39. $\frac{25}{60} \cdot \frac{30}{75} \cdot \frac{15}{10}$

NAME ______________________

CLASS ______________ SECTION ______________ DATE ______________

ANSWERS

40. ______________

41. ______________

42. ______________

43. ______________

44. ______________

45. ______________

46. ______________

47. ______________

48. ______________

49. ______________

50. ______________

51. ______________

52. ______________

53. ______________

54. ______________

40. $\frac{24}{90} \cdot \frac{80}{72} \cdot \frac{120}{200}$

41. $\frac{3}{8} \cdot \frac{4}{7} \cdot \frac{8}{3} \cdot \frac{14}{24}$

42. $\frac{3}{4} \cdot \frac{4}{5} \cdot \frac{10}{21} \cdot \frac{7}{2}$

43. $\frac{12}{20} \cdot \frac{35}{24}$

44. $\frac{36}{55} \cdot \frac{33}{54}$

45. $\frac{7}{12} \cdot \frac{9}{14} \cdot \frac{15}{21} \cdot \frac{7}{10}$

46. $\frac{24}{35} \cdot \frac{40}{44} \cdot \frac{77}{96}$

47. $\frac{56}{65} \cdot \frac{39}{48} \cdot \frac{18}{25}$

48. $\frac{30}{48} \cdot \frac{5}{14} \cdot \frac{7}{25}$

49. $\frac{3}{18} \cdot \frac{9}{10} \cdot \frac{5}{14} \cdot \frac{8}{21}$

50. $\frac{32}{39} \cdot \frac{24}{96} \cdot \frac{52}{72}$

51. $20 \cdot \frac{3}{5} \cdot \frac{7}{12} \cdot \frac{1}{9}$

52. $\frac{22}{25} \cdot \frac{9}{10} \cdot \frac{75}{33}$

53. $\frac{243}{1000} \cdot \frac{25}{81} \cdot \frac{8}{9}$

54. $\frac{27}{45} \cdot \frac{144}{100} \cdot \frac{10}{36}$

D

55. ______________

55. Frank's favorite cake recipe calls for $\frac{3}{4}$ cup of sugar and $\frac{1}{2}$ cup of milk. How much sugar and how much milk will he need to make $\frac{2}{3}$ of the recipe?

56. ______

57. ______

58. ______

59. ______

60. ______

61. ______

62. ______

56. A container holds $\frac{3}{4}$ gallons. What part of a gallon does it contain when it is $\frac{3}{4}$ full?

57. A rectangle is $\frac{5}{8}$ as wide as it is long. How wide is the rectangle if its length is $\frac{4}{5}$ inch?

58. An article was priced to sell for \$6. During a sale the price was reduced $\frac{1}{3}$. By how many dollars was the price reduced?

59. A shirt normally sells for \$20. What is the sale price if it is marked $\frac{1}{5}$ off?

60. In a certain school $\frac{3}{8}$ of the student body take a foreign language. Of those who take a foreign language, $\frac{1}{2}$ takes French. What part of the student body takes French?

61. Lois spends $\frac{1}{2}$ of the family income on rent, utilities, and food. She pays $\frac{3}{7}$ of this amount for rent. What fraction of the family income goes for rent?

62. A building materials business is stocking up on $\frac{5}{8}$-inch plywood for a sale. They have 145 sheets. How high will the sheets reach if they are stacked?

NAME ____________________

CLASS ____________ SECTION ____________ DATE ____________

ANSWERS

63. ____________

63. The width of a door is one-third the height. If the height is 7 feet, find the width.

64. ____________

64. The radius of a circle is one-half the diameter. If the diameter is 9 inches, find the radius.

65. ____________

65. The gasoline tank of an automobile holds 18 gallons. The gauge registers full at the beginning of a trip, and at the end of the trip it indicates that $\frac{1}{4}$ tank was used. How many gallons of gasoline were used on the trip?

66. ____________

66. One-third of Sue's paycheck is deducted for taxes and insurance. If $\frac{1}{4}$ of the deductions are for insurance, what fractional part of her paycheck is deducted for insurance?

E *Maintain your skills* (Sections 1.2, 1.5, 2.6)

67. ____________

67. Find the LCM of 24, 36, and 48.

68. ____________

68. Find the LCM of 12, 15, and 18.

69. ____________

69. Find the LCM of 16, 17, and 18.

ANSWERS

70. ______________

71. ______________

72. ______________

73. ______________

74. ______________

75. ______________

76. ______________

70. Find the LCM of 18, 20, and 22.

71. Multiply and round to the nearest hundred: (781)(34)

72. Multiply and round to the nearest thousand: (655)(477)

73. Divide and round to the nearest ten: $66\overline{)8118}$

74. Divide 172,676 by 49 and round to the nearest hundred.

75. Gerri needs 55 building blocks, each of which weighs 18 lb. Her pick-up can safely carry 1000 lb. Will she be able to haul all the blocks in one load? That is, do the 55 blocks weigh less than 1000 lb? If so, how much less?

76. Kevin needs a load of gravel for a drainage field. His truck can safely carry a load of one ton (2000 lb). If the gravel is sold in 222-lb scoops, how many scoops can Kevin safely haul?

Division of Fractions

OBJECTIVES

33. Find the reciprocal of a number.
34. Divide two fractions.

APPLICATION

If the distance a nut moves on a bolt with one turn is $\frac{3}{16}$ inch, how many turns will it take to move the nut $\frac{3}{4}$ inch?

VOCABULARY

If two fractions have a product of 1, either fraction is called the *reciprocal* of the other. For example, $\frac{2}{3}$ is the reciprocal of $\frac{3}{2}$.

HOW AND WHY

Writing a reciprocal is often called "inverting" the fraction. For instance, the reciprocal of $\frac{3}{7}$ is $\frac{7}{3}$. We check by showing that the product is 1:

$$\frac{3}{7} \cdot \frac{7}{3} = \frac{21}{21} = 1$$

PROCEDURE

The reciprocal of a fraction is found by interchanging the numerator and the denominator.

The reciprocal of a whole number is found by first writing the whole number as a fraction: the reciprocal of 21 $\left(21 = \frac{21}{1}\right)$ is $\frac{1}{21}$.

CAUTION

The number zero, 0, does not have a reciprocal.

It is pointed out in Chapter 1 that division is the inverse of multiplication. That is, the answer to a division problem is the number that is multiplied times the divisor (second number), which will give the first number as an answer. Another way of thinking of division is to ask, "How many groups of a certain size are contained in a number?"

	THINK	ANSWER
$6 \div 2$	How many twos in six?	3
$\frac{4}{5} \div \frac{1}{10}$	How many one-tenths in four-fifths?	See Figure 3.8.

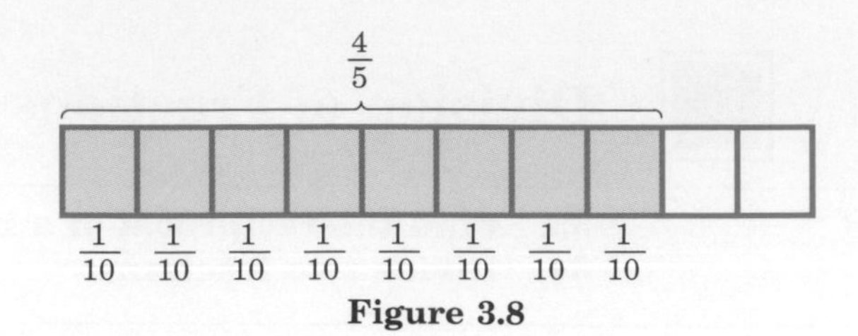

Figure 3.8

In Figure 3.8 we see that there are eight one-tenths in four-fifths. Therefore, we can say:

$$\frac{4}{5} \div \frac{1}{10} = 8$$

Since $8 \cdot \frac{1}{10} = \frac{8}{10} = \frac{4}{5}$, we know that the answer is correct.

The answer can also be obtained from the fractions by multiplying $\frac{4}{5}$ by the reciprocal of $\frac{1}{10}$:

$$\frac{4}{5} \div \frac{1}{10} = \frac{4}{5} \cdot \frac{10}{1} = \frac{40}{5} = 8$$

PROCEDURE

To divide two fractions, multiply the first fraction by the reciprocal of the divisor; that is, invert the divisor and multiply.

CAUTION

Do not reduce the fractions before changing the division to a multiplication, that is, invert the divisor before reducing.

MODEL PROBLEM-SOLVING

Examples | **Strategy**

a. What is the reciprocal of $\frac{7}{10}$?

The reciprocal of $\frac{7}{10}$ is $\frac{10}{7}$. — **Invert $\frac{7}{10}$ to get $\frac{10}{7}$.**

Check: $\frac{7}{10} \cdot \frac{10}{7} = \frac{70}{70} = 1$

Warm Up a. What is the reciprocal of $\frac{8}{11}$?

ANSWERS TO WARM UPS (3.5)

a. $\frac{11}{8}$

b. What is the reciprocal of $1\frac{4}{5}$?

The mixed number $1\frac{4}{5}$ is:

$1\frac{4}{5} = \frac{9}{5}$

Change the mixed number to an improper fraction.

The reciprocal of $\frac{9}{5}$ is $\frac{5}{9}$.

Invert the fraction.

Therefore, the reciprocal of $1\frac{4}{5}$ is $\frac{5}{9}$.

Check: $\frac{9}{5} \cdot \frac{5}{9} = \frac{45}{45} = 1$

Warm Up b. What is the reciprocal of $1\frac{5}{6}$?

c. Divide and reduce if possible:

$$\frac{7}{23} \div \frac{16}{23} = \frac{7}{\cancel{23}} \cdot \frac{\cancel{23}}{16}$$

$$= \frac{7}{16}$$

Invert the divisor and multiply. The like factors of 23 are eliminated.

Warm Up c. Divide and reduce if possible: $\frac{11}{19} \div \frac{15}{19}$

d. Divide and reduce if possible:

$$\frac{8}{3} \div \frac{4}{5} = \frac{\overset{2}{\cancel{8}}}{3} \cdot \frac{5}{\underset{1}{\cancel{4}}}$$

$$= \frac{10}{3} \text{ or } 3\frac{1}{3}$$

Invert the divisor and multiply. Divide out like factors of 4.

Warm Up d. Divide and reduce if possible: $\frac{9}{2} \div \frac{6}{5}$

ANSWERS TO WARM UPS (3.5) **b.** $\frac{6}{11}$ **c.** $\frac{11}{15}$ **d.** $\frac{15}{4}$ or $3\frac{3}{4}$

e. Divide and reduce if possible:

$$\frac{1}{12} \div \frac{3}{5} = \frac{1}{12} \cdot \frac{5}{3}$$

$$= \frac{5}{36}$$

Invert the divisor and multiply. There are no common factors so the quotient (answer) is in lowest terms.

CAUTION

Do not reduce before changing to multiplication.

Warm Up e. Divide and reduce if possible: $\frac{5}{11} \div \frac{2}{3}$

APPLICATION SOLUTION

f. If the distance a nut moves on a bolt with one turn is $\frac{3}{16}$ inch, how many turns will it take to move the nut $\frac{3}{4}$ inch?

$$\frac{3}{4} \div \frac{3}{16}$$

To find the number of turns needed to move the nut the required distance, divide the required distance by the distance the nut moves in one turn.

$$\frac{3}{4} \div \frac{3}{16} = \frac{\cancel{3}^{1}}{\cancel{4}_{1}} \cdot \frac{\cancel{16}^{4}}{\cancel{3}_{1}}$$

$$= 4$$

Invert the divisor and multiply.

Therefore, it takes 4 turns to move the nut $\frac{3}{4}$ inch.

Warm Up f. If in Example g. the distance the nut moves on the bolt with one turn is $\frac{3}{32}$ inch, how many turns will it take to move the nut $\frac{3}{4}$ inch?

ANSWERS TO WARM UPS (3.5)

e. $\frac{15}{22}$ **f.** 8 turns

NAME ______________________

CLASS ____________ SECTION ____________ DATE ____________

Exercises 3.5

A

ANSWERS

1. ________
2. ________
3. ________
4. ________
5. ________
6. ________
7. ________
8. ________
9. ________
10. ________
11. ________
12. ________
13. ________
14. ________
15. ________
16. ________
17. ________
18. ________
19. ________
20. ________
21. ________
22. ________

1. What is the reciprocal of $\frac{3}{4}$?

2. What is the reciprocal of $\frac{9}{7}$?

3. What is the reciprocal of 5?

4. What is the reciprocal of $3\frac{2}{3}$?

5. What is the reciprocal of $1\frac{1}{2}$?

6. What is the reciprocal of 0?

Divide. Reduce to the lowest terms:

7. $\frac{3}{7} \div \frac{5}{6}$

8. $\frac{9}{8} \div \frac{3}{4}$

9. $\frac{7}{10} \div \frac{14}{15}$

10. $\frac{8}{9} \div \frac{2}{9}$

11. $\frac{8}{9} \div \frac{8}{3}$

12. $\frac{5}{6} \div \frac{5}{3}$

13. $\frac{6}{35} \div \frac{3}{7}$

14. $\frac{8}{9} \div \frac{5}{9}$

15. $\frac{2}{5} \div \frac{3}{5}$

16. $\frac{3}{16} \div \frac{15}{16}$

17. $\frac{8}{9} \div \frac{4}{5}$

18. $\frac{6}{7} \div \frac{3}{14}$

19. $\frac{2}{3} \div \frac{24}{36}$

20. $\frac{12}{16} \div \frac{9}{8}$

21. $\frac{24}{36} \div \frac{3}{4}$

22. $\frac{6}{7} \div \frac{18}{21}$

B

ANSWERS

23. ______

24. ______

25. ______

26. ______

27. ______

28. ______

29. ______

30. ______

31. ______

32. ______

33. ______

34. ______

35. ______

36. ______

37. ______

38. ______

39. ______

40. ______

41. ______

42. ______

23. What is the reciprocal of $\frac{13}{20}$?

24. What is the reciprocal of $\frac{5}{12}$?

25. What is the reciprocal of $3\frac{6}{11}$?

26. What is the reciprocal of $5\frac{11}{25}$?

27. What is the reciprocal of $\frac{17}{24}$?

28. What is the reciprocal of 18?

Divide. Reduce to the lowest terms:

29. $\frac{14}{25} \div \frac{21}{40}$

30. $\frac{30}{49} \div \frac{20}{21}$

31. $\frac{5}{18} \div \frac{10}{27}$

32. $\frac{4}{15} \div \frac{5}{8}$

33. $\frac{9}{50} \div \frac{3}{7}$

34. $\frac{7}{32} \div \frac{2}{3}$

35. $\frac{30}{55} \div \frac{3}{5}$

36. $\frac{32}{45} \div \frac{8}{9}$

37. $\frac{7}{20} \div \frac{7}{5}$

38. $\frac{7}{15} \div \frac{14}{15}$

39. $\frac{20}{21} \div \frac{7}{10}$

40. $\frac{14}{15} \div \frac{3}{7}$

41. $\frac{25}{40} \div \frac{5}{8}$

42. $\frac{12}{15} \div \frac{20}{25}$

NAME ______________________

CLASS ______________ SECTION ______________ DATE ______________

C

ANSWERS

43. ______
44. ______
45. ______
46. ______
47. ______
48. ______
49. ______
50. ______
51. ______
52. ______
53. ______
54. ______
55. ______
56. ______
57. ______
58. ______
59. ______
60. ______

43. $\frac{10}{16} \div \frac{5}{9}$ **44.** $\frac{11}{30} \div \frac{1}{6}$ **45.** $\frac{35}{48} \div \frac{7}{8}$ **46.** $\frac{7}{18} \div \frac{14}{27}$

47. $\frac{13}{22} \div \frac{26}{33}$ **48.** $\frac{51}{10} \div \frac{14}{15}$ **49.** $\frac{25}{55} \div \frac{75}{10}$ **50.** $\frac{8}{9} \div \frac{24}{36}$

51. $\frac{70}{63} \div \frac{154}{33}$ **52.** $\frac{81}{100} \div \frac{27}{75}$ **53.** $\frac{25}{64} \div \frac{4}{16}$ **54.** $\frac{120}{45} \div \frac{40}{15}$

55. $\frac{72}{90} \div \frac{20}{75}$ **56.** $\frac{1}{96} \div \frac{3}{50}$ **57.** $\frac{14}{40} \div \frac{7}{30}$ **58.** $\frac{16}{81} \div \frac{8}{108}$

59. $\frac{35}{60} \div \frac{25}{48}$ **60.** $\frac{75}{90} \div \frac{50}{72}$

ANSWERS

61. ______

61. Robin buys $\frac{3}{4}$ yard of rope at the hardware store. If she needs $\frac{1}{8}$ yard for each ring she is making, how many rings can she make from this piece of rope?

62. ______

62. If the distance a nut moves on a bolt with one turn is $\frac{3}{8}$ inch, how many turns are needed to make the nut move $\frac{7}{8}$ inch?

63. ______

63. Chang has $\frac{7}{8}$ lb of cheese. If an omelet recipe calls for $\frac{1}{4}$ lb of cheese, how many omelets can he make?

64. ______

64. The specialty at the El-Tiviola restaurant calls for $\frac{1}{16}$ cup of chopped exotic mushrooms per serving. How many orders for the specialty can be filled if the chef has $\frac{5}{8}$ cup of the mushrooms?

65. ______

65. If the head of a pin is $\frac{1}{20}$ inch wide, how many pinheads would it take to form a line that is $\frac{4}{5}$ inch long?

66. ______

66. Exact Electronics manufactures copper wire that has a diameter of $\frac{3}{32}$ inch. How many turns of the wire will fill the first level on a spool of length $\frac{21}{32}$ inch?

NAME ______________________

CLASS ______________ SECTION ______________ DATE ______________

ANSWERS

67. ______________

68. ______________

69. ______________

70. ______________

71. ______________

72. ______________

73. ______________

74. ______________

67. As part of her job at a pet store, Becky feeds each gerbil $\frac{1}{8}$ cup of seeds each day. If the seeds come in packages of $\frac{3}{4}$ cup, how many gerbils can be fed from one package?

68. The Green Thumb Nursery advises that when planting carrots you should use $\frac{1}{16}$ cup of seed for a 50-foot row. How many rows can be planted using $\frac{3}{4}$ cup of seed?

E *Maintain your skills* (Sections 2.6, 3.2, 3.3, 3.4)

69. Find the LCM of 24, 15, and 10.

70. Find the LCM of 24, 15, and 14.

Change to a mixed number.

71. $\frac{45}{13}$

72. $\frac{45}{16}$

Change to an improper fraction.

73. $17\frac{2}{3}$

74. $13\frac{3}{7}$

ANSWERS

75. ____________

76. ____________

77. ____________

78. ____________

Reduce to the lowest terms.

75. $\frac{28}{42}$

76. $\frac{30}{42}$

77. In Mr. Smart's math class 12 of 32 students did not have their textbooks. What fraction (in lowest terms) represents the part of the class that did not have its books?

78. Roberta is taking classes at a community college. She has budgeted $17 per week for transportation for the school year. What is her total transportation budget for the year if there are three terms of 10 weeks each with an average of two weeks between terms? After five weeks she notes that there is $488 left in her transportation budget. If her travel cost remains the same, will she be over or under her budget at the end of the year? By how much?

Multiplication and Division of Mixed Numbers

OBJECTIVES

35. Multiply two or more mixed numbers.
36. Divide two mixed numbers.

APPLICATION

The Elko Aluminum Company produces ingots that are $6\frac{3}{4}$ inches thick. What is the height in feet of a stack of these ingots that is 15 ingots high?

VOCABULARY

No new words.

HOW AND WHY

PROCEDURE

To multiply or divide whole numbers and/or mixed numbers:

1. **Write them as improper fractions.**
2. **Multiply or divide, following the procedures for fractions.**

MODEL PROBLEM-SOLVING

Examples

a. Multiply:

$$\left(1\frac{1}{2}\right) \cdot \frac{3}{4} = \frac{3}{2} \cdot \frac{3}{4}$$

$$= \frac{9}{8} = 1\frac{1}{8}$$

Strategy

Change the mixed number to an improper fraction, then multiply.

The answer can be left as an improper fraction or a mixed number. Either answer is acceptable.

Warm Up a. Multiply: $\left(1\frac{1}{4}\right) \cdot \frac{3}{7}$

ANSWERS TO WARM UPS (3.6)

a. $\frac{15}{28}$

b. Multiply:

$$5\left(2\frac{1}{4}\right) = \frac{5}{1} \cdot \frac{9}{4}$$

Change the whole number to an improper fraction, change the mixed number to an improper fraction, and then multiply.

$$= \frac{5 \cdot 9}{1 \cdot 4}$$

$$= \frac{45}{4} = 11\frac{1}{4}$$

Change the improper fraction to a mixed number. Either answer is acceptable.

Warm Up b. Multiply: $6\left(3\frac{1}{5}\right)$

c. Multiply:

$$\left(3\frac{3}{4}\right) \cdot \left(2\frac{2}{5}\right) = \frac{15}{4} \cdot \frac{12}{5}$$

Change the mixed numbers to improper fractions and multiply.

$$= \frac{\overset{3}{\cancel{15}}}{\underset{1}{\cancel{4}}} \cdot \frac{\overset{3}{\cancel{12}}}{\underset{1}{\cancel{5}}}$$

Reduce.

$$= \frac{9}{1}$$

$$= 9$$

Change the improper fraction to a whole number.

Warm Up c. Multiply: $\left(2\frac{2}{3}\right) \cdot \left(2\frac{1}{4}\right)$

d. Multiply:

$$6\left(\frac{3}{4}\right)\left(2\frac{1}{3}\right) = \frac{6}{1} \cdot \frac{3}{4} \cdot \frac{7}{3}$$

Change the whole number and the mixed number to improper fractions.

$$= \frac{\overset{3}{\cancel{6}}}{1} \cdot \frac{\overset{1}{\cancel{3}}}{\underset{2}{\cancel{4}}} \cdot \frac{7}{\underset{1}{\cancel{3}}}$$

Reduce and multiply.

$$= \frac{21}{2} = 10\frac{1}{2}$$

Change the improper fraction to a mixed number, if desired.

ANSWERS TO WARM UPS (3.6) **b.** $\frac{96}{5}$ or $19\frac{1}{5}$ **c.** 6

Warm Up d. Multiply: $\left(3\frac{3}{5}\right)\left(\frac{5}{9}\right)4$

e. Divide:

$$3\frac{1}{3} \div 6\frac{7}{8} = \frac{10}{3} \div \frac{55}{8}$$

Change the mixed numbers to improper fractions.

$$= \frac{10}{3} \cdot \frac{8}{55}$$

Invert the divisor and multiply.

$$= \frac{\overset{2}{\cancel{10}}}{3} \cdot \frac{8}{\underset{11}{\cancel{55}}}$$

Reduce.

$$= \frac{16}{33}$$

Multiply.

Warm Up e. Divide: $4\frac{5}{6} \div 2\frac{11}{12}$

f. Divide:

$$12\frac{6}{10} \div 21\frac{3}{5} = \frac{126}{10} \div \frac{108}{5}$$

Change the mixed numbers to improper fractions.

$$= \frac{126}{10} \cdot \frac{5}{108}$$

Invert the divisor and multiply.

$$= \frac{\cancel{2} \cdot \cancel{3} \cdot \cancel{3} \cdot 7 \cdot \cancel{5}}{\cancel{2} \cdot \cancel{5} \cdot 2 \cdot 2 \cdot \cancel{3} \cdot \cancel{3} \cdot 3}$$

Prime factor and eliminate all common factors, then multiply.

$$= \frac{7}{12}$$

Warm Up f. Divide: $15\frac{6}{8} \div 12\frac{1}{2}$

ANSWERS TO WARM UPS (3.6)

d. 8 **e.** $\frac{58}{35}$ or $1\frac{23}{35}$ **f.** $\frac{63}{50}$ or $1\frac{13}{50}$

g. Divide:

$$4 \div 3\frac{1}{3} = \frac{4}{1} \div \frac{10}{3}$$

Change the whole number and the mixed number to improper fractions.

$$= \frac{\overset{2}{\cancel{4}}}{1} \cdot \frac{3}{\underset{5}{\cancel{10}}}$$

Invert the divisor and multiply. Eliminate common factors.

$$= \frac{6}{5} = 1\frac{1}{5}$$

Change the improper fraction to a mixed number, if desired.

Warm Up g. Divide: $12 \div 7\frac{1}{2}$

h. Divide:

$$7\frac{3}{4} \div 5 = \frac{31}{4} \div \frac{5}{1}$$

Change the mixed number and the whole number to improper fractions.

$$= \frac{31}{4} \cdot \frac{1}{5}$$

Invert the divisor and multiply.

$$= \frac{31}{20} = 1\frac{11}{20}$$

Change to a mixed number. Either answer is acceptable.

Warm Up h. Divide: $8\frac{5}{6} \div 17$

APPLICATION SOLUTION

i. The Elko Aluminum Company produces ingots that are $6\frac{3}{4}$ inches thick. What is the height in feet of a stack of these ingots that is 15 ingots high?

To find the height of the stack of ingots in feet, first multiply the height of one ingot $\left(6\frac{3}{4}\text{ inches}\right)$ by the number of ingots (15).

ANSWERS TO WARM UPS (3.6)

g. $\frac{8}{5}$ or $1\frac{3}{5}$ **h.** $\frac{53}{102}$

$$\left(6\frac{3}{4}\right)(15) = \frac{27}{4} \cdot \frac{15}{1}$$

Write the mixed number and the whole number as improper fractions, then multiply.

$$= \frac{405}{4}$$

The stack is $\frac{405}{4}$ inches high.

To change this to height in feet, divide by 12:

$$\frac{405}{4} \div 12 = \frac{405}{4} \div \frac{12}{1}$$

Write the whole number as an improper fraction.

$$= \frac{405}{4} \cdot \frac{1}{12}$$

Invert the divisor and multiply.

$$= \frac{\overset{135}{\cancel{405}}}{4} \cdot \frac{1}{\underset{4}{\cancel{12}}}$$

Reduce.

$$= \frac{135}{16}$$

Multiply.

$$= 8\frac{7}{16}$$

Change to a mixed number.

The stack is $8\frac{7}{16}$ feet high.

Warm Up i. The Elko Aluminum Company also produces ingots that are $7\frac{1}{2}$ inches thick. What is the height in feet of a stack of these ingots that is 28 ingots high?

ANSWERS TO WARM UPS (3.6)

i. $17\frac{1}{2}$ ft

NAME ______

CLASS ______ SECTION ______ DATE ______

Exercises 3.6

A

ANSWERS

1. ______
2. ______
3. ______
4. ______
5. ______
6. ______
7. ______
8. ______
9. ______
10. ______
11. ______
12. ______
13. ______
14. ______
15. ______
16. ______
17. ______
18. ______

Multiply. Where possible, write the answers as mixed numbers:

1. $\left(\frac{3}{4}\right)\left(1\frac{3}{4}\right)$ **2.** $\left(\frac{3}{7}\right)\left(2\frac{5}{7}\right)$ **3.** $\left(8\frac{1}{2}\right)\left(2\frac{2}{3}\right)$

4. $\left(3\frac{1}{3}\right)\left(1\frac{1}{5}\right)$ **5.** $3\left(2\frac{1}{3}\right)$ **6.** $4\left(2\frac{3}{4}\right)$

7. $\left(3\frac{1}{3}\right)\left(1\frac{1}{4}\right)(5)$ **8.** $\left(\frac{3}{5}\right)\left(2\frac{1}{8}\right)(3)$ **9.** $\left(7\frac{1}{2}\right)\cdot\left(3\frac{1}{5}\right)$

10. $\left(\frac{25}{36}\right)\left(2\frac{2}{15}\right)\left(2\frac{7}{10}\right)$ **11.** $\left(3\frac{4}{7}\right)(4)\left(\frac{14}{15}\right)$ **12.** $\left(4\frac{4}{9}\right)(6)\left(\frac{12}{25}\right)$

Divide. Reduce to lowest terms and write as mixed numbers where possible:

13. $3 \div 1\frac{1}{2}$ **14.** $4 \div 2\frac{3}{4}$ **15.** $3\frac{7}{8} \div 5\frac{1}{6}$

16. $8\frac{2}{5} \div 2\frac{1}{3}$ **17.** $2\frac{1}{2} \div \frac{5}{8}$ **18.** $2\frac{1}{2} \div 1\frac{1}{4}$

ANSWERS

19. ____

20. ____

21. ____

22. ____

23. ____

24. ____

19. $1\frac{5}{8} \div \frac{3}{4}$

20. $\frac{5}{8} \div 1\frac{3}{4}$

21. $3\frac{1}{3} \div 4\frac{1}{2}$

22. $5\frac{1}{3} \div 2\frac{1}{4}$

23. $4\frac{4}{15} \div 6\frac{2}{5}$

24. $6\frac{1}{4} \div 7\frac{1}{2}$

B

25. ____

26. ____

27. ____

28. ____

29. ____

30. ____

31. ____

32. ____

33. ____

34. ____

35. ____

36. ____

Multiply. Where possible, write the answers as mixed numbers:

25. $7\left(1\frac{5}{6}\right)$

26. $8\left(1\frac{4}{5}\right)$

27. $\left(1\frac{2}{3}\right)\left(1\frac{3}{4}\right)$

28. $\left(1\frac{1}{2}\right)\left(2\frac{3}{4}\right)$

29. $\left(\frac{7}{8}\right)\left(1\frac{5}{6}\right)$

30. $\left(\frac{15}{8}\right)\left(1\frac{3}{5}\right)$

31. $\left(4\frac{1}{2}\right)\left(\frac{2}{9}\right)(0)$

32. $\left(2\frac{1}{4}\right)\left(2\frac{2}{3}\right)(0)$

33. $\left(3\frac{1}{3}\right)(6)\left(5\frac{1}{10}\right)$

34. $\left(4\frac{1}{5}\right)(5)\left(2\frac{7}{9}\right)$

35. $\left(3\frac{2}{3}\right)\left(\frac{15}{22}\right)\left(7\frac{1}{2}\right)$

36. $\left(4\frac{3}{4}\right)\left(3\frac{1}{5}\right)\left(5\frac{5}{8}\right)$

NAME ______________________

CLASS ______________ SECTION ______________ DATE ______________

ANSWERS

37. ______________

38. ______________

39. ______________

40. ______________

41. ______________

42. ______________

43. ______________

44. ______________

45. ______________

46. ______________

47. ______________

48. ______________

Divide. Reduce to the lowest terms and write as mixed numbers where possible:

37. $3\frac{1}{5} \div 1\frac{1}{5}$ **38.** $3\frac{2}{3} \div 2\frac{2}{3}$ **39.** $3\frac{7}{8} \div 7\frac{3}{4}$

40. $2\frac{1}{5} \div 3\frac{1}{7}$ **41.** $\frac{7}{8} \div 3\frac{3}{4}$ **42.** $\frac{11}{15} \div 2\frac{4}{5}$

43. $1\frac{1}{2} \div 10$ **44.** $6\frac{1}{3} \div 17$ **45.** $3\frac{2}{3} \div \frac{1}{5}$

46. $4\frac{3}{4} \div 7\frac{3}{5}$ **47.** $3\frac{3}{4} \div \frac{7}{15}$ **48.** $5\frac{5}{6} \div \frac{20}{9}$

C

49. ______________

50. ______________

51. ______________

Multiply. Where possible, write the answers as mixed numbers:

49. $\left(4\frac{1}{5}\right)\left(1\frac{1}{3}\right)\left(6\frac{2}{7}\right)$ **50.** $\left(12\frac{1}{4}\right)\left(1\frac{1}{7}\right)\left(2\frac{1}{3}\right)$ **51.** $\left(14\right)\left(6\frac{1}{2}\right)\left(1\frac{2}{13}\right)$

ANSWERS

52. ______

53. ______

54. ______

55. ______

56. ______

57. ______

58. ______

59. ______

60. ______

61. ______

62. ______

63. ______

64. ______

65. ______

66. ______

67. ______

68. ______

52. $(12)\left(3\frac{4}{15}\right)\left(4\frac{4}{7}\right)$

53. $\left(5\frac{2}{3}\right)\left(1\frac{1}{2}\right)\left(8\frac{3}{17}\right)$

54. $\left(2\frac{4}{8}\right)\left(3\frac{1}{5}\right)\left(1\frac{5}{12}\right)(6)$

55. $\left(5\frac{1}{3}\right)\left(2\frac{3}{14}\right)(7)\left(2\frac{1}{4}\right)$

56. $\left(3\frac{3}{5}\right)\left(4\frac{1}{3}\right)\left(\frac{6}{7}\right)\left(3\frac{1}{9}\right)$

57. $\left(\frac{8}{11}\right)\left(3\frac{2}{3}\right)\left(4\frac{1}{2}\right)\left(1\frac{1}{5}\right)$

Divide. Reduce to the lowest terms and write as mixed numbers where possible:

58. $31\frac{1}{3} \div 1\frac{1}{9}$

59. $21\frac{3}{7} \div 8\frac{1}{3}$

60. $10\frac{2}{3} \div 100\frac{2}{3}$

61. $11\frac{2}{3} \div 11\frac{1}{4}$

62. $22\frac{2}{3} \div 6\frac{4}{7}$

63. $33\frac{1}{3} \div 11\frac{1}{9}$

64. $16\frac{2}{3} \div 12\frac{1}{12}$

65. $21\frac{5}{6} \div 5\frac{3}{8}$

66. $18\frac{2}{5} \div 17\frac{1}{4}$

67. $15\frac{3}{7} \div 14$

68. $13 \div 7\frac{3}{7}$

NAME ______

CLASS ______ SECTION ______ DATE ______

D

ANSWERS

69. ______

69. Thirty-two metal bars that are each $1\frac{1}{8}$ inch thick are to be stacked. How high will the stack reach?

70. If a steel bar weighs $\frac{3}{4}$ pound per foot, what is the weight of a bar that is $20\frac{1}{2}$ feet in length?

70. ______

71. How many boards, each with a length of $3\frac{1}{3}$ feet, can Norma cut from a board that is $13\frac{1}{3}$ feet long?

71. ______

72. How many metal pins weighing $\frac{1}{16}$ lb each are there in a sack weighing $3\frac{1}{4}$ lb?

72. ______

73. Each student in the television repair class is allowed $6\frac{3}{4}$ inches of wire solder. How many inches of wire must be ordered for a class of 32 students?

73. ______

74. The water pressure during a bad fire is reduced to $\frac{5}{9}$ its original pressure at the hydrant. What is the reduced pressure if the original pressure was $73\frac{2}{3}$ lb/in.^{2}?

74. ______

75. ______

76. ______

77. ______

78. ______

79. ______

80. ______

75. The height of an I-beam is $2\frac{1}{4}$ times the width. What is the height of an I-beam that is $3\frac{1}{2}$ inches wide?

76. If a woodcutter chops $1\frac{3}{4}$ cords of wood in one day, how many cords can he chop in $3\frac{1}{2}$ days?

77. A machinist takes $74\frac{1}{2}$ minutes to machine 5 pins. How long will it take to machine one pin?

78. A metal ingot weighs $3\frac{1}{4}$ lb. How many ingots will weigh 26 lb?

79. The formula for the area of a triangle is $A = \frac{1}{2}b \cdot h$, where b is the base and h is the height of the triangle:

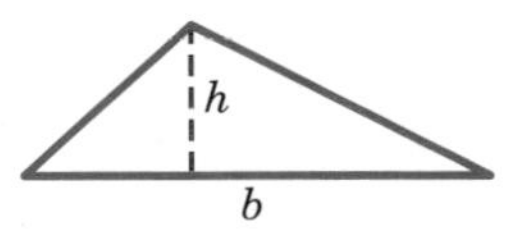

What is the area of a triangle that has a height of $3\frac{3}{4}$ inches and a base of $5\frac{1}{2}$ inches? Give the answer in square inches.

80. What is the area of a triangle whose base is $12\frac{3}{4}$ inches and whose height is $15\frac{1}{2}$ inches? Give the answer in square inches.

NAME ______________________

CLASS ______________ SECTION ______________ DATE ______________

E *Maintain your skills* (Sections 1.3, 1.4, 1.5, 2.6, 3.2)

ANSWERS

81. ______________

82. ______________

83. ______________

84. ______________

85. ______________

86. ______________

87. ______________

88. ______________

81. Add: $2 + 157 + 9854 + 765 + 19 + 4356$

82. Subtract: $73{,}021 - 56{,}489$

83. Find the product of 180 and 231.

84. Find the quotient of 187701 and 267.

85. Find the LCM of 18, 20, and 24.

86. Find the LCM of 22, 24, and 25.

Change to a mixed number:

87. $\frac{553}{12}$

88. $\frac{628}{19}$

ANSWERS

89. ______________________

89. For you to be eligible for a drawing at the Flick movie house, your ticket stub number must be a multiple of three. If Jean's ticket number is 234572, is she eligible for the drawing?

90. The sales of the Goodstone Company totaled $954,000 last year. During the first six months of last year the monthly sales were $72,400, $68,200, $85,000, $89,500, $92,700, and $87,200. What were the average monthly sales for the rest of the year?

90. ______________________

▲▲ *Getting Ready for Algebra*

We have previously solved equations in which variables (letters) were either multiplied or divided by whole numbers. We performed the inverse operations to solve for the variable. To eliminate multiplication, we divided by the number being multiplied. To eliminate division, we multiplied by the number that is the divisor. Now we solve some equations in which variables are multiplied by fractions. Recall from Chapter 1 that if a number is multiplied times a variable, there is usually no multiplication sign between them. That is, $2x$ is understood to mean 2 times x, and $\frac{2}{3}x$ means $\frac{2}{3}$ times x. However, we usually do not write $\frac{2}{3}x$. Instead, we write this as $\frac{2x}{3}$. We can do this because:

$$\frac{2}{3}x = \frac{2}{3} \cdot x = \frac{2}{3} \cdot \frac{x}{1} = \frac{2x}{3}$$

Therefore, we will write $\frac{2}{3}x$ as $\frac{2x}{3}$. Remember, however, that if for convenience we need to use any one of the three forms shown above, they all name the same number. Recall that $\frac{2x}{3}$ means 2 times x and that product divided by 3.

▲▲ MODEL PROBLEM-SOLVING

Examples	Strategy
a. Solve:	
$\frac{3x}{4} = 2$	
$4\left(\frac{3x}{4}\right) = 4 \cdot 2$	**We first eliminate the division by multiplying both sides by 4.**
$3x = 8$	**Multiply.**
$\frac{3x}{3} = \frac{8}{3}$	
$x = \frac{8}{3}$	**To eliminate the multiplication we divide both sides by 3.**
Check:	
$\frac{3x}{4} = 2$	**Substitute $\frac{8}{3}$ for x in the original equation, recall that $\frac{3x}{4} = \frac{3}{4} \cdot x$.**
$\frac{\cancel{3}^{1}}{\cancel{4}_{1}} \cdot \frac{\cancel{8}^{2}}{\cancel{3}_{1}} = 2$	

(Continued on next page)

$\frac{2}{1} = 2$ **Simplify.**

$2 = 2$

The statement is true, so the solution is $\frac{8}{3}$ or $2\frac{2}{3}$.

Warm Up a. $\frac{7x}{8} = 14$

b. Solve:

$\frac{3}{4} = \frac{5x}{4}$

$4 \cdot \frac{3}{4} = 4\left(\frac{5x}{4}\right)$ **To eliminate the division by 4, multiply both sides by 4.**

$3 = 5x$

$\frac{3}{5} = \frac{5x}{5}$ **To eliminate the multiplication by 5, divide both sides of the equation by 5.**

$\frac{3}{5} = x$

Check:

$\frac{3}{4} = \frac{5x}{4}$ **Substitute $\frac{3}{5}$ for x in the original equation, recall that $\frac{5x}{4} = \frac{5}{4} \cdot x$.**

$\frac{3}{4} = \frac{5}{4} \cdot \frac{3}{5}$

$\frac{3}{4} = \frac{\overset{1}{\cancel{5}}}{4} \cdot \frac{3}{\underset{1}{\cancel{5}}}$ **Reduce.**

$\frac{3}{4} = \frac{3}{4}$ **Multiply.**

The statement is true, so the solution is $\frac{3}{5}$.

Warm Up b. $\frac{19x}{8} = \frac{7}{8}$

GETTING READY FOR ALGEBRA **a.** 16 **b.** $\frac{7}{19}$

NAME ____________________

CLASS ____________ SECTION ____________ DATE ____________

▲▲ EXERCISES

ANSWERS

1. ____________
2. ____________
3. ____________
4. ____________
5. ____________
6. ____________
7. ____________
8. ____________
9. ____________
10. ____________
11. ____________
12. ____________
13. ____________
14. ____________
15. ____________
16. ____________

1. $\frac{2x}{3} = \frac{1}{2}$ **2.** $\frac{2x}{5} = \frac{2}{3}$ **3.** $\frac{3y}{4} = \frac{4}{5}$ **4.** $\frac{7y}{8} = \frac{5}{6}$

5. $\frac{4z}{5} = \frac{3}{4}$ **6.** $\frac{5z}{4} = \frac{8}{9}$ **7.** $\frac{17}{9} = \frac{8x}{9}$ **8.** $\frac{29}{10} = \frac{9x}{5}$

9. $\frac{7a}{4} = \frac{5}{2}$ **10.** $\frac{15a}{4} = \frac{24}{5}$ **11.** $\frac{47}{4} = \frac{47b}{6}$ **12.** $\frac{13}{3} = \frac{52b}{9}$

13. $\frac{41z}{6} = \frac{41}{3}$ **14.** $\frac{9b}{23} = \frac{23}{3}$ **15.** $\frac{2a}{15} = \frac{11}{4}$ **16.** $\frac{119x}{12} = \frac{119}{8}$

Building Fractions

OBJECTIVE

37. Build a fraction by finding the missing numerator.

APPLICATION

The applications of building are in comparing, adding, and subtracting fractions.

VOCABULARY

Equivalent fractions are fractions that are different names for the same number. For instance, $\frac{3}{6}$ and $\frac{4}{8}$ are equivalent, since both represent one-half $\left(\frac{1}{2}\right)$ a unit.

HOW AND WHY

This process of renaming fractions (see the description that follows) is sometimes referred to as "building" fractions. It is the opposite of the process called "reducing" fractions, which we studied in Section 3.3.

$$\frac{3}{5} = \frac{3}{5} \cdot 1 = \frac{3}{5} \cdot \frac{2}{2} = \frac{6}{10}$$

We "build" fractions by multiplying by a fraction that is equal to one.

	MULTIPLY BY: $\frac{2}{2}$	$\frac{3}{3}$	$\frac{4}{4}$	$\frac{5}{5}$	$\frac{10}{10}$	$\frac{15}{15}$
$\frac{3}{5}$ =	$\frac{6}{10}$	$\frac{9}{15}$	$\frac{12}{20}$	$\frac{15}{25}$	$\frac{30}{50}$	$\frac{45}{75}$
$\frac{1}{2}$ =	$\frac{2}{4}$	$\frac{3}{6}$	$\frac{4}{8}$	$\frac{5}{10}$	$\frac{10}{20}$	$\frac{15}{30}$
$\frac{3}{8}$ =	$\frac{6}{16}$	$\frac{9}{24}$	$\frac{12}{32}$	$\frac{15}{40}$	$\frac{30}{80}$	$\frac{45}{120}$
$\frac{2}{9}$ =	$\frac{4}{18}$	$\frac{6}{27}$	$\frac{8}{36}$	$\frac{10}{45}$	$\frac{20}{90}$	$\frac{30}{135}$
$\frac{7}{15}$ =	$\frac{14}{30}$	$\frac{21}{45}$	$\frac{28}{60}$	$\frac{35}{75}$	$\frac{70}{150}$	$\frac{105}{225}$

To find a missing numerator such as:

$$\frac{3}{5} = \frac{?}{60}$$

divide 60 by 5 to find out what fraction to multiply by:

$60 \div 5 = 12$

The correct multiplier is $\frac{12}{12}$:

$$\frac{3}{5} = \frac{3}{5} \cdot \frac{12}{12} = \frac{36}{60}$$

The shortcut is to multiply the numerator (3) by 12:

$$\frac{3}{5} = \frac{3 \cdot 12}{60} = \frac{36}{60}$$

The fractions $\frac{3}{5}$ and $\frac{36}{60}$ are equivalent. Either fraction can be used in place of the other.

PROCEDURE

To find a missing numerator when building fractions:

1. **Divide the larger denominator by the smaller denominator.**
2. **Multiply the answer by the given numerator. This product is the missing numerator.**

MODEL PROBLEM-SOLVING

Examples | **Strategy**

a. Rename $\frac{3}{10}$ using $\frac{7}{7}$ to represent 1:

$$\frac{3}{10} \cdot \frac{7}{7} = \frac{21}{70}$$

The new fraction, $\frac{21}{70}$, is equivalent to the original fraction.

Warm Up a. Rename $\frac{8}{9}$ using $\frac{6}{6}$ to represent 1.

b. Find the missing numerator: $\frac{3}{5} = \frac{?}{80}$

$80 \div 5 = 16$

Divide the larger denominator, 80, by the smaller denominator, 5.

$$\frac{3}{5} = \frac{3 \cdot 16}{80} = \frac{48}{80}$$

Multiply the answer, 16, times the given numerator, 3. The product, 48, is the missing numerator.

Warm Up b. Find the missing numerator: $\frac{6}{7} = \frac{?}{28}$

ANSWERS TO WARM UPS (3.7)

a. $\frac{48}{54}$ **b.** 24

c. Find the missing numerator: $\frac{2}{7} = \frac{?}{56}$

$56 \div 7 = 8$

Divide the larger denominator, 56, by the smaller denominator, 7.

$\frac{2}{7} = \frac{2 \cdot 8}{56} = \frac{16}{56}$

Multiply the answer, 8, times the given numerator, 2. The product, 16, is the missing numerator.

Warm Up c. Find the missing numerator: $\frac{5}{8} = \frac{?}{64}$

d. Find the missing numerator: $\frac{13}{10} = \frac{?}{120}$

$120 \div 10 = 12$

$\frac{13}{10} = \frac{13 \cdot 12}{120} = \frac{156}{120}$

Building is the opposite of reducing. A built-up fraction can be reduced to the original fraction.

Warm Up d. Find the missing numerator: $\frac{22}{15} = \frac{?}{120}$

ANSWERS TO WARM UPS (3.7)

c. 40 **d.** 176

NAME ______________________________

CLASS ______________ SECTION ______________ DATE ______________

Exercises 3.7

A

ANSWERS

1. ______
2. ______
3. ______
4. ______
5. ______
6. ______
7. ______
8. ______
9. ______
10. ______
11. ______
12. ______
13. ______
14. ______
15. ______
16. ______
17. ______
18. ______
19. ______
20. ______
21. ______
22. ______
23. ______
24. ______

Write four fractions equivalent to each of the given fractions by multiplying by $\frac{2}{2}$, $\frac{3}{3}$, $\frac{4}{4}$, and $\frac{5}{5}$:

1. $\frac{2}{3}$ **2.** $\frac{3}{5}$ **3.** $\frac{1}{4}$

4. $\frac{1}{7}$ **5.** $\frac{5}{4}$ **6.** $\frac{9}{8}$

7. $\frac{11}{12}$ **8.** $\frac{15}{16}$ **9.** $\frac{3}{4}$

10. $\frac{7}{8}$ **11.** $\frac{5}{6}$ **12.** $\frac{7}{10}$

Find the missing numerator:

13. $\frac{1}{2} = \frac{?}{10}$ **14.** $\frac{3}{4} = \frac{?}{16}$ **15.** $\frac{2}{3} = \frac{?}{15}$

16. $\frac{4}{7} = \frac{?}{14}$ **17.** $\frac{4}{5} = \frac{?}{20}$ **18.** $\frac{7}{8} = \frac{?}{32}$

19. $\frac{11}{13} = \frac{?}{52}$ **20.** $\frac{13}{14} = \frac{?}{56}$ **21.** $\frac{?}{24} = \frac{7}{8}$

22. $\frac{?}{30} = \frac{5}{6}$ **23.** $\frac{5}{12} = \frac{?}{48}$ **24.** $\frac{7}{10} = \frac{?}{60}$

B

ANSWERS

25. ______
26. ______
27. ______
28. ______
29. ______
30. ______
31. ______
32. ______
33. ______
34. ______
35. ______
36. ______
37. ______
38. ______
39. ______
40. ______
41. ______
42. ______

Write four fractions equivalent to each of the given fractions by multiplying by $\frac{2}{2}, \frac{3}{3}, \frac{4}{4}$, and $\frac{5}{5}$:

25. $\frac{3}{7}$ **26.** $\frac{1}{8}$ **27.** $\frac{4}{10}$

28. $\frac{3}{9}$ **29.** $\frac{7}{3}$ **30.** $\frac{6}{5}$

Find the missing numerator:

31. $\frac{7}{8} = \frac{?}{16}$ **32.** $\frac{3}{4} = \frac{?}{36}$ **33.** $\frac{2}{3} = \frac{?}{24}$ **34.** $\frac{6}{7} = \frac{?}{28}$

35. $\frac{5}{6} = \frac{?}{30}$ **36.** $\frac{1}{2} = \frac{?}{30}$ **37.** $\frac{1}{5} = \frac{?}{75}$ **38.** $\frac{5}{9} = \frac{?}{45}$

39. $\frac{?}{24} = \frac{3}{8}$ **40.** $\frac{?}{48} = \frac{5}{8}$ **41.** $\frac{?}{72} = \frac{5}{6}$ **42.** $\frac{?}{36} = \frac{11}{12}$

C

43. ______
44. ______
45. ______
46. ______
47. ______
48. ______

43. $\frac{3}{4} = \frac{?}{100}$ **44.** $\frac{7}{4} = \frac{?}{16}$ **45.** $\frac{?}{12} = \frac{2}{3}$

46. $\frac{?}{66} = \frac{3}{11}$ **47.** $\frac{23}{6} = \frac{?}{12}$ **48.** $\frac{9}{5} = \frac{?}{100}$

NAME ______________________

CLASS ____________ SECTION ____________ DATE ____________

ANSWERS

49. ______________

50. ______________

51. ______________

52. ______________

53. ______________

54. ______________

55. ______________

56. ______________

57. ______________

58. ______________

59. ______________

60. ______________

49. $\frac{?}{90} = \frac{5}{6}$

50. $\frac{?}{80} = \frac{12}{16}$

51. $\frac{?}{300} = \frac{7}{15}$

52. $\frac{3}{4} = \frac{?}{68}$

53. $\frac{6}{9} = \frac{?}{108}$

54. $\frac{11}{15} = \frac{?}{90}$

55. $\frac{?}{126} = \frac{19}{42}$

56. $\frac{?}{147} = \frac{16}{7}$

57. $\frac{15}{18} = \frac{?}{144}$

58. $\frac{11}{16} = \frac{?}{144}$

59. Build the three fractions $\frac{4}{5}$, $\frac{1}{2}$, and $\frac{3}{10}$ so that each has a denominator of 30.

$\frac{4}{5} = \frac{?}{30}$ $\quad \frac{1}{2} = \frac{?}{30}$ $\quad \frac{3}{10} = \frac{?}{30}$

60. Build the three fractions $\frac{10}{7}$, $\frac{3}{4}$, and $\frac{15}{14}$ so that each has a denominator of 28.

61. ______

61. Find the LCM of the denominators of $\frac{1}{2}$, $\frac{2}{3}$, $\frac{1}{6}$, and $\frac{5}{8}$. Build the four fractions so that each has the LCM as the denominator.

62. ______

62. Find the LCM of the denominators of $\frac{1}{4}$, $\frac{4}{13}$, and $\frac{5}{26}$. Build the fractions so that each has the LCM as the denominator.

D

63. ______

63. Janie answered $\frac{4}{5}$ of the problems correctly on her Chapter I test. If there are 40 problems on the Chapter III test, how many must she get correct to answer the same fractional amount?

64. ______

64. Five of every eight packages of gum sold at the Maxi-Mart are bubble gum. At that rate, how many packages of bubble gum were sold if a total of 136 packages of gum were sold?

65. ______

65. Ruth has $\frac{5}{6}$ of a cake left from yesterday's dinner. She wants to cut the pieces into eighteenths. How many pieces will she have?

66. ______

66. Liz wants to cut $\frac{3}{4}$ of her pie into twelfths. How many pieces will there be?

NAME ______________________________

CLASS ______________ SECTION ______________ DATE ______________

E. *Maintain your skills* (Sections 2.6, 3.2, 3.3)

ANSWERS

67. ______________

68. ______________

69. ______________

70. ______________

71. ______________

72. ______________

73. ______________

74. ______________

67. Find the LCM of 30, 20, and 15.

68. Find the LCM of 25, 35, and 45.

Reduce to the lowest terms:

69. $\frac{124}{160}$

70. $\frac{812}{928}$

Change to an improper fraction:

71. $22\frac{5}{7}$

72. $34\frac{8}{9}$

Change to a mixed number:

73. $\frac{125}{9}$

74. $\frac{125}{13}$

ANSWERS

75. ______

76. ______

75. A-Best Construction Company charges 15 dollars per foot to pour a cement walkway. If they poured 125 feet, what did they charge?

76. Ms. Wallington is taking one capsule containing 250 mg of a drug every 8 hours. Beginning next week her doctor's instructions are to increase the dosage to 500 mg every 6 hours. How many 250-mg capsules should the pharmacist give her for the following week (seven days)?

3.8 Listing Fractions in Order of Value

OBJECTIVES

38. List a group of fractions from the smallest to largest.
39. Determine whether an inequality statement involving fractions is true or false.

APPLICATION

The Acme Hardware Store sells bolts with the following diameters: $\frac{5}{16}$, $\frac{3}{8}$, $\frac{1}{2}$, $\frac{5}{8}$, $\frac{1}{4}$, and $\frac{7}{16}$ inch. List the diameters from the smallest to largest.

VOCABULARY

When two or more fractions have the same denominator, we say they have a *common denominator.*
Recall that the symbol "<" is read "is less than" and that the symbol ">" is read "is greater than."

HOW AND WHY

We see from Figure 3.9 that $\frac{4}{5}$ is larger than $\frac{2}{5}$:

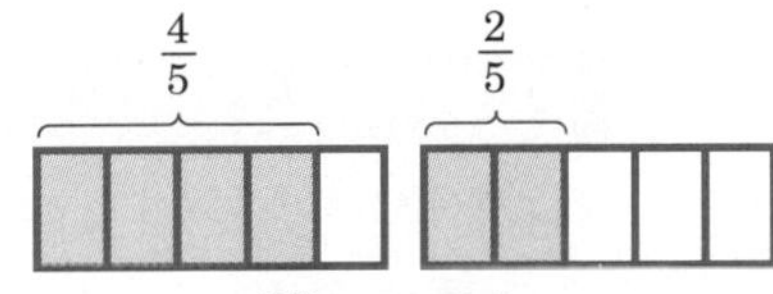

Figure 3.9

If two fractions have the same denominator, the one with the smaller numerator is the smaller.

The symbols "<" (less than) and ">" (greater than) are used as in Chapter 1:

$\frac{2}{5} < \frac{4}{5}$ means "$\frac{2}{5}$ is less than $\frac{4}{5}$"

$\frac{4}{5} > \frac{1}{5}$ means "$\frac{4}{5}$ is greater than $\frac{1}{5}$"

If fractions to be compared do not have a common denominator, then one or more must be renamed so that all have a common denominator. The preferred common denominator is the least common multiple (LCM) of all the denominators.

To list, $\frac{5}{8}$, $\frac{7}{16}$, $\frac{1}{2}$, and $\frac{9}{16}$ from the smallest to largest, we write each with a common denominator and then compare the numerators. We note that the LCM of all the denominators is 16. Therefore, we build each fraction so that it has a denominator of 16:

$\frac{5}{8} = \frac{10}{16}$ $\quad \frac{7}{16} = \frac{7}{16}$ $\quad \frac{1}{2} = \frac{8}{16}$ $\quad \frac{9}{16} = \frac{9}{16}$ **Each fraction now has a denominator of 16.**

We arrange those fractions whose denominator is 16 in order from the smallest to largest:

$\frac{7}{16} < \frac{8}{16} < \frac{9}{16} < \frac{10}{16}$ **They are now listed in order from the smallest to largest with a common denominator of 16.**

We replace each fraction by the original, so:

$\frac{7}{16} < \frac{1}{2} < \frac{9}{16} < \frac{5}{8}$ **They are now listed in order from the smallest to largest.**

To determine whether an inequality statement is true or false, use the preceding method to arrange the fractions from the smallest to largest. The smaller of the two fractions is less than the other. The larger of the two fractions is greater than the other.

We saw in the previous problem that $\frac{7}{16}$ was smaller than $\frac{1}{2}$, so we can write the true inequality statement:

$\frac{7}{16} < \frac{1}{2}$

If $\frac{7}{16}$ is smaller than $\frac{1}{2}$, then $\frac{1}{2}$ is larger than $\frac{7}{16}$, and we can write the true inequality:

$\frac{1}{2} > \frac{7}{16}$

PROCEDURE

To list fractions from the smallest to largest:

1. **Build the fractions so that they have a common denominator. Use the LCM of the denominators.**
2. **List the fractions (with common denominators) with numerators from the smallest to largest.**
3. **Replace each fraction by the original.**

MODEL PROBLEM-SOLVING

Examples **Strategy**

a. Which fraction is larger, $\frac{6}{11}$ or $\frac{8}{11}$?

Since $6 < 8$, then $\frac{6}{11} < \frac{8}{11}$.

Since both fractions have a common denominator, we compare the numerators.

Therefore, $\frac{8}{11}$ is larger.

Another way of writing the relationship is $\frac{8}{11} > \frac{6}{11}$.

Warm Up a. Which fraction is larger, $\frac{8}{15}$ or $\frac{10}{15}$?

ANSWERS TO WARM UPS (3.8)

a. $\frac{10}{15}$

b. Which fraction is larger, $\frac{6}{11}$ or $\frac{1}{2}$?

$\frac{6}{11} = \frac{12}{22}$ and $\frac{1}{2} = \frac{11}{22}$

The LCM of 11 and 2 is 22. Build each fraction to have 22 for a common denominator.

Therefore, $\frac{6}{11}$ is larger than $\frac{1}{2}$.

That is, $\frac{6}{11} > \frac{1}{2}$

Warm Up b. Which fraction is larger, $\frac{8}{13}$ or $\frac{3}{5}$?

c. List the fractions $\frac{2}{3}$, $\frac{3}{8}$, and $\frac{3}{4}$ from the smallest to largest:

$\frac{2}{3} = \frac{16}{24}$ $\quad \frac{3}{8} = \frac{9}{24}$ $\quad \frac{3}{4} = \frac{18}{24}$

The LCM of 3, 8, and 4 is 24. With denominators of 24, the fractions from the smallest to largest are:

$\frac{9}{24} < \frac{16}{24} < \frac{18}{24}$

$\frac{9}{24}$, $\frac{16}{24}$, and $\frac{18}{24}$

From the smallest to largest the list is $\frac{3}{8} < \frac{2}{3} < \frac{3}{4}$.

Replace each fraction by the original.

Warm Up c. List the fractions $\frac{5}{6}$, $\frac{7}{8}$, and $\frac{4}{5}$ from the smallest to largest.

d. List the mixed numbers $3\frac{1}{2}$, $3\frac{5}{6}$, and $3\frac{5}{8}$ from the smallest to largest.

The LCM of 2, 6, and 8 is 24.

Since the whole number is the same in each mixed number, rewrite each fraction part with a common denominator.

$3\frac{1}{2} = 3\frac{12}{24}$ $\quad 3\frac{5}{6} = 3\frac{20}{24}$

$3\frac{5}{8} = 3\frac{15}{24}$

$3\frac{12}{24} < 3\frac{15}{24} < 3\frac{20}{24}$

Since each of them has a common denominator, they can be listed from the smallest to largest.

ANSWERS TO WARM UPS (3.8)

b. $\frac{8}{13}$ **c.** $\frac{4}{5} < \frac{5}{6} < \frac{7}{8}$

From the smallest to largest the list is $3\frac{1}{2} < 3\frac{5}{8} < 3\frac{5}{6}$.

Replace each fraction by its original.

Warm Up d. List the mixed numbers $4\frac{3}{5}$, $4\frac{4}{9}$, and $4\frac{2}{3}$ from the smallest to largest.

e. True or false: $\frac{3}{4} > \frac{7}{9}$

$\frac{3}{4} = \frac{27}{36}$ $\quad \frac{7}{9} = \frac{28}{36}$

The inequality symbol is "greater than." To tell if the statement is true or false, we build each fraction to have a common denominator of 36.

Since 27 is not greater than 28, the statement is false.

Warm Up e. True or false: $\frac{11}{16} > \frac{13}{20}$

APPLICATION SOLUTION

f. The Acme Hardware Store sells bolts with the following diameters: $\frac{5}{16}, \frac{3}{8}, \frac{1}{2}, \frac{5}{8}, \frac{1}{4}$, and $\frac{7}{16}$ inch. List the diameters from the smallest to largest:

$\frac{5}{16} = \frac{5}{16}$ $\quad \frac{3}{8} = \frac{6}{16}$ $\quad \frac{1}{2} = \frac{8}{16}$

$\frac{5}{8} = \frac{10}{16}$ $\quad \frac{1}{4} = \frac{4}{16}$ $\quad \frac{7}{16} = \frac{7}{16}$

To list the diameters of the bolts from the smallest to largest, first list the diameters using a common denominator. The LCM of 16, 8, 2, and 4 is 16.

$\frac{4}{16}, \frac{5}{16}, \frac{6}{16}, \frac{7}{16}, \frac{8}{16}, \frac{10}{16}$

List the fractions with common denominators in order of size.

The list is $\frac{1}{4}, \frac{5}{16}, \frac{3}{8}, \frac{7}{16}, \frac{1}{2}$, and $\frac{5}{8}$ inch.

Replace each fraction by the original to get the list of diameters from the smallest to largest.

Warm Up f. The Acme Hardware Store also sells "rebar" with the following diameters: $\frac{3}{4}, \frac{7}{8}, \frac{7}{16}, \frac{15}{32}, \frac{15}{64}$, and $\frac{7}{12}$ inch. List the diameters from the smallest to largest.

ANSWERS TO WARM UPS (3.8)

d. $4\frac{4}{9} < 4\frac{3}{5} < 4\frac{2}{3}$ **e.** true **f.** $\frac{15}{64}, \frac{7}{16}, \frac{15}{32}, \frac{7}{12}, \frac{3}{4}, \frac{7}{8}$

NAME ______________________

CLASS ______________ SECTION ______________ DATE ______________

Exercises 3.8

A

ANSWERS

1. ______
2. ______
3. ______
4. ______
5. ______
6. ______
7. ______
8. ______
9. ______
10. ______
11. ______
12. ______
13. ______
14. ______
15. ______
16. ______
17. ______
18. ______
19. ______
20. ______
21. ______
22. ______

Which fraction is larger?

1. $\frac{4}{9}, \frac{7}{9}$ **2.** $\frac{9}{12}, \frac{8}{12}$ **3.** $\frac{1}{3}, \frac{2}{9}$

4. $\frac{2}{3}, \frac{1}{2}$ **5.** $\frac{3}{5}, \frac{3}{4}$ **6.** $\frac{3}{4}, \frac{7}{8}$

7. $\frac{1}{2}, \frac{4}{6}$ **8.** $\frac{3}{10}, \frac{1}{5}$

List these fractions from the smallest to largest:

9. $\frac{4}{7}, \frac{5}{7}, \frac{3}{7}$ **10.** $\frac{5}{11}, \frac{4}{11}, \frac{2}{11}$ **11.** $\frac{1}{2}, \frac{1}{4}, \frac{3}{8}$

12. $\frac{1}{2}, \frac{3}{5}, \frac{7}{10}$ **13.** $\frac{1}{2}, \frac{3}{8}, \frac{1}{4}$ **14.** $\frac{2}{3}, \frac{8}{15}, \frac{3}{5}$

15. $\frac{2}{3}, \frac{3}{4}, \frac{7}{12}$ **16.** $\frac{17}{18}, \frac{8}{9}, \frac{5}{6}$

Are the following statements true or false?

17. $\frac{1}{4} < \frac{3}{4}$ **18.** $\frac{5}{9} > \frac{7}{9}$ **19.** $\frac{11}{16} > \frac{7}{8}$

20. $\frac{9}{16} < \frac{5}{8}$ **21.** $\frac{5}{6} > \frac{25}{30}$ **22.** $\frac{12}{24} > \frac{6}{12}$

B

ANSWERS

23. ____
24. ____
25. ____
26. ____
27. ____
28. ____
29. ____
30. ____
31. ____
32. ____
33. ____
34. ____
35. ____
36. ____
37. ____
38. ____
39. ____
40. ____
41. ____
42. ____
43. ____
44. ____
45. ____
46. ____

Which fraction is larger?

23. $\frac{2}{3}, \frac{4}{5}$ **24.** $\frac{8}{9}, \frac{9}{10}$ **25.** $\frac{5}{4}, \frac{13}{10}$ **26.** $\frac{5}{3}, \frac{3}{2}$

27. $\frac{7}{6}, \frac{11}{6}$ **28.** $\frac{6}{7}, \frac{6}{11}$ **29.** $2\frac{3}{8}, 2\frac{5}{16}$ **30.** $5\frac{4}{9}, 5\frac{3}{7}$

31. $\frac{5}{11}, \frac{3}{9}$ **32.** $\frac{5}{6}, \frac{3}{4}$ **33.** $\frac{2}{3}, \frac{5}{8}$ **34.** $\frac{1}{2}, \frac{3}{7}$

List these fractions from the smallest to largest:

35. $\frac{2}{5}, \frac{2}{3}, \frac{4}{7}$ **36.** $\frac{4}{5}, \frac{2}{3}, \frac{3}{4}$ **37.** $\frac{5}{8}, \frac{7}{10}, \frac{3}{4}$

38. $\frac{5}{7}, \frac{7}{9}, \frac{9}{11}$ **39.** $\frac{13}{15}, \frac{4}{5}, \frac{5}{6}, \frac{9}{10}$ **40.** $\frac{7}{9}, \frac{2}{3}, \frac{3}{4}, \frac{5}{6}$

41. $2\frac{3}{4}, 2\frac{7}{8}, 2\frac{5}{6}$ **42.** $1\frac{3}{8}, 1\frac{5}{16}, 1\frac{1}{4}$ **43.** $\frac{3}{4}, \frac{11}{12}, \frac{5}{8}, \frac{1}{2}$

44. $\frac{5}{6}, \frac{5}{8}, \frac{11}{12}, \frac{19}{24}$ **45.** $7\frac{5}{6}, 7\frac{2}{3}, 7\frac{3}{4}$ **46.** $5\frac{3}{5}, 5\frac{1}{2}, 5\frac{7}{10}$

NAME ______________________

CLASS ______________ SECTION ______________ DATE ______________

ANSWERS

47. ______________

48. ______________

49. ______________

50. ______________

Are the following statements true or false?

47. $\frac{3}{10} > \frac{7}{15}$ **48.** $\frac{5}{8} > \frac{13}{12}$ **49.** $\frac{5}{9} < \frac{2}{7}$ **50.** $\frac{6}{7} > \frac{7}{8}$

C

51. ______________

52. ______________

53. ______________

54. ______________

55. ______________

56. ______________

57. ______________

58. ______________

List these fractions from smallest to largest:

51. $\frac{11}{24}, \frac{17}{36}, \frac{35}{72}$ **52.** $\frac{3}{5}, \frac{8}{25}, \frac{31}{50}, \frac{59}{100}$ **53.** $\frac{13}{28}, \frac{17}{35}, \frac{6}{14}$

54. $\frac{11}{15}, \frac{17}{20}, \frac{9}{12}$ **55.** $\frac{7}{18}, \frac{2}{5}, \frac{11}{30}, \frac{17}{45}$ **56.** $\frac{47}{80}, \frac{9}{16}, \frac{13}{20}, \frac{5}{8}$

57. $\frac{29}{30}, \frac{14}{15}, \frac{19}{20}, \frac{11}{12}$ **58.** $\frac{7}{8}, \frac{1}{2}, \frac{2}{3}, \frac{5}{6}$

59. ______________

60. ______________

61. ______________

62. ______________

63. ______________

64. ______________

Are the following statements true or false?

59. $\frac{5}{8} < \frac{47}{80}$ **60.** $\frac{8}{25} < \frac{59}{100}$ **61.** $\frac{15}{20} < \frac{55}{75}$

62. $\frac{19}{40} > \frac{31}{60}$ **63.** $\frac{11}{30} < \frac{7}{18}$

64. $\frac{11}{27} > \frac{29}{36}$

ANSWERS

65. ____________

66. ____________

67. ____________

68. ____________

69. ____________

65. The night nurse at Malcolm X Community Hospital found bottles containing codeine tablets out of the usual order. The bottles contained tablets having the following strengths of codeine: $\frac{1}{8}$, $\frac{3}{32}$, $\frac{5}{16}$, $\frac{3}{8}$, $\frac{9}{16}$, $\frac{1}{2}$, and $\frac{1}{4}$ grain, respectively. Arrange the bottles in order of the strength of codeine from the smallest to largest.

66. Joe, an apprentice, was given the task of sorting a bin of bolts according to their diameters. The bolts had the following diameters: $\frac{11}{16}$, $\frac{7}{8}$, $1\frac{1}{16}$, $\frac{3}{4}$, $1\frac{1}{8}$, and $1\frac{3}{32}$ inches. How should he list the diameters from the smallest to largest?

67. Four partly filled cans of milk were delivered to the local dairy. They contained $\frac{3}{8}$ gallon, $\frac{7}{16}$ gallon, $\frac{1}{4}$ gallon, and $\frac{1}{2}$ gallon. Which is the largest amount and which is the smallest amount of milk?

68. Four pick-up trucks were advertised in the local car ads. The load capacities listed were $\frac{3}{4}$ ton, $\frac{5}{8}$ ton, $\frac{7}{16}$ ton, and $\frac{1}{2}$ ton. Which capacity is the smallest and which is the largest?

69. A container of a chemical was weighed by three people. Mary recorded the weight as $3\frac{1}{8}$ lb. George read the weight as $3\frac{3}{16}$ lb. Chang read the weight as $3\frac{1}{4}$ lb. Whose measurement was heaviest?

NAME ______________________

CLASS ______________ SECTION ______________ DATE ______________

ANSWERS

70. ______________

70. Three rulers are marked in inches. On the first ruler the spaces are divided into tenths, on the second they are divided into sixteenths, and on the third they are divided into eighths. All are used to measure a line on a scale drawing. The nearest mark on the first ruler is $5\frac{7}{10}$, the nearest mark on the second is $5\frac{11}{16}$, and the nearest mark on the third is $5\frac{6}{8}$. Which is the largest (longest) measurement?

E *Maintain your skills* (Sections 1.2, 1.3, 2.1, 2.2, 2.4)

71. ______________

71. Find the sum of 3796, 34, 296, 4099, and 4568.

72. ______________

72. Find the sum of 624, 9, 758, 19, 902, and 83.

73. ______________

73. Is 611 a prime number or a composite number?

74. ______________

74. Is 151 a prime number or a composite number?

75. ______________

75. List the first five multiples of 13.

76. ______________

76. List all the factors of 882.

ANSWERS

77. ______________

78. ______________

79. ______________

80. ______________

77. Round 65,374,182 to the nearest ten million.

78. Round 65,374,182 to the nearest ten thousand.

79. On Saturday the Hot-Wire Appliance Company advertised a 19-inch TV for $349. The usual price for this TV is $419. How much could be saved if the TV was bought at the reduced price?

80. The Forest Service rents a two-engine plane at $625 per hour and a single-engine plane at $365 per hour to drop fire retardant. During a forest fire the two-engine plane was used for four hours and the single engine plane was used for two hours. What was the cost of using the two planes?

Addition of Like Fractions

OBJECTIVE

40. Add like fractions.

APPLICATION

The stock of the Wesin Corporation rose $\frac{1}{8}$ point on Monday, $\frac{3}{8}$ point on Tuesday, $\frac{1}{8}$ point on Wednesday, $\frac{5}{8}$ point on Thursday, and $\frac{5}{8}$ point on Friday. What was the total rise of the stock for the week?

VOCABULARY

Like fractions are fractions with common denominators.

HOW AND WHY

What is the sum of $\frac{1}{5} + \frac{2}{5}$? The denominators tell the number of parts in the unit. The numerator tells us how many of these parts are shaded. By adding the numerators we find the total number of shaded parts. The common denominator keeps track of the size of the parts. (See Fig. 3.10.)

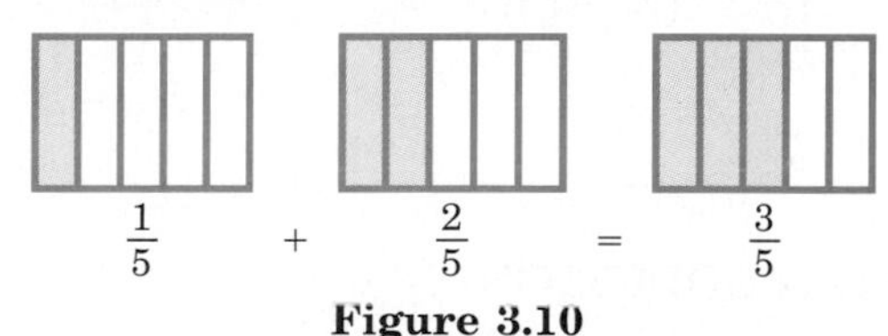

Figure 3.10

PROCEDURE

To add like fractions, add the numerators and write the sum over the common denominator.

MODEL PROBLEM-SOLVING

Examples | **Strategy**

a. Add:

$$\frac{3}{8} + \frac{4}{8} = \frac{7}{8}$$

The fractions have common denominators. Add the numerators and write the sum over the common denominator.

Warm Up a. Add: $\frac{2}{7} + \frac{4}{7}$

ANSWERS TO WARM UPS (3.9)

a. $\frac{6}{7}$

b. Add:

$$\frac{3}{7}+\frac{2}{7}=\frac{5}{7}$$

Warm Up b. Add: $\frac{4}{9}+\frac{3}{9}$

c. Add:

$$\frac{1}{3}+\frac{2}{3}+\frac{1}{3}=\frac{4}{3}$$

Note that sometimes it is appropriate to leave an answer as an improper fraction.

Warm Up c. Add: $\frac{1}{6}+\frac{5}{6}+\frac{5}{6}$

d. Add:

$$\frac{2}{6}+\frac{1}{6}+\frac{1}{6}=\frac{4}{6}=\frac{2}{3}$$

Add the numerators and reduce to the lowest terms.

Warm Up d. Add: $\frac{3}{10}+\frac{2}{10}+\frac{1}{10}$

APPLICATION SOLUTION

e. The stock of the Wesin Corporation rose $\frac{1}{8}$ point on Monday, $\frac{3}{8}$ point on Tuesday, $\frac{1}{8}$ point on Wednesday, $\frac{5}{8}$ point on Thursday, and $\frac{5}{8}$ point on Friday. What was the total rise of the stock for the week?

To find the total number of points that the stock of the Wesin Corporation rose during the week, add the gains of each day.

$$\frac{1}{8}+\frac{3}{8}+\frac{1}{8}+\frac{5}{8}+\frac{5}{8}=\frac{15}{8}$$

The fractions have common denominators, so add numerators. Write the sum over the common denominator.

$$=1\frac{7}{8}$$

The answer may be left as an improper fraction or changed to a mixed number.

The stock rose $1\frac{7}{8}$ points during the week.

Warm Up e. The previous week the Wesin Corporation stock prices rose $\frac{1}{8}$ point on Monday, $\frac{7}{8}$ point on Tuesday, $\frac{1}{8}$ point on Wednesday, $\frac{3}{8}$ point on Thursday, and $\frac{1}{8}$ point on Friday. What was the total rise for the week?

ANSWERS TO WARM UPS (3.9)

b. $\frac{7}{9}$ **c.** $\frac{11}{6}$ **d.** $\frac{3}{5}$ **e.** $\frac{13}{8}$ or $1\frac{5}{8}$

NAME ______________________

CLASS ______________ SECTION ______________ DATE ______________

Exercises 3.9

A

ANSWERS

1. ______
2. ______
3. ______
4. ______
5. ______
6. ______
7. ______
8. ______
9. ______
10. ______
11. ______
12. ______
13. ______
14. ______
15. ______
16. ______

Add. Reduce to the lowest terms:

1. $\frac{4}{11} + \frac{5}{11}$

2. $\frac{5}{12} + \frac{2}{12}$

3. $\frac{1}{9} + \frac{4}{9} + \frac{1}{9}$

4. $\frac{3}{8} + \frac{1}{8} + \frac{2}{8}$

5. $\frac{1}{2} + \frac{1}{2}$

6. $\frac{1}{4} + \frac{3}{4}$

7. $\frac{2}{5} + \frac{2}{5}$

8. $\frac{4}{13} + \frac{4}{13}$

9. $\frac{3}{10} + \frac{4}{10} + \frac{1}{10}$

10. $\frac{5}{12} + \frac{4}{12} + \frac{1}{12}$

11. $\frac{5}{11} + \frac{2}{11} + \frac{1}{11}$

12. $\frac{3}{7} + \frac{1}{7} + \frac{2}{7}$

13. $\frac{3}{6} + \frac{1}{6} + \frac{2}{6}$

14. $\frac{4}{10} + \frac{5}{10} + \frac{1}{10}$

15. $\frac{4}{12} + \frac{3}{12} + \frac{8}{12}$

16. $\frac{7}{16} + \frac{8}{16} + \frac{5}{16}$

B

17. ______
18. ______
19. ______

17. $\frac{2}{15} + \frac{4}{15} + \frac{3}{15}$

18. $\frac{5}{24} + \frac{2}{24} + \frac{1}{24}$

19. $\frac{5}{7} + \frac{3}{7}$

ANSWERS

20. ______

21. ______

22. ______

23. ______

24. ______

25. ______

26. ______

27. ______

28. ______

29. ______

30. ______

31. ______

32. ______

20. $\frac{14}{20} + \frac{9}{20}$

21. $\frac{5}{48} + \frac{7}{48} + \frac{3}{48}$

22. $\frac{3}{16} + \frac{2}{16} + \frac{5}{16}$

23. $\frac{8}{30} + \frac{9}{30} + \frac{1}{30}$

24. $\frac{13}{50} + \frac{7}{50} + \frac{2}{50}$

25. $\frac{9}{2} + \frac{7}{2}$

26. $\frac{7}{5} + \frac{8}{5}$

27. $\frac{5}{24} + \frac{7}{24} + \frac{9}{24}$

28. $\frac{3}{20} + \frac{9}{20} + \frac{3}{20}$

29. $\frac{7}{36} + \frac{8}{36} + \frac{5}{36}$

30. $\frac{7}{36} + \frac{7}{36} + \frac{1}{36}$

31. $\frac{5}{18} + \frac{7}{18} + \frac{3}{18}$

32. $\frac{7}{40} + \frac{1}{40} + \frac{7}{40}$

C

33. ______

34. ______

35. ______

36. ______

37. ______

38. ______

39. ______

40. ______

33. $\frac{1}{12} + \frac{1}{12} + \frac{1}{12} + \frac{1}{12} + \frac{1}{12}$

34. $\frac{3}{12} + \frac{3}{12} + \frac{3}{12} + \frac{3}{12}$

35. $\frac{17}{100} + \frac{31}{100} + \frac{9}{100} + \frac{3}{100}$

36. $\frac{29}{50} + \frac{3}{50} + \frac{3}{50}$

37. $\frac{7}{120} + \frac{9}{120} + \frac{25}{120} + \frac{15}{120}$

38. $\frac{29}{144} + \frac{17}{144} + \frac{3}{144} + \frac{25}{144}$

39. $\frac{7}{25} + \frac{3}{25} + \frac{2}{25} + \frac{8}{25}$

40. $\frac{23}{150} + \frac{31}{150} + \frac{17}{150} + \frac{9}{150}$

NAME ______________________

CLASS ______________ SECTION ______________ DATE ______________

ANSWERS

41. ______________

42. ______________

43. ______________

44. ______________

41. $\frac{21}{40} + \frac{15}{40} + \frac{4}{40}$

42. $\frac{27}{60} + \frac{18}{60} + \frac{5}{60}$

43. $\frac{8}{75} + \frac{9}{75} + \frac{8}{75}$

44. $\frac{30}{90} + \frac{22}{90} + \frac{20}{90}$

D

45. ______________

45. The stock of the Eastern Corporation rose $\frac{5}{8}$ point on Monday, $\frac{7}{8}$ point on Tuesday, $\frac{3}{8}$ point on Wednesday, $\frac{5}{8}$ point on Thursday, and $\frac{7}{8}$ point on Friday. What was the total rise for the week?

46. ______________

46. The stock of the Northern Corporation dropped $\frac{4}{8}$ point on Monday, $\frac{9}{8}$ point on Tuesday, $\frac{5}{8}$ point on Wednesday, 0 points on Thursday, and $\frac{3}{8}$ point on Friday. What was the total decline for the week?

47. Chef Ramon prepares a punch for the stockholders' meeting of the Northern Corporation. The punch calls for $\frac{1}{4}$ gallon lemon juice, $\frac{3}{4}$ gallon raspberry juice, $\frac{2}{4}$ gallon cranberry juice, $\frac{1}{4}$ gallon lime juice, $\frac{3}{4}$ gallon 7-Up, and $\frac{3}{4}$ gallon vodka. How many gallons of punch does the recipe make?

47. ______________

ANSWERS

48. ______

49. ______

50. ______

51. ______

52. ______

53. ______

48. A physical therapist prescribes that Belinda swim $\frac{5}{16}$ mile on Monday and increase the distance by $\frac{1}{16}$ mile each day from Tuesday through Friday. How much total swimming was prescribed for the five days?

49. Mary spends $\frac{3}{16}$ of her income for rent, $\frac{1}{16}$ for entertainment, and $\frac{3}{16}$ for food. How much of her income goes for rent, entertainment, and food?

50. At Hilhi school $\frac{1}{9}$ of the students take algebra, $\frac{5}{9}$ take arithmetic, $\frac{2}{9}$ take geometry, and the rest take no math at all. What part of the students take math?

51. To find the perimeter of (distance around) a rectangle, add the two lengths and the two widths together. What is the perimeter of a rectangle if each of the lengths is $\frac{7}{10}$ meter and each width is $\frac{1}{10}$ meter?

52. What is the perimeter of a rectangle if each of the lengths is $\frac{9}{10}$ meter and each width is $\frac{3}{10}$ meter?

53. In order to make a certain project, Charles needs $\frac{1}{10}$ inch of foam, $\frac{3}{10}$ inch of metal, $\frac{4}{10}$ inch of wood, $\frac{7}{10}$ inch of fabric. What will be the total thickness of this project when these materials are piled up?

NAME ______________________________

CLASS ______________ SECTION ______________ DATE ______________

ANSWERS

54. ______________

54. Marilyn is collecting buttons for a certain project. The first button is $\frac{5}{12}$ inch thick, the second is $\frac{7}{12}$ inch thick, and the last one is $\frac{4}{12}$ inch thick. They will be piled up for this project. What is the total height?

E *Maintain your skills* (Sections 1.2, 1.8, 2.6, 3.2, 3.6)

55. ______________

55. Find the LCM of 6, 8, 10, and 12.

56. ______________

56. Find the LCM of 8, 9, 12, and 20.

57. ______________

57. Round 678,223,500,000 to the nearest million.

58. ______________

58. Round 678,223,500,000 to the nearest hundred thousand.

59. ______________

Change to an improper fraction.

60. ______________

59. $13\frac{3}{4}$

60. $17\frac{7}{12}$

ANSWERS

61. ______________________

62. ______________________

63. ______________________

64. ______________________

61. Find the average of 256, 432, 367, 575, and 5.

62. Find the average of 1803, 72, 630, 1433, 1009, and 753.

63. How many 12-foot pieces of pipe are needed to plumb a house that requires a total of 288 feet of pipe?

64. In a metal bench work class that has 42 students each student is allowed $8\frac{1}{4}$ inches of wire solder. How many inches of wire must the instructor provide for the class?

Addition of Unlike Fractions

OBJECTIVE

41. Add unlike fractions.

APPLICATION

Sheila Frankowski is assembling a tricycle for her son's birthday. She needs a bolt that must reach through a $\frac{1}{32}$-inch thick washer, a $\frac{3}{16}$-inch thick plastic bushing, a $\frac{3}{4}$-inch piece of steel tubing, a second $\frac{1}{32}$-inch thick washer, and a $\frac{1}{4}$-inch thick nut. How long a bolt does she need?

VOCABULARY

Unlike fractions are fractions with different denominators.

HOW AND WHY

The sum $\frac{1}{2} + \frac{1}{5}$ cannot be worked in this form. A look at Figure 3.11 shows that the parts are not the same size:

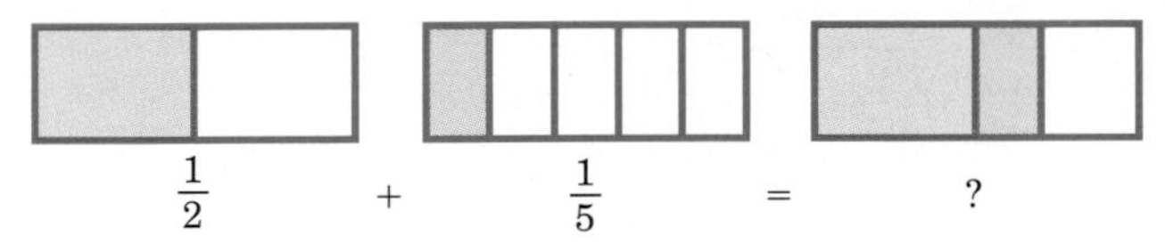

$\frac{1}{2} \quad + \quad \frac{1}{5} \quad = \quad ?$

Figure 3.11

To add, rename $\frac{1}{2}$ and $\frac{1}{5}$ as like fractions.

The LCM (Least Common Multiple) of the two denominators serves as the least common denominator. The LCM of 2 and 5 is 10. We can now write:

$$\frac{1}{2} = \left(\frac{1}{2}\right)\left(\frac{5}{5}\right) = \frac{5}{10} \quad \text{and} \quad \frac{1}{5} = \left(\frac{1}{5}\right)\left(\frac{2}{2}\right) = \frac{2}{10}$$

and the problem now becomes:

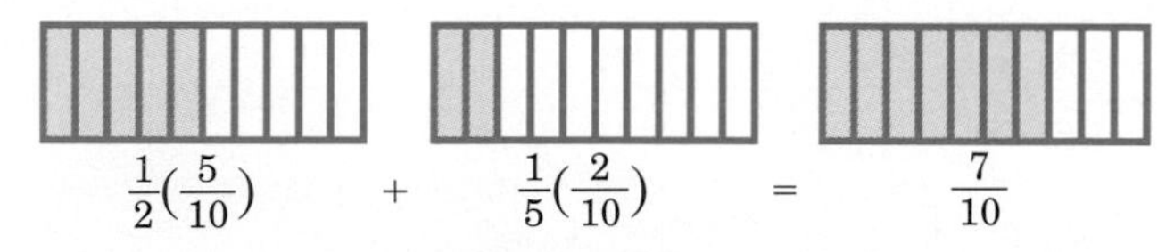

$\frac{1}{2}\left(\frac{5}{10}\right) \quad + \quad \frac{1}{5}\left(\frac{2}{10}\right) \quad = \quad \frac{7}{10}$

Figure 3.12

PROCEDURE

To add unlike fractions:

1. **Build the fractions so that they have a common denominator.**
2. **Add the numerators and write the sum over the common denominator.**
3. **Reduce, if possible.**

MODEL PROBLEM-SOLVING

Examples	Strategy
a. Add:	
$\frac{3}{8} + \frac{1}{4}$	**The fractions do not have common denominators.**
$\frac{3}{8} = \frac{3}{8} \qquad \frac{1}{4} = \frac{2}{8}$	**Build the fraction $\frac{1}{4}$ to have a denominator of 8, which is the LCM of 8 and 4.**
$\frac{3}{8} + \frac{1}{4} = \frac{3}{8} + \frac{2}{8}$	**Replace the original fractions with their equivalents that have a common denominator.**
$= \frac{5}{8}$	**Add.**

Warm Up a. Add: $\frac{2}{3} + \frac{1}{6}$

Examples	Strategy
b. Add:	
$\frac{5}{12} + \frac{2}{9}$	**The fractions do not have a common denominator.**
$\frac{5}{12} = \frac{15}{36} \qquad \frac{2}{9} = \frac{8}{36}$	**Build each to have a denominator of 36, which is the LCM of 12 and 9.**
$\frac{5}{12} + \frac{2}{9} = \frac{15}{36} + \frac{8}{36}$	**Replace the original fractions with their equivalents that have a common denominator.**
$= \frac{23}{36}$	**Add.**

Warm Up b. Add: $\frac{5}{8} + \frac{1}{6}$

ANSWERS TO WARM UPS (3.10) **a.** $\frac{5}{6}$ **b.** $\frac{19}{24}$

c. Add: $\frac{3}{4} + \frac{7}{10}$

$\frac{3}{4} = \frac{15}{20} \qquad \frac{7}{10} = \frac{14}{20}$

Build each fraction to have a denominator of 20.

$\frac{3}{4} + \frac{7}{10} = \frac{15}{20} + \frac{14}{20}$

Replace the original fractions with their equivalents that have a common denominator.

$= \frac{29}{20}$

Add.

or

$1\frac{9}{20}$

The answer can be left as an improper fraction or rewritten as a mixed number.

Warm Up c. Add: $\frac{5}{8} + \frac{5}{12}$

d. Add:

$\frac{11}{96} + \frac{35}{72}$

We see that $96 = 2^5 \cdot 3$ and $72 = 2^3 \cdot 3^2$, so the LCM is $2^5 \cdot 3^2 = 288$.

$\frac{11}{96} = \frac{3 \cdot 11}{288} = \frac{33}{288}$

$\frac{35}{72} = \frac{4 \cdot 35}{288} = \frac{140}{288}$

Build each fraction to have a denominator of 288.

$\frac{33}{288} + \frac{140}{288} = \frac{173}{288}$

Replace the original fractions with their equivalents that have a common denominator and add.

Warm Up d. Add: $\frac{13}{45} + \frac{28}{75}$

ANSWERS TO WARM UPS (3.10) **c.** $\frac{25}{24}$ or $1\frac{1}{24}$ **d.** $\frac{149}{225}$

APPLICATION SOLUTION

e. Sheila Frankowski is assembling a tricycle for her son's birthday. She needs a bolt that must reach through a $\frac{1}{32}$-inch thick washer, a $\frac{3}{16}$-inch thick plastic bushing, a $\frac{3}{4}$-inch piece of steel tubing, a second $\frac{1}{32}$-inch thick washer, and a $\frac{1}{4}$-inch thick nut. How long a bolt does she need?

$$\frac{1}{32}+\frac{3}{16}+\frac{3}{4}+\frac{1}{32}+\frac{1}{4}$$

The bolt must be long enough to reach through all five pieces to hold them together. Add all the thicknesses together.

$$\frac{1}{32}=\frac{1}{32} \qquad \frac{3}{16}=\frac{6}{32} \qquad \frac{3}{4}=\frac{24}{32}$$

Since the fractions are unlike, we find a common denominator.

$$\frac{1}{32}=\frac{1}{32} \qquad \frac{1}{4}=\frac{8}{32}$$

The LCM of 32, 16, and 4 is 32.

$$\frac{1}{32}+\frac{6}{32}+\frac{24}{32}+\frac{1}{32}+\frac{8}{32}=\frac{40}{32}$$

Write the fractions so they all have a common denominator of 32, and add.

$$=\frac{5}{4}$$

Reduce.

$$=1\frac{1}{4}$$

Change to a mixed number.

So, the bolt must be $1\frac{1}{4}$ inches long.

Warm Up e. A nail must reach through three thicknesses of wood and penetrate the fourth thickness $\frac{1}{4}$ inch. If the first piece of wood is $\frac{5}{16}$ inch, the second is $\frac{3}{8}$ inch, and the third is $\frac{9}{16}$ inch, how long must the nail be?

ANSWERS TO WARM UPS (3.10)

e. $1\frac{1}{2}$ inches

NAME ______________________

CLASS ____________ SECTION ____________ DATE ____________

Exercises 3.10

A

ANSWERS

1. ______
2. ______
3. ______
4. ______
5. ______
6. ______
7. ______
8. ______
9. ______
10. ______
11. ______
12. ______
13. ______
14. ______
15. ______
16. ______
17. ______
18. ______
19. ______
20. ______

Add. Reduce to the lowest terms:

1. $\frac{4}{9}+\frac{1}{6}$ **2.** $\frac{1}{3}+\frac{1}{4}$ **3.** $\frac{1}{2}+\frac{1}{4}$ **4.** $\frac{2}{3}+\frac{1}{6}$

5. $\frac{1}{8}+\frac{7}{24}$ **6.** $\frac{7}{15}+\frac{1}{3}$ **7.** $\frac{3}{16}+\frac{5}{8}$ **8.** $\frac{2}{9}+\frac{5}{18}$

9. $\frac{1}{2}+\frac{1}{3}+\frac{1}{6}$ **10.** $\frac{1}{3}+\frac{1}{4}+\frac{1}{12}$ **11.** $\frac{2}{5}+\frac{3}{10}$ **12.** $\frac{4}{15}+\frac{3}{5}$

13. $\frac{2}{7}+\frac{3}{14}$ **14.** $\frac{3}{4}+\frac{1}{5}$ **15.** $\frac{1}{2}+\frac{3}{4}+\frac{3}{16}$ **16.** $\frac{1}{6}+\frac{3}{8}+\frac{1}{2}$

17. $\frac{3}{15}+\frac{7}{12}$ **18.** $\frac{5}{9}+\frac{5}{12}$ **19.** $\frac{1}{6}+\frac{7}{15}+\frac{3}{5}$ **20.** $\frac{1}{4}+\frac{1}{6}+\frac{1}{3}$

B

21. ______
22. ______
23. ______

21. $\frac{7}{16}+\frac{3}{20}+\frac{1}{5}$ **22.** $\frac{5}{12}+\frac{9}{16}+\frac{7}{24}$ **23.** $\frac{3}{35}+\frac{8}{21}$

ANSWERS

24. ____________________

25. ____________________

26. ____________________

27. ____________________

28. ____________________

29. ____________________

30. ____________________

31. ____________________

32. ____________________

33. ____________________

34. ____________________

35. ____________________

36. ____________________

37. ____________________

38. ____________________

39. ____________________

40. ____________________

24. $\frac{9}{14} + \frac{5}{21}$

25. $\frac{3}{10} + \frac{7}{20} + \frac{11}{30}$

26. $\frac{5}{12} + \frac{3}{16} + \frac{7}{20}$

27. $\frac{5}{8} + \frac{7}{12} + \frac{1}{6}$

28. $\frac{5}{12} + \frac{7}{16} + \frac{1}{24}$

29. $\frac{1}{10} + \frac{2}{5} + \frac{5}{6} + \frac{1}{15}$

30. $\frac{1}{2} + \frac{3}{10} + \frac{3}{5} + \frac{1}{4}$

31. $\frac{3}{4} + \frac{5}{8} + \frac{7}{16}$

32. $\frac{15}{36} + \frac{7}{9} + \frac{5}{12}$

33. $\frac{9}{16} + \frac{13}{20} + \frac{1}{10}$

34. $\frac{5}{6} + \frac{2}{3} + \frac{4}{15}$

35. $\frac{5}{6} + \frac{7}{8} + \frac{3}{4} + \frac{1}{2}$

36. $\frac{9}{10} + \frac{4}{5} + \frac{7}{15} + \frac{11}{18}$

37. $\frac{13}{24} + \frac{17}{36} + \frac{11}{48} + \frac{21}{72}$

38. $\frac{12}{27} + \frac{5}{9} + \frac{32}{81} + \frac{2}{3}$

39. $\frac{7}{9} + \frac{4}{5} + \frac{4}{15} + \frac{11}{30}$

40. $\frac{11}{15} + \frac{7}{12} + \frac{9}{10} + \frac{17}{20}$

NAME ______________________

CLASS ______________ SECTION ______________ DATE ______________

C

ANSWERS

41. ______________

42. ______________

43. ______________

44. ______________

45. ______________

46. ______________

47. ______________

48. ______________

49. ______________

50. ______________

51. ______________

52. ______________

53. ______________

54. ______________

55. ______________

56. ______________

57. ______________

58. ______________

41. $\frac{11}{20} + \frac{2}{5} + \frac{13}{40}$

42. $\frac{21}{25} + \frac{11}{50} + \frac{7}{75}$

43. $\frac{1}{12} + \frac{1}{36} + \frac{1}{54}$

44. $\frac{23}{96} + \frac{14}{64} + \frac{7}{80}$

45. $\frac{117}{240} + \frac{41}{270}$

46. $\frac{7}{18} + \frac{5}{27} + \frac{4}{36}$

47. $\frac{27}{72} + \frac{19}{48}$

48. $\frac{32}{75} + \frac{41}{90}$

49. $\frac{19}{54} + \frac{7}{48}$

50. $\frac{7}{96} + \frac{37}{84}$

51. $\frac{29}{100} + \frac{7}{50} + \frac{33}{75}$

52. $\frac{10}{33} + \frac{9}{22} + \frac{4}{55}$

53. $\frac{25}{36} + \frac{19}{48}$

54. $\frac{72}{85} + \frac{69}{102}$

55. $\frac{39}{102} + \frac{14}{17} + \frac{18}{51}$

56. $\frac{25}{72} + \frac{9}{108} + \frac{19}{144}$

57. $\frac{41}{48} + \frac{23}{30}$

58. $\frac{29}{50} + \frac{62}{75}$

ANSWERS

59. ____________

60. ____________

59. $\frac{23}{30} + \frac{17}{20} + \frac{57}{75}$ **60.** $\frac{11}{18} + \frac{29}{30} + \frac{13}{24}$

D

61. ____________

61. Jonnie Lee Simms is assembling a rocking horse for his granddaughter. He needs a bolt to reach through a $\frac{7}{8}$-inch piece of steel tubing, a $\frac{1}{16}$-inch bushing, a $\frac{1}{2}$-inch piece of tubing, a $\frac{1}{8}$-inch thick washer, and a $\frac{1}{4}$-inch thick nut. How long a bolt does he need?

62. ____________

62. On the American Stock Exchange, Joan's stock rose $\frac{1}{8}$ of a point the first hour and an additional $\frac{3}{16}$ of a point during the remainder of the day. What was the total rise for the day?

63. What is the total distance (perimeter) around this triangle?

63. ____________

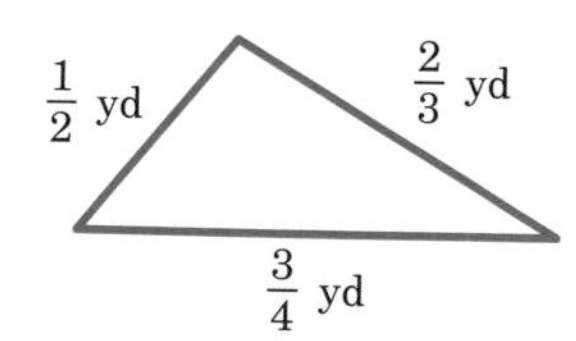

64. Jim ran $\frac{7}{8}$ mile on Monday, $\frac{11}{12}$ mile on Tuesday, and $\frac{5}{6}$ mile each on Wednesday and Thursday. What was the total distance he ran?

64. ____________

NAME ______________________________

CLASS ______________ SECTION ______________ DATE ______________

ANSWERS

65. ______________

65. Find the length of this pin:

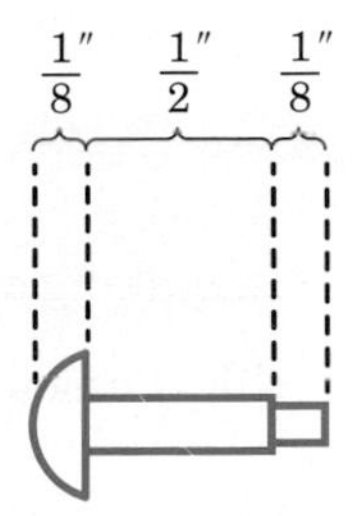

66. ______________

66. An elephant-ear bamboo grew $\frac{1}{2}$ inch on Tuesday, $\frac{3}{8}$ inch on Wednesday, and $\frac{1}{4}$ inch on Thursday. How much did it grow in the three days?

67. ______________

67. Mike needs $\frac{1}{2}$ cup of milk for the biscuit mix, $\frac{3}{4}$ cup of milk for the pancake mix, and $\frac{1}{3}$ cup of milk for the scrambled eggs. How much milk does he need in all?

68. If one board that is $\frac{7}{16}$ inch thick is nailed to another board that is $\frac{5}{8}$ inch thick, what is the total thickness of the boards?

68. ______________

ANSWERS

69. ______

70. ______

69. Jonathan purchased the following remnants of fabric one Saturday: $\frac{1}{2}$ yard, $\frac{3}{8}$ yard, $\frac{2}{3}$ yard, $\frac{3}{4}$ yard, and $\frac{7}{8}$ yard. What was the total amount of fabric he bought?

70. In making a certain piece of furniture, you glue together boards of various thicknesses. Suppose you use a $\frac{3}{4}$-inch board, a $\frac{5}{8}$-inch board, a $\frac{11}{16}$-inch board, and a $\frac{5}{6}$-inch board. What is the total thickness?

E *Maintain your skills* (Sections 1.8, 2.5, 2.6)

71. ______

72. ______

73. ______

74. ______

75. ______

76. ______

77. ______

78. ______

79. ______

80. ______

71. Prime factor 132.

72. Prime factor 248.

73. Find the LCM of 12, 16, and 24.

74. Find the LCM of 12, 14, 16, and 20.

Perform the indicated operations.

75. $38 - 5 \cdot 7 + 6 \cdot 5 - 5$

76. $(38 - 5)7 + 6(5 - 5)$

77. $4^6 + 2^2 - 7^3$

78. $(6 + 2^3)^2 - 6 \cdot 7$

79. Mrs. Teech has five classes to teach this term. The enrollment in each class is as follows:

8 am MWF	32	9 am TT	33
10 am MWF	36	11 am TT	29
1 pm MWF	40		

What is the average enrollment of her classes?

80. A retail furniture store buys seven dinette sets for \$214 each. If the store wants to make a total profit of \$756 on the dinette sets, how much should each set sell for?

3.11 Addition of Mixed Numbers

OBJECTIVE

42. Add mixed numbers.

APPLICATION

Whitney works part-time and takes two or three classes per term. His employer allows him to choose the number of hours he works each day. He worked the following schedule one week: Monday, $6\frac{2}{3}$ hours; Tuesday, $1\frac{3}{4}$ hours; Wednesday, $10\frac{5}{6}$ hours; Thursday, $6\frac{1}{2}$ hours; Friday, $7\frac{1}{4}$ hours. How many hours did he work that week?

VOCABULARY

No new words.

HOW AND WHY

What is the answer to $3\frac{1}{4} + 5\frac{1}{6} = ?$ This can also be written $\left(3 + \frac{1}{4}\right) + \left(5 + \frac{1}{6}\right)$.

$$\left(3 + \frac{1}{4}\right) + \left(5 + \frac{1}{6}\right) = (3 + 5) + \left(\frac{1}{4} + \frac{1}{6}\right)$$
$$= 8 + \left(\frac{3}{12} + \frac{2}{12}\right) = 8\frac{5}{12}$$

If we write the problem vertically, this grouping takes place naturally:

$3\frac{1}{4} - 3\frac{3}{12}$ **Rewrite the fractional parts with a common denominator.**

$\underline{5\frac{1}{6}} = \underline{5\frac{2}{12}}$

$8\frac{5}{12}$ **Add the whole-number parts and add the fractional parts.**

When mixed numbers are added, it is possible for the sum of the fraction parts to be greater than 1. In this case the fraction part of the answer can be changed to a mixed number and added to the whole-number part. For example:

$11\frac{7}{10} = 11\frac{21}{30}$ **Write the fractional parts with a common denominator.**

$\underline{23\frac{8}{15}} = \underline{23\frac{16}{30}}$

$34\frac{37}{30}$ **Add the whole-number parts and add the fraction parts.**

$= 34 + 1\frac{7}{30}$ **The fraction is improper, so rewrite it as a mixed number.**

$= 35\frac{7}{30}$ **Add the mixed number to the whole number.**

PROCEDURE

To add mixed numbers:

1. **Add the whole numbers.**
2. **Add the fractions. If the sum of the fractions is more than 1, change the fraction to a mixed number and add again.**

MODEL PROBLEM-SOLVING

Examples | **Strategy**

a. Add:

$$5\frac{5}{12} = 5\frac{5}{12}$$

$$\underline{1\frac{3}{4}} = \underline{1\frac{9}{12}}$$

$$6\frac{14}{12}$$

Write the fractions with common denominators. The LCM of 12 and 4 is 12.

$$= 6 + 1\frac{2}{12}$$

Write the improper fraction as a mixed number and add to the whole number.

$$= 7\frac{2}{12}$$

$$= 7\frac{1}{6}$$

Reduce.

Warm Up a. Add: $8\frac{5}{16}$

$$\underline{3\frac{1}{2}}$$

ANSWERS TO WARM UPS (3.11)

a. $11\frac{13}{16}$

b. Add:

$$\begin{aligned} 25\frac{7}{8} &= 25\frac{63}{72} \\ 13\frac{5}{9} &= 13\frac{40}{72} \\ 7\frac{1}{6} &= 7\frac{12}{72} \\ \hline &\,45\frac{115}{72} \end{aligned}$$

Write the fractions with common denominators. The LCM of 8, 9, and 6 is 72.

$$= 45 + 1\frac{43}{72}$$

Write the improper fraction as a mixed number and add to the whole number.

$$= 46\frac{43}{72}$$

Warm Up b. Add: $12\frac{8}{9}$, $14\frac{3}{4}$, $6\frac{2}{3}$

c. Add:

$$\begin{aligned} 8 &= 8\frac{0}{3} \\ 5\frac{2}{3} &= 5\frac{2}{3} \\ \hline &\,13\frac{2}{3} \end{aligned}$$

Recall that any whole number can be written as a mixed number by writing the fraction part with 0 as a numerator and any nonzero number for a denominator. The denominator 3 is selected so that the denominators will be common.

Warm Up c. Add: 16, $12\frac{4}{9}$

ANSWERS TO WARM UPS (3.11)

b. $34\frac{11}{36}$ **c.** $28\frac{4}{9}$

APPLICATION SOLUTION

d. Whitney works part-time and takes two or three classes per term. His employer allows him to choose the number of hours he works each day. He worked the following schedule one week: Monday, $6\frac{2}{3}$ hours; Tuesday, $1\frac{3}{4}$ hours; Wednesday, $10\frac{5}{6}$ hours; Thursday, $6\frac{1}{2}$ hours; Friday, $7\frac{1}{4}$ hours. How many hours did he work that week?

$$6\frac{2}{3} = 6\frac{8}{12}$$
$$1\frac{3}{4} = 1\frac{9}{12}$$
$$10\frac{5}{6} = 10\frac{10}{12}$$
$$6\frac{1}{2} = 6\frac{6}{12}$$
$$7\frac{1}{4} = 7\frac{3}{12}$$

To find out how many hours he worked that week, find the sum of the hours he worked. Write the fraction parts with a common denominator. The LCM of 2, 3, 4, and 6 is 12.

$$30\frac{36}{12}$$

Change the improper fraction to a whole number.

$$= 30 + 3 = 33$$

Find the sum of the two whole numbers.

Therefore, Whitney worked a total of 33 hours that week.

Warm Up d. The next week Whitney worked the following schedule: Monday, $5\frac{3}{4}$ hours; Tuesday, $6\frac{11}{12}$ hours; Wednesday, $4\frac{1}{2}$ hours; Thursday, $8\frac{1}{4}$ hours; Friday, $9\frac{5}{6}$ hours. How many hours did he work that week?

ANSWERS TO WARM UPS (3.11)

d. $35\frac{1}{4}$ hours

NAME ______________________

CLASS ____________ SECTION ____________ DATE ____________

Exercises 3.11

A

ANSWERS

1. ______
2. ______
3. ______
4. ______
5. ______
6. ______
7. ______
8. ______
9. ______
10. ______
11. ______
12. ______
13. ______
14. ______
15. ______
16. ______
17. ______
18. ______

Add. Write answers as mixed numbers where possible:

1. $1\frac{3}{7}$ $+\ 2\frac{2}{7}$

2. $1\frac{1}{3}$ $+\ 3\frac{1}{3}$

3. $6\frac{4}{5}$ $+\ 1\frac{2}{5}$

4. $5\frac{3}{8}$ $+\ 6\frac{7}{8}$

5. $3\frac{2}{3}$ $+\ 2\frac{1}{4}$

6. $6\frac{1}{6}$ $+\ 8\frac{2}{3}$

7. $5\frac{5}{9}$ $+\ \frac{2}{3}$

8. $8\frac{5}{12}$ $+\ 6\frac{1}{12}$

9. $2\frac{4}{7}$ $+\ 3\frac{11}{14}$

10. $17\frac{5}{12}$ $+\ 1\frac{5}{6}$

11. $2\frac{7}{15}$ $+\ 13\frac{2}{5}$

12. $11\frac{2}{3}$ $+\ 5\frac{1}{4}$

13. $6\frac{3}{8}$ $+\ 14\frac{5}{12}$

14. $6\frac{2}{9}$ $+\ 12\frac{1}{6}$

15. $3\frac{2}{3}$ $+\ 2\frac{1}{2}$ $+\ 4\frac{5}{6}$

16. $7\frac{3}{10}$ $+\ 8\frac{2}{5}$ $+\ 6\frac{1}{2}$

17. $4\frac{1}{2} + 8\frac{3}{4} + 6 + 7\frac{2}{3}$

18. $5\frac{2}{5} + 4 + 3\frac{3}{10} + 9\frac{7}{15}$

B

ANSWERS

19. ____________

20. ____________

21. ____________

22. ____________

23. ____________

24. ____________

25. ____________

26. ____________

27. ____________

28. ____________

29. ____________

30. ____________

19. $\begin{array}{r} 7\frac{3}{8} \\ 4\frac{2}{8} \\ \hline \end{array}$

20. $\begin{array}{r} 15\frac{3}{5} \\ 17\frac{1}{5} \\ \hline \end{array}$

21. $\begin{array}{r} 1\frac{1}{5} \\ 2\frac{4}{5} \\ 3\frac{2}{5} \\ \hline \end{array}$

22. $\begin{array}{r} 5\frac{1}{4} \\ 2\frac{3}{4} \\ 7\frac{1}{4} \\ \hline \end{array}$

23. $\begin{array}{r} 14\frac{7}{10} \\ 11\frac{3}{5} \\ \hline \end{array}$

24. $\begin{array}{r} 9\frac{5}{24} \\ 101\frac{7}{12} \\ \hline \end{array}$

25. $21\frac{5}{7} + 15\frac{9}{14} + 12\frac{10}{21}$

26. $18\frac{3}{4} + 17\frac{7}{8} + 23\frac{1}{6}$

27. $\begin{array}{r} 213\frac{5}{18} \\ 506\frac{7}{12} \\ \hline \end{array}$

28. $\begin{array}{r} 213\frac{5}{6} \\ 347\frac{3}{10} \\ \hline \end{array}$

29. $47\frac{1}{5} + 23\frac{2}{3} + 15\frac{1}{2}$

30. $47\frac{3}{8} + 23 + 42\frac{5}{12}$

NAME ______________________________

CLASS ______________ SECTION ______________ DATE ______________

ANSWERS

31. ______________

32. ______________

33. ______________

34. ______________

35. ______________

36. ______________

37. ______________

38. ______________

39. ______________

31. $25\frac{2}{3} + 16\frac{1}{2} + 18\frac{3}{4}$

32. $29\frac{7}{8} + 19\frac{5}{6} + 32\frac{3}{4}$

33. $62 + 18\frac{5}{9} + 37\frac{1}{5}$

34. $41\frac{5}{7} + 29\frac{9}{14} + 3\frac{1}{6}$

35. $41\frac{5}{7} + 29\frac{9}{14} + 36$

36. $12\frac{11}{12} + 22\frac{5}{8} + 8$

37. $26 + 18\frac{6}{7} + 11\frac{15}{28}$

38.

$15\frac{5}{6}$

$12\frac{9}{10}$

16

$\underline{17\frac{1}{6}}$

39.

$22\frac{7}{8}$

19

$16\frac{5}{9}$

$\underline{10\frac{5}{12}}$

ANSWERS

40. ______

41. ______

42. ______

43. ______

44. ______

45. ______

46. ______

47. ______

48. ______

49. ______

50. ______

51. ______

40. $$\begin{array}{r} 7\frac{7}{20} \\ \frac{3}{4} \\ 8\frac{9}{40} \\ \hline \end{array}$$

41. $$\begin{array}{r} 6\frac{2}{9} \\ \frac{1}{4} \\ 12\frac{5}{6} \\ \hline \end{array}$$

42. $2\frac{2}{5} + 7\frac{1}{6} + 1\frac{4}{15} + 3\frac{1}{10}$

43. $1\frac{1}{5} + 3\frac{7}{10} + \frac{1}{2} + 7\frac{16}{25}$

44. $$\begin{array}{r} 14\frac{13}{18} \\ 22\frac{23}{27} \\ \hline \end{array}$$

45. $$\begin{array}{r} 28\frac{7}{15} \\ 19\frac{13}{20} \\ \hline \end{array}$$

46. $$\begin{array}{r} 82\frac{41}{45} \\ 97\frac{25}{27} \\ \hline \end{array}$$

47. $$\begin{array}{r} 74\frac{23}{42} \\ 57\frac{7}{18} \\ \hline \end{array}$$

48. $$\begin{array}{r} 82\frac{7}{24} \\ 16\frac{5}{12} \\ 5\frac{7}{36} \\ \hline \end{array}$$

49. $$\begin{array}{r} 100\frac{37}{100} \\ 31\frac{7}{50} \\ 15\frac{17}{20} \\ \hline \end{array}$$

50. $$\begin{array}{r} 17\frac{2}{5} \\ 28\frac{7}{10} \\ 18\frac{5}{6} \\ 22\frac{2}{3} \\ \hline \end{array}$$

51. $$\begin{array}{r} 57\frac{19}{24} \\ 23\frac{7}{12} \\ 32\frac{5}{18} \\ 46\frac{25}{36} \\ \hline \end{array}$$

NAME ______________________________

CLASS ______________ SECTION ______________ DATE ______________

ANSWERS

52. ______________

53. ______________

54. ______________

55. ______________

56. ______________

57. ______________

52. $16\frac{7}{16}$

$72\frac{8}{9}$

$\underline{52\frac{5}{144}}$

53. $17\frac{3}{35}$

$8\frac{1}{5}$

$9\frac{5}{7}$

$\underline{12\frac{1}{2}}$

54. $15\frac{3}{8} + 22\frac{1}{2} + 19\frac{5}{9} + 36\frac{2}{3}$

55. $14\frac{3}{5} + 29\frac{1}{8} + 7\frac{3}{4} + 12\frac{9}{10}$

56. $13\frac{11}{12} + 3\frac{17}{18} + \frac{19}{24} + 15$

57. $41\frac{17}{25} + 34\frac{11}{15} + 16 + 25\frac{3}{5}$

D

58. ______________

58. Maria worked the following schedule in her part-time job. How many hours did she work during the month?

Oct 1–Oct 7	$25\frac{1}{2}$ hours
Oct 8–Oct 14	$16\frac{2}{3}$ hours
Oct 15–Oct 21	10 hours
Oct 22–Oct 28	$19\frac{5}{6}$ hours
Oct 29–Oct 31	$4\frac{3}{4}$ hours

59. ______________________

59. Michelle worked $4\frac{1}{2}$ hours on both Monday and Wednesday, $6\frac{3}{4}$ hours on Tuesday, and 8 hours on Saturday. How many hours did she work during the week?

60. ______________________

60. On successive days Wendy picked $21\frac{3}{4}$, $31\frac{5}{8}$, and $27\frac{3}{16}$ pounds of beans. How many pounds of beans did she pick during the three days?

61. ______________________

61. John rode his bicycle $2\frac{3}{8}$ miles on Tuesday, $7\frac{3}{4}$ miles on Wednesday, $10\frac{2}{3}$ miles on Thursday, and $4\frac{5}{6}$ miles on Friday. What was his total mileage during the four days?

62. In a recent road test three cars used the following amounts of gasoline: $9\frac{3}{10}$ gal, $10\frac{1}{2}$ gal, and $9\frac{3}{4}$ gal. How much gasoline was used?

62. ______________________

NAME ______________________________

CLASS ______________ SECTION ______________ DATE ______________

ANSWERS

63. ______________

63. What is the overall length of the bolt in the drawing?

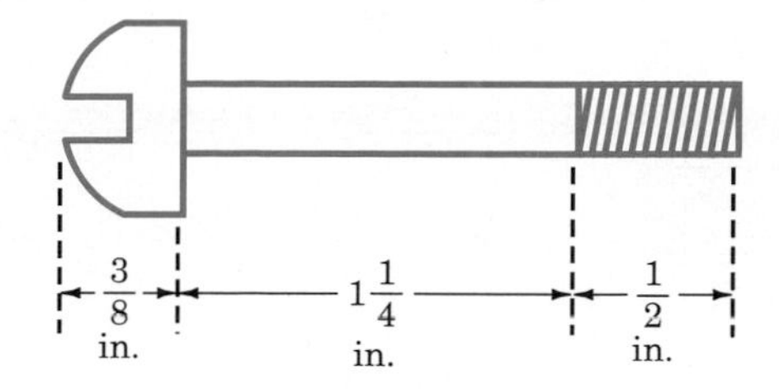

Recall that to find the perimeter of any figure, you must find the distance around the figure. That is, add the lengths of all the sides together.

64. ______________

64. Find the perimeter of a rectangle that has two lengths of $24\frac{1}{2}$ ft and two widths of $17\frac{3}{4}$ ft.

65. ______________

65. Find the perimeter of a triangle (three sides) that has sides of $12\frac{5}{6}$ ft, $16\frac{3}{4}$ ft, and $14\frac{5}{12}$ ft.

E *Maintain your skills* (Sections 2.5, 3.3, 3.4)

ANSWERS

66. ______

67. ______

68. ______

69. ______

70. ______

71. ______

72. ______

73. ______

74. ______

75. ______

Multiply.

66. $\frac{7}{8} \cdot \frac{8}{9} \cdot \frac{3}{7}$

67. $\frac{18}{5} \cdot \frac{15}{4} \cdot \frac{7}{9}$

68. $\frac{12}{15} \cdot \frac{14}{33} \cdot \frac{22}{42}$

69. $\frac{35}{36} \cdot \frac{16}{21} \cdot \frac{38}{57}$

70. Prime factor 111.

71. Prime factor 1323.

Reduce.

72. $\frac{1950}{4095}$

73. $\frac{385}{847}$

74. Par on the first nine holes of the Richet Country Club golf course is 36. If Millie recorded scores of 5, 4, 6, 2, 3, 3, 3, 5, and 4 on those nine holes, what was her total score? Is she under or over par for the first nine holes?

75. Dried prunes weigh one-third the weight of fresh prunes. How many pounds of fresh prunes must be dried to make up 124 half-pound packages of dried prunes?

Subtraction of Fractions

OBJECTIVE

43. Subtract fractions.

APPLICATION

Lumber mill operators must plan for the shrinkage of "green" (wet) boards when they cut logs. If the shrinkage for a $\frac{5}{8}$-inch thick board is expected to be $\frac{1}{16}$ inch, what will be the thickness of the dried board?

VOCABULARY

No new words.

HOW AND WHY

What is the answer to $\frac{2}{3} - \frac{1}{3} = ?$ Looking at Figure 3.13, we can see that we subtract the numerators and keep the common denominator (subtract the yellow region from the blue region).

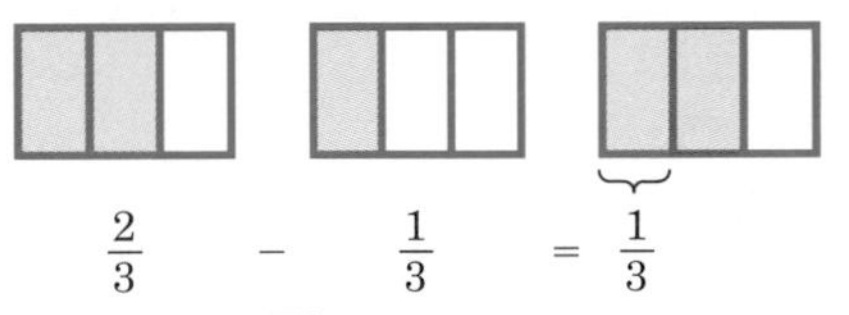

Figure 3.13

What is the answer to $\frac{3}{4} - \frac{1}{3} = ?$ (See Figure 3.14.)

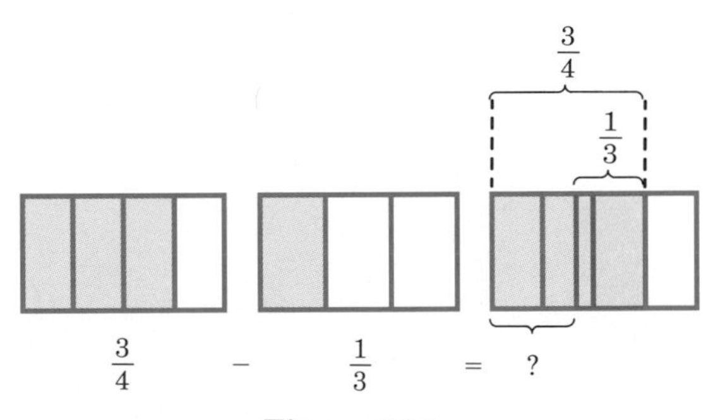

Figure 3.14

The blue shaded area with the question mark cannot be named immediately, because the original parts are not the same size. If the fractions had common denominators, we could subtract as before. Using the common denominator 12, we see in Figure 3.15 that the answer is $\frac{5}{12}$:

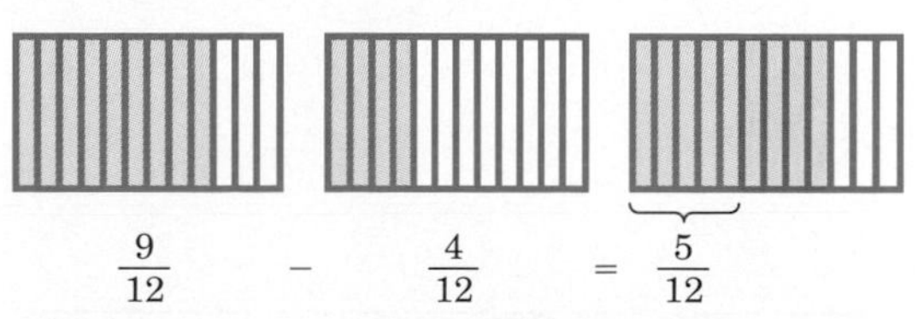

Figure 3.15

$$\frac{3}{4} - \frac{1}{3} = \frac{9}{12} - \frac{4}{12} = \frac{5}{12}$$

The process for subtracting fractions is similar to that for adding fractions. We subtract instead of adding.

PROCEDURE

To subtract fractions:

1. **Build the fractions so that they have a common denominator.**
2. **Subtract the numerators and write the difference over the common denominator.**
3. **Reduce, if possible.**

MODEL PROBLEM-SOLVING

Examples | **Strategy**

a. Subtract:

$$\frac{11}{20} - \frac{5}{20} = \frac{6}{20}$$

Since the fractions have common denominators, subtract the numerators and write the answer over the common denominator.

$$= \frac{3}{10}$$

Reduce.

Warm Up a. Subtract: $\frac{15}{16} - \frac{11}{16}$

b. Subtract:

$$\frac{7}{8} - \frac{2}{3} = \frac{21}{24} - \frac{16}{24}$$

The fractions do not have common denominators. Build each fraction to have 24, the LCM of 8 and 3, as a denominator.

$$= \frac{5}{24}$$

Subtract the numerators and put the answer over the common denominator.

Warm Up b. Subtract: $\frac{11}{12} - \frac{4}{5}$

ANSWERS TO WARM UPS (3.12) a. $\frac{1}{4}$ b. $\frac{7}{60}$

c. Subtract:

$$\frac{7}{15} - \frac{1}{4} = \frac{28}{60} - \frac{15}{60}$$

The fractions do not have common denominators. Build each to have 60, the LCM of 15 and 4, for a denominator.

$$= \frac{13}{60}$$

Subtract the numerators and put the answer over the common denominator.

Warm Up c. Subtract: $\frac{15}{32} - \frac{1}{3}$

d. Subtract:

$$\frac{25}{48} - \frac{15}{64}$$

Find the LCM of 48 and 64: $48 = 2^4 \cdot 3$ and $64 = 2^6$. The LCM is $2^6 \cdot 3 = 192$.

$$\frac{25}{48} = \frac{4 \cdot 25}{192} = \frac{100}{192}$$

$$\frac{15}{64} = \frac{3 \cdot 15}{192} = \frac{45}{192}$$

Build each fraction so that the denominator is 192.

$$= \frac{55}{192}$$

Subtract the numerators and put the answer over the common denominator.

Warm Up d. Subtract: $\frac{71}{72} - \frac{28}{45}$

ANSWERS TO WARM UPS (3.12)

c. $\frac{13}{96}$ **d.** $\frac{131}{360}$

APPLICATION SOLUTION

e. Lumber mill operators must plan for the shrinkage of "green" (wet) boards when they cut logs. If the shrinkage for a $\frac{5}{8}$-inch thick board is expected to be $\frac{1}{16}$ inch, what will be the thickness of the dried board?

$$\frac{5}{8} = \frac{10}{16}$$
$$\frac{1}{16} = \frac{1}{16}$$
$$\frac{9}{16}$$

Subtract the shrinkage from the thickness of the "green" board to find the thickness of the dry board. Since the denominators are not common, build the fraction $\frac{5}{8}$ so that it has a denominator of 16, which is the LCM of 8 and 16. Subtract in the usual way.

The dried board will be $\frac{9}{16}$ inch thick.

Warm Up e. Mike must plane $\frac{3}{32}$ inch from the thickness of a board. If the board is now $\frac{3}{4}$ inch thick, how thick will it be after he has planed it?

ANSWERS TO WARM UPS (3.12)

e. $\frac{21}{32}$ inch

NAME ________________

CLASS ____________ SECTION ____________ DATE ____________

Exercises 3.12

A

ANSWERS

1. ________
2. ________
3. ________
4. ________
5. ________
6. ________
7. ________
8. ________
9. ________
10. ________
11. ________
12. ________
13. ________
14. ________
15. ________
16. ________
17. ________
18. ________
19. ________
20. ________

Subtract. Reduce to the lowest terms:

1. $\frac{3}{8} - \frac{1}{8}$ **2.** $\frac{8}{9} - \frac{3}{9}$ **3.** $\frac{3}{4} - \frac{2}{4}$ **4.** $\frac{5}{7} - \frac{3}{7}$

5. $\frac{17}{30} - \frac{7}{30}$ **6.** $\frac{14}{15} - \frac{11}{15}$ **7.** $\frac{5}{7} - \frac{3}{14}$ **8.** $\frac{11}{15} - \frac{2}{5}$

9. $\frac{3}{4} - \frac{5}{16}$ **10.** $\frac{8}{9} - \frac{5}{18}$ **11.** $\frac{3}{15} - \frac{2}{45}$ **12.** $\frac{5}{18} - \frac{2}{9}$

13. $\frac{5}{6} - \frac{1}{3}$ **14.** $\frac{5}{6} - \frac{1}{2}$ **15.** $\frac{17}{18} - \frac{2}{3}$ **16.** $\frac{19}{24} - \frac{3}{8}$

17. $\frac{3}{5} - \frac{1}{3}$ **18.** $\frac{3}{4} - \frac{1}{3}$ **19.** $\frac{5}{8} - \frac{1}{3}$ **20.** $\frac{9}{10} - \frac{1}{3}$

B

ANSWERS

21. ______
22. ______
23. ______
24. ______
25. ______
26. ______
27. ______
28. ______
29. ______
30. ______
31. ______
32. ______
33. ______
34. ______
35. ______
36. ______
37. ______
38. ______
39. ______
40. ______

21. $\frac{7}{8} - \frac{5}{6}$ **22.** $\frac{2}{3} - \frac{3}{8}$ **23.** $\frac{9}{20} - \frac{3}{10}$ **24.** $\frac{7}{8} - \frac{1}{4}$

25. $\frac{6}{14} - \frac{5}{21}$ **26.** $\frac{5}{6} - \frac{7}{15}$ **27.** $\frac{9}{16} - \frac{1}{6}$ **28.** $\frac{5}{6} - \frac{4}{5}$

29. $\frac{8}{9} - \frac{5}{6}$ **30.** $\frac{12}{21} - \frac{5}{14}$ **31.** $\frac{5}{8} - \frac{1}{12}$ **32.** $\frac{7}{10} - \frac{7}{15}$

33. $\frac{9}{10} - \frac{3}{4}$ **34.** $\frac{11}{15} - \frac{7}{12}$ **35.** $\frac{8}{9} - \frac{3}{4}$ **36.** $\frac{11}{12} - \frac{11}{15}$

37. $\frac{23}{25} - \frac{13}{15}$ **38.** $\frac{19}{32} - \frac{17}{48}$ **39.** $\frac{13}{16} - \frac{11}{24}$ **40.** $\frac{13}{18} - \frac{7}{12}$

C

41. ______
42. ______
43. ______
44. ______

41. $\frac{17}{18} - \frac{11}{12}$ **42.** $\frac{9}{10} - \frac{7}{8}$ **43.** $\frac{19}{50} - \frac{13}{40}$ **44.** $\frac{33}{35} - \frac{17}{20}$

NAME ____________________

CLASS ____________ SECTION ____________ DATE ____________

ANSWERS

45. ____________
46. ____________
47. ____________
48. ____________
49. ____________
50. ____________
51. ____________
52. ____________
53. ____________
54. ____________
55. ____________
56. ____________
57. ____________
58. ____________
59. ____________
60. ____________

45. $\frac{3}{10} - \frac{1}{15}$ **46.** $\frac{7}{16} - \frac{1}{6}$ **47.** $\frac{7}{24} - \frac{5}{18}$ **48.** $\frac{23}{48} - \frac{21}{80}$

49. $\frac{19}{24} - \frac{17}{36}$ **50.** $\frac{14}{15} - \frac{11}{20}$ **51.** $\frac{29}{54} - \frac{13}{36}$ **52.** $\frac{47}{60} - \frac{17}{24}$

53. $\frac{5}{9} - \frac{2}{75}$ **54.** $\frac{22}{33} - \frac{9}{44}$ **55.** $\frac{23}{56} - \frac{25}{72}$ **56.** $\frac{47}{64} - \frac{53}{80}$

57. $\frac{121}{144} - \frac{13}{36}$ **58.** $\frac{153}{175} - \frac{12}{25}$ **59.** $\frac{101}{240} - \frac{101}{360}$ **60.** $\frac{133}{200} - \frac{133}{300}$

D

61. ____________

61. If the shrinkage of a $\frac{5}{4}$-inch thick "green" board is $\frac{1}{8}$ inch, what will be the thickness of the dried board?

62. ____________

62. Mary fills a pitcher with $\frac{11}{4}$ inches of milk. Due to a small leak, she loses $\frac{3}{16}$ inch. How much milk is left?

63. ______

64. ______

65. ______

66. ______

67. ______

68. ______

63. A cake recipe calls for $\frac{3}{4}$ cup of milk. Jorge has $\frac{1}{3}$ cup of milk. How much milk will he need to borrow from his neighbor in order to make the cake?

64. Dan has a bolt that is $\frac{3}{8}$ inch in length. He found that it is $\frac{3}{16}$ inch too long. What is the length of the bolt Dan needs?

65. Mary is given $\frac{7}{8}$ lb of dried fruit. She uses $\frac{2}{3}$ lb to make a fruit cake. How much dried fruit does she have left?

66. The diameter at the large end of a tapered pin is $\frac{7}{8}$ inch and at the smaller end it is $\frac{3}{16}$ inch. What is the difference between the diameters?

67. A carpenter planed the thickness of a board from $\frac{13}{16}$ inch to $\frac{5}{8}$ inch. How much was removed?

68. A machinist needs a bar that is $\frac{5}{8}$ inches thick. If he cuts it from a bar that is $\frac{27}{32}$ inches thick, how much must be cut off?

NAME ______________________________

CLASS ______________ SECTION ______________ DATE ______________

ANSWERS

69. ______________

70. ______________

69. Melissa needs $\frac{5}{16}$ yard of ribbon for a project. How much ribbon will be left if she cuts it from a piece of ribbon that is $\frac{7}{8}$ yard?

70. Anne requested $\frac{5}{8}$ yard of fabric from each of her students. Jon brought in $\frac{11}{16}$ yard. How much more or less does he have?

E *Maintain your skills* (Sections 1.3, 1.4, 1.5, 1.6, 3.3, 3.5)

71. ______________

72. ______________

73. ______________

74. ______________

75. ______________

76. ______________

77. ______________

71. Add: $483 + 2031 + 3450 + 8002 + 5998$

72. Subtract: $30{,}102 - 2178$

73. Divide: $123{,}567 \div 121$

74. Multiply: $(3080)(104)$

75. Reduce: $\frac{65}{91}$

76. Reduce: $\frac{85}{119}$

77. Divide: $\frac{2}{7} \div \frac{8}{9}$

ANSWERS

78. ______

79. ______

80. ______

78. Divide: $\frac{2}{7} \div 3$

79. Three bricklayers can lay 800 bricks per person per day, on the average. How many bricks can they lay in four days?

80. If a retaining wall requires 21,600 bricks, how many days will it take the three bricklayers in Exercise 79 to build the wall?

Subtraction of Mixed Numbers

OBJECTIVES

44. Subtract mixed numbers (no borrowing).
45. Subtract mixed numbers (with borrowing).

APPLICATION

Shawn McCord bought a roast for Sunday dinner that weighed 7 pounds. He cut off some fat and took out a bone. The meat left weighed $4\frac{1}{3}$ pounds. How many pounds of bone and fat did he trim off?

VOCABULARY

No new words.

HOW AND WHY

A subtraction problem may be written in horizontal or vertical form.

Horizontally:

$$5\frac{7}{8} - 3\frac{3}{8} = (5 - 3) + \left(\frac{7}{8} - \frac{3}{8}\right)$$

$$= 2 + \frac{4}{8} = 2\frac{1}{2}$$

Since the denominators are the same, subtract the whole-number parts and then subtract the fraction parts.

Vertically:

$$\begin{array}{r} 5\frac{7}{8} \\ 3\frac{3}{8} \\ \hline \end{array}$$

The process is similar to that for adding mixed numbers. We subtract instead of adding.

$$2\frac{4}{8} = 2\frac{1}{2}$$

Reduce.

When subtracting mixed numbers, we sometimes need to rename, or "borrow," first:

$$8\frac{2}{5} - 3\frac{3}{4} = ?$$

$$8\frac{2}{5} = 8\frac{8}{20}$$

$$\underline{3\frac{3}{4}} = \underline{3\frac{15}{20}}$$

Since the denominators are not the same, build the fractions to have a common denominator of 20.

Notice that $\frac{15}{20}$ is larger than $\frac{8}{20}$, so we cannot subtract. We will rename $8\frac{8}{20}$:

$$8\frac{8}{20} = 7 + 1\frac{8}{20}$$

$$= 7 + \frac{28}{20}$$

$$= 7\frac{28}{20}$$

Borrow 1 from the 8 and add to the fraction part. We now have a whole number plus a mixed number. Change the mixed number to an improper fraction.

CAUTION

Do not write $\frac{18}{20}$. If we "borrow" 1 from 8, we must add 1 to $\frac{8}{20}$.

The example can now be completed.

$$8\frac{2}{5} = 8\frac{8}{20} = 7\frac{28}{20}$$

Rename or "borrow" 1.

$$3\frac{3}{4} = 3\frac{15}{20} = 3\frac{15}{20}$$

$$4\frac{13}{20}$$

Do the subtraction.

PROCEDURE

To subtract mixed numbers:

1. **Build the fraction parts so that they have a common denominator.**
2. **If the second fraction part is larger than the first fraction part, rename the first mixed number by "borrowing" 1 from the whole-number part and adding it to the fraction part.**
3. **Subtract the fraction parts and then subtract the whole-number parts.**

MODEL PROBLEM-SOLVING

Examples | **Strategy**

a. Subtract:

$$47\frac{5}{8} = 47\frac{35}{56}$$

Write the fractions with common denominators.

$$36\frac{3}{7} = 36\frac{24}{56}$$

$$11\frac{11}{56}$$

Subtract the fraction parts and the whole-number parts.

Warm Up a. Subtract: $48\frac{5}{9}$

$22\frac{2}{5}$

ANSWERS TO WARM UPS (3.13)

a. $26\frac{7}{45}$

b. Subtract:

$$\begin{array}{rl} 15\frac{3}{10} = 14 + 1\frac{3}{10} = & 14\frac{13}{10} \\ \underline{7\frac{7}{10}} \qquad\qquad\quad = & \underline{7\frac{7}{10}} \\ & 7\frac{6}{10} \\ = & 7\frac{3}{5} \end{array}$$

Since we cannot subtract $\frac{7}{10}$ from $\frac{3}{10}$, we borrow 1 from 15. We then change the mixed number to an improper fraction and subtract.

Reduce.

Warm Up b. Subtract: $\begin{array}{r} 18\frac{11}{15} \\ \underline{8\frac{12}{15}} \end{array}$

c. Subtract:

$$\begin{array}{rl} 32\frac{1}{3} = 32\frac{4}{12} = 31 + 1\frac{4}{12} = & 31\frac{16}{12} \\ \underline{27\frac{3}{4}} = \underline{27\frac{9}{12}} \qquad\qquad\quad = & \underline{27\frac{9}{12}} \\ & 4\frac{7}{12} \end{array}$$

We first write the fractions with a common denominator. We cannot subtract the larger fraction from the smaller fraction. We borrow 1 from 32. We then change the mixed number to an improper fraction and subtract.

Warm Up c. Subtract: $\begin{array}{r} 47\frac{3}{8} \\ \underline{32\frac{5}{6}} \end{array}$

d. Subtract:

$$\begin{array}{rl} 4\frac{2}{3} = & 4\frac{2}{3} \\ \underline{3} \;\; = & \underline{3\frac{0}{3}} \\ & 1\frac{2}{3} \end{array}$$

To subtract a whole number from a mixed number, write the whole number as a mixed number by adding on a fraction with numerator zero. We choose 3 as the denominator because that gives us a common denominator.

Warm Up d. Subtract: $\begin{array}{r} 8\frac{2}{5} \\ \underline{6\phantom{\frac{2}{5}}} \end{array}$

ANSWERS TO WARM UPS (3.13) **b.** $9\frac{14}{15}$ **c.** $14\frac{13}{24}$ **d.** $2\frac{2}{5}$

e. Subtract:

$$\begin{array}{r} 11 \\ 2\frac{2}{9} \\ \hline \end{array}$$

> **CAUTION**
>
> **Do not bring down the fraction $\frac{2}{9}$ to the answer. It must be subtracted also.**

$$\begin{array}{rcl} 11 = 11\frac{0}{9} = 10 + 1\frac{0}{9} &=& 10\frac{9}{9} \\ 2\frac{2}{9} = 2\frac{2}{9} &=& 2\frac{2}{9} \\ \hline && 8\frac{7}{9} \end{array}$$

The whole number is changed to a mixed number. We see that we cannot subtract the larger fraction from the smaller fraction. We borrow 1 from 11. The 1 is added to $\frac{0}{9}$. The mixed number $1\frac{0}{9}$ is changed to an improper fraction. Then we can subtract.

Warm Up e. Subtract:

$$\begin{array}{r} 33 \\ 11\frac{5}{7} \\ \hline \end{array}$$

APPLICATION SOLUTION

f. Shawn McCord bought a roast for Sunday dinner that weighed 7 pounds. He cut off some fat and took out a bone. The meat left weighed $4\frac{1}{3}$ pounds. How many pounds of bone and fat did he trim off?

$$\begin{array}{rcl} 7 = 7\frac{0}{3} = 6 + 1\frac{0}{3} &=& 6\frac{3}{3} \\ 4\frac{1}{3} = 4\frac{1}{3} &=& 4\frac{1}{3} \\ \hline && 2\frac{2}{3} \end{array}$$

To determine the number of pounds that Shawn trimmed off, we must subtract the amount that remains from the starting weight.
We follow the same steps as in example f.

So, $2\frac{2}{3}$ pounds were trimmed off.

Warm Up f. Jamie weighed $138\frac{1}{2}$ pounds and decided to lose some weight. She lost a total of $5\frac{3}{4}$ pounds last week. What is her weight after the loss?

ANSWERS TO WARM UPS (3.13)

e. $21\frac{2}{7}$ **f.** $132\frac{3}{4}$ pounds

NAME ____________________

CLASS ____________ SECTION ____________ DATE ____________

Exercises 3.13

A

ANSWERS

1. ________
2. ________
3. ________
4. ________
5. ________
6. ________
7. ________
8. ________
9. ________
10. ________
11. ________
12. ________
13. ________
14. ________
15. ________
16. ________
17. ________
18. ________

Subtract. Reduce to the lowest terms:

1. $13\frac{6}{7} - 8\frac{4}{7}$

2. $27\frac{5}{9} - 23\frac{2}{9}$

3. $212\frac{37}{80} - 109\frac{21}{80}$

4. $205\frac{6}{11} - 112\frac{4}{11}$

5. $10\frac{7}{8} - 5\frac{3}{4}$

6. $11\frac{4}{5} - 7\frac{3}{10}$

7. $12 - 11\frac{4}{7}$

8. $21 - 20\frac{5}{6}$

9. $145\frac{2}{3} - 27\frac{1}{2}$

10. $6\frac{5}{6} - 3\frac{3}{10}$

11. $5\frac{1}{4} - 2\frac{3}{4}$

12. $7\frac{3}{8} - 4\frac{5}{8}$

13. $28\frac{3}{10} - 14\frac{7}{10}$

14. $16\frac{1}{8} - 3\frac{5}{8}$

15. $212\frac{1}{9} - 57$

16. $400\frac{3}{8} - 62$

17. $7\frac{1}{4} - 5\frac{5}{8}$

18. $9\frac{9}{16} - 2\frac{5}{6}$

B

19. ______

20. ______

21. ______

22. ______

23. ______

24. ______

25. ______

26. ______

27. ______

28. ______

29. ______

30. ______

31. ______

32. ______

33. ______

34. ______

35. ______

36. ______

19. $17\frac{15}{16}$ − $13\frac{9}{16}$

20. $21\frac{11}{12}$ − $10\frac{5}{12}$

21. $310\frac{23}{24}$ − $254\frac{5}{8}$

22. $118\frac{7}{12}$ − $93\frac{1}{4}$

23. $63\frac{8}{15}$ − $43\frac{1}{12}$

24. $5\frac{11}{16}$ − $\frac{5}{12}$

25. $37\frac{2}{3}$ − $15\frac{11}{12}$

26. $28\frac{1}{3}$ − $15\frac{7}{9}$

27. $29\frac{5}{16}$ − $10\frac{5}{6}$

28. $35\frac{11}{30}$ − $22\frac{7}{9}$

29. 45 − $16\frac{2}{3}$

30. 76 − $26\frac{2}{3}$

31. $5\frac{31}{32}$ − $\frac{3}{16}$

32. $8\frac{11}{20}$ − $\frac{9}{10}$

33. $7\frac{13}{18}$ − $\frac{5}{12}$

34. $9\frac{1}{6}$ − $\frac{5}{8}$

35. $37\frac{2}{5}$ − 28

36. $43\frac{7}{9}$ − 39

NAME ______________________

CLASS ____________ SECTION ____________ DATE ____________

ANSWERS

37. ____________

38. ____________

39. ____________

40. ____________

37. $45\frac{11}{15}$
$\underline{37\frac{31}{40}}$

38. $68\frac{13}{18}$
$\underline{58\frac{37}{45}}$

39. $75\frac{7}{12}$
$\underline{47\frac{13}{18}}$

40. $82\frac{4}{15}$
$\underline{56\frac{7}{12}}$

C

41. ____________

42. ____________

43. ____________

44. ____________

45. ____________

46. ____________

47. ____________

48. ____________

49. ____________

41. $27\frac{7}{39}$
$\underline{2\frac{3}{26}}$

42. $18\frac{5}{24}$
$\underline{11\frac{3}{40}}$

43. $25\frac{5}{12}$
$\underline{14\frac{11}{15}}$

44. $37\frac{1}{9}$
$\underline{22\frac{5}{12}}$

45. $113\frac{29}{36}$
$\underline{85\frac{13}{24}}$

46. $47\frac{8}{15}$
$\underline{29\frac{7}{20}}$

47. $143\frac{19}{24}$
$\underline{58\frac{7}{16}}$

48. $178\frac{11}{15}$
$\underline{97\frac{5}{9}}$

49. $17\frac{1}{12}$
$\underline{9\frac{5}{18}}$

ANSWERS

50. ____________

51. ____________

52. ____________

53. ____________

54. ____________

55. ____________

56. ____________

57. ____________

58. ____________

59. ____________

60. ____________

50. $\begin{array}{r} 22\frac{1}{8} \\ \underline{19\frac{3}{5}} \end{array}$

51. $14 - 5\frac{3}{14}$

52. $11 - 5\frac{11}{12}$

53. $6\frac{2}{3} - 1\frac{5}{6}$

54. $3\frac{4}{15} - 2\frac{7}{10}$

55. $\begin{array}{r} 16 \\ \underline{14\frac{17}{20}} \end{array}$

56. $\begin{array}{r} 41 \\ \underline{29\frac{17}{36}} \end{array}$

57. $47\frac{24}{35} - 19$

58. $78\frac{43}{58} - 39$

59. $\begin{array}{r} 131\frac{43}{45} \\ \underline{99\frac{27}{60}} \end{array}$

60. $\begin{array}{r} 103\frac{29}{30} \\ \underline{84\frac{75}{75}} \end{array}$

D

61. ____________

61. Han Kwong trimmed bone and fat from a $6\frac{3}{4}$-pound roast. The meat left weighed $5\frac{1}{4}$ pounds. How many pounds did she trim off?

62. Patti has a piece of lumber that measures $10\frac{7}{12}$ ft, and it is to be used in a spot that calls for a length of $8\frac{5}{12}$ ft. How much of the board must be cut off?

62. ____________

NAME ____________________

CLASS __________ SECTION __________ DATE __________

ANSWERS

63. ____________

64. ____________

65. ____________

66. ____________

67. ____________

68. ____________

63. Dick harvested $30\frac{3}{4}$ tons of wheat. He sold $12\frac{3}{10}$ tons to the Cartwright Flour Mill. How many tons of wheat does he have left?

64. A $14\frac{3}{4}$-inch casting shrinks $\frac{3}{16}$ inch on cooling. Find the size when the casting is cold.

65. Larry and Greg set out to hike 42 miles in two days. At the end of the first day they had covered $24\frac{3}{10}$ miles. How many miles do they have yet to go?

66. Frank poured $8\frac{3}{10}$ yards of cement for a fountain. Another fountain took $5\frac{3}{8}$ yards. How much more cement was needed for the larger fountain than for the smaller one?

67. From a tank containing $9\frac{3}{4}$ gallons of gas, Charlie filled a can that holds $2\frac{1}{2}$ gallons. How much gas is still in the tank?

68. How much liquid must be drained from a can containing $2\frac{4}{5}$ qt so that the amount remaining will be $1\frac{1}{2}$ qt?

E *Maintain your skills* (Sections 3.2, 3.3, 3.4)

ANSWERS

69. ______

70. ______

71. ______

72. ______

73. ______

74. ______

75. ______

76. ______

77. ______

78. ______

Reduce to the lowest terms:

69. $\frac{52}{68}$

70. $\frac{286}{429}$

71. Change $\frac{125}{36}$ to a mixed number.

72. Change $25\frac{5}{9}$ to an improper fraction.

Multiply:

73. $\frac{15}{24} \cdot \frac{9}{25} \cdot \frac{16}{27}$

74. $\frac{16}{25} \cdot \frac{24}{27} \cdot \frac{9}{15}$

75. $\frac{1}{2} \cdot \frac{4}{7} \cdot \frac{4}{5} \cdot 7$

76. $\frac{2}{3} \cdot \frac{6}{11} \cdot \frac{1}{4} \cdot 22$

77. Last week when Karla filled the tank of her car with gasoline, the odometer read 57,832 miles. Yesterday when she filled the tank with 18 gallons of gasoline, the odometer read 58,426 miles. How many miles to the gallon was the car averaging?

78. A pet-food canning company packs Feelein Cat Food in cans, of which each contains $7\frac{3}{4}$ ounces of cat food. Each empty can weighs $1\frac{1}{2}$ ounces. Twenty-four cans are packed in a case that weighs 10 ounces empty. What is the shipping weight of five cases of the cat food?

▲▲ *Getting Ready for Algebra*

We have solved equations in which whole numbers were either added to or subtracted from a letter. Let's solve some equations in which fractions or mixed numbers are either added to or subtracted from a variable. To eliminate addition, subtract the number being added from both sides of the equation. To eliminate subtraction, add the number being subtracted to both sides of the equation.

▲▲ MODEL PROBLEM-SOLVING

Examples | **Strategy**

a. Solve: $x - \frac{4}{5} = 3\frac{1}{2}$

$$x - \frac{4}{5} + \frac{4}{5} = 3\frac{1}{2} + \frac{4}{5}$$

Since $\frac{4}{5}$ is subtracted from x, we eliminate the subtraction by adding $\frac{4}{5}$ to both sides of the equation.

$$x = 3\frac{1}{2} + \frac{4}{5}$$

$$x = 3\frac{5}{10} + \frac{8}{10}$$

We now do the addition on the right side.

$$x = 3\frac{13}{10}$$

$$x = 4\frac{3}{10}$$

Check:

$$4\frac{3}{10} - \frac{4}{5} = 3\frac{1}{2}$$

Substitute $4\frac{3}{10}$ for x in the original equation.

$$3\frac{1}{2} = 3\frac{1}{2}$$

We get a true statement, so the solution is $4\frac{3}{10}$.

Warm Up a. Solve: $x - \frac{5}{8} = 3\frac{7}{8}$

GETTING READY FOR ALGEBRA

a. $x = 4\frac{1}{2}$

b. Solve: $4\frac{1}{2} = x - 2\frac{2}{3}$

$4\frac{1}{2} + 2\frac{2}{3} = x - 2\frac{2}{3} + 2\frac{2}{3}$

$4\frac{1}{2} + 2\frac{2}{3} = x$

$7\frac{1}{6} = x$

Since $2\frac{2}{3}$ is subtracted from x, we eliminate the subtraction by adding $2\frac{2}{3}$ to both sides of the equation and doing the addition on the right side.

Check:

$4\frac{1}{2} = 7\frac{1}{6} - 2\frac{2}{3}$

$4\frac{1}{2} = 4\frac{1}{2}$

Substitute $7\frac{1}{6}$ for x in the original equation.

Do the subtraction on the left side of the equation.

The statement is true, so the solution is $7\frac{1}{6}$.

Warm Up b. Solve: $5\frac{1}{2} = a - 5\frac{5}{8}$

GETTING READY FOR ALGEBRA

b. $11\frac{1}{8} = a$

NAME ______________________

CLASS ____________ SECTION ____________ DATE ____________

▲▲ EXERCISES

ANSWERS

1. ____________
2. ____________
3. ____________
4. ____________
5. ____________
6. ____________
7. ____________
8. ____________
9. ____________
10. ____________
11. ____________
12. ____________
13. ____________
14. ____________
15. ____________

Solve:

1. $a + \frac{3}{8} = \frac{7}{8}$

2. $y - \frac{1}{8} = \frac{5}{8}$

3. $c - \frac{3}{16} = \frac{7}{16}$

4. $w + \frac{5}{12} = \frac{11}{12}$

5. $x + \frac{5}{8} = \frac{7}{9}$

6. $x - \frac{7}{8} = \frac{3}{2}$

7. $y - \frac{5}{7} = \frac{8}{9}$

8. $y + \frac{5}{9} = \frac{9}{10}$

9. $a + \frac{9}{8} = \frac{12}{5}$

10. $a - \frac{7}{4} = \frac{5}{8}$

11. $c - 1\frac{1}{2} = 2\frac{2}{3}$

12. $c + 2\frac{4}{5} = 3\frac{5}{8}$

13. $x + 4\frac{3}{4} = 7\frac{8}{9}$

14. $x - 7\frac{8}{9} = 5\frac{7}{8}$

15. $12 = w + 8\frac{5}{6}$

ANSWERS

16. ______________

17. ______________

18. ______________

19. ______________

20. ______________

16. $25 = m + 15\frac{5}{8}$

17. $a - 13\frac{5}{6} = 22\frac{11}{18}$

18. $b + 23\frac{11}{12} = 34\frac{1}{3}$

19. $c + 44\frac{13}{21} = 65\frac{5}{7}$

20. $x - 27\frac{5}{8} = 48\frac{2}{3}$

Order of Operations and Average

OBJECTIVES

46. Do any combination of operations with fractions.
47. Find the average of a group of fractions.

APPLICATION

Gwen, Sam, Carlos, and Sari have equal shares in a florist shop. Sam decides to sell his share. He sells $\frac{3}{8}$ of his share to Gwen, $\frac{1}{2}$ to Carlos, and the rest to Sari. What is Gwen's share of the florist shop now?

VOCABULARY

No new words.

HOW AND WHY

The order of operations for fractions is the same as for whole numbers. (See Section 1.7.) The method for finding the average of a group of fractions is the same as for the whole numbers. (See Section 1.8.)

The following chart summarizes some of the processes that need to be remembered when working with fractions:

OPERATION	FIND THE LCM OR LEAST COMMON DENOMINATOR AND BUILD	CHANGE MIXED NUMBER TO IMPROPER FRACTION	INVERT DIVISOR AND MULTIPLY	REDUCE ANSWER
Add	Yes	No	No	Yes
Subtract	Yes	No	No	Yes
Multiply	No	Yes	No	Yes
Divide	No	Yes	Yes	Yes

MODEL PROBLEM-SOLVING

Examples | **Strategy**

a. Perform the indicated operations:

$$\frac{5}{6} - \frac{1}{2} \cdot \frac{2}{3} = \frac{5}{6} - \frac{1}{3}$$

Multiplication is performed first.

$$= \frac{5}{6} - \frac{2}{6}$$

Write each fraction with a common denominator and subtract.

$$= \frac{3}{6}$$

$$= \frac{1}{2}$$

Reduce to the lowest terms.

Warm Up a. Perform the indicated operations: $\frac{7}{8} - \frac{3}{4} \cdot \frac{8}{9}$

ANSWERS TO WARM UPS (3.14)

a. $\frac{5}{24}$

b. Perform the indicated operations:

$$\frac{1}{2} \div \frac{2}{3} \cdot \frac{1}{4} = \frac{3}{4} \cdot \frac{1}{4}$$

$$= \frac{3}{16}$$

Division is performed first, because as we read from left to right we come to division first. We then do the multiplication.

Warm Up b. Perform the indicated operations: $\frac{5}{8} \div \frac{9}{16} \cdot \frac{7}{5}$

c. Perform the indicated operations:

$$\left(\frac{3}{4}\right)^2 \cdot \frac{2}{5} - \frac{1}{5} = \frac{9}{\underset{8}{\cancel{16}}} \cdot \frac{\overset{1}{\cancel{2}}}{5} - \frac{1}{5}$$

Exponentiation is done first then reduce.

$$= \frac{9}{40} - \frac{1}{5}$$

Multiply.

$$= \frac{9}{40} - \frac{8}{40}$$

$$= \frac{1}{40}$$

Subtract.

Warm Up c. Perform the indicated operations: $\left(\frac{5}{8}\right)^2 \cdot \frac{8}{15} - \frac{1}{8}$

d. Find the average of $\frac{1}{2}$, $\frac{1}{3}$, and $\frac{3}{4}$.

$$\left(\frac{1}{2} + \frac{1}{3} + \frac{3}{4}\right) \div 3$$

To find the average of three numbers, add the three numbers and divide their sum by three.

$$\frac{1}{2} + \frac{1}{3} + \frac{3}{4} = \frac{6}{12} + \frac{4}{12} + \frac{9}{12}$$

$$= \frac{19}{12}$$

We do the addition first.

ANSWERS TO WARM UPS (3.14)

b. $\frac{14}{9}$ or $1\frac{5}{9}$ **c.** $\frac{1}{12}$

$\frac{19}{12} \div 3 = \frac{19}{12} \cdot \frac{1}{3} = \frac{19}{36}$ **Divide the sum by three.**

The average is $\frac{19}{36}$.

Warm Up d. Find the average of $\frac{5}{6}$, $\frac{7}{8}$, and $\frac{3}{4}$.

e. A class of 10 students took a 12-problem test. Their results are listed in the following table.

NUMBER OF STUDENTS	FRACTION OF PROBLEMS CORRECT
1	$\frac{12}{12}$
2	$\frac{11}{12}$
3	$\frac{10}{12}$
4	$\frac{9}{12}$

What was the average?

$\left[\frac{12}{12} + 2\left(\frac{11}{12}\right) + 3\left(\frac{10}{12}\right) + 4\left(\frac{9}{12}\right)\right] \div 10$

To find the class average, add all the grades together and divide by 10. There were two scores of $\frac{11}{12}$, three scores of $\frac{10}{12}$, and four scores of $\frac{9}{12}$.

$\left(\frac{12}{12} + \frac{22}{12} + \frac{30}{12} + \frac{36}{12}\right) \div 10$ **Multiply.**

$\frac{100}{12} \div 10$ **Add.**

$\frac{10}{12}$ **Divide.**

The class average was $\frac{10}{12}$ correct.

CAUTION

Do not reduce the answer, since the test scores are based on 12.

ANSWERS TO WARM UPS (3.14)

d. $\frac{59}{72}$

Warm Up e. A class of 12 students took a 20-problem test. Their results are listed in the following table:

NUMBER OF STUDENTS	FRACTION OF PROBLEMS CORRECT
1	$\frac{20}{20}$
2	$\frac{19}{20}$
4	$\frac{16}{20}$
5	$\frac{14}{20}$

What was the average?

f. Find the average of $7\frac{1}{2}$, $6\frac{1}{4}$, and $4\frac{5}{8}$.

$$\left(7\frac{1}{2}\right) + \left(6\frac{1}{4}\right) + \left(4\frac{5}{8}\right) = 18\frac{3}{8}$$

Add the mixed numbers.

$$18\frac{3}{8} \div 3 = \frac{147}{8} \div 3$$

Since there are three mixed numbers, divide the sum by 3. Change the mixed number to an improper fraction.

$$= \frac{147}{8} \cdot \frac{1}{3}$$

Invert the divisor.

$$= \frac{49}{8}$$

$$= 6\frac{1}{8}$$

Warm Up f. Find the average of $3\frac{5}{6}$, $4\frac{1}{2}$, and $2\frac{2}{3}$.

ANSWERS TO WARM UPS (3.14)

e. $\frac{16}{20}$ **f.** $3\frac{2}{3}$

APPLICATION SOLUTION

g. Gwen, Sam, Carlos, and Sari have equal shares in a florist shop. Sam decides to sell his share. He sells $\frac{3}{8}$ of his share to Gwen, $\frac{1}{2}$ to Carlos, and the rest to Sari. What is Gwen's share of the florist shop now?

$\frac{1}{4}+\frac{3}{8}\left(\frac{1}{4}\right)$

Since each of the four had equal shares, each of them owned $\frac{1}{4}$ of the business.

Gwen now owns her original $\frac{1}{4}$ plus $\frac{3}{8}$ of Sam's share, which was $\frac{1}{4}$.

$\frac{1}{4}+\frac{3}{8}\left(\frac{1}{4}\right)=\frac{1}{4}+\frac{3}{32}$ **Do the multiplication first.**

$=\frac{8}{32}+\frac{3}{32}$ **Write each with a common denominator.**

$=\frac{11}{32}$ **Add.**

Gwen's share is now $\frac{11}{32}$ of the shop.

Warm Up g. Jill, Jenny, and Joan have equal shares in a gift shop. Jill decides to sell out her share. She sells $\frac{1}{4}$ to Jenny and the remainder to Joan. What is Joan's share of the gift shop?

ANSWERS TO WARM UPS (3.14)

g. $\frac{7}{12}$

NAME ______________________

CLASS ______________ SECTION ______________ DATE ______________

Exercises 3.14

A

ANSWERS

1. ______________
2. ______________
3. ______________
4. ______________
5. ______________
6. ______________
7. ______________
8. ______________
9. ______________
10. ______________
11. ______________
12. ______________
13. ______________
14. ______________
15. ______________
16. ______________

Perform the indicated operations:

1. $\frac{6}{7} - \frac{3}{7} - \frac{1}{7}$ **2.** $\frac{6}{7} + \frac{3}{7} - \frac{1}{7}$ **3.** $\frac{1}{2} \cdot \left(\frac{3}{7} - \frac{1}{7}\right)$ **4.** $\frac{1}{3} \div \frac{1}{2} \cdot \frac{1}{3}$

5. $\frac{2}{6} - \frac{1}{2} \cdot \frac{1}{3}$ **6.** $\frac{11}{6} + \frac{1}{2} \div \frac{3}{2}$ **7.** $\frac{1}{4} + \frac{3}{8} \div \frac{1}{2}$ **8.** $\frac{1}{4} \div \frac{3}{8} + \frac{1}{2}$

9. $\frac{3}{4} \cdot \frac{1}{2} - \frac{3}{8}$ **10.** $\frac{5}{6} \cdot \frac{1}{3} - \frac{5}{18}$ **11.** $\frac{3}{8} \div \frac{3}{4} + \frac{1}{2}$ **12.** $\frac{2}{3} \div \frac{5}{6} + \frac{1}{5}$

13. $\frac{2}{3} + \left(\frac{1}{2}\right)^2$ **14.** $\frac{4}{5} - \frac{2}{3}\left(\frac{1}{2}\right)^2$

15. Find the average of $\frac{1}{9}$ and $\frac{7}{9}$.

16. Find the average of $\frac{1}{5}$, $\frac{2}{5}$, and $\frac{3}{5}$.

ANSWERS

17. ______________________

17. Find the average of $\frac{2}{7}, \frac{3}{7}$, and $\frac{5}{7}$.

18. ______________________

18. Find the average of $\frac{3}{11}, \frac{5}{11}$, and $\frac{9}{11}$.

19. ______________________

19. Find the average of $\frac{2}{7}, \frac{3}{7}$, and $\frac{5}{7}$.

B

20. ______________________

21. ______________________

22. ______________________

23. ______________________

24. ______________________

25. ______________________

26. ______________________

27. ______________________

28. ______________________

29. ______________________

Perform the indicated operations:

20. $\frac{3}{4} \div \frac{1}{3} \cdot \frac{1}{6}$

21. $\frac{3}{4} \div \left(\frac{1}{3} \cdot \frac{1}{6}\right)$

22. $\frac{1}{3} \cdot \frac{1}{6} \div \frac{3}{4}$

23. $\frac{4}{5} - \frac{2}{5} \cdot \frac{1}{2}$

24. $\frac{4}{5} - \frac{2}{5} + \frac{1}{2}$

25. $\frac{1}{12} - \frac{3}{4} \cdot \frac{1}{18}$

26. $\frac{5}{6} - \frac{3}{4} \div \frac{3}{2} + \frac{1}{2}$

27. $\frac{7}{8} - \frac{5}{6} \div \frac{4}{3} + \frac{5}{6}$

28. $\frac{5}{12} \cdot \frac{4}{5} + \frac{1}{3} \div \frac{1}{2} - \frac{5}{6}$

29. $\frac{2}{9} \div \frac{1}{3} - \frac{3}{4} \cdot \frac{6}{12} + \frac{5}{6}$

NAME ______________________________

CLASS ______________ SECTION ______________ DATE ______________

ANSWERS

30. ______________

31. ______________

32. ______________

33. ______________

34. ______________

35. ______________

36. ______________

37. ______________

38. ______________

39. ______________

30. $\frac{7}{12} + \left(\frac{3}{4}\right)^2 - \frac{1}{2} \cdot \frac{3}{4}$

31. $\frac{3}{4} \cdot \frac{4}{5} - \frac{1}{3} + \left(\frac{2}{3}\right)^2$

32. $\frac{15}{16} - \left(\frac{3}{8}\right)^2 + \frac{7}{8} \div \frac{2}{3}$

33. $\frac{21}{25} - \frac{2}{5} + \left(\frac{4}{5}\right)^2 \div \frac{3}{4}$

34. $\frac{3}{8} - \left(\frac{2}{3} \div \frac{4}{5} - \frac{1}{2}\right)$

35. $\frac{3}{4} + \left(\frac{4}{5} \cdot \frac{5}{8} + \frac{2}{3}\right)$

36. Find the average of $\frac{1}{3}$, $\frac{3}{4}$, and $\frac{1}{2}$.

37. Find the average of $\frac{2}{3}$, $1\frac{2}{3}$, and $3\frac{2}{3}$.

38. Find the average of $2\frac{1}{3}$, $3\frac{3}{4}$, and $6\frac{1}{2}$.

39. Find the average of $1\frac{2}{3}$, $3\frac{2}{3}$, and $6\frac{2}{3}$.

ANSWERS

40. ______

41. ______

40. Find the average of $\frac{3}{4}$, $\frac{4}{3}$, $\frac{3}{2}$, and $\frac{2}{3}$.

41. Find the average of $\frac{5}{6}$, $\frac{11}{12}$, $\frac{17}{24}$, and $\frac{3}{2}$.

C

42. ______

43. ______

44. ______

45. ______

46. ______

47. ______

48. ______

49. ______

Perform the indicated operations:

42. $\frac{7}{9} - \frac{1}{3} + \frac{1}{2} \div \frac{2}{5} \cdot \frac{4}{5}$

43. $\left(\frac{7}{9} - \frac{1}{3} + \frac{1}{2}\right) \div \frac{2}{5} \cdot \frac{4}{5}$

44. $\frac{17}{9} - \left(\frac{1}{3} + \frac{1}{2} \div \frac{5}{2}\right) \cdot \frac{5}{4}$

45. $\frac{1}{2} \cdot \frac{2}{3} \div 3 \cdot \frac{3}{4} \div \frac{1}{2}$

46. $\left(\frac{5}{8} - \frac{1}{2} \cdot \frac{3}{4}\right) \div \frac{1}{2} + \frac{1}{2}$

47. $\left(\frac{29}{12} - \frac{5}{4} \cdot \frac{3}{2}\right) \div \frac{5}{6} + \frac{3}{4}$

48. $\left(\frac{3}{4} + \frac{7}{8} \cdot \frac{2}{3}\right) \cdot \left(\frac{3}{2}\right)^3$

49. $\frac{7}{8} \div \frac{5}{4} \cdot \frac{5}{14} \div \left(\frac{1}{2}\right)^3$

NAME ______________________________

CLASS ______________ SECTION ______________ DATE ______________

ANSWERS

50. ______________

50. Find the average of $\frac{3}{8}$, $\frac{1}{4}$, $\frac{1}{2}$, and $\frac{3}{4}$.

51. ______________

51. Find the average of $\frac{2}{3}$, $\frac{5}{12}$, $\frac{1}{2}$, $\frac{3}{4}$, and $\frac{5}{6}$.

52. ______________

52. Find the average of $\frac{4}{15}$, $\frac{7}{10}$, $\frac{18}{25}$, and $\frac{7}{30}$.

53. ______________

53. Find the average of $3\frac{2}{3}$, $4\frac{5}{6}$, and $2\frac{5}{9}$.

54. ______________

54. Find the average of $6\frac{7}{8}$, $8\frac{3}{4}$, and $8\frac{1}{2}$.

ANSWERS

55. ______

56. ______

57. ______

55. Find the average of $5\frac{1}{3}$, $6\frac{2}{5}$, and $9\frac{13}{15}$.

56. Find the average of $\frac{3}{4}$, $1\frac{5}{6}$, $\frac{8}{9}$, and $5\frac{5}{8}$.

57. Find the average of $\frac{15}{32}$, $1\frac{13}{16}$, $\frac{17}{4}$, and $9\frac{3}{8}$.

D

58. ______

58. Anita found the following lengths of boards in her workshop: $3\frac{3}{8}$ inch, $4\frac{1}{2}$ inch, $5\frac{3}{4}$ inch and $6\frac{2}{3}$ inch. What was the average length of these pieces?

59. ______

59. Georgia started with 22 yards of fabric. She made shirts for her nephews that required the following lengths: $2\frac{3}{8}$ yd, $3\frac{1}{4}$yd, $4\frac{7}{8}$ yd. How much fabric is left for Georgia and her brother if they split the remaining fabric?

NAME ______________________

CLASS ____________ SECTION ____________ DATE ____________

ANSWERS

60. ____________

61. ____________

62. ____________

60. Three people measured the diameter of a pipe. One individual said his reading was $7\frac{1}{2}$ in., another reported $7\frac{7}{10}$ in., and the third got $7\frac{2}{5}$ in. What is the average of the measurements?

61. Nurse Louise weighed five new babies at General Hospital. They weighed $6\frac{1}{2}$ lb, $7\frac{3}{4}$ lb, $10\frac{3}{8}$ lb, $7\frac{1}{2}$ lb, and $8\frac{1}{2}$ lb. What was the average weight?

62. A class of fifteen students took a 10-problem quiz. Their results were as follows:

NUMBER OF STUDENTS	FRACTION OF PROBLEMS CORRECT
1	$\frac{10}{10}$ (all correct)
2	$\frac{9}{10}$
3	$\frac{8}{10}$
5	$\frac{7}{10}$
1	$\frac{6}{10}$
2	$\frac{5}{10}$
1	$\frac{2}{10}$

What was the class average?

ANSWERS

63. ______________

64. ______________

63. On the second quiz the class in Exercise 53 scored as follows:

NUMBER OF STUDENTS	FRACTION OF PROBLEMS CORRECT
2	$\frac{10}{10}$ (all correct)
1	$\frac{9}{10}$
2	$\frac{8}{10}$
2	$\frac{7}{10}$
4	$\frac{5}{10}$
3	$\frac{3}{10}$
1	$\frac{2}{10}$

What was the class average?

64. During a recent 100-yd dash three sprinters were clocked as follows: $9\frac{9}{10}$ sec, $10\frac{1}{10}$ sec, and $10\frac{1}{2}$ sec. What was the average time of the sprinters?

E *Maintain your skills* (Sections 2.5, 3.4, 3.5)

65. ______________

66. ______________

Divide:

65. $\frac{7}{9} \div \frac{14}{3}$

66. $\frac{7}{3} \div \frac{14}{9}$

NAME ______________________________

CLASS ______________ SECTION ______________ DATE ______________

ANSWERS

67. ______________

68. ______________

69. ______________

70. ______________

71. ______________

72. ______________

73. ______________

74. ______________

67. $\frac{25}{32} \div \frac{15}{36}$

68. $\frac{25}{36} \div \frac{15}{32}$

Multiply:

69. $\frac{15}{28} \cdot \frac{21}{45} \cdot \frac{16}{9}$

70. $\frac{9}{15} \cdot \frac{21}{28} \cdot \frac{45}{16}$

71. Prime factor 325.

72. Prime factor 975.

73. A coffee table is made of a piece of maple that is $\frac{3}{4}$ in. thick, a piece of chipboard that is $\frac{3}{8}$ in. thick, and a veneer that is $\frac{1}{8}$ in. thick. How thick is the tabletop?

74. A woman worked a five-day week for the following hours: $6\frac{3}{4}$ hours, $7\frac{1}{4}$ hours, $6\frac{2}{3}$ hours, $9\frac{1}{3}$ hours, $6\frac{1}{2}$ hours. How many hours did she work for the week? What was her pay if the rate was $\$5\frac{1}{2}$ per hour?

NAME ______________________

CLASS ______________ SECTION ______________ DATE ______________

ANSWERS

1. ______________
2. ______________
3. ______________
4. ______________
5. ______________
6. ______________
7. ______________
8. ______________
9. ______________
10. ______________
11. ______________
12. ______________
13. ______________
14. ______________
15. ______________
16. ______________
17. ______________
18. ______________

CHAPTER 3 TRUE–FALSE CONCEPT REVIEW

Check your understanding of the language of basic mathematics. Tell whether each of the following statements is True (always true) or False (not always true).

1. It is not possible to picture an improper fraction using unit regions.

2. The fraction, $\frac{2}{3}$, written as a mixed number, is $0\frac{2}{3}$.

3. The whole number, 1, can also be written as a proper fraction.

4. A fraction is another way of writing a division problem.

5. When a fraction is reduced to lowest terms, its value remains the same.

6. Every improper fraction can be reduced.

7. There are some fractions with large numerators and denominators that cannot be reduced to lowest terms.

8. Two mixed numbers can be subtracted without first changing them to improper fractions.

9. The reciprocal of a mixed number greater than 1, is a proper fraction.

10. The quotient of two non-zero fractions can always be found by multiplication.

11. Building fractions is the opposite of reducing fractions.

12. The primary reason for building fractions is so that they will have a common denominator.

13. Unlike fractions have different numerators.

14. Mixed numbers must be changed to improper fractions before adding them.

15. It is sometimes necessary to use "borrowing" to subtract mixed numbers as is used to subtract whole numbers.

16. The order of operations for fractions is the same as the order of operations for whole numbers.

17. The average of three different fractions is larger than at least one of the fractions.

18. The quotient of two fractions can be larger than either of the two fractions.

NAME ______________________

CLASS ______________ SECTION ______________ DATE ______________

ANSWERS

1. ______
2. ______
3. ______
4. ______
5. ______
6. ______
7. ______
8. ______
9. ______
10. ______
11. ______
12. ______
13. ______
14. ______
15. ______
16. ______
17. ______

CHAPTER 3 POST-TEST

1. **(Section 3.2, Obj. 29)** Change to a mixed number: $\frac{49}{16}$

2. **(Section 3.10, Obj. 41)** Add: $\frac{3}{8} + \frac{5}{12}$

3. **(Section 3.2, Obj. 30)** Change to an improper fraction: $8\frac{9}{10}$

4. **(Section 3.8, Obj. 38)** List these fractions from the smallest to largest: $\frac{3}{10}, \frac{3}{8}, \frac{2}{5}$

5. **(Section 3.2, Obj. 30)** Change to an improper fraction: 3

6. **(Section 3.7, Obj. 37)** Find the missing numerator: $\frac{7}{8} = \frac{?}{64}$

7. **(Section 3.11, Obj. 42)** Add:
$$\begin{array}{r} 5\frac{5}{6} \\ 3\frac{3}{10} \\ \hline \end{array}$$

8. **(Section 3.6, Obj. 35)** Multiply. Write the result as a mixed number: $\left(2\frac{3}{4}\right)\left(3\frac{3}{5}\right)$

9. **(Section 3.14, Obj. 46)** Perform the indicated operations: $\frac{3}{8} \div \frac{3}{4} + \frac{1}{4}$

10. **(Section 3.3, Obj. 31)** Reduce to the lowest terms: $\frac{36}{54}$

11. **(Section 3.13, Obj. 44)** Subtract: $11\frac{1}{8} - 6$

12. **(Section 3.4, Obj. 32)** Multiply: $\frac{3}{16} \cdot \frac{8}{9} \cdot \frac{15}{4}$

13. **(Section 3.4, Obj. 32)** Multiply: $\frac{3}{8} \cdot \frac{3}{2}$

14. **(Section 3.12, Obj. 43)** Subtract: $\frac{3}{4} - \frac{3}{10}$

15. **(Section 3.6, Obj. 36)** Divide: $3\frac{5}{9} \div 1\frac{1}{3}$

16. **(Section 3.13, Obj. 45)** Subtract:
$$\begin{array}{r} 12\frac{3}{10} \\ 2\frac{7}{15} \\ \hline \end{array}$$

17. **(Section 3.3, Obj. 31)** Reduce to the lowest terms: $\frac{44}{66}$

ANSWERS

18. ____

19. ____

20. ____

21. ____

22. ____

23. ____

24. ____

25. ____

26. ____

27. ____

28. ____

29. ____

30. ____

31. ____

32. ____

18. **(Section 3.10, Obj. 41)** Add: $\frac{3}{35}+\frac{5}{14}+\frac{3}{10}$

19. **(Section 3.5, Obj. 33)** What is the reciprocal of $3\frac{5}{8}$?

20. **(Section 3.5, Obj. 33)** What is the reciprocal of $\frac{7}{8}$?

21. **(Section 3.1, Obj. 27)** Which of these fractions are proper?

$$\frac{7}{8}, \frac{8}{8}, \frac{9}{8}, \frac{7}{9}, \frac{9}{7}, \frac{8}{9}, \frac{9}{9}$$

22. **(Section 3.5, Obj. 34)** Divide: $\frac{8}{9} \div \frac{10}{3}$

23. **(Section 3.13, Obj. 44)** Subtract:

$$\begin{array}{r} 11\frac{7}{10} \\ 2\frac{3}{8} \\ \hline \end{array}$$

24. **(Section 3.1, Obj. 26)** Write the fraction for the following figure:

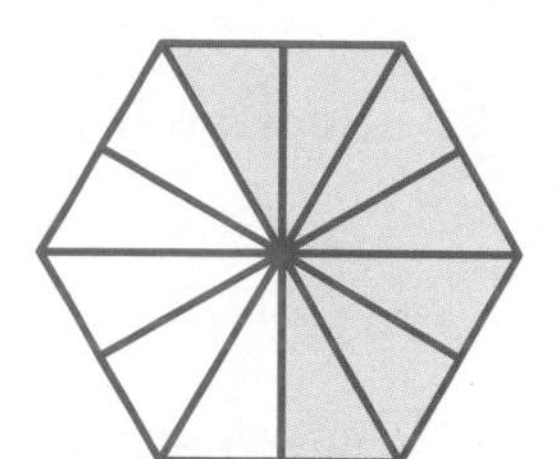

25. **(Section 3.13, Obj. 45)** Subtract: $10 - 8\frac{5}{9}$

26. **(Section 3.9, Obj. 40)** Add: $\frac{5}{12}+\frac{5}{12}$

27. **(Section 3.14, Obj. 47)** Find the average: $1\frac{1}{2}, \frac{3}{4}, 2\frac{1}{8}, 1\frac{1}{8}$

28. **(Section 3.1, Obj. 28)** Which of these fractions represent the number one?

$$\frac{6}{5}, \frac{5}{5}, \frac{7}{6}, \frac{6}{6}, \frac{7}{7}, \frac{6}{7}, \frac{5}{7}$$

29. **(Section 3.4, Obj. 32)** Multiply: $\frac{8}{9} \cdot \frac{3}{10}$

30. **(Section 3.8, Obj. 39)** True or false? $\frac{4}{5} > \frac{8}{9}$

31. A railroad car contains $126\frac{1}{2}$ tons of baled hay. A truck that is being used to unload the hay can haul $5\frac{3}{4}$ tons. How many truckloads of hay are in the railroad car?

32. Jill buys a board that is 16 ft long. She needs a board that is $14\frac{1}{2}$ ft long. How much will she need to cut off the board that she bought?

CHAPTER 4

Decimals

NAME ____________________

CLASS ____________ SECTION ____________ DATE ____________

ANSWERS

1. ____________

2. ____________

3. ____________

4. ____________

5. ____________

6. ____________

7. ____________

8. ____________

9. ____________

10. ____________

11. ____________

12. ____________

13. ____________

14. ____________

15. ____________

CHAPTER 4 PRE-TEST

The problems in the following pre-test are a sample of the material in the chapter. You may already know how to work some of these. If so, this will allow you to spend less time on those parts. As a result, you will have more time to give to the sections that gave you difficulty. Please work the following problems. The answers are in the back of the text.

1. **(Section 4.1, Obj. 48)** What is the place value of the 9 in 3.0795?

2. **(Section 4.1, Obj. 48)** What is the place value of the 6 in 0.649?

3. **(Section 4.1, Obj. 49)** Write the place value name for "three thousand six ten-thousandths."

4. **(Section 4.1, Obj. 50)** Write the word name for 203.4075.

5. **(Section 4.2, Obj. 51)** Write in expanded form: 13.0108

6. **(Section 4.2, Obj. 52)** Write the place value name for $5 + \frac{3}{10} + \frac{0}{100} + \frac{0}{1000} + \frac{7}{10000}$.

7. **(Section 4.3, Obj. 53)** Change to a fraction and reduce to the lowest terms: 0.135

8. **(Section 4.4, Obj. 55)** List the following decimals from the smallest to largest: 2.651, 2.6499, 2.6509, 2.65099

9. **(Section 4.5, Obj. 56)** Round to the nearest hundredth: 2.6352

10. **(Section 4.5, Obj. 56)** Round to the nearest hundred: 249.352

11. **(Section 4.6, Obj. 57)** Add: $6.3 + 10.075 + 2.36 + 1.7905$

12. **(Section 4.7, Obj. 58)** Subtract: $\begin{array}{r} 14.073 \\ \underline{9.875} \end{array}$

13. **(Section 4.7, Obj. 58)** Subtract: $9 - 3.652$

14. **(Section 4.8, Obj. 59)** Multiply: $\begin{array}{r} 2.035 \\ \underline{2.5} \end{array}$

15. **(Section 4.8, Obj. 59)** Multiply: $(8.16)(0.009)$

ANSWERS

16. ______

17. ______

18. ______

19. ______

20. ______

21. ______

22. ______

23. ______

24. ______

25. ______

26. ______

27. ______

28. ______

16. (Section 4.8, Obj. 60) Multiply: 78.59(1000)

17. (Section 4.9, Obj. 61) Divide: $31.5 \div 10000$

18. (Section 4.9, Obj. 62) Write in scientific notation: 340,000

19. (Section 4.9, Obj. 62) Write in scientific notation: 0.000000345

20. (Section 4.9, Obj. 63) Change to place value form: 1.73×10^6

21. (Section 4.10, Obj. 64) Divide: $8\overline{)2.3}$

22. (Section 4.10, Obj. 65) Divide. Round answer to the nearest hundredth: $2.3\overline{)86.35}$

23. (Section 4.11, Obj. 66) Write as a decimal: $\frac{9}{40}$

24. (Section 4.11, Obj. 67) Write as an approximate decimal to the nearest hundredth: $\frac{5}{17}$

25. (Section 4.11, Obj. 67) Write as an approximate decimal to the nearest thousandth: $11\frac{7}{9}$

26. (Section 4.12, Obj. 68) Perform the indicated operations: $0.2 \div 0.02 + 0.01$

27. (Section 4.12, Obj. 69) Gene drove 31.6 miles on Monday, 110.5 miles on Tuesday, 56.8 miles on Wednesday, 157.1 miles on Thursday, 96.3 miles on Friday, and 55.7 miles on Saturday. What was his average mileage per day (to the nearest tenth)?

28. What is the cost per ounce of tortilla chips if an 11-oz bag costs $2.22?

Decimal Numbers: Place Value and Word Names

OBJECTIVES

48. Write the place value of any digit, given the place value name for a decimal number.
49. Write place value names from word names.
50. Write word names from place value names.

APPLICATION

Robert bought a new tire for his car. The tire cost $48.52. He wrote a check to the Low-Pressure Tire Co. What word name will he write on the check?

753
76-231/1234
May 1 19 90
PAY TO THE ORDER OF Low Pressure Tire Co. $ 48.52
DOLLARS
SECOND CHANCE BANK AND TRUST CO.
ANYTOWN, USA 00000
FOR
Robert E. Lee
0005⑆03103⑆ 510 143 8⑈ 0110

VOCABULARY

The digits, and a period called a *decimal point,* are used to write place value names for fractions and mixed numbers. These names are called *decimals.*
The number of digits to the right of the decimal point is the *number of decimal places;* for example, 3.47 has two decimal places.

HOW AND WHY

Decimals are written by using a standard place value in the same way as for whole numbers. The place value for decimals is:

1. The same as whole numbers for digits to the left of the decimal point, and
2. A fraction whose denominator is 10, 100, 1000, and so on, for digits to the right of the decimal point.

The digits to the right of the decimal point have place values of:

$$\frac{1}{10^1} = \frac{1}{10} = 0.1$$

$$\frac{1}{10^2} = \frac{1}{10 \cdot 10} = \frac{1}{100} = 0.01$$

$$\frac{1}{10^3} = \frac{1}{10 \cdot 10 \cdot 10} = \frac{1}{1000} = 0.001$$

$$\frac{1}{10^4} = \frac{1}{10 \cdot 10 \cdot 10 \cdot 10} = \frac{1}{10{,}000} = 0.0001$$

and so on, in that order from left to right.

Using the ones place as the central position (the place value 10^0), the place value of a decimal looks like the following:

	Whole-number part				Fraction part				
ten thousands	thousands	hundreds	tens	ones . decimal point	tenths	hundredths	thousandths	ten-thousandths	hundred-thousandths
10,000	1000	100	10	1	$\frac{1}{10}$	$\frac{1}{100}$	$\frac{1}{1000}$	$\frac{1}{10,000}$	$\frac{1}{100,000}$
10^4	10^3	10^2	10^1	10^0	$\frac{1}{10^1}$	$\frac{1}{10^2}$	$\frac{1}{10^3}$	$\frac{1}{10^4}$	$\frac{1}{10^5}$

. . . (the place values continue in both directions)

Note that the decimal point separates the whole-number part from the fraction part:

$$26.573 = \underbrace{26}_{\text{WHOLE-NUMBER PART}} + \underbrace{.573}_{\text{FRACTION PART}}$$

If the decimal point is not written, as in the case of a whole number, it is understood to follow the ones place. Thus:

23 = 23. 8 = 8. 415 = 415.

Consider 226.35 and 0.127:

	NUMBER TO LEFT OF DECIMAL POINT	DECIMAL POINT	NUMBER TO RIGHT OF DECIMAL POINT	PLACE VALUE OF LAST DIGIT
PLACE VALUE NAME	226	.	35	$\frac{1}{100}$
WORD NAME OF EACH	two hundred twenty-six	and	thirty-five	hundredths
WORD NAME OF DECIMAL	Two hundred twenty-six and thirty-five hundredths			

	NUMBER TO LEFT OF DECIMAL POINT	DECIMAL POINT	NUMBER TO RIGHT OF DECIMAL POINT	PLACE VALUE OF LAST DIGIT
PLACE VALUE NAME	0	.	127	$\frac{1}{1000}$
WORD NAME OF EACH	omit	omit	one hundred twenty-seven	thousandths
WORD NAME OF DECIMAL	One hundred twenty-seven thousandths			

For numbers greater than zero and less than one (such as 0.127) the digit 0 is written in the ones place.

PROCEDURE

To write the word name for a decimal:

1. **Write the name for the whole number to the left of the decimal point.**
2. **Write the word "and" for the decimal point.**
3. **Write the whole-number name for the number to the right of the decimal point.**
4. **Write the place value of the digit farthest to the right.**

If the decimal has only zero or no digit to the left of the decimal point, omit steps 1 and 2.
Work backward to find the place value name.

MODEL PROBLEM-SOLVING

Examples	Strategy
a. Write the place value of the digit 4 in 0.21345:	
ten-thousandths	**The fourth place to the right of the decimal point is the ten-thousandths place.**

Warm Up a. Write the place value of the digit 3 in 0.21345.

b. Write the word name for 0.58:	
fifty-eight hundredths	**First, write the word name for the whole number 58. Second, write the word name for the place value of the digit 8. "Zero and fifty-eight hundredths" is also correct but unnecessary.**

Warm Up b. Write the word name for 0.31.

c. Write the word name for 0.003:	
three thousandths	**Write the word name for three and then the word name for the place value of the digit 3.**

Warm Up c. Write the word name for 0.031.

d. Write the word name for 2.24:	
two	**First, write the word name for the whole number.**
two and	**Second, write "and" for the decimal point.**

ANSWERS TO WARM UPS (4.1) **a.** thousandths **b.** thirty-one hundredths **c.** thirty-one thousandths

two and twenty-four — **Third, write the word name for the whole number to the right of the decimal, 24.**

two and twenty-four hundredths — **Fourth, write the place value of the digit 4.**

Warm Up d. Write the word name for 7.06.

e. Write the place value name for fifteen ten-thousandths:

0.0015 — **First, write the numeral for fifteen. The place value "ten-thousandths" indicates four decimal places, so write two zeros *before* the numeral 15 and then a decimal point. This places the numeral 5 in the ten-thousandths place. Since fifteen ten-thousandths is greater than zero and less than one, we write "0" in the ones place.**

Warm Up e. Write the place value name for twenty-nine thousandths.

f. Write the place value name for four hundred five and four hundred five thousandths:

405 — **The whole-number part is 405.**

405. — **Write a decimal point for "and."**

405.405 — **The decimal part is also 405. Thousandths indicates three decimal places.**

Warm Up f. Write the place value name for seven hundred three and three hundred seven thousandths.

g. Write the place value name for four hundred five and four hundred five ten-thousandths:

405 — **The whole-number part is 405.**

405. — **Write a decimal point for "and."**

405.0405 — **The decimal part is also 405. Ten-thousandths indicates four decimal places.**

Warm Up g. Write the place value name for seven hundred three and three hundred seven ten-thousandths.

h. Write the place value names for three hundred-thousandths and for three hundred thousandths:

0.00003 — **The word name "hundred-thousandths" indicates the fifth place to the right of the decimal point. The hyphen in the name "hundred-thousandths" is part of the word name. So, three hundred-thousandths has five decimal places.**

ANSWERS TO WARM UPS (4.1)

d. seven and six hundredths **e.** 0.029

f. 703.307 **g.** 703.0307

0.300 — **Three hundred thousandths has three decimal places, since the place value is "thousandths." Note that three hundred thousandths is equal to three tenths, 0.3.**

Warm Up h. Write the place value names for thirty hundredths and for three hundredths.

APPLICATION SOLUTION

i. Robert bought a new tire for his car. The tire cost $48.52. He wrote a check to the Low-Pressure Tire Co. What word name will he write on the check?

753
76-231/1234
May 1 1990
PAY TO THE ORDER OF Low Pressure Tire Co. $ 48.52
Forty-eight and fifty-two hundredths DOLLARS
SECOND CHANCE BANK AND TRUST CO.
ANYTOWN, USA 00000
FOR ________ Robert E. Lee
0005⑆031103⑆ 510 143 8⑈ 0110

(*Note:* Many people use a combination of word and numeral names on checks, such as Forty-eight and 52/100.)

Warm Up i. Freda bought a chain saw for cutting firewood. The saw cost $64.49. If she wrote a check for the saw, what word name did she write on the check?

ANSWERS TO WARM UPS (4.1) **h.** 0.30, 0.03 **i.** sixty-four and forty-nine hundredths

NAME ______________________

CLASS ______________ SECTION ______________ DATE ______________

Exercises 4.1

A

ANSWERS

1. ______
2. ______
3. ______
4. ______
5. ______
6. ______
7. ______
8. ______
9. ______
10. ______
11. ______
12. ______
13. ______
14. ______
15. ______
16. ______
17. ______
18. ______
19. ______
20. ______
21. ______
22. ______

Write the place value of the digit 2 in each of the following:

1. 0.23 **2.** 0.0032 **3.** 17.10432

4. 0.527 **5.** 8.0325 **6.** 2.0355

7. 203.309 **8.** 142.891

Write the word names for these numbers:

9. 0.5 **10.** 0.7 **11.** 0.12

12. 0.34 **13.** 0.67 **14.** 0.94

15. 0.267 **16.** 0.712 **17.** 0.4865

18. 0.8975 **19.** 8.7543 **20.** 5.3113

For problems 21 to 24 use this number: 7,182.65409:

21. What digit is in the hundredths place?

22. What digit is in the ten-thousandths place?

ANSWERS

23. ______

24. ______

25. ______

26. ______

27. ______

28. ______

29. ______

30. ______

23. What digit is in the tenths place?

24. What digit is in the hundreds place?

Write the place value names for these numbers:

25. six tenths **26.** nine tenths

27. eleven hundredths **28.** forty-five hundredths

29. one hundred eleven thousandths

30. five hundred fourteen thousandths

B

31. ______

32. ______

33. ______

34. ______

35. ______

36. ______

37. ______

38. ______

39. ______

40. ______

41. ______

42. ______

43. ______

44. ______

What is the place value of the 6 in each of the following?

31. 0.486 **32.** 0.2746 **33.** 0.0632 **34.** 2.3269

35. 164.7829 **36.** 216.95173

What is the place value of the 8 in each of the following?

37. 0.082 **38.** 0.834 **39.** 3.278 **40.** 4.7938

Write the word names for these numbers:

41. 0.504 **42.** 5.04 **43.** 50.4 **44.** 500.4

NAME ______

CLASS ______ SECTION ______ DATE ______

ANSWERS

45. ______
46. ______
47. ______
48. ______
49. ______
50. ______
51. ______
52. ______
53. ______
54. ______
55. ______
56. ______
57. ______
58. ______
59. ______
60. ______
61. ______
62. ______

45. 50.04 **46.** 5.004 **47.** 8.0205 **48.** 45.0051

49. 384.0 **50.** 45. **51.** 405.05 **52.** 48.055

Write the place value name for each of the following:

53. fifteen hundredths **54.** five thousand

55. eighteen ten-thousandths **56.** twelve thousandths

57. Which digit is in the hundredths place? 1435.268

58. Which digit is in the tenths place? 2164.8935

59. Write the place value of the digit 3 in 1435.268.

60. Write the place value of the digit 6 in 4156.239

61. Which digit is in the hundreds place? 7239.564

62. Which digit is in the thousandths place? 8036.7519

C

ANSWERS

63. ______________

64. ______________

65. ______________

66. ______________

67. ______________

68. ______________

69. ______________

70. ______________

71. ______________

72. ______________

73. ______________

74. ______________

75. ______________

76. ______________

Write the word names for these numbers:

63. 2.0202

64. 1.23456

65. 121.00023

66. 75.00203

67. 72.1653

68. 311.0615

Write the place value names for these numbers:

69. seven hundred and ninety-six thousandths

70. six hundred and seven thousandths

71. five hundred and five thousandths

72. five thousand and five hundredths

73. five hundred-thousandths

74. one thousand and five thousandths

75. one thousand five and five thousandths

76. two hundred and thirty-one thousandths

NAME ____________________

CLASS ____________ SECTION ____________ DATE ____________

ANSWERS

77. ____________

77. two hundred thirty-one thousandths

78. ____________

78. eighty-nine thousand, fifty-nine and forty-six thousandths

79. ____________

79. eighty-nine thousand, fifty and nine hundred forty-six thousandths

80. ____________

80. two hundred-thousandths

81. ____________

81. two hundred thousandths

82. ____________

82. two hundred thousand

D

83. ____________

83. Dan Ngo bought a toaster-oven that had a marked price of \$53.98. What word name will he write on the check?

84. ____________

84. Fari Alhadet bought a pickup truck load of organic fertilizer for her garden. The price of the load was \$83.45. What word name will she write on the check?

85. ____________

85. Alton Jewett made a 35% down payment on a tiller. If the down payment was \$77.15, what word name will she write on the check?

ANSWERS

86. ______________

87. ______________

88. ______________

89. ______________

90. ______________

91. ______________

92. ______________

93. ______________

94. ______________

95. ______________

96. ______________

86. Harlan Niehauss sent a check to his bank for his safety deposit box and some printed checks. If the amount was $104.08, what word name did he write on the check?

E *Maintain your skills* (Sections 3.4 3.5, 3.6)

Perform the indicated operations:

87. $\frac{25}{27} \cdot \frac{18}{35} \cdot \frac{7}{15}$

88. $\frac{36}{75} \cdot \frac{15}{16} \cdot \frac{40}{27}$

89. $4\frac{2}{5} \cdot 2\frac{4}{5}$

90. $5\frac{7}{10} \cdot 1\frac{7}{10}$

91. $\frac{85}{48} \div \frac{51}{32}$

92. $\frac{95}{72} \div \frac{38}{64}$

93. $4\frac{2}{5} \div 2\frac{4}{5}$

94. $5\frac{7}{10} \div 1\frac{7}{10}$

95. Gordon is driving to his parents' home, which is 432 miles away. He plans to drive three-fourths of the distance the first day. How far will he travel the first day?

96. Berta plans to drive $1127\frac{1}{4}$ miles to see her brother, visiting friends along the way and doing some sightseeing. If her trip covers an average of $125\frac{1}{4}$ miles each day, how many days will the trip take?

4.2 Decimals: Expanded Form

OBJECTIVES

51. Write the expanded form from the place value name.
52. Write the place value name from the expanded form.

APPLICATION

Expanded form helps one to understand the methods we use to add, subtract, multiply, and divide decimals.

VOCABULARY

No new words.

HOW AND WHY

The expanded form shows the place value of each digit:

PLACE VALUE NAME	EXPANDED FORM INDICATED SUM OF VALUES
0.346	$\frac{3}{10} + \frac{4}{100} + \frac{6}{1000}$
32.9	$30 + 2 + \frac{9}{10}$
8.6421	$8 + \frac{6}{10} + \frac{4}{100} + \frac{2}{1000} + \frac{1}{10,000}$

The number 0.346 can be written: 3 tenths + 4 hundredths + 6 thousandths.

PROCEDURE

To change from place value name to expanded form, write the indicated sum of the values of each of the digits.
To change from expanded form to place value name, add.

MODEL PROBLEM-SOLVING

Examples | **Strategy**

a. Write 0.31 in expanded form:

$$0.31 = \frac{3}{10} + \frac{1}{100}$$

Write the indicated sum of the fractions for three tenths and for one hundredth.

Warm Up a. Write 0.79 in expanded form.

ANSWERS TO WARM UPS (4.2)

a. $\frac{7}{10} + \frac{9}{100}$

b. Write 0.6006 in expanded form:

$$0.6006 = \frac{6}{10} + \frac{0}{100} + \frac{0}{1000} + \frac{6}{10{,}000}$$

Write the fractions for six tenths, zero hundredths, zero thousandths, and six ten-thousandths.

Warm Up b. Write 0.2035 in expanded form.

c. Write 182.56 in expanded form:

$$182.56 = 100 + 80 + 2 + \frac{5}{10} + \frac{6}{100}$$

The expanded form of the whole number (to the left of the decimal) is the same as in Chapter 1. The decimal point does not show in expanded form since the fractions give the place values.

Warm Up c. Write 37.09 in expanded form.

d. Write the place value name for $\frac{2}{10} + \frac{7}{100} + \frac{3}{1000}$:

0.273

It is also correct to write .273 but it is standard practice to write the zero.

Warm Up d. Write the place value name for $\frac{8}{10} + \frac{0}{100} + \frac{6}{1000}$.

e. Write the place value name for $700 + 50 + 0 + \frac{9}{10} + \frac{8}{100}$:

750.98

Warm Up e. Write the place value name for $100 + 0 + 7 + \frac{2}{10} + \frac{5}{100}$.

ANSWERS TO WARM UPS (4.2)

b. $\frac{2}{10} + \frac{0}{100} + \frac{3}{1000} + \frac{5}{10{,}000}$

c. $30 + 7 + \frac{0}{10} + \frac{9}{100}$

d. 0.806

e. 107.25

NAME ______________________

CLASS ______________ SECTION ______________ DATE ______________

Exercises 4.2

A

ANSWERS

1. ______
2. ______
3. ______
4. ______
5. ______
6. ______
7. ______
8. ______
9. ______
10. ______
11. ______
12. ______
13. ______
14. ______
15. ______
16. ______
17. ______
18. ______
19. ______
20. ______
21. ______
22. ______

Write each of the following in expanded form:

1. 0.2 **2.** 0.7 **3.** 0.21 **4.** 0.45 **5.** 0.61

6. 0.83 **7.** 0.257 **8.** 0.682 **9.** 0.314 **10.** 0.561

11. 5.3 **12.** 4.71

Write the place value name for each:

13. $\frac{5}{10}$

14. $\frac{3}{10}$

15. $\frac{1}{10} + \frac{6}{100}$

16. $\frac{2}{10} + \frac{3}{100}$

17. $\frac{3}{10} + \frac{8}{100}$

18. $\frac{4}{10} + \frac{7}{100}$

19. $\frac{1}{10} + \frac{2}{100} + \frac{1}{1000}$

20. $\frac{3}{10} + \frac{5}{100} + \frac{7}{1000}$

21. $\frac{9}{10} + \frac{3}{100} + \frac{8}{1000}$

22. $\frac{8}{10} + \frac{1}{100} + \frac{5}{1000}$

ANSWERS

23. ______

24. ______

23. $\frac{4}{10} + \frac{0}{100} + \frac{3}{1000}$

24. $\frac{0}{10} + \frac{5}{100} + \frac{8}{1000}$

B

25. ______
26. ______
27. ______
28. ______
29. ______
30. ______
31. ______
32. ______
33. ______
34. ______
35. ______
36. ______
37. ______
38. ______
39. ______
40. ______
41. ______
42. ______
43. ______
44. ______
45. ______
46. ______

Write each of the following in expanded form:

25. 0.421 **26.** 0.682 **27.** 0.025 **28.** 0.302 **29.** 0.908

30. 0.301 **31.** 0.0509 **32.** 0.0102 **33.** 0.0011 **34.** 0.0023

35. 14.723 **36.** 62.081

Write the place value name for each:

37. $\frac{0}{10} + \frac{3}{100}$

38. $\frac{0}{10} + \frac{5}{100}$

39. $\frac{0}{10} + \frac{1}{100} + \frac{1}{1000}$

40. $\frac{0}{10} + \frac{0}{100} + \frac{2}{1000}$

41. $\frac{3}{10} + \frac{0}{100} + \frac{2}{1000}$

42. $\frac{5}{10} + \frac{0}{100} + \frac{1}{1000}$

43. $\frac{7}{10} + \frac{8}{100} + \frac{2}{10000}$

44. $\frac{2}{10} + \frac{5}{1000} + \frac{9}{10000}$

45. $\frac{4}{10} + \frac{0}{100} + \frac{0}{1000} + \frac{3}{10000}$

46. $\frac{0}{10} + \frac{0}{100} + \frac{0}{1000} + \frac{8}{10000}$

NAME ______________________

CLASS ____________ SECTION ____________ DATE ____________

ANSWERS

47. ______

48. ______

49. ______

50. ______

51. ______

52. ______

53. ______

54. ______

55. ______

56. ______

57. ______

58. ______

59. ______

47. $\frac{4}{10} + \frac{5}{100} + \frac{0}{1000} + \frac{0}{10000} + \frac{8}{100000}$

48. $\frac{0}{10} + \frac{0}{100} + \frac{1}{1000} + \frac{0}{10000} + \frac{2}{100000}$

49. $70 + 3 + \frac{0}{10} + \frac{2}{100}$

50. $800 + 60 + 5 + \frac{6}{10} + \frac{0}{100}$

C

Write each of the following in expanded form:

51. 2.3 **52.** 12.82 **53.** 91.321

54. 128.025 **55.** 625.0031 **56.** 102.0007

Write the place value name for each:

57. $2 + \frac{1}{10} + \frac{3}{100} + \frac{5}{1000}$ **58.** $10 + 3 + \frac{8}{10} + \frac{9}{100} + \frac{7}{1000}$

59. $100 + 0 + 5 + \frac{5}{10} + \frac{0}{100} + \frac{8}{1000}$

ANSWERS

60. ______

61. ______

62. ______

60. $5000 + 200 + 30 + 1 + \frac{0}{10} + \frac{0}{100} + \frac{8}{1000}$

61. $5000 + \frac{8}{10} + \frac{0}{100} + \frac{9}{1000} + \frac{6}{10000}$

62. $3000 + 100 + 0 + 2 + \frac{0}{10} + \frac{0}{100} + \frac{3}{1000} + \frac{5}{10000}$

E *Maintain your skills* (Sections 3.3, 3.5, 3.7, 3.8)

63. ______

64. ______

65. ______

66. ______

67. ______

68. ______

69. ______

70. ______

71. ______

72. ______

Build the fractions as indicated:

63. $\frac{11}{12} = \frac{?}{36}$

64. $\frac{7}{8} = \frac{?}{48}$

Reduce these fractions:

65. $\frac{75}{80}$

66. $\frac{75}{120}$

Write the reciprocal:

67. $\frac{22}{25}$

68. $4\frac{7}{8}$

True or false?

69. $\frac{2}{3} < \frac{7}{9}$

70. $1\frac{1}{5} > 1\frac{2}{7}$

71. Trina bought $\frac{1}{2}$ lb of chocolates, $\frac{1}{4}$ lb of peanut brittle, and $\frac{1}{2}$ lb of gumdrops. How many pounds of candy did she buy?

72. Saul needs eight and one-half ounces of tomato paste for a recipe. He already has three and seven-eighths ounces left over from an earlier meal. How many more ounces of tomato paste does he need?

Decimals to Fractions

OBJECTIVE

53. Change decimals to fractions.

APPLICATION

The chance (probability) that a coin will land heads up when flipped once is 0.5. Express this decimal as a fraction in the lowest terms.

VOCABULARY

No new words.

HOW AND WHY

Consider 0.625:

READ: six hundred twenty-five thousandths

WRITE: $\frac{625}{1000}$

So, $0.625 = \frac{625}{1000} = \frac{5}{8}$

PROCEDURE

To change a decimal to a fraction:

1. **Read the decimal word name.**
2. **Write the fraction that has the same name.**
3. **Reduce if possible.**

Notice that because of place value, the number of decimal places in a decimal tells us the number of zeros in the denominator of the fraction. This fact can be used as another way to write the fraction or to check that the fraction is correct:

$$\underbrace{2.78}_{\text{TWO DECIMAL PLACES}} = 2\frac{78}{\underbrace{100}_{\text{TWO ZEROS}}} = 2\frac{39}{50} \text{ or } \frac{278}{100} = 2\frac{39}{50}$$

MODEL PROBLEM-SOLVING

Examples	Strategy
a. Change 0.75 to a fraction: $0.75 = \frac{75}{100} = \frac{3}{4}$	**READ: "seventy-five hundredths," then write the fraction name and reduce.**

Warm Up a. Change 0.12 to a fraction.

ANSWERS TO WARM UPS (4.3)

a. $\frac{3}{25}$

b. Change 12.3 to a mixed number:

$12.3 = 12\frac{3}{10}$

READ: "twelve and three tenths," then write the mixed number.

The denominator (in this case, 10) gives the place value of the last place, so we can also write:

$\mathbf{12.3 = \frac{123}{10} = 12\frac{3}{10}}$

Warm Up b. Change 1.3 to a mixed number.

c. Change 5.055 and 0.0075 to fractions or mixed numbers and reduce:

$5.055 = 5\frac{55}{1000} = 5\frac{11}{200}$

$0.0075 = \frac{75}{10{,}000} = \frac{3 \cdot 25}{400 \cdot 25} = \frac{3}{400}$

Warm Up c. Change 3.25 and 0.050 to fractions or mixed numbers and reduce.

APPLICATION SOLUTION

d. The chance (probability) that a coin will land heads up when flipped once is 0.5. Express this decimal as a fraction in the lowest terms.

$0.5 = \frac{5}{10} = \frac{1}{2}$

READ: "five tenths," then write the fraction for five tenths and reduce.

The chance of getting "heads" is $\frac{1}{2}$, or one in every two flips.

Warm Up d. The chance that a pair of dice will land with a sum of five or six is 0.25. Write this as a fraction in the lowest terms.

ANSWERS TO WARM UPS (4.3)

b. $1\frac{3}{10}$ **c.** $3\frac{1}{4}, \frac{1}{20}$ **d.** $\frac{1}{4}$

NAME ______________________________

CLASS ______________ SECTION ______________ DATE ______________

Exercises 4.3

A

ANSWERS

1. ____
2. ____
3. ____
4. ____
5. ____
6. ____
7. ____
8. ____
9. ____
10. ____
11. ____
12. ____
13. ____
14. ____
15. ____
16. ____
17. ____
18. ____
19. ____
20. ____
21. ____
22. ____
23. ____
24. ____

Change each of the following decimals to fractions or mixed numbers:

1. 0.7 **2.** 0.3 **3.** 0.9 **4.** 0.1

5. 0.17 **6.** 0.23 **7.** 0.07 **8.** 0.03

9. 0.13 **10.** 0.31 **11.** 0.22 **12.** 0.18

13. 0.24 **14.** 0.36 **15.** 0.003 **16.** 0.001

17. 0.085 **18.** 0.024 **19.** 0.135 **20.** 0.214

21. 0.028 **22.** 0.055 **23.** 3.5 **24.** 4.8

B

ANSWERS

25. ______
26. ______
27. ______
28. ______
29. ______
30. ______
31. ______
32. ______
33. ______
34. ______
35. ______
36. ______
37. ______
38. ______
39. ______
40. ______
41. ______
42. ______
43. ______
44. ______

25. 1.85 **26.** 2.17 **27.** 0.0004 **28.** 0.0008

29. 3.002 **30.** 4.005 **31.** 0.875 **32.** 0.382

33. 3.95 **34.** 6.49 **35.** 2.10 **36.** 2.01

37. 0.305 **38.** 0.032 **39.** 0.444 **40.** 0.555

41. 8.55 **42.** 8.055 **43.** 15.65 **44.** 25.25

C

45. ______
46. ______
47. ______
48. ______

45. 68.5 **46.** 21.88 **47.** 25.675 **48.** 112.08

NAME ______________________

CLASS ____________ SECTION ____________ DATE ____________

ANSWERS

49. ________
50. ________
51. ________
52. ________
53. ________
54. ________
55. ________
56. ________
57. ________
58. ________

49. 700.007 **50.** 305.305 **51.** 233.1234 **52.** 890.0204

53. 2.00001 **54.** 3.000025 **55.** 0.7152 **56.** 0.834

57. 7.21605 **58.** 8.47985

D

59. ________

59. The probability (or chance) that a flipped coin will come up "heads" three times in a row is 0.125. Write this as a fraction in the lowest terms.

60. ________

60. The probability (or chance) that a flipped coin will come up "heads" twice and "tails" once out of three flips is 0.375. Write this as a fraction in the lowest terms.

61. ________

61. The micrometer reading for the diameter of a piece of stock is 0.875 inch. What fraction of an inch is the diameter?

62. ________

62. The diameter of a piston pin measured with a micrometer is 0.6875 inch. What fraction of an inch is this?

63. ________

63. Nike stock rose 0.625 in one day. What fraction represents this rise?

ANSWERS

64. ____________

65. ____________

66. ____________

67. ____________

64. Nordstrom stock fell 0.75 during a recent day. Represent this fall as a fraction.

65. Maria may choose 0.725 yard or $\frac{3}{4}$ yard in remnants for the same price. Which should she choose to get the most fabric? Compare as fractions.

66. How much more fabric will Maria get? (See Exercise 65.)

67. Gerald may choose a 0.055 raise in pay or $\frac{1}{20}$ increase. Which will give him more money? Compare using fractions.

E *Maintain your skills* (Section 3.6)

68. ____________

69. ____________

70. ____________

71. ____________

72. ____________

73. ____________

74. ____________

75. ____________

76. ____________

77. ____________

Multiply or divide as indicated:

68. $\left(3\frac{1}{2}\right)\left(6\frac{3}{4}\right)$

69. $\left(3\frac{3}{4}\right)\left(6\frac{1}{2}\right)$

70. $\left(7\frac{3}{4}\right)\left(8\frac{2}{3}\right)$

71. $\left(7\frac{2}{3}\right)\left(8\frac{3}{4}\right)$

72. $\left(16\frac{2}{3}\right) \div \left(4\frac{1}{6}\right)$

73. $\left(16\frac{1}{6}\right) \div \left(4\frac{2}{3}\right)$

74. $\left(12\frac{1}{2}\right) \div \left(1\frac{1}{4}\right)$

75. $\left(12\frac{1}{4}\right) \div \left(1\frac{1}{2}\right)$

76. Sheela drove 385 miles on seventeen and one-half gallons of gasoline. How many miles per gallon did she average?

77. Jorge drove 660 miles on fourteen and two-thirds gallons of diesel fuel. How many miles per gallon did he average?

Listing Decimals in Order of Value

OBJECTIVES

54. Determine if an inequality statement is true or false.
55. List a group of decimals in order from the smallest to largest.

APPLICATION

Using a micrometer to measure the diameter of a piece of metal stock, Mike measured 0.8247 and Mildred measured 0.825. Which measurement is larger? (A micrometer is a device used to make measurements that are more precise than those made with a ruler or tape.)

VOCABULARY

The symbols for less than, "<", and greater than, ">", are used the same way as in Chapters 1 and 3.

HOW AND WHY

Fractions can be listed in order, when they have a common denominator, by ordering the numerators. This idea can be extended to decimals when they have the same number of decimal places. For instance, $0.26 = \frac{26}{100}$ and $0.37 = \frac{37}{100}$ have a common denominator when written in fraction form. So, 0.26 is less than 0.37: $0.26 < 0.37$.

We can see that $1.5 < 3.6$ because 1.5 is to the left of 3.6 on the number line:

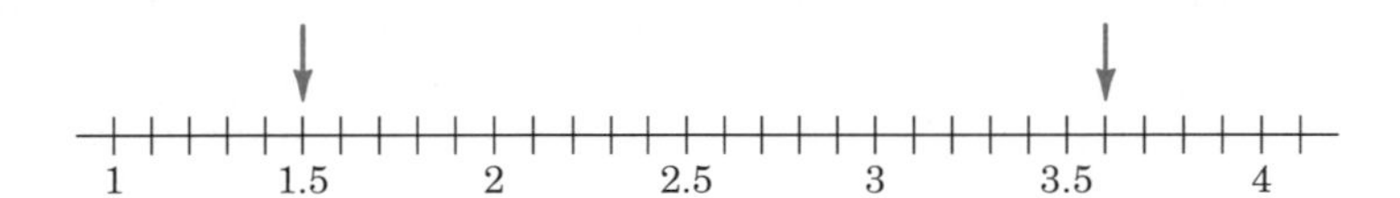

The decimals 0.3 and 0.15 have a common denominator when zeros are placed after the 3. Thus:

$$0.3 = \frac{3}{10} \quad \text{and} \quad \frac{3}{10} = \frac{3}{10} \cdot \frac{10}{10} = \frac{30}{100}$$

so that:

$$0.3 = \frac{30}{100} \quad \text{and} \quad 0.15 = \frac{15}{100}$$

Then, since $\frac{15}{100} < \frac{30}{100}$, we conclude that $0.15 < 0.3$.

PROCEDURE

Decimals can be listed in order by the following procedure:

1. **Make sure that all numbers have the same number of decimal places to the right of the decimal point by placing zeros to the right of the last digit when necessary.**
2. **Write the numbers in order from the smallest to largest, ignoring the decimal point.**
3. **Remove the extra zeros.**

MODEL PROBLEM-SOLVING

Examples	Strategy

a. Determine whether the following inequalities are true or false: $0.45 > 0.41$, $0.92 < 0.919$, $1.003 > 1.01$, $53.0001 < 53.009$:

$0.45 > 0.41$ true	**The decimal points can be ignored since both numbers have the same number of decimal places. Since $45 > 41$, the statement is true.**
$0.92 < 0.919$ false	**Write 0.92 as 0.920. Since $920 < 919$ is false, the statement is false.**
$1.003 > 1.01$ false	**Write 1.01 as 1.010. Since $1003 > 1010$ is false, the statement is false.**
$53.0001 < 53.009$ true	**True, since $530001 < 530090$.**

Warm Up a. Determine whether the following inequalities are true or false: $2.09 > 2.001$, $0.5 < 0.499$, $33.7 > 33.6989$

b. List these decimals from the smallest to largest: 0.62, 0.637, 0.6159, 0.621:

0.62	0.6200	**First, give all the numbers the same number of decimal places by writing zeros on the right where needed.**
0.637	0.6370	
0.6159	0.6159	
0.621	0.6210	

0.6159, 0.6200, 0.6210, 0.6370 — **Second, write the numbers in order, ignoring the decimal points.**

0.6159, 0.62, 0.621, 0.637 — **Third, remove the extra zeros.**

Warm Up b. List these decimals from the smallest to largest: 0.75, 0.692, 0.748

c. List these decimals from the smallest to largest: 1.357, 1.361, 1.3534, 1.358:

1.357	1.3570	**Step 1**
1.361	1.3610	
1.3534	1.3534	
1.358	1.3580	

ANSWERS TO WARM UPS (4.4) **a.** true, false, true **b.** 0.692, 0.748, 0.75

1.3534, 1.3570, 1.3580, 1.3610 **Step 2**

1.3534, 1.357, 1.358, 1.361 **Step 3**

Warm Up c. List these decimals from the smallest to largest: 0.03, 0.0033, 0.0333, 0.0303

APPLICATION SOLUTION

d. Using a micrometer to measure the diameter of a piece of metal stock, Mike measured 0.8247 and Mildred measured 0.825. Which measurement is larger?

0.8247, 0.8250

Since 0.8250 is larger than 0.8247 ($0.8250 > 0.8247$), we conclude that 0.825 is larger than 0.8247, so Mildred's measurement is larger.

Warm Up d. Robert and Roberta measured the same stock. Robert measured 0.824 in. and Roberta measured 0.8237 in. Which measurement is larger?

ANSWERS TO WARM UPS (4.4)

c. 0.0033, 0.03, 0.0303, 0.0333

d. Robert's (0.824)

NAME ____________________

CLASS ____________ SECTION ____________ DATE ____________

Exercises 4.4

A

ANSWERS

1. ______
2. ______
3. ______
4. ______
5. ______
6. ______
7. ______
8. ______
9. ______
10. ______
11. ______
12. ______
13. ______
14. ______
15. ______
16. ______
17. ______
18. ______

Determine whether the following statements are true or false:

1. $0.2 < 0.7$ **2.** $0.5 < 0.4$ **3.** $0.12 > 0.11$ **4.** $0.27 > 0.26$

5. $0.2 < 0.22$ **6.** $0.23 < 0.021$ **7.** $0.19 > .019$ **8.** $0.6 > 0.62$

Insert the appropriate inequality symbol, $>$ or $<$, between the given numbers to make a true statement:

9. 0.86 ____ 0.93 **10.** 0.212 ____ 0.202

11. 0.4 ____ 0.04 **12.** 82.16 ____ 82.2

13. 0.751 ____ 0.571 **14.** 0.1234 ____ 0.124

List the decimals from the smallest to largest:

15. 0.2, 0.9, 0.7 **16.** 0.02, 0.09, 0.07

17. 1.4, 1.7, 1.3 **18.** 7.3, 7.7, 7.0

ANSWERS

19. ______

20. ______

21. ______

22. ______

19. 0.05, 0.6, 0.07

20. 34.01, 34.1, 34.09

21. 6.14, 6.141, 6.139

22. 5.18, 5.183, 5.179

B

23. ______

24. ______

25. ______

26. ______

27. ______

28. ______

29. ______

30. ______

31. ______

32. ______

33. ______

34. ______

35. ______

36. ______

37. ______

38. ______

Determine whether the following statements are true or false:

23. $0.31 < 0.3$

24. $0.32 > 0.315$

25. $1.099 < 1.09$

26. $13.1 < 13.099$

27. $7.86 > 7.68$

28. $2.42 > 2.24$

29. $0.112 < 0.1119$

30. $0.335 < 0.3351$

Place the appropriate inequality symbol, $>$ or $<$, between the given numbers to make a true statement:

31. 7.261 ____ 7.61

32. 85.91 ____ 95.9

33. 4.316 ____ 4.32

34. 22.71 ____ 22.701

35. 0.8216 ____ 0.821

36. 0.931 ____ 0.9316

List the decimals from the smallest to largest:

37. 7.61, 7.59, 7.6, 7.62, 7.63

38. 1.67, 1.65, 1.6, 1.7

NAME ______________________________

CLASS ______________ SECTION ______________ DATE ______________

ANSWERS

39. ______________

40. ______________

41. ______________

42. ______________

43. ______________

44. ______________

39. 0.565, 0.556, 0.566, 0.555

40. 3.2, 3.19, 3.21, 3.27

41. 0.899, 0.86, 0.91, 0.903, 0.9

42. 0.1163, 0.116, 0.1159, 0.117

43. 17.05, 17.16, 17.0506, 17.057

44. 1.999, 2, 1, 1.5, 2.006

C

45. ______________

46. ______________

47. ______________

48. ______________

49. ______________

50. ______________

51. ______________

52. ______________

53. ______________

Determine whether the following statements are true or false:

45. $3.274 < 3.3$

46. $3.3 > 3.269$

47. $3.269 < 3.369$

48. $3.369 > 3.274$

49. $0.1231 < 0.1243$

50. $0.1243 > 0.124$

51. $0.124 < 0.123$

52. $0.123 > 0.1231$

List the decimals from the smallest to largest:

53. 0.0729, 0.073001, 0.072, 0.073, 0.073015

ANSWERS

54. ______

55. ______

56. ______

57. ______

58. ______

59. ______

60. ______

61. ______

62. ______

63. ______

54. 3.0009, 0.301, 0.3008, 0.30101

55. 0.888, 0.88799, 0.8881, 0.88759

56. 8.36, 8.2975, 8.3599, 8.3401

57. 50.004, 50.04, 50.039, 50.093

58. 22.4506, 22.0456, 22.0546, 22.6405

Place the appropriate inequality symbol, > or < between the given numbers to make a true statement:

59. 184.841 ____ 184.481

60. 611.2 ____ 611.1875

61. $6\frac{1}{4}$ ____ 6.27*

62. 7.53* ____ $7\frac{1}{2}$

63. 2.625* ____ $2\frac{9}{16}$

D

64. ______

*Change to a fraction!

64. Greg and Larry each measured the diameter of a coin and got 0.916 centimeter and 0.921 centimeter, respectively. Which measurement is smaller?

NAME ____________________

CLASS ____________ SECTION ____________ DATE ____________

ANSWERS

65. ____________

66. ____________

67. ____________

68. ____________

69. ____________

70. ____________

71. ____________

65. The Davis Meat Company bid 98.375 cents per pound to provide meat to the Ajax Grocery. Circle K Meats put in a bid of 98.35 cents, and J & K Meats made a bid of 98.3801 cents. Which is the best bid for the Ajax Grocery?

66. Three service stations advertise gasoline. The first offers 1.879 dollars per gallon; the second, 1.919 dollars per gallon; and the third, 1.819 dollars per gallon. Which is the most economical?

67. A vial contains 2.3059 grams of charcoal. Dan weighs the charcoal and determines that it weighs 2.31 grams. Is Dan's weighing too light or too heavy?

68. Elaine got 32.783 miles per gallon in her new car. Jeff got 32.837 in his new truck. Who has the least gas mileage?

69. Charles lost 2.165 pounds during the week Karla lost 2.203 pounds. Who lost the most weight that week?

70. A certain chemistry class required 0.2178 mg of soap per student. Hoa has only 0.21 mg of soap. Does she need more or less soap?

71. Ilga needs 0.3426 g of cereal for a certain recipe. She has 0.342 g. Does she have more or less than she needs?

E *Maintain your skills* (Sections 3.3, 3.7, 3.8)

ANSWERS

72. ____________________

73. ____________________

74. ____________________

75. ____________________

76. ____________________

77. ____________________

78. ____________________

79. ____________________

80. ____________________

81. ____________________

List the fractions and mixed numbers from the smallest to largest:

72. $\frac{3}{4}, \frac{7}{8}, \frac{5}{6}, \frac{2}{3}$ **73.** $\frac{9}{10}, \frac{17}{20}, \frac{3}{4}, \frac{4}{5}$ **74.** $2\frac{1}{2}, 2\frac{13}{20}, 3\frac{1}{15}, 2\frac{7}{10}$

Reduce to the lowest terms:

75. $\frac{128}{192}$ **76.** $\frac{136}{192}$ **77.** $\frac{144}{192}$

Find the missing numerator:

78. $\frac{15}{16} = \frac{?}{128}$ **79.** $\frac{15}{16} = \frac{?}{192}$

80. The Oak Lumber Company ordered 128 sheets of $\frac{5}{16}$-inch paneling. If the sheets are stacked in one pile, how high will the stack be?

81. Three tape measures are marked in inches. On the first tape measure the spaces are divided into tenths, on the second they are divided into eighths, and on the third they are divided into thirty-seconds. They are all used to measure the length of a 1″ × 6″ board. The nearest mark on the first tape measure is $38\frac{7}{10}$, the nearest mark on the second is $38\frac{6}{8}$, and the nearest mark on the third is $38\frac{25}{32}$. Which is the shortest measurement?

Approximation by Rounding

OBJECTIVE

56. Round off a given decimal.

APPLICATION

Sam is reading the instructions about listing deductions on his income tax return. The instructions are to round off each deduction to the nearest whole dollar. If he is going to deduct $832.57 for mortgage interest he paid, what will he enter on his return?

VOCABULARY

Decimals are either *exact* or *approximate.* For example, decimals that count money are exact. The figure $14.95 shows an exact amount. Most decimals are approximations of measurements. For example, 5.9 ft shows a person's height to the nearest tenth of a foot, and 1.8 m shows the height to the nearest tenth of a meter.

If 15.38 is written as 15.4, it has been *rounded off* to the nearest tenth. Recall that 15.38 ≈ 15.4 is read, "15.38 is approximately equal to 15.4."

HOW AND WHY

Look at this ruler:

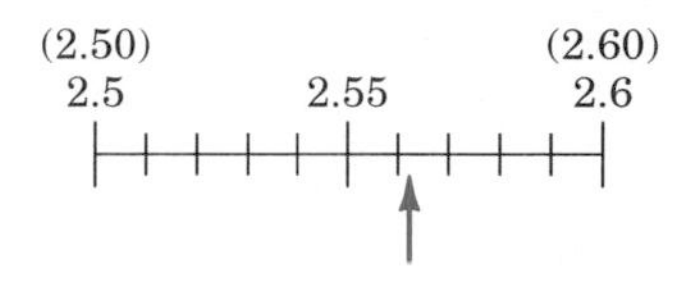

To the nearest tenth, the arrow shows the larger number, 2.6, because it is closer to 2.6 than to 2.5. Rounded off to the nearest hundredth, the arrow shows the smaller number, 2.56, because it is closer to 2.56 than 2.57.

To round off 6.3265 to the nearest hundredth, draw an arrow under the hundredths place to identify the round-off position:

6.3265

We must choose between 6.32 and 6.33. Since the digit to the right of the round-off position is 6, the number is more than halfway to 6.33. So, we choose the larger number:

6.3265 ≈ 6.33

PROCEDURE

To round off a decimal number to a given place value:

1. **Until you can round off mentally, draw an arrow under the given place value. (After enough practice, you will not need the arrow.)**
2. **If the digit to the right of the arrow is 5, 6, 7, 8, or 9, add 1 to the digit above the arrow. That is, round off to the larger number.**
3. **If the digit to the right of the arrow is 0, 1, 2, 3, or 4, keep the digit above the arrow. That is, round off to the smaller number.**
4. **Write whatever zeros are necessary after the arrow so that the number above the arrow has the same place value as the original. See Example b.**

This is sometimes called the "four-five" rule. Although this rounding procedure is the most commonly used, it is not the only way to round off. Many government agencies round by *truncation,* that is, by dropping the digits after the decimal point. Thus, $56.65 ≈ $56. It is common for retail stores to round up for any amounts larger than one cent. Thus, $1.333 ≈ $1.34. There is also a rule for rounding numbers in science, which is sometimes referred to as the "even/odd" rule. You might need to learn and use a different round-off rule depending on what kind of work you are doing.

MODEL PROBLEM-SOLVING

Examples | **Strategy**

a. Round off 0.3582 to the nearest hundredth.

$0.3582 \approx 0.36$
↑ (under the 5)

Draw an arrow under the digit 5 in the hundredths place.
The digit to the right of the arrow is 8, so we round "up" by adding 5 + 1.
The reason for this is that the digit 8, in the thousandths place, is between 0 and 10 thousandths:

$$\frac{0}{1000} < \frac{8}{1000} < \frac{10}{1000}$$

but closer to ten thousandths, so we round "up."
Note that 0.36 > 0.3582.

Warm Up a. Round off 0.3548 to the nearest hundredth.

b. Round off 3582.9 to the nearest thousand:

$3582.9 \approx 4000$
↑ (under the 3)

Draw an arrow under the digit 3 in the thousands place.
The digit to the right of the arrow is 5, so we round "up" by adding 3 + 1. In this case three zeros must be written after the digit 4 to keep it in the thousands place.

Note that 4000 > 3582.9.

Warm Up b. Round off 3489.5 to the nearest thousand.

ANSWERS TO WARM UPS (4.5) **a.** 0.35 **b.** 3000

c. Round off 16.3499 to the nearest tenth:

$16.3499 \approx 16.3$
↑

Draw an arrow under the digit 3 in the tenths place.
The digit to the right of the arrow is 4, so we round "down."

CAUTION

Do not round off by working from right to left. Here, the "99" is ignored since only the digit 4, to the right of the arrow is used for rounding

Note that $16.3499 > 16.3$.

Warm Up c. Round off 16.7506 to the nearest tenth.

d. Round 249.7 to the nearest unit:

$249.7 \approx 250$
↑

Draw an arrow under the digit 9 in the units or ones place.
The digit to the right of the arrow is 7, so we round "up" by adding $9 + 1$ or $249 + 1 = 250$.

Note that $250 > 249.7$.

Warm Up d. Round off 370.2 to the nearest unit.

e. Round off 37.2828 and 3.9964 to the nearest unit, the nearest tenth, the nearest hundredth, and the nearest thousandth:

	UNIT	TENTH	HUNDREDTH	THOUSANDTH
37.2828 ≈	37 ≈	37.3 ≈	37.28 ≈	37.283
3.9964 ≈	4 ≈	4.0 ≈	4.00 ≈	3.996

Warm Up e. Round off 12.8947 to the nearest unit, the nearest tenth, the nearest hundredth, and the nearest thousandth.

ANSWERS TO WARM UPS (4.5) **c.** 16.8 **d.** 370 **e.** 13, 12.9, 12.89, 12.895

APPLICATION SOLUTION

f. Sam is reading the instructions about listing deductions on his income tax return. The instructions are to round off each deduction to the nearest whole dollar. If he is going to deduct $832.57 for mortgage interest he paid, what will he enter on his return?

832.57
↑

Draw an arrow under the digit 2 in the units place.

833

The digit after the arrow is 5, so we round "up" by adding 2 + 1.

Therefore, Sam will enter $833 on his return.

Warm Up f. If Sam can deduct $204.48 for charitable contributions, what will he enter on his income tax return?

ANSWERS TO WARM UPS (4.5)

f. $204

NAME ______________________

CLASS ______________ SECTION ______________ DATE ______________

Exercises 4.5

A

ANSWERS

1. ______
2. ______
3. ______
4. ______
5. ______
6. ______
7. ______
8. ______
9. ______
10. ______
11. ______
12. ______
13. ______
14. ______
15. ______
16. ______

Round off each of the following to the nearest unit, tenth, and hundredth:

	UNIT	TENTH	HUNDREDTH
23.461	23	23.5	23.46
1. 321.222			
2. 46.777			
3. 529.655			
4. 8.496			
5. 0.6493			

Round off each of the following to the nearest ten, tenth, and thousandth:

	TEN	TENTH	THOUSANDTH
6. 32.4643			
7. 55.6767			
8. 10.0752			
9. 53.3125			
10. 321.9896			

Round off each of the following to the nearest penny:

11. \$75.6182 **12.** \$8.4167 **13.** \$123.4136

14. \$28.1625 **15.** \$18.915 **16.** \$20.813

B

17. ______
18. ______
19. ______
20. ______
21. ______
22. ______
23. ______
24. ______

Round off each of the following decimals to the nearest tenth, hundredth, and thousandth:

17. 2.6532 **18.** 0.8359 **19.** 12.3015 **20.** 1.3347

21. 9.9892 **22.** 10.0752 **23.** 53.3125 **24.** 9.7765

ANSWERS

25. ____

26. ____

27. ____

28. ____

29. ____

30. ____

31. ____

32. ____

33. ____

34. ____

35. ____

36. ____

25. 2.1789 **26.** 3.0007 **27.** 0.7934 **28.** 14.5553

29. 0.7891 **30.** 1.1139 **31.** 0.9999 **32.** 129.6788

Round off each of the following to the nearest dime:

33. \$75.6182 **34.** \$8.4167 **35.** \$123.4936 **36.** \$28.1625

C

37. ____

38. ____

39. ____

40. ____

41. ____

42. ____

43. ____

44. ____

45. ____

46. ____

Round off each of the following to the nearest hundred and to the nearest hundredth:

37. 543.543 **38.** 756.348 **39.** 3971.244 **40.** 3700.891

41. 49.506

Round off each of the following to the nearest ten and to the nearest tenth:

42. 379.153 **43.** 396.51 **44.** 14.95 **45.** 19.54

46. 4.28

NAME ____________________

CLASS ____________ SECTION ____________ DATE ____________

ANSWERS

47. ____________

48. ____________

49. ____________

50. ____________

51. ____________

52. ____________

53. ____________

54. ____________

55. ____________

56. ____________

57. ____________

58. ____________

59. ____________

60. ____________

61. ____________

62. ____________

63. ____________

64. ____________

Round off each of the following to the nearest thousand and to the nearest thousandth:

47. 23,786.2234 **48.** 4,978.5638 **49.** 965.0348 **50.** 432.1577

51. 9501.9936 **52.** 4198.0955 **53.** 7,688.2673

Round off each of the following to the nearest whole number:

54. 36.58 **55.** 100.495 **56.** 15.50 **57.** 5.399 **58.** 78.905

59. 6234.8167 **60.** 255.7923

Round off to the nearest dollar:

61. $10.89 **62.** $216.27 **63.** $3167.17 **64.** $543.76

D

65. ____________

65. What is the position of the arrow to the nearest tenth?

3.7 3.75 3.8

66. ____________

66. What is the position of the arrow in Exercise 65 to the nearest hundredth?

ANSWERS

67. ______

68. ______

69. ______

70. ______

71. ______

72. ______

73. ______

74. ______

75. ______

Exercises 67 through 70 refer to the figure that follows:

1.62 1.625 1.63

67. What is the position of the arrow to the nearest hundredth?

68. What is the position of the arrow to the nearest thousandth?

69. What is the position of the arrow to the nearest tenth?

70. What is the position of the arrow to the nearest unit?

71. Helen is filing her income tax and wishes to deduct \$150.60 for union dues. If the instructions are to round off to the nearest dollar, how much will she deduct?

72. Michelle plans to deduct the cost of her uniforms when she files her tax return. The total cost of her uniforms was \$662.48. The instructions are to round off to the nearest dollar. How much will she deduct?

73. When the cost of material needed to lay a cement sidewalk is estimated, all measurements are rounded off to the nearest tenth of a foot. What estimation will be used for the length if the rod person measures 482.65 feet?

74. The computer shows that Fran's savings account, with the interest she has earned, has a value of \$145.7888345. Round off the value to the nearest cent.

75. The computer shows that Frank's savings account, with the interest he has earned, has a value of \$132.2506681. Round off the value to the nearest cent.

NAME ______________________________

CLASS ______________ SECTION ______________ DATE ______________

ANSWERS

76. ______________

76. If the Argentine peso is valued at \$0.8051, what is its value to the nearest cent?

77. ______________

77. If an Australian dollar is worth \$0.7157, what is its value to the nearest cent?

78. ______________

78. An ounce of gold is selling on the exchange for \$368.947. What is the price of an ounce, rounded off to the nearest dollar?

79. ______________

79. What is the price of gold in Exercise 78, rounded off to the nearest cent?

E *Maintain your skills* (Sections 3.2, 3.5, 3.9)

80. ______________

81. ______________

Change to a mixed number:

80. $\frac{48}{13}$

81. $\frac{53}{13}$

82. ______________

83. ______________

Change to an improper fraction:

82. $15\frac{5}{7}$

83. $17\frac{2}{3}$

84. ______________

85. ______________

Divide:

84. $\frac{15}{4} \div \frac{5}{8}$

85. $\frac{14}{15} \div \frac{35}{40}$

ANSWERS

86. ________

87. ________

88. ________

89. ________

Add and reduce:

86. $\frac{12}{57} + \frac{7}{57}$

87. $\frac{19}{68} + \frac{13}{68} + \frac{19}{68}$

88. A spring has 12 coils, and each coil requires $1\frac{1}{4}$ in. of wire. How many inches of wire does it take to make one spring?

89. How many springs like those in Exercise 88 can be made from 300 inches of wire?

Addition of Decimals

OBJECTIVE

57. Add decimals.

APPLICATION

What is the total cost of an automobile tire if the retail price is $57.95, the federal excise tax is $2.05, the state sales tax is $3.48, and the local sales tax is $1.16?

VOCABULARY

No new words.

HOW AND WHY

What is the answer to 6.3 + 2.5 = ? We make use of the expanded form to explain addition:

$$\begin{array}{l} 6.3 = 6 \text{ ones} + 3 \text{ tenths} \\ \underline{2.5} = \underline{2 \text{ ones} + 5 \text{ tenths}} \\ \quad\quad\; 8 \text{ ones} + 8 \text{ tenths} = 8.8 \end{array}$$

The vertical form gives us a natural grouping of the ones and tenths. The addition problem 2.8 + 13.4 + 6.22 can be written 2.80 + 13.40 + 6.22, and then:

$$\begin{array}{r} 2.80 \\ 13.40 \\ \underline{6.22} \\ 22.42 \end{array}$$

PROCEDURE

Decimals can be added as if they were whole numbers by writing them in vertical columns. Align the decimal points so that the columns have the same place value.

MODEL PROBLEM-SOLVING

Examples

a. Add: 1.3 + 21.41 + 32 + 0.05

$$\begin{array}{r} 1.30 \\ 21.41 \\ 32.00 \\ \underline{0.05} \\ 54.76 \end{array}$$

Strategy

The numerals 21.41 and 0.05 each have two decimal places, the largest number of decimal places in the four numerals. So, we give each decimal two decimal places by writing zeros where needed. The extra zeros help in lining up the place values.

CAUTION

Be sure to align the decimal points so that the place values will be lined up.

The whole number 32 is the same as 32. or 32.0 or 32.00 because the decimal point always follows the units place.

Warm Up a. Add: 2.4 + 37.52 + 19 + 0.08

b. Add: 1.05 + 0.723 + 72.6 + 8

$$\begin{array}{r} 1.050 \\ 0.723 \\ 72.600 \\ \underline{8.000} \\ 82.373 \end{array}$$

Write each decimal with three decimal places.

The whole number 8 can be written as 8 or 8. or 8.0 or 8.00 or 8.000 because the decimal point follows the units place.

Warm Up b. Add: 7.09 + 0.385 + 37.7 + 12

c. Calculator example

Add: 6.3975 + 0.0116 + 3.410 + 18.624

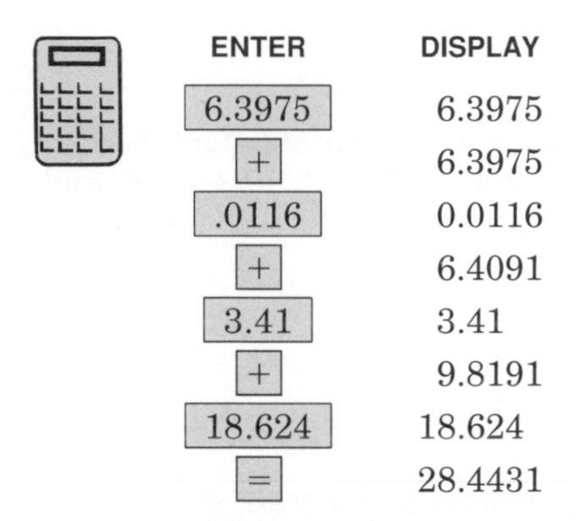

ENTER	DISPLAY
6.3975	6.3975
+	6.3975
.0116	0.0116
+	6.4091
3.41	3.41
+	9.8191
18.624	18.624
=	28.4431

Extra zeros do not need to be written or entered on the calculator. The place values will be added correctly by the calculator.

The sum is 28.4431.

Warm Up c. Add: 8.4086 + 0.0229 + 4.56 + 34.843

APPLICATION SOLUTION

d. What is the total cost of an automobile tire if the retail price is $57.95, the federal excise tax is $2.05, the state sales tax is $3.48, and the local sales tax is $1.16?

Retail price	57.95
Excise tax	2.05
State sales tax	3.48
Local sales tax	1.16
Total	64.64

To find the total cost of the tire, add the retail price and the taxes.

Since the prices are in dollars and cents, every decimal already has the same number of decimal places.

The cost of the tire is $64.64.

Warm Up d. What is the total cost of a pair of emerald earrings if the retail price is $83.95, the federal tax is $3.06, the state sales tax is $2.78, and the city sales tax is $0.67?

ANSWERS TO WARM UPS (4.6) **a.** 59 **b.** 57.175 **c.** 47.8345 **d.** $90.46

NAME ______________________________

CLASS ____________ SECTION ____________ DATE ____________

Exercises 4.6

A

ANSWERS

1. ______
2. ______
3. ______
4. ______
5. ______
6. ______
7. ______
8. ______
9. ______
10. ______
11. ______
12. ______
13. ______
14. ______
15. ______
16. ______
17. ______
18. ______
19. ______
20. ______

Add:

1. $\begin{array}{r} 0.3 \\ \underline{0.4} \end{array}$

2. $\begin{array}{r} 0.5 \\ \underline{0.4} \end{array}$

3. $\begin{array}{r} 0.6 \\ 0.5 \\ \underline{0.7} \end{array}$

4. $\begin{array}{r} 0.8 \\ 0.7 \\ \underline{0.4} \end{array}$

5. $\begin{array}{r} 0.12 \\ \underline{0.34} \end{array}$

6. $\begin{array}{r} 0.86 \\ \underline{0.25} \end{array}$

7. $\begin{array}{r} 0.31 \\ 0.49 \\ \underline{0.28} \end{array}$

8. $\begin{array}{r} 0.11 \\ 0.67 \\ \underline{0.74} \end{array}$

9. $\begin{array}{r} 0.113 \\ 0.121 \\ \underline{0.314} \end{array}$

10. $\begin{array}{r} 0.217 \\ 0.384 \\ \underline{0.276} \end{array}$

11. $\begin{array}{r} 0.194 \\ 0.631 \\ 0.823 \\ \underline{0.654} \end{array}$

12. $\begin{array}{r} 0.0816 \\ 0.5329 \\ 0.1283 \\ \underline{0.0015} \end{array}$

13. $4 + 8.2$

14. $9.1 + 5$

15. $4.68 + 9$

16. $7.23 + 8$

17. $2.61 + 7 + 0.8 + 0.404$

18. $11.34 + 16 + 0.16 + 0.016$

19. $\begin{array}{r} 7.210 \\ 8.600 \\ \underline{16.502} \end{array}$

20. $\begin{array}{r} 23.060 \\ 19.200 \\ \underline{7.000} \end{array}$

B

ANSWERS

21. ______
22. ______
23. ______
24. ______
25. ______
26. ______
27. ______
28. ______
29. ______
30. ______
31. ______
32. ______
33. ______
34. ______
35. ______
36. ______
37. ______
38. ______
39. ______
40. ______

21.
$$\begin{array}{r} 7.1 \\ \underline{2.653} \end{array}$$

22.
$$\begin{array}{r} 4.249 \\ \underline{6.55} \end{array}$$

23.
$$\begin{array}{r} 9.19 \\ 0.38 \\ 12.7 \\ \underline{1.32} \end{array}$$

24.
$$\begin{array}{r} 2.295 \\ 5.84 \\ 4.4 \\ \underline{0.963} \end{array}$$

25. 0.85 + 0.15 + 0.23

26. 7.3 + 8.1 + 6.9

27. 7.5 + 8.3 + 9.2

28. 0.08 + 1.58 + 3.47

29. 0.002 + 0.022 + 0.222 + 2.222

30. 7.007 + 0.707 + 0.077 + 0.007

31. 0.0015 + 1.005 + 5 + 1.051

32. 0.0308 + 0.0805 + 0.5003 + 2

33. 33.368 + 32.411 + 29 + 3.007

34. 54.778 + 57 + 55.0062 + 4.708

35. 0.00067 + 0.0035 + 0.034 + 0.1007

36. 0.00704 + 0.0003 + 0.0886 + 0.041

37.
$$\begin{array}{r} 7.5000 \\ 14.3870 \\ \underline{22.6385} \end{array}$$

38.
$$\begin{array}{r} 10.0100 \\ 122.2210 \\ \underline{6057.7506} \end{array}$$

39.
$$\begin{array}{r} 32.615 \\ 41.82 \\ \underline{17.921} \end{array}$$

40.
$$\begin{array}{r} 52.316 \\ 18.792 \\ \underline{25.364} \end{array}$$

NAME ____________________

CLASS ____________ SECTION ____________ DATE ____________

C

ANSWERS

41. ____________

42. ____________

43. ____________

44. ____________

45. ____________

46. ____________

47. ____________

48. ____________

49. ____________

50. ____________

51. ____________

52. ____________

53. ____________

54. ____________

55. ____________

41.
```
0.8966
4.804
5.61
3.545
```

42.
```
10.78
 8.2
 0.479
 7.15
```

43.
```
6.2921
0.297
0.0599
2.5
```

44.
```
 5.302
 9.01
 0.9479
14.85
```

45.
```
 87.5521
572.63
 98.007
113.98
```

46.
```
213.6
 48.089
117.35
 92.91
```

47. 21.75 + 8 + 6.5 + 1.034

48. 52 + 9.631 + 12.23 + 0.6

49. 9.2 + 5.69 + 4.143 + 1.7

50. 1.904 + 3.33 + 7.9 + 16.63

51. 0.995 + 2.7 + 0.065 + 7.84 + 8.004

52. 126.5 + 8.75 + 13.007 + 15.9

53. 12.2 + 3.72 + 5.14 (Round off the sum to the nearest tenth.)

54. 0.0075 + 0.012 + 0.0009 + 0.003 (Round off the sum to the nearest thousandth.)

55. 32.168 + 75.216 + 12 (Round off the sum to the nearest hundredth.)

ANSWERS

56. ______

57. ______

58. ______

59. ______

60. ______

56. \$75.42 + \$16.76 + \$81.79 (Round off the sum to the nearest dollar.)

57. 26.35 + 72.09 + 8.7 + 61.05 + 142.007 (Round off the sum to the nearest whole number.)

58. 217.3 + 25.97 + 6.935 + 21.64 + 18.5 (Round off the sum to the nearest whole number.)

59. 42.16 + 81.88 + 7.316 (Round off the sum to the nearest tenth.)

60. 7.1653 + 8.3783 + 9.2165 (Round off the sum to the nearest thousandth.)

D

61. ______

62. ______

61. Russ was hitchhiking across the city. The first car to offer him a ride took him 3.8 miles. The second car took him 8.5 miles, and the third car went 7.4 miles before he got out. How many miles did he ride altogether?

62. On a short vacation trip Paul stopped for gas four times. The first time he bought 8.6 gallons. At the second station he bought 14.9 gallons, and at the third he bought 15.4 gallons. At the last stop he bought 13.5 gallons. How much gas did he buy on the trip?

NAME ______________________

CLASS ____________ SECTION ____________ DATE ____________

ANSWERS

63. ____________

63. Heather wrote five checks in the amounts of \$34.85, \$17.37, \$67.48, \$21.10, and \$8.10. She has \$152.29 in her checking account. Does she have enough money deposited to cover the five checks?

64. ____________

64. Mr. Jones made the following purchases: one shirt for \$18.98, one pair of slacks for \$39.95, and one sport coat for \$88. What was the total price he paid for the clothes?

65. ____________

65. Find the total cost of an automobile tire that has a retail price of \$75.84 if the federal excise tax is \$2.45, the cost for balancing is \$4, and the sales tax is \$2.29.

66. ____________

66. What is the total cost of a bag of groceries that contains the following?

Bread	\$0.99
Margarine	\$0.68
Hamburger (1 lb)	\$1.98
Eggs (1 doz)	\$1.02
Coffee (1 lb)	\$3.98

67. ____________

67. Bev received her paycheck. Her "take-home" pay was \$285.64. The check stub indicated that a total of \$184.32 had been deducted from her check. What was the total amount of money she earned that pay period?

68. ____________

68. Sam just bought an advanced electron scale to weigh his ore samples. The quartz weighed 4.1675 g, the coal weighed 5.2166 g, and the tin weighed 2.1679 g. What is the total weight in this collection?

ANSWERS

69. ______

70. ______

69. A trio is driving to their grandmother's house for a holiday gathering. Jasmine drives 100 miles, Juniper drives 125.316 miles, and Jupiter drives 127.8163 miles. How far is it to grandmother's house?

70. On that same trip to grandmother's house (see Exercise 69) Jupiter used 4.7892 gallons of gas, Juniper used 4.817 gallons, and Jasmine used 4.2 gallons. How many gallons of gas did they use?

E *Maintain your skills* (Sections 3.10, 3.12)

71. ______

72. ______

73. ______

74. ______

75. ______

76. ______

77. ______

78. ______

79. ______

80. ______

Add:

71. $\frac{1}{2} + \frac{3}{4} + \frac{1}{8}$

72. $\frac{2}{3} + \frac{5}{12} + \frac{5}{6}$

73. $\frac{11}{16} + \frac{11}{24} + \frac{11}{12}$

74. $\frac{11}{18} + \frac{11}{30} + \frac{11}{20}$

Subtract:

75. $\frac{15}{16} - \frac{3}{8}$

76. $\frac{25}{32} - \frac{5}{8}$

77. $\frac{13}{18} - \frac{7}{12}$

78. $\frac{17}{20} - \frac{7}{12}$

79. How many $1\frac{1}{2}$ in.-long machine bolt blanks can be cut from a piece of stock that is 36 in. long? Make no allowance for waste.

80. A board is $3\frac{5}{6}$ in. thick. If $\frac{1}{16}$ in. is sanded from each side, how thick will the board be?

Subtraction of Decimals

OBJECTIVE

58. Subtract decimals.

APPLICATION

Mark purchased a small radio for $13.89. He gave the clerk a $20 bill to pay for the radio. How much change should he get?

VOCABULARY

No new words.

HOW AND WHY

Consider 6.59 − 2.34. Written vertically in expanded form, we have:

6.59 = 6 ones + 5 tenths + 9 hundredths
2.34 = 2 ones + 3 tenths + 4 hundredths
4.25 = 4 ones + 2 tenths + 5 hundredths

When necessary, we can regroup, or "borrow," as with whole numbers. What is the answer to 6.271 − 3.845 = ? In expanded form we have:

6.271 = 6 ones + 2 tenths + 7 hundredths + 1 thousandth
3.845 = 3 ones + 8 tenths + 4 hundredths + 5 thousandths

By borrowing 1 one (1 one = 10 tenths) and adding it to the tenths and then borrowing 1 hundredth (1 hundredth = 10 thousandths) and adding it to the thousandths, we have:

6.271 = 5 ones + 12 tenths + 6 hundredths + 11 thousandths
3.845 = 3 ones + 8 tenths + 4 hundredths + 5 thousandths
2.426 = 2 ones + 4 tenths + 2 hundredths + 6 thousandths

The examples show the common shortcuts.

PROCEDURE

Decimals can be subtracted as if they were whole numbers by writing them in vertical columns. Align the decimal points; the columns will then be aligned so that they have the same place value.

MODEL PROBLEM-SOLVING

Examples | **Strategy**

a. Subtract: 5.831 − 0.287

5.831
0.287

We need to regroup since we cannot subtract 7 thousandths from 1 thousandth.

$$\begin{array}{r} {\scriptstyle 2}\;{\scriptstyle 11} \\ 5.8\not{3}\,\not{1} \\ \underline{0.2\,8\,7} \end{array}$$

We borrow 1 hundredth from the "3" in the hundredths place to add to the "1" in the thousandths place (1 hundredth = 10 thousandths).

$$\begin{array}{r} \overset{7}{\cancel{8}}\ \overset{\overset{12}{\cancel{2}}}{\cancel{3}}\ \overset{11}{\cancel{1}} \\ 5.\ \ \ \ \ \ \ \ \ \ \\ -\ 0.2\ 8\ 7 \\ \hline 5.5\ 4\ 4 \end{array}$$

Now we cannot subtract 8 hundredths from 2 hundredths, so we borrow 1 tenth from the "8" in the tenths place to add to the "2" in the hundredths place (1 tenth = 10 hundredths).

$$\begin{array}{r} 0.287 \\ 5.544 \\ \hline 5.831 \end{array}$$

Check by addition.

Warm Up a. Subtract: 7.946 − 0.378

b. Subtract: 6.271 − 3.845

$$\begin{array}{r} \overset{5}{\cancel{6}}.\overset{12}{\cancel{2}}\ \overset{6}{\cancel{7}}\ \overset{11}{\cancel{1}} \\ 3.8\ 4\ 5 \\ \hline 2.4\ 2\ 6 \end{array}$$

Each of these numbers has three decimal places, so when we write one under the other, the place values line up.

Warm Up b. Subtract: 9.382 − 5.736

c. Subtract: 6 − 2.94

$$\begin{array}{r} 6.00 \\ 2.94 \\ \hline \end{array}$$

We write the "6" as 6.00 so that both numbers will have the same number of decimal places.

$$\begin{array}{r} \overset{5}{\cancel{6}}.\overset{10}{\cancel{0}}\ \cancel{0} \\ 2.9\ 4 \\ \hline \end{array}$$

We need to borrow to subtract in the hundredths place. Since there is a "0" in the tenths place, we start by borrowing "1" from the ones place (1 one = 10 tenths).

$$\begin{array}{r} \overset{5}{\cancel{6}}.\overset{\overset{9}{\cancel{10}}}{\cancel{0}}\ \overset{10}{\cancel{0}} \\ 2.9\ 4 \\ \hline 3.0\ 6 \end{array}$$

Now borrow 1 tenth to add to the hundredths place, (1 tenth = 10 hundredths) and then subtract.

$$\begin{array}{r} 2.94 \\ 3.06 \\ \hline 6.00 \end{array}$$

Check by addition.

Warm Up c. Subtract: 13 − 7.88

ANSWERS TO WARM UPS (4.7) **a.** 7.568 **b.** 3.646 **c.** 5.12

d. Calculator example

Subtract: 127.9635 − 96.9358

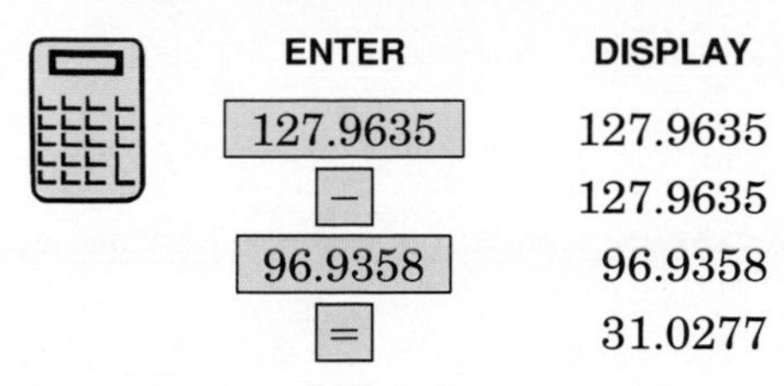

ENTER	DISPLAY
127.9635	127.9635
−	127.9635
96.9358	96.9358
=	31.0277

The difference is 31.0277.

Warm Up d. Subtract: 340.7225 − 278.8269

APPLICATION SOLUTION

e. Mark purchased a small radio for $13.89. He gave the clerk a $20 bill to pay for the radio. How much change should he get?

$$\begin{array}{r} 20.00 \\ \underline{13.89} \\ 6.11 \end{array}$$

To find out how much change Mark should get, we subtract.

Clerks often make change by counting backward, that is, by adding to 13.89 the amount necessary to equal 20:

$13.89 + a penny = $13.90
$13.90 + a dime = $14.00
$14.00 + 6 dollars = $20.00,
or 0.01 + 0.10 + 6.00

Mark should get $6.11 in change.

Warm Up e. Micky purchased a prerecorded video cassette for $18.69. She gave the clerk a $20 bill to pay for the cassette. How much change should she get?

ANSWERS TO WARM UPS (4.7)

d. 61.8956 **e.** $1.31

NAME ______________________________

CLASS ______________ SECTION ______________ DATE ______________

Exercises 4.7

A

ANSWERS

1. ____________
2. ____________
3. ____________
4. ____________
5. ____________
6. ____________
7. ____________
8. ____________
9. ____________
10. ____________
11. ____________
12. ____________
13. ____________
14. ____________
15. ____________
16. ____________
17. ____________
18. ____________
19. ____________
20. ____________
21. ____________
22. ____________

Subtract:

1. $0.9 - 0.6$ **2.** $0.8 - 0.3$ **3.** $2.7 - 2.2$

4. $5.7 - 2.3$ **5.** $0.25 - 0.12$ **6.** $0.89 - 0.63$

7. $\begin{array}{r} 8.31 \\ \underline{2.11} \end{array}$ **8.** $\begin{array}{r} 16.37 \\ \underline{5.12} \end{array}$ **9.** $\begin{array}{r} 9.7 \\ \underline{1.9} \end{array}$

10. $\begin{array}{r} 7.63 \\ \underline{1.49} \end{array}$ **11.** $\begin{array}{r} 13.7 \\ \underline{2.6} \end{array}$ **12.** $\begin{array}{r} 6.28 \\ \underline{1.19} \end{array}$

13. $\begin{array}{r} 2.299 \\ \underline{1.210} \end{array}$ **14.** $\begin{array}{r} 19.05 \\ \underline{12.64} \end{array}$

15. Subtract 8.11 from 16.20. **16.** Subtract 1.22 from 4.63.

17. Subtract 5.29 from 7.73. **18.** Subtract 1.14 from 2.71.

19. $8 - 2.3$ **20.** $5 - 4.16$

21. $23.6 - 17.89$ **22.** $21.3 - 18.62$

B

ANSWERS

23. ____________

24. ____________

25. ____________

26. ____________

27. ____________

28. ____________

29. ____________

30. ____________

31. ____________

32. ____________

33. ____________

34. ____________

35. ____________

36. ____________

37. ____________

38. ____________

39. ____________

40. ____________

Subtract:

23. $\begin{array}{r} 0.612 \\ \underline{0.155} \end{array}$ **24.** $\begin{array}{r} 0.56 \\ \underline{0.17} \end{array}$ **25.** $\begin{array}{r} 3.457 \\ \underline{2.509} \end{array}$

26. $\begin{array}{r} 2.712 \\ \underline{1.148} \end{array}$ **27.** $\begin{array}{r} 7.29 \\ \underline{4.4} \end{array}$ **28.** $\begin{array}{r} 19.01 \\ \underline{2.2} \end{array}$

29. $\begin{array}{r} 9.531 \\ \underline{0.709} \end{array}$ **30.** $\begin{array}{r} 7.303 \\ \underline{0.178} \end{array}$

31. Subtract 9.34 from 12.1.

32. Subtract 0.176 from 8.4.

33. Subtract 3 from 4.08.

34. Subtract 4 from 6.295.

35. 8.164 − 7.215

36. 23.175 − 18.627

37. 6.21 − 4.5

38. 17.86 − 13.92

39. 7.5 − 2.18

40. 18.21 − 15.753

C

41. ____________

42. ____________

43. ____________

44. ____________

45. ____________

46. ____________

Subtract:

41. 9 − 3.257

42. 10 − 3.582

43. $300 − $148.76

44. $500 − $347.23

45. $\begin{array}{r} 121.8 \\ \underline{16.047} \end{array}$

46. $\begin{array}{r} 64.21 \\ \underline{8.0635} \end{array}$

NAME ______________________

CLASS ______________ SECTION ______________ DATE ______________

ANSWERS

47. ______________

48. ______________

49. ______________

50. ______________

51. ______________

52. ______________

53. ______________

54. ______________

55. ______________

56. ______________

57. ______________

58. ______________

59. ______________

60. ______________

47. $\begin{array}{r} 30.97 \\ \underline{6.582} \end{array}$

48. $\begin{array}{r} 11.49 \\ \underline{5.0693} \end{array}$

49. $\begin{array}{r} 7.416 \\ \underline{3.2189} \end{array}$

50. $\begin{array}{r} 72.815 \\ \underline{37.9272} \end{array}$

51. $\begin{array}{r} 32.0405 \\ \underline{17.9999} \end{array}$

52. $\begin{array}{r} 789.0003 \\ \underline{599.8787} \end{array}$

53. 10.9 − 8.359

54. 3.18 − 0.0835

55. 162.365 − 87.496 (Round off the difference to the nearest hundredth.)

56. 569.843 − 378.294 (Round off the difference to the nearest hundredth.)

57. Subtract 2.651 from 3 and round off the difference to the nearest tenth.

58. Subtract 0.0632 from 1 and round off the difference to the nearest tenth.

59. Find the change from a $50 bill if a person spent $18.92.

60. Find the change from a $100 bill after spending $73.27.

D

61. ______________

61. At the beginning of a chemistry experiment Ross had 10.3 cubic centimeters (cc) of a solution. He used 7.6 cc. How much of the solution did he have left?

ANSWERS

62. ______________

63. ______________

64. ______________

65. ______________

66. ______________

67. ______________

68. ______________

69. ______________

70. ______________

62. Jack went shopping with \$50 in cash. He bought a tape for \$8.79 and a sweater for \$21.88. On the way home he bought \$14.75 worth of gas. How much cash did he have left?

63. If \$8.80 is deducted from a marked price of \$88, what is the new price?

64. What is the change from a \$20 bill if a purchase of \$7.68 is made?

65. A mason charges \$802.74 for the completion of a job. His material cost was \$389.61. How much does he receive for his labor?

66. The Build-Em-Up Construction Company makes a bid of \$425,625.62 on a job. If materials and labor will cost a total of \$362,979.84, what profit is the company expecting to make on the project?

67. Donald started his trip with a full tank of gas at 18 gallons. If he used 14.126 gallons on his trip, how many gallons were left?

68. Suppose Donald (see Exercise 67) filled up again for the return trip and had 7.168 gallons left once he got home. How many gallons did he use on the return trip?

69. Muthoni ran a race in 12.16 seconds, whereas Sera ran the same race in 11.382 seconds. How much faster was Sera?

70. Ilga used 7.216 kg of flour in making bread for the school. Jon used 6.8383 kg to make the same number of loaves. How much more did Ilga use?

NAME ______________________________

CLASS ______________ SECTION ______________ DATE ______________

E *Maintain your skills* (Sections 3.6, 3.11, 3.13)

ANSWERS

71. ______________

72. ______________

73. ______________

74. ______________

75. ______________

76. ______________

77. ______________

78. ______________

79. ______________

80. ______________

Add:

71. $3\frac{3}{4} + 2\frac{1}{2} + 8\frac{5}{6} + 4\frac{7}{8}$

72. $5\frac{1}{4} + 1\frac{5}{8} + \frac{1}{2} + 6\frac{7}{8}$

Subtract:

73. $15\frac{2}{3} - 7\frac{5}{6}$

74. $15\frac{5}{6} - 7\frac{2}{3}$

Multiply:

75. $\left(5\frac{2}{5}\right)\left(2\frac{2}{9}\right)$

76. $\left(5\frac{5}{9}\right)\left(9\frac{2}{5}\right)$

Divide:

77. $5\frac{5}{6} \div 4\frac{4}{9}$

78. $4\frac{5}{6} \div 5\frac{4}{9}$

79. A new spool of wire contains $41\frac{1}{2}$ lb of wire. The spool costs $249. What is the cost per pound?

80. Shana uses $\frac{3}{8}$ of her income for payments on her student loan, car, and furniture. If she pays half this amount for her car payment, what fraction of her income goes for the car payment?

▲▲ *Getting Ready for Algebra*

We can now solve equations that involve addition and subtraction of decimals. Again, we perform the inverse operation to solve for the variable. To eliminate addition, subtract the number being added from both sides of the equation. To eliminate subtraction, add the number being subtracted to both sides of the equation.

▲▲ MODEL PROBLEM-SOLVING

Examples / **Strategy**

a. Solve:

$8.6 = x + 3.5$

$8.6 - 3.5 = x + 3.5 - 3.5$

Since 3.5 is added to x, we eliminate the addition by subtracting 3.5 from both sides of the equation. Recall that subtraction is the inverse of addition.

$5.1 = x$

Simplify.

Check:

$8.6 = 5.1 + 3.5$

Substitute 5.1 for x in the original equation and simplify.

$8.6 = 8.6$

The statement is true.

The solution is 5.1.

Warm Up a. Solve: $15.4 = p + 2.9$

b. Solve:

$z - 12.6 = 27.7$

$z - 12.6 + 12.6 = 27.7 + 12.6$

Since 12.6 is subtracted from z, we eliminate the subtraction by adding 12.6 to both sides of the equation. Recall that addition is the inverse of subtraction.

$z = 40.3$

Simplify.

Check:

$40.3 - 12.6 = 27.7$

Substitute 40.3 for z in the original equation and simplify.

$27.7 = 27.7$

The statement is true.

The solution is 40.3.

Warm Up b. Solve: $t - 9.37 = 6.5$

GETTING READY FOR ALGEBRA **a.** $12.5 = p$ **b.** $t = 15.87$

c. Solve:

$$b + 12.875 = 22.4$$
$$b + 12.875 - 12.875 = 22.4 - 12.875$$ **Subtract 12.875 from both sides.**
$$b = 9.525$$ **Simplify.**

Check:

$9.525 + 12.875 = 22.4$ **Substitute 9.525 for b in the original equation and simplify.**

$22.4 = 22.4$ **The statement is true.**

The solution is 9.525.

Warm Up c. Solve: $c + 45.349 = 47$

d. Solve:

$$y - 3.947 = 7.0721$$
$$y - 3.947 + 3.947 = 7.0721 + 3.947$$ **Add 3.947 to both sides.**
$$y = 11.0191$$ **Simplify.**

Check:

$11.0191 - 3.947 = 7.0721$ **Substitute 11.0191 for y in the original equation and simplify.**

$7.0721 = 7.0721$ **The statement is true.**

The solution is 11.0191.

Warm Up d. Solve: $w - 34.87 = 19.6641$

c. $c = 1.651$ **d.** $w = 54.5341$

NAME ________________________________

CLASS ______________ SECTION ______________ DATE ______________

▲▲ EXERCISES

ANSWERS

1. ____________
2. ____________
3. ____________
4. ____________
5. ____________
6. ____________
7. ____________
8. ____________
9. ____________
10. ____________
11. ____________
12. ____________
13. ____________
14. ____________
15. ____________
16. ____________
17. ____________
18. ____________

Solve:

1. $13.8 = x + 2.3$

2. $2.408 = x + 1.7$

3. $y - 0.7 = 14.28$

4. $w - 0.03 = 0.378$

5. $t + 0.05 = 0.123$

6. $x + 11.6 = 367.72$

7. $x - 2.6 = 9.4$

8. $y - 8.1 = 0.33$

9. $2.66 = w + 0.04$

10. $12 = x + 5.3$

11. $t - 7.14 = 0.09$

12. $w - 0.06 = 0.235$

13. $1.56 = a + 0.78$

14. $24.8 = w - 0.65$

15. $2.2 = x - 4.36$

16. $4 = 2.3 + x$

17. $a + 78.4 = 100$

18. $b + 29.76 = 45$

ANSWERS

19. ______________

20. ______________

21. ______________

22. ______________

19. $s - 2.4 = 3.889$

20. $r - 3.8 = 5.6231$

21. $c + 567.8 = 1043.82$

22. $d + 234.87 = 505.1$

Multiplication of Decimals

OBJECTIVE

59. Multiply decimals.

APPLICATION

If exactly eight strips, each 3.875 inches wide, are to be cut from a piece of sheet metal, what is the smallest (in width) piece of sheet metal that can be used?

VOCABULARY

No new words.

HOW AND WHY

The "multiplication table" for decimals is the same as for whole numbers. In fact, decimals are multiplied the same way as whole numbers, with one exception. The exception is in locating the decimal point. To locate the decimal point in (0.3)(0.8), change the decimals to fractions:

DECIMALS	FRACTIONS	PRODUCT OF FRACTIONS	PRODUCT OF DECIMALS
0.3×0.8	$\frac{3}{10} \times \frac{8}{10}$	$\frac{24}{100}$	0.24
11.2×0.07	$\frac{112}{10} \times \frac{7}{100}$	$\frac{784}{1000}$	0.784
7.2×0.13	$\frac{72}{10} \times \frac{13}{100}$	$\frac{936}{1000}$	0.936

The shortcut is to multiply the numbers and insert the decimal point. If necessary, insert zeros so that there are enough decimal places. The product 0.2×0.3 has two decimal places, since tenths multiplied by tenths yields hundredths:

$0.2 \times 0.3 = 0.06$ $\left(\text{Note that } \frac{2}{10} \times \frac{3}{10} = \frac{6}{100}.\right)$

PROCEDURE

To multiply two decimals:

1. **Multiply the numbers as if they were whole numbers.**
2. **Locate the decimal point by counting the number of decimal places (to the right of the decimal point) in both factors. The total of these two counts is the number of decimal places the product must have.**
3. **If necessary, zeros are inserted at the *left of the numeral* so there are enough decimal places (see Example g).**

MODEL PROBLEM-SOLVING

Examples	Strategy
a. Multiply: 0.7×6	
$0.7 \times 6 = 4.2$	**Multiply as if they were whole numbers. The total count of decimal places in both factors is one (1), so there will be one decimal place in the product. Note that tenths times ones is tenths. In this case 7 tenths times 6 ones is 42 tenths (or 4 and 2 tenths).**
Warm Up a. Multiply: 0.8×7	
b. Multiply: 7×0.6	
$7 \times 0.6 = 4.2$	**Again, there is one decimal place.**
Warm Up b. Multiply: 8×0.7	
c. Multiply: 0.7×0.6	
$0.7 \times 0.6 = 0.42$	**This time there are two decimal places, since each factor contains one decimal place. Note that tenths times tenths equals hundredths. In this case 7 tenths times 6 tenths is 42 hundredths.**
Warm Up c. Multiply: 0.8×0.7	
d. Multiply: 11×0.33	
$\begin{array}{r} 0.33 \\ \underline{11} \\ 33 \\ \underline{3\,3} \\ 3.63 \end{array}$	**Since 0.33 has two decimal places and 11 has none, the product has two decimal places.**
Warm Up d. Multiply: 14×0.27	

ANSWERS TO WARM UPS (4.8)
a. 5.6 **b.** 5.6 **c.** 0.56 **d.** 3.78

c. Multiply: 0.53×15

```
0.53
  15
----
2 65
5 3
----
7.95
```

Note that the product has the same number of decimal places as Example d.

Warm Up e. Multiply: 0.23×66

f. Multiply: 2.31(3.4)

```
 2.31
  3.4
-----
  924
6 93
-----
7.854
```

There are three decimal places in the factors (two in 2.31 and one in 3.4), so the product has three decimal places.

Warm Up f. Multiply: 6.2(5.78)

g. Multiply: 0.21(0.14)

```
  0.21
  0.14
------
    84
   21
------
0.0294
```

There are four decimal places in the factors. The product must also have four decimal places.

CAUTION

We must insert a 0 before the 2 because the product must have four decimal places.

Warm Up g. Multiply: 0.09(0.35)

ANSWERS TO WARM UPS (4.8) **e.** 15.18 **f.** 35.836 **g.** 0.0315

h. Calculator example

Multiply: (62.75)(136.492)

ENTER	DISPLAY
62.75	62.75
×	62.75
136.492	136.492
=	8564.873

The product is 8564.873.

Warm Up h. Multiply: (3.456)(82.74)

APPLICATION SOLUTION

i. If exactly eight strips, each 3.875 inches wide, are to be cut from a piece of sheet metal, what is the smallest (in width) piece of sheet metal that can be used?

$$\begin{array}{r} 3.875 \\ 8 \\ \hline 31.000 \end{array}$$

To find the width of the blank piece of metal, we multiply.
Drop the zeros.

Therefore, the blank must be at least 31 inches wide.

Warm Up i. If exactly 12 strips, each 6.45 centimeters wide, are to be cut from a piece of sheet metal, what is the smallest (in width) piece of sheet metal that can be used?

ANSWERS TO WARM UPS (4.8)

h. 285.94944 **i.** 77.4 inches wide

NAME ______________________

CLASS ______________ SECTION ______________ DATE ______________

Exercises 4.8

A

ANSWERS

1. ______
2. ______
3. ______
4. ______
5. ______
6. ______
7. ______
8. ______
9. ______
10. ______
11. ______
12. ______
13. ______
14. ______
15. ______
16. ______
17. ______
18. ______
19. ______
20. ______
21. ______
22. ______
23. ______
24. ______
25. ______

Multiply:

1. $\begin{array}{r} 0.3 \\ \underline{8} \end{array}$ **2.** $\begin{array}{r} 0.8 \\ \underline{3} \end{array}$ **3.** $\begin{array}{r} 0.7 \\ \underline{6} \end{array}$ **4.** $\begin{array}{r} 0.6 \\ \underline{7} \end{array}$

5. $\begin{array}{r} 0.9 \\ \underline{2} \end{array}$ **6.** $\begin{array}{r} 0.2 \\ \underline{9} \end{array}$ **7.** 3×0.09 **8.** 0.03×9

9. $\begin{array}{r} 0.5 \\ \underline{0.7} \end{array}$ **10.** $\begin{array}{r} 0.2 \\ \underline{0.6} \end{array}$ **11.** 0.2×0.3 **12.** 0.4×0.2

13. $\begin{array}{r} 0.21 \\ \underline{4} \end{array}$ **14.** $\begin{array}{r} 0.13 \\ \underline{2} \end{array}$ **15.** $\begin{array}{r} 2.7 \\ \underline{5} \end{array}$ **16.** $\begin{array}{r} 6.4 \\ \underline{7} \end{array}$

17. $\begin{array}{r} 0.05 \\ \underline{0.3} \end{array}$ **18.** $\begin{array}{r} 0.14 \\ \underline{0.7} \end{array}$ **19.** $\begin{array}{r} 1.3 \\ \underline{0.2} \end{array}$ **20.** $\begin{array}{r} 2.4 \\ \underline{0.3} \end{array}$

21. $\begin{array}{r} 3.18 \\ \underline{0.7} \end{array}$ **22.** $\begin{array}{r} 4.12 \\ \underline{0.4} \end{array}$ **23.** 0.26×0.3 **24.** 0.38×0.4

25. 0.56×0.32

B

ANSWERS

26. ________
27. ________
28. ________
29. ________
30. ________
31. ________
32. ________
33. ________
34. ________
35. ________
36. ________
37. ________
38. ________
39. ________
40. ________
41. ________
42. ________
43. ________

26. $\begin{array}{r} 7.38 \\ \underline{0.002} \end{array}$ **27.** $\begin{array}{r} 1.11 \\ \underline{0.023} \end{array}$ **28.** $\begin{array}{r} 1.45 \\ \underline{2.1} \end{array}$ **29.** $\begin{array}{r} 8.97 \\ \underline{0.61} \end{array}$

30. $\begin{array}{r} 34.6 \\ \underline{6.5} \end{array}$ **31.** $\begin{array}{r} 0.272 \\ \underline{3.1} \end{array}$ **32.** $\begin{array}{r} 9.04 \\ \underline{0.47} \end{array}$ **33.** $\begin{array}{r} 7.41 \\ \underline{0.206} \end{array}$

34. $\begin{array}{r} 8.2 \\ \underline{0.68} \end{array}$ **35.** $\begin{array}{r} 7.21 \\ \underline{0.88} \end{array}$ **36.** $\begin{array}{r} 3.32 \\ \underline{45} \end{array}$ **37.** $\begin{array}{r} 6.005 \\ \underline{75} \end{array}$

38. $\begin{array}{r} 0.059 \\ \underline{6.4} \end{array}$ **39.** $\begin{array}{r} 9.8 \\ \underline{0.079} \end{array}$ **40.** $\begin{array}{r} 0.345 \\ \underline{4.8} \end{array}$ **41.** $\begin{array}{r} 5.31 \\ \underline{5.9} \end{array}$

42. 7.02×3.5 **43.** 8.06×5.3

C

44. ________
45. ________
46. ________
47. ________
48. ________
49. ________

44. (0.011)(0.032) **45.** (2.52)(1.37)

46. (162.5)(6.37) **47.** (89.6)(5.37)

48. (99.63)(30.07) **49.** (12.62)(760.02)

NAME ______

CLASS ______ SECTION ______ DATE ______

ANSWERS

50. ______

51. ______

52. ______

53. ______

54. ______

55. ______

56. ______

57. ______

58. ______

59. ______

60. ______

61. ______

50. (0.7)(0.3)(0.4)

51. (0.8)(0.6)(0.2)

52. (1.6)(7)(2.03)

53. (5.06)(3)(2.06)

54. (2.13)(4.06)(3.2)

55. (7.05)(5.21)(4.6)

56. (18)(5.09)(9.3)

57. (3.99)(9.4)(2.1)

58. (8.95)(16.5)(37.65)

59. (1.98)(127.95)(12.88)

60. (2.15)(1.2)(0.32)(12.2)

61. (6.9)(0.69)(6.09)(60.9)

D

62. ______

62. Joe earns $4.97 an hour. How much did he earn during the week if he worked 30.25 hours? (Round off to the nearest hundredth of a dollar.)

63. ______

63. If gasoline costs $1.839 per gallon, how much must Faye pay for 12.7 gallons? (Round off to the nearest cent.)

64. ______

64. Melanie bought six records on sale. The advertised sale price was two records for $15.75. How much did she pay for the six records?

ANSWERS

65. ____________

66. ____________

67. ____________

68. ____________

69. ____________

70. ____________

71. ____________

72. ____________

65. The monthly payment for a car is $125.95. How much is paid in 15 months?

66. Upholstery fabric costs $13.98 a yard. What is the cost of 8.3 yards? (Round off to the nearest penny.)

67. What is the width of a piece of iron if exactly 12 strips, each 2.625 inches wide, can be cut from it with no iron left over?

68. Mr. Bee has a diesel-powered automobile. The state in which he lives charges 7 cents per gallon tax on diesel fuel. Mr. Bee chooses to pay this tax at the end of the year. If during the year he bought 501 gallons of fuel, how much state tax must he pay?

69. George is making uniforms for his son's team. Each shirt takes 3.375 yards of fabric. How much fabric will he need for 18 shirts?

70. A monthly mortgage payment is $623.89. How much is paid in one year?

71. The quarterly (3 months worth) sewer bill is $89.27. How much does the sewer cost for the entire year?

72. A monthly car payment is $249.77. How much is paid in 4 years?

NAME ____________________

CLASS ____________ SECTION ____________ DATE ____________

E *Maintain your skills* (Sections 3.6, 3.11, 3.13, 3.14)

ANSWERS

73. ____________

73. Find the sum of $4\frac{4}{7}$, $8\frac{2}{3}$, and $1\frac{1}{6}$.

74. ____________

74. Find the total of $4\frac{7}{8}$ and $8\frac{4}{7}$.

75. ____________

75. Find the product of $16\frac{2}{3}$ and $4\frac{1}{5}$.

76. ____________

76. Multiply $4\frac{3}{8}$ by 6.

77. ____________

77. Find the difference between $32\frac{7}{8}$ and $16\frac{9}{10}$.

78. ____________

78. Subtract $\frac{13}{18}$ from $3\frac{1}{6}$.

79. ____________

79. Perform the indicated operations: $\frac{5}{3} \cdot \frac{3}{4} - \frac{7}{8} \cdot \frac{4}{35}$

ANSWERS

80. ______

81. ______

82. ______

80. Perform the indicated operations: $\frac{5}{3} - \frac{3}{4} \cdot \frac{7}{8} - \frac{7}{12}$

81. An automobile is $18\frac{1}{2}$ ft long. How much clearance space is left in a garage that is 25 ft long inside, if a bench that is $2\frac{1}{4}$ ft deep is in front of the car?

82. Transit buses $5\frac{3}{4}$ feet wide are to be parked in a car barn so that there is $2\frac{1}{4}$ feet between the buses and the same distance between the end buses and the walls. How many buses can be parked in a row if the barn is $50\frac{1}{4}$ feet wide?

Multiplication and Division by Powers of Ten

OBJECTIVES

60. Multiply a decimal by a power of ten.
61. Divide a decimal by a power of ten.
62. Write a number in scientific notation.
63. Change a number written in scientific notation to place value form.

APPLICATION

A stack of sheets of metal contains 100 sheets. The stack is 6.25 inches high. How thick is each sheet of metal?

VOCABULARY

A *power of 10* is a number that can be written as a product of tens; 10, 100, 1000, 10,000 are powers of ten. In exponent form these are 10^1, 10^2, 10^3, and 10^4. A power of ten can be recognized by looking for the number 10 written with an exponent or a single "1" followed by zeros.

Scientific notation is a special way to write numbers as a product using a number between one and ten and a power of ten. This notation is especially useful for writing very large and very small numbers in a small space. Many numbers are displayed on a calculator using scientific notation.

HOW AND WHY

The shortcut used in Section 1.7 for multiplying and dividing by ten or a power of ten works in a similar way with decimals. Consider the following products:

0.5	0.23	5.67
10	10	10
0	0	0
5 0	2 30	56 70
5.0 = 5	2.30 = 2.3	56.70 = 56.7

Notice in each case that multiplying a decimal by 10 has the effect of moving the decimal point one place to the right.

Since $100 = 10 \cdot 10$, multiplying by 100 is the same as multiplying by 10 two times in succession. So, multiplying by 100 has the effect of moving the decimal point two places to the right. For instance:

$(0.53)(100) = 0.53(10 \cdot 10) = (0.53 \cdot 10) \cdot 10 = 5.3 \cdot 10 = 53$

Since $1000 = 10 \cdot 10 \cdot 10$, the decimal point will move three places to the right when multiplying by 1000. Since $10{,}000 = 10 \cdot 10 \cdot 10 \cdot 10$, the decimal point will move four places to the right when multiplying by 10,000, and so on in the same pattern:

$(0.08321)(10{,}000) = 832.1$

Zeros may have to be placed on the right in order to move the correct number of decimal places:

$(2.3)(1000) = 2.300 = 2300$

Two zeros are placed on the right.

Since multiplying a decimal by 10 has the effect of moving the decimal point one place to the right, dividing a number by 10 must move the decimal point one place to the left. Again, we are using the fact that multiplication and division are inverse operations. Division by 100 will move the decimal point two places to the left, and so on. Thus:

347.1 ÷ 100 = 347.1 = 3.471, and

0.763 ÷ 1000 = 0.000763

Three zeros are placed on the left so that the decimal point may be moved three places to the left.

PROCEDURE

To multiply a number by a power of ten, move the decimal point to the right. The number of places to move is shown by the number of zeros in the power of ten.

To divide a number by a power of ten, move the decimal point to the left. The number of places to move is shown by the number of zeros in the power of ten.

Scientific notation is widely used in science, technology, and industry to write large and small numbers. Every "scientific calculator" has a key for entering numbers in scientific notation. This notation makes it possible for a calculator or computer to deal with much larger or smaller numbers than those that take up eight, nine, or ten spaces on the display of a calculator.

PROPERTY

A number in scientific notation is written as the product of two numbers. The first number is between 1 and 10 (including 1 but not 10) and the second number is a power of ten.

For example:

WORD FORM	PLACE VALUE (NUMERAL FORM)	SCIENTIFIC NOTATION
one million	1,000,000	1×10^6
five billion	5,000,000,000	5×10^9
one trillion, three billion	1,003,000,000,000	1.003×10^{12}

Small numbers are shown by writing the power of ten using a negative exponent. (You will learn more about this when you take a course in algebra.) For now, remember that multiplying by a negative power of ten is the same as *dividing* by a power of ten, which means you will be moving the decimal point to the left:

WORD FORM	PLACE VALUE (NUMERAL FORM)	SCIENTIFIC NOTATION
seven thousandths	0.007	7×10^{-3}
six ten-millionths	0.0000006	6×10^{-7}
fourteen hundred-billionths	0.00000000014	1.4×10^{-10}

The shortcut for multiplying by a power of ten is to move the decimal to the right, and the shortcut for dividing by a power of ten is to move the decimal to the left.

PROCEDURE

To write a number in scientific notation:

1. **Move the decimal point right or left so that only one digit remains to the left of the decimal point. The result will be a number between 1 and 10. If the choice is 1 or 10 itself, use 1.**
2. **Multiply the decimal found in step 1 by a power of ten. The exponent of ten to use is one that will make the new product equal to the original number.**
   a. **If you had to move the decimal to the left, multiply by the same number of tens as the number of places moved.**
   b. **If you had to move the decimal to the right, divide (by writing a negative exponent) by the same number of tens as the number of places moved.**

PROCEDURE

To change from scientific notation to place value (numeral) form:

1. **If the exponent of ten is positive, multiply by as many tens (move the decimal to the right as many places) as the exponent shows.**
2. **If the exponent of ten is negative, divide by as many tens (move the decimal to the left as many places) as the exponent shows.**

For numbers larger than 1:

PLACE VALUE FORM:	12,000	3,400,000	12,300,000,000,000	
NUMBER BETWEEN 1 and 10:	1.2	3.4	1.23	**Move the decimal (which is after the units place) to the left until the number is between 1 and 10 (one digit to the left of the decimal).**
SCIENTIFIC NOTATION:	1.2×10^4	3.4×10^6	1.23×10^{13}	**Multiply each by a power of ten that shows how many places left the decimal moved, or how many places you would have to move to the right to recover the original number.**

For numbers smaller than 1:

PLACE VALUE FORM:	0.000033	0.00000007	0.0000000000345	
NUMBER BETWEEN 1 and 10:	3.3	7.	3.45	**Move the decimal to the right until the number is between 1 and 10.**
SCIENTIFIC NOTATION:	3.3×10^{-5}	7×10^{-8}	3.45×10^{-11}	**Divide each by the power of ten that shows how many places right the decimal moved. Show this division by a negative power of 10.**

It is important to note that scientific notation is not rounding off. The scientific notation has exactly the same value as the original name:

$13.7 = 1.37 \times 10^1$ but $13.7 \neq 1.37 \times 10^{-1}$ because $1.37 \times 10^{-1} = 1.37 \div 10 = 0.137$, and $0.065 = 6.5 \times 10^{-2}$, but $0.065 \neq 6.5 \times 10^2$ because $6.5 \times 10^2 = 6.5 \times 100 = 650$.

MODEL PROBLEM-SOLVING

Examples	Strategy
a. Multiply: 3.828(10)	
3.828(10) = 38.28	**Multiplying by 10 moves the decimal point one place to the right.**
Warm Up a. Multiply: 22.58(10)	
b. Multiply: 0.482(100)	
0.482(100) = 48.2	**Multiplying by 100 moves the decimal point two places to the right.**
Warm Up b. Multiply: 0.017(100)	
c. Multiply: 264.3(1000)	
264.3(1000) = 264,300	**Multiplying by 1000 moves the decimal point three places to the right. Two zeros must be placed on the right.**
Warm Up c. Multiply: 31.45(1000)	
d. Divide: $27.3 \div 10$	
$27.3 \div 10 = 2.73$	**Dividing by 10 moves the decimal one place to the left.**
Warm Up d. Divide: $334 \div 10$	
e. Divide: $3.12 \div 100$	
$3.12 \div 100 = 0.0312$	**Dividing by 100 moves the decimal two places to the left.**
Warm Up e. Divide: $5.267 \div 100$	

ANSWERS TO WARM UPS (4.9)

a. 225.8 **b.** 1.7 **c.** 31,450
d. 33.4 **e.** 0.05267

f. Divide: $47.8 \div 10000$

$47.8 \div 10000 = 0.00478$

Dividing by 10000 moves the decimal four places to the left.

Warm Up f. Divide: $68.935 \div 10000$

g. Write 782,000,000 in scientific notation:

7.82 is between 1 and 10

Move the decimal to the left eight places so the number is between 1 and 10.

$7.82 \times 100{,}000{,}000$ is 782,000,000

Moving the decimal point to the left is the same as dividing by 10 for each place. To keep the values the same, we now multiply by 10 eight times.

$782{,}000{,}000 = 7.82 \times 10^8$

Warm Up g. Write 13,000,000 in scientific notation.

h. Write 0.000000092 in scientific notation:

9.2 is between 1 and 10.

Move the decimal to the right eight places so the number is between 1 and 10.

$9.2 \div 100{,}000{,}000$ is 0.000000092.

Moving the decimal point to the right is the same as multiplying by 10 for each place. To keep the values the same, we now divide by 10 eight times.

$0.000000092 = 9.2 \times 10^{-8}$

Warm Up h. Write 0.00000774 in scientific notation.

APPLICATION SOLUTION

i. A stack of sheets of metal contains 100 sheets. The stack is 6.25 inches high. How thick is each sheet of metal?

$6.25 \div 100 = 0.0625$

To find the thickness of each sheet, divide the height by the number of sheets. To divide by 100, move the decimal two places to the left.

Each sheet is 0.0625 inch thick.

Warm Up i. One hundred sheets of clear plastic would be 0.05 inch thick. How thick is each sheet? (This is the thickness of some household plastic wrap.)

ANSWERS TO WARM UPS (4.9)

f. 0.0068935 **g.** 1.3×10^7 **h.** 7.74×10^{-6} **i.** 0.0005 inch

NAME ______________________

CLASS ______________ SECTION ______________ DATE ______________

Exercises 4.9

A

ANSWERS

1. ______
2. ______
3. ______
4. ______
5. ______
6. ______
7. ______
8. ______
9. ______
10. ______
11. ______
12. ______
13. ______
14. ______
15. ______
16. ______
17. ______
18. ______
19. ______
20. ______
21. ______
22. ______
23. ______
24. ______

Multiply or divide as indicated:

1. $42.5 \div 10$ **2.** $4.67 \div 10$ **3.** $1.83(100)$

4. $6.275(100)$ **5.** $0.8214(1000)$ **6.** $682.34(10)$

7. $276 \div 100$ **8.** $2195 \div 100$ **9.** $2143.61 \div 1000$

10. $217.16 \div 1000$ **11.** $1.86 \div 100$ **12.** $6.72 \div 100$

Write these numbers in scientific notation:

13. 1000 **14.** 100,000 **15.** 0.0001

16. 0.00001 **17.** 1 **18.** 10

Change these numbers to place value form:

19. 1×10^4 **20.** 1×10^7 **21.** 1×10^{-2}

22. 1×10^{-7} **23.** 1×10^2 **24.** 1×10^{-4}

B

ANSWERS

25. ______
26. ______
27. ______
28. ______
29. ______
30. ______
31. ______
32. ______
33. ______
34. ______
35. ______
36. ______
37. ______
38. ______
39. ______
40. ______
41. ______
42. ______
43. ______
44. ______
45. ______
46. ______
47. ______
48. ______
49. ______
50. ______
51. ______
52. ______
53. ______
54. ______

25. 6.274(1000) **26.** 6.72(100) **27.** $1.85 \div 10$

28. $768.2 \div 1000$ **29.** 487(100) **30.** 487(1000)

31. $876.21 \div 1000$ **32.** $149718 \div 100000$ **33.** 4.2756(10000)

34. 2.75(1000) **35.** $36.95 \div 100$ **36.** $756 \div 10000$

37. 26(100000) **38.** 314.5(10000) **39.** $122.31 \div 100$

40. $41.62 \div 100$ **41.** 9.87×10 **42.** 16.75×10

Write these numbers in scientific notation:

43. 700 **44.** 9500 **45.** 0.078

46. 0.002 **47.** 15,000 **48.** 24,000

Change these numbers to place value form:

49. 6×10^4 **50.** 9×10^5 **51.** 1.22×10^{-1}

52. 5.7×10^{-2} **53.** 2.34×10^3 **54.** 4.57×10^2

NAME ______________________________

CLASS ____________ SECTION ____________ DATE ____________

C

ANSWERS

55. ____________
56. ____________
57. ____________
58. ____________
59. ____________
60. ____________
61. ____________
62. ____________
63. ____________
64. ____________
65. ____________
66. ____________
67. ____________
68. ____________
69. ____________
70. ____________
71. ____________
72. ____________
73. ____________
74. ____________
75. ____________
76. ____________
77. ____________
78. ____________

55. $3589.6 \div 10000$ **56.** $0.21 \div 10$ **57.** $62.3 \div 1000$

58. $1.23(10000)$ **59.** $82.5 \div 10000$ **60.** $182.14 \div 1000$

61. $0.00214(10000)$ **62.** $0.000214(100)$ **63.** $0.832 \div 100$

64. $0.068 \div 10000$ **65.** 0.08216×100 **66.** $0.2317 \div 10$

Write these numbers in scientific notation:

67. 70,000 **68.** 7,500,000 **69.** 0.00816

70. 0.00000000044 **71.** 627,000 **72.** 0.0000461

Change these numbers to place value form:

73. 6×10^9 **74.** 3×10^{15} **75.** 4.44×10^{-5}

76. 2.3×10^{-7} **77.** 7.851×10^5 **78.** 3.492×10^{-3}

ANSWERS

79. ______

80. ______

81. ______

82. ______

83. ______

84. ______

85. ______

86. ______

87. ______

88. ______

79. Ken's Shoe Store bought 100 pairs of shoes that cost \$22.29 per pair. What was the total cost of the shoes?

80. If Mae's Shoe Store bought 100 pairs of shoes and the total cost was \$4897.50, what was the cost of each pair of shoes?

81. If Jean traveled 8275 feet in 100 seconds, how far did she travel in 1 second?

82. If Norman drove at the rate of 68.24 feet per second for 100 seconds, how many feet did he travel?

83. Ms. James bought 100 acres of land at a cost of \$985 per acre. What was the total cost of her land?

84. Jamie bought 100 pieces of candy and paid \$29 for them. What was the cost of each piece?

85. If 1000 bricks weigh 5900 pounds, how much does each brick weigh?

86. Mr. Tuck loaded 1000 boxes of paper on his trailer. If each box weighed 12.5 pounds, what was the total weight of the boxes that he loaded?

87. A standard brick together with its mortar joint measures 2.75 inches high. If one brick and one mortar joint make up one course, how high will a wall be that has 100 courses?

88. If 100 courses of brick measure 287.5 inches high, how high is one course?

NAME ______________________

CLASS ____________ SECTION ____________ DATE ____________

ANSWERS

89. ____________

90. ____________

91. ____________

92. ____________

93. ____________

94. ____________

95. ____________

96. ____________

97. ____________

89. A wire of a certain size has a resistance of 1.82 ohms per one thousand feet. What is the resistance in 1 foot of such wire?

90. If a certain wire size has a resistance of 0.00235 ohms per foot, what will be the total resistance of 1000 feet of such wire?

91. The total land area of the Earth is approximately 52,000,000 square miles. What is the total area written in scientific notation?

92. The local computer store offers a small computer with 1152K (1,152,000) bytes of memory. Write the number of bytes in scientific notation.

93. The length of a red light ray is 0.000000072 cm. Write this length in scientific notation.

94. The time it takes light to travel 1 kilometer is approximately 0.0000033 seconds. Write this time in scientific notation.

95. The speed of light is approximately 1.116×10^7 miles per minute. Write this speed in place value form.

96. The Earth is approximately 1.5×10^8 kilometers from the sun. Write this distance in place value form.

97. The shortest wavelength of visible light is approximately 4×10^{-5} centimeters. Write this length in place value form.

E *Maintain your skills* (Sections 3.9, 3.13, 3.14)

ANSWERS

98. ____________

99. ____________

100. ____________

101. ____________

102. ____________

103. ____________

104. ____________

105. ____________

106. ____________

107. ____________

Perform the indicated operations:

98. $\frac{4}{5} + \frac{3}{5} + \frac{7}{5} + \frac{2}{5} + \frac{4}{5}$

99. $\frac{7}{12} + \frac{7}{12} + \frac{1}{12} + \frac{5}{12} + \frac{11}{12}$

100. $\left(12\frac{1}{2}\right)\left(2\frac{3}{4}\right) - \left(9\frac{1}{2}\right)\left(2\frac{3}{4}\right)$

101. $\left(11\frac{3}{8}\right)\left(1\frac{5}{6}\right) - \left(5\frac{3}{8}\right)\left(2\frac{1}{6}\right)$

102. $4\frac{7}{8} \div 2\frac{1}{4} - \frac{2}{3} \cdot \frac{1}{2}$

103. $4\frac{7}{8} + 2\frac{1}{4} \cdot \frac{2}{3} \div \frac{1}{2}$

104. $14\frac{9}{10} - 6\frac{19}{20}$

105. $14\frac{17}{20} - 6\frac{9}{10}$

106. A workman is allowed $18\frac{3}{4}$ hours to complete a job. He works on the job four different times, as follows: $1\frac{3}{4}$ hr, $7\frac{1}{2}$ hr, $4\frac{1}{2}$ hr, and completes the job in another $4\frac{1}{2}$ hr. Was he able to complete the job in the allotted time?

107. A carpet layer is given $5\frac{7}{8}$ hours to complete a job. She has already worked three different times, as follows: $2\frac{1}{4}$ hr, $1\frac{1}{4}$ hr, and $\frac{1}{2}$ hr. How much time does she have to complete the job?

Division of Decimals

OBJECTIVES

64. Divide a decimal by a whole number.
65. Divide two decimals.

APPLICATION

What is the cost per ounce of a 12-ounce can of root beer that costs 45¢? This is called the "unit price" and is used for comparing prices. Many stores are required to show this price for the food they sell.

VOCABULARY

No new words.

HOW AND WHY

Division of decimals is the same as division of whole numbers, with one exception. The exception is in locating the decimal point.

```
    0.0019
20)0.0380
      20
      180
      180
        0
```

When dividing by a whole number, you can correctly place the decimal point for the quotient by writing it above the decimal point in the dividend.

Check:

```
 0.0019
     20
 0.0380
```

It may be necessary to insert zeros to do the division. See Example b.

When a decimal is divided by 7, the division process may not have a remainder of zero at any step:

```
  0.33
7)2.34
  2 1
    24
    21
     3
```

At this step we can write zeros to the right of the digit 4, since:

2.34 = 2.340 = 2.3400 = 2.34000 = 2.340000

```
  0.33428
7)2.34000
  2 1
    24
    21
     30
     28
      20
      14
       60
       56
        4
```

It appears that we might go on inserting zeros and continue endlessly. This is indeed what happens. Such decimals are called "nonterminating, repeating decimals." For example, this one is sometimes written:

$0.33428571428571\ldots$ or $0.33\overline{428571}$

The bar written above the sequence of digits, 428571, indicates that these digits are repeated endlessly. Whenever a fraction is changed to a decimal, we have one of two possibilities:

1. The quotient is exact, as in $\frac{2}{5} = 0.4$. This decimal is a "terminating decimal."
2. The quotient is a nonterminating, repeating decimal, as in $\frac{2}{3} = 0.6666666\ldots = 0.\overline{6}$.

In practical applications we stop the division process one place value beyond the accuracy required by the situation and then round off. Therefore:

$2.34 \div 7 \approx 0.33$ to the nearest hundredth

$2.34 \div 7 \approx 0.3343$ to the nearest ten-thousandth

If the divisor contains a decimal point, we change the problem:

$0.7\overline{)2.338} = 7\overline{)23.38}$ because $\frac{2.338}{0.7} \cdot \frac{10}{10} = \frac{23.38}{7}$

$0.014\overline{)7.8} = 14\overline{)7800}$ because $\frac{7.8}{0.014} \cdot \frac{1000}{1000} = \frac{7800}{14}$

$0.23\overline{)4.865} = 23\overline{)486.5}$

PROCEDURE

To divide two numbers:

1. **If the divisor is not a whole number, move both decimal points to the right the same number of decimal places until the divisor is a whole number. In other words, multiply both the divisor and the dividend by the same power of ten so the divisor is a whole number.**
2. **Place the decimal point in the quotient above the decimal point in the dividend.**
3. **Divide as if both numbers were whole numbers.**
4. **Round off to the given place value. (If no round-off place is given, divide until the remainder is 0 or round off as appropriate in the problem. For instance, in problems with money, round off to the nearest cent.)**

MODEL PROBLEM-SOLVING

Examples | **Strategy**

a. Divide: $13\overline{)13.026}$

```
       1.002
  13)13.026
      13
       0 0
       0 0
        02
        00
         26
         26
          0
```

CAUTION

Write the decimal point for the quotient above the decimal point in the dividend.

The numerals in the answer are lined up in columns, which have the same place value.

```
   1.002
      13
   3 006
  10 02
  13.026
```

Check by multiplying 13 × 1.002.

Warm Up a. Divide: $15\overline{)45.105}$

b. Divide: 1.88 ÷ 8

```
     0.23
  8)1.88
    1 6
      28
      24
       4
```

Here the remainder is not zero, so the division is not complete. We can place a zero on the right (1.880) without changing the value of the dividend and continue dividing.

```
     0.235
  8)1.880
    1 6
      28
      24
       40
       40
        0
```

Note that the quotient (0.235) and the rewritten dividend (1.880) both have three decimal places.

```
  0.235
      8
  1.880
```

Check by multiplying 8 × 0.235.

Warm Up b. Divide: 2.16 ÷ 16

ANSWERS TO WARM UPS (4.10) **a.** 3.007 **b.** 0.135

c. Divide and round off to the nearest hundredth: 486.5 ÷ 23

```
      21.152
23)486.500
    46
     26
     23
      3 5
      2 3
      1 20
      1 15
         50
         46
          4
```

It is necessary to place two zeros on the right in order to round off to the hundredths place.

Hence, 486.5 ÷ 23 ≈ 21.15.

Warm Up c. Divide and round off to the nearest hundredth: 241.3 ÷ 21

d. Divide: $0.7\overline{)1.32}$

$7\overline{)13.2}$

First, multiply both divisor and dividend by 10 or (to achieve the same result) move both decimal places one place to the right:

$$\frac{1.32}{0.7} \times \frac{10}{10} = \frac{13.2}{7}$$

```
   1.8
7)13.2
  7
  6 2
  5 6
    6
```

```
   1.885
7)13.200
  7
  6 2
  5 6
    60
    56
     40
     35
      5
```

The number of zeros you place on the right depends on either the directions for rounding or your own choice of the number of decimal places. Here we find the approximate quotient rounded off to the nearest hundredth.

Hence, 1.32 ÷ 0.7 ≈ 1.89.

Warm Up d. Divide. Round off the quotient to the nearest hundredth: $0.7\overline{)2.48}$

ANSWERS TO WARM UPS (4.10) **c.** 11.49 **d.** 3.54

e. Divide: $82 \div 0.16$

```
0.16)82

    512.5
16)8200.0
   80
    20
    16
     40
     32
      8 0
      8 0
```

Move the decimal two places to the right. Remember that 82 = 82. = 82.00.

We write a zero on the right (8200 = 8200.0) to complete the division.

The quotient is 512.5.

Warm Up e. Divide: $82 \div 0.8$

f. Find the quotient correct to the nearest thousandth: $0.47891 \div 0.072$

```
0.072)0.47891

      6.6515
72)478.9100
   432
    46 9
    43 2
     3 71
     3 60
       110
        72
        380
        360
         20
```

Move both decimals three places to the right; that is, multiply both by 1000.

Since we will round off to the thousandth place, we carry the division to one place past thousandths, that is, to four places.

Hence, $0.47891 \div 0.072 \approx 6.652$

Warm Up f. Find the quotient correct to the nearest thousandth:

$0.75593 \div 0.043$

ANSWERS TO WARM UPS (4.10) **e.** 102.5 **f.** 17.580

g. Calculator example

Divide and round off to the nearest thousandth: $78.1936 \div 8.705$

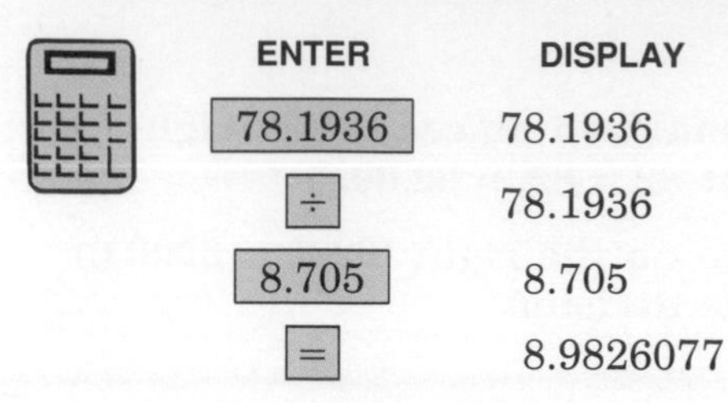

ENTER	DISPLAY
78.1936	78.1936
÷	78.1936
8.705	8.705
=	8.9826077

The quotient is 8.983, to the nearest thousandth.

Warm Up g. Divide and round off to the nearest thousandth: $103.843 \div 4.088$

APPLICATION SOLUTION

h. What is the cost per ounce of a 12-ounce can of root beer that costs 45¢? This is called the "unit price" and is used for comparing prices. Many stores are required to show this price for the food they sell.

$$\begin{array}{r} 3.75 \\ 12\overline{)45.00} \\ \underline{36} \\ 9\,0 \\ \underline{8\,4} \\ 60 \\ \underline{60} \end{array}$$

To find the "unit price" (cost per ounce), we divide the cost by the number of ounces.

The root beer costs 3.75 cents per ounce.

Warm Up h. What is the unit price of potato chips if a 7-ounce bag costs $1.26?

ANSWERS TO WARM UPS (4.10)

g. 25.402 **h.** 18 cents per ounce

NAME ______

CLASS ______ SECTION ______ DATE ______

Exercises 4.10

A

ANSWERS

1. ______
2. ______
3. ______
4. ______
5. ______
6. ______
7. ______
8. ______
9. ______
10. ______
11. ______
12. ______
13. ______
14. ______
15. ______
16. ______

Divide:

1. $7\overline{)3.5}$ **2.** $6\overline{)4.8}$ **3.** $2\overline{)53.28}$

4. $4\overline{)14.88}$ **5.** $4\overline{)0.544}$ **6.** $4\overline{)2.95}$

7. $7.21 \div 0.4$ **8.** $9.21 \div 0.3$

9. $8 \div 0.05$ **10.** $6 \div 0.25$

Divide and round off to the nearest tenth:

11. $7\overline{)8.96}$ **12.** $8\overline{)0.896}$ **13.** $0.02\overline{)0.0034}$

14. $0.03\overline{)0.8226}$ **15.** $1.2\overline{)10.872}$ **16.** $0.2\overline{)9.19}$

ANSWERS

17. ______

18. ______

19. ______

20. ______

21. ______

22. ______

23. ______

24. ______

25. ______

26. ______

17. $0.1725 \div 0.15$

18. $0.6426 \div 0.03$

Divide and round off to the nearest hundredth:

19. $0.21\overline{)9.19}$

20. $0.94\overline{)0.0481}$

21. $3\overline{)13.7}$

22. $9\overline{)2.202}$

23. $1.3\overline{)534}$

24. $1.5\overline{)7.36}$

25. $17.7 \div 7.4$

26. $0.039 \div 7.5$

B

27. ______

28. ______

29. ______

30. ______

31. ______

32. ______

33. ______

34. ______

Divide:

27. $12\overline{)38.4}$

28. $25\overline{)12.3}$

29. $80\overline{)1008}$

30. $80\overline{)104.8}$

31. $16.64 \div 32$

32. $3.936 \div 32$

33. $0.1664 \div 3.2$

34. $0.3936 \div 3.2$

NAME ____________________

CLASS ____________ SECTION ____________ DATE ____________

ANSWERS

35. ____________

36. ____________

37. ____________

38. ____________

39. ____________

40. ____________

41. ____________

42. ____________

43. ____________

44. ____________

45. ____________

46. ____________

47. ____________

48. ____________

49. ____________

50. ____________

35. $0.5418 \div 4.3$

36. $14.42 \div 3.5$

37. $32.43 \div 4.6$

38. $35.09 \div 5.8$

Divide and round off to the nearest thousandth:

39. $1.09\overline{)6.48}$

40. $0.28\overline{)0.11247}$

41. $2.2\overline{)34.22}$

42. $24.9\overline{)60.363}$

43. $131\overline{)29}$

44. $36\overline{)1.975}$

45. $1.0374 \div 0.23$

46. $3.433 \div 1.45$

47. $0.12\overline{)64}$

48. $0.17\overline{)22}$

49. $17 \div 0.043$

50. $29 \div 0.081$

C

ANSWERS

51. ____________

52. ____________

53. ____________

54. ____________

55. ____________

56. ____________

57. ____________

58. ____________

59. ____________

60. ____________

61. ____________

62. ____________

63. ____________

64. ____________

65. ____________

Divide:

51. $64\overline{)6211.84}$

52. $32\overline{)201.824}$

53. $16\overline{)1921.096}$

54. $1.9656 \div 84$

55. $5201.6 \div 320$

56. $3610.32 \div 112$

57. $2.2905 \div 0.45$

58. $2.8872 \div 0.36$

59. $0.68685 \div 0.057$

Divide and round off to the nearest tenth:

60. $0.029\overline{)0.226}$

61. $3.46\overline{)56.99}$

62. $67\overline{)983.7}$

63. $0.347\overline{)15}$

64. $33 \div 0.092$

65. $16 \div 0.176$

NAME ______________________

CLASS ______________ SECTION ______________ DATE ______________

ANSWERS

66. ______________

67. ______________

68. ______________

69. ______________

70. ______________

71. ______________

72. ______________

73. ______________

74. ______________

75. ______________

66. $18.3 \div 0.21$

67. $27.8 \div 0.35$

68. $0.12 \div 0.007$

69. $0.15 \div 0.008$

Divide and round off to the nearest unit (one).

70. $0.142\overline{)3.6039}$

71. $0.033\overline{)1.1982}$

72. $0.0046\overline{)0.08919}$

73. $0.389\overline{)11.098}$

74. $0.452 \div 0.0281$

75. $0.365 \div 0.0321$

D

76. ______________

76. Find the unit price (the cost per pound) of 2.3 pounds of chuck steak that cost $3.63. Round off to the nearest cent.

77. ______________

77. Find the unit price (the cost per ounce) of a 16-ounce package of tortilla chips that costs $1.49. Round off to the nearest tenth of a cent.

ANSWERS

78. ____________________

79. ____________________

80. ____________________

81. ____________________

82. ____________________

83. ____________________

84. ____________________

78. Find the unit price (the cost per pound) of a 3-pound can of coffee that costs $8.29. Round off to the nearest cent.

79. Find the unit price (the cost per ounce) of a 32-ounce block of cheese that costs $5.49. Round off to the nearest tenth of a cent.

80. Eighty alumni of Tech U. donated $3561.60 to the university. What was the average donation?

81. If you drove a car 1363.2 miles on 32 gallons of gas, what was the average number of miles per gallon?

82. Vern bought a pair of red socks. The socks were on sale at three pairs for $12.50. How much did he pay for the socks?

83. Ross drove 214.6 miles on 11.3 gallons of gas. What was his mileage (miles per gallon)? Round off to the nearest mile per gallon.

84. A 55-gallon drum of cleaning solvent in an auto repair shop is being used at the rate of 1.5 gallons per day. At this rate, how many days will the drum last? (Round off to the nearest day.)

NAME ______________________________

CLASS ______________ SECTION ______________ DATE ______________

ANSWERS

85. ______________

86. ______________

87. ______________

88. ______________

89. ______________

90. ______________

85. It is known that a motor-run electric generator uses 0.75 gallon of gasoline per hour. How many hours can the generator be used if the gasoline tank will hold 13 gallons? (Round off to the nearest tenth.)

86. What is the cost per pound of a 5-pound box of candy that costs $12.45?

87. If 19 dry-cell batteries connected in series (so that their voltages add) yield 28.5 volts, what is the yield of each battery?

88. The Williams Construction Company uses cable that weighs 3.5 pounds per foot. A partly filled spool of the cable is weighed. It is found that the cable itself weighs 813 pounds after taking off for the weight of the spool. To the nearest foot, how many feet of cable are on the spool?

89. A partly filled spool contains 1348 pounds of cable. If the cable weighs 2.3 pounds per foot, how many feet of cable remain on the spool? (Round off to the nearest foot.)

90. If a box of 144 pencils costs $19.20, what is the cost of one pencil (rounded off to the nearest cent)?

ANSWERS

91. ______

92. ______

93. ______

94. ______

95. ______

96. ______

97. ______

98. ______

99. ______

100. ______

101. ______

91. How many auto repair stalls can be put into a space that is 102 feet long if each stall requires 8.5 feet?

E *Maintain your skills* (Sections 3.11, 3.13, 3.14)

Perform the indicated operations:

92. $2\frac{1}{4} - \frac{3}{4} + 1\frac{5}{8}$

93. $3 + 2\frac{3}{4} - 1\frac{5}{16}$

94. $7\frac{5}{8} + 12\frac{1}{2} - 8\frac{3}{4}$

95. $5\frac{5}{8} - 2\frac{3}{16} + 4\frac{1}{4}$

96. $17 - 5\frac{7}{9}$

97. $23 - 19\frac{15}{16}$

98. Find the average of $\frac{2}{3}$, $\frac{5}{8}$, and $\frac{2}{9}$.

99. Find the average of $1\frac{1}{4}$, $2\frac{2}{5}$, $3\frac{3}{8}$, and $4\frac{7}{10}$.

100. What is the total length of angle iron needed to make 12 braces that each use $8\frac{3}{4}$ in. of metal?

101. Mr. Lewis bought 350 books for \$60 at an auction. He sold two-fifths of them for \$25, 25 books at \$1.50 each, and 45 books at \$1 each and gave away the rest. How many books did he give away? What was his total profit if his handling cost was \$15?

▲▲ *Getting Ready for Algebra*

We can now solve equations that involve multiplication and division of decimals. Again, we perform the inverse operation to solve for the variable. To eliminate division (which usually occurs in fraction form), multiply by the number, which is the divisor, on both sides of the equation. To eliminate multiplication, divide both sides of the equation by the number being multiplied.

▲▲ MODEL PROBLEM-SOLVING

Examples	Strategy
a. Solve:	
$1.5x = 43.5$ $\frac{1.5x}{1.5} = \frac{43.5}{1.5}$	**Since 1.5 is multiplied times *x*, we eliminate the multiplication by dividing both sides by 1.5. The division is traditionally written in fraction form.**
$x = 29$	**Simplify.**
Check:	
$1.5(29) = 43.5$	**Substitute 29 for *x* in the original equation and simplify.**
$43.5 = 43.5$	**The statement is true.**
The solution is 29.	

Warm Up a. Solve: $1.2t = 84$

b. Solve:	
$18.2 = \frac{a}{12.7}$	
$12.7(18.2) = (12.7)\frac{a}{12.7}$	**Since *a* is divided by 12.7, we eliminate the division by multiplying both sides by 12.7.**
$231.14 = a$	**Simplify.**
Check:	
$18.2 = \frac{231.14}{12.7}$	**Substitute 231.14 for *a* in the original equation and simplify.**
$18.2 = 18.2$	**The statement is true.**
The solution is 231.14.	

GETTING READY FOR ALGEBRA

a. $t = 70$

Warm Up b. Solve: $10.9 = \frac{r}{6.8}$

c. Solve: $22.3y = 62.44$

$$\frac{22.3y}{22.3} = \frac{62.44}{22.3}$$ **Divide both sides by 22.3 to eliminate the multiplication.**

$$y = 2.8$$

Check:

$(22.3)(2.8) = 62.44$ **Substitute 2.8 for y in the original equation.**

$62.44 = 62.44$ **The statement is true.**

The solution is 2.8.

Warm Up c. Solve: $0.075a = 1.065$

d. Solve: $\frac{c}{0.234} = 1.2$

$$(0.234)\frac{c}{0.234} = (0.234)(1.2)$$ **Multiply both sides by 0.234 to eliminate the division.**

$$c = 0.2808$$ **Simplify.**

Check:

$\frac{0.2808}{0.234} = 1.2$ **Substitute 0.2808 for c in the original equation and simplify.**

$1.2 = 1.2$ **The statement is true.**

The solution is 0.2808.

Warm Up d. Solve: $\frac{x}{0.508} = 3.5$

b. $74.12 = r$ **c.** $a = 14.2$ **d.** $x = 1.778$

NAME ______________________

CLASS ____________ SECTION ____________ DATE ____________

▲▲ EXERCISES

ANSWERS

1. ________
2. ________
3. ________
4. ________
5. ________
6. ________
7. ________
8. ________
9. ________
10. ________
11. ________
12. ________
13. ________
14. ________
15. ________

Solve:

1. $2.3x = 13.8$ **2.** $1.7x = 0.408$ **3.** $0.07y = 14.28$

4. $0.03w = 0.378$ **5.** $0.123 = 2.05t$ **6.** $367.72 = 11.6x$

7. $1.1m = 0.044$ **8.** $0.004p = 8$ **9.** $0.016q = 7$

10. $6 = 0.004w$ **11.** $8 = 0.016h$ **12.** $\frac{x}{2.6} = 9.4$

13. $\frac{y}{8.1} = 0.33$ **14.** $0.03 = \frac{b}{0.23}$ **15.** $0.215 = \frac{c}{0.48}$

ANSWERS

16. ______

17. ______

18. ______

19. ______

20. ______

21. ______

22. ______

16. $\frac{w}{0.04} = 2.66$

17. $0.0425 = \frac{x}{23}$

18. $0.09 = \frac{t}{7.14}$

19. $\frac{w}{0.06} = 0.235$

20. $\frac{y}{12.3} = 1.07$

21. $\frac{z}{14.5} = 2.08$

22. $\frac{c}{9.07} = 1.003$

Fractions to Decimals

OBJECTIVES

66. Change fractions and mixed numbers to decimals when the denominator of the fraction contains only 2's and/or 5's as prime factors.

67. Write approximate decimals for fractions.

APPLICATION

Jan needs to make a pattern of the shape shown below. Her ruler is marked in tenths. Change all the measurements to tenths so that she can make an accurate pattern:

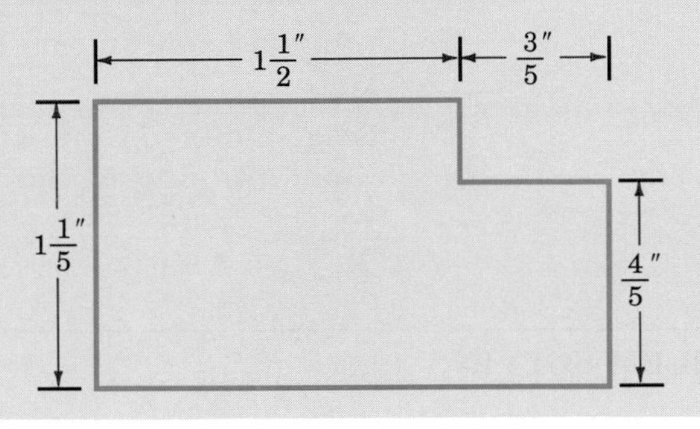

VOCABULARY

No new words.

HOW AND WHY

Every decimal can be written as a whole number times the place value of the last digit on the right:

$$0.63 = 63 \times \frac{1}{100} = \frac{63}{100}$$

The fraction has a power of ten for the denominator. Any fraction that has only prime factors of 2 and 5 in the denominator can be written as a decimal:

$$\frac{2}{5} = \frac{2}{5} \cdot \frac{2}{2} = \frac{4}{10} = 4 \times \frac{1}{10} = 0.4$$

$$\frac{1}{20} = \frac{1}{20} \cdot \frac{5}{5} = \frac{5}{100} = 5 \times \frac{1}{100} = 0.05$$

The power of ten in the denominator tells the number of decimal places in the result:

$\frac{1}{10}$ One decimal place (0.1)

$\frac{1}{100}$ Two decimal places (0.01)

$\frac{1}{1000}$ Three decimal places (0.001)

Every fraction can be thought of as a division problem $\left(\frac{2}{5} = 2 \div 5\right)$.

Therefore, a second method for changing fractions to decimals is division. As you discovered in the previous section, many division problems with decimals do not "divide evenly." If the denominator of the fraction has prime factors other than 2's or 5's, it is probable that the quotient is a nonterminating decimal. The fraction one-third is an example:

$$\frac{1}{3} = 0.3333333\ldots \quad \text{or} \quad 0.\overline{3}$$

The bar over the 3 indicates that the decimal repeats the number 3 forever. In the exercises for this section, round off the division to the desired decimal place.

PROCEDURE

To change a fraction to a decimal, divide the numerator by the denominator.

To change a mixed number to a decimal, change the fraction part to a decimal and add to the whole-number part.

MODEL PROBLEM-SOLVING

Examples | **Strategy**

a. Change $\frac{7}{20}$ to a decimal:

$$\begin{array}{r} 0.35 \\ 20\overline{)7.00} \\ \underline{6\,0} \\ 1\,00 \\ \underline{1\,00} \\ 0 \end{array}$$

Divide the numerator, 7, *by* the denominator, 20.

$$\frac{7}{20} = 0.35$$

Warm Up a. Change $\frac{9}{20}$ to a decimal.

b. Change the mixed number $7\frac{3}{50}$ to a decimal:

$$\frac{3}{50} = \frac{3}{50} \cdot \frac{2}{2} = \frac{6}{100} = 0.06$$

So, $7\frac{3}{50} = 7 + 0.06 = 7.06$

A fraction with a denominator that has only 2's and 5's for prime factors can be changed easily to a decimal by building (instead of dividing).

ANSWERS TO WARM UPS (4.11)

a. 0.45

Warm Up b. Change $3\frac{4}{25}$ to a decimal.

c. Change the fraction $\frac{9}{16}$ to a decimal:

$$
\begin{array}{r}
0.5625 \\
16\overline{)9.0000} \\
\underline{8\ 0} \\
1\ 00 \\
\underline{96} \\
40 \\
\underline{32} \\
80 \\
\underline{80} \\
0
\end{array}
$$

Divide the numerator by the denominator. (This fraction can also be changed by building, but we usually divide when the decimal is not easily recognized.)

$$\frac{9}{16} = \frac{9}{16} \cdot \frac{625}{625} = \frac{5625}{10{,}000}$$

$\frac{9}{16} = 0.5625$

Warm Up c. Change $\frac{9}{40}$ to a decimal.

d. Write $\frac{7}{12}$ as a decimal correct to the nearest hundredth:

$$
\begin{array}{r}
0.583 \\
12\overline{)7.000} \\
\underline{6\ 0} \\
1\ 00 \\
\underline{96} \\
40 \\
\underline{36} \\
4
\end{array}
$$

Divide 7 by 12. Carry out the division to three decimal places and then round off to the nearest hundredth.

$\frac{7}{12} \approx 0.58$ to the nearest hundredth.

Warm Up d. Write $\frac{2}{7}$ as a decimal correct to the nearest hundredth.

ANSWERS TO WARM UPS (4.11) **b.** 3.16 **c.** 0.225 **d.** 0.29

e. Write $\frac{49}{55}$ as a decimal correct to the nearest tenth:

$$\begin{array}{r} 0.89 \\ 55\overline{)49.00} \\ \underline{44\ 0} \\ 5\ 00 \\ \underline{4\ 95} \\ 5 \end{array}$$

Divide 49 by 55. Carry out the division to two decimal places and then round off to the nearest tenth.

$\frac{49}{55} \approx 0.9$ to the nearest tenth.

Warm Up e. Write $\frac{7}{8}$ as a decimal correct to the nearest tenth.

f. Calculator example

Find $\frac{49}{69}$ to the nearest thousandth:

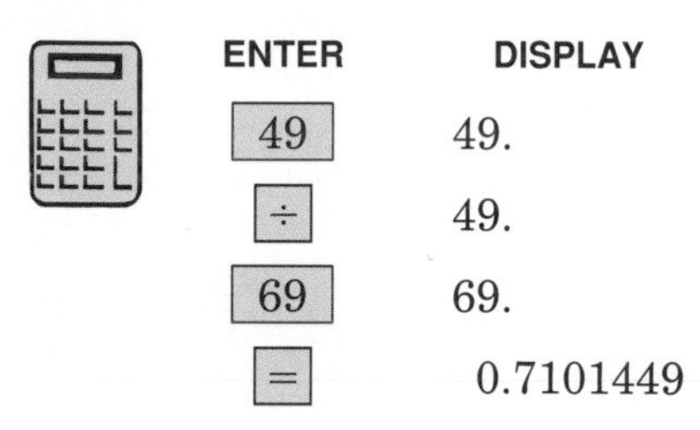

ENTER	DISPLAY
49	49.
÷	49.
69	69.
=	0.7101449

So, $\frac{49}{69} \approx 0.710$ to the nearest thousandth.

Warm Up f. Write $\frac{1}{7}$ as a decimal correct to the nearest thousandth.

APPLICATION SOLUTION

g. Jan needs to make a pattern of the shape shown below. Her ruler is marked in tenths. Change all the measurements to tenths so that she can make an accurate pattern:

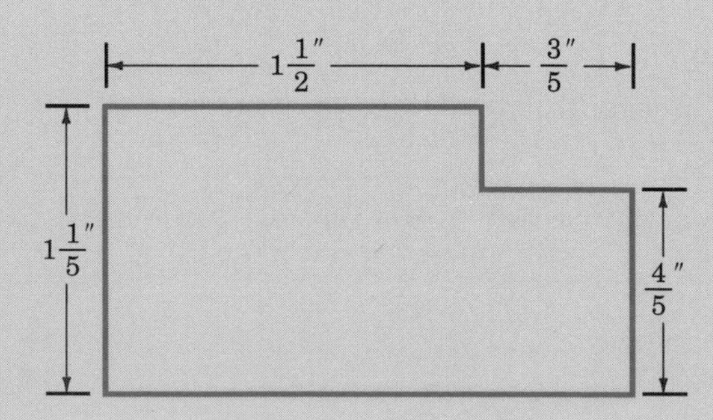

ANSWERS TO WARM UPS (4.11) **e.** 0.9 **f.** 0.143

So that Jan can measure more accurately, each fraction is changed to a decimal to the nearest tenth. So:

$1\frac{1}{2} = 1\frac{5}{10} = 1.5$

$\frac{3}{5} = \frac{6}{10} = 0.6$

$1\frac{1}{5} = 1\frac{2}{10} = 1.2$

$\frac{4}{5} = \frac{8}{10} = 0.8$

Each fraction or mixed number can be changed to a decimal by either building each fraction to a denominator of 10 (as shown at the left) or by dividing the numerator by the denominator. The measurements on the drawing can be labeled:

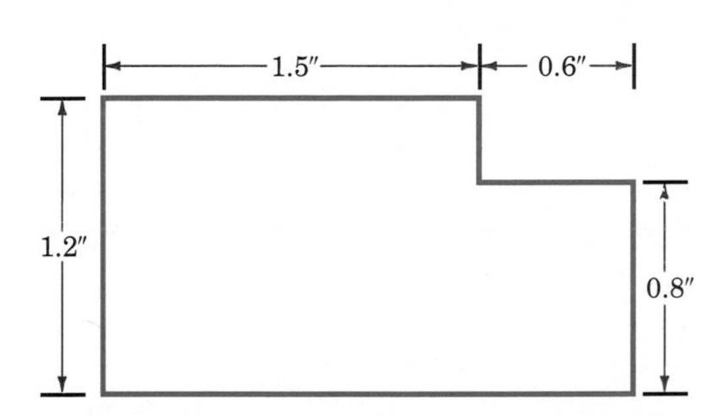

Warm Up g. Change the measurements shown on the shape below to tenths for use with a ruler marked in tenths:

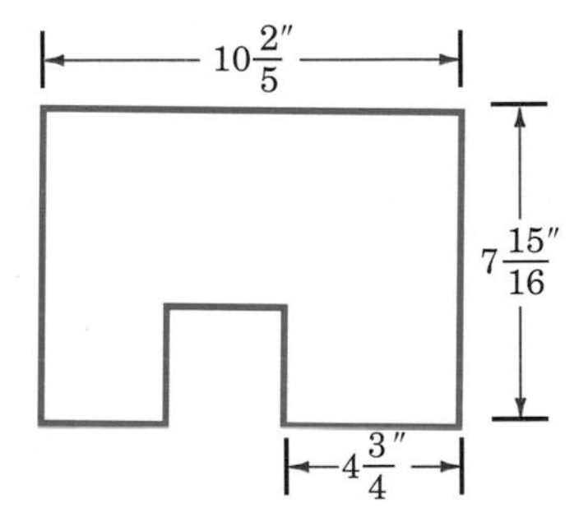

ANSWERS TO WARM UPS (4.11)

g. $10\frac{2}{5}$ in. = 10.4 in.; $7\frac{15}{16}$ in. = 7.9375 in. ≈ 7.9 in.; $4\frac{3}{4}$ in. = 4.75 in. ≈ 4.8 in.

NAME ______

CLASS ______ SECTION ______ DATE ______

Exercises 4.11

A

ANSWERS

1. ______
2. ______
3. ______
4. ______
5. ______
6. ______
7. ______
8. ______
9. ______
10. ______
11. ______
12. ______
13. ______
14. ______
15. ______
16. ______
17. ______

Change each of the following to decimals:

1. $\frac{1}{8}$ **2.** $\frac{1}{4}$ **3.** $\frac{3}{8}$

4. $\frac{5}{8}$ **5.** $\frac{11}{16}$ **6.** $\frac{13}{16}$

7. $\frac{9}{20}$ **8.** $\frac{3}{4}$ **9.** $\frac{1}{32}$

10. $\frac{1}{16}$ **11.** $\frac{7}{16}$

Find the approximate decimal to the nearest indicated place value:

	TENTH	HUNDREDTH
12. $\frac{3}{7}$		
13. $\frac{1}{9}$		
14. $\frac{2}{9}$		
15. $\frac{4}{7}$		
16. $\frac{1}{11}$		
17. $\frac{5}{11}$		

B

ANSWERS

18. ______

19. ______

20. ______

21. ______

22. ______

23. ______

24. ______

25. ______

26. ______

27. ______

28. ______

29. ______

30. ______

31. ______

32. ______

33. ______

34. ______

35. ______

36. ______

37. ______

38. ______

Change each of the following to decimals:

18. $\frac{11}{20}$ **19.** $\frac{13}{20}$ **20.** $\frac{3}{125}$

21. $\frac{99}{125}$ **22.** $\frac{9}{40}$ **23.** $\frac{19}{40}$

24. $\frac{11}{50}$ **25.** $\frac{51}{60}$ **26.** $\frac{57}{80}$ **27.** $\frac{80}{25}$ **28.** $\frac{12}{5}$

Find the approximate decimal to the nearest indicated place value:

	TENTH	HUNDREDTH
29. $\frac{5}{6}$		
30. $\frac{7}{9}$		
31. $\frac{4}{11}$		
32. $\frac{2}{13}$		
33. $\frac{3}{14}$		
34. $\frac{17}{18}$		
35. $\frac{8}{15}$		
36. $\frac{9}{17}$		
37. $\frac{11}{19}$		
38. $\frac{18}{23}$		

NAME ______________________________

CLASS ______________ SECTION ______________ DATE ______________

C

ANSWERS

39. ________
40. ________
41. ________
42. ________
43. ________
44. ________
45. ________
46. ________
47. ________
48. ________
49. ________
50. ________
51. ________
52. ________
53. ________
54. ________
55. ________
56. ________
57. ________
58. ________
59. ________
60. ________

Change each of the following to decimals:

39. $\frac{187}{200}$ **40.** $\frac{333}{400}$ **41.** $\frac{189}{200}$ **42.** $9\frac{4}{5}$

43. $3\frac{5}{8}$ **44.** $17\frac{13}{40}$ **45.** $15\frac{82}{125}$ **46.** $22\frac{147}{250}$

47. $33\frac{313}{400}$ **48.** $6\frac{293}{500}$ **49.** $\frac{213}{500}$ **50.** $\frac{393}{400}$

Find the approximate decimal to the nearest indicated place value:

	TENTH	HUNDREDTH	THOUSANDTH
51. $\frac{15}{43}$			
52. $\frac{28}{33}$			
53. $\frac{12}{31}$			
54. $\frac{72}{79}$			
55. $11\frac{17}{21}$			
56. $18\frac{19}{30}$			
57. $2\frac{7}{9}$			
58. $4\frac{7}{23}$			
59. $21\frac{45}{52}$			
60. $33\frac{78}{81}$			

ANSWERS

61. ____________

62. ____________

63. ____________

64. ____________

65. ____________

66. ____________

61. A piece of blank stock of metal is $\frac{3}{4}$ inch in diameter. A micrometer measures in decimal units. If the stock is measured with the micrometer, what will the reading be?

62. A wrist pin is $\frac{3}{8}$ inch in diameter. What micrometer reading would this be?

63. The diameter of a wrist pin is $\frac{15}{16}$ inch. What reading is this on a micrometer?

64. Convert the measurements in the following drawing to decimals:

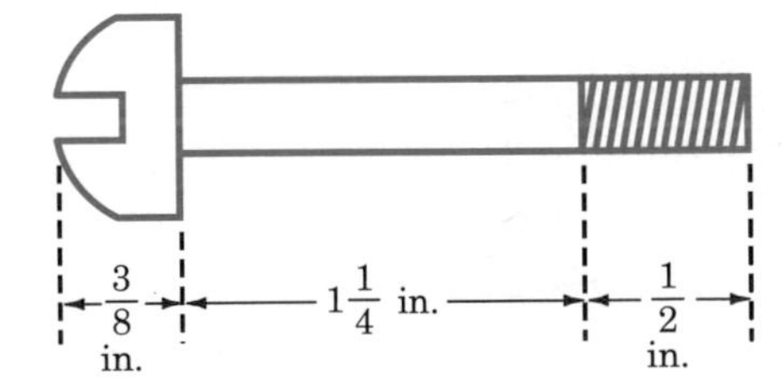

65. What reading, to the nearest thousandth, will show on a micrometer for a measure of $\frac{25}{32}$ inch?

66. What reading, to the nearest hundredth, will show on a micrometer for a measure of $\frac{15}{16}$ inch?

NAME ______________________________

CLASS ______________ SECTION ______________ DATE ______________

ANSWERS

67. ______________

68. ______________

69. ______________

70. ______________

67. Fran purchases $\frac{5}{16}$ yard of fabric. What is the decimal equivalent?

68. Stephen needs $\frac{31}{40}$ inches of rope. What is the decimal equivalent?

69. A certain recipe calls for $4\frac{7}{8}$ cups of flour. What is this as a decimal?

70. A remnant of fabric measures $3\frac{5}{12}$ yard. What is the decimal equivalent to the nearest hundredth?

E *Maintain your skills* (Sections 4.6, 4.7)

71. ______________

72. ______________

73. ______________

74. ______________

75. ______________

76. ______________

Add:

71. $0.32 + 0.024 + 3.1 + 7.214$

72. $375.3 + 2.8041 + 1984.7$

73. $3475 + 3.475 + 34.75 + 377.05$

74. $314.2 + 2.314 + 42.13 + 4312$

Subtract:

75. $48 - 27.83$

76. $49.721 - 36$

ANSWERS

77. ______________

78. ______________

79. ______________

80. ______________

77. 17.66 − 16

78. 17 − 16.66

79. A flight of stairs has five risers. What is the height of the flight of stairs if each riser is 7.5 in. high?

80. Katya made the following transactions in her checking account for June. With a starting balance of $65.87, she made a deposit of $562.34, wrote checks for $255, $34.22, and $75.29, and made another deposit of $281.17. What is the balance of her account now?

Order of Operations and Average

OBJECTIVES

68. Do any combination of operations with decimals.
69. Find the average of a group of decimals.

APPLICATION

Wanda has a morning and an evening rural paper route. She must drive her car to deliver the papers, so she uses a lot of gasoline. During one week she made the following "fill-ups":

Monday	12.8 gallons
Tuesday	16.3 gallons
Wednesday	13.2 gallons
Thursday	15.6 gallons
Friday	18.3 gallons
Saturday	17.5 gallons
Sunday	8.8 gallons (morning only)

What was her average daily purchase, rounded off to the nearest tenth of a gallon?

VOCABULARY

No new words.

HOW AND WHY

The order of operations for decimals and the method for finding the average of a group of decimals are the same as those for whole numbers.

MODEL PROBLEM-SOLVING

Examples | **Strategy**

a. Perform the indicated operations: $0.45 - (0.32)(0.75)$

$0.45 - (0.32)(0.75) = 0.45 - 0.24$ — **Multiplication is performed first.**

$= 0.21$ — **Subtraction is performed next.**

Warm Up a. Perform the indicated operations: $0.67 - (0.26)(0.5)$

b. Perform the indicated operations: $0.75 \div (0.03)(1.6)$

$0.75 \div (0.03)(1.6) = (25)(1.6)$ — **Division is performed first, since it occurs first.**

$= 40$ — **Multiplication is performed next.**

Warm Up b. Perform the indicated operations: $0.08 \div (0.05)(2.8)$

ANSWERS TO WARM UPS (4.13) **a.** 0.54 **b.** 4.48

c. Perform the indicated operations: $(3.2)^2 - (0.4)^3$

$(3.2)^2 - (0.4)^3 = 10.24 - 0.064$
$= 10.176$

Exponents are done first, as in Chapter 1.
Subtraction is performed next.

Warm Up c. Perform the indicated operations: $(4.5)^2 - (0.2)^4$

d. Find the average of 0.1, 0.27, 0.48, and 0.03.

$0.1 + 0.27 + 0.48 + 0.03 = 0.88$
$0.88 \div 4 = 0.22$

First, add the numbers.
Second, divide by 4 (the number of numbers).

The average is 0.22.

Warm Up d. Find the average of 0.7, 5.2, 1.18, 0.5, and 2.4

e. Calculator example

Perform the indicated operations: $2.55 \div 17 + (2.3)(4.5) + 1.37$

[2.55] [÷] [17] [+] [17] [+] [2.3] [×]
[4.5] [+] [1.37] [=] 11.87

If your calculator has algebraic logic, the numbers and operations can be entered in the same order as the exercise.

If your calculator is a four-function (basic), nonalgebraic model:

ENTER	DISPLAY	SAVE FOR LATER USE
2.55	2.55	
÷	2.55	
17	17.	
=	0.15	**0.15**
AC	0.	
2.3	2.3	
×	2.3	
4.5	4.5	
=	10.35	**10.35**
AC	0.	
0.15	0.15	**Recall the value saved.**
+	0.15	
10.35	10.35	**Recall the value saved.**

ANSWERS TO WARM UPS (4.13) **c.** 20.2484 **d.** 1.996

+	10.5
1.37	1.37
=	11.87

The answer is 11.87.

Warm Up e. Perform the indicated operations: $1.82 \div 1.3 + (0.3)(14.4) + 3.79$

APPLICATION SOLUTION

f. Wanda has a morning and an evening rural paper route. She must drive her car to deliver the papers, so she uses a lot of gasoline. During one week she made the following "fill-ups":

Monday	12.8 gallons
Tuesday	16.3 gallons
Wednesday	13.2 gallons
Thursday	15.6 gallons
Friday	18.3 gallons
Saturday	17.5 gallons
Sunday	8.8 gallons

What was her average daily purchase, rounded off to the nearest tenth of a gallon?

$12.8 + 16.3 + 13.2 + 15.6 + 18.3 + 17.5 + 8.8 = 102.5$

$102.5 \div 7 \approx 14.64$

To find the daily average, find the sum and divide by 7.

Wanda's average daily purchase, to the nearest tenth, is 14.6 gallons.

Warm Up f. During another week Wanda made the following "fill-ups":

Monday	13.7 gallons
Tuesday	12.6 gallons
Wednesday	13.9 gallons
Thursday	14.4 gallons
Friday	27.7 gallons
Saturday	12.8 gallons
Sunday	0 gallons (typesetters' strike)

What was her average daily purchase? (Sunday, of course, does not count.)

ANSWERS TO WARM UPS (4.13)

e. 9.51

f. 15.85 gallons

NAME ______

CLASS ______ SECTION ______ DATE ______

Exercises 4.13

A

ANSWERS

1. ______
2. ______
3. ______
4. ______
5. ______
6. ______
7. ______
8. ______
9. ______
10. ______
11. ______
12. ______
13. ______
14. ______
15. ______
16. ______

Perform the indicated operations:

1. $0.6 + 0.2 - 0.3$ **2.** $0.9 - 0.4 + 0.3$ **3.** $0.21 \div 7 + 0.02$

4. $1.8 - 2(0.8)$ **5.** $(0.02)(10) - 0.15 \div 5$ **6.** $(1.4)(3) - (2.1)(2)$

7. $(0.8)(0.4) \div (0.5)(0.2)$ **8.** $(0.6)(0.8) \div (0.4)(0.2)$

9. $(0.2)^3 + 0.08$ **10.** $(0.3)^2 + 0.12$

Find the average of the following:

11. 0.32, 0.16, 0.21 **12.** 0.23, 0.56, 0.77

13. 1.3, 3.1, 2.5 **14.** 4.7, 5.3, 3.5

15. 2.8, 1.5, 1.4, 2.3 **16.** 5.1, 2.6, 3.4, 6.2, 1.7

B

17. ______
18. ______
19. ______
20. ______

Perform the indicated operations:

17. $5.03 - 2.19 + 0.75 - 1.17$ **18.** $0.08 + 2.37 - 1.6 + 0.98$

19. $4.97 \div (0.07)(3.1)$ **20.** $(52.5) \div (0.05)(0.2)$

ANSWERS

21. ______

22. ______

23. ______

24. ______

25. ______

26. ______

27. ______

28. ______

29. ______

30. ______

31. ______

32. ______

33. ______

34. ______

35. ______

36. ______

21. $(25.6)(0.08) \div 0.02$

22. $(6.5)(2.14) \div (0.13)$

23. $(1.62 + 0.87) \div 0.05$

24. $7.2 + 6.15 - 2.34 \times 0.2 \div 0.05$

25. $3.67 + 8.3 - 1.27 \times 0.3 \div 0.06$

26. $1.62 + 0.87 \div 0.05$

27. $(6.3)^2 - 12.02$

28. $(0.5)^2(2.1 \div 1.05)$

29. $(4.6)(0.3)^3 \div 2.07$

30. $(1.4)^3 - (0.2)^2$

Find the average:

31. 8, 7.5, 2.13, 4.05

32. 19, 20.16, 17.85, 16.4

33. 6.2, 5.73, 6.01, 5.9, 5.81

34. 0.863, 0.605, 0.431, 0.501

35. 0.071, 0.065, 0.058, 0.069, 0.061, 0.054

36. 0.113, 0.217, 0.202, 0.178, 0.247, 0.135

NAME ______________________

CLASS ______________ SECTION ______________ DATE ______________

C

ANSWERS

37. ______________

38. ______________

39. ______________

40. ______________

41. ______________

42. ______________

43. ______________

44. ______________

45. ______________

46. ______________

Perform the indicated operations:

37. $(5.5)(2.4) \div (0.22)(0.35)$

38. $9 \div (1.5)(2.4)(0.05)$

39. $(6.75)(1.3) - 2.61 + (3.2)(0.5)$

40. $0.0033 \div 0.88 + 1.075 - 0.0976$

41. $4.067 - [(3.7)(0.33) + 1.108]$

42. $(7.86)(5.06 - 3.4 + 7.09)$

43. $(6.157 - 3.05)(1.6 + 0.09)$

44. $6.157 - [(3.05)(1.6) + 0.09]$

45. $3.62 \div [0.02 + (72.3)(0.2)]$

46. $(3.62 \div 0.02 + 8.6)(0.51) - 82.6$

ANSWERS

47. ____________

48. ____________

49. ____________

50. ____________

51. ____________

52. ____________

53. ____________

54. ____________

55. ____________

56. ____________

47. $(4.2 - 3.17)^2 + 4.61$

48. $0.5 - (2.4 - 1.68)^3$

49. $[3.6 + (2.1)^2] - [(0.5)^2 + 0.3]$

50. $6.3 - [(1.2)^2 - 0.72]$

Find the average:

51. 21.98, 36.02, 25.05, 31.11

52. 6.05, 5.08, 9.06, 1.6, 0.035

53. 15.61, 7.8, 22.11, 7.826

54. 0.1896, 7.83, 25, 0.998

55. 1.6, 0.85, 3.06, 0.91, 4, 1.1

56. 16.8, 17.2, 36.02, 1.494, 38.2, 47

NAME ______________________________

CLASS ______________ SECTION ______________ DATE ______________

D

ANSWERS

57. ______________

58. ______________

59. ______________

60. ______________

57. The common stock of the Acme Corporation closed at $16.58 on Wednesday, $18.09 on Thursday, and $17.31 on Friday. What was the average closing price of the stock (to the nearest cent)?

58. Rita took a trip and needed to fill the gas tank four times. If the prices of a gallon of gasoline were $1.299, $1.349, $1.379, and $1.429, what was the average price per gallon?

59. The Vernon family's natural gas bills for last year were:

January	$125.64	July	$10.45
February	142.63	August	9.75
March	83.91	September	11.84
April	65.64	October	25.32
May	41.23	November	35.72
June	20.31	December	80.16

The gas company will allow them to make equal monthly payments this year equal to the monthly average of last year. How much would the monthly payment be?

60. The price in dollars per gallon of the same grade of gasoline at eight different service stations is 1.299, 1.315, 1.419, 1.259, 1.249, 1.309, 1.319, and 1.339. What is the average price per gallon of the eight stations?

ANSWERS

61. ______________

62. ______________

63. ______________

64. ______________

65. ______________

61. Rock City police reported the following numbers of automobile accidents for a week:

Monday	28
Tuesday	31
Wednesday	34
Thursday	40
Friday	70
Saturday	63
Sunday	75

What was the average number of accidents reported each day? (Round off to the nearest tenth.)

62. Roxanne sent each of her five children to five different stores to buy 1 dozen large grade A eggs. Find the average cost of a dozen eggs if the kids paid these prices: \$1.21, \$1.01, \$1.15, \$1.07, and \$1.18. Round off your answer to the nearest cent.

63. Pens cost \$0.89 for two, while pencils cost 25 cents for four. What is the total cost of 20 of each?

64. Shampoo costs \$2.15, while conditioner costs \$1.98. What is the total cost of 8 bottles of each.

65. Find the change from a \$20 bill if Ross purchases shampoo, conditioner, 10 pens, and 20 pencils. (See Exercises 63 and 64.)

NAME ____________________

CLASS ____________ SECTION ____________ DATE ____________

ANSWERS

66. ____________

66. A shopper buys toothpaste for \$0.89, a toothbrush for \$1.19, and two pair of pantyhose at \$1.59 each. What is the change from a \$10 bill?

E *Maintain your skills* (Sections 4.3, 4.11)

67. ____________

Change to a decimal:

67. $\frac{15}{16}$ **68.** $\frac{25}{32}$ **69.** $\frac{17}{80}$ **70.** $\frac{59}{80}$

68. ____________

69. ____________

70. ____________

Change to a fraction:

71. 0.82 **72.** 0.724 **73.** 0.085 **74.** 0.0045

71. ____________

72. ____________

73. ____________

75. The sale price of a hand-held calculator is \$18.95. If the sale price was marked down \$5.84 from the original price, what was the original price?

74. ____________

75. ____________

76. The sale price on a Tomaya VCR is \$239.95. If the discount was \$49.85, what was the original price?

76. ____________

▲▲ *Getting Ready for Algebra*

PROCEDURE

To solve an equation in which more than one operation is involved:

1. **Eliminate the addition or subtraction by performing the inverse operation.**
2. **Eliminate the multiplication by dividing both sides; that is, perform the inverse operation.**

▲▲ MODEL PROBLEM-SOLVING

Examples / **Strategy**

a. Solve: $2.5x - 3.7 = 18.8$

$2.5x - 3.7 + 3.7 = 18.8 + 3.7$ — **Eliminate the subtraction by adding 3.7 to both sides.**

$2.5x = 22.5$

$\frac{2.5x}{2.5} = \frac{22.5}{2.5}$ — **Eliminate the multiplication by dividing both sides by 2.5.**

$x = 9$

Check:

$2.5(9) - 3.7 = 18.8$ — **Substitute 9 for x in the original equation.**

$22.5 - 3.7 = 18.8$

$18.8 = 18.8$ — **The statement is true.**

The solution is 9.

Warm Up a. Solve: $0.06y - 2.8 = 0.5$

GETTING READY FOR ALGEBRA

a. $y = 55$

b. Solve: $6.3 = 1.75x + 1.05$

$6.3 - 1.05 = 1.75x + 1.05 - 1.05$ **Subtract 1.05 from both sides.**

$5.25 = 1.75x$

$\frac{5.25}{1.75} = \frac{1.75x}{1.75}$ **Divide both sides by 1.75.**

$3 = x$

Check:

$6.3 = 1.75(3) + 1.05$ **Substitute 3 for x in the original equation.**

$6.3 = 5.25 + 1.05$

$6.3 = 6.3$

The solution is 3.

Warm Up b. Solve: $2.9494 = 2.49t + 2.8$

GETTING READY FOR ALGEBRA

b. $0.06 = t$

NAME ______________________________

CLASS ______________ SECTION ______________ DATE ______________

▲▲ EXERCISES

ANSWERS

1. ______________
2. ______________
3. ______________
4. ______________
5. ______________
6. ______________
7. ______________
8. ______________
9. ______________
10. ______________
11. ______________
12. ______________
13. ______________
14. ______________
15. ______________
16. ______________

Solve:

1. $2.5x - 8.9 = 13.6$

2. $0.25x - 2.2 = 0.47$

3. $2.4x + 2.6 = 4.04$

4. $14w + 0.004 = 43.404$

5. $4.115 = 2.15t + 3.9$

6. $10.175 = 1.25y + 9.3$

7. $0.03x - 13.5 = 2.22$

8. $0.07r - 2.35 = 61.7$

9. $7x + 9.06 = 11.3$

10. $13x + 14.66 = 15.7$

11. $3.45m - 122 = 109.15$

12. $11.6t - 398 = 193.6$

13. $2000 = 92y + 482$

14. $1500 = 48w + 516$

15. $50p - 149 = 1.1$

16. $14.4 = 0.44y + 5.6$

ANSWERS

17. ______

18. ______

19. ______

20. ______

21. ______

17. $7.5 = 2.3 + 0.13x$

18. $7 = 0.25w - 3.6$

19. $9 = 1.25h - 0.1$

20. $1000 = 90y + 415$

21. $1250 = 80c - 130$

NAME ______________________________

CLASS ______________ SECTION ______________ DATE ______________

ANSWERS

1. ______
2. ______
3. ______
4. ______
5. ______
6. ______
7. ______
8. ______
9. ______
10. ______
11. ______
12. ______
13. ______
14. ______
15. ______
16. ______
17. ______
18. ______

CHAPTER 4 TRUE–FALSE CONCEPT REVIEW

Check your understanding of the language of basic mathematics. Tell whether or not each of the following is true (always true) or false (not always true).

1. The word name for 0.502 is "five hundred and two thousandths."
2. .502 and 0.502 name the same number.
3. To write 0.75 in expanded form we write $\frac{75}{100}$.
4. Since 0.145 is read "one hundred forty five thousandths", we write $\frac{145}{1000}$ and reduce to change the decimal to a fraction.
5. True or false: $0.821597 > 0.84$.
6. Since $2.3 > 1.5$ is true, 1.5 is to the left of 2.3 on the number line.
7. To list a group of decimals in order, we need to write or think of all of the numbers as having the same number of decimal places.
8. Decimals are either exact or approximate.
9. To round 123.3477 to the nearest tenth, we write 123.4 since the four in the hundredths place rounds up to 5 since it is followed by a 7.
10. The sum of 0.5 and 0.25 is 0.75.
11. $8.5 - 0.2$ is 6.5.
12. The final answer of a multiplication problem will always contain the same number of decimal places as the total number of places in the two numbers being multiplied.
13. To multiply a number by a positive power of 10, move the decimal point in the number the same number of places as the number of zeros in the power of 10.
14. To divide a number by a positive power of 10 which is written in exponent form, move the decimal the same number of places to the right as the exponent indicates.
15. To change 1.56×10^{-5} to place value form, move the decimal five places to the right.
16. To divide a number by a decimal, first change the decimal to a whole number by moving the decimal point to the right.
17. All fractions can be changed to exact decimals.
18. The decimal 0.5649, rounds to 0.57 to the nearest hundredth.

ANSWERS

19. ______________

20. ______________

19. The order of operations for decimals is the same as for whole numbers.

20. To find the average of a group of decimals, find their sum and divide by the number of decimals in the group.

NAME ______________________________

CLASS ______________ SECTION ______________ DATE ______________

ANSWERS

1. ______________

2. ______________

3. ______________

4. ______________

5. ______________

6. ______________

7. ______________

8. ______________

9. ______________

10. ______________

11. ______________

12. ______________

13. ______________

14. ______________

15. ______________

CHAPTER 4 POST-TEST

1. **(Section 4.10, Obj. 65)** Divide. Round off the answer to the nearest thousandth: $0.98\overline{)0.0394}$

2. **(Section 4.4, Obj. 55)** List the following decimals from the smallest to largest: 0.728, 0.731, 0.7279, 0.7308, 0.7299

3. **(Section 4.1, Obj. 48)** What is the place value of the 3 in 89.438?

4. **(Section 4.1, Obj. 50)** Write the word name for 27.027.

5. **(Section 4.8, Obj. 59)** Multiply: $\begin{array}{r} 5.36 \\ \underline{7.8} \end{array}$

6. **(Section 4.1, Obj. 48)** What is the place value of the 4 in 0.0394?

7. **(Section 4.11, Obj. 66)** Write as a decimal: $\frac{3}{125}$

8. **(Section 4.5, Obj. 56)** Round off to the nearest hundredth: 8.996

9. **(Section 4.7, Obj. 58)** Subtract: $11 - 6.789$

10. **(Section 4.3, Obj. 53)** Change to a mixed number with the fraction part reduced to the lowest terms: 16.925

11. **(Section 4.2, Obj. 51)** Write in expanded form: 702.305

12. **(Section 4.9, Obj. 62)** Write in scientific notation: 0.0024

13. **(Section 4.11, Obj. 67)** Write as an approximate decimal to the nearest hundredth: $\frac{4}{13}$

14. **(Section 4.5, Obj. 56)** Round off to the nearest hundred: 47,495.6

15. **(Section 4.12, Obj. 68)** Perform the indicated operations: $4.56 \div 0.6(1.03) + 7.5$

ANSWERS

16. ______

17. ______

18. ______

19. ______

20. ______

21. ______

22. ______

23. ______

24. ______

25. ______

26. ______

27. ______

28. ______

16. **(Section 4.7, Obj. 58)** Subtract: $\begin{array}{r} 7.846 \\ \underline{2.948} \end{array}$

17. **(Section 4.9, Obj. 63)** Change to place value form: 2.66×10^{-1}

18. **(Section 4.1, Obj. 49)** Write the place value name for "two hundred twelve and sixty-three thousandths."

19. **(Section 4.9, Obj. 60)** Multiply: 0.00216(10000)

20. **(Section 4.2, Obj. 52)** Write the place value name for:

$$20 + 5 + \frac{7}{10} + \frac{0}{100} + \frac{9}{1000}$$

21. **(Section 4.9, Obj. 62)** Write in scientific notation: 32,750

22. **(Section 4.11, Obj. 67)** Write as an approximate decimal to the nearest thousandth:

$$7\frac{3}{11}$$

23. **(Section 4.9, Obj. 61)** Divide: $0.08 \div 100$

24. **(Section 4.8, Obj. 59)** Multiply: (7.6)(0.0018)

25. **(Section 4.12, Obj. 69)** For each of the four Sundays of February, the offering at the Chapel on the Hill was $68.25, $76.55, $82.76, and $71.33. What was the average Sunday offering?

26. **(Section 4.6, Obj. 57)** Add: 1.07 + 0.659 + 12.36 + 8.9

27. **(Section 4.10, Obj. 65)** Divide: $25\overline{)1.375}$

28. During a canned vegetable sale Ted bought 14 cans of various vegetables. If the sale price was 4 cans for $1.79, how much did Ted pay for the canned vegetables?

CHAPTER 5

Ratio and Proportion

NAME ______________________________

CLASS ______________ SECTION ______________ DATE ______________

ANSWERS

1. ______________

2. ______________

3. ______________

4. ______________

5. ______________

6. ______________

7. ______________

8. ______________

CHAPTER 5 PRE-TEST

The problems in the following pre-test are a sample of the material in the chapter. You may already know how to work some of these. If so, this will allow you to spend less time on those parts. As a result, you will have more time to give to the sections that gave you difficulty. Please work the following problems. The answers are in the back of the text.

1. **(Section 5.1, Obj. 70)** Write a ratio to compare 3 miles to 8 miles.

2. **(Section 5.2, Obj. 70)** Write a ratio to compare 3 quarters to 6 dollars (in quarters) and reduce to the lowest terms.

3. **(Section 5.2, Obj. 71)** Is the following proportion true or false?

$$\frac{16}{30} = \frac{24}{45}$$

4. **(Section 5.2, Obj. 71)** Is the following proportion true or false?

$$\frac{7.8}{0.5} = \frac{0.39}{0.025}$$

5. **(Section 5.2, Obj. 72)** Solve the proportion: $\frac{5}{6} = \frac{x}{33}$

6. **(Section 5.2, Obj. 72)** Solve the proportion: $\frac{\frac{1}{3}}{7} = \frac{2}{x}$

7. If 30 lb of beef contains 6 lb of bones, how many pounds of bones may be expected in 100 lb of beef?

8. There is a canned food sale at the supermarket. A case of 24 cans of peas is priced at $10.50. To the nearest cent, what is the price of 10 cans of peas?

Ratio and Proportion

OBJECTIVES

70. Write a fraction that shows a ratio comparison of two numbers or two measurements.
71. Determine whether a proportion is true or false.

APPLICATION

In the parking lot at the Rural Community Center there are 48 parking spaces for compact cars and 32 parking spaces for larger cars:

1. What is the ratio of the number of compact spaces to the number of larger spaces?
2. What is the ratio of the number of compact spaces to the total number of spaces?

VOCABULARY

A *ratio* is a comparison of a pair of numbers by division $\left(3 \div 4 = \frac{3}{4}\right)$.
A ratio is commonly written as a fraction.

A *proportion* is a statement that says two ratios are equal; $\frac{3}{4} = \frac{6}{8}$ is an example of a proportion. *Cross multiplication* is often used when working with proportions. This involves multiplying the numerator of one ratio times the denominator of the other ratio:

$$\frac{3}{4} = \frac{6}{8}$$

The products $3 \cdot 8$ and $4 \cdot 6$ are called the *cross products.*
A *measurement* is written with a number and a unit of measure (3 pounds).

HOW AND WHY

Two numbers can be compared by subtraction or by division. Compare 12 dollars and 3 dollars. Since $12 - 3 = 9$, we say:

\$12 is nine dollars more than \$3.

And since $12 \div 3$ is 4, we say:

\$12 is four times larger than \$3.

The indicated division is called a ratio. These are common ways to write the ratio of 12 and 3:

$12:3 \qquad 12 \div 3 \qquad 12 \text{ to } 3 \qquad \frac{12}{3}$

Here we write ratios as fractions.

Ratios are used to compare both like and unlike measurements. The ratio $\frac{31 \text{ children}}{10 \text{ families}}$ compares the unlike measurements "31 children" and "10 families."

The ratio $\frac{\$3}{\$100}$ compares the like measurements "\$3" and "\$100."

If a car runs 208 miles on 8 gallons of gas, we compare miles to gallons by writing $\frac{208 \text{ miles}}{8 \text{ gallons}}$. This symbol can be reduced in the same way as a fraction, as long as the units are stated:

$$\frac{208 \text{ miles}}{8 \text{ gallons}} = \frac{104 \text{ miles}}{4 \text{ gallons}} = \frac{26 \text{ miles}}{1 \text{ gallon}} = 26 \text{ miles per gallon}$$

Reducing ratios leads to statements such as, "There are 3.1 children to a family," since:

$$\frac{31 \text{ children}}{10 \text{ families}} = \frac{3.1 \text{ children}}{1 \text{ family}}$$

The last ratio is a comparison, not a fact, since no family has 3.1 children.

If two measurements have the same units, then the units may be dropped:

$\frac{\$3}{\$100} = \frac{3}{100}$ **Since the units are the same, they are dropped.**

If two measurements do not have the same units, as in $\frac{26 \text{ miles}}{1 \text{ gallon}}$, the units must be written.

Statements such as the following are called proportions:

$\frac{62 \text{ miles}}{1 \text{ hour}} = \frac{100 \text{ kilometers}}{1 \text{ hour}}$ **In each ratio the units of measurement are different, so they are not dropped.**

This is read "62 miles is to 1 hour as 100 kilometers is to 1 hour." To be true, the ratios must be equivalent fractions when the units are the same.

The test to tell whether a proportion is true or false is called "cross multiplication."

$\frac{14}{8} = \frac{35}{20}$ is true because

CROSS MULTIPLICATION

$$\frac{14}{8} = \frac{35}{20}$$

$8 \times 35 = 280$ and $20 \times 14 = 280$

so, $20 \times 14 = 8 \times 35$

This test is based on building the two ratios or fractions to a common denominator of 8×20 or 160:

$$\frac{14}{8} = \frac{35}{20} \quad \text{or} \quad \frac{280}{160} = \frac{280}{160}$$

In the cross-multiplying test we take a shortcut by comparing only the two numerators ($280 = 280$). The proportion $\frac{4}{9} = \frac{2}{3}$ is not true because 3×4 is not equal to 9×2:

PROCEDURE

To tell whether a proportion is true or false:

1. **Check that the ratios have the same units.**
2. **Cross multiply.**
3. **If the products are equal, the proportion is true.**

MODEL PROBLEM-SOLVING

Examples | **Strategy**

a. Express the ratio of 5 chairs to 6 chairs:

$$\frac{5 \text{ chairs}}{6 \text{ chairs}} = \frac{5}{6}$$

The first measure (5 chairs) is written in the numerator. Since the units are the same, they are dropped.

The ratio of 5 chairs to 6 chairs is $\frac{5}{6}$.

Warm Up a. Express the ratio of 8 tables to 11 tables.

b. Express the ratio of 10 chairs to 8 people:

$$\frac{10 \text{ chairs}}{8 \text{ people}} = \frac{5 \text{ chairs}}{4 \text{ people}}$$

The first measure (10 chairs) is written in the numerator.
In this case the units must be kept since they are different. The fraction $\frac{10}{8}$ can be reduced.

The ratio of 10 chairs to 8 people is the same as 5 chairs to 4 people.

Warm Up b. Express the ratio of 15 people to 8 tables.

c. Express the ratio of the length of a room to its width, if the room measures 24 feet by 18 feet:

$$\frac{24 \text{ feet}}{18 \text{ feet}} = \frac{24}{18} = \frac{4}{3}$$

The length is longer than the width, so the length is 24 feet and the width is 18 feet. Since the units are the same they are dropped. The fraction is reduced.

The ratio of the length to the width is $\frac{4}{3}$.

Warm Up c. Express the ratio of the width of a room to its length, if the room measures 24 feet by 18 feet.

ANSWERS TO WARM UPS (5.1) **a.** $\frac{8}{11}$ **b.** $\frac{15 \text{ people}}{8 \text{ tables}}$ **c.** $\frac{3}{4}$

d. Express the ratio of 6 dimes to 14 nickels (compare in dimes).

$$\frac{6 \text{ dimes}}{14 \text{ nickels}} = \frac{6 \text{ dimes}}{7 \text{ dimes}}$$

$$= \frac{6}{7}$$

It takes two nickels to make a dime, so 14 nickels are equal to 7 dimes. This allows us to drop the measures since they are the same.

The ratio of 6 dimes to 14 nickels is $\frac{6}{7}$.

Warm Up d. Express the ratio of 10 dimes to 18 nickels (compare in dimes).

e. Is the proportion $\frac{6}{5} = \frac{72}{60}$ true?

$$\frac{6}{5} \times \frac{72}{60}$$

$6 \cdot 60 = 360$ and $5 \cdot 72 = 360$

If the proportion is true, the cross products will be equal.

The cross products are equal; therefore, the proportion is true.

Warm Up e. Is the proportion $\frac{8}{7} = \frac{48}{42}$ true?

f. Is the proportion $\frac{2.1}{3.1} = \frac{2}{3}$ true?

$$\frac{2.1}{3.1} \times \frac{2}{3}$$

$(2.1) \cdot (3) = 6.3$ and $(3.1) \cdot (2) = 6.2$.
The products are not equal, so the proportion is not true.

If the proportion is true, the cross products will be equal.

Warm Up f. Is the proportion $\frac{4.2}{6.3} = \frac{4}{5}$ true?

g. Is the proportion $\frac{1 \text{ dollar}}{3 \text{ quarters}} = \frac{8 \text{ dimes}}{12 \text{ nickels}}$ true?

$$\frac{20 \text{ nickels}}{15 \text{ nickels}} = \frac{16 \text{ nickels}}{12 \text{ nickels}}$$

$$\frac{20}{15} = \frac{16}{12}$$

First, change each unit to nickels. There are 20 nickels in 1 dollar, 16 nickels in 8 dimes, and 15 nickels in 3 quarters. Since the units are the same, they can be dropped.

ANSWERS TO WARM UPS (5.1) **d.** $\frac{10}{9}$ **e.** true **f.** false

$20 \cdot 12 = 240$ and $15 \cdot 16 = 240$ **Check the cross products.**

Since the cross products are equal, the original proportion is true.

Warm Up g. Is the proportion $\frac{1 \text{ dollar}}{2 \text{ quarters}} = \frac{16 \text{ nickels}}{4 \text{ dimes}}$ true?

h. Calculator example

The population density of a region is defined as the ratio of the number of people to the number of square miles of area. Find the population density of Stone County if the population is 12550 and the area is 1700 square miles. Reduce to a 1 square mile of area comparison, rounded off to the nearest tenth.

$$\text{Density} = \frac{12{,}550 \text{ people}}{1700 \text{ square miles}}$$

ENTER	DISPLAY
12550	12550.
÷	12550.
1700	1700.
=	7.3823529

The density to the nearest tenth is 7.4 people per square mile.

Warm Up h. The population of Washington County is 145,989 and the area is 1800 square miles. What is the population density reduced to 1 square mile of area comparison, rounded off to the nearest tenth?

ANSWERS TO WARM UPS (5.1)

g. true

h. 81.1 people per square mile

APPLICATION SOLUTION

i. In the parking lot at the Rural Community Center there are 48 parking spaces for compact cars and 32 parking spaces for larger cars:

1. What is the ratio of the number of compact spaces to the number of larger spaces?
2. What is the ratio of the number of compact spaces to the total number of spaces?

1. $\frac{48 \text{ compact spaces}}{32 \text{ large spaces}} =$

To compare, write the first unit (48 compact spaces) in the numerator.

$\frac{3 \text{ compact}}{2 \text{ large}}$

Reduce the number part of the ratio.

The ratio of compact spaces to the number of large spaces is 3 to 2. There are 3 compact spaces for each 2 large spaces.

2. Total spaces = 48 + 32 = 80

To find the total number of spaces, add the number of compact spaces to the number of large spaces.

$\frac{48 \text{ compact spaces}}{80 \text{ total spaces}} =$

We compare the number of compact spaces to the total number of spaces. Again, we write the first measure in the numerator.

$\frac{3 \text{ compact}}{5 \text{ total}}$

The ratio of compact spaces to total spaces is 3 to 5. That is, 3 of every 5 is a compact space.

Warm Up i. In the parking lot at the Downtown Theatre there are 112 spaces for compact cars and 182 spaces for larger cars. What is the ratio of the number of compact spaces to larger spaces? What is the ratio of the number of compact spaces to the total number of spaces?

ANSWERS TO WARM UPS (5.1)

i. $\frac{8 \text{ compact}}{13 \text{ large}}, \frac{8 \text{ compact}}{21 \text{ total}}$

NAME ______

CLASS ______ SECTION ______ DATE ______

Exercises 5.1

A

ANSWERS

1. ______
2. ______
3. ______
4. ______
5. ______
6. ______
7. ______
8. ______
9. ______
10. ______
11. ______
12. ______
13. ______
14. ______
15. ______
16. ______
17. ______
18. ______
19. ______

Write a ratio and reduce:

1. 8 people to 11 chairs

2. 6 families to 18 children

3. 3 nickels to 16 cents (Compare in cents.)

4. 4 dimes to 3 nickels (Compare in pennies.)

5. 12 meters to 10 meters

6. 110 miles to (per) 2 hours

7. 16 families to 48 children

8. 48 miles to 3 gallons

9. 75 pounds to 15 feet

10. 40 buttons to 16 bows

Are the following proportions true or false?

11. $\frac{2}{4} = \frac{12}{24}$

12. $\frac{6}{8} = \frac{9}{12}$

13. $\frac{4}{5} = \frac{5}{6}$

14. $\frac{4}{6} = \frac{7}{10}$

15. $\frac{4}{10} = \frac{20}{50}$

16. $\frac{1}{3} = \frac{33}{100}$

17. $\frac{12}{5} = \frac{24}{10}$

18. $\frac{3}{4} = \frac{8}{11}$

19. $\frac{7}{6} = \frac{6}{5}$

ANSWERS

20. ______

21. ______

22. ______

20. $\frac{10}{15} = \frac{6}{9}$ **21.** $\frac{3}{8} = \frac{8}{22}$ **22.** $\frac{7}{5} = \frac{8}{6}$

B

23. ______

24. ______

25. ______

26. ______

27. ______

28. ______

29. ______

30. ______

31. ______

32. ______

33. ______

34. ______

Write a ratio and reduce:

23. 1 quarter to 1 dime (Compare in cents.)

24. 3 dimes to 7 nickels (Compare in nickels.)

25. 600 miles to (per) 8 hours.

26. $2.37 to (per) 3 pounds of tomatoes

27. 220 miles to (per) 12 gallons

28. 400 feet to (per) 25 seconds

Are the following proportions true or false?

29. $\frac{2}{4} = \frac{14}{28}$ **30.** $\frac{7}{5} = \frac{84}{60}$ **31.** $\frac{7}{8} = \frac{11}{12}$

32. $\frac{8}{34} = \frac{9}{32}$ **33.** $\frac{14}{16} = \frac{21}{24}$ **34.** $\frac{12}{17} = \frac{204}{289}$

NAME ______________________

CLASS ______________ SECTION ______________ DATE ______________

ANSWERS

35. ______________

36. ______________

37. ______________

38. ______________

39. ______________

40. ______________

41. ______________

42. ______________

43. ______________

35. $\frac{16}{21} = \frac{176}{231}$ **36.** $\frac{17}{15} = \frac{289}{255}$ **37.** $\frac{60}{45} = \frac{72}{54}$

38. $\frac{480}{90} = \frac{80}{15}$ **39.** $\frac{20}{13} = \frac{158}{103}$ **40.** $\frac{15}{23} = \frac{224}{343}$

41. $\frac{83}{125} = \frac{16}{24}$ **42.** $\frac{104}{216} = \frac{13}{27}$ **43.** $\frac{120}{232} = \frac{15}{29}$

C

44. ______________

Write a ratio and reduce:

44. 18 cups of sugar for 12 cakes

45. ______________

45. 650 tickets for 250 people

46. ______________

46. 21 cups of flour to 6 loaves of bread (Reduce to a one-loaf comparison.)

47. ______________

47. 1300 television sets to 1000 houses (Reduce to a one-house comparison.)

48. ______________

48. The low gear ratio in a truck's transmission, if the large gear has 189 teeth and the small gear has 14 teeth.

ANSWERS

49. ______

50. ______

51. ______

52. ______

53. ______

54. ______

55. ______

56. ______

57. ______

58. ______

59. ______

60. ______

61. ______

62. ______

63. ______

49. 385 miles per 14 gallons (Reduce to a one-gallon comparison and the nearest tenth of a mile.)

Are the following proportions true or false?

50. $\frac{224}{100} = \frac{305}{125}$

51. $\frac{150}{360} = \frac{100}{240}$

52. $\frac{15.2}{35.1} = \frac{27.1}{61.5}$

53. $\frac{21.5}{18.3} = \frac{19.3}{15.4}$

54. $\frac{16.24}{24.08} = \frac{406}{602}$

55. $\frac{0.006}{0.024} = \frac{0.04}{0.16}$

56. $\frac{4\frac{1}{2}}{3} = \frac{9}{6}$

57. $\frac{5}{2\frac{2}{3}} = \frac{2}{1\frac{1}{10}}$

58. $\frac{2.1}{3.2} = \frac{1.2}{2.3}$

59. $\frac{2.6}{4.8} = \frac{3.9}{7.2}$

60. $\frac{5}{8} = \frac{\frac{1}{5}}{\frac{1}{4}}$

61. $\frac{\frac{1}{3}}{\frac{1}{2}} = \frac{\frac{7}{9}}{1\frac{1}{6}}$

62. $\frac{\text{8 inches}}{\text{2 feet}} = \frac{\text{6 inches}}{\text{18 inches}}$

63. $\frac{\text{5 pounds}}{\$1.30} = \frac{\text{8 ounces}}{\text{14 cents}}$

NAME ______________________

CLASS ______________ SECTION ______________ DATE ______________

D

ANSWERS

64. ______________

65. ______________

66. ______________

67. ______________

68. ______________

64. The parking lot in the lower level of the Knew Office Building has 18 spaces for compact cars and 24 spaces for larger cars.

a. What is the ratio of compact spaces to larger spaces?

b. What is the ratio of compact spaces to the total number of spaces?

65. The Reliable Auto Repair Service building has eight stalls for repairing automobiles and four stalls for repairing small trucks.

a. What is the ratio of the number of stalls for small trucks to the number of stalls for automobiles?

b. What is the ratio of the number of stalls for small trucks to the total?

66. One section of the country has 3500 TV sets per 1000 houses. A second section has 500 TV sets per 150 houses. Are the ratios of the TV sets to the number of houses the same in both parts of the country?

67. In City A there are 5000 automobiles per 3000 households. In City B there are 8000 automobiles per 4800 households. Are the ratios of the number of automobiles to the number of households the same?

68. What is the population density of the city of Dryton if 22,450 people live there and the area is 230 square miles? Reduce to a 1-square-mile comparison, rounded off to the nearest tenth.

ANSWERS

69. ______________________

70. ______________________

71. ______________________

72. ______________________

73. ______________________

74. ______________________

75. ______________________

69. What is the population density of Struvaria if there are 950,000 people and the area is 18,000 square miles? Reduce to a 1-square-mile comparison, rounded off to the nearest tenth.

70. What was the population density of your city in 1980? your county? your state?

71. A store bought a sofa for a cost of $175 and sold it for $300. What is the ratio of the cost to the selling price?

72. In Exercise 71, what is the ratio of the markup (the difference between the cost and the selling price) to the cost?

73. In Exercise 71, what is the ratio of the markup to the selling price?

74. A coat is regularly priced at $99.99, but during a sale its price is $66.66. What is the ratio of the sale price to the regular price?

75. In Exercise 74, what is the ratio of the discount (difference between the sale price and the regular price) to the regular price?

NAME ______________________________

CLASS ______________ SECTION ______________ DATE ______________

ANSWERS

76. ______________

76. In accounting the current ratio is defined as:

$$\text{Current ratio} = \frac{\text{current assets}}{\text{current liabilities}}$$

What is the current ratio for a business if the current assets are $208,000 and the current liabilities are $1,600,000?

77. ______________

77. Using the formula in Exercise 76, what is the current ratio for a business if the current assets are $8,250,000 and the current liabilities are $2,500,000?

E *Maintain your skills* (Sections 4.1, 4.3, 4.4, 4.5, 4.8)

78. ______________

78. Write the word name for 2,300.03.

79. ______________

79. Change 0.234 to a fraction.

80. ______________

80. True or false? $0.04 > 0.03988$

81. ______________

81. List these decimals from the smallest to largest: 7.05, 7.005, 7.095, 7.009, 7.059

82. Round off 0.6847 to the nearest hundredth.

82. ____________________

83. Round off 0.6847 to the nearest unit.

83. ____________________

84. Multiply: (13.275)(0.32)

84. ____________________

85. Multiply: (13.075)(0.032)

85. ____________________

86. Jill bought a pair of shoes. The cost of the shoes was $38.49. She gave the salesperson two $20 bills to pay for the shoes. How much change did she get?

86. ____________________

87. A hospital patient is scheduled to be given 1250 cc of an IV solution during the next eight hours. How many cubic centimeters per minute is this, rounded off to the nearest tenth?

87. ____________________

Solving Proportions

OBJECTIVE

72. Solve a proportion.

APPLICATION

Applications of this objective come from a variety of fields. These applications are so important that an entire section (5.3) is given to them.

VOCABULARY

In proportions we use a letter to hold the place of a missing number: The letter that is used is called an unknown or a *variable*. Finding that replacement for the missing number that will make the proportion true is called *solving the proportion*.

HOW AND WHY

Proportions are used to solve many problems in science, technology, and business. There are four numbers in a proportion. If three numbers are given, we can find the missing number:

What number is to 5 as 15 is to 25? To answer this, we use x to hold the place of the missing number:

$$\frac{x}{5} = \frac{15}{25}$$

Since the cross products are equal, we have:

$$\frac{x}{5} \nearrow\!\!\!\!\swarrow \frac{15}{25}$$

$25 \cdot x = 5 \cdot 15$

$25 \cdot x = 75$

Any multiplication problem has two related division problems. The product divided by either factor gives the other factor. So:

$x = 75 \div 25$ or $x = 3$

The missing number is 3.

PROCEDURE

To solve a proportion:

1. **Cross multiply.**
2. **Do the related division problem to find the missing number.**

MODEL PROBLEM-SOLVING

Examples	Strategy
a. Solve the proportion: $\frac{4}{9} = \frac{8}{x}$	
$4x = 9 \cdot 8$	**Cross multiply.**
$4x = 72$	**Simplify the right-hand side.**
$x = 72 \div 4$	**Do the related division problem.**
$x = 18$	
Check:	
$\frac{4}{9} = \frac{8}{18}$	**Substitute 18 for x in the original proportion.**
$4 \cdot 18 = 9 \cdot 8$	**To check, cross multiply.**
$72 = 72$	**The statement is true.**
The missing number is 18.	

Warm Up a. Solve: $\frac{5}{9} = \frac{10}{x}$

Examples	Strategy
b. Solve the proportion: $\frac{5}{3} = \frac{y}{8}$	
$5 \cdot 8 = 3y$	**Cross multiply.**
$40 = 3y$	**Simplify the right-hand side.**
$40 \div 3 = y$	**Do the related division problem.**
$13\frac{1}{3} = y$	**The check is left for the student.**

Warm Up b. Solve: $\frac{7}{5} = \frac{a}{4}$

ANSWERS TO WARM UPS (5.2)

a. $x = 18$ **b.** $a = \frac{28}{5}$ or $5\frac{3}{5}$

c. Solve the proportion: $\frac{0.6}{x} = \frac{1.2}{0.84}$

$(0.6)(0.84) = (1.2)x$	**Cross multiply.**
$0.504 = (1.2)x$	**Simplify the left-hand side.**
$0.504 \div 1.2 = x$	**Do the related division problem.**
$0.42 = x$	**The check is left for the student.**

Warm Up c. Solve: $\frac{0.5}{x} = \frac{1.5}{0.75}$

d. Solve: $\frac{\frac{3}{4}}{1\frac{2}{3}} = \frac{\frac{1}{2}}{x}$

$\frac{3}{4}x = 1\frac{2}{3} \cdot \frac{1}{2}$	**Cross multiply.**
$\frac{3}{4}x = \frac{5}{3} \cdot \frac{1}{2}$	**Change the mixed number to an improper fraction.**
$\frac{3}{4}x = \frac{5}{6}$	**Simplify.**
$x = \frac{5}{6} \div \frac{3}{4}$	**Do the related division problem.**
$x = \frac{5}{\not{6}_{3}} \cdot \frac{\not{4}^{2}}{3}$	**Change division to multiplication, then reduce and multiply.**
$x = \frac{10}{9} = 1\frac{1}{9}$	**The check is left to the student.**

Warm Up d. Solve: $\frac{\frac{3}{4}}{\frac{5}{8}} = \frac{\frac{1}{2}}{x}$

ANSWERS TO WARM UPS (5.2)

c. $x = 0.25$ **d.** $x = \frac{5}{12}$

e. Calculator example

Proportions with whole numbers or decimals can be solved with a calculator.

For instance, if:

$$\frac{3}{x} = \frac{9.6}{7.32}$$

$3(7.32) = 9.6x$

$[3(7.32)] \div 9.6 = x$

Cross multiply. Write the related division problem. Now we are ready to use the calculator.

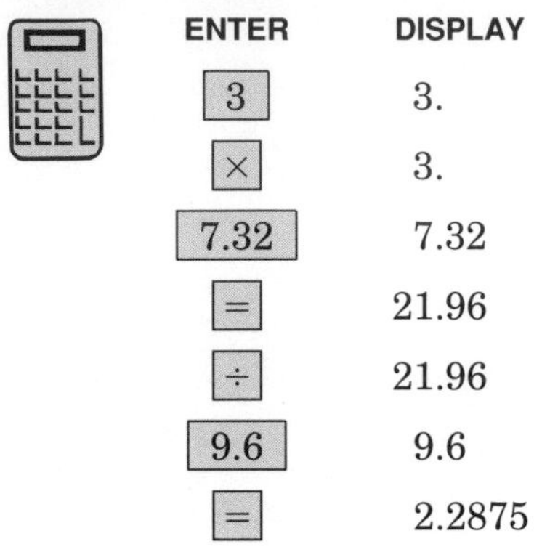

ENTER	DISPLAY
3	3.
×	3.
7.32	7.32
=	21.96
÷	21.96
9.6	9.6
=	2.2875

Therefore, $x \approx 2.29$ to the nearest hundredth.

Warm Up e. Solve and round answer to the nearest tenth: $\frac{8}{x} = \frac{1.82}{21.24}$

ANSWERS TO WARM UPS (5.2)

e. $x \approx 93.4$

NAME ______________________

CLASS ____________ SECTION ____________ DATE ____________

Exercises 5.2

A

ANSWERS

1. ________
2. ________
3. ________
4. ________
5. ________
6. ________
7. ________
8. ________
9. ________
10. ________
11. ________
12. ________
13. ________
14. ________
15. ________
16. ________
17. ________
18. ________

Solve these proportions:

1. $\frac{x}{1} = \frac{14}{2}$ **2.** $\frac{2}{x} = \frac{6}{9}$ **3.** $\frac{4}{6} = \frac{x}{9}$

4. $\frac{3}{6} = \frac{4}{x}$ **5.** $\frac{y}{10} = \frac{3}{5}$ **6.** $\frac{5}{y} = \frac{4}{8}$

7. $\frac{9}{6} = \frac{y}{2}$ **8.** $\frac{11}{2} = \frac{44}{y}$ **9.** $\frac{u}{72} = \frac{1}{9}$

10. $\frac{72}{v} = \frac{4}{1}$ **11.** $\frac{10}{7} = \frac{20}{x}$ **12.** $\frac{9}{y} = \frac{3}{11}$

13. $\frac{w}{4} = \frac{25}{10}$ **14.** $\frac{5}{w} = \frac{50}{20}$ **15.** $\frac{0.1}{0.2} = \frac{w}{10}$

16. $\frac{0.3}{x} = \frac{0.4}{8}$ **17.** $\frac{7}{42} = \frac{\frac{1}{2}}{w}$ **18.** $\frac{a}{30} = \frac{2}{\frac{3}{4}}$

B

ANSWERS

19. ____________

20. ____________

21. ____________

22. ____________

23. ____________

24. ____________

25. ____________

26. ____________

27. ____________

28. ____________

29. ____________

30. ____________

31. ____________

32. ____________

33. ____________

34. ____________

35. ____________

36. ____________

37. ____________

38. ____________

Solve these proportions:

19. $\frac{2}{x} = \frac{10}{15}$ **20.** $\frac{x}{7} = \frac{21}{49}$ **21.** $\frac{6}{8} = \frac{12}{x}$ **22.** $\frac{x}{42} = \frac{5}{7}$

23. $\frac{9}{3} = \frac{n}{2}$ **24.** $\frac{8}{4} = \frac{b}{3}$ **25.** $\frac{15}{8} = \frac{4}{x}$ **26.** $\frac{18}{5} = \frac{12}{x}$

27. $\frac{36}{y} = \frac{45}{25}$ **28.** $\frac{y}{25} = \frac{18}{30}$ **29.** $\frac{5}{2} = \frac{w}{9}$ **30.** $\frac{35}{42} = \frac{7}{w}$

31. $\frac{3}{2} = \frac{R}{100}$ **32.** $\frac{5}{8} = \frac{R}{100}$ **33.** $\frac{0.2}{0.3} = \frac{8}{x}$ **34.** $\frac{0.3}{0.4} = \frac{12}{x}$

35. $\frac{8}{9} = \frac{\frac{1}{3}}{y}$ **36.** $\frac{x}{40} = \frac{\frac{3}{4}}{5}$ **37.** $\frac{w}{2.5} = \frac{3}{5}$ **38.** $\frac{6.5}{26} = \frac{y}{2}$

C

39. ____________

40. ____________

41. ____________

42. ____________

Solve these proportions:

39. $\frac{418}{154} = \frac{w}{7}$ **40.** $\frac{T}{6} = \frac{207}{138}$ **41.** $\frac{y}{3} = \frac{15}{16}$ **42.** $\frac{25}{36} = \frac{a}{20}$

NAME ______________________________

CLASS ______________ SECTION ______________ DATE ______________

ANSWERS

43. ______________
44. ______________
45. ______________
46. ______________
47. ______________
48. ______________
49. ______________
50. ______________
51. ______________
52. ______________
53. ______________
54. ______________
55. ______________
56. ______________
57. ______________
58. ______________
59. ______________
60. ______________
61. ______________
62. ______________
63. ______________

43. $\frac{40}{55} = \frac{60}{c}$ **44.** $\frac{30}{8} = \frac{y}{6}$ **45.** $\frac{x}{19} = \frac{60}{57}$ **46.** $\frac{90}{w} = \frac{18}{45}$

47. $\frac{37}{7.4} = \frac{m}{15}$ **48.** $\frac{0.014}{x} = \frac{7}{50}$ **49.** $\frac{0.05}{0.9} = \frac{y}{4.5}$ **50.** $\frac{1.2}{2.7} = \frac{3.4}{w}$

51. $\frac{3\frac{1}{2}}{10\frac{1}{2}} = \frac{8}{x}$ **52.** $\frac{2\frac{1}{2}}{3\frac{1}{3}} = \frac{4\frac{1}{4}}{x}$

Solve to the nearest tenth:

53. $\frac{\frac{1}{2}}{0.8} = \frac{c}{0.4}$ **54.** $\frac{\frac{2}{3}}{m} = \frac{1.5}{\frac{3}{4}}$

55. $\frac{8}{21} = \frac{10}{x}$ **56.** $\frac{9}{50} = \frac{w}{16}$

Solve these proportions. Round off to the nearest hundredth:

57. $\frac{7}{11} = \frac{y}{15}$ **58.** $\frac{6}{z} = \frac{7}{17}$ **59.** $\frac{4}{3.7} = \frac{a}{0.53}$

60. $\frac{4.4}{0.23} = \frac{9}{b}$ **61.** $\frac{341}{202} = \frac{c}{8}$ **62.** $\frac{76}{d} = \frac{314}{49}$

63. $\frac{44.5}{p} = \frac{41}{80}$

E *Maintain your skills* (Sections 4.5, 4.8, 4.10)

ANSWERS

64. ______

65. ______

66. ______

67. ______

68. ______

69. ______

70. ______

71. ______

72. ______

73. ______

Multiply:

64. (0.758)(36)

65. (2.004)(0.076)

66. (14.25)(9.09)

67. (300.7)(2.06)

Divide:

68. $0.23\overline{)0.46023}$

69. $0.46\overline{)2.3}$

70. Divide 3.3 by 77 and round off to the nearest thousandth.

71. Divide 77 by 3.3 and round off to the nearest thousandth.

72. The records show that the tax bill on John and Freda's house was really $872.3449. What were the taxes to the nearest cent?

73. A keg of rivets weighs 500 pounds. If the empty keg weighs 11.4 lb and the average rivet weighs 0.78 lb, approximately how many rivets are in the keg?

Word Problems

OBJECTIVE

73. Solve word problems using proportions.

APPLICATION

Paula decided to start a savings account. Her weekly take-home pay (without overtime) is \$278. She decides to save \$13.90 of this each week. One week she works overtime and her take-home pay is \$320. If she wants to save from this check in the same ratio, how much should she save?

VOCABULARY

No new words.

HOW AND WHY

If the ratio of two quantities is constant, the ratio is used to find the missing part of a second ratio. For instance, if two pounds of grapes cost \$0.98, what will 12 pounds of grapes cost?

	CASE I	CASE II
Pounds of grapes	2	12
Cost in dollars	0.98	

In the chart the cost in Case II is missing. Call the missing value y.

	CASE I	CASE II
Pounds of grapes	2	12
Cost in dollars	0.98	y

Write the proportion using the ratios as shown in the chart.

$$\frac{2 \text{ lb of grapes}}{\$0.98} = \frac{12 \text{ lb of grapes}}{\$y}$$

Cross multiplying gives us:

$(2 \text{ lb of grapes})(\$y) = (12 \text{ lb of grapes})(\$0.98)$

The units are the same on each side, so we can drop them and have:

$$2y = 12(0.98)$$
$$2y = 11.76$$
$$y = 11.76 \div 2$$
$$y = 5.88$$

So, 12 pounds of grapes will cost \$5.88.

MODEL PROBLEM-SOLVING

Examples **Strategy**

a. On a road map of Oregon, $\frac{1}{4}$ inch represents 50 miles. How many miles are represented by $1\frac{1}{2}$ inches?

	CASE I	CASE II
Inches	$\frac{1}{4}$	$1\frac{1}{2}$
Miles	50	N

Make a chart in which the columns are labeled Case I and Case II. The first row is labeled "inches" and the second row is labeled "miles." Enter the appropriate measure in each box.

The letter N in the box is a place holder for the number of miles represented by $1\frac{1}{2}$ inches.

$$\frac{\frac{1}{4}}{50} = \frac{1\frac{1}{2}}{N}$$

A proportion to solve the problem is in the boxes. Write this proportion.

$$\frac{1}{4}N = 1\frac{1}{2} \cdot 50$$

Cross multiply. The units can be dropped, since they are the same on each side of the equal sign after cross multiplying.

$$\frac{1}{4}N = \frac{3}{2} \cdot \frac{50}{1}$$

Change the mixed number and the whole number to improper fractions and multiply.

$$\frac{1}{4}N = \frac{150}{2} = 75$$

$$N = 75 \div \frac{1}{4}$$

Do the related division problem.

$$N = \frac{75}{1} \cdot \frac{4}{1}$$

Change the division problem to a multiplication problem.

$$N = 300$$

Therefore, $1\frac{1}{2}$ inches on the map represents 300 miles.

Warm Up a. On a road map of Jackson County, $\frac{1}{4}$ inch represents 25 miles. How many miles are represented by $2\frac{1}{2}$ inches?

ANSWERS TO WARM UPS (5.3)

a. 250 miles

APPLICATION SOLUTION

b. Paula decided to start a savings account. Her weekly take-home pay (without overtime) is $278. She decides to save $13.90 of this each week. One week she works overtime and her take-home pay is $320. If she wants to save from this check in the same ratio, how much should she save?

	CASE I	CASE II
Pay	278	320
Savings	13.90	x

Draw a chart.

The letter x is the placeholder for the amount of savings from the larger check.

$$\frac{278}{13.90} = \frac{320}{x}$$

$$278x = 13.90(320)$$

Cross multiply. The units can be ignored, since they will be the same on each side of the equal sign after cross multiplying.

$$278x = 4448$$

Simplify the right-hand side.

$$x = 4448 \div 278$$

Do the related division problem.

$$x = 16$$

If she wants to save from this check in the same ratio, she should save $16.

Warm Up b. Paula receives a raise and is now taking home $330 (without overtime). If she wants to save in the same ratio as in Example b, how much should she now save each week?

ANSWERS TO WARM UPS (5.3)

b. $16.50

NAME ______________________

CLASS ____________ SECTION ____________ DATE ____________

Exercises 5.3

A

ANSWERS

1. ____________
2. ____________
3. ____________
4. ____________
5. ____________
6. ____________
7. ____________
8. ____________
9. ____________
10. ____________
11. ____________
12. ____________

A photograph that measures 3 inches wide and 5 inches high is to be enlarged so that the height will be 10 inches. What will be the width of the enlargement?

	CASE I	CASE II
Width	(a)	(c)
Height	(b)	(d)

1. What goes in box (a)?

2. What goes in box (b)?

3. What goes in box (c)?

4. What goes in box (d)?

5. What is the proportion for the problem?

6. What is the width of the enlargement?

If a fir tree that is 20 feet tall casts a shadow of 12 feet, how tall is a tree that casts a shadow of 18 feet at the same time and location?

	FIRST TREE	SECOND TREE
Height	(1)	(3)
Shadow	(2)	(4)

7. What goes in box (1)?

8. What goes in box (2)?

9. What goes in box (3)?

10. What goes in box (4)?

11. What is the proportion for the problem?

12. How tall is the second tree?

ANSWERS

13. ____________________

14. ____________________

15. ____________________

16. ____________________

17. ____________________

18. ____________________

Three women can assemble 17 television sets in 34 days. How many days will it take them to assemble 68 sets?

	CASE I	CASE II
TV sets	(5)	(7)
Number of days	(6)	(8)

13. What goes in box (5)?

14. What goes in box (6)?

15. What goes in box (7)?

16. What goes in box (8)?

17. What is the proportion for the problem?

18. How many days does it take to assemble 68 sets?

B

19. ____________________

20. ____________________

21. ____________________

22. ____________________

23. ____________________

The Mudville Elementary School expects an enrollment of 980 students. The district assigns teachers at the rate of 3 teachers for every 70 students. The district now employs 32 teachers. How many additional teachers does the district need to hire?

	CASE I	CASE II
Teachers	3	(e)
Students	70	(f)

19. What goes in box (e)?

20. What goes in box (f)?

21. What is the proportion for the problem?

22. How many teachers are needed next year?

23. How many additional teachers must be hired?

NAME ____________________

CLASS ____________ SECTION ____________ DATE ____________

ANSWERS

24. ____________

25. ____________

26. ____________

27. ____________

28. ____________

29. ____________

30. ____________

Marilee's recipe calls for 8 cups of flour for every 3 cups of sugar. For a large batch if she uses 120 cups of flour, how many cups of sugar will she need?

	CASE I	CASE II
Flour	8	g
Sugar	3	h

24. What goes into box (g)?

25. What goes into box (h)?

26. What is the original ratio?

27. What is the proportion?

28. How much sugar will be used?

The average restaurant in Universeville produces 30 pounds of garbage in $1\frac{1}{2}$ days. How many pounds of garbage do they produce in two weeks?

	CASE I	CASE II
Days		
Garbage		

29. What goes in each of the four boxes?

30. What is the proportion for the problem?

ANSWERS

31. ______

32. ______

33. ______

34. ______

35. ______

36. ______

37. ______

31. How many pounds of garbage do they have at the end of two weeks?

32. Merle is knitting a sweater. The knitting gauge is 8 rows to the inch. How many rows must she knit to complete $12\frac{1}{2}$ inches of the sweater?

	CASE I	CASE II
Rows		
Inches		

33. Floyd has 75 yards of fabric to make shirts. Every 2 shirts take 5 yards of fabric. How many shirts can he make from this fabric.

34. For every 2 hours a week that Helen is in class, she plans to spend 5 hours a week doing her homework. If she is in class 15 hours each week, how many hours will she plan to study each week?

35. It takes 50 hours of work to pay the tuition for 4 college credits at the local university. If John is going to take 15 credits in the fall, how many hours will he need to work to pay for his tuition?

36. If John (see Exercise 35) works 40 hours per week, how many weeks will he need to work to pay for this tuition? (Any part of a week counts as a full week.)

37. If 16 lb of fertilizer will cover 1500 square feet of lawn, how much fertilizer is needed to cover 2500 square feet?

NAME ______________________

CLASS ____________ SECTION ____________ DATE ____________

ANSWERS

38. ____________

39. ____________

40. ____________

41. ____________

42. ____________

43. ____________

44. ____________

38. The Joneses pay \$2100 in taxes on their home, which has an assessed value of \$64,000. How much will the taxes be on a \$96,000 home in the same district?

39. If 18 ounces of soap powder costs \$3.50, what will 45 ounces cost?

40. It takes 4 secretaries to type 85 pages of a certain book. At this rate, how many pages (to the nearest page) can 9 secretaries type?

41. A certain detergent is 28 ounces for \$4.20. What would 52 ounces cost at this rate?

42. At a local community college 2 credits cost \$47. What will be the cost of 15 credits at this college?

43. A certain new car traveled 300 miles in 4.2 hours. How long will it take to go 500 miles? (Assume no speeding tickets!)

44. Twenty-five pounds of tomatoes cost \$20.50 at a local market. At this rate, what would be the cost of 10 pounds?

ANSWERS

45. ____________

46. ____________

47. ____________

48. ____________

45. A baseball team won 85 out of the first 150 games. If the team is playing 180 total games this season, how many more should it win?

46. If 72 ounces of a soft drink cost $1.50, how many ounces will $10 buy?

47. James earns a salary of $800 a month, from which he saves $50. If his salary is increased to $900 a month, how much must he save each month to save at the same rate?

48. A room that contains 28 square yards was carpeted at a cost of $322. If the same kind of carpet is used in a room that contains 20 square yards, what will be the cost?

C

49. ____________

50. ____________

51. ____________

49. The counter on a tape recorder registers 640 after the recorder has been running for 20 minutes. What would the counter register after half an hour?

50. During the first 720 miles on their vacation trip the Scaberys used 35 gallons of gasoline. At this rate, to the nearest tenth of a gallon, how many gallons will be needed to finish the remaining 580 miles of their trip?

51. The Utah Construction Co. has a job that takes 5 people 18 hours to do. How many of these jobs could they do in 90 hours?

NAME ______________________________

CLASS ______________ SECTION ______________ DATE ______________

ANSWERS

52. ______________

53. ______________

54. ______________

55. ______________

56. ______________

57. ______________

52. A map of the western United States is scaled so that $\frac{3}{4}$ inch represents 100 miles. How many miles is it between San Diego and Seattle if the distance on the map is 9 inches?

53. If Wayne receives \$650 for $\frac{3}{4}$ ton of strawberries, how much will he receive for $1\frac{4}{5}$ ton?

54. If Nora receives \$800 for $\frac{2}{3}$ ton of raspberries, how much will 3.5 tons yield?

55. Eighty gallons of paint will cover 32,000 square feet of surface. How many gallons of paint should be purchased for 15,000 square feet. (Any part of a gallon means purchasing the entire gallon.)

56. Twelve cups of flour will make 900 cookies in a certain recipe. How much flour is needed for 1500 cookies?

57. Eight double rolls of wallpaper will cover 320 square feet of wall. How many square feet will 11 double rolls cover?

ANSWERS

58. ______

59. ______

60. ______

61. ______

62. ______

58. In the first 12 games of a 22-game schedule Mavis's basketball team scored a total of 990 points. At this rate, how many points can the team expect to score in its remaining games?

59. A 16-ounce can of pears costs $0.59 and a 29-ounce can costs $0.99. Is the price per ounce the same in both cases? If not, then to the nearest cent, what should be the price of the 29-ounce can to equalize the price per ounce?

60. A doctor requires that Ida, the nurse, give 9 milligrams of a certain drug to a patient. The drug is in a solution that contains 30 milligrams in one cubic centimeter. How many cubic centimeters should Ida use for the injection?

61. If a 24-foot beam of structural steel contracts 0.0064 inch for each drop of 5 degrees in temperature, then, to the nearest ten-thousandth of an inch, how much would a 16-foot beam of structural steel contract for a drop of 5 degrees in temperature?

62. If a box of mints weighing 1.5 ounces costs 40¢, what is the cost of one pound (16 ounces) of the mints?

D

63. ______

63. The ratio of boys to girls taking chemistry is 4 to 3. How many boys are there in a chemistry class of 84 students? (*Hint:* Fill in the rest of the table.)

	CASE I	CASE II
Number of boys		
Number of students	7	84

NAME ______________________________

CLASS ______________ SECTION ______________ DATE ______________

ANSWERS

64. ______________

64. Betty prepares a mixture of nuts that has cashews and peanuts in a ratio of 2 to 7. How many pounds of each will she need to make 54 pounds of the mixture?

65. ______________

65. Marcia's cereal mix has nuts to cereal in a ratio of 2 to 5. If she uses 98 ounces of mix, how many ounces of nuts will she need?

66. ______________

66. Debra is mixing green paint by using 2 quarts of blue for every 3 quarts of yellow. How much blue paint should she buy if she needs 100 quarts of green paint?

67. ______________

67. A concrete mix takes 3 bags of cement for every 2 bags of sand and every 2 bags of gravel. How many bags of cement are necessary if 70 bags of concrete are needed?

68. ______________

68. Mario makes meatballs for his famous spaghetti by using 5 pounds of ground round to 1.5 pounds of additives. How many pounds of ground round should he buy for 130 pounds of meatballs?

69. ______________

69. Martha makes her own glue by mixing 5 parts flour with 2 parts water. How much flour will she need for 3 quarts of glue?

ANSWERS

70. ______________

71. ______________

72. ______________

73. ______________

74. ______________

75. ______________

76. ______________

70. The estate of the late Ms. June Redgrave is to be divided among her 3 nephews in the ratio of 4 to 3 to 3. How much of the $84,930 estate will each nephew receive?

71. A brass alloy is 4 parts copper and 3 parts zinc. How many kilograms of copper are needed to make 200 kilograms of the alloy (to the nearest tenth of a kg)?

72. A yard (cubic yard) of concrete will make a 4-inch-thick slab that is 81 ft^2. How many yards of concrete (to the nearest tenth of a yard) are needed to pour a 4-inch-thick garage floor of 440 ft^2?

73. Jean paid a total of $848 to the bank when she borrowed $800 for one year. The bank still charges the same rate when she wants to borrow money again. What is the total amount she will pay to the bank if she borrows $950 and keeps it for one year?

74. If George pays back $1008 when he borrows $900 for one year, how much will he pay back if he borrows $750 and keeps it for one year?

75. The Corner Grocery Store bought $6875 worth of goods and sold them for $7700. At the same rate, what would goods cost that were sold for $5600?

76. The Slightly Used Auto Company bought an automobile for $7500 and sold it for $9000. If the company bought another automobile for $8200 at the same rate, for what price would the company expect to sell it?

NAME ______________________

CLASS ______________ SECTION ______________ DATE ______________

ANSWERS

77. ______________

78. ______________

79. ______________

80. ______________

81. ______________

77. When \$1 was worth ¥280 (Japanese yen) and a fan cost ¥37800, what was the cost in dollars?

78. When \$1 (U.S. dollars) was worth \$1.35 (Canadian dollars) and an automobile cost \$14,850 Canadian, what was the cost in U.S. dollars?

79. When \$1 was worth 130 drachma (Greek currency) and a refrigerator cost \$399, what was the cost in drachmas?

80. When \$1 was worth £0.8 (British pound) and a computer cost \$2595, what was the cost in pounds?

81. When \$1 was worth 1,874 lire (Italian currency) and a pair of shoes cost 89,952 lire, what was the cost in dollars?

E *Maintain your skills* (Sections 4.3, 4.5, 4.10, 4.11, 4.12)

82. ______________

83. ______________

Change each decimal to a fraction:

82. 0.554

83. 0.08875

ANSWERS

Change each fraction to a decimal to the nearest hundredth:

84. $\frac{9}{16}$

85. $\frac{17}{32}$

84. ________

85. ________

Perform the indicated operations:

86. $0.8(0.21) + (0.67)(9) - (0.3)(0.2)$

86. ________

87. $(0.094)(0.0008) \div (0.04) + 0.2(0.5)$

87. ________

88. Divide 38 by 0.67 and round off to the nearest tenth.

88. ________

89. Divide 49 by 27 and round off to the nearest hundredth.

89. ________

90. What is the cost of 15.8 gallons of gasoline that costs $1.419 per gallon? Round off your answer to the nearest cent.

90. ________

91. What is the cost per quart of a package of drink mix if a 2.3-oz package costs $2.59 and makes eight quarts of drink?

91. ________

NAME ____________________

CLASS ____________ SECTION ____________ DATE ____________

ANSWERS

1. ____________

2. ____________

3. ____________

4. ____________

5. ____________

6. ____________

7. ____________

8. ____________

9. ____________

10. ____________

CHAPTER 5 TRUE–FALSE CONCEPT REVIEW

Check your understanding of the language of basic mathematics. Tell whether or not each of the following is true (always true) or false (not always true).

1. A ratio is a comparison of two fractions.

2. A proportion is a comparison of two ratios.

3. The cross products of a proportion are always equal.

4. 5 people is an example of a measurement.

5. To determine whether a proportion is true or false, the ratios must have the same units.

6. To solve a proportion one needs to find the replacement for the missing number which will make the proportion true.

Use the following information to answer questions 7–10.

If a fir tree that is 18 ft tall casts a shadow of 17 ft, how tall is a tree that casts a shadow of 25 ft?

7. The following chart can be used to solve the given problem.

	FIRST TREE	SECOND TREE
Height	17	18
Shadow	x	25

8. The following chart can be used to solve the given problem.

	FIRST TREE	SECOND TREE
Shadow	17	25
Height	18	x

9. The proportion $\frac{18}{17} = \frac{x}{25}$ can be used to solve the given problem.

10. The height of the second tree will be less than the length of its shadow.

NAME ______________________

CLASS ______________ SECTION ______________ DATE ______________

ANSWERS

1. ______________

2. ______________

3. ______________

4. ______________

5. ______________

6. ______________

7. ______________

8. ______________

CHAPTER 5 POST-TEST

1. **(Section 5.1, Obj. 70)** Write a ratio to compare 16 pounds to 9 pounds.

2. On a test Ken answered 36 of 40 questions correctly. At the same rate, how many would he answer correctly if there were 100 questions on a test?

3. **(Section 5.2, Obj. 72)** Solve the proportion: $\frac{2.1}{9} = \frac{0.35}{w}$

4. **(Section 5.2, Obj. 71)** Is the following proportion true or false?
$$\frac{16}{34} = \frac{24}{51}$$

5. **(Section 5.2, Obj. 71)** Is the following proportion true or false?
$$\frac{9 \text{ inches}}{2 \text{ feet}} = \frac{6 \text{ inches}}{16 \text{ inches}}$$

6. **(Section 5.2, Obj. 72)** Solve the proportion: $\frac{9}{24} = \frac{y}{14}$

7. If Mary is paid $46.06 for 7 hours of work, how much should she expect to earn for 12 hours of work?

8. **(Section 5.1, Obj. 70)** Write a ratio to compare 6 hours to 2 days and reduce. (Compare in hours.)

CHAPTER 6

Percent

NAME ______________________

CLASS ______________ SECTION ______________ DATE ______________

ANSWERS

1. ______________

2. ______________

3. ______________

4. ______________

5. ______________

6. ______________

7. ______________

8. ______________

9. ______________

10. ______________

11. ______________

12. ______________

13. ______________

14. ______________

CHAPTER 6 PRE-TEST

The problems in the following pre-test are a sample of the material in the chapter. You may already know how to work some of these. If so, this will allow you to spend less time on those parts. As a result, you will have more time to give to the sections that gave you difficulty. Please work the following problems. The answers are in the back of the text.

1. **(Section 6.1, Obj. 74)** The latest graduation class at City U had 35 women out of every 100 graduates. What percent of the class were women?

2. **(Section 6.2, Obj. 75)** Write as a percent: 1.7

3. **(Section 6.2, Obj. 75)** Write as a percent: 0.357

4. **(Section 6.3, Obj. 76)** Write as a decimal: 0.3%

5. **(Section 6.3, Obj. 76)** Write as a decimal: 112%

6. **(Section 6.4, Obj. 77)** Change to a percent: $1\frac{5}{8}$

7. **(Section 6.4, Obj. 77)** Change to a percent (give answer to the nearest tenth of a percent): $\frac{7}{9}$

8. **(Section 6.5, Obj. 78)** Change to a fraction: 7.6%

9. **(Section 6.5, Obj. 78)** Change to a mixed number: $102\frac{2}{9}\%$

10. **(Section 6.7, Obj. 82)** What percent of 21 is 28? (To the nearest tenth of a percent.)

11. **(Section 6.7, Obj. 82)** 26% of what number is 11.7?

12. **(Section 6.7, Obj. 82)** What number is 73% of 73?

13. There were 2375 students enrolled at the local community college last fall. If 412 of the students dropped out, what percent of the students dropped out? (To the nearest whole percent.)

14. A TV set is listed at $679.95. During a sale it sold for a 15% discount. What was the sale price?

ANSWERS

15. ____________

16. ____________

17. ____________

15. What is the selling price of a radio if the markup is 27% of the cost and the cost is \$18.36? (To the nearest cent.)

16. **(Section 6.10, Obj. 85)** Given the following graph showing the grade distribution in an economics class:

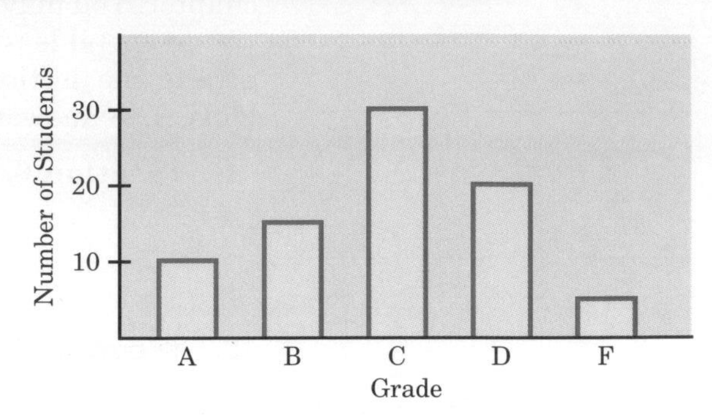

a. What was the grade given most often?
b. What was the total of the D and F grades?
c. How many more students received C grades than received A grades?

17. **(Section 6.11, Obj. 87)** Given the following table:

NUTRITIONAL INFORMATION PER SERVING SALAD DRESSINGS		
	Brand A	Brand B
Calories	30	16
Fat	3 g	0 g

a. Which dressing has the least calorie content?
b. How many grams of fat would be consumed in 5 servings of Brand A?
c. How many calories are saved by using 6 servings of Brand B over Brand A?

18. Construct a bar graph to display the career choices of students entering vocational education:

Aviation Maintenance	125
Auto Body Repair	75
Welding	50
Electronics	175
Diesel Services	100

What Is Percent?

OBJECTIVE

74. Write a percent to express a comparison of two numbers.

APPLICATION

Donna has $300 in her savings account. Last year she received a total of $24 in interest. The interest is what percent of the total in her account?

VOCABULARY

When ratios are used to compare numbers, the denominator is called the *base unit.* In comparing 80 to 100 $\left(\text{as the ratio } \frac{80}{100}\right)$, 100 is the base unit. The *percent comparison,* or just the *percent,* is a ratio with a base unit of 100. The percent $\frac{80}{100} = (80)\left(\frac{1}{100}\right)$ is usually written 80%. The symbol "%" is read "percent," and $\% = \frac{1}{100} = 0.01$.

HOW AND WHY

The word "percent" means "by the hundred." It is from the Roman word "percentum." In Rome taxes were collected by the hundred. For example, if you had 100 cattle, the tax collector might take 14 of them to pay your taxes. Hence, 14 per one hundred, or 14 percent, would be the tax rate.

Look at Figure 6.1 to see an illustration of the concept of "by the hundred." The base unit is 100, and 24 of the 100 parts are shaded. The ratio of shaded parts to total parts is $\frac{24}{100} = 24\left(\frac{1}{100}\right) = 24\%$. We say that 24% of the unit is shaded.

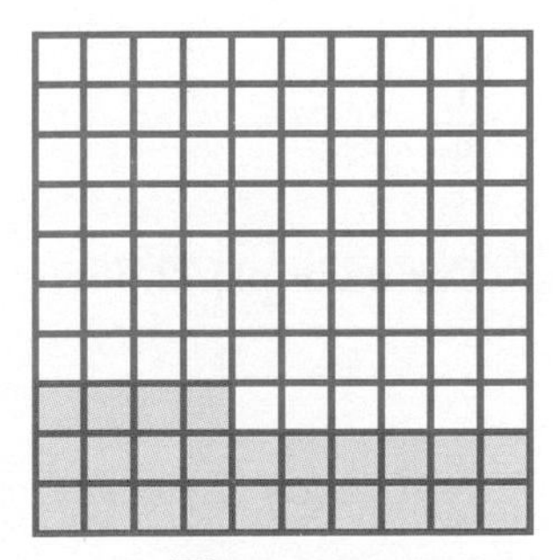

Figure 6.1

Figure 6.1 also illustrates that if the numerator is smaller than the denominator, then not all of the base unit will be shaded, and hence the comparison will be less than 100%. If the numerator equals the denominator, the entire unit will be shaded and the comparison will be 100%. If the numerator is larger than the denominator, more than one entire unit will be shaded, and the comparison will be more than 100%.

The ratio of two numbers can be used to find the percent when the base unit is not 100. Compare 7 to 20. The ratio is $\frac{7}{20}$. Now find the equivalent ratio with a denominator of 100:

$$\frac{7}{20} = \frac{35}{100} = 35 \cdot \frac{1}{100} = 35\%$$

If the equivalent ratio with a denominator of 100 cannot be found easily, solve as a proportion. See Example f.

PROCEDURE

To find the percent comparison of two numbers:

1. **Write the ratio of the first number to the base number.**
2. **Find the equivalent ratio with denominator 100.**
3. $\dfrac{\textbf{numerator}}{\textbf{100}} = \textbf{numerator} \cdot \dfrac{\textbf{1}}{\textbf{100}} = \textbf{numerator \%}$

The following chart shows some common fractions and their decimal equivalents. Some of the decimals are repeating decimals. Remember that a repeating decimal is shown by the bar over the digits that repeat. These fractions occur often in applications of percents. They should be memorized so that you can recall the patterns when they appear.

$\frac{1}{2} = 0.5$

$\frac{1}{3} = 0.33\overline{3}$ $\quad \frac{2}{3} = 0.66\overline{6}$

$\frac{1}{4} = 0.25$ $\quad \frac{3}{4} = 0.75$

$\frac{1}{5} = 0.2$ $\quad \frac{2}{5} = 0.4$ $\quad \frac{3}{5} = 0.6$ $\quad \frac{4}{5} = 0.8$

$\frac{1}{6} = 0.166\overline{6}$ $\quad \frac{5}{6} = 0.833\overline{3}$

$\frac{1}{8} = 0.125$ $\quad \frac{3}{8} = 0.375$ $\quad \frac{5}{8} = 0.625$ $\quad \frac{7}{8} = 0.875$

For example, $2.66\overline{6} = 2\frac{2}{3}$, $8.166\overline{6} = 8\frac{1}{6}$, and $17.125 = 17\frac{1}{8}$

MODEL PROBLEM-SOLVING

Examples | **Strategy**

a. What percent of the following region is shaded?

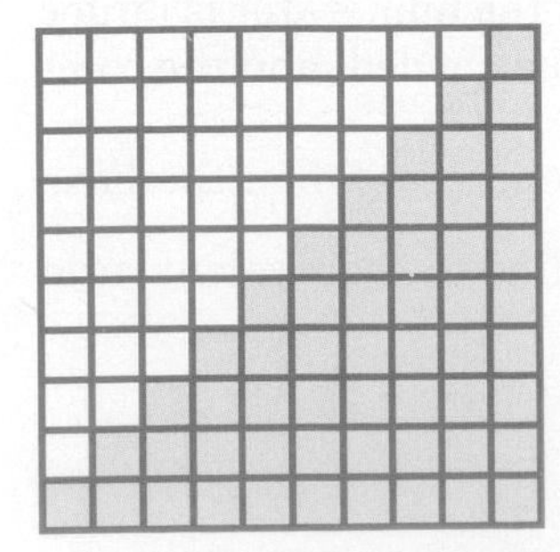

There are 100 small squares in the base unit. There are 55 shaded squares.

55 out of 100 parts or:

$$\frac{55}{100} = 55 \cdot \frac{1}{100} = 55\%.$$

The shaded portion represents 55% of the base unit.

Warm Up a. What percent of the following region is shaded?

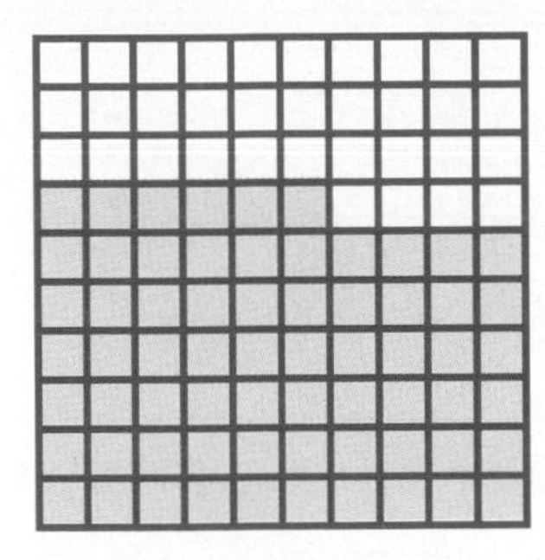

b. What percent of the following region is shaded?

Four out of 4 parts are shaded.

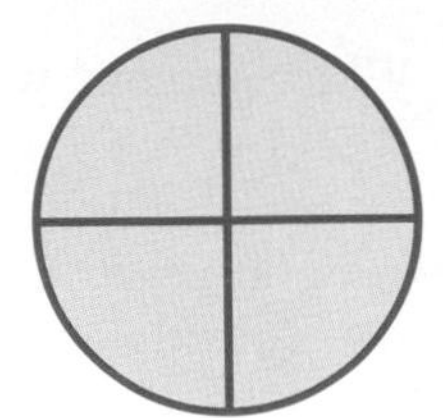

$$4 \text{ out of } 4 \text{ or } \frac{4}{4} = \frac{100}{100}$$

Change to a fraction with a denominator of 100.

$$= 100 \cdot \left(\frac{1}{100}\right)$$

$$= 100\%$$

The shaded portion represents 100% of the figure.

Warm Up b. What percent of the following region is shaded?

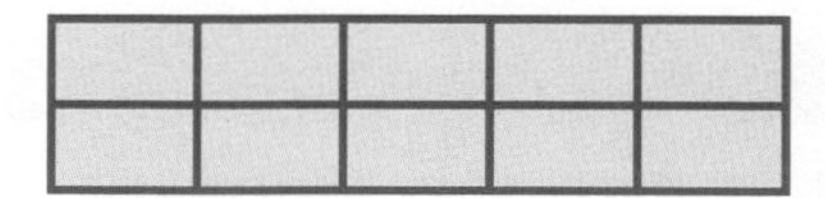

c. What percent of the figure is shaded?

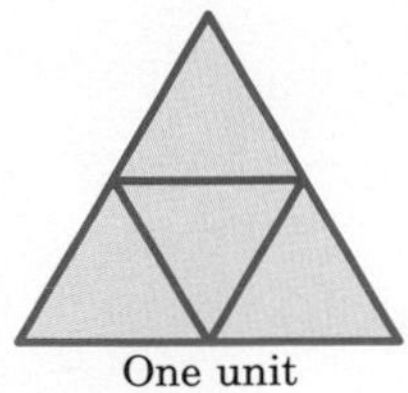
One unit

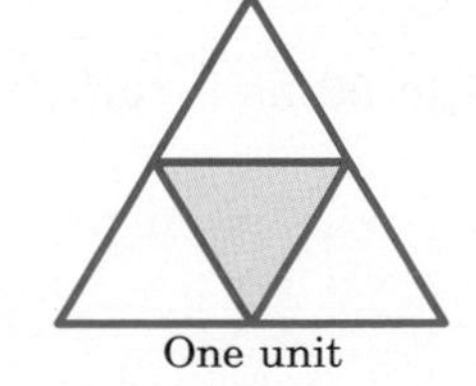
One unit

Each unit is divided into four parts. All four parts in one unit and one part in the second unit are shaded. Since each unit is divided into 4 units, the number of shaded parts must be compared to 4. Therefore, more than 100% is shaded.

ANSWERS TO WARM UPS (6.1) **a.** 66% **b.** 100%

$$\frac{5}{4} = \frac{125}{100}$$ **Change to a fraction with a denominator of 100.**

$$= 125 \cdot \left(\frac{1}{100}\right)$$

$$= 125\%$$

Therefore, 125% of the figure is shaded.

Warm Up c. What percent of the figure is shaded?

One unit

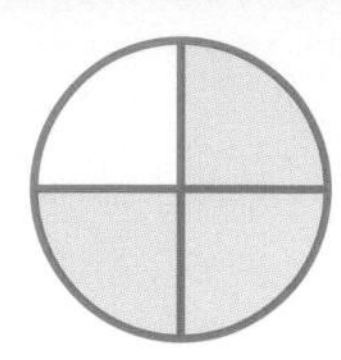

One unit

d. At a recent sporting event 30 women were among the first 100 people to enter. What percent were women?

$$\frac{30}{100} = 30 \cdot \left(\frac{1}{100}\right)$$ **The ratio of women to people is 30 to 100. Write the ratio and change to percent.**

$$= 30\%$$

Warm Up d. At the last basketball game of the season for Local High School, of the first 100 tickets sold, 61 were student tickets. What percent were student tickets?

e. What percent corresponds to the comparison of 42 to 25 as a ratio?

$$\frac{42}{25} = \frac{168}{100}$$ **Write the ratio and then change this fraction to one with a denominator of 100.**

$$= 168 \cdot \left(\frac{1}{100}\right)$$

CAUTION

The base unit for percent is *always* 100

$$= 168\%$$ **Change this fraction to a percent.**

The ratio of 42 to 25 is the same as 168%.

Warm Up e. What percent corresponds to the comparison of 80 to 50 as a ratio?

ANSWERS TO WARM UPS (6.1) **c.** 175% **d.** 61% **e.** 160%

f. Write the comparison of 12 to 18 as a percent.

$$\frac{12}{18} = \frac{R}{100}$$ **Write a proportion to solve the problem.**

$$12 \cdot 100 = 18 \cdot R$$ **Cross multiply.**

$$1200 \div 18 = R$$

$$66\frac{2}{3} = R$$ **We can write the remainder in fraction form or as a repeating decimal.**

$$66.66\overline{6} = R$$

So, $$\frac{12}{18} = \frac{66\frac{2}{3}}{100}$$ **Replace R in the original proportion with $66\frac{2}{3}$**

$$= 66\frac{2}{3} \cdot \frac{1}{100}$$ **Change to a percent.**

$$= 66\frac{2}{3}\%$$

Warm Up f. Write a comparison of 10 to 12 as a percent.

g. Calculator example

Find the percent comparison of 35 to 280.

$$\frac{35}{280} = \frac{R}{100}$$ **Percents can be found by proportions, so a calculator can be used in the same way as in Chapter 5.**

$$280 \cdot R = 35 \cdot 100$$ **Cross multiply and solve.**

$$R = (35 \cdot 100) \div 280$$

ENTER	DISPLAY
35	35.
×	35.
100	100.
÷	100.
280	280.
=	12.5

The comparison is 12.5%

ANSWERS TO WARM UPS (6.1)

f. $83\frac{1}{3}\%$

ENTER	DISPLAY
35	35.
÷	35.
280	280.
%	12.5

If your calculator has a "%" key, you can save two steps.

We read the percent from the display. Some calculators require pressing the = key also.

So, 35 is 12.5% of 280.

Warm Up g. Find the percent comparison of 45 to 240.

APPLICATION SOLUTION

h. Donna has \$300 in her savings account. Last year she received a total of \$24 in interest. The interest is what percent of the total in her account?

$$\frac{24}{300} = \frac{R}{100}$$

We want to compare \$24 to \$300. Write the proportion that expresses the ratio of 24 to 300 and R to 100.

$300 \cdot R = 100 \cdot 24$

Cross multiply.

$300 \cdot R = 2400$

$R = 2400 \div 300$

$R = 8$

or

We can also reduce the ratio and get the percent:

$$\frac{24}{300} = \frac{8}{100} = 8\left(\frac{1}{100}\right) = 8\%$$

Therefore, the interest was 8% of the total in her account.

If \$450 in a savings account earns \$27 interest in a year, what percent is the interest of the total in the account?

ANSWERS TO WARM UPS (6.1)

g. 18.75% **h.** 6%

NAME ______________________________

CLASS ______________ SECTION ______________ DATE ______________

Exercises 6.1

A

ANSWERS

1. ______________

2. ______________

3. ______________

4. ______________

5. ______________

6. ______________

7. ______________

8. ______________

9. ______________

10. ______________

11. ______________

12. ______________

13. ______________

14. ______________

15. ______________

16. ______________

What percent of each of the following regions are shaded?

1.

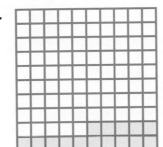

2.

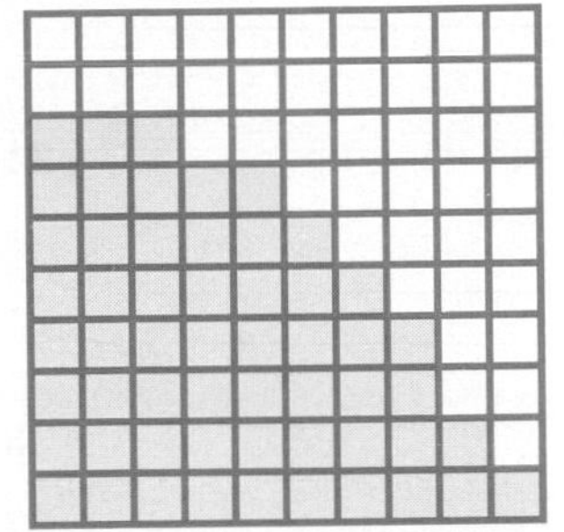

3.

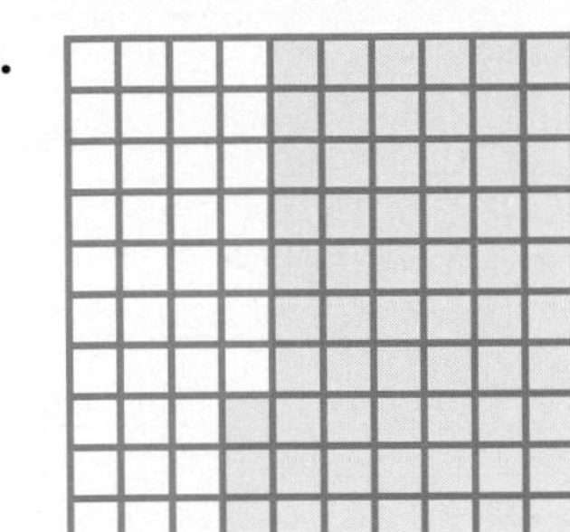

4.

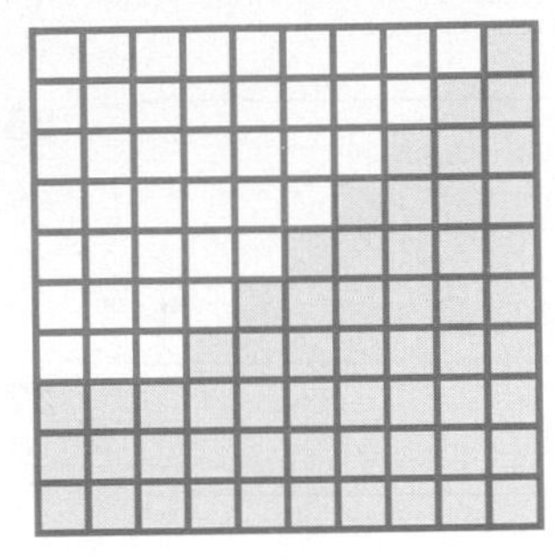

Write an exact percent for these comparisons; use fractions when necessary:

5. 17 out of 100

6. 78 per 100

7. 6 to 100

8. 7 out of 50

9. 10 per 50

10. 45 to 50

11. 11 out of 25

12. 24 per 25

13. 7 out of 20

14. 15 to 20

15. 30 per 25 parts

16. 33 per 20

B

ANSWERS

17. ______

18. ______

19. ______

20. ______

21. ______

22. ______

23. ______

24. ______

25. ______

26. ______

27. ______

28. ______

29. ______

30. ______

31. ______

32. ______

33. ______

34. ______

35. ______

36. ______

37. ______

38. ______

39. ______

40. ______

17. 78 to 50

18. 3 per 5

19. 68 parts per 200 parts

20. 120 per 300

21. 16 to 10

22. 300 to 120

23. 12 to 25

24. 160 out of 160

25. 62 to 62

26. 500 to 400

27. 215 to 100

28. 8 to 4

29. 30 to 10

30. 24 to 16

31. 30 to 25

32. 70 to 200

33. 65 to 50

34. 120 to 80

35. 20 to 60

36. 80 to 120

37. 9 per 15

38. 53 per 500

39. 93 to 80

40. 75 to 80

NAME ____________________

CLASS ______________ SECTION ______________ DATE ______________

C

ANSWERS

41. ______
42. ______
43. ______
44. ______
45. ______
46. ______
47. ______
48. ______
49. ______
50. ______
51. ______
52. ______
53. ______
54. ______
55. ______
56. ______
57. ______
58. ______
59. ______
60. ______
61. ______
62. ______

41. 106 to 1000

42. 279 to 240

43. 7 parts per 16 parts

44. 27 per 45

45. 95 to 114

46. 28 to 42

47. 49 to 56

48. 11 to 12

49. 7 to 32

50. 11 to 15

51. 62 to 93

52. 78 to 117

53. 98 to 21

54. 102 to 16

55. 105 to 25

56. 200 to 64

57. 200 to 48

58. 300 to 96

59. 37 to 75

60. 61 to 160

61. 105 to 15

62. 245 to 35

ANSWERS

63. ______

64. ______

65. ______

66. ______

67. ______

68. ______

69. ______

70. ______

71. ______

63. Carol spent $65 on a new outfit. If she had $100, what percent of her money did she spend on the outfit?

64. The fact that 10% of all women are blonde indicates that __?__ out of 100 women are blonde.

65. In a recent election, out of every 100 eligible voters, 62 cast their ballots. What percent of the eligible voters exercised the right to vote?

66. Of the people who use Shiny toothpaste, 37 out of 100 report that they have fewer cavities. Out of every 100 people who report, what percent does not report fewer cavities?

67. If the telephone tax rate is 8 cents per dollar, what percent is this?

68. For every $100 spent on gasoline, the state receives $6 tax. What percent of the price of gasoline is the state tax?

69. A bank pays $5.25 interest per year for every $100 in savings. What is the annual interest rate?

70. An electronic calculator originally priced at $100 is on sale for $93. What is the percent of discount? (Discount is the difference between the original price and the sale price.)

71. James has $500 in his savings account. Of that amount, $25 was interest that was paid to him. What percent of the total amount is interest?

NAME ______________________

CLASS ____________ SECTION ____________ DATE ____________

ANSWERS

72. ____________

73. ____________

74. ____________

72. Mickie bought a TV and makes monthly payments to pay for it. Last year she paid a total of $600. Of the total that she paid, $90 was interest. What percent of the total was the interest?

73. Last year Mr. and Mrs. Average were informed that the property tax rate on their house was $3 per $100 of the house's assessed value. What percent is the tax rate?

74. Salespeople are often paid by commission, which is expressed as a percent. If Pat is told that she will receive $8 for every $100 worth of merchandise she sells, what is her rate of commission?

E *Maintain your skills* (Sections 4.9, 4.11, 4.12)

75. ____________

76. ____________

77. ____________

78. ____________

79. ____________

80. ____________

81. ____________

82. ____________

Change to a decimal:

75. $\frac{15}{32}$

76. $\frac{35}{64}$

Change to a decimal rounded off to the hundredths place:

77. $\frac{17}{18}$

78. $\frac{31}{24}$

Find the average:

79. 16.2, 13.9, 15.6, 20.8, 17.5

80. 216.7, 345.77, 256.03, 198.54

Divide:

81. $456.003 \div 10^4$

82. $55{,}000 \div 10^7$

ANSWERS

83. ______

84. ______

83. Bill went to the store with $10. He used his calculator to keep track of the money he was spending. He decided that he could make the following purchases. Was he correct?

ARTICLE	COST
2 loaves of bread	$0.89 each
5 cans of soup	$0.38 each
1 box of crackers	$0.97
2 lb hamburger	$1.19 per lb
6 cans of root beer	6 cans for $2.49

84. Ms. Henderson earns $5.62 per hour and works the following hours during one month. How much are her monthly earnings?

WEEK	HOURS
1	$28\frac{1}{2}$
2	$32\frac{1}{4}$
3	33
4	$30\frac{1}{6}$
5	$6\frac{3}{4}$

Changing Decimals to Percents

OBJECTIVE

75. Write a given decimal as a percent.

APPLICATION

If the tax rate on a building lot is given as 0.03, what is the tax rate expressed as a percent?

VOCABULARY

No new words.

HOW AND WHY

In multiplication, where one factor is $\frac{1}{100}$, the indicated multiplication can be read as a percent. That is, $75\left(\frac{1}{100}\right) = 75\%$, $0.8\left(\frac{1}{100}\right) = 0.8\%$, and $\frac{3}{4}\left(\frac{1}{100}\right) = \frac{3}{4}\%$.

To write a number as a percent, multiply by $100 \cdot \frac{1}{100}$, a name for one. This is shown in the following table:

NUMBER	MULTIPLY BY 1 $100\left(\frac{1}{100}\right) = 1$	MULTIPLY BY 100	PERCENT
0.45	$0.45\ (100)\left(\frac{1}{100}\right)$	$45.\left(\frac{1}{100}\right)$	45%
0.2	$0.2\ (100)\left(\frac{1}{100}\right)$	$20.\left(\frac{1}{100}\right)$	20%
5	$5\ (100)\left(\frac{1}{100}\right)$	$500.\left(\frac{1}{100}\right)$	500%

In each case the decimal point is moved two places to the right and the percent symbol (%) is inserted.

PROCEDURE

To change a decimal to a percent:

1. **Move the decimal point two places to the right. (Write zeros on the right if necessary.)**
2. **Write the percent symbol (%) on the right.**

MODEL PROBLEM-SOLVING

Examples | **Strategy**

a. Change 0.35 to a percent:

0.35 = 35%

Move the decimal point two places to the right and write the percent symbol on the right.

Warm Up a. Change 0.24 to a percent.

b. Change 0.04 to a percent:

0.04 = 004% = 4%

Move the decimal point two places to the right and write the percent symbol on the right. Since the zeros are to the left of the "4," we can drop them.

Warm Up b. Change 0.09 to a percent.

c. Change 0.217 to a percent:

0.217 = 21.7%

Move the decimal point two places to the right and write the percent symbol on the right.

Warm Up c. Change 0.324 to a percent.

d. Change 0.003 to a percent:

0.003 = 000.3% = 0.3%

Warm Up d. Change 0.007 to a percent.

e. Change 9 to a percent:

9 = 9.00 = 900%

The decimal point follows the 9. Two zeros must be written on the right so we can move the decimal point two places.

Warm Up e. Change 4 to a percent.

ANSWERS TO WARM UPS (6.2)

a. 24% **b.** 9% **c.** 32.4% **d.** 0.7% **e.** 400%

f. Change 0.7 to a percent:

$0.7 = 0.70 = 70\%$

Warm Up f. Change 0.8 to a percent.

g. Change $0.2533\overline{3}$ to a percent:

$0.2533\overline{3} = 25.33\overline{3}\%$ — **Move the decimal two places to the right and write the percent symbol on the right.**

$= 25\frac{1}{3}\%$ — **Recall that the repeating decimal $0.33\overline{3} = \frac{1}{3}$.**

Warm Up g. Change $0.2266\overline{6}$ to a percent.

APPLICATION SOLUTION

h. If the tax rate on a building lot is given as 0.03, what is the tax rate expressed as a percent?

$0.03 = 003\% = 3\%$ — **To change 0.03 to a percent, move the decimal point two places to the right and write the percent symbol on the right.**

Warm Up h. If the tax rate on a building lot is given as 0.025, what is the tax rate expressed as a percent?

ANSWERS TO WARM UPS (6.2) **f.** 80% **g.** $22\frac{2}{3}\%$ **h.** 2.5%

NAME ______________________________

CLASS ______________ SECTION ______________ DATE ______________

Exercises 6.2

A

ANSWERS

1. ______
2. ______
3. ______
4. ______
5. ______
6. ______
7. ______
8. ______
9. ______
10. ______
11. ______
12. ______
13. ______
14. ______
15. ______
16. ______
17. ______
18. ______
19. ______
20. ______

Write each decimal as a percent:

1. 0.36 **2.** 0.24 **3.** 4.76 **4.** 8.33

5. 0.08 **6.** 0.06 **7.** 1.6 **8.** 2.4

9. 12 **10.** 11 **11.** 0.009 **12.** 0.003

13. 0.531 **14.** 0.555 **15.** 0.29 **16.** 0.74

17. 1 **18.** 2 **19.** 0.1 **20.** 0.5

B

21. ______
22. ______
23. ______
24. ______
25. ______
26. ______
27. ______
28. ______

21. 0.214 **22.** 0.083 **23.** 7 **24.** 27

25. 13.21 **26.** 1.27 **27.** 0.005 **28.** 0.745

ANSWERS

29. ____________
30. ____________
31. ____________
32. ____________
33. ____________
34. ____________
35. ____________
36. ____________
37. ____________
38. ____________
39. ____________
40. ____________

29. 0.7 **30.** 0.4 **31.** 3.2 **32.** 1.85

33. 0.0317 **34.** 0.265 **35.** 2.84 **36.** 1.23

37. 0.008 **38.** 0.007 **39.** 0.0015 **40.** 0.026

C

41. ____________
42. ____________
43. ____________
44. ____________
45. ____________
46. ____________
47. ____________
48. ____________
49. ____________
50. ____________
51. ____________
52. ____________
53. ____________
54. ____________
55. ____________
56. ____________

41. 5.75 **42.** 7.36 **43.** 0.5625 **44.** 4.23

45. $0.7416\overline{6}$ **46.** $0.66\overline{6}$ **47.** 0.2051 **48.** 0.3618

49. 0.1025 **50.** 0.205 **51.** 0.052 **52.** $0.033\overline{3}$

53. $0.276\overline{6}$ **54.** 0.27375 **55.** 0.0009 **56.** 0.0001

NAME ______

CLASS ______ SECTION ______ DATE ______

ANSWERS

57. ______

58. ______

59. ______

60. ______

61. ______

62. ______

57. 10 **58.** 12 **59.** 1.234 **60.** 2.145

61. $0.2033\overline{3}$ **62.** 0.156

D

63. ______

63. If the tax rate on a person's income is 0.22, what is that rate expressed as a percent?

64. ______

64. The completion rate in a certain math class is 0.85. What is this rate as a percent?

65. ______

65. Oregon had 0.37 of a month in sunny days. Express this as a percent.

66. ______

66. During that same month Buffalo had 0.26 of the month in sunny days. Express this as a percent.

67. ______

67. A certain class has completed 0.62 of its class work. Express this as a percent.

68. ______

68. The sales tax in a certain state is 0.055. Express this as a percent.

ANSWERS

69. ____________

70. ____________

71. ____________

72. ____________

73. ____________

69. If 0.375 of the contestants in a race withdraw, what is this expressed as a percent?

70. During one August the price of fans increased by 1.25. Express this as a percent.

71. During the blizzard of 1977 the price of a snow blower increased by 1.87. Express this as a percent.

72. The taxes on the McNamara home were raised 0.012 this year. Express this as a percent.

73. Mary Ellen measured the July rainfall. She found it was 0.235 of the year's total. Express this as a percent.

E *Maintain your skills* (Sections 4.5, 4.12)

74. ____________

75. ____________

76. ____________

77. ____________

74. Round off to the nearest thousandth: 3.87245

75. Round off to the nearest thousand: 3872.45

Perform the indicated operations:

76. $(0.37)(0.4) + 2.5 - (0.04)(0.02)$

77. $(0.75) - (0.5)(0.3) + (1.8)(0.2)$

NAME ____________________

CLASS ____________ SECTION ____________ DATE ____________

ANSWERS

78. ____________

79. ____________

80. ____________

81. ____________

82. ____________

83. ____________

78. $(0.18)(3.6) \div 18 - (1.2)(0.01)$

79. $0.5^2 - 0.2^3$

80. $[2.3^2 + 7.3(0.1)](0.3)^2 - 0.041$

81. $23 - 3.7^3(0.1)$

82. Marilyn needs to buy four textbooks for this semester. She goes to the bookstore to find out how much they cost:

Algebra	$27.45
Chemistry	$31.50
Psychology	$26.85
American history	$34.35

What is the total cost of the books that she needs? What is the average cost of the books?

83. Bill is making a table top. The two ends of the piece of wood he is using need to be trimmed and sanded. The piece of wood is 52.25 inches long. If he requires a length of 48.5 inches, the saw removes $\frac{1}{8}$ inch during each cut, and he wants to allow $\frac{1}{16}$ inch for finishing sanding on each end, how much should he cut from each end?

6.3 Changing Percents to Decimals

OBJECTIVE

76. Write a given percent as a decimal.

APPLICATION

When ordering bricks, a contractor orders 3% more than is needed to account for breakage. What decimal is entered into the computer to calculate the extra number of bricks to be added to the order?

VOCABULARY

No new words.

HOW AND WHY

The percent symbol indicates multiplication by $\frac{1}{100}$, so:

$$17\% = 17 \cdot \frac{1}{100} = \frac{17}{100} = 17 \div 100$$

To divide a number by 100, move the decimal point two places to the left:

$$17\% = 17 \div 100 = 0.17$$

PROCEDURE

To change a percent to a decimal

1. **Move the decimal point two places to the left. (Write zeros on the left if necessary.)**
2. **Drop the percent symbol (%).**

MODEL PROBLEM-SOLVING

Examples

Strategy

a. Change 14.5% to a decimal:

14.5% = 0.145

Move the decimal point two places to the left and drop the percent symbol.

Warm Up a. Change 92.3% to a decimal.

b. Change 35% to a decimal:

35% = 0.35

Warm Up b. Change 68% to a decimal.

ANSWERS TO WARM UPS (6.3)

a. 0.923 **b.** 0.68

c. Change 295% to a decimal:

295% = 2.95

Warm Up c. Change 127% to a decimal.

d. Change $83\frac{1}{2}\%$ to a decimal:

$83\frac{1}{2}\% = 83.5\%$ — **Change the fraction part of the percent to a decimal.**

$= 0.835$ — **Move the decimal point two places to the left and drop the percent symbol.**

Warm Up d. Change $75\frac{1}{5}\%$ to a decimal.

e. Change $29\frac{3}{8}\%$ to a decimal:

$29\frac{3}{8}\% = 29.375\%$ — **Recall that $\frac{3}{8} = 0.375$**

$= 0.29375$ — **Move the decimal point two places to the left and drop the percent symbol.**

Warm Up e. Change $32\frac{3}{4}\%$ to a decimal.

f. Change $83\frac{1}{6}\%$ to a decimal rounded off to the nearest thousandth:

$83\frac{1}{6}\% = 83.166\overline{6}\%$ — **Recall that $\frac{1}{6} = 0.166\overline{6}$**

$= 0.83166\overline{6}$ — **Move the decimal point two places to the left and drop the percent symbol.**

≈ 0.832 — **Round to the nearest thousandth.**

Warm up f. Change $9\frac{5}{6}\%$ to a decimal rounded off to the nearest thousandth.

ANSWERS TO WARM UPS (6.2) **c.** 1.27 **d.** 0.752 **e** 0.3275 **f.** 0.098

g. Change 0.5% to a decimal:

$0.5\% = 00.5\%$	**To move the decimal point two places to the left, we need two zeros to the left of the decimal point.**
$= 0.005$	**Move the decimal point two places to the left and drop the percent symbol.**

Warm Up g. Change 0.6% to a decimal.

h. Change $\frac{3}{4}\%$ to a decimal:

$\frac{3}{4}\% = 0.75\%$	**Change the fraction to a decimal:** $\frac{3}{4} = 0.75$
$= 00.75\%$	
$= 0.0075$	**Move the decimal point two places to the left and drop the percent symbol.**

Warm Up h. Change $\frac{3}{5}\%$ to a decimal.

APPLICATION SOLUTION

i. When ordering bricks, a contractor orders 3% more than is needed to account for breakage. What decimal is entered into the computer to calculate the extra number of bricks to be added to the order?

$3\% = 03\% = 0.03$	**To change the percent to a decimal, write a zero to the left of the 3 so there will be two places.**

Therefore, 0.03 is the number that must be entered into the computer.

Warm Up i. When ordering cement, a contractor orders 2% more than is needed. She does this to make sure that enough is on hand when a job is to be done. What decimal is entered into the computer to calculate the extra amount of cement to be added to the order?

ANSWERS TO WARM UPS (6.2) **g.** 0.006 **h.** 0.006 **i.** .02

NAME ______________________

CLASS ______________ SECTION ______________ DATE ______________

Exercises 6.3

A

ANSWERS

1. ______
2. ______
3. ______
4. ______
5. ______
6. ______
7. ______
8. ______
9. ______
10. ______
11. ______
12. ______
13. ______
14. ______
15. ______
16. ______
17. ______
18. ______
19. ______
20. ______

Write each of the following as a decimal:

1. 16% **2.** 59% **3.** 82% **4.** 36%

5. 73% **6.** 57% **7.** 2.15% **8.** 3.75%

9. 312% **10.** 563% **11.** 110.6% **12.** 53.7%

13. 0.04% **14.** 0.08% **15.** 135% **16.** 112%

17. 2.79% **18.** 1.79% **19.** 17.9% **20.** 179%

B

21. ______
22. ______
23. ______
24. ______
25. ______
26. ______
27. ______
28. ______

21. 314.7% **22.** 261.3% **23.** 0.12% **24.** 0.52%

25. $\frac{1}{2}\%$ **26.** $\frac{1}{4}\%$ **27.** 0.25% **28.** 0.75%

ANSWERS

29. ____
30. ____
31. ____
32. ____
33. ____
34. ____
35. ____
36. ____
37. ____
38. ____
39. ____
40. ____

29. 1% **30.** 100% **31.** 200% **32.** 0.082%

33. 0.058% **34.** 0.002% **35.** 125% **36.** 234%

37. $\frac{5}{8}\%$ **38.** $\frac{4}{5}\%$ **39.** 0.009% **40.** 0.006%

C

41. ____
42. ____
43. ____
44. ____
45. ____
46. ____
47. ____
48. ____
49. ____
50. ____
51. ____
52. ____
53. ____
54. ____
55. ____
56. ____
57. ____
58. ____
59. ____
60. ____

41. $29\frac{3}{4}\%$ **42.** $22\frac{2}{5}\%$ **43.** $475\frac{1}{2}\%$ **44.** $325\frac{1}{5}\%$

45. $\frac{7}{8}\%$ **46.** $\frac{1}{8}\%$ **47.** $1\frac{1}{4}\%$ **48.** $2\frac{7}{8}\%$

49. $\frac{7}{5}\%$ **50.** $\frac{11}{8}\%$ **51.** 72.61% **52.** 81.94%

53. 29.3468% **54.** 46.0134%

Change to a decimal rounded off to the nearest thousandth.

55. $\frac{1}{6}\%$ **56.** $\frac{2}{3}\%$ **57.** $35\frac{5}{6}\%$ **58.** $48\frac{1}{3}\%$

59. $\frac{11}{6}\%$ **60.** $\frac{5}{3}\%$

NAME ______________________________

CLASS ______________ SECTION ______________ DATE ______________

D

ANSWERS

61. ______________

62. ______________

63. ______________

64. ______________

65. ______________

66. ______________

67. ______________

68. ______________

61. A contractor orders 5% more sand than is needed to allow for shrinkage. What decimal part is this?

62. Employees just settled their new contract and got a 4.7% raise. Express this as a decimal.

63. When bidding for a job, an estimator adds 10% to cover unexpected expenses. What decimal part is this?

64. Interest rates are expressed as percents. The Last Federal Bank charges 16% interest on auto loans. What decimal will they use to compute the interest?

65. What decimal is used to compute the interest on a mortgage that has an interest rate of 14.25%?

66. Marc got a 15.75% bonus over last year. What decimal is this?

67. James is paid a 7.5% rate of commission. What decimal is used to compute the amount of his commission?

68. Mary is paid a 9% rate of commission. What will be the decimal used to compute the amount of her commission?

ANSWERS

69. ______

70. ______

69. Unemployment is down 0.2%. Express this as a decimal.

70. The cost of living rose 0.7% during June. Express this as a decimal.

E *Maintain your skills* (Sections 4.3, 4.11)

71. ______

72. ______

73. ______

74. ______

75. ______

76. ______

77. ______

78. ______

79. ______

80. ______

Change to a decimal:

71. $\frac{7}{8}$

72. $\frac{9}{64}$

73. $\frac{19}{16}$

74. $\frac{24}{75}$

75. $\frac{27}{15}$

76. $\frac{117}{65}$

Change to a fraction:

77. 0.715

78. 0.1025

79. In one week George earns $245. His deductions (income tax, Social Security, and so on) total $38.45. What is his "take-home" pay?

80. The cost of gasoline is reduced from $0.419 per liter to $0.379 per liter. How much money is saved on an automobile trip that requires 230 liters?

6.4 Fractions to Percent

OBJECTIVE

77. Change a fraction or mixed number to a percent.

APPLICATION

A motor that needs repair is only turning $\frac{3}{4}$ the number of revolutions per minute that it should. What percent of the normal rate is this?

VOCABULARY

No new words.

HOW AND WHY

We already know how to change fractions to decimals and decimals to percent. We combine the two ideas to change fractions to a percent.

PROCEDURE

To change a fraction or mixed number to a percent:

1. **Change to a decimal with two decimal places. The decimal is rounded off or carried out as directed.**
2. **Change the decimal to percent.**

MODEL PROBLEM-SOLVING

Examples	Strategy
a. Change $\frac{4}{5}$ to a percent:	
$\frac{4}{5} = 0.8$	**Divide 4 by 5 to change the fraction to a decimal.**
$= 80\%$	**Move the decimal point two places to the right and write the percent symbol on the right.**

Warm Up a. Change $\frac{3}{5}$ to a percent.

ANSWERS TO WARM UPS (6.4)

a. 60%

b. Change $\frac{5}{8}$ to a percent:

$\frac{5}{8} = 0.625$

$= 62.5\%$

Change to a decimal. Then move the decimal point two places to the right and place the percent symbol on the right.

Warm Up b. Change $\frac{1}{8}$ to a percent.

c. Change $\frac{5}{6}$ to a percent:

$\frac{5}{6} = 0.833\overline{3}$

Write the fraction, $\frac{5}{6}$, as a repeating decimal.

$= 83.3\overline{3}\%$

Move the decimal point two places to the right and write the percent symbol on the right.

$= 83\frac{1}{3}\%$

The repeating decimal $0.3\overline{3} = \frac{1}{3}$

Warm Up c. Change $\frac{1}{6}$ to a percent.

d. Change $\frac{3}{7}$ to a percent rounded off to the nearest tenth of a percent.

$\frac{3}{7} \approx 0.429$

$\approx 42.9\%$

Change $\frac{3}{7}$ to a decimal rounded to the nearest thousandth.

CAUTION

One tenth of a percent, that is $\frac{1}{10}$ of $\frac{1}{100}$, is $\frac{1}{1000} = 0.001$.

Warm Up d. Change $\frac{4}{7}$ to a percent rounded off to the nearest tenth of a percent.

ANSWERS TO WARM UPS (6.4)

b. 12.5% **c.** $16\frac{2}{3}\%$ **d.** 57.1%

e. Change $\frac{1}{40}$ to a percent:

$\frac{1}{40} = 0.025 = 002.5\% = 2.5\%$

Warm Up e. Change $\frac{3}{40}$ to a percent:

f. Change $2\frac{1}{2}$ to a percent:

$2\frac{1}{2} = 2.5$ — Change the fraction part to a decimal.

$= 2.50$

$= 250\%$

Warm Up f. Change $2\frac{3}{4}$ to a percent.

g. Change $\frac{7}{320}$ to a percent rounded off to the nearest tenth of a percent.

$\frac{7}{320} \approx 0.022$ — Change $\frac{7}{320}$ to a decimal rounded to the nearest thousandth.

$\approx 2.2\%$ — Change the decimal to percent.

Warm Up g. Change $\frac{7}{160}$ to a percent rounded off to the nearest tenth of a percent.

h. Calculator example

Change $\frac{3}{80}$ to a percent:

ENTER	DISPLAY
3	3.
÷	3.
80	80.
%	3.75

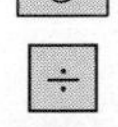
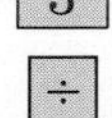

Press the keys indicated in the "enter" column. The numbers should show on the display as indicated.

The "%" key moves the decimal two places to the right.

Therefore, $\frac{3}{80} = 3.75\%$

ANSWERS TO WARM UPS (6.4) **e.** 7.5% **f.** 275% **g.** 4.4%

Warm Up h. Change $\frac{23}{400}$ to a percent.

i. Calculator example

Change $2\frac{9}{40}$ to a percent:

$2\frac{9}{40} = \frac{89}{40}$

Change the mixed number to an improper fraction. Change this fraction to a percent.

ENTER	DISPLAY
89	89.
÷	89.
40	40.
%	222.5

Therefore, $2\frac{9}{40} = 222.5\%$

Warm Up i. Change $3\frac{7}{80}$ to a percent.

APPLICATION SOLUTION

j. A motor that needs repair is only turning $\frac{3}{4}$ the number of revolutions per minute that it should. What percent of the normal rate is this?

$\frac{3}{4} = 0.75$

We need to change $\frac{3}{4}$ to a percent.

$= 75\%$

First change the fraction to a decimal, then change the decimal to a percent.

Therefore, the motor is turning at 75% of its normal rate.

Warm Up j. An 8-cylinder motor has only 7 of its cylinders firing. What percent of its cylinders are firing?

ANSWERS TO WARM UPS (6.4) **h.** 5.75% **i.** 308.75% **j.** 87.5% or $87\frac{1}{2}\%$

NAME ______________________

CLASS ______________ SECTION ______________ DATE ______________

Exercises 6.4

A

ANSWERS

1. ______
2. ______
3. ______
4. ______
5. ______
6. ______
7. ______
8. ______
9. ______
10. ______
11. ______
12. ______
13. ______
14. ______
15. ______
16. ______
17. ______
18. ______
19. ______
20. ______

Change each fraction to a percent:

1. $\frac{75}{100}$ **2.** $\frac{50}{100}$ **3.** $\frac{11}{50}$ **4.** $\frac{3}{10}$

5. $\frac{17}{20}$ **6.** $\frac{4}{25}$ **7.** $\frac{1}{2}$ **8.** $\frac{3}{5}$

9. $\frac{7}{25}$ **10.** $\frac{7}{50}$ **11.** $\frac{3}{20}$ **12.** $\frac{3}{25}$

13. $\frac{21}{20}$ **14.** $\frac{13}{10}$ **15.** $2\frac{1}{2}$ **16.** $4\frac{1}{4}$

17. $\frac{72}{200}$ **18.** $\frac{24}{300}$ **19.** $\frac{3}{8}$ **20.** $\frac{5}{16}$

B

21. ______
22. ______
23. ______
24. ______
25. ______
26. ______
27. ______
28. ______

Change each fraction or mixed number to a percent.

21. $\frac{2}{3}$ **22.** $\frac{2}{12}$ **23.** $\frac{7}{6}$ **24.** $\frac{8}{3}$

25. $1\frac{5}{6}$ **26.** $2\frac{1}{3}$ **27.** $\frac{21}{400}$ **28.** $\frac{17}{200}$

ANSWERS

29. ______
30. ______
31. ______
32. ______
33. ______
34. ______
35. ______
36. ______
37. ______
38. ______

Change each fraction or mixed number to a percent. Round to the nearest tenth of a percent.

29. $\frac{14}{15}$ **30.** $\frac{15}{28}$ **31.** $\frac{4}{9}$ **32.** $\frac{1}{6}$

33. $\frac{5}{6}$ **34.** $\frac{7}{9}$ **35.** $1\frac{5}{6}$ **36.** $2\frac{7}{9}$

37. $\frac{33}{400}$ **38.** $\frac{57}{200}$

C

39. ______
40. ______
41. ______
42. ______
43. ______
44. ______
45. ______
46. ______
47. ______
48. ______
49. ______
50. ______
51. ______
52. ______
53. ______
54. ______
55. ______
56. ______
57. ______
58. ______

39. $\frac{1}{3}$ **40.** $\frac{1}{6}$ **41.** $\frac{7}{13}$ **42.** $\frac{8}{15}$

43. $1\frac{3}{7}$ **44.** $2\frac{5}{9}$ **45.** $\frac{8}{21}$ **46.** $\frac{5}{6}$

47. $\frac{1}{400}$ **48.** $\frac{37}{400}$ **49.** $\frac{23}{6000}$ **50.** $\frac{17}{1200}$

51. $\frac{93}{5000}$ **52.** $\frac{121}{4000}$ **53.** $\frac{79}{8000}$ **54.** $\frac{153}{1600}$

55. $4\frac{8}{9}$ **56.** $3\frac{5}{6}$ **57.** $5\frac{4}{7}$ **58.** $6\frac{2}{3}$

NAME ______________________________

CLASS ______________ SECTION ______________ DATE ______________

D

ANSWERS

59. ______________

60. ______________

61. ______________

62. ______________

63. ______________

64. ______________

65. ______________

59. In a certain algebra class 32 out of 40 students took arithmetic before taking algebra. What percent took arithmetic?

60. Maureen got 19 problems correct on a 25-problem test. What percent was correct?

61. Tasha applied for a loan to buy a motorcycle. The annual interest charged was 35¢ for each \$2 of the loan $\left(\frac{35}{200}\right)$. What was the annual rate (percent) of interest?

62. Four-sevenths of the eligible voters turned out for the recent Hillsboro city elections. What percent of the voters turned out, to the nearest tenth of a percent?

63. In a supermarket 2 eggs out of 9 dozen are lost because of cracks. What percent of eggs must be discarded, to the nearest tenth of a percent?

64. The owner of a small business pays 17 cents for insurance for every \$100 of insurance coverage. What is the percent rate of the insurance?

65. The tachometer on an automobile can register 6000 revolutions per minute (rpm). If the reading on the tachometer when the motor is idling is 900, what percent of the maximum reading is the motor turning?

ANSWERS

66. ______________

67. ______________

68. ______________

66. In one day the Wilted Produce Store sold 18 of the 45 boxes of apples they had in stock. This means that they sold $\frac{18}{45}$ of their stock. What percent of their stock did they sell?

67. On the same day as in Exercise 66 the Wilted Produce Store sold 46 of the 200 sacks of onions it had in stock. What percent of the stock of onions did the store sell?

68. The price of gasoline at the 4-Corner Service Station currently is three times the price it was 10 years ago. Today's price is what percent of the price 10 years ago?

E *Maintain your skills* (Sections 4.8, 4.10)

69. ______________

70. ______________

71. ______________

72. ______________

73. ______________

74. ______________

75. ______________

76. ______________

77. ______________

78. ______________

Multiply:

69. (8.003)(0.87)

70. (19)(0.0115)

71. (0.02)(0.2)(2.02)

72. (1.45)(4.05)(1.4)

Divide:

73. $0.38\overline{)5.738}$

74. $6.22\overline{)0.202772}$

75. Divide 48 by 6.2 and round off to the nearest hundredth.

76. Divide 62 by 480 and round to the nearest thousandth.

77. If Abel worked 37 hours and earned a total of $197.95, what was his hourly rate?

78. If Spencer lost 11.6 lb in two weeks, what was his rate of weight loss per day (that is, what is the average loss per day) to the nearest hundredth of a pound?

Percents to Fractions

OBJECTIVE

78. Change percents to fractions or mixed numbers.

APPLICATION

It is estimated that the spraying for the gypsy moth in Salemtown was 92% successful. What fraction (ratio comparison) of the moths was eliminated?

VOCABULARY

No new words.

HOW AND WHY

The expression $6.5\% = 6.5 \times \frac{1}{100}$. This gives a very efficient method for changing a percent to a fraction. Change 6.5 to a fraction and do the multiplication. See Example c.

PROCEDURE

To change a percent to a fraction or a mixed number:

1. **Replace the percent symbol (%) with the fraction $\left(\frac{1}{100}\right)$.**
2. **Rewrite the other factor (if necessary) as a fraction.**
3. **Multiply.**

MODEL PROBLEM-SOLVING

Examples / **Strategy**

a. Change 35% to a fraction:

$$35\% = 35 \times \frac{1}{100}$$

$$= \frac{35}{100}$$

Replace the percent symbol (%) with $\frac{1}{100}$ and multiply.

CAUTION

You need to multiply by $\frac{1}{100}$, not just write it down.

$$= \frac{7}{20}$$

Reduce.

Warm Up a. Change 25% to a fraction.

ANSWERS TO WARM UPS (6.5)

a. $\frac{1}{4}$

b. Change 312% to a fraction:

$312\% = 312 \times \frac{1}{100}$ — **Replace the percent symbol (%) with $\frac{1}{100}$ and multiply.**

$= \frac{312}{100}$

$= 3\frac{12}{100}$ — **Change the improper fraction to a mixed number.**

$= 3\frac{3}{25}$ — **Reduce the fraction part.**

Warm Up b. Change 120% to a fraction.

c. Change 6.5% to a fraction:

$6.5\% = 6.5 \times \frac{1}{100}$ — **Replace the percent symbol with the fraction $\frac{1}{100}$.**

$= 6\frac{1}{2} \times \frac{1}{100}$ — **Change the decimal to a mixed number.**

$= \frac{13}{2} \times \frac{1}{100}$ — **Change the mixed number to an improper fraction.**

$= \frac{13}{200}$ — **Multiply.**

Warm Up c. Change 8.2% to a fraction.

d. Change $12\frac{2}{3}\%$ to a fraction:

$12\frac{2}{3}\% = 12\frac{2}{3} \times \frac{1}{100}$ — **Replace the percent symbol (%) with $\frac{1}{100}$ and multiply.**

$= \frac{38}{3} \times \frac{1}{100}$ — **Change the mixed number to an improper fraction.**

ANSWERS TO WARM UPS (6.5)

b. $\frac{6}{5}$ or $1\frac{1}{5}$ **c.** $\frac{41}{500}$

$= \dfrac{\overset{19}{\cancel{38}}}{3} \times \dfrac{1}{\underset{50}{\cancel{100}}}$ **Reduce.**

$= \dfrac{19}{150}$ **Multiply.**

Warm Up d. Change $5\frac{5}{6}\%$ to a fraction.

e. Change $5\frac{5}{8}\%$ to a fraction:

$5\frac{5}{8}\% = \dfrac{45}{8} \times \dfrac{1}{100}$ **Change the mixed number to a fraction and replace the percent symbol (%) with $\frac{1}{100}$.**

$= \dfrac{9}{160}$ **Multiply.**

Warm Up e. Change $12\frac{2}{3}\%$ to a fraction.

APPLICATION SOLUTION

f. It is estimated that the spraying for the gypsy moth in Salemtown was 92% successful. What fraction (ratio comparison) of the moths was eliminated?

$92\% = 92 \times \dfrac{1}{100}$ **Replace the percent symbol with the fraction $\frac{1}{100}$ and multiply.**

$= \dfrac{92}{100}$

$= \dfrac{23}{25}$

Therefore, $\frac{23}{25}$ of the moths were destroyed.

Warm Up f. Gail scored 84% on a math test. What fraction of the questions did she answer correctly?

ANSWERS TO WARM UPS (6.5) d. $\frac{7}{120}$ e. $\frac{19}{150}$ f. $\frac{21}{25}$

NAME ______________________

CLASS ______________ SECTION ______________ DATE ______________

Exercises 6.5

A

ANSWERS

1. ______
2. ______
3. ______
4. ______
5. ______
6. ______
7. ______
8. ______
9. ______
10. ______
11. ______
12. ______
13. ______
14. ______
15. ______
16. ______
17. ______
18. ______
19. ______
20. ______

Change each of the following percents to fractions or mixed numbers:

1. 5% **2.** 8% **3.** 35% **4.** 65%

5. 125% **6.** 175% **7.** 400% **8.** 600%

9. 70% **10.** 20% **11.** 56% **12.** 82%

13. 75% **14.** 80% **15.** 90% **16.** 15%

17. 100% **18.** 150% **19.** 112% **20.** 285%

B

ANSWERS

21. ______
22. ______
23. ______
24. ______
25. ______
26. ______
27. ______
28. ______
29. ______
30. ______
31. ______
32. ______
33. ______
34. ______
35. ______
36. ______
37. ______
38. ______
39. ______
40. ______
41. ______
42. ______
43. ______
44. ______

21. 27.5% **22.** 38.4% **23.** 6.3% **24.** 2.5%

25. 0.045% **26.** 0.072% **27.** 20.5% **28.** 40.6%

29. $\frac{1}{3}\%$ **30.** $\frac{2}{3}\%$ **31.** $5\frac{1}{2}\%$ **32.** $7\frac{3}{4}\%$

33. $10\frac{1}{2}\%$ **34.** $16\frac{1}{2}\%$ **35.** $\frac{3}{4}\%$ **36.** $\frac{3}{5}\%$

37. 32.6% **38.** 72.7% **39.** $50\frac{1}{2}\%$ **40.** $27\frac{3}{4}\%$

41. 0.85% **42.** 0.68% **43.** 0.004% **44.** 0.008%

NAME ______________________________

CLASS ______________ SECTION ______________ DATE ______________

C

ANSWERS

45. ______
46. ______
47. ______
48. ______
49. ______
50. ______
51. ______
52. ______
53. ______
54. ______
55. ______
56. ______
57. ______
58. ______
59. ______
60. ______
61. ______
62. ______

45. $16\frac{5}{7}\%$ **46.** $27\frac{3}{7}\%$ **47.** $9\frac{1}{5}\%$

48. $16\frac{1}{4}\%$ **49.** 0.0016% **50.** 0.0024%

51. $2\frac{41}{80}\%$ **52.** $1\frac{1}{9}\%$ **53.** 0.375%

54. 0.125% **55.** $116\frac{2}{3}\%$ **56.** $183\frac{1}{3}\%$

57. 0.006% **58.** 0.004% **59.** 0.0008%

60. 0.0004% **61.** 0.0375% **62.** 0.0128%

ANSWERS

63. ______________

64. ______________

63. $266\frac{2}{3}\%$ **64.** $283\frac{5}{6}\%$

D

65. ______________

65. George and Ethel paid 16% of their annual income in taxes last year. What fractional part of their income went to taxes?

66. ______________

66. One summer Judy earned 25% of her college expenses working at a local restaurant. What fractional part of her expenses did she earn that summer?

67. ______________

67. The Kilroys spend 35% of their monthly income on their mortgage payment. What fractional part goes toward the mortgage?

68. ______________

68. George spends 42% of his salary on his car each month. What fractional part is spent on George's car?

69. ______________

69. The enrollment at City Community College this year is 118% of last year's enrollment. What fractional increase in enrollment took place this year?

NAME ______________________________

CLASS ______________ SECTION ______________ DATE ______________

ANSWERS

70. ______________

70. The city budget is 132% over last year's budget. What fraction does this represent?

71. ______________

71. A census determined that $37\frac{1}{2}\%$ of the residents of a certain city was of age 40 or over and that 50% was of age 25 or under. What fraction of the residents is between the ages of 25 and 40?

72. ______________

72. Brooks and Foster form a partnership. If Brooks' investment is $56\frac{1}{4}\%$ of the total, what fraction of the total is Foster's share?

73. ______________

73. The spraying for the med fly was found to be 88% successful. What fraction (ratio comparison) of the med flies were eliminated?

74. ______________

74. Roger got a 92% on his algebra test. What fractional part did he get correct?

E *Maintain your skills* (Section 5.1)

ANSWERS

75. ______

76. ______

77. ______

78. ______

79. ______

80. ______

81. ______

82. ______

83. ______

84. ______

Are the following proportions true or false?

75. $\dfrac{\frac{2}{3}}{\frac{4}{5}} = \dfrac{500}{600}$

76. $\dfrac{\frac{1}{2}}{\frac{7}{8}} = \dfrac{50}{87}$

77. $\dfrac{36}{12.5} = \dfrac{81}{35}$

78. $\dfrac{25.75}{12} = \dfrac{103}{48}$

79. Write a ratio to compare $39,000,000 to 75,000,000 people.

80. Reduce Exercise 79 to a one-person comparison.

81. Write a ratio to compare 14,765,000 gallons of water to 100,000 people.

82. Reduce Exercise 81 to a one-person comparison.

83. The taxes on a home valued at $89,000 are $2225. At the same rate, what are the taxes on a house valued at $75,000?

84. Five measurements of the diameter of a wire are taken with a micrometer screw gauge. The five estimated measurements were 2.31 mm, 2.32 mm, 2.30 mm, 2.34 mm, and 2.30 mm. What is the average estimate? (Note that although 2.3 = 2.30, writing 2.30 mm shows greater precision than writing 2.3 mm.)

Fractions, Decimals, and Percents (Review)

OBJECTIVES

79. Write a percent as a decimal and as a fraction.
80. Write a fraction as a percent and as a decimal.
81. Write a decimal as a percent and as a fraction.

VOCABULARY

No new words.

HOW AND WHY

Decimals, fractions, and percents can each be expressed in terms of the others:

$$50\% = 50 \cdot \frac{1}{100} = \frac{50}{100} = \frac{1}{2} \quad \text{and} \quad 50\% = 0.50$$

$$\frac{3}{4} = 3 \div 4 = 0.75 \quad \text{and} \quad \frac{3}{4} = 0.75 = 75\%$$

$$0.65 = 65\% \quad \text{and} \quad 0.65 = \frac{65}{100} = \frac{13}{20}$$

MODEL PROBLEM-SOLVING

Examples | **Strategy**

In the following table fill in the empty spaces with the related fraction, decimal, or percent.

FRACTION	DECIMAL	PERCENT
		30%
$\frac{7}{8}$		
	0.62	

FRACTION	DECIMAL	PERCENT
$\frac{3}{10}$	0.30	30%
$\frac{7}{8}$	0.875	87.5% or $87\frac{1}{2}\%$
$\frac{31}{50}$	0.62	62%

$$30\% = 0.30 = \frac{30}{100} = \frac{3}{10}$$

$$\frac{7}{8} = 0.875 = 87.5\% = 87\frac{1}{2}\%$$

$$0.62 = 62\% = \frac{62}{100} = \frac{31}{50}$$

Warm Up. In the following table fill in the empty spaces with the related fraction, decimal or percent.

FRACTION	DECIMAL	PERCENT
$\frac{2}{3}$		
		27%
	0.72	
$\frac{73}{100}$		
		160%
	1.3	

ANSWERS TO WARM UPS (6.6)

FRACTION	DECIMAL	PERCENT
$\frac{2}{3}$	$0.66\overline{6}$	$66\frac{2}{3}\%$
$\frac{27}{100}$	0.27	27%
$\frac{18}{25}$	0.72	72%
$\frac{73}{100}$	0.73	73%
$1\frac{3}{5}$	1.6	160%
$1\frac{3}{10}$	1.3	130%

NAME ______________________________

CLASS ______________ SECTION ______________ DATE ______________

Exercises 6.6

A–D

Fill in the empty spaces with the related percent, decimal, or fraction:

	FRACTION	DECIMAL	PERCENT
*	$\frac{1}{10}$		
*			30%
*		0.75	
*	$\frac{9}{10}$		
			145%
*	$\frac{3}{8}$		
		0.001	
*		1	
	$2\frac{1}{4}$		
*		0.8	
			$5\frac{1}{2}\%$
*		0.875	
			$\frac{1}{2}\%$
*		0.6	
*			$62\frac{1}{2}\%$
			50%

*If you are going to be working with problems that will involve percent in your job or in your personal finances (loans, savings, insurance, and so on), it is advisable to know (memorize) these special relationships.

	FRACTION	DECIMAL	PERCENT
		0.86	
	$\frac{5}{6}$		
		0.08	
	$\frac{2}{3}$		
*			25%
*		0.20	
*			40%
*			$33\frac{1}{3}\%$
*		0.125	
*	$\frac{7}{10}$		

1. ______________

1. Louis went to buy new tires for his truck. He found them on sale for $\frac{1}{4}$ off. What percent is this?

2. ______________

2. George bought a new VCR at a 50% off sale. What fraction is this?

3. Michael went on a diet. He now weighs $66\frac{2}{3}\%$ of his original weight. What fraction is this?

3. ______________

NAME ______

CLASS ______ SECTION ______ DATE ______

ANSWERS

4. ______

4. Melinda is researching the best place to buy a computer. Family Computers offers $\frac{1}{8}$ off, Computer will give a 12% discount, and Machines ETC will allow a 0.13 discount. Where does she get the best deal?

5. ______

5. Teresa is negotiating a business deal. The client has offered a 5% increase in price, while her boss has authorized up to 0.0526 more. A competitor has offered a deal that is $\frac{1}{18}$ more. Who has offered the least?

6. ______

6. A local department store is having its red tag sale. All merchandise will now be 20% off. What fraction is this?

7. ______

7. Randy is trading in his swimming pool for a larger model. Prices are the same for Model PS+ and Model PT. PS+ has 11% more water while PT has $\frac{1}{9}$ more water. Which should he choose to get the most value for his money?

8. ______

8. During the month of August the grocery store has a special on sweet corn: $\frac{3}{8}$ or 35% more corn for 1 cent. Which is the better deal?

E *Maintain your skills* (Section 5.2)

ANSWERS

9. ____________________

10. ____________________

11. ____________________

12. ____________________

13. ____________________

14. ____________________

15. ____________________

16. ____________________

17. ____________________

18. ____________________

Solve the following proportions:

9. $\frac{25}{30} = \frac{x}{45}$

10. $\frac{45}{81} = \frac{16}{y}$

11. $\frac{x}{72} = \frac{38}{9}$

12. $\frac{x}{9.5} = \frac{125}{250}$

13. $\frac{\frac{1}{2}}{100} = \frac{A}{40}$

14. $\frac{50}{100} = \frac{70}{B}$

15. $\frac{R}{100} = \frac{8}{27}$ Round off R to the nearest tenth.

16. $\frac{17}{100} = \frac{A}{22.3}$ Round off A to the nearest tenth.

17. Sean drove 413 miles and used 11.8 gallons of gasoline. At that rate, how many gallons will he need to drive 1032.5 miles?

18. The Bacons' house is worth $78,000 and is insured so that the Bacons will be paid four-fifths of the value for any damage. One-third of the house is totally destroyed by fire. How much insurance should they collect?

Solving Percent Problems

OBJECTIVE

82. Solve problems written in the form "A is $R\%$ of B" or "$R\%$ of B is A."

APPLICATION

Applications of this objective come from a variety of fields. These applications are so important that two entire sections (6.8 and 6.9) are given to them.

VOCABULARY

To *solve* a percent problem means to do one of the following:

1. Find A, given R and B.
2. Find B, given R and A.
3. Find R, given A and B.

In the statement "$R\%$ of B is A,"

R is the *rate* of percent and is followed by the "%" symbol.

B is the *base* unit and follows the word "of."

A is the *amount* that is compared to B.*

HOW AND WHY

We show two methods for solving percent problems. Both methods give the same answer so either can be used. A few examples show the use of method two, but most are done using method one.

To help determine A, R, and B, keep the following in mind. R is followed by the percent symbol (%), B follows the word(s) "of" or "percent of," and A, sometimes called the percentage, is that which is compared to B.

METHOD ONE

Since $R\%$ is a comparison of A to B and we have seen that this comparison can be thought of as a ratio, the following proportion can be formed:

$$\frac{R}{100} = \frac{A}{B}$$

Now if any one of A, B, or R is missing, it can be found by solving the proportion:

25% of 60 is what number?

Here $R = 25$, $B = 60$, and $A = ?$
So:

$$\frac{25}{100} = \frac{A}{60}$$

$$(25)(60) = 100(A)$$

$$1500 = 100(A)$$

$$1500 \div 100 = A$$

$$15 = A$$

So, 25% of 60 is 15.

METHOD TWO

The word "of" in the statement above and in other places in mathematics indicates multiplication. The word "is" describes the relationship "is equal to" or "=". Thus, we may write $R\%$ of B is A in the form:

$R\%$ of B is A

$\downarrow \quad \downarrow$

$R\% \cdot B = A$

When any two of R, B, and A are known, we can find the other value.

25% of 60 is what number?
$25\% \cdot 60 = A$

Since $25\% = 0.25$:

$0.25(60) = A$

$15.00 = A$

So, 25% of 60 is 15.

*A is also called the *percentage*.

MODEL PROBLEM-SOLVING

Examples | **Strategy**

a. Solve: 45% of what number is 9?

METHOD ONE

$$\frac{R}{100} = \frac{A}{B}$$

$$\frac{45}{100} = \frac{9}{B}$$

$$45B = 9(100)$$

$$B = 900 \div 45$$

$$B = 20$$

So, 45% of 20 is 9.

METHOD TWO

45% of B is 9.

$$45\%(B) = 9$$

$$0.45B = 9$$

$$B = 9 \div 0.45$$

$$B = 20$$

45 is followed by "%," so $R = 45$. The base B (follows "of") is unknown. $A = 9$.

Warm Up a. Solve: 75% of what number is 9?

b. Solve: 5 is what percent of 20?

METHOD ONE

$$\frac{R}{100} = \frac{A}{B}$$

R is unknown.

$$\frac{R}{100} = \frac{5}{20}$$

$B = 20$ since it follows "of." $A = 5$.

$$20(R) = 5(100)$$

$$R = 500 \div 20$$

$$R = 25$$

So, 5 is 25% of 20.

Warm Up b. Solve: 3 is what percent of 20?

ANSWERS TO WARM UPS (6.7) **a.** 12 **b.** 15%

c. Solve: 135% of ___?___ is 54.

METHOD ONE

$$\frac{R}{100} = \frac{A}{B}$$

$$\frac{135}{100} = \frac{54}{B}$$

Since "?" follows "of," we need to find B. $R = 135$ and $A = 54$.

$$135(B) = 54(100)$$

$$B = 5400 \div 135$$

$$B = 40$$

So, 135% of 40 is 54.

Warm Up c. Solve: 140% of ___?___ is 105.

d. Solve: 78% of 36 is ___?___ (to the nearest tenth).

METHOD ONE

$$\frac{R}{100} = \frac{A}{B}$$

$$\frac{78}{100} = \frac{A}{36}$$

$$78(36) = 100(A)$$

$$2808 = 100(A)$$

$$2808 \div 100 = A$$

$$28.08 = A$$

$$28.1 \approx A$$

METHOD TWO

78% of 36 is A.

$$78(0.01)(36) = A$$

$$0.78(36) = A$$

$$28.08 = A$$

$$28.1 \approx A$$

$R = 78$ and $B = 36$. Find A. Note that % = 0.01.

So, 78% of 36 is 28.1, to the nearest tenth.

Round off the answer to the nearest tenth.

Warm Up d. Solve: 48% of 98 is ___?___ (to the nearest tenth).

ANSWERS TO WARM UPS (6.7)

c. 75 **d.** 47.0

e. Solve: 50 is ___?___% of 180 (to the nearest tenth of 1 percent).

METHOD ONE

$$\frac{R}{100} = \frac{A}{B}$$

$$\frac{R}{100} = \frac{50}{180}$$

We need to find R.
$A = 50$ and $B = 180$.

$$180(R) = (100)(50)$$

$$R = 5000 \div 180$$

$$R \approx 27.77$$

$$R \approx 27.8$$

Round off answer to the nearest tenth.

So, 50 is 27.8% of 180 to the nearest tenth of 1 percent.

Warm Up e. Solve: 78 is ___?___% of 162 (to the nearest tenth of 1 percent).

f. Solve: $27\frac{2}{3}$% of 60 is ___?___.

METHOD ONE

$$\frac{R}{100} = \frac{A}{B}$$

A is unknown.

$$\frac{27\frac{2}{3}}{100} = \frac{A}{60}$$

$R = 27\frac{2}{3}$ and $B = 60$.

$$\left(27\frac{2}{3}\right)(60) = (100)(A)$$

$$1660 = (100)(A)$$

$$1660 \div 100 = A$$

$$16.6 = A$$

Since $16\frac{3}{5} = 16.6$, we have:

$27\frac{2}{3}$% of 60 is $16\frac{3}{5}$, or 16.6.

Warm Up f. Solve: $42\frac{1}{6}$% of 120 is ___?___.

ANSWERS TO WARM UPS (6.7)

e. 48.1%

f. 50.6

NAME ______

CLASS ______ SECTION ______ DATE ______

Exercises 6.7

A

ANSWERS

1. ______
2. ______
3. ______
4. ______
5. ______
6. ______
7. ______
8. ______
9. ______
10. ______
11. ______
12. ______
13. ______
14. ______

Solve:

1. 9 is 50% of __?__.

2. 9 is 90% of __?__.

3. What is 75% of 80?

4. What is 45% of 70?

5. 3 is __?__% of 1.

6. 8 is __?__% of 4.

7. __?__% of 60 is 30.

8. __?__% of 65 is 13.

9. 70% of __?__ is 28.

10. 80% of __?__ is 28.

11. 80% of 45 is __?__.

12. __?__ is 80% of 25.

13. 64 is __?__% of 80.

14. __?__% of 56 is 14.

ANSWERS

15. ______________

16. ______________

17. ______________

18. ______________

15. 19% of ___?___ is 19.

16. 16 is ___?___% of 16.

17. $\frac{1}{2}$% of 200 is ___?___.

18. $\frac{1}{4}$% of 800 is ___?___.

B

19. ______________

20. ______________

21. ______________

22. ______________

23. ______________

24. ______________

25. ______________

26. ______________

27. ______________

28. ______________

29. ______________

30. ______________

19. 67% of 40 is ___?___.

20. 38% of 70 is ___?___.

21. 48 is ___?___% of 36.

22. 56 is ___?___% of 48.

23. 140% of ___?___ is 35.

24. 175% of ___?___ is 52.5.

25. 9.3% of 60 is ___?___.

26. 36.75% of 28 is ___?___.

27. 0.4 is ___?___% of 20.

28. 78 is 52% of ___?___.

29. 45 is 36% of ___?___.

30. 1 is ___?___% of 1000.

NAME ________________________________

CLASS ____________ SECTION ____________ DATE ____________

ANSWERS

31. ____________

32. ____________

33. ____________

34. ____________

35. ____________

36. ____________

37. ____________

38. ____________

39. ____________

40. ____________

31. 48% of 40 is ___?___.

32. 72% of 80 is ___?___.

33. 76% of ___?___ is 152.

34. 27% of ___?___ is 162.

35. 135% of ___?___ is 48.6.

36. 165% of ___?___ is 112.2.

37. 76 is ___?___% of 125.

38. 84 is ___?___% of 160.

39. 15.6% of 80 is ___?___.

40. 24.3% of 90 is ___?___.

C

41. ____________

42. ____________

43. ____________

41. 3.2% of 0.7 is ___?___.

42. ___?___ is $\frac{1}{2}$% of 0.5.

43. 28% of ___?___ is 36. (To the nearest tenth.)

ANSWERS

44. ______

45. ______

46. ______

47. ______

48. ______

49. ______

50. ______

51. ______

52. ______

53. ______

44. 219 is 12% of __?__.

45. What percent of 75 is 8? (To the nearest tenth of a percent.)

46. What percent of 92 is 56? (To the nearest tenth of a percent.)

47. $11\frac{1}{9}\%$ of 1845 is __?__.

48. $16\frac{2}{3}\%$ of 3522 is __?__.

49. $83\frac{1}{3}\%$ of __?__ is 1040.

50. $57\frac{1}{7}\%$ of __?__ is 1008.

51. 47 is __?__% of 30. (To the nearest tenth of one percent.)

52. 82.5 is __?__% of 37.6. (To the nearest whole number percent.)

53. __?__ is 31.6% of 57.8. (To the nearest tenth.)

NAME ______________________

CLASS ____________ SECTION ____________ DATE ____________

ANSWERS

54. ____________

55. ____________

56. ____________

57. ____________

58. ____________

54. $5\frac{1}{3}\%$ of $6\frac{1}{2}$ is ___?___. (As a fraction.)

55. ___?___ is 53% of $15\frac{2}{3}$. (Write as a mixed number.)

56. 1.25% of 1250 is ___?___.

57. ___?___% of 82 is 105.5. (To the nearest tenth of one percent.)

58. $5\frac{1}{2}\%$ of ___?___ is 34.5. (To the nearest hundredth.)

E *Maintain your skills* (Sections 5.2, 5.3)

59. ____________

60. ____________

Solve the following proportions:

59. $\frac{14}{24} = \frac{x}{30}$

60. $\frac{4.8}{2.5} = \frac{96}{y}$

ANSWERS

61. ____________

62. ____________

63. ____________

64. ____________

65. ____________

66. ____________

67. ____________

68. ____________

61. $\dfrac{a}{\frac{5}{8}} = \dfrac{1\frac{1}{2}}{3\frac{3}{4}}$

62. $\dfrac{1\frac{1}{2}}{t} = \dfrac{5\frac{5}{8}}{1\frac{2}{3}}$

63. $\dfrac{1.3}{0.07} = \dfrac{w}{3.01}$

64. $\dfrac{1.3}{0.07} = \dfrac{5.59}{t}$

If one and one-half inches on a map represent 60 miles:

65. How many miles are represented by two and seven-eighths inches?

66. How many miles are represented by three and three-sixteenths inches?

67. How many inches are needed to represent 820 miles?

68. How many inches are needed to represent 22 miles?

Word Problems

OBJECTIVE **83. Solve word problems involving percent.**

APPLICATION

In a recent survey it was found that, of the 285 people surveyed, 114 people preferred eating whole wheat bread. What percent of the people surveyed preferred eating whole wheat bread?

VOCABULARY No new words.

HOW AND WHY

When a word problem is translated into the simpler word form, "What percent of what is what?" such as:

"What percent of a loan is the interest?" (See Example a.)

the answers can be found by the procedures described in the last section. Either Method One or Method Two can be used. The two methods are used alternately in the examples.

MODEL PROBLEM-SOLVING

Examples **Strategy**

a. Rod bought a car with a 15% annual loan. If the interest payment was $75 per year, how much was his loan? (Method One)

$$\frac{15}{100} = \frac{75}{B}$$

$$15B = 7500$$

$$B = 500$$

The $75 interest is 15% of the loan, so the problem translates into the simpler word form "15% of what is 75?"

CAUTION

Do not confuse the amount with the base. The amount is some percent of the base so the base is the total amount or value.

So, Rod's loan was for $500.

Warm Up a. Jamie bought a car with a 14% annual loan. If the interest payment was $112 per year, how much was the loan?

ANSWERS TO WARM UPS (6.8)

a. $800

b. The population of Century County is now 130% of its population ten years ago. The population ten years ago was 117,000. What is the present population? (Method One)

$$\frac{130}{100} = \frac{A}{117{,}000}$$

The simpler word form is "130% of 117,000 is what number?"

$$15{,}210{,}000 = 100A$$

$$15{,}210{,}000 \div 100 = A$$

$$152{,}100 = A$$

The population is now 152,100.

Warm Up b. The cost of an automobile is now 140% of what it was five years ago. If the cost of an automobile five years ago was $4850, what would be its cost today?

c. A car has depreciated to 65% of its original cost. If the value of the car is now $5070, what did it cost originally? (Method Two)

$$0.65B = 5070$$
$$B = 5070 \div 0.65$$
$$B = 7800$$

The simpler word form is "65% of what is 5070?"

The original cost was $7800.

Warm Up c. A tractor has depreciated to 52% of its original cost. If the value of the tractor is now $8216, what did it cost originally?

d. The Goliath Bakery has 500 loaves of day-old bread they want to sell. If the price was originally 92¢ a loaf and they sell it for 64¢ a loaf, what percent discount, based on the original price, should the bakery advertise? (Method One)

Note: 92 − 64 = 28, which is the amount of discount.

The discount is 28¢, so the simpler word form is "What percent of 92 is 28?"

$$\frac{R}{100} = \frac{28}{92}$$

$$92R = (100)(28)$$

$$R = 2800 \div 92$$

$$R \approx 30.4$$

Since the discount is about 30.4%, the bakery will probably advertise "over 30% off."

ANSWERS TO WARM UPS (6.8)   **b.** $6790   **c.** $15,800

Warm Up d. The Goliath Bakery also has 200 packages of day-old buns they want to sell. If the price was originally \$1.08 per package and they sell it for 81¢ per package, what percent discount, based on the original price, should the bakery advertise?

e. In a poll taken among a group of students 2 said they walked to school, 7 said they rode the bus, 10 drove in car pools, and 3 drove their own cars. What percent of the group rode the bus? (Method Two)

$$R\%(22) = 7$$
$$R(0.01)(22) = 7$$
$$0.22R = 7$$
$$R = 7 \div 0.22$$
$$R \approx 31.8.$$

There are a total of 22 students in the group, so the simpler word form is "What percent of 22 is 7?"

So, approximately 32%, take the bus.

Warm Up e. A listing of the grades of a certain class indicated that 7 students received a grade of A, 15 received a grade of B, 23 received a grade of C, and 5 received a grade of D. What percent of the students received a grade of B?

APPLICATION SOLUTION

f. In a recent survey it was found that, of the 285 people surveyed, 114 people preferred eating whole wheat bread. What percent of the people surveyed preferred eating whole wheat bread? (Method One)

$$\frac{R}{100} = \frac{114}{285}$$
$$285R = 114(100)$$
$$285(R) = 11400$$
$$R = 11400 \div 285$$
$$R = 40$$

The simpler word form of the problem is "114 is what percent of 285?"

Therefore, 40% of the people surveyed preferred eating whole wheat bread.

Warm Up f. In a recent survey it was found that, of the 372 people surveyed, 93 jogged for exercise. What percent of the people surveyed jogged for exercise?

ANSWERS TO WARM UPS (6.8) **d.** 25% **e.** 30% **f.** 25%

NAME

CLASS SECTION DATE

Exercises 6.8

A

ANSWERS

1. ______

2. ______

3. ______

4. ______

5. ______

6. ______

1. If there is a 4% sales tax on a television set costing $119.95, how much is the tax?

2. Dan bought a used motorcycle for $955. He made a down payment of 18%. How much cash did he pay as a down payment?

3. Frank is selling magazine subscriptions. He keeps 16% of the cost of each subscription he sells as his salary. How many dollars' worth of subscriptions must he sell to earn $125?

4. Last year Joan had 14% of her salary withheld for taxes. If the total amount withheld was $2193.10, what is Joan's yearly salary?

5. The manager of a fruit stand lost $16\frac{2}{3}\%$ of his bananas to spoilage and sold the rest. He discarded 3 boxes of bananas in two weeks. How many boxes did he have in stock at the beginning of the two weeks?

6. A state sales tax is 7%. Rich paid $513.60 for a new TV, which included the sales tax. Find the cost of the TV.

ANSWERS

7. ______________________

8. ______________________

9. ______________________

10. ______________________

7. Ilga bought a new van for \$15,980. She paid 12% down. How much does she still owe?

8. Janet paid \$225 as a down payment on her new appliances. The appliances cost \$2,500. What percent did she pay as a down payment? How much does she still owe?

9. It rained only 6 days during June, 7 days in July, and 2 days in August. What percent of the days during these 3 months were rainy? Round off to the nearest tenth of a percent.

10. The manager of a fruit stand lost 25% of his bananas to spoilage and sold the rest. He discarded 3 boxes of bananas in two weeks. How many boxes did he have in stock at the beginning of the two weeks?

B

11. ______________________

12. ______________________

11. John got 26 problems correct on a 30-problem test. What was his percent score (to the nearest whole-number percent)?

12. To pass a test to qualify for a job interview, Carol must score at least 70%. If there are 40 questions on the test, how many must Carol get correct to score 70%?

NAME ______________________________

CLASS ______________ SECTION ______________ DATE ______________

ANSWERS

13. ______________

14. ______________

15. ______________

16. ______________

17. ______________

13. Vera's house is valued at $65,000 and rents for $4875 per year. What percent of the value of the house is the annual income from rent? (To the nearest tenth of a percent.)

14. Eddie and his family went to a restaurant for dinner. The dinner check was $19.75. He left the waiter a tip of $3. What percent of the check was the tip (to the nearest whole-number percent)?

15. Floyd earns a monthly salary of $805. He spends $130 a month at the supermarket. What percent of his salary is spent at the supermarket (to the nearest whole-number percent)?

16. VegiMart receives 78 heads of lettuce during one week. If $16\frac{2}{3}\%$ of all lettuce spoils and is discarded, how many heads of lettuce will the store discard that week?

The following five problems refer to the same family. You will need to use the information from previous problems to answer some of them.

17. Marilyn and George Harris spend 35% of their combined net income on a mortgage payment. What is their mortgage payment if they net $2,500 per month?

ANSWERS

18. ______

19. ______

20. ______

21. ______

C

22. ______

23. ______

24. ______

18. The Harris family (see Exercise 17) will spend \$180 on property taxes per month. What percent of their monthly income is this?

19. George donates \$71.25 to a local charity one month. This is 5% of his monthly net income. What is his monthly net income?

20. What percent of their combined net income is Marilyn's net income?

21. George Harris got a 4.5% raise in his net income. How much was this? (round off to the nearest cent).

C

22. Adams High School's basketball team finished the season with a record of 15 wins and 9 losses. What percent of the games played were won?

23. The Top Company offered a 6% rebate on the purchase of its best model of canopy. If the regular price is \$398.98, to the nearest cent, what is the amount of the rebate?

24. A supplier increased the price of a dishwasher 8% to its dealers. If the dishwasher now costs \$270, what was the former price?

NAME ____________________

CLASS ____________ SECTION ____________ DATE ____________

ANSWERS

25. ____________

26. ____________

27. ____________

28. ____________

29. ____________

30. ____________

25. A state charges a gasoline tax of 9% of the cost. The federal tax is 4¢ per gallon. If gasoline costs $1.30 per gallon before taxes, what is the total price per gallon, including both taxes?

26. The local police department set up a vehicle inspection station at the high school parking lot. Of the 128 cars inspected on a particular day, 6.25% was found to have faulty brakes. How many vehicles did not have faulty brakes?

27. Prices at the local supermarket rose 6.3% last month. Chicken now costs $1.15 per pound. What was last month's price to the nearest cent?

28. The cost of dairy items increased an average of 2.3% during the month of February. If the price of eggs on February 1 was 94¢ per dozen, what was the price on March 1? (Assume that the rate of increase for eggs was close to the average increase.)

29. Gene bought 125 shares of IWW Inc. for $21.75 a share and sold them for $29.33 a share. If the brokerage fee on both transactions together totaled $39.50, what percent profit did he make on his investment? (To the nearest whole percent.)

30. The town of Verboort has a population of 15,560, which is 45% male. Of the men, 32% is of age 40 or older. How many men are there in Verboort who are younger than 40?

ANSWERS

31. ______

32. ______

33. ______

34. ______

35. ______

36. ______

31. Jill's weekly salary of $297.50 is increased by 6%. What is the amount of increase in her weekly salary?

32. George's weekly salary increased to $315.35 per week. If this resulted from a 6% increase, find his former salary.

33. On a test given by Mr. Jones each question missed caused a loss of 4%. How many questions were on the test?

34. If the population of Sparse County decreased from 2305 to 2289, what is the percent of decrease? (To the nearest tenth.)

35. After an 8% increase in population the city of Croburg had 1998 inhabitants. What was the former population?

36. James received a 5% decrease in his hourly wages. If after the decrease he receives $4.75 an hour, what was his wage before the decrease?

D

37. ______

37. The population of Port City increased 15% since the last census. If the former population was 124,000, what is the present population?

NAME ____________________

CLASS ____________ SECTION ____________ DATE ____________

ANSWERS

38. ____________

39. ____________

40. ____________

41. ____________

42. ____________

43. ____________

38. For customers who use a bank's credit card, there is a $1\frac{3}{4}\%$ finance charge on monthly accounts that have a balance of \$400 or less. Merle's finance charge for August was \$2.80. What was the amount of her account for that month?

39. In preparing a mixture of concrete, Susan uses 300 pounds of gravel, 100 pounds of cement, and 200 pounds of sand. What percent of the mixture is gravel?

40. St. Joseph's Hospital has 8 three-bed wards, 20 four-bed wards, 12 two-bed wards, and 10 private rooms. What percent of the capacity of St. Joseph's Hospital is in private rooms? (To the nearest tenth of a percent.)

41. A discount house advertises that it sells all appliances at cost plus 10%. If Kathy buys a TV set for \$605, what is the profit for the store?

42. It is claimed that in 15,000 hours, or six years, a gasoline engine will be down 32 days for routine maintenance, whereas a diesel engine will be down only 13 days. What percent less time is the diesel engine down compared with the gasoline engine? (To the nearest whole percent.)

43. During a six-year period the cost of maintaining a diesel engine averages \$2650 and the cost of maintaining a gasoline engine averages \$4600. What is the percent of saving of the diesel compared with the gasoline engine? (To the nearest whole percent.)

44. ____________

44. Carol's baby weighed $7\frac{1}{2}$ lb when he was born. On his first birthday he weighed $23\frac{3}{4}$ lb. What was the percent of increase during the year? (To the nearest whole percent.)

E *Maintain your skills* (Sections 6.1, 6.2)

45. ____________

45. There are 100 true-false problems on a history test. What is the percent score if a student gets 78 correct?

46. ____________

46. There are 50 questions on a psychology final exam. What is the percent score for 43 questions correct?

47. ____________

47. There are 20 problems on an algebra test. What is the percent score for 17 problems correct?

48. ____________

48. There are 23 questions on a test for volleyball rules. What is the percent score for 18 correct answers, to the nearest percent?

49. ____________

50. ____________

Change the following decimals to percents:

49. 0.47

50. 0.69

51. ____________

51. 3.34

52. 0.1555

52. ____________

53. The sales tax rate in a certain state is 0.057. Express this as a percent.

53. ____________

54. A neighboring state has a sales tax rate of 0.045. Express this as a percent.

54. ____________

Business-Related Problems

OBJECTIVE **84. Solve business-related problems that involve percent.**

VOCABULARY No new words.

HOW AND WHY Business uses percent in many ways. Among these are percent of markup, percent of discount, percent of profit, interest rates, taxes, salary increases, and commissions. These terms are explained in the examples.

In this section we work with word problems that involve these terms.

MODEL PROBLEM-SOLVING

Examples / **Strategy**

a. The cost of an electric iron is $18.50. The markup is 30% of the cost. What is the selling price of the iron?

Markup **is the amount added to the cost of an article so the store can pay its expenses and also make a profit.**

SIMPLER WORD FORM

Markup is 30% of cost.

Recall that the word "of" can be thought of as "times" and that the word "is" can mean "equal."

$M = 30\%(\$18.50)$ — **Let M represent the markup.**

$M = 0.30(\$18.50)$ — **Change the percent to a decimal.**

$M = \$5.55$

The markup is $5.55.

$$
\begin{aligned}
\text{S.P.} &= \text{cost} + \text{markup} \\
&= \$18.50 + \$5.55 \\
&= \$24.05
\end{aligned}
$$

To find the selling price, add the cost and the markup. Let "S.P." represent the selling price.

The selling price of the iron is $24.05.

Warm Up a. The cost of a coffee brewer is $28.50. The markup is 40% of the cost. What is the selling price of the coffee brewer?

ANSWERS TO WARM UPS (6.9)

a. $39.90

b. A toaster-oven is priced to sell at \$29.45. If the markup is 40% of the selling price, what is the cost of the article?

SIMPLER WORD FORM

We want to find the markup.
Let M represent the markup.

M is 40% of selling price:

$$\frac{40}{100} = \frac{M}{29.45}$$

$$(40)(29.45) = 100(M)$$
$$M = 1178 \div 100$$
$$M = 11.78$$

The markup is \$11.78.

$$\begin{aligned} \text{Cost} &= \text{S.P.} - \text{markup} \\ &= 29.45 - 11.78 \\ &= 17.67 \end{aligned}$$

The cost of the toaster-oven is \$17.67.

Warm Up b. A radio is priced to sell at \$38.80. If the markup is 60% of the selling price, what is the cost of the article?

c. The regular price of a radio is \$29.95. What is the sale price of the radio if it is discounted 30%? (To the nearest cent.)

The *amount of discount* is the amount deducted from the regular selling price. Therefore, the percent of discount will be the amount deducted from the selling price compared to the selling price.
Let S.P. represent the selling price.

SIMPLER WORD FORM

Amount of discount is 30% of S.P.:

$$\begin{aligned} \text{Amount of discount} &= 0.30(\$29.95) \\ &= \$8.985 \\ &= \$8.99 \end{aligned}$$

Round off to the nearest cent.

$$\begin{aligned} \text{Sale price} &= \$29.95 - \$8.99 \\ &= \$20.96 \end{aligned}$$

The sale price is the regular price minus the amount of discount.

The sale price is \$20.96.

Warm Up c. The regular price of a tape recorder is \$79.95. What is the sale price of the tape recorder if it is discounted 28%? (To the nearest cent.)

ANSWERS TO WARM UPS (6.9) **b.** \$15.52 **c.** \$57.56

d. A company bought a new automobile that cost \$12,000. During the first year it will depreciate $12\frac{1}{2}\%$ of its original value. What will be its value at the end of the first year?

SIMPLER WORD FORM

$\begin{pmatrix}\text{amount of}\\ \text{depreciation}\end{pmatrix}$ is $12\frac{1}{2}\%$ of $\begin{pmatrix}\text{original}\\ \text{price}\end{pmatrix}$:

Depreciation is the name given to decline in value, which is caused by age or use.

$$\begin{pmatrix}\text{amount of}\\ \text{depreciation}\end{pmatrix} = 12\frac{1}{2}\%\ (\$12000)$$

$$= (0.125)(\$12000)$$

$$= \$1500$$

The depreciation is \$1500:

$$\text{Value} = \$12000 - \$1500$$
$$= \$10500$$

The value at the end of the first year is the original value minus the depreciation.

The value of the automobile at the end of the first year is \$10,500.

Warm Up d. A company bought a new truck that cost \$28,530. During the first year it will depreciate $11\frac{1}{9}\%$ of its original value. What will be its value at the end of the first year?

e. Jill, Bob, and Sally pay a total of \$140,000 for a business. Jill invested \$56,000, Bob \$49,000, and Sally \$35,000. They share in the profits according to the percent of each partner's investment in the business. The profits of the business for one month were \$6000. What was each partner's share of the profits?

Jill's percent of the profits:

Determine each partner's percent of the business. This is his or her percent of the profits. Compare each person's investment to the total investment.

$$\frac{56{,}000}{140{,}000} = \frac{R}{100}$$

$$5{,}600{,}000 = 140{,}000(R)$$

$$5{,}600{,}000 \div 140{,}000 = R$$

$$40 = R$$

Jill's share is 40% of the total profits
Bob's percent of the profits:

$$\frac{49{,}000}{140{,}000} = \frac{R}{100}$$

$$49{,}000 = 140{,}000(R)$$

$$49{,}000 \div 140{,}000 = R$$

$$35 = R$$

Bob's share is 35% of the total profits.

ANSWERS TO WARM UPS (6.9)

d. \$25,360

Sally's percent of the profits:

Since Jill receives 40% and Bob receives 35%, Sally must receive 25% (100% − 40% − 35%) of the total profits.

$$\begin{aligned}\text{Jill's share} &= 40\% \text{ of } \$6000\\ &= 0.4(6000)\\ &= 2400\\ \text{Bob's share} &= 35\% \text{ of } \$6000\\ &= 0.35(6000)\\ &= 2100\\ \text{Sally's share} &= 25\% \text{ of } \$6000\\ &= 0.25(6000)\\ &= 1500\end{aligned}$$

Jill's share is $2,400. Bob's share is $2,100. Sally's share is $1,500.

Warm Up e. Joe, James, and John pay a total of $180,000 for a business. Joe invested $108,000 and James and John each invested $36,000. They share in the profits according to the percent of each partner's investment in the business. The profits for the business for one month were $8,000. What was each partner's share of the profits?

f. Jean's base salary rate of pay is $6.48 per hour. She gets time and one-half for each hour over 40 hours worked in one week. What are her earnings if one week she works 46.5 hours?

Time and one-half means that she will earn 1.5 times or 150% of her regular hourly wage each hour.

She has 40 hours at $6.48 per hour. She has 6.5 hours at 150% of her regular wage.

SIMPLER WORD FORM

Overtime is 150% of regular wage:

Total of earnings is regular pay plus overtime pay (OT).

$$\begin{aligned}\frac{150}{100} &= \frac{OT}{6.48}\\ 100(OT) &= (150)(6.48)\\ OT &= 972 \div 100\\ &= 9.72\\ \text{Earnings} &= 40(6.48) + 6.5(9.72)\\ &= 259.20 + 63.18\\ &= 322.38\end{aligned}$$

Jean earned a total of $322.38 for the week.

ANSWERS TO WARM UPS (6.9)

e. Joe $4,800, James $1,600, John $1,600

Warm Up f. What are Jean's earnings if one week she worked 49.75 hours, her rate of pay is the same as in Example f, and she gets time and one-half for each hour over 40 hours worked in one week?

g. Mr. Jordan sells men's clothing. He receives a salary of $150 per week plus a commission of 7.5% of his total sales. One week he sold $1,250 worth of clothing. What were his total earnings for the week?

***Commission* is a name for pay that depends on how much one sells.**

SIMPLER WORD FORM

Commission is 7.5% of sales:

$$\begin{aligned} \text{Commission} &= 7.5\%(\$1250) \\ &= 0.075(1250) \\ &= 93.75 \end{aligned}$$

His total commission was $93.75.

SIMPLER WORD FORM

$$\begin{aligned} \text{Total earnings} &= \text{salary} + \text{commission} \\ &= \$150 + \$93.75 \\ &= \$243.75 \end{aligned}$$

To find the total earnings, add the salary and the commission.

Mr. Jordan's total earnings for the week were $243.75.

Warm Up g. If Mr. Janes receives the same salary and commission as Mr. Jordan, what were his total earnings if he sold $950 worth of clothing in one week?

ANSWERS TO WARM UPS (6.9)

f. $353.97 **g.** $221.25

NAME ______________________

CLASS ____________ SECTION ____________ DATE ____________

Exercises 6.9

A–D

ANSWERS

1. ____________

2. ____________

3. ____________

4. ____________

5. ____________

6. ____________

1. An article that costs the store owner \$8.43 is to be marked up \$2.81. What is the percent of markup based on the cost?

2. In Exercise 1, what is the percent of markup based on the selling price?

3. A screwdriver that costs the merchant \$8.40 is to be marked up 30% of the cost. What is the selling price?

4. In Exercise 3, to the nearest percent, the markup is what percent of the selling price?

5. An article is priced to sell at \$39.99. If the markup is 28% of the selling price, to the nearest cent, how much is the markup?

6. In Exercise 5 the markup is what percent of the cost (to the nearest percent)?

ANSWERS

7. ______

7. Martin earns $12\frac{1}{2}\%$ commission on all his sales. Find the amount of sales if his commission was $3,000 last month.

8. ______

8. Matthew receives a weekly salary of $85 plus 9% of everything he sells. Last week he earned $226.30. What was his sales total for that week?

9. ______

9. The Bright TV Store regularly sells a television set for $319.99. An advertisement in the paper indicated that it is on sale and is being discounted 25%. What is the sale price (to the nearest cent)?

10. ______

10. A competitor of the Bright TV Store in Exercise 9 has the same television set on sale. The competitor normally sells the television set for $325.99 and has it advertised at a 27% discount. To the nearest cent, what is the sale price of this set? Which is the better buy and by how much?

11. ______

11. People pay 7.51% of their salary as social security. What will Jon pay to social security if he made $1,576 last month (to the nearest cent).

12. ______

12. Jill gives 0.5% of her salary to United Way every week. Last week she gave $9.25. Find her salary last week.

NAME ______________________________

CLASS ______________ SECTION ______________ DATE ______________

ANSWERS

13. ______________

13. Susan makes $1,900 each month. She pays 12% toward retirement and $1\frac{1}{4}$% to United Way. How much does she give to each?

14. ______________

14. Recently Susan (See Exercise 13) increased her contribution to the United Way by $\frac{1}{2}$% of her salary and decreased her retirement by 1.5% of her salary. How much does she give to each now?

15. ______________

15. A salesman earns a 9% commission on all his sales. How much did he earn last week if his total sales were $5482?

16. ______________

16. If the salesman in Exercise 15 received a 12% commission on all sales and his total sales for one week were $4,725, what were his earnings for the week?

17. ______________

17. Mary's base rate of pay is $7.82 per hour. She receives time and one-half for all hours over 40 that she works in one week. What were her total earnings if last week she worked a total of 47 hours?

18. ______________

18. James' base rate of pay is $6.72 per hour. He receives time and one-half for all hours over 40 that he works in one week. What were his earnings if last week he worked a total of 42 hours?

ANSWERS

19. ______________

20. ______________

21. ______________

22. ______________

23. ______________

24. ______________

25. ______________

19. Last year the president of the Sweet Tooth Candy Company received a bonus of $12,824. If the president receives 4% of the profits as a bonus, what were the company's profits last year?

20. The president of the Big Boy Bicycle Company is to receive a bonus of 3% of the profits. The profits for last quarter were $18,275. What will be the president's bonus for last quarter?

The following four problems refer to the same company.

21. O'HARA CLOCKS marks up all merchandise 22% above cost. Recently they bought 30 clocks for $9 each. Find the selling price of each clock.

22. A year later O'HARA CLOCKS has a sale on the remaining 12 clocks at 8% off. Find the sale price of each clock.

23. Three months later 5 clocks still haven't sold. O'HARA CLOCKS decide to give an additional 13% off. What is this sale price?

24. How much profit did they make on the entire 30 clocks?

25. A new automobile depreciated 14% during the first year. If its original cost was $9850, what is its depreciated value at the end of the first year? (*Hint:* the original value minus the depreciation.)

NAME ______________________________

CLASS ______________ SECTION ______________ DATE ______________

ANSWERS

26. ______________

27. ______________

28. ______________

29. ______________

30. ______________

26. A new machine that was purchased by the Hot-Blast Foundry had a new cost of $11,728. At the end of the first year it had depreciated to a value of $10,262. What was the rate of depreciation for the first year?

27. The cost of a new truck is $48,500. The depreciated value at the end of the first year is $39,770. What is the rate of depreciation for the first year?

28. If the truck in Exercise 27 had cost $36,800 new and the rate of depreciation for the first year is 18.5%, what was the depreciated value at the end of the first year?

29. Four people decide to become partners and buy a business. The first partner buys 25% of the business, the second partner buys 20% of the business, the third partner buys 32% of the business, and the fourth partner buys the remaining 23% of the business. The profits are to be divided among the four according to their share of the ownership. The company's profits for last month were $18,256. How much did each partner receive?

30. Three people decide to become partners and buy a business. The first partner buys 36% of the business, the second partner buys 29% of the business, and the third partner buys the remaining 35%. The profits are to be divided among the three according to each partner's share of the ownership. The profits for last month were $5,274. How much did each partner receive?

ANSWERS

31. ______

32. ______

33. ______

34. ______

35. ______

36. ______

37. ______

31. During one week Ms. James sold a total of $12,875 worth of hardware to the stores in her territory. She receives a 3% commission on sales of $2,000 or less, 4% on that portion of her sales from $2,001 to $8,000, and 5% on all sales over $8,000. What was her total commission for the week?

32. Referring to Exercise 31, if the week before Ms. James sold $14,280, how much commission did she receive?

33. The retail price (selling price) of a dining room table is $489. If the markup is 50% of the cost, what is the cost? (*Hint:* The selling price is cost + 50% of the cost; that is, the selling price is 150% of the cost.)

34. The retail price of a china hutch is $624. If the markup is 30% of the cost, what is the cost? (Refer to the hint in Exercise 33.)

35. Mr. Johnson had total sales of $5,240 in one week. His weekly quota is $4800. What percent were his sales above his quota? Round off to the nearest tenth of a percent.

36. If in Exercise 35 Mr. Johnson was paid an additional commission of 5% on all sales over his quota, what amount of additional commission did he receive?

37. The base price of an automobile is $8400. The optional equipment that had been installed on the car increased the price to $11,200. What percent of the original base price is the price of the optional equipment?

NAME ______________________

CLASS ______________ SECTION ______________ DATE ______________

ANSWERS

38. ______________

39. ______________

40. ______________

41. ______________

42. ______________

43. ______________

38. In Exercise 37, what percent of the total price is the optional equipment?

39. The tax rate in Rural Township is \$3.20 per \$100 of assessed value. What is the tax rate expressed as a percent?

40. Using the tax rate from Exercise 39, what would be the taxes on a house that has an assessed value of \$68,950?

41. The taxes levied on a piece of property are \$2238.30, and the property has an assessed value of \$82,900. What is the tax rate in the tax district in which the property is located?

42. A labor union renegotiated a contract with the Bobcat Manufacturing Company. The average wage under the new contract will increase 8.25%. If the average wage under the old contract were \$8.52 per hour, what will be the average wage under the new contract?

43. A labor union renegotiated a contract with the Copycat Manufacturing Company. The average wage under the new contract will increase $7\frac{3}{4}\%$. The average wage under the old contract was \$9.24 per hour. What will be the average wage under the new contract?

ANSWERS

44. ______

44. A retailer bought a small black-and-white TV for \$56.56 and sold it for \$98.98. The cost to the retailer increased to \$60; the retailer in turn increased the selling price to \$109.98. Will the retailer make a larger or smaller percent of markup?

45. ______

45. A radio shop buys a small radio for \$25.80 and sells it for \$38.70. The cost to the radio shop is increased 12%, and it, in turn, increases the selling price to \$41.80. Is the percent of markup more or less after the increase?

46. ______

46. The markup on furniture at the New Furniture Mart is 48% based on cost. The markup on a lounge chair is \$32.64. What are the cost and the selling price?

47. ______

47. In Exercise 46 the Mart marks up a sofa \$48.36. This markup is 62% based on cost. What are the cost and the selling price of the sofa?

48. ______

48. The local department store sells a flower pot for \$3.98. The store operates with a 25% markup on the cost. What is the most the store can afford to pay for the pot?

49. ______

49. A toaster sells for \$28.96. If the markup is $33\frac{1}{3}\%$ of the cost, what is the most the store can afford to pay for the toaster?

NAME ______________________

CLASS ______________ SECTION ______________ DATE ______________

ANSWERS

50. ______________

50. The Corner Department is having a sale. In the newspaper all merchandise is advertised at 25% off. When you go in to buy a set of golf clubs that originally sold for \$279.95, you find that the store is giving an additional 10% discount off the original price. What is the price you will pay for the set of golf clubs?

51. ______________

51. A furniture store has a sale on sofas. Every sofa is marked 20% off. What is the sale price of a sofa that was priced at \$549.95?

52. ______________

52. For customers who use a bank's credit card, there is a $1\frac{1}{4}\%$ finance charge on monthly accounts that have a balance of \$400 or less. Merle's finance charge for August was \$2.80. What was the balance of her account for that month?

53. ______________

53. A discount house advertises that it sells all appliances at cost plus 10%. If Kathy buys a TV set for \$847, what is the profit for the store?

E *Maintain your skills* (Sections 5.3, 6.1, 6.2, 6.3)

54. ______________

54. Peter attended 18 of the 20 G.E.D. classes held last month. What percent of the classes did he attend?

ANSWERS

55. ______

56. ______

57. ______

58. ______

59. ______

60. ______

61. ______

62. ______

63. ______

55. A family spends $120 for food out of a budget of $500. What percent goes for food?

Change to percent:

56. 0.035

57. 2.34

Write as a decimal:

58. 1.8%

59. 0.6%

60. 20.5%

61. 400%

62. An engine with a displacement of 400 cubic inches develops 260 horsepower. How much horsepower is developed by an engine with a displacement of 175 cubic inches?

63. Sally and Rita are partners. How much does each receive of the income if they are to share $8,200 in a ratio of 6 to 4 (Sally 6, Rita 4)?

Draw and Interpret Graphs

OBJECTIVES

85. Read data from bar, circle, pictorial, and line graphs.
86. Construct a bar, circle, pictorial, or line graph.

APPLICATION

The sources of City Community College's revenue are given in the following graph:

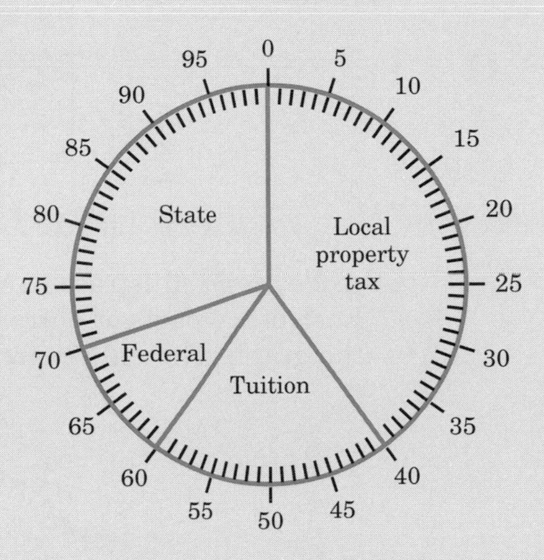

1. What percent of the revenue is from the federal government?
2. What percent of the revenue is from tuition and property taxes?
3. What percent of the revenue is from the federal and state government?

VOCABULARY

Graphs are used to illustrate sets of numerical information.

Bar Graph

A *bar graph* uses solid lines or heavy bars of fixed length to represent numbers from a set. Bar graphs contain two *scales,* a *vertical scale* and *horizontal scale.* The *vertical scale* represents one set of values and the *horizontal scale* represents a second set of values. These values depend on the information to be presented. The following bar graph illustrates four types of cars (first set of values) and the number of each type of car sold (second set of values):

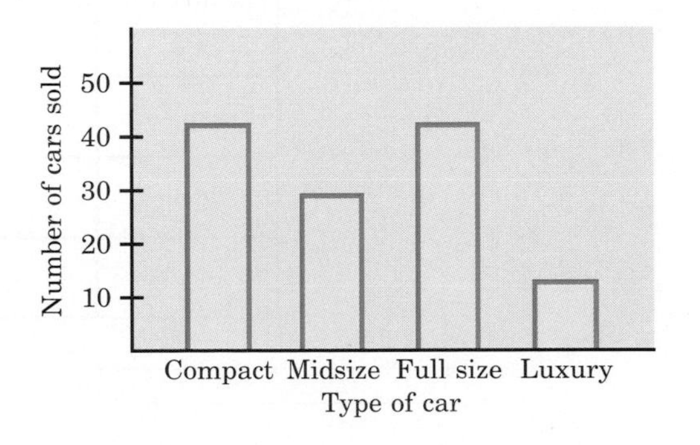

Line Graph

A *line graph* uses lines connecting points to represent numbers from a set. A line graph has a vertical and a horizontal scale like a bar graph:

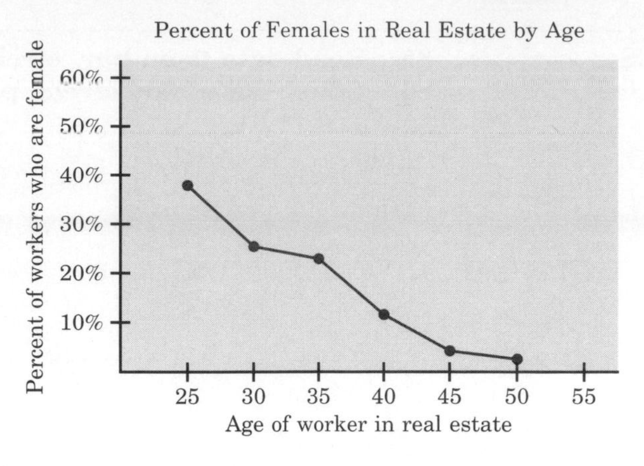

Circle Graph

A *circle graph* illustrates a whole unit or total divided into parts or percents. Each of the parts or percents is represented by a sector (pie shaped piece) of the graph. The entire circle represents 100%:

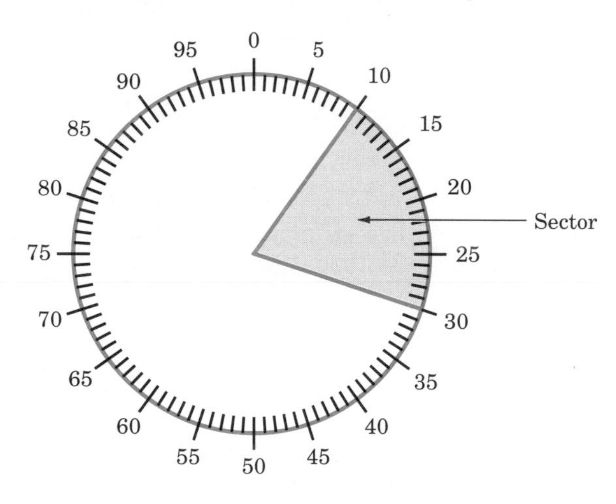

Pictograph

A *pictograph* uses symbols or simple drawings to represent numbers from a set:

Mathematics Class	Distribution of Mathematics Students ♀ = 20 students
Prealgebra	♀ ♀ ♀ ♀ ♀
Algebra	♀ ♀ ♀ ♀ ♀ ♀ ♀
Calculus	♀ ♀ ♀ ♀

HOW AND WHY

A graph is a picture used for presenting data for the purpose of comparison. To "read a graph" means to find values from the graph.

The first graph we consider is a bar graph. It contains a vertical and a horizontal scale. Examine the following bar graph:

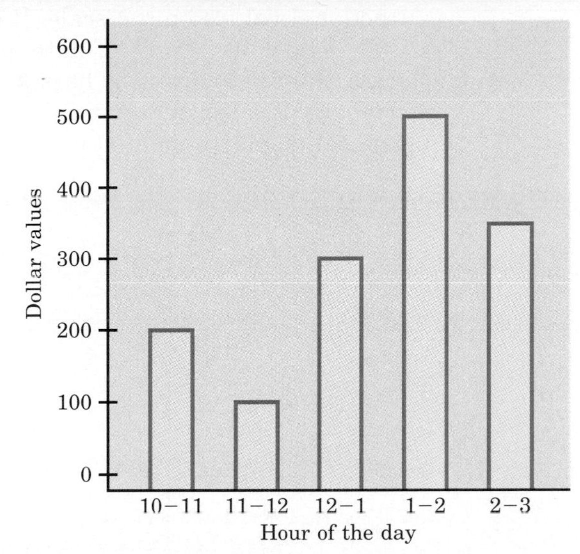

The vertical scale shows dollar values and is divided into units of $100. The horizontal scale shows time values and is divided into units of one hour periods from 10:00 A.M. to 3:00 P.M. From the graph we can see that;

1. The hour of greatest sales (highest bar) was 1 to 2 P.M. with $500 in sales.
2. The hour of least sales (lowest bar) was 11 A.M. to 12 noon with sales of $100.
3. The total sales during the 2 P.M. to 3 P.M. time period is estimated to be $350, as the top of the bar falls between the scale divisions.
4. The total of the morning sales was $300.
5. The total of the afternoon sales was $1150.

Other observations may be made by studying the graph.

Some advantages of displaying data with a graph:

1. Each person can easily find the data most useful to him or her.
2. The visual display is easier for most people to read.
3. Some questions can be answered by a quick look at the graph. For example, "What time does the store need the most sales clerks?"

Let us construct a bar graph to show the variation in used car sales at Oldies but Goldies used car lot. The data are shown in the following table:

MONTH	JAN.	FEB.	MAR.	APR.	MAY	JUN.
Cars sold	30	15	25	20	15	35

To draw and label the bar graph for these data, we show the number of cars sold on the vertical scale and the months on the horizontal scale. This is a logical display as we will most likely be asked to find the highest and lowest months of car sales and a vertical display of numbers is easier to read than a horizontal display of numbers. This is the typical way bar graphs are displayed. Be sure to write the names on the vertical and horizontal scales as soon as you have chosen how the data will be displayed. Now title the graph so that the reader will recognize the data it contains.

The next step is to construct the two scales of the graph. Since each monthly total is divisible by 5, we choose multiples of 5 for the vertical scale. We could have chosen 1 for the vertical scale, but the bars would be very long and the graph would take up a lot of space. We could also have chosen a larger scale, say, 10, then the graph might be too compact and we would need to find fractional values on the scale. It is easier to draw the graph if we use a scale that divides each unit of data. The months are displayed on the horizontal scale. Be sure to draw the bars with uniform width, for each of them represents a month of sales. A vertical display of between 5 and 12 units is typical. The vertical display should start with 0:

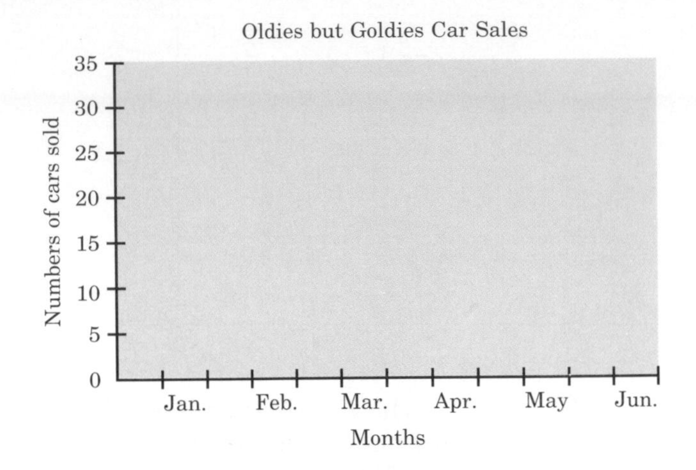

We stop the vertical scale at 35, since that is the maximum number of cars to be displayed. The next step is to draw the bars. Start by finding the number of cars for January. Since 30 cars were sold in January, we draw the bar for January until the height of 30 is reached. This is the top of the bar. Now draw the solid bar for January:

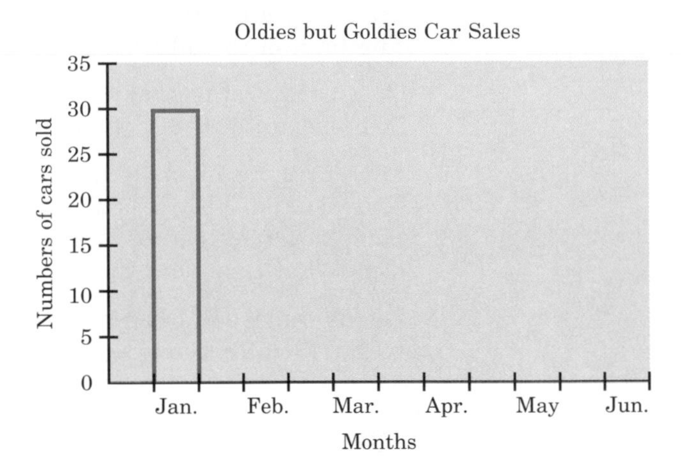

Complete the graph by drawing the bars for the other months:

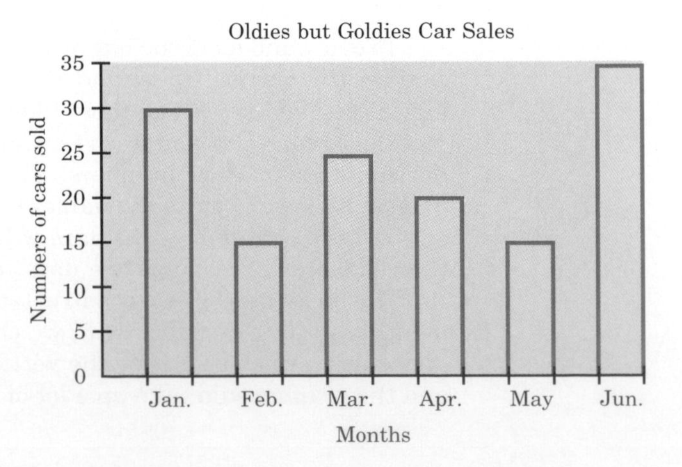

A line graph is similar to a bar graph in that it has vertical and horizontal scales. The data are represented by points rather than bars and the points are connected by line segments. We use a line graph to display the following data:

PROPERTY TAXES

Year	Tax Rate (per $1000)
1970	12.00
1975	15.00
1980	14.00
1985	16.00
1990	20.00

We will let the vertical scale represent the tax rate and each unit represent $2. This requires using a half space for the $15.00 rate, for using $1 would make the graph too tall. The horizontal scale represents the years:

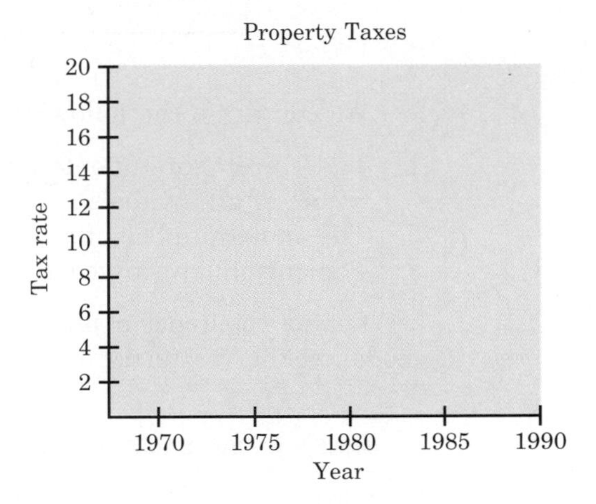

To find the points that represent the data, find the points that are at the intersection of the horizontal line through the tax rate and the vertical line through the corresponding year. Once all the points have been located, connect them with line segments:

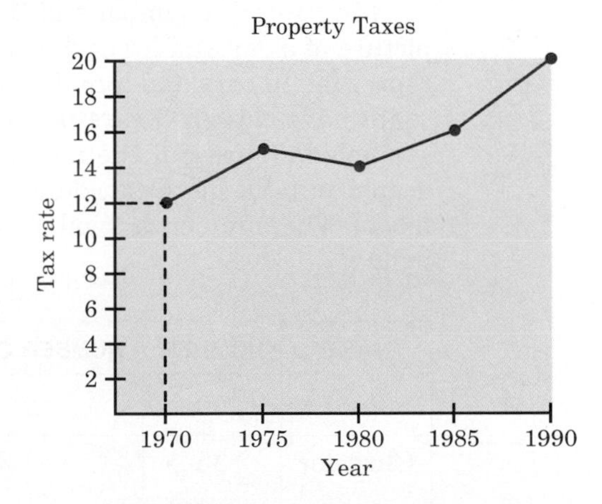

From the graph we can conclude:

1. Only during one five-year period (1975–1980) did the tax rate decline.
2. The largest increase in the tax rate took place from 1985 to 1990.
3. The tax rate has increased $8 per thousand from 1970 to 1990.

The third type of graph we consider is a pictograph. A pictograph is similar to a bar graph where symbols replace the bar to represent the data. Consider the following graph:

Class	Mathematics Students at City College = 50 students = 25 students
Prealgebra	
Algebra	
Geometry	
Calculus	
Statistics	

We can draw the following conclusions from the graph:

1. There are 525 students enrolled in mathematics courses at City College (value of all symbols).
2. The same number of students takes prealgebra and calculus.
3. The enrollment in geometry is 25 more than the enrollment in statistics.

Let us construct a pictorial graph to show the number of cars sold, by model, at the Western Car Corral. The data are shown in the following table:

MAKE	CAR SALES AT WESTERN CAR CORRAL
Ford	50
Chrysler	35
Honda	60
Toyota	25
Pontiac	20

First select a symbol and the number of cars it represents. Here we use a picture of a car and let each symbol represent 10 cars. By letting each symbol represent 10 cars and a half a symbol represent 5 cars, we can save space. We could have chosen 5 cars per symbol, but then we would need to display 12 symbols to represent the number of Hondas, and 12 symbols would make the graph quite large. Next, determine the number of symbols we need for each model. The number of symbols can be found by dividing each number of cars by 10:

MAKE	NUMBER	NUMBER OF SYMBOLS
Ford	50	5
Chrysler	35	$3\frac{1}{2}$
Honda	60	6
Toyota	25	$2\frac{1}{2}$
Pontiac	20	2

Now draw the graph using the symbols (pictures) to represent the data:

Model	Car Sales at Western Car Corral = 10 cars = 5 cars
Ford	
Chrysler	
Honda	
Toyota	
Pontiac	

The circle graph is used to show how a whole unit is divided into parts, or in this case percents. Consider the following graph:

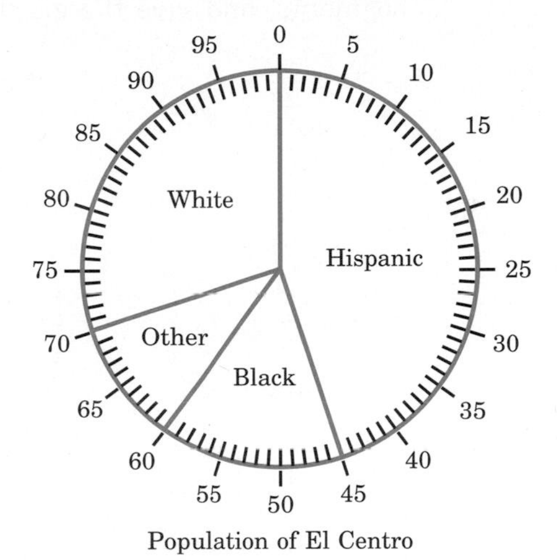

Population of El Centro

From the graph we can conclude:

1. The largest ethnic group in El Centro is Hispanic.
2. The white population is twice the black population.
3. Blacks and Hispanics make up 60% of the population.

If the population of El Centro is 125,000, we can also compute the approximate number of each group. For instance, the number of blacks is found by:

$$RB = A$$

$$15\% \ (125{,}000) = A$$

$$0.15 \ (125{,}000) = A$$

$$18{,}750 = A$$

So, there are approximately 18,750 blacks in El Centro.

Construct a circle graph to show the percent of age groups in Central City. The data are listed in the following chart:

Age Groups	0–21	22–50	Over 50
Population	14,560	29,120	14,560

The first step is to find the size of the whole unit or total population. The sum of the population by age groups is 58,240. The second step is to determine the percent of the population in each age group. Since the total population is 58,240, the percents are:

0–21 age group:

$$\frac{14560}{58240} = 0.25 = 25\%$$

22–50 age group:

$$\frac{29120}{58240} = 0.5 = 50\%$$

Over 50 age group:

$$\frac{14560}{58240} = 0.25 = 25\%$$

From the percents we see that we will need to divide the circle in half (each half is 50%), and then divide one of the halves. Draw the graph, label the segments, and give the graph a title:

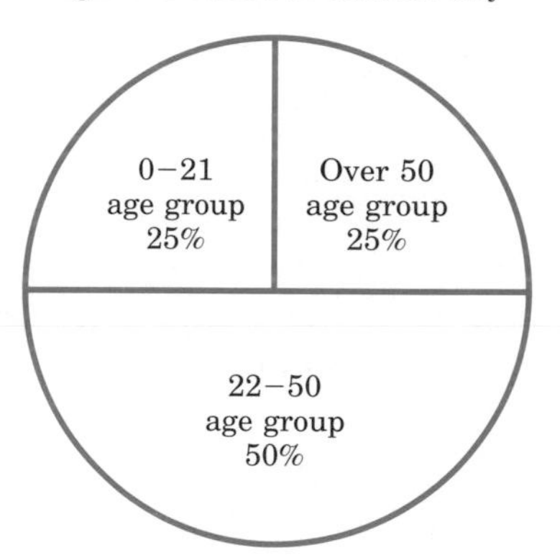

Many circle graphs are divided into sectors using a degree as the unit of measurement. We do not discuss degrees in this text but instead divide the circumference of the circle using increments of 1%. A more accurate graph can be drawn when you have learned about degree measurement.

MODEL PROBLEM-SOLVING

Examples | **Strategy**

a. The following highway department graph shows the number of cars that used highway 37 during a one-week period:

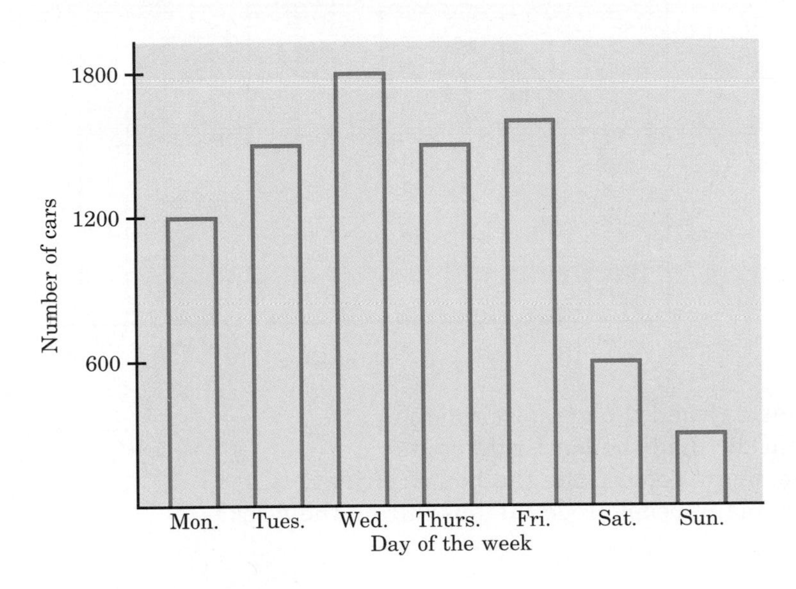

1. What day was the highest traffic day?
2. What day was the lowest traffic day?
3. How many cars used the highway on Tuesday?
4. How many cars used the highway on the weekend?

1. Wednesday

To find the highest traffic day, identify the tallest bar.

2. Sunday

To find the lowest traffic day, identify the shortest bar.

3. 1500

Read the vertical scale at the top of the bar for Tuesday. The value is estimated since the top of the bar is not on a scale line.

4. 900

Add the number of cars for Saturday and Sunday.

(Warm Up continued on next page)

Warm Up a. The number of bus riders for each day of a week is given in the following bar graph:

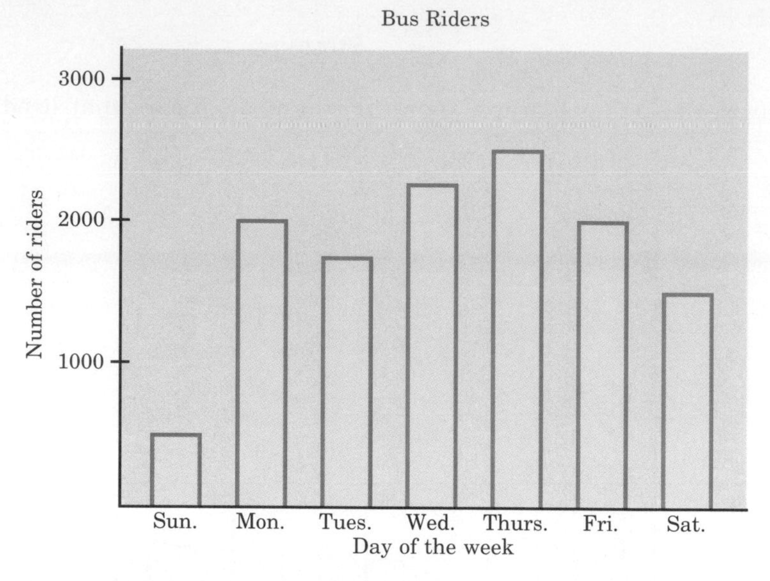

1. What day had the greatest ridership?
2. What day had the least ridership?
3. How many people rode the bus on Friday?
4. How many people rode the bus during the week?

b. Construct a bar graph to display the following data:

The number of babies born during the first six months at Tuality Hospital: January, 15; February, 9; March, 7; April, 18; May, 10; June, 6.

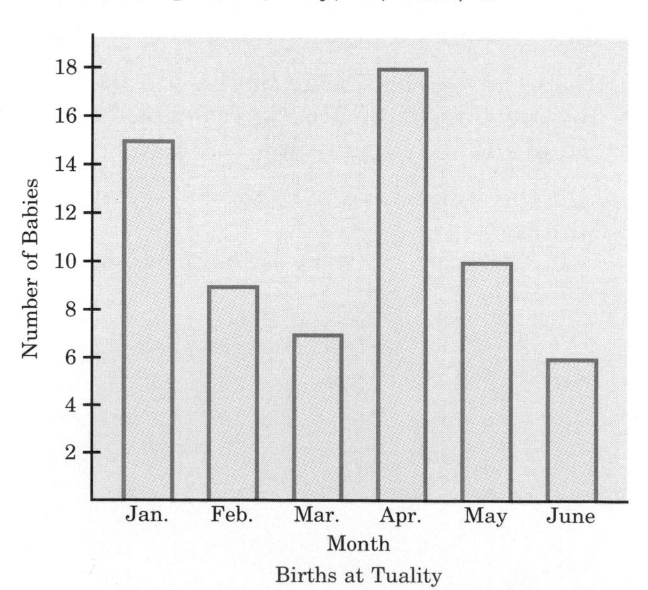

Choose a scale of 1 unit = 2 babies for the vertical scale.
Divide the horizontal scale so that it will accommodate six months with a common unit between them.
Construct the graph, label the scales, and give the graph a title.

ANSWER TO WARM UPS

a. (1) Thursday (2) Sunday (3) 2000 (4) 12,500

Warm Up b. Construct a bar graph to display the following data: The number of plants sold at the Pick-a-Posey Nursery: geraniums, 45; fuchsias, 60; marigolds, 150; impatiens, 90; daisies, 120.

ANSWERS TO WARM UPS (6.10)

b.

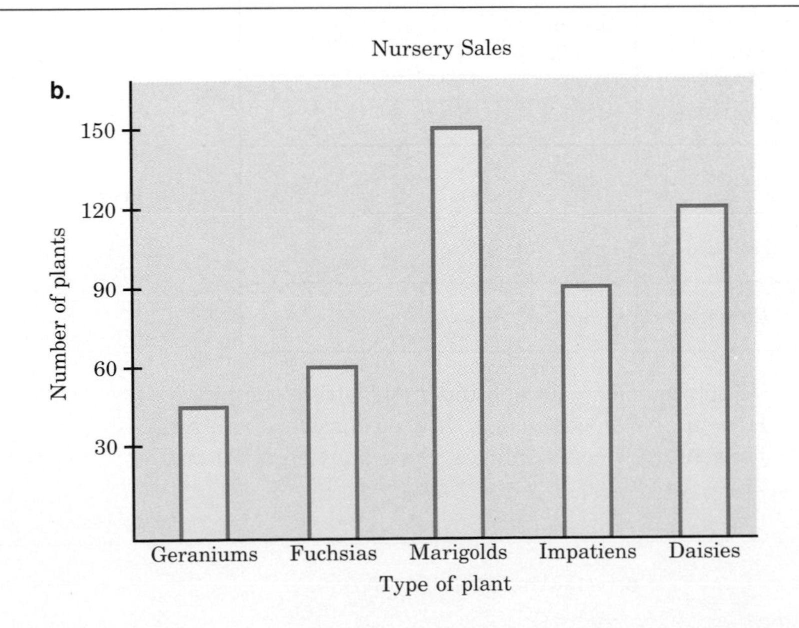

c. The total sales from hot dogs, soda, T-shirts, and buttons during the local air show are given in the following graph:

Item	Sales at Air Show = \$1000 = \$500
Hotdogs	
Soda	
T-shirts	
Buttons	

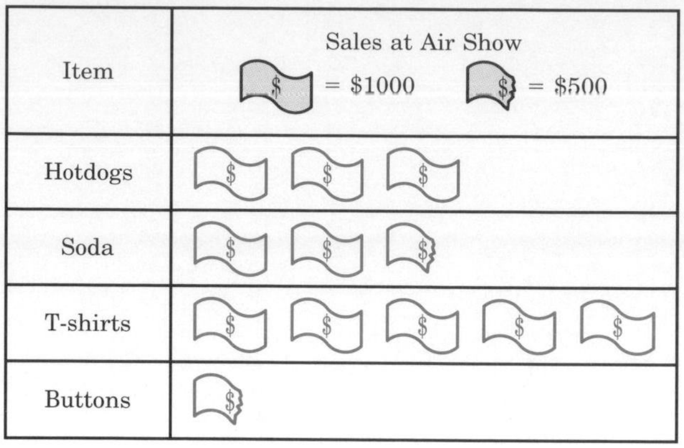

1. What item had the highest dollar sales?
2. What were the total sales from hot dogs and buttons?
3. How many more dollars were realized from the sale of T-shirts than from buttons?

1. T-shirts	**There are more bills representing dollar sales in the T-shirt row than any other row.**
2. Hot dogs = \$3000	**The three bills represents \$3000 of sales for hot dogs.**
Buttons = \$500	**The half bill represents \$500 of sales for buttons.**
Total sales of hotdogs and buttons is \$3500.	**Add the sales.**
3. T-shirts = \$5000	**The five bills represent \$5000 in sales.**
\$4500 more was earned from T-shirt sales.	**Subtract the sales of buttons from the sales of T-shirts.**

Warm Up c. The number of birds spotted during a recent expedition of the Huntsville Bird Society is given in the following graph:

Species	Birds Spotted = 50 = 25
Crows	
Woodpeckers	
Wrens	
Canaries	

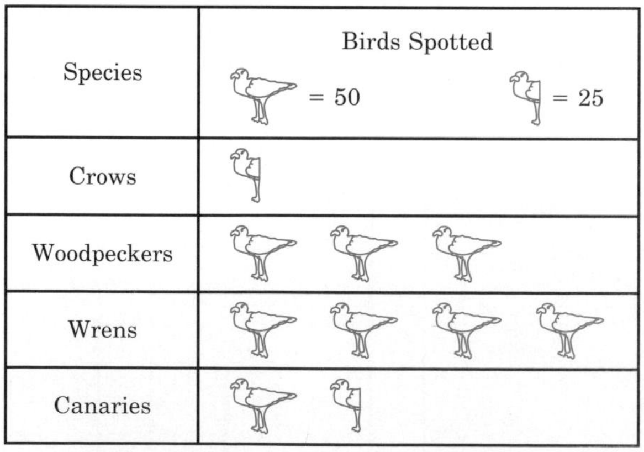

1. Which species was spotted most often?
2. How many woodpeckers and wrens were spotted?
3. How many more canaries were spotted than crows?

ANSWER TO WARM UPS

c. (1) Wrens (2) 350 (3) 50

APPLICATION SOLUTION

d. The sources of City Community College's revenue are given in the following graph:

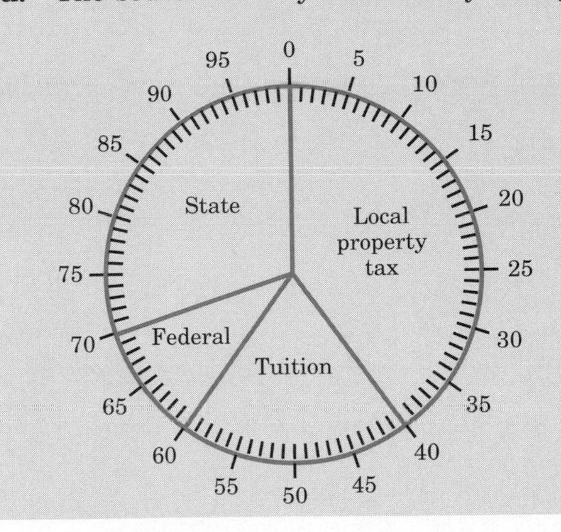

1. What percent of the revenue is from the federal government?
2. What percent of the revenue is from tuition and property taxes?
3. What percent of the revenue is from the federal and state government?

1. 10% — **Read directly from the graph.**
2. 60% — **Add the amounts from tuition and property taxes.**
3. 40% — **Add the amounts from state and federal sources.**

Warm Up d. The percent of sales from food, sundries, drugs, and hardware at the local Mart are given in the following graph:

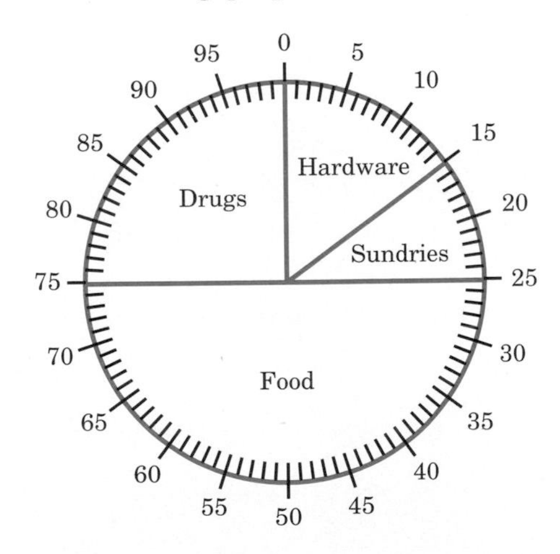

1. What was the area of highest sales?
2. What percent of the total sales was from sundries and drugs?
3. What percent of the total sales was from food and hardware?

ANSWERS TO WARM UPS (6.10)

d. (1) Food (2) 35% (3) 65%

NAME ______________________

CLASS ____________ SECTION ____________ DATE ____________

Exercises 6.10

A

ANSWERS

1. ____________

2. ____________

3. ____________

4. ____________

5. ____________

The following graph shows the variation in the number of phone calls during normal business hours:

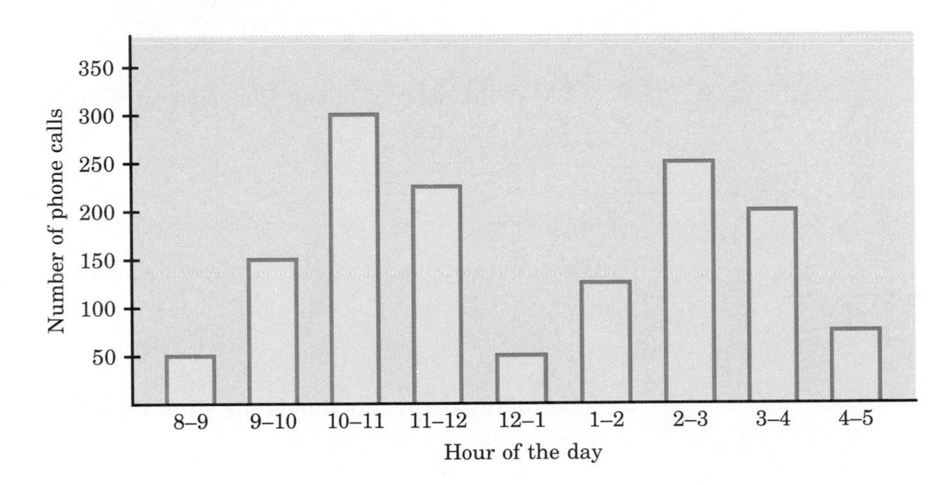

1. What hour of the day was the number of phone calls greatest?

2. What hour of the day was the number of phone calls least?

3. What was the number of phone calls made between 2 and 3?

4. What was the number of phone calls made between 8 and 12?

5. What was the total number of phone calls made during the times listed?

The following graph displays the way the Andrews spend their monthly income:

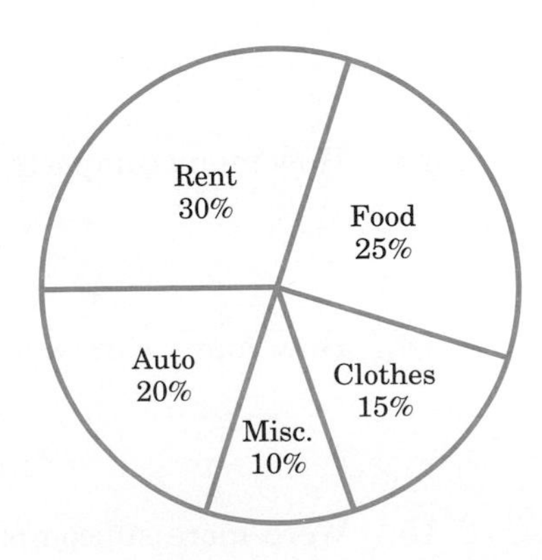

ANSWERS

6. ______

7. ______

8. ______

9. ______

10. ______

11. ______

12. ______

13. ______

14. ______

15. ______

6. What percent of their income is spent on food?

7. What percent of their income is spent on rent?

8. Do the Andrews spend more money per month on automobile expenses or clothes?

9. What percent of the Andrews' income is spent on clothes and miscellaneous items?

10. What percent of the Andrews' income is spent on food and rent?

The following graph shows the number of cars in the shop for repair during a given year:

Type of car	1988 Repair Intake Record (one car symbol = 20 cars)
Compact	4 car symbols
Full size	8 car symbols
Van	2 car symbols
Subcompact	6 car symbols

11. How many vans were in the shop for repair during the year?

12. What type of car had the most cars in for repair?

13. How many compacts and subcompacts were in for repair during the year?

14. How many cars were in for repair during the year?

15. Were more subcompacts or compacts in for service during the year?

NAME ______________________

CLASS ____________ SECTION ____________ DATE ____________

B

ANSWERS

16. ____________

17. ____________

18. ____________

19. ____________

20. ____________

The following graph shows the number of production units at What-Co during the last five years:

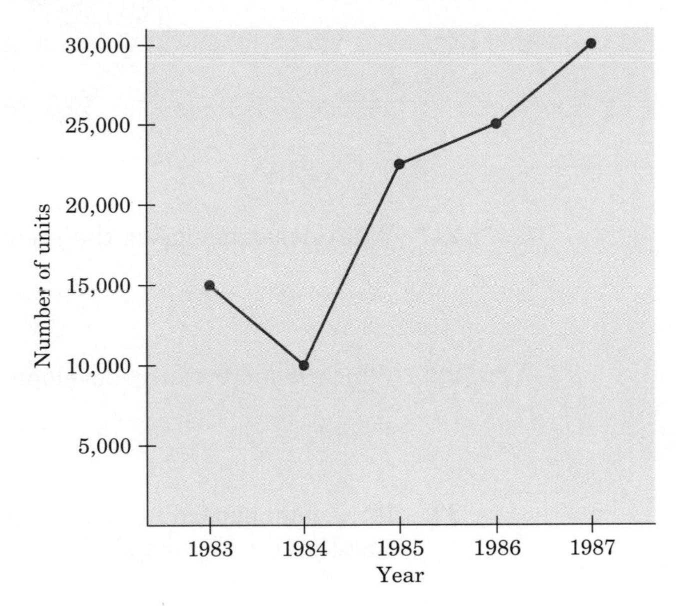

16. What was the greatest production year?

17. What was the least production year?

18. What was the increase in production between 1984 and 1985?

19. What was the decrease in production between 1983 and 1984?

20. What was the percent increase in production from 1983 to 1987?

ANSWERS

21. ______________________

22. ______________________

23. ______________________

24. ______________________

25. ______________________

The following circle graph shows how dollars are spent in a particular industry:

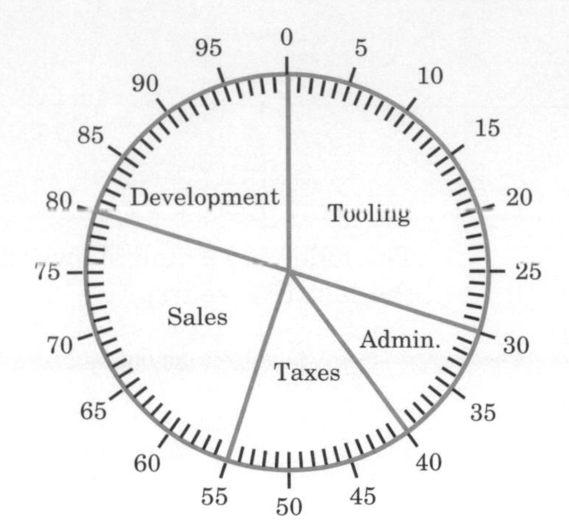

21. What department has the greatest expenditure of funds?

22. Which is more costly, development or taxes?

23. If the total expenditures of the industry were $2,500,000, how much was spent on development?

24. Using the total expenditures in Exercise 23, how much was spent on tooling?

The following graph shows the amounts paid for raw materials:

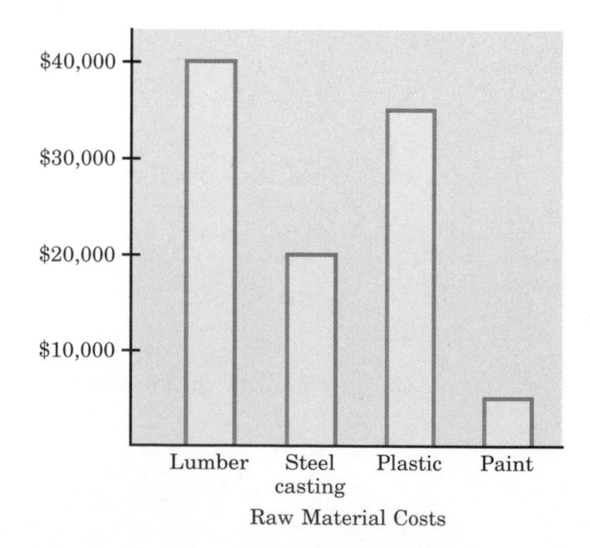

25. What was the total paid for paint and lumber?

NAME ______________________________

CLASS ______________ SECTION ______________ DATE ______________

ANSWERS

26. ______________

27. ______________

28. ______________

29. ______________

26. How much less was paid for steel castings than for plastics?

27. What was the total amount paid for raw materials?

28. What percent of the total cost of raw materials was paid for lumber?

29. What two raw materials account for 45% of the total cost?

C

In Exercises 30 to 42 be sure to name the graph and label the parts. Draw bar graphs to display the following data:

30. Distribution of grades in an algebra class: A—8, B—6, C—15, D—8, F—4.

31. Dinner choices at the LaPlante restaurant in one week: Steak—45, Salmon—80, Chicken—60, Lamb—10, Others—25.

32. The distribution of monthly income: Rent—$350, Automobile—$300, Taxes—$250, Clothes—$100, Food—$250, Miscellaneous—$100.

33. The type of television sets sold in one week at an appliance store: Black and White—15, Color—25, Color with Stereo—5, Miniature—10.

34. Career preference as expressed by a senior class: Business—120, Law—20, Medicine—40, Science—100, Public Service—80, Armed Service—40.

Draw circle graphs to display the following data:

35. The distribution of male and female workers in the Acme Corporation: Males—75%, Females—25%.

36. The type of fish caught during the Far West Angling Tournament: Salmon—44, Trout—22, Sturgeon—11, Sea Bass—11.

NAME ______________________________

CLASS ____________ SECTION ____________ DATE ____________

Draw line graphs to display the following data:

37. Daily sales at the local men's store: Monday—$1500, Tuesday—$2500, Wednesday—$1500, Thursday—$3500, Friday—$4000, Saturday—$6000, Sunday—$4500.

38. The gallons of water used each quarter of the year by a certain city:

Jan.–Mar. 20,000,000 gallons
Apr.–June 30,000,000 gallons
July–Aug. 45,000,000 gallons
Sept.–Dec. 25,000,000 gallons

39. Income from various sources for a given year: Wages—$32,000, Interest—$2,000, Dividends—$4,000, Sale of Property—$16,000.

Draw pictorial graphs to display the following data:

40. The cost of an average three-bedroom house in Oregon:

YEAR	COST
1960	$45,000
1965	$55,000
1970	$60,000
1975	$70,000
1980	$75,000
1985	$70,000

41. The population of Wilsonville over 20 years: 1965—15,000; 1970—17,500; 1975—22,500; 1980—30,000; 1985—32,500.

42. The oil production from a local well over a five-year period;

YEAR	BARRELS PRODUCED
1980	500
1981	1500
1982	3000
1983	2250
1984	1250

NAME ______________________

CLASS ____________ SECTION ____________ DATE ____________

E *Maintain your skills* (Sections 4.9, 4.12, 5.2)

ANSWERS

43. ____________

44. ____________

45. ____________

46. ____________

47. ____________

48. ____________

49. ____________

50. ____________

51. ____________

52. ____________

43. $\dfrac{30}{36} = \dfrac{x}{42}$

44. $\dfrac{72}{80} = \dfrac{y}{220}$

45. $\dfrac{12}{x} = \dfrac{102}{17}$

46. $\dfrac{52}{a} = \dfrac{286}{44}$

47. $\dfrac{8.4}{3.5} = \dfrac{12}{b}$

48. $\dfrac{4}{x} = \dfrac{8}{21}$

49. $\dfrac{3.4}{5} = \dfrac{8.5}{x}$

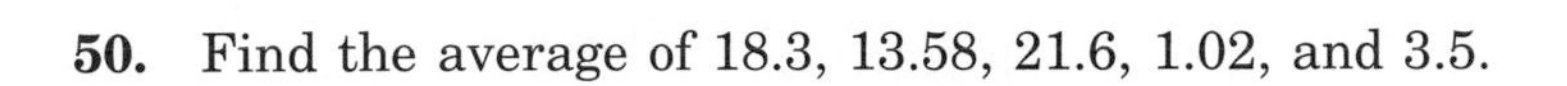

50. Find the average of 18.3, 13.58, 21.6, 1.02, and 3.5.

51. Multiply: 8614.371×10^2.

52. Divide: $2.384 \div 10^3$.

Read and Interpret Tables

OBJECTIVE

87. Read and interpret information given in a table.

APPLICATION

The following table displays nutritional information about four breakfast cereals:

NUTRITIONAL VALUE PER 1-OZ SERVING				
Ingredient	Oat Bran	Rice Puffs	Raisin Bran	Wheat Flakes
Calories	90	110	120	110
Protein	6 g	2 g	3 g	3 g
Carbohydrate	17 g	25 g	31 g	23 g
Fat	0 g	0 g	1 g	1 g
Sodium	5 mg	290 mg	230 mg	270 mg
Potassium	180 mg	35 mg	260 mg	4 mg

1. Which cereal has the most calories per serving?
2. How many grams (g) of carbohydrates are there in 5 ounces of Rice Puffs?
3. How many more milligrams (mg) of sodium are there in a one-ounce serving of Wheat Flakes as compared to Raisin Bran?
4. How many ounces of Oat Bran can one eat before consuming the same amount of sodium as in one ounce of Rice Puffs?

VOCABULARY

Row
5 17 23 9

Column
5
17
23
9

A *table* is a method of displaying data in an array using a horizontal and vertical arrangement to distinguish the type of data. A *row* of a table is a horizontal line of a table and reads left or right across the page. A *column* of a table is a vertical line of a table and reads up or down the page. For example, in the table,

Column

	134	**56**	89	102
	14	**116**	7	98
Row	**65**	**45**	**12**	**67**
	23	**32**	7	213

The number "45" is in row 3 and column 2.

HOW AND WHY

Data are often displayed in the form of a table. We see tables in the print media, in advertisements, and in business presentations. Reading a table involves finding the correct column and row that describes the needed information and then reading the data at the intersection of that column and that row. For example:

STUDENT COURSE ENROLLMENT				
Class	Mathematics	English	Science	Humanities
Freshman	950	1500	500	1200
Sophomore	600	700	650	1000
Junior	450	200	950	1550
Senior	400	250	700	950

To find the number of sophomores who take English, find the column headed English and the row headed sophomore and read the number at the intersection. The number of sophomores taking English is 700.

We can use the table for predicting by scaling the values in the table upward or downward. If the number of seniors doubles next year, we can assume the number of seniors taking humanities will also double. So, the number of seniors taking humanities next year will be 1900, since $2 \cdot 950 = 1900$.

We can also find the percent of the enrollments listed by subject area. For instance, to find the percent of those enrolled in mathematics that are freshmen, we divide the freshman enrollment in mathematics by the total mathematics enrollment:

$$\frac{950}{950 + 600 + 450 + 400} = \frac{950}{2400} \approx 0.39583 \approx 39.6\%$$

So, about 39.6% of the mathematics enrollment is freshmen.

Other ways to interpret data from a table are shown in examples.

MODEL PROBLEM-SOLVING

Examples Strategy

a. The following table shows the decline in the number of railroad workers in four western states:

RAILROAD WORKERS		
State	1980	1988
Oregon	2991	1338
Idaho	3368	1748
Wyoming	3416	1486
Utah	3046	1717

1. Which state had the most railroad workers in 1980?
2. Which state had the least number of railroad workers in 1988?
3. Which state suffered the largest loss in the number of railroad workers?
4. What was the percent of decrease in railroad workers in Wyoming from 1980 to 1988?

1. Wyoming

Read down the column headed "1980" to locate the largest number of workers, 3416. Now read across to find the state, Wyoming.

2. Oregon

Read down the column headed "1988" to find the least number of workers. Now read across to find the state, Oregon.

3. Oregon:

$2991 - 1338 = 1653$

Idaho:

$3368 - 1748 = 1620$

Wyoming:

$3416 - 1486 = 1930$

Utah:

$3046 - 1717 = 1329$

Find the difference in the number of workers for each state.

Wyoming suffered the greatest loss.

1930 is greater than 1653, 1620, or 1329.

4. $\dfrac{1930}{3416} \approx 0.56498829$

Divide the loss in workers by the number in 1980.

≈ 0.565

Round off to three decimal places to find the percent to the nearest tenth.

$\approx 56.5\%$

Write as a percent.

The percent of decrease was approximately 56.5%.

Warm Up a. Using the table in Example a, find the following:

1. What was the total number of railroad workers in 1980 in the four states?
2. How many more railroad workers did Idaho have than Oregon in 1988?
3. What was the total number of railroad workers in Wyoming and Utah in 1988?
4. The railroad workers in Idaho represent what percent of the total railroad workers in 1988? Round off to the nearest percent.

ANSWER TO WARM UPS

a. (1) 12,821 (2) 410 (3) 3,203 (4) 28%

b. The following table shows the value of homes sold in the Portland metropolitan area for a given month in 1989:

VALUE OF HOUSES SOLD			
	Price		
Location	Lowest	Highest	Average
North Portland	\$16,000	\$ 58,500	\$ 34,833
N.E. Portland	\$18,000	\$120,000	\$ 47,091
S.E. Portland	\$18,000	\$114,000	\$ 51,490
Lake Oswego	\$40,000	\$339,000	\$121,080
West Portland	\$29,500	\$399,000	\$112,994
Beaverton	\$20,950	\$165,000	\$ 78,737

1. In which section of town was the highest priced home sold?
2. What was the price difference between the average cost of a house and the lowest cost of a house in Lake Oswego?
3. The lowest priced home is what percent of the highest priced home in West Portland? (To the nearest percent.)
4. What percent of the average priced home in Beaverton is the average priced home in S.E. Portland? (To the nearest percent.)

1. West Portland — **Read down the highest cost column and find the largest price, \$399,000.**

2. $\begin{array}{r} \$121{,}080 \\ -\ 40{,}000 \\ \hline \$\ 81{,}080 \end{array}$ — **Subtract the lowest cost from the average cost for Lake Oswego to get the price difference.**

   The price difference is \$81,080.

3. $\dfrac{\$29{,}500}{\$399{,}000} \approx 0.073934$ — **Divide the lowest price by the highest price.**

   $\approx 7.3934\%$ — **Change to a percent.**

   $\approx 7\%$ — **Round off to the nearest percent.**

The lowest priced house is 7% of the highest priced house.

4. $\dfrac{\$51{,}490}{\$78{,}737} \approx 0.65394$ — **Divide the average cost in S.E. Portland by the average cost in Beaverton.**

   $\approx 65.394\%$ — **Change to percent.**

   $\approx 65\%$ — **Round off to the nearest percent.**

The average priced S.E. home is 65% of the average priced Beaverton home.

Warm Up b. Use the table in Example b to find the following:

1. Which area has the highest average sale price?
2. What is the difference between the highest and lowest priced house in Beaverton?
3. What is the percent of increase from the average priced house in N.E. Portland to the highest priced house in N.E. Portland? (To the nearest percent.)
4. What is the percent of increase from the cost of the lowest priced house in North Portland to the highest priced house in West Portland? (To the nearest percent.)

ANSWERS TO WARM UPS (6.11)

b. (1) Lake Oswego (2) \$144,050 (3) 155% (4) 2,394%

APPLICATION SOLUTION

c. The following table displays nutritional information about four breakfast cereals:

NUTRITIONAL VALUE PER 1-OZ SERVING				
Ingredient	Oat Bran	Rice Puffs	Raisin Bran	Wheat Flakes
Calories	90	110	120	110
Protein	6 g	2 g	3 g	3 g
Carbohydrate	17 g	25 g	31 g	23 g
Fat	0 g	0 g	1 g	1 g
Sodium	5 mg	290 mg	230 mg	270 mg
Potassium	180 mg	35 mg	260 mg	4 mg

1. Which cereal has the most calories per serving?
2. How many grams (g) of carbohydrates are there in 5 ounces of Rice Puffs?
3. How many more milligrams (mg) of sodium are there in a one-ounce serving of Wheat Flakes as compared to Raisin Bran?
4. How many ounces of Oat Bran can one eat before consuming the same amount of sodium as in one ounce of Rice Puffs?

1. Raisin Bran

Find the largest value in the calorie row, 120, and then read the cereal heading over that column, Raisin Bran.

2. $5(25\text{ g}) = 125\text{ g}$

Multiply the grams of carbohydrates in Rice Puffs by 5.

3. $(270 - 230)\text{ mg} = 40\text{ mg}$

Subtract the milligrams of sodium in Raisin Bran from that of Wheat Flakes.

4. $\dfrac{290\text{ mg}}{5\text{ mg}} = 58$

Divide the milligrams of sodium in Rice Puffs by those in Raisin Bran.

Fifty-eight servings of Oat Bran have the same sodium as one serving of Rice Puffs.

Warm Up c. Use the table in Example c to find the following:

1. Which cereal has the most sodium per one-ounce serving?
2. How many milligrams (mg) of potassium are there in six ounces of Wheat Flakes?
3. Mary's doctor has urged her to eat 18 grams (g) of protein for breakfast. How many servings of Raisin Bran would she need to eat to meet the requirement?
4. How many ounces of Wheat Flakes can one eat before consuming the same amount of potassium as in one ounce of Rice Puffs?

ANSWER TO WARM UPS

c. (1) Rice Puffs (2) 24 mg (3) 6 servings (4) 8.75 ounces

NAME ________________________________

CLASS ______________ SECTION ______________ DATE ______________

Exercises 6.11

A

ANSWERS

1. ______________

2. ______________

3. ______________

4. ______________

5. ______________

6. ______________

7. ______________

Use the following table to answer Exercises 1 to 10:

TELEPHONE NETWORK MONTHLY ACCESS CHARGES			
City	1-Party Line	2-Party Line	4-Party Line
Portland	$18.00	$15.69	$13.84
Seattle	$18.31	$16.00	$14.15
Tampa	$17.33	$15.02	$13.17
Dallas	$19.34	$17.03	$15.18
Boston	$18.61	$16.30	$14.45

1. What is the cost of a one-party line in Seattle?

2. What is the cost of a four-party line in Boston?

3. Which city has the highest two-party rate?

4. Which city has the lowest one-party rate?

5. What are the monthly savings of a two-party line over a one-party line in Portland?

6. How much more does it cost a person in Dallas to have a four-party line than it does a person in Tampa?

7. How much more does it cost to have a one-party line than a two-party line in Tampa?

ANSWERS

8. ______

9. ______

10. ______

11. ______

12. ______

8. How much can be saved on a yearly basis by using a two-party line instead of a one-party line in Seattle?

9. How much will be saved yearly when José moves from Dallas to Tampa and keeps a two-party line?

10. What is the percent of increase in the cost from a four-party line to a one-party line in Boston? (To the nearest percent.)

11. What is the percent of increase in the cost of a one-party line in Dallas as compared to Portland? (To the nearest percent.)

12. How much monthly income is realized by the phone company in Seattle from 3500 four-party lines?

B

13. ______

14. ______

15. ______

Use the following table to answer Exercises 13 to 24:

NUTRITIONAL INFORMATION PER SERVING OF ENTREE				
Ingredient	Fish Cakes	Veal Chops	Chicken Dijon	Pepper Steak
Calories	259	421	247	240
Protein	28 g	34 g	31 g	28 g
Fat	12 g	24 g	12 g	10 g
Carbohydrate	7 g	15 g	2 g	9 g
Sodium	783 mg	687 mg	649 mg	820 mg
Cholesterol	147 mg	115 mg	99 mg	76 mg

13. Which entrée has the highest level of cholesterol per serving?

14. How much less cholesterol does one consume when ordering chicken dijon as opposed to fish cakes?

15. How much more sodium is consumed when eating pepper steak as opposed to veal chops?

NAME ______________________________

CLASS ______________ SECTION ______________ DATE ______________

ANSWERS

16. ______________

17. ______________

18. ______________

19. ______________

20. ______________

21. ______________

22. ______________

23. ______________

24. ______________

16. Jerry's doctor put him on a 900-calorie diet. What percent of his total calorie intake is in one serving of chicken dijon? (To the nearest tenth of a percent.)

17. How much fat is contained in 3 servings of veal chops?

18. How many grams of carbohydrate are there in four servings of chicken dijon and two servings of fish cakes?

19. Jessica is restricted to 1200 mg of sodium per day. What percent of Jessica's sodium intake for the day is in one serving of pepper steak? (To the nearest tenth of a percent.)

20. What is the minimum number of servings of chicken dijon needed to equal the fat content of veal chops?

21. At a buffet Dan eats one serving each of veal chops, chicken dijon, and pepper steak. How many calories did he consume?

22. At a buffet Susan eats two servings of fish cakes and one serving of veal chops. How many milligrams (mg) of cholesterol did she consume?

23. What percent of the cholesterol in a serving of fish cakes is in a serving of chicken dijon? (To the nearest tenth of a percent.)

24. Which entrée has the lowest ratio of cholesterol to calories?

ANSWERS

25. ______

26. ______

27. ______

28. ______

29. ______

30. ______

31. ______

32. ______

33. ______

Use the following table to answer Exercises 25 to 36:

FLEXIBLE LIFE INSURANCE POLICY			
Age	Death Benefit	Account Value	Cash Surrender Value
57	$400,000	$144,276	$128,677
59	400,000	165,219	150,928
61	400,000	189,522	178,494
63	400,000	217,865	217,865
65	400,000	251,206	251,206

25. What is the difference between the account value and the cash surrender value at age 57?

26. How much did the cash surrender value increase between the ages of 61 and 65?

27. What is the difference between the death benefit and the account value at age 65?

28. What percent of the death benefit is the cash surrender value at age 59? (To the nearest tenth of a percent.)

29. What was the percent of increase in the cash surrender value from age 57 to age 61? (To the nearest tenth of a percent.)

30. Between what two consecutive ages in the table did the largest increase in the account value occur?

31. What was the percent of increase in the account value from age 57 to 65? (To the nearest tenth of a percent.)

32. Between what two ages was the greatest percent of increase in the cash surrender value?

33. At what age is the difference between the account value and the cash surrender value the least?

NAME ______________________________

CLASS ______________ SECTION ______________ DATE ______________

ANSWERS

34. ______________

35. ______________

36. ______________

34. At what age is the cash surrender value the greatest percent of the account value?

35. Assuming the same ratio, what would be the account value of a $300,000 death benefit policy at age 63? (To the nearest dollar.)

36. Assuming the same ratio, what would be the cash surrender value of a $500,000 death benefit policy at age 59? (To the nearest dollar.)

D

37. ______________

38. ______________

39. ______________

40. ______________

Use the following table to answer Exercises 37 to 40:

TRUCK SHIPPING RATES BY STATE

State	Minimum Charge	Price per Pound	Tax Rate
Montana	$58.37	34¢	None
Maine	31.22	15¢	5%
Michigan	44.79	23¢	4%
Missouri	35.13	15¢	4.225%

37. What does it cost to ship a 300-pound object within the state of Michigan?

38. What does it cost to ship a 1200-pound object within the state of Missouri?

39. How much more does it cost to ship a 600-pound object in Montana than it does in Maine?

40. What is the heaviest object, in pounds, that can be shipped in Michigan for $44.90 plus tax? (To the nearest pound.)

ANSWERS

41. ______

42. ______

43. ______

44. ______

45. ______

46. ______

47. ______

48. ______

Use the following table to answer Exercises 41 to 44:

A midwestern state travel bureau made up a table showing the number of people using the facilities of Lizard Lake State Park for May through September.

VISITORS AT LIZARD LAKE STATE PARK					
	May	June	July	August	September
Overnight Camping	231	378	1104	1219	861
Picnics	57	265	2371	2873	1329
Boat Rental	29	45	147	183	109
Hiking/Climbing	48	72	178	192	56
Horse Rental	22	29	43	58	27

41. How many overnight campers used these facilities during these months?

42. What month did the park have the greatest use of facilities?

43. The horse rentals were what percent of the boat rentals during these months? (To the nearest percent.)

44. Which month was the ratio of the number of hikers to the other facility users the greatest?

Use the following table to answer Exercises 45 to 48:

ZOO ATTENDANCE IN MILLIONS					
	1985	1986	1987	1988	1989
Fisher Zoo	2.30	2.25	2.71	2.74	2.72
Delaney Zoo	1.06	1.11	1.31	1.15	1.29
Shefford Garden	2.19	2.25	2.27	2.27	2.31
Utaki Park	0.35	0.39	0.47	0.52	0.65

45. Which zoo had the greatest increase in attendance from 1985 to 1987?

46. Which zoo had the least increase in attendance from 1985 to 1989?

47. Which zoo had the greatest percent of increase in attendance from 1988 to 1989?

48. What was the percent of increase in attendance at Utaki Park from 1985 to 1989? Round off to the nearest tenth of a percent.

NAME ____________________

CLASS ____________ SECTION ____________ DATE ____________

E *Maintain your skills* (Sections 4.12, 5.1, 5.3)

ANSWERS

49. ____________

50. ____________

51. ____________

52. ____________

53. ____________

54. ____________

55. ____________

56. ____________

57. ____________

58. ____________

49. Write the ratio of 5 dimes to 12 nickels in cents.

50. Write the ratio of 15 people to 35 chairs.

51. Billie earned \$39.20 working for 5 hours. At the same rate how much will she earn in 8 hours?

52. The cost of 24 pounds of nails cost \$21.36. At the same rate what will 18 pounds cost?

53. If Tang earns \$45 in interest on an investment of \$1200 for one year, at the same rate how much interest would hc carn on an investment of \$5760?

54. The cost to rent a car is \$10 per day plus 12 cents per mile. Jenny wants to rent a car to drive a distance of 50 miles and return in the same day. How much will the car cost her?

55. Bob earns \$6.72 per hour. He works 6 hours on Monday, 8 hours on Tuesday, 7.5 hours on Wednesday, 3.5 hours on Thursday, and 7.75 hours on Friday. How much did he earn in those five days?

Perform the indicated operations:

56. $(0.25)(3.6) \div 0.03 - (3.5)(1.2)$

57. $[3.5^2 + 6.5(0.3)](0.4)^2 - 0.234$

58. $4.45 - (2.1)^2(0.2)$

NAME ____________________

CLASS ____________ SECTION ____________ DATE ____________

ANSWERS

1. ____________
2. ____________
3. ____________
4. ____________
5. ____________
6. ____________
7. ____________
8. ____________
9. ____________
10. ____________
11. ____________
12. ____________
13. ____________
14. ____________
15. ____________
16. ____________

CHAPTER 6 TRUE–FALSE CONCEPT REVIEW

Check your understanding of the language of basic mathematics. Tell whether or not each of the following is true (always true) or false (not always true).

1. Percent means per one-hundred.

2. Percent is a ratio.

3. The symbol "%" is read "percent."

4. In percent the base unit can be more than 100.

5. To change a decimal to a percent, move the decimal point 2 places to the left.

6. To change a percent to a decimal, drop the percent symbol and move the decimal two places to the left.

7. To change a fraction to a percent, move the decimal in the numerator two places to the left and add the percent sign.

8. A percent can be equal to a whole number.

9. To solve a problem written in the form A is R of B, we can use the proportion $\frac{B}{A} = \frac{R}{100}$

10. To solve the problem:

If there is a 5% sales tax on a radio costing $49.95, how much is the tax?

The simpler word form could be, 5% of $49.95 is what?

11. A line graph has at least two scales.

12. In a pictograph simple drawings are used as the unit of measure.

13. If ☺ represents 50 people, then ☺ ☺ ☺ ☺ represents 80 people.

14. A table is a method of displaying data in an array using a horizontal and vertical arrangement to distinguish the type of data.

15. A graph is a visual display of data that is easier for most people to read.

16. In a circle graph the circle is usually divided into 80 equal parts called sectors (pie shaped pieces).

NAME ____________________

CLASS ____________ SECTION ____________ DATE ____________

ANSWERS

1. ____________

3. ____________

4. ____________

5. ____________

6. ____________

7. ____________

8. ____________

9. ____________

10. ____________

11. ____________

12. ____________

13. ____________

14. ____________

CHAPTER 6 POST-TEST

1. A TV set regularly sells for \$450. During a sale the dealer discounts the price \$67.50. What is the percent of the discount?

2. **(Section 6.4, Obj. 77)** Change to a percent: $\frac{5}{16}$

3. **(Section 6.1, Obj. 74)** Write as a percent: 0.07125

4. **(Section 6.7, Obj. 82)** One hundred seventeen percent of what number is 65.52?

5. **(Section 6.1, Obj. 74)** If 12 out of every 100 people are left-handed, what percent of the population is left-handed?

6. **(Section 6.5, Obj. 78)** Change to a fraction: $71\frac{3}{7}\%$

7. **(Section 6.7, Obj. 82)** What percent of $5\frac{1}{2}$ is $3\frac{1}{8}$? (To the nearest tenth of a percent.)

8. **(Section 6.2, Obj. 75)** Write as a percent: 2.3

9. **(Section 6.7, Obj. 82)** What number is 6.3% of 570?

10. **(Section 6.4, Obj. 77)** Change to a percent (give answer to the nearest tenth of a percent): $1\frac{1}{7}$

11. **(Section 6.5, Obj. 78)** Change to a fraction: 230%

12. **(Section 6.3, Obj. 76)** Write as a decimal: 5%

13. The Mitchells spend 23.6% of their monthly income on food. If their monthly income is \$1753.35, how much money do they spend on food?

14. **(Section 6.3, Obj. 76)** Write as a decimal: 0.2%

ANSWERS

15. ______________________

16. ______________________

17. ______________________

15. If a tire cost the dealer \$48.52 and the markup is 30% of the cost, what is the selling price of the tire?

16. (Section 6.10, Obj. 85) Given the following graph showing auto sales distribution for a local dealer:

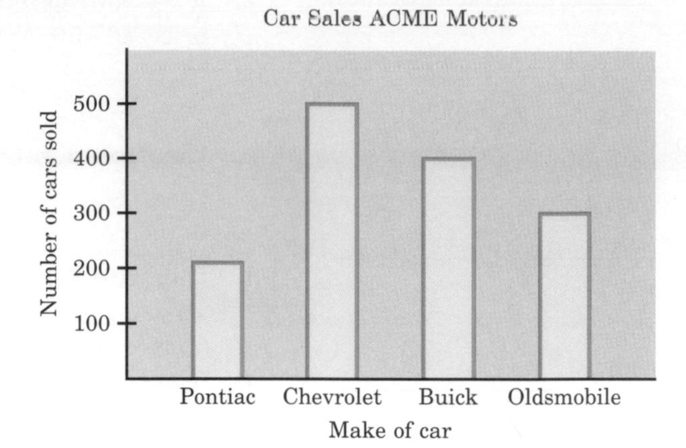

a. What make of auto had the greatest sales?
b. What was the total number of Pontiacs and Chevrolets sold?
c. How many more Buicks were sold than Oldsmobiles?

17. (Section 6.11, Obj. 87) Given the following table:

EMPLOYEES BY DIVISION EXACTO ELECTRONICS		
Division	Day Shift	Swing Shift
A	350	175
B	400	125
C	125	25

a. Which division has the greatest number of employees?
b. How many more employees are in the day shift in Division A as compared to Division C?
c. What percent of Exacto's employees are in Division B?

18. Construct a bar graph to display the type of lunch that was purchased at the local fast food bar over one week: Hamburger—1100, Fishburger—300, Chef Salad—500, Roll and Soup—400, Omelet—300.

CHAPTER 7

Measurement

NAME ______________________

CLASS ______________ SECTION ______________ DATE ______________

ANSWERS

1. ______________

2. ______________

3. ______________

4. ______________

5. ______________

6. ______________

7. ______________

8. ______________

9. ______________

10. ______________

11. ______________

CHAPTER 7 PRE-TEST

The problems in the following pre-test are a sample of the material in the chapter. You may already know how to work some of these. If so, this will allow you to spend less time on those parts. As a result, you will have more time to give to the sections that gave you difficulty. Please work the following problems. The answers are in the back of the text.

1. Ken needs 4 pieces of wire, each 8.3 in. long. What is the total length of wire he needs?

2. A board that is 18 ft long is to be cut into 4 pieces, each the same length. What will be the length of each piece?

3. **(Section 7.1, Obj. 89)** Change $6\frac{2}{3}$ hours to minutes.

4. **(Section 7.1, Obj. 90)** Add:

$$\begin{array}{r} 6 \text{ ft } 8 \text{ in.} \\ \underline{4 \text{ ft } 5 \text{ in.}} \end{array}$$

5. Norm uses 22 liters of gasoline in his automobile to drive to and from work each day. At this rate, how many liters does he use each week if he works 5 days each week?

6. Four candy bars weigh a total of 458 g. What is the weight of each candy bar?

7. **(Section 7.2, Obj. 92)** 2.7 km = ? m

8. **(Section 7.2, Obj. 93)** Subtract:

$$\begin{array}{r} 8\ \ell\ 4\ \text{d}\ell \\ \underline{2\ \ell\ 5\ \text{d}\ell} \end{array}$$

9. **(Section 7.3, Obj. 94)** Convert 54 in. to yd.

10. **(Section 7.3, Obj. 94)** Convert 45 miles per hour to feet per second.

11. **(Section 7.4, Obj. 95)** Convert 45 kilograms to pounds.

ANSWERS

12. ______

13. ______

14. ______

15. ______

16. ______

17. ______

18. ______

19. ______

12. **(Section 7.5, Obj. 96)** Find the perimeter of the following geometric figure:

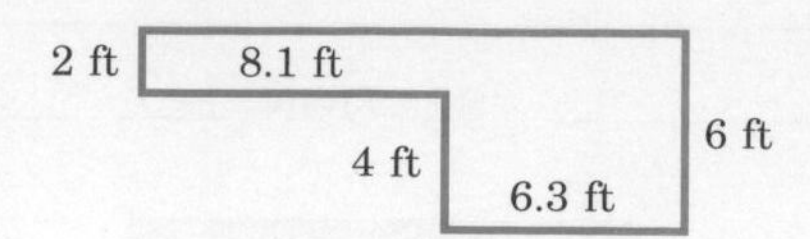

13. **(Section 7.5, Obj. 96)** Find the circumference of a circle whose radius is 6 cm. (Let $\pi \approx 3.14$.)

14. **(Section 7.6, Obj. 97)** Find the area of the following geometric figure:

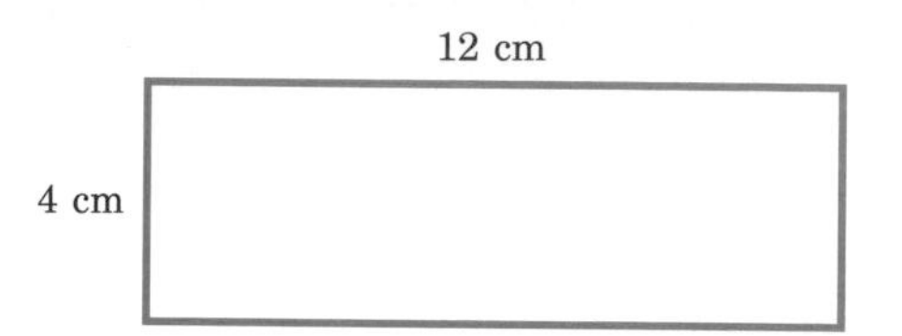

15. **(Section 7.6, Obj. 97)** Find the area of the following circle. (Let $\pi \approx 3.14$.)

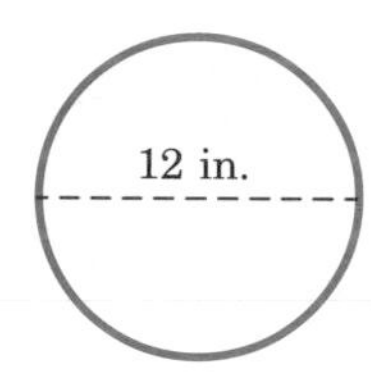

16. **(Section 7.7, Obj. 98)** Find the area of the following geometric figure:

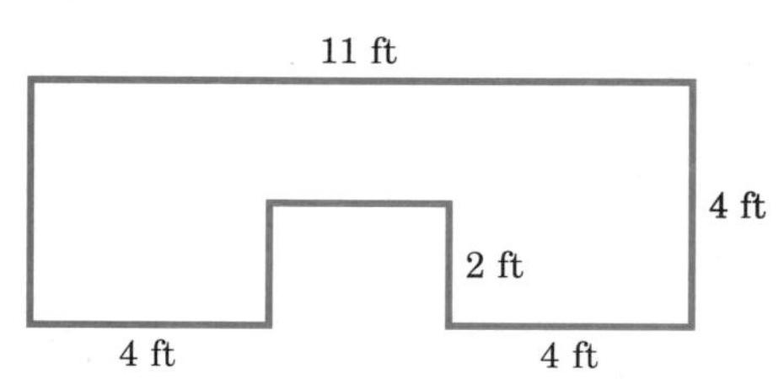

17. **(Section 7.8, Obj. 99)** Find the volume of a beachball whose radius is 9 in. (Let $\pi \approx 3.14$.)

18. **(Section 7.9, Obj. 100)** Predict what number seems most likely to appear next in the following sequence.
5, 3, 8, 6, 11, 9, __

19. **(Section 7.9, Obj. 100)** State whether the following is true or false.
All houses have roofs. This building has a roof, therefore it is a house.

7.1 English Measurements

OBJECTIVES

88. Multiply or divide an English measurement and a number.
89. Change the units of an English measurement.
90. Add or subtract English measurements.

APPLICATION

To run a mile race on the indoor track at Low High School, the runners must go around the track 8 times. If Tom runs around the track once, how many feet has he gone?

VOCABULARY

A *measurement* is written with a number and a unit of measure. It shows "how many" or "how much." For example, *5 miles* is a measurement.

English measurements are those whose units came from English-speaking nations. Examples of English measures are 1 pound (weight), 1 foot (length), 1 quart (volume).

Equivalent measurements are different measures of the same thing using a different unit (for example, 12 inches = 1 foot).

HOW AND WHY

Any unit of measure can be chosen arbitrarily. Multiples of this unit are used to measure objects. For example, 1 inch measures length and is shown here:

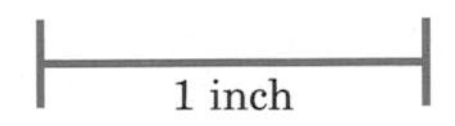

To measure a longer object, we count how many of these units are needed to measure the length. Let's try it on the following line:

It takes 4 of the 1-inch units to measure the line. We say the line has length $4 \cdot (1 \text{ inch})$ or just 4 inches. "$4 \cdot (1 \text{ inch}) = 4$ inches" is a way of writing a measurement as a number times a unit of measure. So:

5 feet = $5 \cdot$ (1 foot)

25 miles = $25 \cdot$ (1 mile)

3 teaspoons = $3 \cdot$ (1 teaspoon)

This way of writing measurements is used to find the multiple of any measurement. Consider 3 boards, each 5 feet in length. Since the boards are of equal length, the total length should be 3 times one length:

$3 \cdot (5 \text{ feet}) = 3 \cdot 5 \cdot (1 \text{ foot}) = 15 \cdot (1 \text{ foot}) = 15 \text{ feet}$

PROCEDURE

To multiply a measurement times a number, multiply the two numbers and write the unit of measure.

If a container holds 6 quarts of water, how many quarts does it contain when it is $\frac{1}{3}$ full?

$\frac{1}{3}$ of 6 quarts $= \frac{1}{3} \cdot 6$ quarts $= 2$ quarts

PROCEDURE

To divide a measurement by a number, divide the two numbers and write the unit of measure.

If you are given 65 dollars and need to divide it into 5 equal parts, how much would be in each part?

(65 dollars) ÷ 5 = (65 ÷ 5) dollars = 13 dollars

English measures and their equivalents that you should know are listed in Table 7.1.

TABLE 7.1 English Measures and Equivalents

LENGTH	TIME
12 inches (in.) = 1 foot (ft) 3 feet (ft) = 1 yard (yd) 5280 feet (ft) = 1 mile (mi)	60 seconds (sec) = 1 minute (min) 60 minutes (min) = 1 hour (hr) 24 hours (hr) = 1 day 7 days = 1 week
LIQUID (VOLUME)	**WEIGHT**
3 teaspoons (tsp) = 1 tablespoon (tbs) 2 cups (c) = 1 pint (pt) 2 pints (pt) = 1 quart (qt) 4 quarts (qt) = 1 gallon (gal)	16 ounces (oz) = 1 pound (lb) 2000 pounds (lb) = 1 ton

Using the equivalent measures in this table, we can change the unit of a measurement. Consider converting 14 feet to inches. Since 12 inches = 1 foot, these two are equivalent measures. The indicated division, (12 inches) ÷ (1 foot), can be interpreted as "how many units of measure 1 foot will it take to make 12 inches?" since they measure the same length, the answer is 1, so

$$\frac{12 \text{ inches}}{1 \text{ foot}} = \frac{1 \text{ foot}}{12 \text{ inches}} = 1$$

Either of these fractions, along with the multiplication property of one can be used to convert from on measure to another. The units of measure are treated like factors and are reduced (canceled) before multiplying.

To convert 14 feet to inches, multiply:

$14 \text{ feet} = (14 \text{ feet}) \cdot 1$ **Multiply by 1.**

$= \frac{14 \cancel{\text{ft}}}{1} \cdot \frac{12 \text{ inches}}{1 \cancel{\text{ft}}}$ **Substitute $\frac{12 \text{ inches}}{1 \text{ ft}}$ for 1 so the ft units will reduce.**

$= 168$ inches

PROCEDURE

To change the unit of a measurement, multiply by a fraction formed by equivalent measurements (name for 1) to get the required units. Then multiply and reduce.

As another example convert 17,160 feet to miles.

$17160 \text{ feet} = (17160 \text{ feet}) \cdot 1$ — **Multiply by 1.**

$= \dfrac{17160 \cancel{\text{feet}}}{1} \cdot \dfrac{1 \text{ mile}}{5280 \cancel{\text{feet}}}$ — **Substitute $\dfrac{1 \text{ mile}}{5280 \text{ feet}}$ for 1 so the feet units will reduce.**

$= 3.25 \text{ miles}$ — **Reduce (divide).**

CAUTION

When choosing the substitution for 1, be sure to select the fraction whose *denominator* includes the units you wish to reduce (cancel).

When 7 ounces is added to 3 pounds, the sum is neither 10 ounces nor 10 pounds. We cannot add unlike measures, so we convert 3 pounds to ounces:

3 pounds = 3 (16 ounces) = 48 ounces

Then we can add 7 ounces and 48 ounces and find the total measure to be 55 ounces.

PROCEDURE

To add or subtract two measurements with common units of measure, add or subtract the numbers and write the unit of measure.

MODEL PROBLEM-SOLVING

Examples	Strategy
a. Multiply 12 pints by 3:	
(12 pints)(3) = (12 · 3)(pints)	Group the numbers.
= 36 pints	Simplify.

Warm Up a. Multiply 32 ounces by 5.

ANSWERS TO WARM UPS (7.1)

a. 160 ounces

b. Divide 55 minutes into 4 equal time periods:

$(55 \text{ minutes}) \div 4 = \frac{55}{4} \text{ minutes}$ **Divide the numbers.**

$= 13\frac{3}{4} \text{ minutes}$ **Simplify.**

Warm Up b. Divide 85 feet into 5 equal lengths.

c. Change $13\frac{3}{4}$ minutes to seconds:

$13\frac{3}{4} \text{ minutes} = \left(13\frac{3}{4} \text{ minutes}\right) \cdot 1$ **Multiply by 1.**

$= \frac{13\frac{3}{4} \cancel{\text{minutes}}}{1} \cdot \frac{60 \text{ seconds}}{1 \cancel{\text{minute}}}$ **Substitute $\frac{60 \text{ seconds}}{1 \text{ minute}}$ for 1 so the minutes units will reduce.**

$= \frac{55}{4} \cdot 60 \text{ seconds}$

$= 825 \text{ seconds}$ **Multiply.**

Warm Up c. Change 230 miles to feet.

d. Add: 6 feet + 7 feet + 11 feet

6 feet + 7 feet + 11 feet = 24 feet **Add the numbers and retain the common measure.**

Warm Up d. Add: 15 lb + 24 lb + 45 lb + 17 lb

ANSWERS TO WARM UPS (7.1) **b.** 17 feet **c.** 1,214,400 feet **d.** 101 lb

e. Add: 3 gal 2 qt + 5 gal 3 qt

3 gal 2 qt
5 gal 3 qt
8 gal 5 qt = 8 gal + 4 qt + 1 qt
= 8 gal + 1 gal + 1 qt
= 9 gal 1 qt

Write in columns, lining up the place values and units.
Add. Since 5 qt is greater than 1 gal (4 qt = 1 gal), we change four qt to 1 gal.

Add the gallons.

Warm Up e. Add: 6 ft 9 in + 4 ft 7 in.

f. If a carpenter cuts a board that is 2 ft 5 in. long from a board that is 8 ft 3 in. long, how much board will be left?

8 ft 3 in.
2 ft 5 in.

Subtract the length of the piece cut off from the length of the original board.

7 ft 1 ft 3 in.
2 ft 5 in.

Since 5 cannot be subtracted from 3, borrow 1 ft from the 8 ft (1 ft = 12 in.).

7 ft 15 in.
2 ft 5 in.
5 ft 10 in.

Subtract.

So, there will be 5 ft 10 in. of board remaining.

Warm Up f. A wine maker draws off 5 gallons 3 quarts of wine from a 15-gallon keg. How much wine is left in the keg?

APPLICATION SOLUTION

g. To run a mile race on the indoor track at Low High School, the runners must go around the track 8 times. If Tom runs around the track once, how many feet has he gone?

$1 \text{ mile} \div 8 = 5280 \text{ ft} \div 8$
$= 660 \text{ ft}$

To find the number of feet Tom has run, we must divide the number of feet in one mile by 8 (8 laps make one mile). Since 1 mile = 5280 ft, change 1 mile to 5280 ft.

Therefore, Tom has run 660 ft around the track.

Warm Up g. Five hundred twenty-five pounds of ground beef are divided evenly among three school lunch programs. How many pounds of ground beef will each lunch program receive?

ANSWERS TO WARM UPS (7.1) **e.** 11 ft 4 in. **f.** 9 gal 1 qt **g.** 175 lb

NAME ______________________________

CLASS ______________ SECTION ______________ DATE ______________

Exercises 7.1

A

ANSWERS

1. ______________
2. ______________
3. ______________
4. ______________
5. ______________
6. ______________
7. ______________
8. ______________
9. ______________
10. ______________
11. ______________
12. ______________
13. ______________
14. ______________
15. ______________
16. ______________
17. ______________
18. ______________
19. ______________
20. ______________

Multiply or divide:

1. (4 feet)(6)

2. (4 cups)(5)

3. (2.3)(7 pounds)

4. (3)(18 yards)

5. (200 feet) ÷ 25

6. (28 days) ÷ 4

7. 30 feet ÷ 6

8. 180 pounds ÷ 15

Change the units as shown:

9. 5 ft = ? inches

10. 2 miles = ? feet

11. 3 lb = ? oz

12. 2 hours = ? min

13. 3 days = ? hours

14. 2 gallons = ? qt

15. 7 yards = ______ inches

16. 9 yards = ______ feet

17. 7 feet = ______ inches

18. 5 pints = ______ cups

19. 7 gallons = ______ quarts

20. 18 pounds = ______ ounces

ANSWERS

21. ______

22. ______

23. ______

24. ______

Add:

21. 2 ft 5 in.
6 ft 4 in.

22. 3 days 4 hours
4 days 6 hours
6 days 9 hours

Subtract:

23. 6 yd 2 ft
3 yd 1 ft

24. 27 gal 3 qt
9 gal 1 qt

B

25. ______

26. ______

27. ______

28. ______

29. ______

30. ______

31. ______

32. ______

33. ______

34. ______

35. ______

36. ______

25. Multiply: $3\frac{1}{2}$ (5 ounces)

26. Multiply: $2\frac{3}{4}$ (8 hours)

27. Divide: (4 hours) $\div 1\frac{1}{2}$

28. Divide: (12 feet) $\div 3\frac{1}{3}$

29. 3 ft 7 in. = ? in.

30. 3.5 gal = ? qt

31. 3 qt 1 pt = ? pints

32. 2 days 7 hours = ? hours

33. 8 yards and 2 feet = ______ feet

34. 4.6 miles = ______ feet

35. 7 days and 8 hours = ______ hours

36. 4 gallons 3 quarts = ______ quarts

NAME ____________________

CLASS ____________ SECTION ____________ DATE ____________

ANSWERS

37. ____________

38. ____________

39. ____________

40. ____________

41. ____________

42. ____________

43. ____________

44. ____________

Add:

37.
6 ft 5 in.
2 ft 7 in.
10 ft 7 in.

38.
9 gal 3 qt 1 pt
4 gal 2 qt 1 pt
3 gal 1 qt 1 pt

39. 4 hr 30 min + 7 hr 45 min + 8 hr 10 min

40. 8 lb 12 oz + 6 lb 10 oz + 4 lb 8 oz

Subtract:

41.
21 min 39 sec
14 min 47 sec

42.
5 min 30 sec
3 min 42 sec

43. 1 day − 10 hr 7 min 18 sec

44. 3 gal − 2 gal 1 qt 1 pt

C

45. ____________

46. ____________

47. ____________

48. ____________

49. ____________

45. Multiply: 2.38 (16 miles)

46. Multiply: 5.73 (3 lb)

47. (7.25)(5.5 hr)

48. (6.18)(4.3 gal)

49. Divide: (38 tons) ÷ 4.63 (to the nearest tenth)

ANSWERS

50. ______

51. ______

52. ______

53. ______

54. ______

55. ______

56. ______

57. ______

58. ______

59. ______

60. ______

61. ______

62. ______

63. ______

64. ______

50. Divide: (2036 ft) ÷ 125 (to the nearest tenth)

51. 38.22 yards ÷ 4.7 (to the nearest tenth)

52. 4.56 gal ÷ 1.3 (to the nearest hundredth)

53. 2 yd 2 ft 11 in. = ? in.

54. $4\frac{3}{16}$ lb = ? ounces

55. 2.13 tons = ? pounds

56. 2 miles = ? yd

57. 3 miles 100 yards = ______ yards

58. 15 days 8 hours = ______ hours

59. $4\frac{2}{3}$ days = ______ hours

60. $7\frac{3}{4}$ gallons = ______ quarts

61. 14.8 yards = ______ inches

62. 7.38 miles = ______ feet

63. 4.65 tons = ______ pounds

64. 7.15 yards = ______ inches

NAME ______________________________

CLASS ______________ SECTION ______________ DATE ______________

D

ANSWERS

65. ______________

66. ______________

67. ______________

68. ______________

69. ______________

70. ______________

65. While golfing, Peg made birdie putts of 3 ft 2 in., 11 ft 10 in., 22 ft 9 in., and 7 ft 4 in. during a round. What was the total length of all the birdie putts?

66. If a car travels 55 miles in one hour, how far will it travel in 7 hours at the same speed?

67. A buyer purchased 52 quarts of a rare wine for four clients. If each client is to share equally, how many quarts of wine will each one get?

68. The Corner Grocery sold 20 lb 6 oz of hamburger on Wednesday, 13 lb 8 oz on Thursday, and 17 lb 10 oz on Friday. How much hamburger was sold during the three days?

69. Paul pledged 7 dollars a month to the United Fund. What was his annual donation?

70. Kevin canned 28 pints of strawberry jam, 18 pints of raspberry jam, 10 pints of blueberry jam, and 12 pints of peach jam. How many cups of jam did Kevin make?

ANSWERS

71. ______________

72. ______________

73. ______________

74. ______________

75. ______________

76. ______________

71. Roger drove 300 miles in 6 hours. What was the average speed per hour?

72. Sue bought $2\frac{1}{2}$ yards of plain blue fabric, $3\frac{1}{8}$ yards of blue plaid, $4\frac{1}{4}$ yards of blue print, and $1\frac{3}{8}$ yards of blue polka dot. How many inches of fabric did she buy?

73. Dennis worked 18 hours during the first week, 27 the second week, 35 the third week, and 29 the last week. What is his average number of hours worked per week?

74. Carol bought one gallon of milk at the local store. When she got home, Dan drank 3 cups of milk, Larry drank $1\frac{1}{2}$ cups, Greg drank 2 cups, and Carol used $\frac{1}{2}$ cup in cooking. How much milk remained at the end of the day?

75. A *mil* is a unit of measure for wire diameter (1 mil = 0.001 inch). If the diameter of a wire is 25 mils, find its diameter in inches.

76. Using the unit of measure in Exercise 75, if the diameter of a wire is 75.5 mils, find its diameter in inches.

NAME ______________________________

CLASS ______________ SECTION ______________ DATE ______________

E *Maintain your skills* (Sections 5.3, 6.2, 6.3, 6.4, 6.5, 6.6)

ANSWERS

77. ______________

78. ______________

79. ______________

80. ______________

81. ______________

82. ______________

83. ______________

84. ______________

85. ______________

86. ______________

Change to a percent:

77. $\frac{9}{32}$

78. $\frac{17}{125}$

79. 0.82

80. 3.775

Change to a fraction:

81. 72%

82. 37.8%

Change to a decimal:

83. 72%

84. 37.8%

85. A map of Canada is scaled so that 2.5 cm represents approximately 64 kilometers. How many kilometers is it between Vancouver, British Columbia, and Regina, Saskatchewan, if the distance on the map is 70 cm?

86. Mrs. Hagen owns an apartment complex. She figures that for each three apartments that are occupied she will pay a monthly water bill of $19.00. At present, 47 apartments are occupied. How much should she plan to spend for water this month, to the nearest cent?

Metric Measurements

OBJECTIVES

91. Multiply or divide a metric measurement and a number.
92. Change the units of a metric measurement.
93. Add or subtract metric measurements.

APPLICATION

Michael DesChamps ran five 2000-meter races in seven days. How many kilometers did he run in the five races?

VOCABULARY

The standard units in the metric system are:

Length: 1 meter or 1 m
Weight: 1 gram or 1 g (See table, page 669)
Volume: 1 liter or 1 ℓ

HOW AND WHY

A line six centimeters long requires six 1-centimeter measurements to show the length:

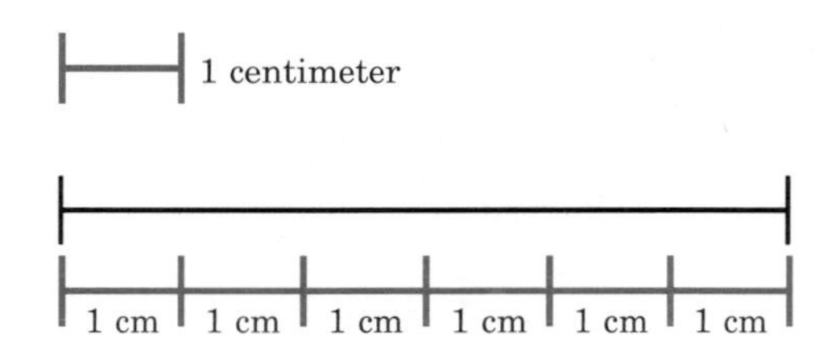

We can say that the line segment is $6 \cdot (1 \text{ cm})$ long, or 6 cm. Three liters of water represent three 1-liter measurements:

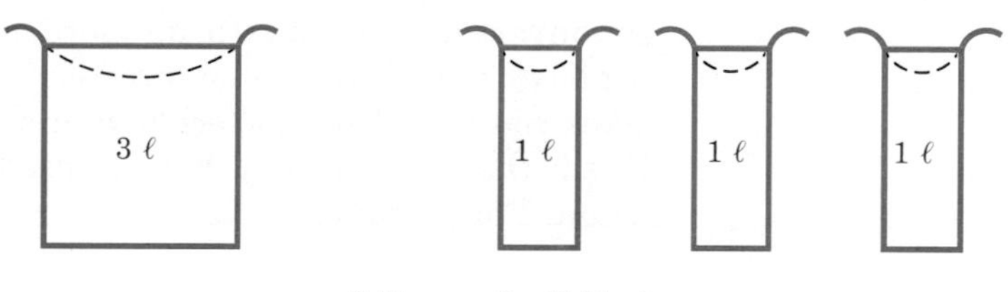

3 liters = $3 \cdot$ (1 liter)

These examples show a useful way of thinking about metric measurements:

1.5 meters = 1.5(1 meter)

15 kilometers = 15(1 kilometer)

0.45 kilograms = 0.45(1 kilogram)

Virtually every country in the world except the United States uses the metric system. Scientists of the world, including those in the United States, use the metric system exclusively.

The metric system was invented to take advantage of our base ten place value system in the same way that our monetary system does. (See Table 7.2.)

TABLE 7.2 U.S. Monetary System

						METRIC PREFIXES
		1 mil	=	0.001	dollar	milli
10 mils	=	1 cent	=	0.01	dollar	centi
10 cents	=	1 dime	=	0.1	dollar	deci
10 dimes	=	1 dollar	=	1	dollar	base unit
10 dollars	=	1 $10 bill	=	10	dollars	deka
10 $10 bills	=	1 $100 bill	=	100	dollars	hecto
10 $100 bills	=	1 $1000 bill	=	1000	dollars	kilo

For most purposes, the units of meter, gram, and liter, along with their multiples, are sufficient.

Some equivalent metric measures are shown in Table 7.3 on the facing page.

Meters, grams, and liters (base units in the metric system) all correspond to DOLLARS. Using this concept, we may think of a $1000 bill as a "kilobuck."

The operations with metric measurements are done in the same way as they are with English measurements.

Since the metric system takes advantage of the base ten place value system, we can convert from one unit to another by moving the decimal point. The following list of the prefixes and the base unit will help.

```
           b
           a
           s
           e
k  h  da       d  c  m
           u
           n
           i
           t
```

To convert to a new metric measure, move the decimal point the same number of places and in the same direction as the number of places and direction it takes you to go from the original prefix to the new one on the preceding chart. This is the same process that was used when dividing or multiplying by powers of ten. For instance:

23 hg = ? dg

```
         base
k h da   unit d c m
2 3 .0    0   0
```

23 hg = 23000 dg

The "d" prefix is three places to the right of the "h" prefix, so move the decimal point three places to the right.

Also:

56 mℓ = ? kℓ
56 mℓ = 0.000056 kℓ

The "k" prefix is six places to the left of the "m" prefix, so move the decimal point six places to the left.

TABLE 7.3 Metric Measures and Equivalents

LENGTH (BASIC UNIT IS 1 METER)				
1 millimeter	(mm)	=	=	0.001 m
1 centimeter	(cm)	= 10 millimeters	=	0.01 m
1 decimeter	(dm)	= 10 centimeters	=	0.1 m
1 METER	(m)	= 10 decimeters	=	1 m
1 dekameter	(dam)	= 10 meters	=	10 m
1 hectometer	(hm)	= 10 dekameters	=	100 m
1 kilometer	(km)	= 10 hectometers	=	1000 m
WEIGHT* (BASIC UNIT IS 1 GRAM)				
1 milligram	(mg)	=	=	0.001 g
1 centigram	(cg)	= 10 milligrams	=	0.01 g
1 decigram	(dg)	= 10 centigrams	=	0.1 g
1 GRAM	(g)	= 10 decigrams	=	1 g
1 dekagram	(dag)	= 10 grams	=	10 g
1 hectogram	(hg)	= 10 dekagrams	=	100 g
1 kilogram	(kg)	= 10 hectograms	=	1000 g
1 metric ton		= 1000 kilograms		
LIQUID AND DRY MEASURE (BASIC UNIT IS 1 LITER)				
1 milliliter	(mℓ)	=	=	0.001 ℓ
1 centiliter	(cℓ)	= 10 milliliters	=	0.01 ℓ
1 deciliter	(dℓ)	= 10 centiliters	=	0.1 ℓ
1 LITER	(ℓ)	= 10 deciliters	=	1 ℓ
1 dekaliter	(daℓ)	= 10 liters	=	10 ℓ
1 hectoliter	(hℓ)	= 10 dekaliters	=	100 ℓ
1 kiloliter	(kℓ)	= 10 hectoliters	=	1000 ℓ

*These units are technically reserved for measuring mass. We will use them for weight or mass. The difference between weight and mass is covered in science classes.

MODEL PROBLEM-SOLVING

Examples | **Strategy**

a. What is the total weight of four packages of cheese if each weighs 0.75 kilogram?

$(4)(0.75 \text{ kg}) = (4)(0.75)(1 \text{ kg})$ — **To find the total weight, multiply the weight of one package by 4.**

$= (3)(1 \text{ kg})$ — **Multiply the numbers.**

$= 3 \text{ kg}$

The total weight of the four packages is 3 kg.

Warm Up a. A package of microwave popcorn weighs 101 grams. What is the weight of 7 packages?

ANSWERS TO WARM UPS (7.2)

a. 707 grams

b. If 400 grams of peanut brittle are divided equally among 6 sacks, how much goes into each sack?

$400 \text{ g} \div 6 = \frac{400}{6}(1 \text{ g})$ **Divide the weight by the number of sacks.**

$= 66.7(1 \text{ g})$ **Divide the numbers, and round off.**

$= 66.7 \text{ g}$

So, each sack contains 66.7 g of peanut brittle.

Warm Up b. If 288 deciliters of soup are to be divided equally among 48 patients at the local hospital, how many deciliters of soup will each receive?

c. Change 5 m to cm.

$5 \text{ m} = ? \text{ cm}$ **Since cm is two places to the right of m in the chart, move the decimal point two places to the right.**

$= 500 \text{ cm}$

Warm Up c. Change 80 g to mg.

d. How many meters are in 0.4 kilometer?

$0.4 \text{ km} = ? \text{ m}$ **Since m is three places to the right of km in the chart, move the decimal point three places to the right.**

$= 400 \text{ m}$

Warm Up d. How many milliliters are in 0.6 liters?

e. What is the measure of 8 meters + 45 cm in meters?

$8 \text{ meters} + 45 \text{ cm} = 8 \text{ m} + 45 \text{ cm}$ **Convert 45 cm to m by moving the decimal point two places to the left.**

$= 8 \text{ m} + 0.45 \text{ m}$

$= 8.45 \text{ m}$ **Add.**

Warm Up e. What is the measure of 8 grams and 15 centigrams in grams?

ANSWERS TO WARM UPS (7.2)

b. 6 deciliters **c.** 80,000 mg **d.** 600 milliliters

e. 8.15 grams

f. What is the measure of 1.3 liters − 135 milliliters in milliliters?

$1.3\ \ell - 135\ \text{m}\ell = 1300\ \text{m}\ell - 135\ \text{m}\ell$
$= 1165\ \text{m}\ell$

Convert ℓ to mℓ by moving the decimal point three places to the right.

Warm Up f. What is the measure of 35 meters and 2.4 centimeters in centimeters?

APPLICATION SOLUTION

g. Michael DesChamps ran five 2000-meter races in seven days. How many kilometers did he run in the five races?

$5(2000\ \text{m}) = 10000\ \text{m}$ — **Multiply the length of one race by five.**

$10000\ \text{m} = 10\ \text{km}$ — **Convert to km by moving the decimal point three places to the left.**

So, Michael ran 10 km in the five races.

Warm Up g. Carol bought four boxes of cereal. The weights of the cereal were 578 g, 645 g, 495 g, and 755 g. How many kilograms of cereal did Carol buy?

ANSWERS TO WARM UPS (7.2)

f. 3502.4 centimeters **g.** 2.473 kilograms

NAME ______________________

CLASS ______________ SECTION ______________ DATE ______________

Exercises 7.2

A

ANSWERS

1. ______
2. ______
3. ______
4. ______
5. ______
6. ______
7. ______
8. ______
9. ______
10. ______
11. ______
12. ______
13. ______
14. ______
15. ______
16. ______

Change the units as shown:

1. 30 m = ? cm **2.** 5 kg = ? g **3.** 100 g = ? kg

4. 30 mm = ? m **5.** 0.2 km = ? m **6.** 10 m = ? cm

7. 7 ℓ to mℓ **8.** 4 km to m **9.** 700 cm to m

10. 18 cm to mm

11. Add: 4.6 m + 2.8 m = ? m

12. Add: 48 mm + 32 mm + 10 mm = ? mm

13. Subtract: 32 g − 12 g = ? g

14. Subtract: 360 kℓ − 155 kℓ = ? kℓ

15. Multiply: 3(120 km) = ? km

16. Divide: 225 mg ÷ 5 = ? mg

B

ANSWERS

17. ______

18. ______

19. ______

20. ______

21. ______

22. ______

23. ______

24. ______

25. ______

26. ______

27. ______

28. ______

29. ______

30. ______

31. ______

32. ______

33. ______

34. ______

35. ______

36. ______

17. 627 cℓ = ? ℓ

18. 7 m = ? cm

19. 1.3 km = ? m

20. 1.3 kg = ? g

21. 1.3 kℓ = ? ℓ

22. 244 mm = ? m

23. 245 mℓ = ? ℓ

24. 246 mg = ? g

25. 3.6 m to cm

26. 0.07 cm to m

27. 0.175 kg to g

28. 52.87 mℓ to ℓ

Multiply:

29. (6)(13 g)

30. (3.5 mℓ)(14)

Divide:

31. (17 km) ÷ 5

32. (187 ℓ) ÷ 17

Add:

33. 35 mℓ + 14 ℓ = ? mℓ

34. 2.6 km + 1900 m = ? m

Subtract:

35. 42 g − 12 cg = ? cg

36. 17 ℓ − 444 mℓ = ? mℓ

NAME ________________

CLASS ________ SECTION ________ DATE ________

C

ANSWERS

37. ________

38. ________

39. ________

40. ________

41. ________

42. ________

43. ________

44. ________

45. ________

46. ________

47. ________

48. ________

49. ________

50. ________

51. ________

52. ________

37. 7.25 km to mm

38. 0.162 kg to cg

39. 6 kℓ 800 ℓ to ℓ

40. 82 cm 7 mm to m

41. 214 m to km

42. 18 mg to g

43. 17,250 mℓ to ℓ

44. 812.3 g to kg

45. Add: 1.3 kℓ + 1200 ℓ = ? kℓ

46. 8 m + 9 cm + 6 mm = ? mm

47. Subtract: 2.6 kg − 1500 g = ? g

48. 45 g − 728 cg = ? cg

49. 11.33 km = ? m

50. 51.395 kg = ? g

Add:

51. 0.4 km + 22 m = ? m

52. 2 kg + 4500 g = ? kg

ANSWERS

53. ____________

54. ____________

55. ____________

56. ____________

53. 3 km 250 m
3 km 900 m
= ? m

54. 2 kg 8 g
14 kg 5 g
= ? g

Subtract:

55. 3 kℓ 500 ℓ
1 kℓ 700 ℓ
= ? kℓ

56. 15 km 700 m
3 km 850 m
= ? m

D

57. ____________

58. ____________

59. ____________

60. ____________

61. ____________

62. ____________

63. ____________

57. Thirty oil drums contain a total of 5.1 kiloliters. How many liters does each drum hold?

58. Twelve hundred milligrams of calcium are recommended as the daily allowance for teenage women. How many grams is this?

59. The U.S. recommended dosage of iron is 18 mg per day for teenage women. How many grams is this?

60. Any raisin weighs about one gram. What part of a raisin would 54 cg be?

61. One cup of nonfat milk will provide 0.3 g of calcium. Change this to milligrams of calcium.

62. A board 2.4 meters long is to be cut into 5 equal parts. How many centimeters long will each piece be?

63. If a can contains 298 grams of soup, how many grams of soup are contained in 7 cans?

NAME ______

CLASS ______ SECTION ______ DATE ______

ANSWERS

64. ______

65. ______

66. ______

67. ______

68. ______

69. ______

70. ______

64. Mia, who is a lab assistant, has 282 mℓ of acid that is to be divided among 24 students. How many mℓ will each student receive?

65. Gayle purchased a package of ground beef that cost $1.43 and weighed 0.5 kg. What was the price per kilogram?

66. Paul used a meter stick to measure the distance from his receiver to the place where he wants to put his stereo speakers. He found that each speaker will be 1 meter and 15 centimeters from the receiver. How many meters of wire does he need to connect the two speakers to the receiver?

67. An electrician uses wire that is 16.5 millimeters in diameter. What is the diameter in centimeters?

68. If the mass of an iron core for a magnet is 525 milligrams, find its mass in grams.

69. The internist at St. Vincent's Medical Center ordered three 0.25-mg tablets of reserpine for a patient. How many milligrams of reserpine did the patient receive?

70. Dr. Mayo at the Children's Clinic orders 0.4 g of Terramycin for Betty Stone. How many milligrams of Terramycin did Betty receive?

E *Maintain your skills* (Sections 5.3, 6.7)

ANSWERS

71. ________________

72. ________________

73. ________________

74. ________________

75. ________________

76. ________________

77. ________________

78. ________________

79. ________________

80. ________________

Solve:

71. 38% of ___?___ is 114.

72. ___?___% of 200 is 126.

73. 48% of 925 is ___?___.

74. ___?___ is 72% of 840.

75. 47% of 48 is ___?___.

76. 832 is ___?___% of 520.

77. 148% of ___?___ is 1332.

78. 168% of 50 is ___?___.

79. Mr. Jensen owns rental properties. He purchases one property that is valued at $45,950. If he pays $1881 in property taxes on one property that is valued at $49,500, how much property tax will he pay on the new purchase?

80. In the first 18 games of a 120-game schedule, the Beavers baseball team had a total of three complete games pitched. At that rate, how many complete games pitched can the team expect in its remaining games?

Conversion of Units Within a System

OBJECTIVE **94. Convert the units within a system of a given measurement.**

APPLICATION

The Sidley Widgit Company wants to change Jean's hourly wage of $8.40 per hour to an equal piecework wage. If Jean averages $2\frac{1}{2}$ widgits in one hour, what would be the equal wage per widgit?

VOCABULARY No new words.

HOW AND WHY

Whenever we can discover equivalent measures we can convert from one unit of measure to another using the procedure of Section 7.1. In some cases, however, it may be necessary to multiply by one more than once using different fractions for one.

How many seconds old is a person aged 22 years? Over one million seconds? Over one billion seconds? Over one trillion? More? In this case it is necessary to multiply by several different names of one.

$$22 \text{ years} = (22 \text{ years}) \cdot 1$$
Multiply by 1.

$$22 \text{ years} = \frac{22 \text{ years}}{1} \cdot \frac{365 \text{ days}}{1 \text{ year}}$$
Substitute $\frac{365 \text{ days}}{1 \text{ year}} = 1$
This gives the number of days (ignoring leap years)

$$22 \text{ years} = \frac{22 \text{ years}}{1} \cdot \frac{365 \text{ days}}{1 \text{ year}} \cdot \frac{24 \text{ hours}}{1 \text{ day}}$$
Now multiply by one $\left(\frac{24 \text{ hours}}{1 \text{ day}} = 1\right)$
This gives the number of hours.

$$22 \text{ years} = \frac{22 \text{ years}}{1} \cdot \frac{365 \text{ days}}{1 \text{ year}} \cdot \frac{24 \text{ hours}}{1 \text{ day}} \cdot \frac{60 \text{ minutes}}{1 \text{ hour}}$$
Now multiply by one $\left(\frac{60 \text{ minutes}}{1 \text{ hour}} = 1\right)$
This gives the number of minutes.

$$22 \text{ years} = \frac{22 \cancel{\text{years}}}{1} \cdot \frac{365 \cancel{\text{days}}}{1 \cancel{\text{year}}} \cdot \frac{24 \cancel{\text{hours}}}{1 \cancel{\text{day}}} \cdot \frac{60 \cancel{\text{minutes}}}{1 \cancel{\text{hour}}} \cdot \frac{60 \text{ seconds}}{1 \cancel{\text{minute}}}$$
Now multiply by one $\left(\frac{60 \text{ seconds}}{1 \text{ minute}} = 1\right)$

$$= 693{,}792{,}000 \text{ seconds}$$
This gives the number of seconds.

A 22-year old is almost 700 million seconds in age.

PROCEDURE

To convert the units of a measurement:

1. **Multiply by fractions formed by equivalent measurements (names for 1) as often as necessary to get the required units of measure**
2. **Multiply and reduce.**

MODEL PROBLEM-SOLVING

Examples | **Strategy**

a. Convert 4 gallons to pints.

$$\begin{aligned} 4 \text{ gallons} &= \frac{4 \text{ gallons}}{1} \cdot \frac{1}{1} \cdot \frac{1}{1} \\ &= \frac{4 \cancel{\text{gallons}}}{1} \cdot \frac{4 \cancel{\text{quarts}}}{1 \cancel{\text{gallon}}} \cdot \frac{2 \text{ pints}}{1 \cancel{\text{quart}}} \\ &= 4 \cdot 4 \cdot 2 \text{ pints} \\ &= 32 \text{ pints} \end{aligned}$$

Multiply by 1 twice, by quarts/gallon to get quarts and then by pints/quart to get pints.

Warm Up a. Convert 4 hours to seconds.

b. Convert 3 ft^2 to in.^2. (See Section 7.6 for an explanation of the units in this example.)

$$\begin{aligned} 3 \text{ ft}^2 &= 3 \cdot 1 \cancel{\text{ft}} \cdot 1 \cancel{\text{ft}} \cdot \frac{12 \text{ in.}}{1 \cancel{\text{ft}}} \cdot \frac{12 \text{ in.}}{1 \cancel{\text{ft}}} \\ &= 3 \cdot 12 \cdot 12 \text{ in.}^2 \\ &= 432 \text{ in.}^2 \end{aligned}$$

Since 1 ft^2 = (1 foot)(1 foot) and 1 in.^2 = (1 inch)(1 inch), we multiply by in./ft twice.

Warm Up b. Convert 6 yd^2 to ft^2.

c. Convert 60 miles per hour to feet per second.

$$\begin{aligned} 60 \text{ mi per hr} &= \frac{60 \cancel{\text{mi}}}{1 \cancel{\text{hr}}} \cdot \frac{1 \cancel{\text{hr}}}{60 \cancel{\text{min}}} \cdot \frac{1 \cancel{\text{min}}}{60 \text{ sec}} \cdot \frac{5280 \text{ ft}}{1 \cancel{\text{mi}}} \\ &= \frac{60 \cdot 5280 \text{ ft}}{60 \cdot 60 \text{ sec}} \\ &= 88 \text{ ft per sec} \end{aligned}$$

First multiply by hr/min and min/sec to get the correct time measure. Then multiply by ft/mi to get the correct length measure. Cancel the common units. If the desired units are left, multiply and divide as indicated.

60 miles per hour is equivalent to 88 feet per second.

ANSWERS TO WARM UPS (7.3) **a.** 14,400 seconds **b.** 54 ft^2

Warm Up c. Convert 8 oz per in. to lb per ft.

d. Convert 480 grams per liter to milligrams per kiloliter.

$$\frac{480 \text{ g}}{1 \ell} = \frac{480000 \text{ mg}}{0.001 \text{ k}\ell}$$

$$= 480{,}000{,}000 \text{ mg/k}\ell$$

Change each measure to the desired units by moving the decimal points. Divide.

Warm Up d. Convert 25 meters per minute to kilometers per hour.

APPLICATION SOLUTION

e. The Sidley Widgit Company wants to change Jean's hourly wage of \$8.40 per hour to an equal piecework wage. If Jean averages $2\frac{1}{2}$ widgits in one hour, what would be the equal wage per widgit?

$$\frac{\$8.40}{1 \text{ hr}} = \frac{\$8.40}{1 \not{\text{hr}}} \cdot \frac{1 \not{\text{hr}}}{2.5 \text{ widgits}}$$

$$= \frac{\$8.40}{2.5 \text{ widgits}}$$

$$= \frac{\$3.36}{1 \text{ widgit}}$$

To change dollars per hour to dollars per widgit, note that Jean makes 2.5 widgits in one hour, so we have:

$$\frac{1 \text{ hour}}{2.5 \text{ widgits}}$$

Multiply and reduce.

So, an equal piecework wage would be \$3.36 per widgit.

Warm Up e. Mary can type an average of 70 words per minute. If a page averages 500 words, how many pages can Mary type in one hour?

ANSWERS TO WARM UPS (7.3)

c. $\frac{6 \text{ lb}}{1 \text{ ft}}$ **d.** $\frac{1.5 \text{ km}}{1 \text{ hr}}$ **e.** 8.4 pages per hour

NAME ____________________

CLASS ____________ SECTION ____________ DATE ____________

Exercises 7.3

A

ANSWERS

1. ____
2. ____
3. ____
4. ____
5. ____
6. ____
7. ____
8. ____
9. ____
10. ____
11. ____
12. ____
13. ____
14. ____
15. ____
16. ____
17. ____
18. ____

Convert the units as shown:

1. 80 ounces = ? pounds

2. 24 quarts = ? gallons

3. 10 cm = ? mm

4. 4 g = ? cg

5. 40 pints = ? gallons

6. 12 feet = ? yards

7. 6000 g = ? kg

8. 7 feet = ? inches

9. 10 kg = ? g

10. 3 pounds − ? ounces

11. 72 inches = ? yards

12. 9 yards = ? feet

13. 550 mℓ = ? ℓ

14. 2 m = ? cm

15. $\dfrac{120 \text{ lb}}{1 \text{ ft}} = \dfrac{? \text{ lb}}{1 \text{ in.}}$

16. $\dfrac{\$330}{\text{hr}} = \dfrac{\$?}{\text{min}}$

17. 1 day = __?__ seconds

18. 1 week = __?__ hours

ANSWERS

19. ______

20. ______

19. 16,560 seconds = __?__ hours

20. 259,200 seconds = __?__ days

B

21. ______
22. ______
23. ______
24. ______
25. ______
26. ______
27. ______
28. ______
29. ______
30. ______
31. ______
32. ______
33. ______
34. ______
35. ______
36. ______

Convert the units as shown:

21. 1080 inches = ? yards

22. 3.6 gallons = ? pints

23. 605 cm = ? m

24. 1000 mm = ? m

25. 1.8 m = ? mm

26. 7.6 kg = ? g

27. 8 inches = ? feet

28. 7.5 pounds = ? ounces

29. 3 yards 1 foot = ? inches

30. 4 feet 6 inches = ? yards

31. 2600 pounds = ? tons

32. 1 mile = ? inches

33. 6800 centimeters = ? meters

34. 10,080 minutes = ? days

35. 13 tons = ? pounds

36. 5 hours = ? seconds

NAME ______________________________

CLASS ______________ SECTION ______________ DATE ______________

ANSWERS

37. ______________

38. ______________

39. ______________

40. ______________

37. 96 cups = __?__ quarts

38. 96 pints = __?__ gallons

39. 3240 in. = __?__ feet

40. 4570 cm = __?__ mm

C

41. ______________

42. ______________

43. ______________

44. ______________

45. ______________

46. ______________

47. ______________

48. ______________

49. ______________

50. ______________

Convert the units as shown:

41. $\frac{30 \text{ miles}}{1 \text{ hour}} = \frac{? \text{ feet}}{1 \text{ hour}}$

42. $\frac{9 \text{ miles}}{1 \text{ hour}} = \frac{? \text{ miles}}{1 \text{ min}}$

43. $\frac{45 \text{ miles}}{1 \text{ hour}} = \frac{? \text{ feet}}{1 \text{ second}}$

44. $\frac{\$180}{1 \text{ ton}} = \frac{? \text{ cents}}{1 \text{ pound}}$

45. $\frac{9 \text{ tons}}{1 \text{ ft}} = \frac{? \text{ pounds}}{1 \text{ in.}}$

46. $\frac{10 \text{ ounces}}{1 \text{ cup}} = \frac{? \text{ pounds}}{1 \text{ gallon}}$

47. $\frac{144 \text{ kilometers}}{1 \text{ hour}} = \frac{? \text{ meters}}{1 \text{ second}}$

48. $\frac{32 \text{ gallons}}{1 \text{ hour}} = \frac{? \text{ quart}}{1 \text{ minute}}$

49. $3540 \text{ cm}^2 = ? \text{ m}^2$

50. 12 feet 9 inches = ? yards

ANSWERS

51. ______

52. ______

53. ______

54. ______

55. ______

56. ______

57. ______

58. ______

59. ______

60. ______

51. $2.5\ \text{m}^2 = ?\ \text{cm}^2$

52. 32 yards = ? feet

53. $\dfrac{25 \text{ pounds}}{1 \text{ foot}} = \dfrac{? \text{ ounces}}{1 \text{ inch}}$

54. $\dfrac{18{,}000 \text{ miles}}{1 \text{ hour}} = \dfrac{? \text{ feet}}{1 \text{ second}}$

55. $\dfrac{\$7.20}{1 \text{ hour}} = \dfrac{? \text{ cents}}{1 \text{ minute}}$

56. $\dfrac{\$120}{1 \text{ day}} = \dfrac{? \text{ cents}}{1 \text{ hour}}$

57. $\dfrac{16 \text{ lb}}{\text{ft}^2} = \dfrac{? \text{ oz}}{\text{in.}^2}$

58. $\dfrac{24 \text{ g}}{\text{m}^2} = \dfrac{? \text{ kg}}{\text{km}^2}$

59. $\dfrac{15 \text{ tons}}{\text{day}} = \dfrac{? \text{ lb}}{\text{min}}$

60. $\dfrac{\$24}{\text{ft}^2} = \dfrac{\$?}{\text{in.}^2}$

D

61. ______

61. If a secretary can type 90 words per minute $\left(\dfrac{90 \text{ words}}{1 \text{ minute}}\right)$, how many words can he type in 1 second?

62. ______

62. Shirley is going on a diet that will cause her to lose 4 ounces every day. At this rate, how many pounds will she lose in 6 weeks?

NAME ______

CLASS ______ SECTION ______ DATE ______

ANSWERS

63. ______

64. ______

65. ______

66. ______

67. ______

68. ______

69. ______

63. A physician orders 0.01 g of Elixir Chlor-Trimeton. The available dose has a label that reads 2 mg per cc (cubic centimeter). How many cubic centimeters are needed to fill the doctor's order?

64. Nurse Paul sees that the doctor has ordered 0.9 g of Tetracyn for his patient. If the available capsules contain 150 mg of Tetracyn, how many capsules will he give the patient?

65. Larry's family normally eats 3 boxes of Yummy Snap And Pops, the latest breakfast craze, in a week. Each box contains 14 ounces of cereal. How many pounds of the cereal will Larry's family eat in one year?

66. If the cereal in Exercise 65 costs $1.12 per box, what was the total cost of the cereal for the year?

67. Carol has a recipe that calls for 120 cubic centimeters of milk. She only has a liter measure. How much of a liter should she add? (1 cubic centimeter = 1 milliliter)

68. If Dan averages 50 miles per hour during an eight-hour day of driving, how many days will it take him to drive 2,000 miles?

69. Greg's Sport Shop decides to donate to charity 1 cent for every yard Joanne jogs during one week. Joanne jogs 2 miles every day. How much does Greg donate to charity?

ANSWERS

70. ______

71. ______

72. ______

70. If a gel weighs 64 pounds per ft^3, how many ounces (to the nearest tenth) does one $in.^3$ weigh? [*Hint:* 1 ft^3 = (1 foot)(1 foot)(1 foot) and 1 $in.^3$ = (1 inch)(1 inch)(1 inch)]

71. What is the equal piecework wage if the hourly wage is $9.80 and the average number of articles made in one hour is 2.8?

72. A certain kind of intravenous (IV) solution contains 20 cc of medicine and 250 cc of solution for a total of 270 cc. There are 10 drops of solution per 1 cc. Dr. Stem orders the IV to run 3 hours for her patient. How many drops per minute should the nurse give the patient?

E *Maintain your skills* (Sections 6.1, 6.7, 6.9)

73. ______

74. ______

75. ______

76. ______

77. ______

78. ______

79. ______

80. ______

81. ______

82. ______

Write as percents:

73. 48 to 60

74. 72 per 90

75. 16 to 48

76. 75 per 60

Write as percents to the nearest percent:

77. 52 to 60

78. 78 per 90

Write as percents to the nearest tenth of a percent:

79. 13 to 48

80. 81 per 60

81. The original price of a 13-in. color television set was $389. The price was discounted (lowered) $58.35 for sale. What was the percent of discount?

82. The Low-Risk Insurance Company pays $8.75 interest per year for every $100 of the cash value of a policy. What annual interest rate is this?

Conversion of Units: English–Metric

OBJECTIVE

95. Convert units of measure from metric to English and English to metric.

APPLICATION

Mary is driving on a road in Canada on which the speed limit is posted as 80 km/hr. The speedometer in her automobile is registering 45 miles/hr. Is she driving within the speed limit?

VOCABULARY

No new words

HOW AND WHY

The world is becoming almost entirely metric and the United States is moving in that direction. We are seeing comparisons of the two systems by double listings. This is apparent on marked boxes of food (ounces and grams) and on the highway (miles per hour and kilometers per hour).

In this discussion we perform some conversions between the systems (English and metric) using the basic units. Since the systems were developed independently there are no exact comparisons. All conversion units are approximate.

ENGLISH–METRIC CONVERSIONS	
1 inch	≈ 2.54 centimeters
1 foot	≈ 0.30 meter
1 yard	≈ 0.91 meter
1 mile	≈ 1.61 kilometers
1 quart	≈ 0.95 liter
1 pound	≈ 0.45 kilograms

METRIC–ENGLISH CONVERSIONS	
1 centimeter	≈ 0.39 inch
1 meter	≈ 3.28 feet
1 meter	≈ 1.09 yards
1 kilometer	≈ 0.62 mile
1 liter	≈ 1.06 quarts
1 kilogram	≈ 2.20 pounds

With these conversions we can convert between the two systems using the procedure of Section 7.3.

CAUTION

1. **Because these conversions are all rounded to the nearest hundredth we cannot expect closer accuracy when using them.**
2. **Since there are two conversion factors that can be used for each conversion it is possible for answers to vary, depending upon which factor is chosen.**

Results in the examples and exercises are rounded to the nearest tenth. Since these conversions are approximate the results may not always be accurate. For greater accuracy, conversion units with more decimal places are required.

MODEL PROBLEM-SOLVING

Examples | **Strategy**

a. Convert 30 inches to centimeters:

$$30 \text{ inches} \approx \frac{30 \cancel{\text{ inches}}}{1} \cdot \frac{2.54 \text{ centimeters}}{1 \cancel{\text{ inch}}}$$

Multiply by the conversion factor 2.54 cm/in. from the table.

$$\approx (30)(2.54 \text{ centimeters})$$

Simplify.

$$\approx 76.2 \text{ centimeters}$$

Multiply.

So, 30 inches is approximately 76.2 centimeters.

Warm Up a. Convert 12 ounces to grams.

b. Convert one pint to liters:

$$\frac{1 \text{ pt}}{1} \approx \frac{1 \cancel{\text{ pt}}}{1} \cdot \frac{1 \cancel{\text{ qt}}}{2 \cancel{\text{ pt}}} \cdot \frac{0.95 \ \ell}{1 \cancel{\text{ qt}}}$$

The second factor of 1 $\left(\frac{0.95 \ \ell}{1 \text{ qt}} = 1\right)$ is needed to convert pints to quarts. We have a conversion from quarts to liters.

$$\approx (1)\left(\frac{1}{2}\right)(0.95)\ell$$

Multiply.

$$\approx 0.475 \ \ell$$

So, one pint is approximately 0.5 liters to the nearest tenth.

Warm Up b. Convert 450 inches to meters.

c. The speed limit used to be 55 miles per hour. Convert the speed limit to kilometers per hour:

$$\frac{55 \text{ miles}}{1 \text{ hour}} \approx \frac{55 \cancel{\text{ mi}}}{1 \text{ hour}} \cdot \frac{1.61 \text{ km}}{1 \cancel{\text{ mi}}}$$

Multiply by the conversion factor $\frac{1.61 \text{ km}}{1 \text{ mi}}$.

$$\approx \frac{88.55 \text{ km}}{1 \text{ hr}}$$

So, 55 miles per hour is approximately 88.6 kilometers per hour.

Warm Up c. Convert 50 miles per hour to meters per second.

ANSWERS TO WARM UPS (7.4) **a.** 337.5 grams **b.** 11.4 meters **c.** 22.4 meters per second

d. Convert 2 kilograms to pounds:

$$\frac{2\text{ kg}}{1} \approx \frac{2\text{ }\cancel{\text{kg}}}{1} \cdot \frac{2.2\text{ lb}}{1\text{ }\cancel{\text{kg}}}$$

Multiply by the conversion factor $\frac{2.2\text{ lb}}{1\text{ kg}}$.

$$\approx 4.4\text{ lb}$$

So, 2 kilograms is approximately 4.4 pounds.

Warm Up d. Convert 2 pounds to kilograms.

e. Convert 6.3 liters to quarts:

$$\frac{6.3\ \ell}{1} \approx \frac{6.3\ \cancel{\ell}}{1} \cdot \frac{1.06\text{ qt}}{1\ \cancel{\ell}}$$

Multiply by the conversion factor $\frac{1.06\text{ qt}}{1\ \ell}$.

$$\approx 6.678\text{ qt}$$

So, 6.3 liters is approximately 6.7 quarts.

Warm Up e. Convert 8.95 kiloliters to gallons (to the nearest hundred gallons).

f. Convert 75 kilometers per hour to miles per hour:

$$\frac{75\text{ km}}{1\text{ hr}} \approx \frac{75\text{ }\cancel{\text{km}}}{1\text{ hr}} \cdot \frac{0.62\text{ mi}}{\cancel{\text{km}}}$$

Multiply by the conversion factor $\frac{0.62\text{ mi}}{1\text{ km}}$.

$$\approx \frac{46.5\text{ mi}}{1\text{ hr}}$$

So, 75 kilometers per hour is approximately 46.5 miles per hour.

Warm Up f. Convert 32 meters per second to feet per second. Round off to the nearest foot.

ANSWERS TO WARM UPS (7.4) **d.** 0.9 kg **e.** 2400 gallons **f.** 105 feet per second

APPLICATION SOLUTION

g. Mary is driving on a road in Canada on which the speed limit is posted as 80 km/hr. The speedometer in her automobile is registering 45 miles/hr. Is she driving within the speed limit?

$$\frac{45 \text{ mi}}{1 \text{ hr}} \approx \frac{45 \text{ }\cancel{\text{mi}}}{1 \text{ hr}} \cdot \frac{1.61 \text{ km}}{1 \text{ }\cancel{\text{mi}}}$$

$$\approx \frac{72.45 \text{ km}}{1 \text{ hr}}$$

We must convert 45 miles per hour to kilometers per hour.
We could also have converted 80 km/hr to miles/hr. 80 km/hr ≈ 49.6 miles/hr.

Mary is traveling approximately 72 km/hr, so she is within the speed limit.

Warm Up g. Mary is traveling on another road in Canada where the speed limit is posted as 115 km/hr. Can she drive 70 miles per hour and be within the speed limit?

ANSWERS TO WARM UPS (7.4)

g. yes

NAME ______________________________

CLASS ______________ SECTION ______________ DATE ______________

Exercises 7.4

A

ANSWERS

1. ______________
2. ______________
3. ______________
4. ______________
5. ______________
6. ______________
7. ______________
8. ______________
9. ______________
10. ______________

Use the conversion tables in 7.1, 7.2, and this section to convert as shown. Round to the nearest tenth where necessary. Your answers may vary slightly from those given, depending on which conversions you use.

1. 75 ft = ? m **2.** 12 lb = ? kg **3.** 5 qt = ? ℓ

4. 1100 yd = ? m **5.** 0.1 ton = ? kg **6.** 4.5 ℓ = ? qt

7. 112 m = ? ft **8.** 31.4 cm = ? in **9.** 5.4 miles = ? km

10. 7 m = ? yd

B

11. ______________
12. ______________
13. ______________
14. ______________
15. ______________
16. ______________

11. 4.26 in = ? cm **12.** 15 kg = ? lb **13.** 345 g = ? lb

14. 3.2 ℓ = ? pints **15.** 15 gallons = ? ℓ **16.** 3560 kg = ? tons

ANSWERS

17. ______________

18. ______________

19. ______________

20. ______________

17. 185 mm = ? in

18. 7.2 km = ? yd

19. 6.3 miles = ? m

20. 0.5 cup = ? mℓ

C

21. ______________

22. ______________

23. ______________

24. ______________

25. ______________

26. ______________

27. ______________

28. ______________

29. ______________

30. ______________

21. 4800 mg = ? oz

22. 27 oz = ? cg (to the nearest thousand)

23. $\dfrac{30 \text{ lb}}{1 \text{ ft}} = \dfrac{? \text{ g}}{1 \text{ m}}$

24. $\dfrac{40 \text{ miles}}{1 \text{ hr}} = \dfrac{? \text{ meters}}{1 \text{ second}}$

25. $\dfrac{80 \text{ km}}{1 \text{ hr}} = \dfrac{? \text{ mi}}{1 \text{ hr}}$

26. $\dfrac{9.6 \text{ g}}{1 \text{ cm}} = \dfrac{? \text{ oz}}{1 \text{ inch}}$

27. $\dfrac{3.5 \text{ lb}}{\text{ft}^2} = \dfrac{? \text{ g}}{\text{cm}^2}$

28. $\dfrac{45 \text{ ft}}{\text{sec}} = \dfrac{? \text{ m}}{\text{sec}}$

29. $\dfrac{525 \text{ g}}{\ell} = \dfrac{? \text{ lb}}{\text{qt}}$

30. $\dfrac{\$1.52}{\text{lb}} = \dfrac{? \ \$}{\text{kg}}$

NAME ______________________

CLASS ______________ SECTION ______________ DATE ______________

D

ANSWERS

31. ______________

32. ______________

33. ______________

34. ______________

35. ______________

36. ______________

37. ______________

31. A box of Whammy-O's cereal weighs 16.5 ounces. What is the weight in grams, to the nearest tenth?

32. A farmer harvested 10 tons of strawberries per acre. How many kilograms per acre did he harvest? Round off to the nearest kilogram.

33. The Cascade Run-Off is a 10-kilometer race. What is the distance expressed in miles? Round off to the nearest tenth of a mile.

34. The tallest player on the all-star basketball team is 7 ft 4 in. Express his height in meters, to the nearest tenth of a meter.

35. A garden pump delivers 15 liters of water per minute. Express this in gallons per hour, to the nearest gallon.

36. The average weight of a man in the town of Golden Rod is 185 pounds. Express this weight in kilograms, to the nearest kilogram.

37. The standard size for a sheet of paper is 8.5 in. by 11 in. Give the dimensions in centimeters, to the nearest tenth of a centimeter.

ANSWERS

38. ______

39. ______

40. ______

41. ______

42. ______

38. An elephant weighs 3.5 tons. What is her weight in kilograms?

39. Soda comes in a 2-liter bottle for 99 cents or half gallon for $1.09. Which is the better buy?

40. Three kilograms of coffee costs $18.95, whereas 6 pounds costs $17.95. Which is the better buy?

41. Laura has two ways to drive to her cousin's house. One route is 85 miles and the other is 130 km. Which way would be shorter?

42. Nancy can purchase 3.5 yards of fabric for $10 or 3 meters for $10. Which gives the most fabric?

E *Maintain your skills* (Sections 6.1, 6.6, 6.9)

43. ______

44. ______

45. ______

46. ______

Write as a percent:

43. 80 to 64 **44.** 78 to 52

Write as a percent to the nearest tenth of a percent:

45. 38 to 64 **46.** 75 to 48

Fill in the blank spaces with the related percent, fraction, or decimal.

	FRACTION	DECIMAL	PERCENT
47.	$\frac{5}{8}$		
48.		$0.33\overline{3}$	
49.			0.8%

50. ______

50. The regular price of a radio is $45.95. What is the sales price of the radio if it is discounted 20%?

Perimeter and Circumference

OBJECTIVE

96. Find the perimeter of geometric figures.

APPLICATION

How long must a hose clamp be to reach around a radiator hose that has an outside diameter of $2\frac{1}{2}$ in.? (Let $\pi \approx 3.14$.)

VOCABULARY

Circles, rectangles, triangles, and *squares* are four examples of *geometric figures.* An example of each is shown in Figure 7.1. Different parts of the figures are labeled:

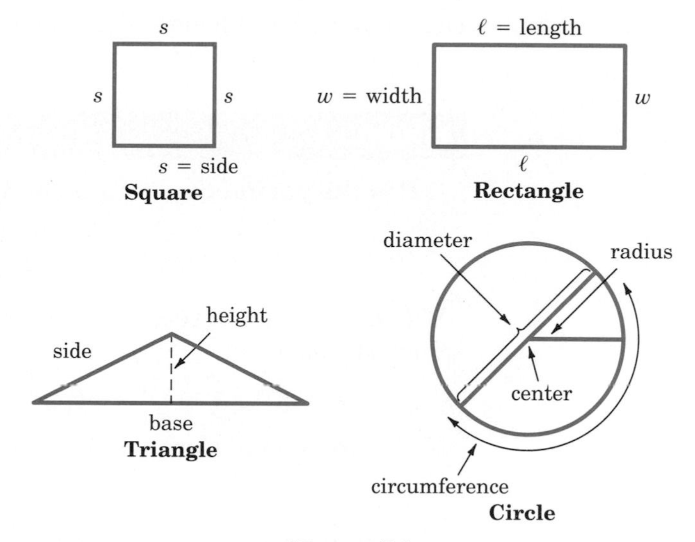

Figure 7.1

The distance around any geometric figure is called the *perimeter* of that figure. The perimeter of a circle is called the *circumference.* The *radius* of the circle is the distance from the center to any point on the circle. The *diameter* is twice the radius.

The number π (read "pi") is the ratio (quotient) of the circumference of a circle to its diameter. This number is the same for *every* circle no matter how large or how small. This remarkable fact was discovered over a long period of time historically and during that time a large number of approximations were used. Here are some of those approximations:

$$3;\ 3\frac{1}{8};\ 3\frac{1}{7}\left(\text{or } \frac{22}{7}\right);\ \frac{355}{113};\ \text{and } 3.1415926$$

In the 1700's mathematicians discovered that the value of π cannot be written as a fraction (ratio of whole numbers) or as an exact or repeating decimal. In this century computers have been used to find the approximate decimal value of π to thousands of decimal places. For our work we will use the approximation to the nearest hundredth: $\pi \approx 3.14$.

HOW AND WHY

The perimeter or circumference can be thought of in terms of building a fence around a figure. The perimeter or circumference is the length of that fence. The perimeter of any figure in which all the sides are straight is the sum of the lengths of the sides.

For example, find the perimeter of Figure 7.2. P (the perimeter) is the sum of the lengths of the four sides; that is:

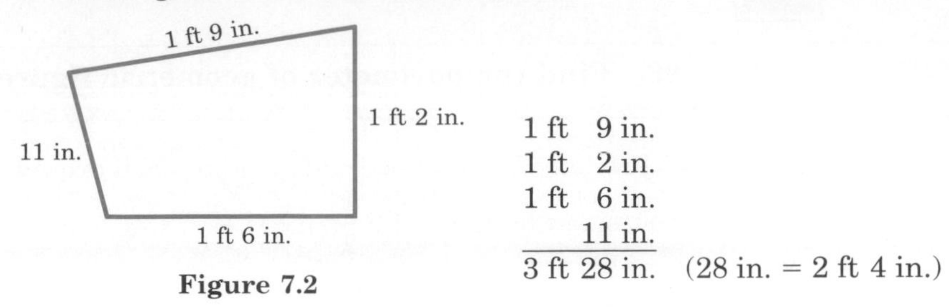

Figure 7.2

$$\begin{array}{r} 1 \text{ ft} \quad 9 \text{ in.} \\ 1 \text{ ft} \quad 2 \text{ in.} \\ 1 \text{ ft} \quad 6 \text{ in.} \\ 11 \text{ in.} \\ \hline 3 \text{ ft } 28 \text{ in.} \end{array} \quad (28 \text{ in.} = 2 \text{ ft } 4 \text{ in.})$$

So, $P = 5$ ft 4 in.

There are special formulas for the perimeters of squares, rectangles, and circles and for the length of a semicircle.

FORMULAS

Square

If P is the perimeter and s is the length of one side, then the formula is:

$$P = 4 \cdot s$$

Rectangle

If P is the perimeter, ℓ is the length, and w is the width, then the formula is:

$$P = 2 \cdot \ell + 2 \cdot w$$

Circle

If C is the circumference and d is the diameter, then the formula is:

$$C = \pi \cdot d = 2 \cdot \pi \cdot r$$

(π is read "pi," and its value is approximately 3.14 $\quad \pi \approx 3.14$)

Semicircle

If L is the length and r is the radius, then the formula is:

$$L = \pi \cdot r = \frac{1}{2} \cdot \pi \cdot d$$

PROCEDURE

To find the perimeter of a geometric figure that is not one of the above, add the lengths of the sides.

MODEL PROBLEM-SOLVING

Examples | **Strategy**

a. Find the perimeter of the following rectangle:

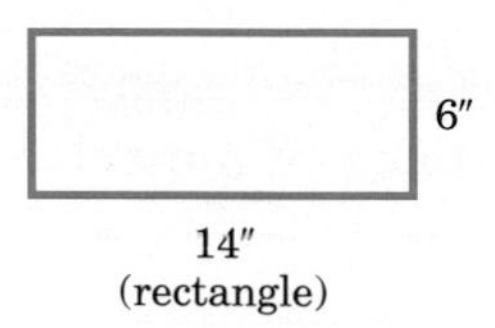

(rectangle)

$P = 2 \cdot \ell + 2 \cdot w$	**Formula for the perimeter of a rectangle.**
$= 2(14'') + 2(6'')$	**Substitute. (1″ = 1 in.)**
$= 28'' + 12''$	**Multiply.**
$= 40''$	**Add.**

Warm Up a. Find the perimeter of the following rectangle:

4 in.

3 in.

b. Find the perimeter of the following square:

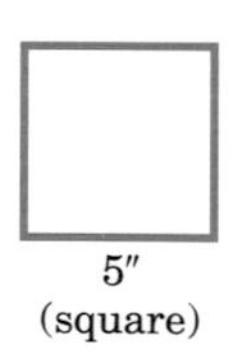

(square)

$P = 4 \cdot s$	**Formula for the perimeter of a square.**
$= 4(5'')$	**Substitute.**
$= 20''$	**Multiply.**

Warm Up b. Find the perimeter of the following hexagon:

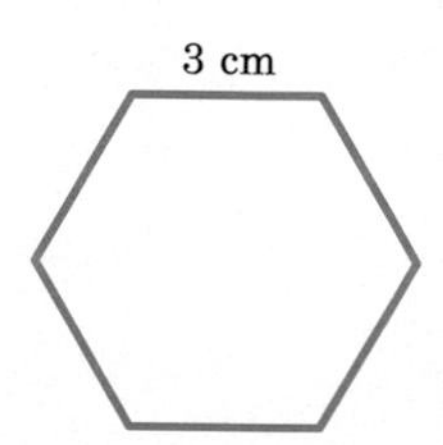

ANSWERS TO WARM UPS (7.5)

a. 14 in. **b.** 18 cm

c. Find the perimeter of the following circle:

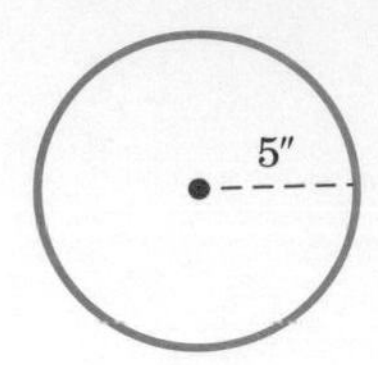

$$\begin{aligned} C &= 2 \cdot \pi \cdot r \\ &\approx 2(3.14)(5'') \\ &\approx 31.4'' \end{aligned}$$

Formula for the circumference of a circle.
Substitute; let $\pi \approx 3.14$.
Multiply.

Warm Up c. Find the perimeter of the following circle:

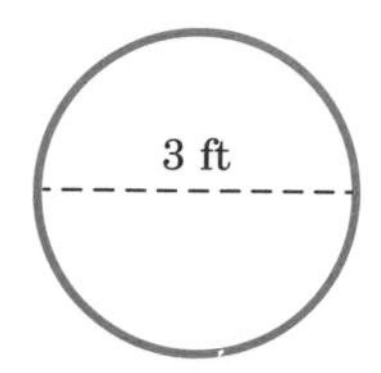

d. Find the perimeter of the following figure:

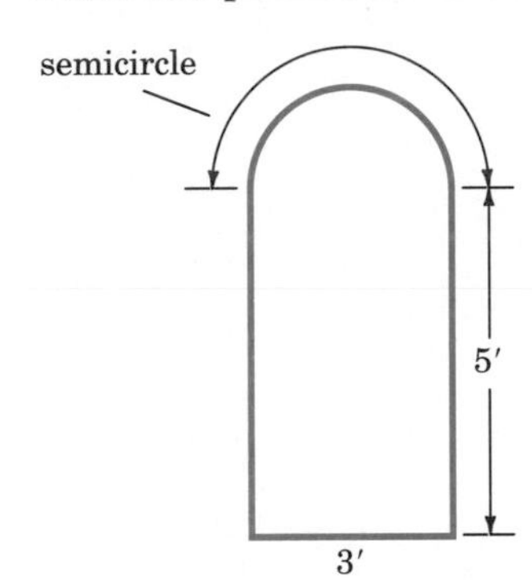

$$\begin{aligned} P &\approx 5' + 3' + 5' + \frac{1}{2}(3.14)(3') \\ &\approx 13' + \frac{1}{2}(9.42') \\ &\approx 13' + 4.71' \\ &\approx 17.71' \end{aligned}$$

This figure is three sides of a rectangle and a semicircle. Add the three sides and the perimeter of the semicircle; let $\pi \approx 3.14$. Note that $1' = 1$ foot.

Warm Up d. Find the perimeter of the following figure:

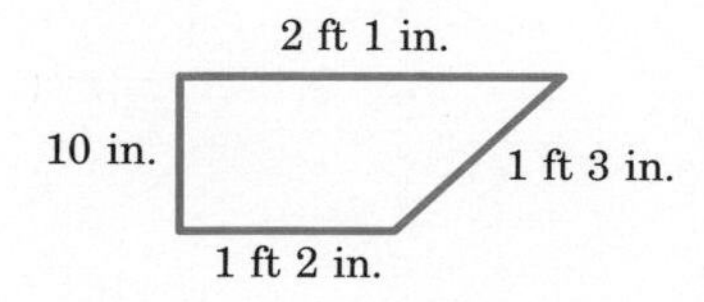

ANSWERS TO WARM UPS (7.5)

c. 9.42 ft

d. 5 ft 4 in.

e. Find the perimeter of the following figure:

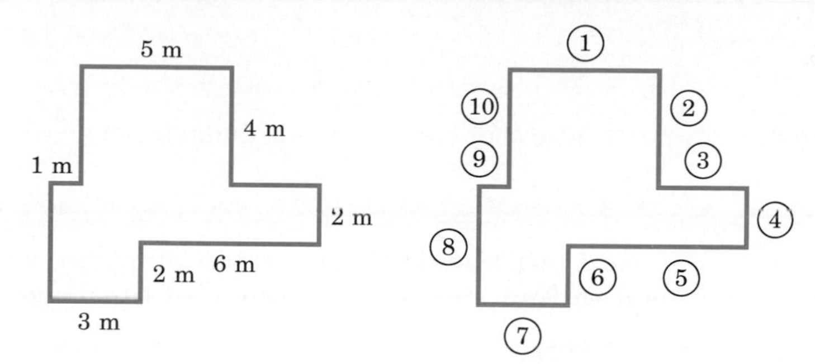

SIDE	LENGTH
1	5 m
2	4 m
3	?
4	2 m
5	6 m
6	2 m
7	3 m
8	?
9	1 m
10	?
3	3 m
8 + 10	8 m

To find the perimeter, number the sides and write down their lengths.

To find the length of side 3, note that:

Side 3 = side 5 + side 7 − side 9 − side 1
= (6 + 3 − 1 − 5) m
= 3 m

Note that the lengths of sides 8 and 10 cannot be found, but their sum can.

Side 8 + side 10 = side 2 + side 4 + side 6
= (4 + 2 + 2) m
= 8 m

So:
$P = 5 \text{ m} + 4 \text{ m} + 3 \text{ m} + 2 \text{ m} + 6 \text{ m} + 2 \text{ m} + 3 \text{ m} + 1 \text{ m} + 8 \text{ m}$
$= 34 \text{ m}$

Now add the lengths.

The perimeter is 34 m.

Warm Up e. Find the perimeter of the following figure:

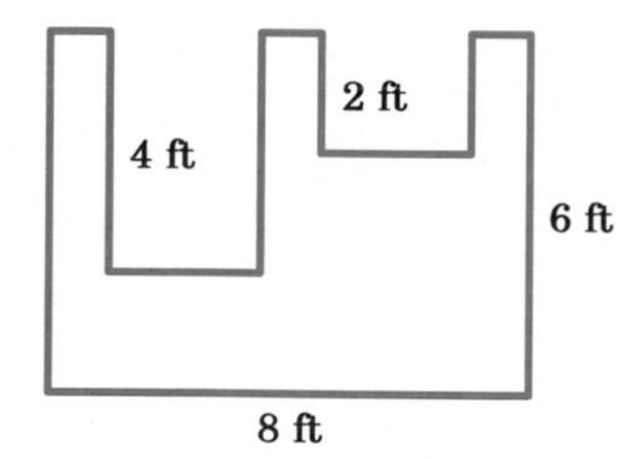

ANSWERS TO WARM UPS (7.5)

e. 40 ft

APPLICATION SOLUTION

f. How long must a hose clamp be to reach around a radiator hose that has an outside diameter of $2\frac{1}{2}$ in.? (Let $\pi \approx 3.14$).

$C = \pi \cdot d$ — **To find out how long the clamp must be, we need to find the circumference of the hose.**

$C \approx (3.14)(2.5 \text{ in.})$ — **Substitute.**
$C \approx 7.85 \text{ in.}$

The clamp needs to be approximately 7.85 in. long to go around the hose.

Warm Up f. What is the circumference of the circle formed by an irrigation sprinkler that has a reach of 45 feet? (Let $\pi \approx 3.14$.)

ANSWERS TO WARM UPS (7.5)

f. 282.6 ft

NAME ______________________

CLASS ____________ SECTION ____________ DATE ____________

Exercises 7.5

A

ANSWERS

1. ____________
2. ____________
3. ____________
4. ____________
5. ____________
6. ____________
7. ____________
8. ____________

1. Find the perimeter of a triangle that has sides of 7 in., 8 in., and 9 in.

2. Find the circumference of a circle that has a diameter of 8 cm. Let $\pi \approx 3.14$.

3. Find the perimeter of a square that is 18 in. on each side.

4. How much fencing will be needed to fence in a rectangular barnyard that is 18 feet by 11 feet?

5. How much fencing is needed for a circular irrigation field if its radius is 15 meters?

Find the perimeter of the following figures. Let $\pi \approx 3.14$.

6.

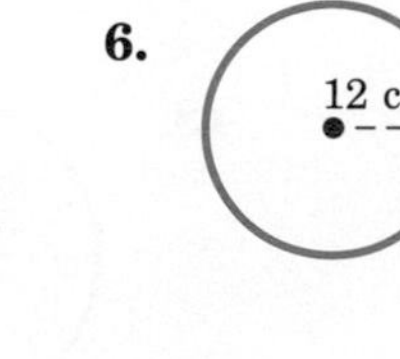

7.

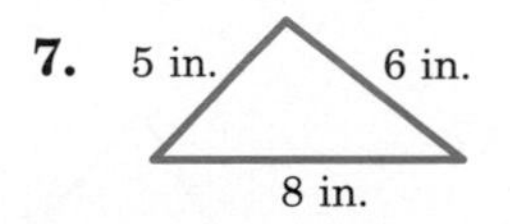

8.

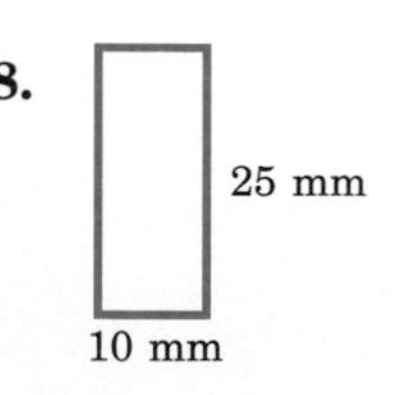

ANSWERS

9. ____________________

10. ____________________

11. ____________________

12. ____________________

13. ____________________

14. ____________________

15. ____________________

16. ____________________

17. ____________________

18. ____________________

9. 3 ft; 10 ft

10. 20 in.; 8 in.; 8 in.; 10 in.

11.

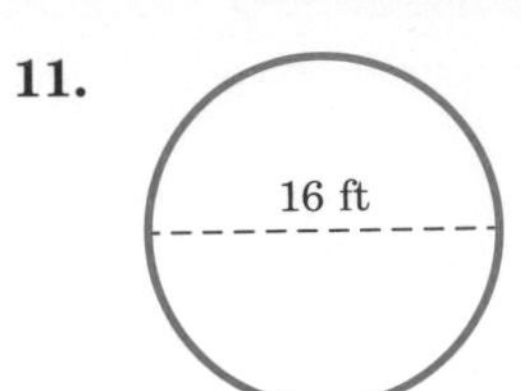

12. 9.5 cm; 15 cm

13. 8″; 8″

14. 6′; 6′; 6′

15. 12″; 4″

16. 4.3″

17.

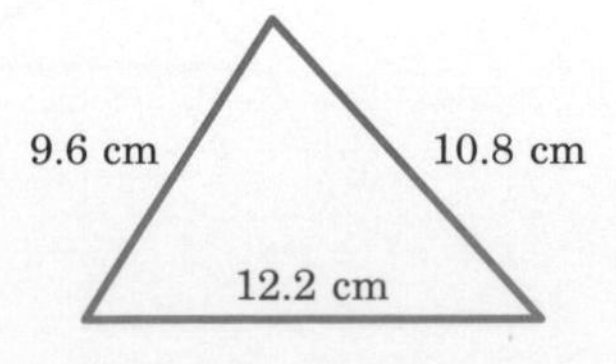

18.

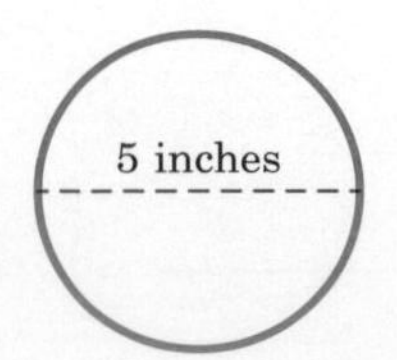

NAME ______________________

CLASS ______________ SECTION ______________ DATE ______________

ANSWERS

19. ______________

19.

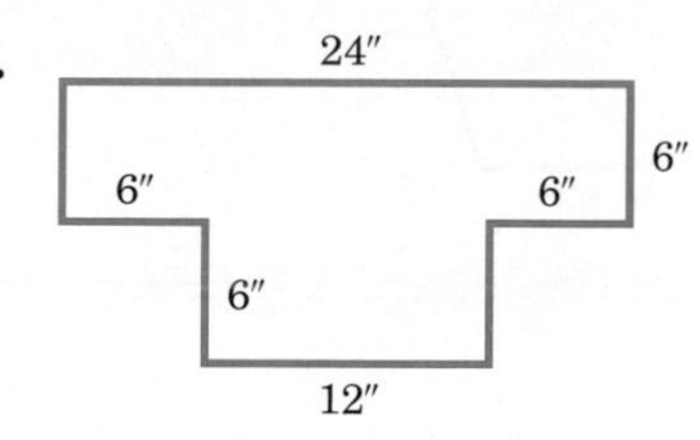

B

20. ______________

21. ______________

22. ______________

23. ______________

24. ______________

25. ______________

Find the perimeter of the following figures. Let $\pi \approx 3.14$.

20.

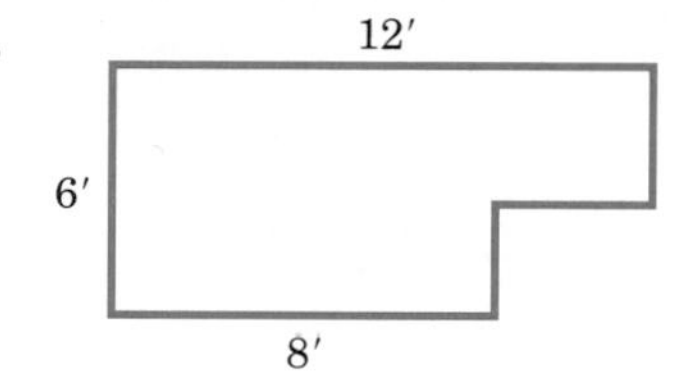

21.

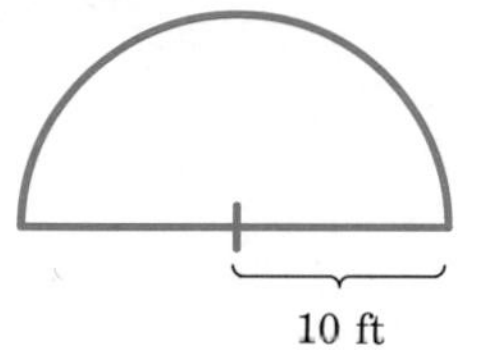

22.

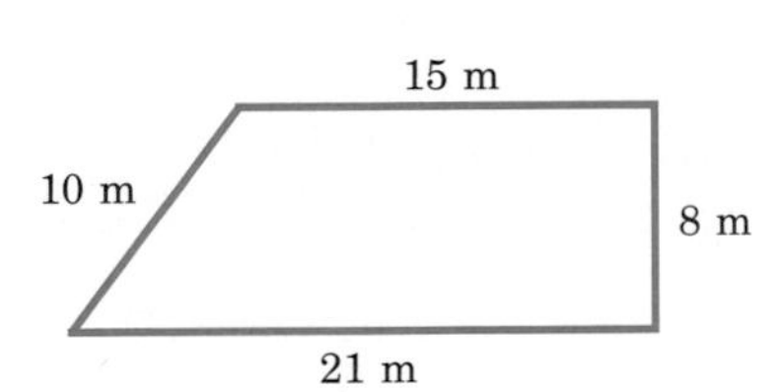

23.

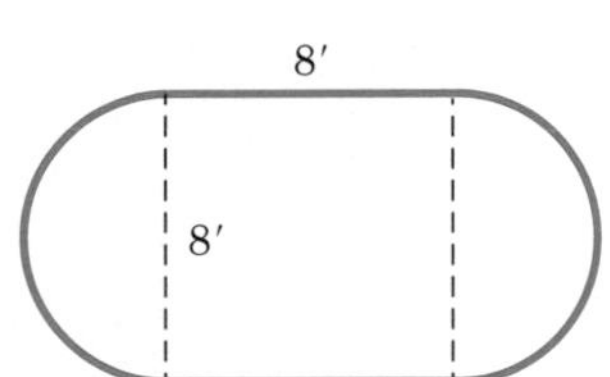

24.

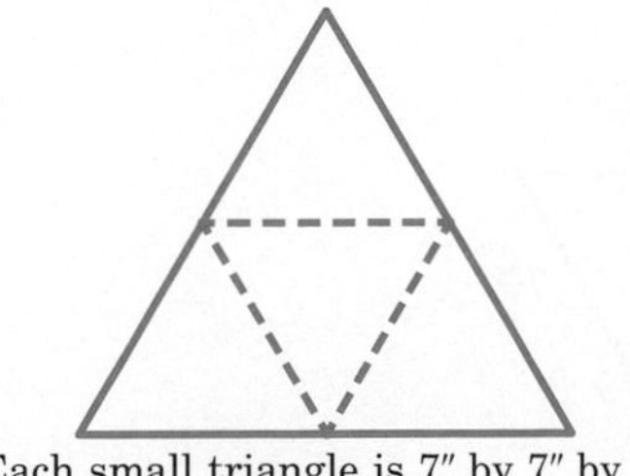

Each small triangle is 7″ by 7″ by 7″

25.

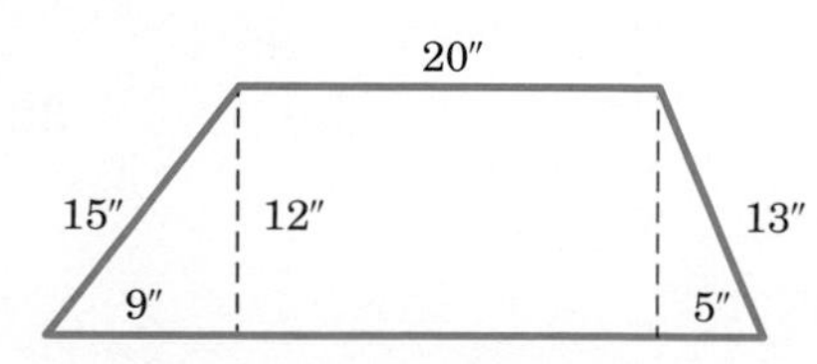

26. ______________________

27. ______________________

28. ______________________

29. ______________________

30. ______________________

31. ______________________

32. ______________________

33. ______________________

26.

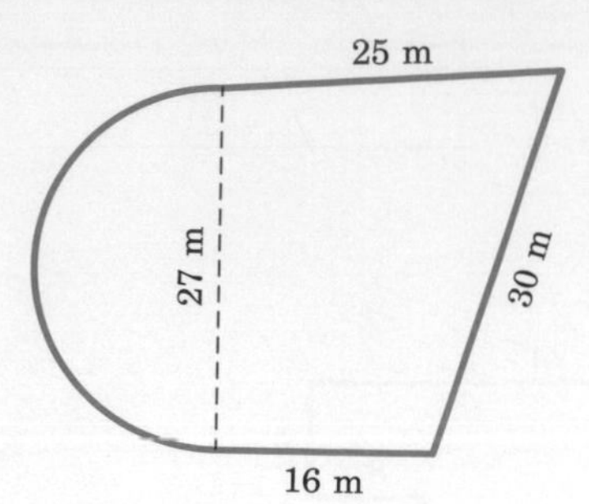

27.

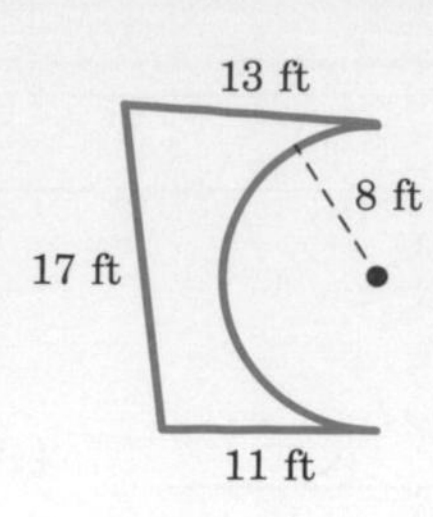

28.

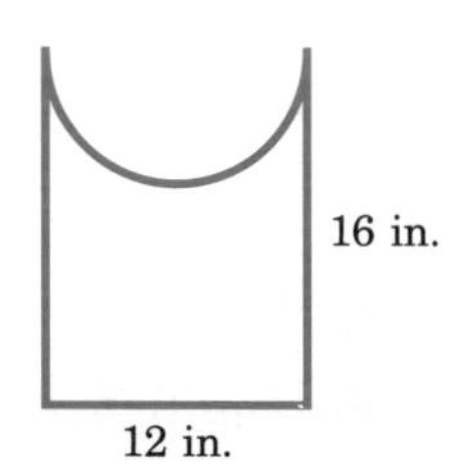

29.

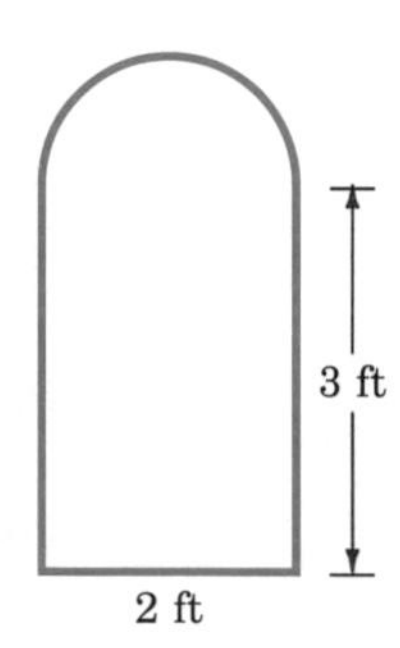

30.

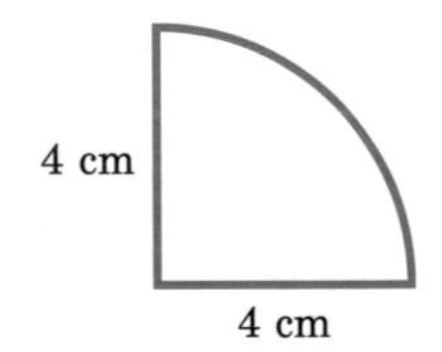

31.

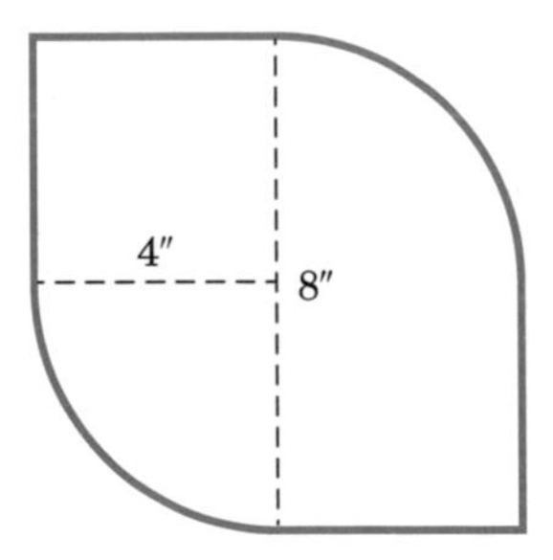

32.

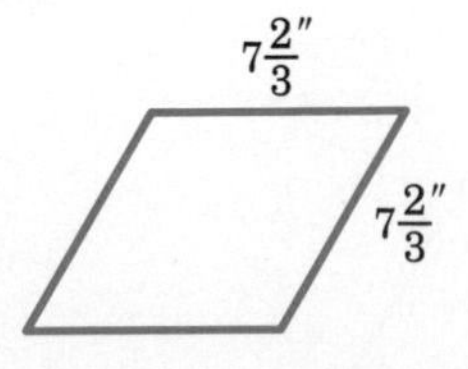

33.

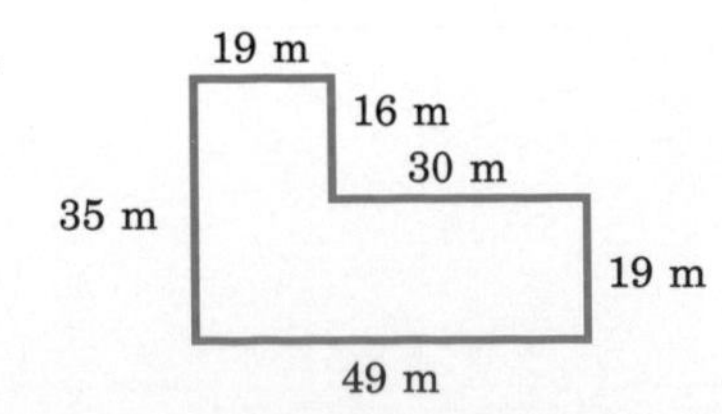

NAME ______________________________

CLASS ______________ SECTION ______________ DATE ______________

ANSWERS

34. ______________

34.

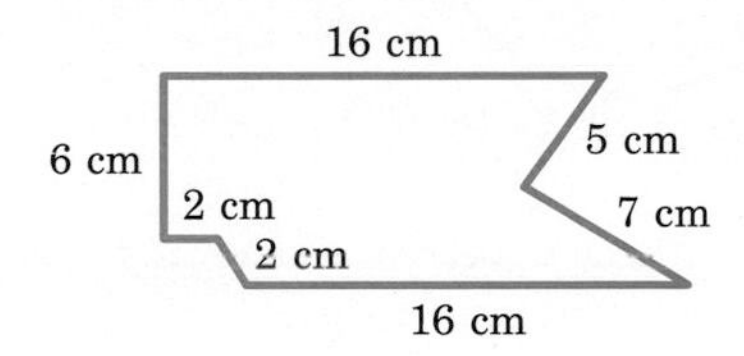

35.

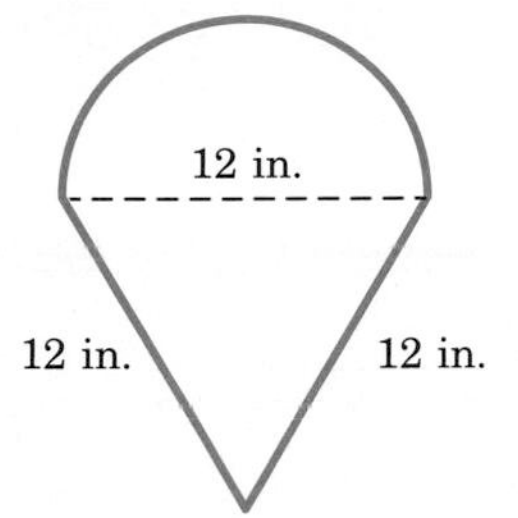

35. ______________

36. ______________

36.

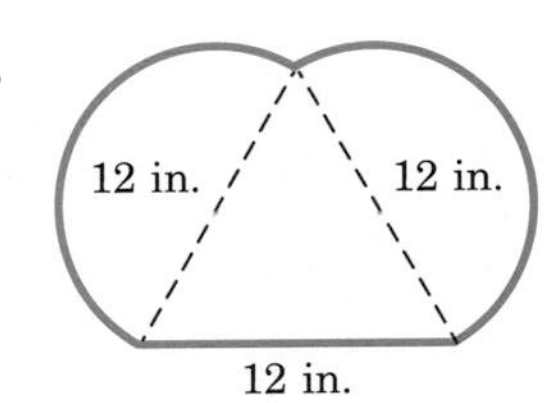

C

37. ______________

38. ______________

Find the perimeter of the following figures. Let $\pi \approx 3.14$.

37.

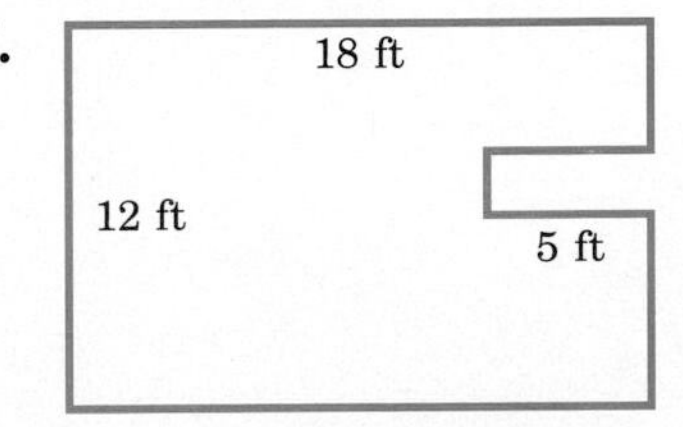

38.

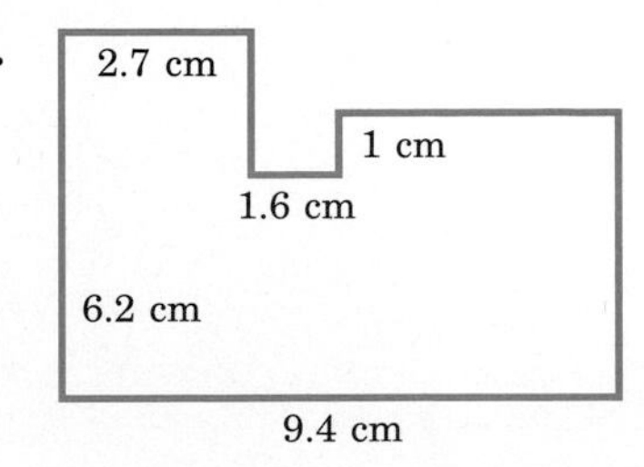

ANSWERS

39. ______

40. ______

41. ______

42. ______

43. ______

44. ______

39.

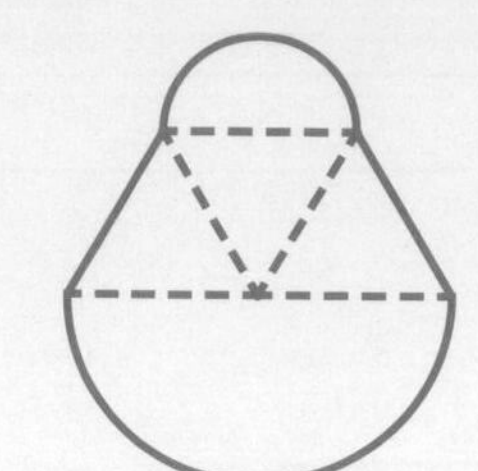

Each triangle is 10″ by 10″ by 10″

40.

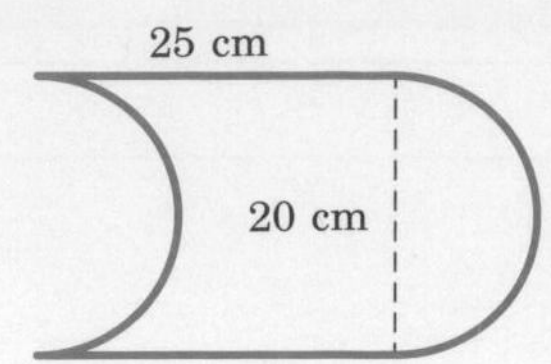

41.

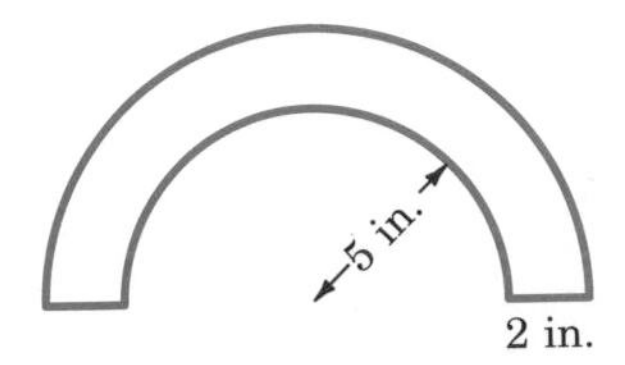

42.

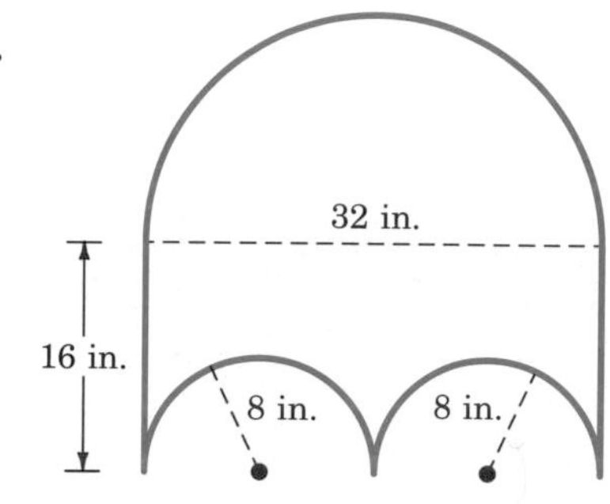

43.

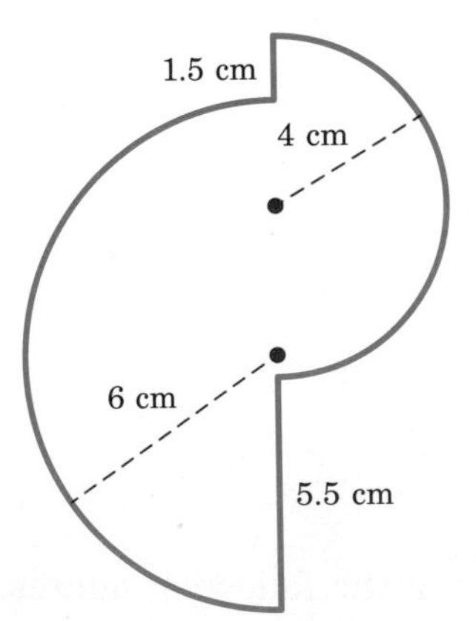

44.

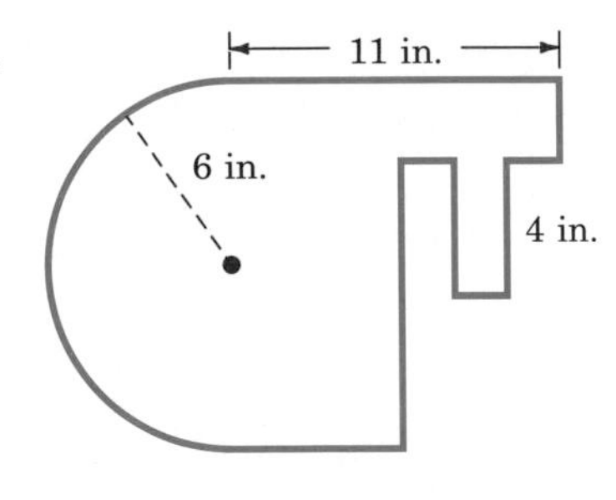

NAME ______________________

CLASS ______________ SECTION ______________ DATE ______________

D

ANSWERS

45. ______________

45. How many feet of picture framing does Bonnie need to frame four pictures, each measuring 8 inches by 10 inches?

46. ______________

46. How far will the tip of the minute hand of a clock travel in 15 minutes if the hand is 3 cm in length?

47. ______________

47. If fencing costs $10 per meter, what will be the cost of fencing a rectangular lot that is 1200 dm long and 240 dm wide?

48. ______________

48. If Atha needs 2 minutes to put one foot of binding on a rug, how long will it take her to put the binding on a rug that is 15 ft by 12 ft?

49. ______________

49. A wheel on Mary's automobile is 16 inches in diameter. To the nearest whole number, how many revolutions will the wheel turn if the automobile travels one mile? ($\pi \approx 3.14$)

50. ______________

50. How much ribbon is needed for the outside edge of a round tablecloth with a 52-inch diameter?

E *Maintain your skills* (Sections 6.6, 6.7, 6.8, 6.9)

ANSWERS

52. ______________

51. Fill in the blank spaces with the related percent, fraction, or decimal:

FRACTION	DECIMAL	PERCENT
		124.5%
$\frac{7}{16}$		
	$0.833\overline{3}$	
		520%

52. In the last election one candidate received 842 votes out of 1236. What percent is this, to the nearest tenth of a percent?

53. Bobbi started out shopping for clothes with $100. She bought a new coat that cost $55. What percent of her money did she have left after she paid for the coat?

53. ______________

Area of Common Geometric Figures

OBJECTIVE

97. Find the area of common geometric figures.

APPLICATION

The normal coverage rate for a gallon of deck paint is known to be 400 ft^2. George has a rectangular-shaped deck that he wants to paint. The length is 26 ft and the width is 15 ft. Will one gallon be enough to paint this deck?

VOCABULARY

Area is a measure of a surface, and surface is measured in square units. Two examples of surface measure are shown in Figure 7.3.

The unit of measure on the right is called a *"square inch,"* since the square is one inch on each side. The "square inch" measures the surface that is contained within the square. "Square inch" is written "in.^{2}."

Note that in.^{2} denotes a unit of measure, not an operation with exponents:

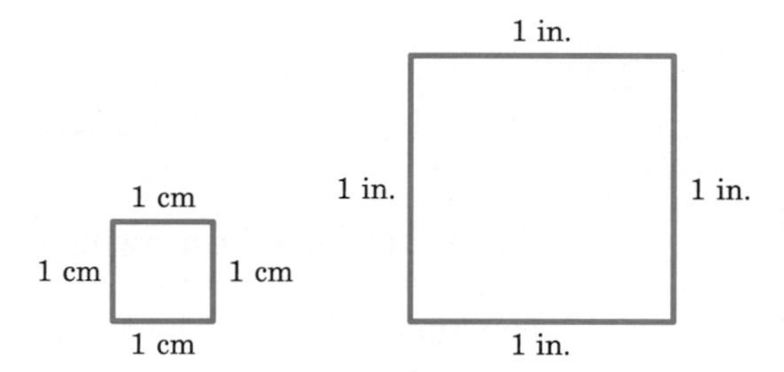

Figure 7.3

The unit of measure on the left is called a *"square centimeter,"* since the square is one centimeter on each side. The "square centimeter" measures the surface that is contained within the square. "Square centimeter" is written "cm^2."

There are other units of surface measure such as *"square foot," "square mile," "square meter,"* and *"square kilometer."*

In Figure 7.4 two different geometric figures are shown: the *parallelogram* and the *trapezoid.* Note that the trapezoid has two bases and one altitude, whereas the parallelogram has one base and one altitude. The altitude of a parallelogram or a trapezoid is always the distance between the parallel sides.

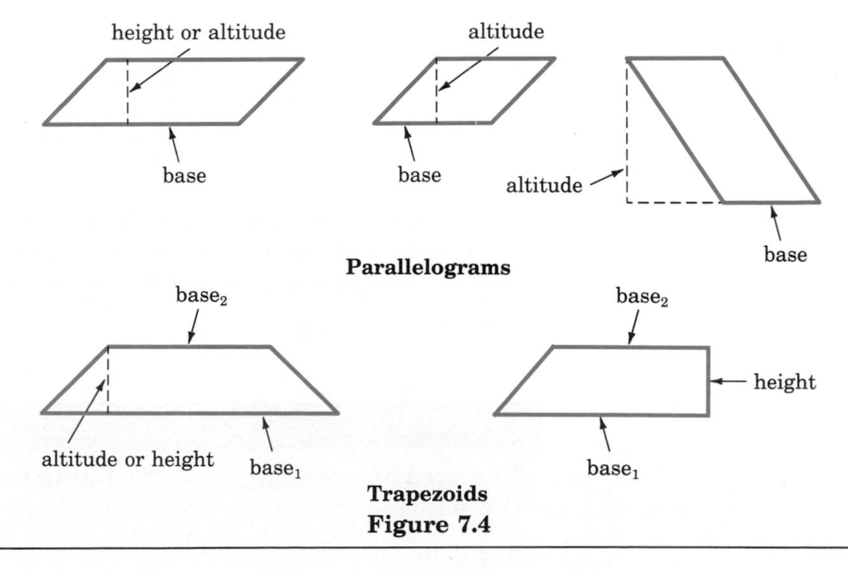

Figure 7.4

HOW AND WHY

Recall that perimeter is the distance around a figure, like building a fence around the figure. Area is the surface of the figure, or like laying carpet tile on a floor. The area is the number of squares (tiles) it takes to cover the floor. So, to find the area of a geometric figure means to determine how many squares are contained within that figure. Each square is the same size and is called a square unit (see Figure 7.3). Consider a square that is 2 inches on each side:

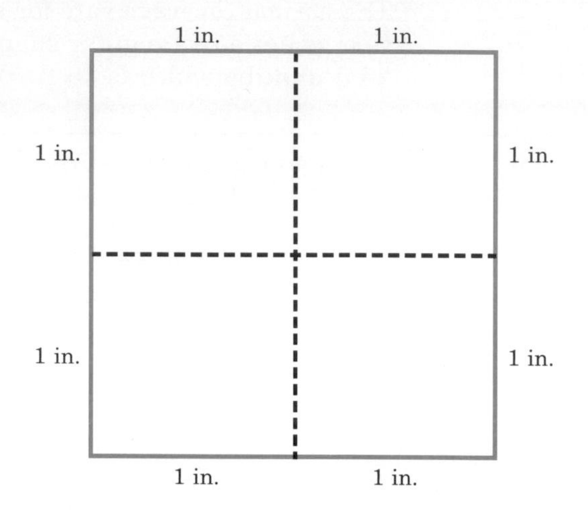

We can see from the drawing that this square can be divided into four squares, each of which is one inch on a side. This shows that the area of the square is four square inches. The area (4 in.^{2}) is the square of the length of a side. That is, $A = (2 \text{ in.})^2 = (2 \text{ in.})(2 \text{ in.}) = 4 \text{ in.}^2$.

PROCEDURE

The area of a square can be found by squaring the length of one of its equal sides.

$A = s^2$

Consider a rectangle that has a length of 3 centimeters and a width of 2 centimeters:

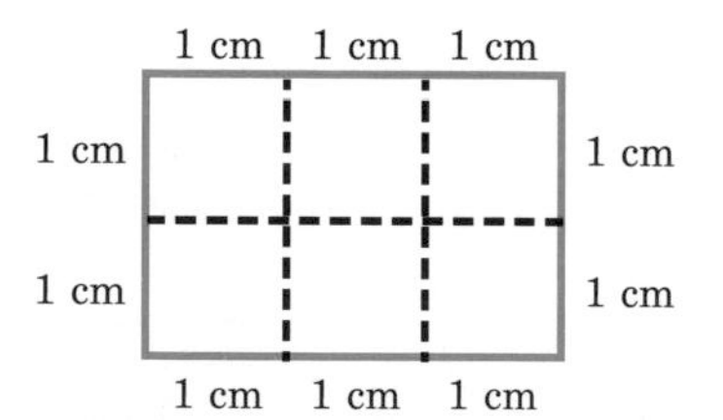

The figure shows that the rectangle can be divided into six squares, each of which is one centimeter on a side. This tells us that the area of the rectangle is 6 cm^2. If we multiply the length (3 cm) times the width (2 cm), we have: $(3 \text{ cm})(2 \text{ cm}) = 6 \text{ cm}^2$

PROCEDURE

The area of a rectangle can be found by multiplying the length times the width:

$A = \ell \cdot w$

The following formulas are used to find the area of a parallelogram, a triangle, a trapezoid, and a circle. See Appendix II.

FORMULAS	
The area of a parallelogram	$A = b \cdot h$
The area of a triangle	$A = \frac{1}{2} b \cdot h$
The area of a trapezoid	$A = \frac{1}{2}(b_1 + b_2) \cdot h$
The area of a circle	$A = \pi r^2$ or $A = \frac{\pi d^2}{4}$

where b indicates the base, h the height, r the radius, and d the diameter of the circle.

CAUTION

Be careful not to confuse the formulas for circumference and area of a circle. $C = 2\pi r$ and $A = \pi r^2$. The formulas can also be written $C = \pi(r + r)$ whereas $A = \pi(r \cdot r)$

MODEL PROBLEM-SOLVING

Examples | **Strategy**

a. Find the area of a square that is 4 in. on each side:

$A = s^2$ — Formula for the area of a square.

$= (4 \text{ in.})^2 = (4 \text{ in.})(4 \text{ in.})$ — Substitute.

$= 16 \text{ in.}^2$ — Multiply.

Warm Up a. Find the area of a square that is 11 cm on each side.

b. Find the area of the rectangle with $\ell = 12$ m and $w = 8$ m.

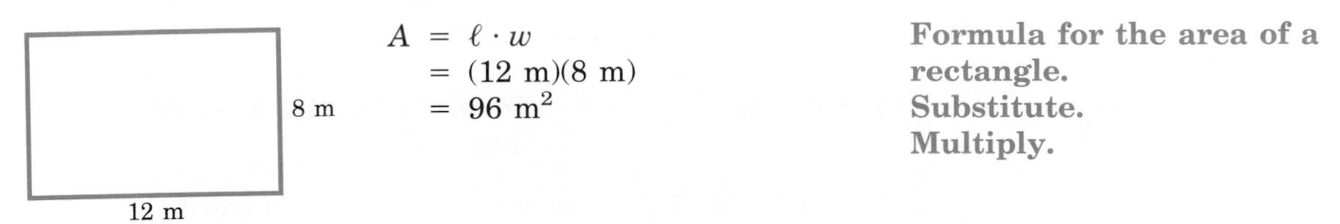

$A = \ell \cdot w$ — Formula for the area of a rectangle.

$= (12 \text{ m})(8 \text{ m})$ — Substitute.

$= 96 \text{ m}^2$ — Multiply.

Warm Up b. Find the area of a rectangle with length of 24 ft and width of 17 ft.

ANSWERS TO WARM UPS (7.6)

a. 121 cm^2 **b.** 408 ft^2

c. Find the area of a triangle with $b = 14.2$ cm and $h = 7$ cm.

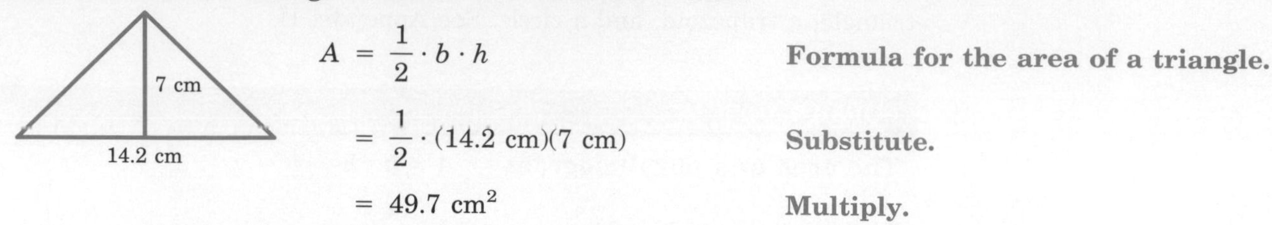

$A = \frac{1}{2} \cdot b \cdot h$ **Formula for the area of a triangle.**

$= \frac{1}{2} \cdot (14.2 \text{ cm})(7 \text{ cm})$ **Substitute.**

$= 49.7 \text{ cm}^2$ **Multiply.**

Warm Up c. Find the area of a triangle with $b = 35.4$ ft and $h = 9.6$ ft.

d. Find the area of a parallelogram with a base of 1 ft and a height of 4 in.

$A = b \cdot h$

$= (1 \text{ ft})(4 \text{ in.})$ or $A = (1 \text{ ft})(4 \text{ in.})$ **We can find the area only if the units are the same, so either change ft to in. or in. to ft.**

$= (12 \text{ in.})(4 \text{ in.})$ $= (1 \text{ ft})\left(\frac{1}{3} \text{ ft}\right)$

$= 48 \text{ in.}^2$ $= \frac{1}{3} \text{ ft}^2$ **The solution is shown both ways.**

Warm Up d. Find the area of a parallelogram with a base of 6 yd and a height of 5 ft.

e. Find the area of a trapezoid with $b_1 = 17$ ft, $b_2 = 13$ ft, and $h = 7$ ft.

13 ft

7 ft

17 ft

$A = \frac{1}{2}(b_1 + b_2) \cdot h$ **Formula for the area of a trapezoid.**

$= \frac{1}{2}(17 \text{ ft} + 13 \text{ ft}) \cdot 7 \text{ ft}$ **Substitute and simplify.**

$= \frac{1}{2}(30 \text{ ft})(7 \text{ ft})$

$= 105 \text{ ft}^2$

ANSWERS TO WARM UPS (7.6)

c. 169.92 ft^2 **d.** 90 ft^2 or 10 yd^2

Warm Up e. Find the area of a trapezoid with bases of 51 m and 36 m and a height of 12 m.

f. Find the area of a circle with $r = 8$ cm.

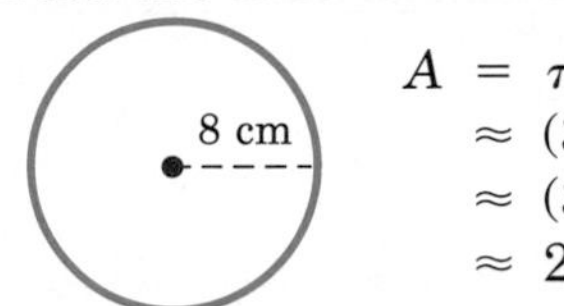

$$\begin{aligned} A &= \pi r^2 \\ &\approx (3.14)(8 \text{ cm})^2 \\ &\approx (3.14)(64 \text{ cm}^2) \\ &\approx 200.96 \text{ cm}^2 \end{aligned}$$

Formula for the area of a circle. Let $\pi \approx 3.14$. Substitute and simplify.

Warm Up f. Find the area of a circle with a diameter of 18 ft. Let $\pi \approx 3.14$.

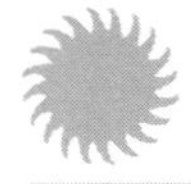

APPLICATION SOLUTION

g. The normal coverage rate for a gallon of deck paint is known to be 400 ft^2. George has a rectangular-shaped deck that he wants to paint. The length is 26 ft and the width is 15 ft. Will one gallon be enough to paint this deck?

$$\begin{aligned} A &= (26 \text{ ft})(15 \text{ ft}) \\ &= 390 \text{ ft}^2 \end{aligned}$$

First find the area of the deck: $A = \ell \cdot w$

Since the area of the deck is less than the coverage of a gallon ($390 < 400$), George has enough paint.

One gallon should be enough to paint the deck.

Warm Up g. Marta found 45 sq yd of plush carpet on sale for \$15 per sq yd. Marta's living room is circular, with a radius of 11 ft. Is there enough carpet to cover her living room?

ANSWERS TO WARM UPS (7.6) **e.** 522 m^2 **f.** 254.34 ft^2 **g.** Yes

NAME ______________________________

CLASS ______________ SECTION ______________ DATE ______________

Exercises 7.6

A

ANSWERS

1. ______________
2. ______________
3. ______________
4. ______________
5. ______________
6. ______________
7. ______________
8. ______________
9. ______________
10. ______________

Find the area of the following. Let $\pi \approx 3.14$.

1.

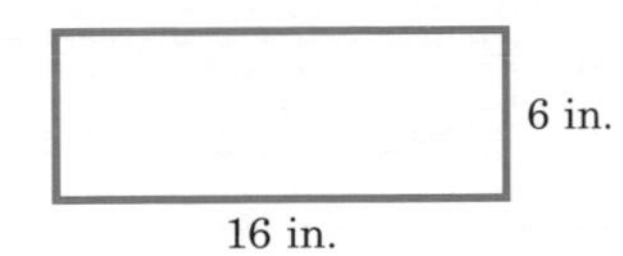

2.

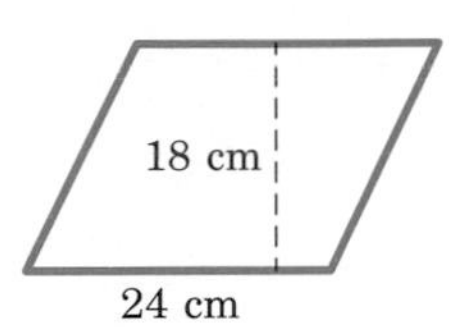

3.

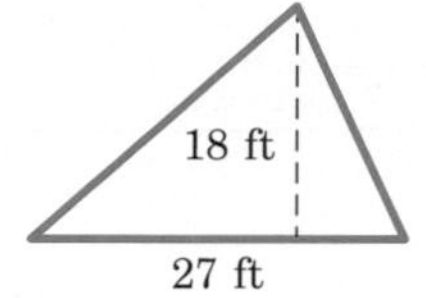

4.

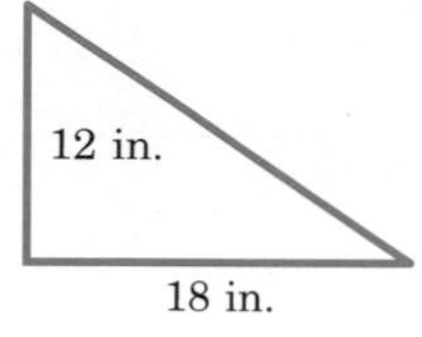

5.

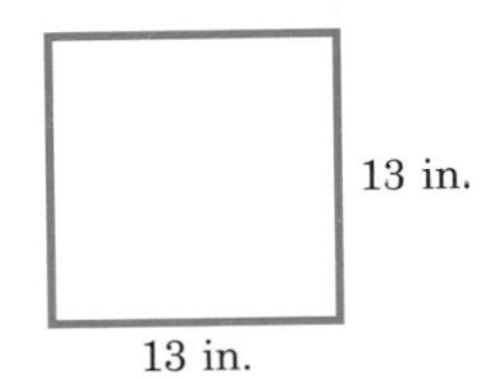

6.

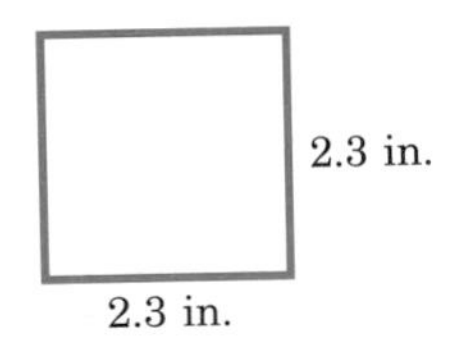

7.

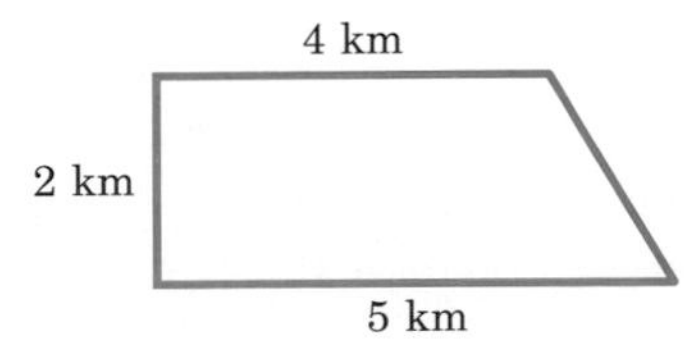

8.

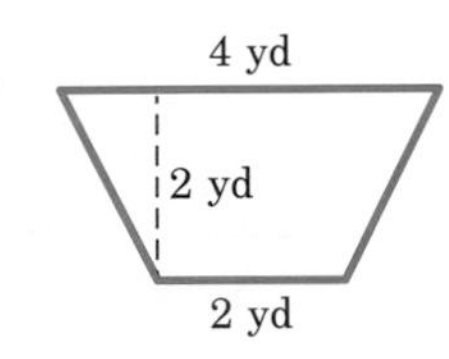

9.

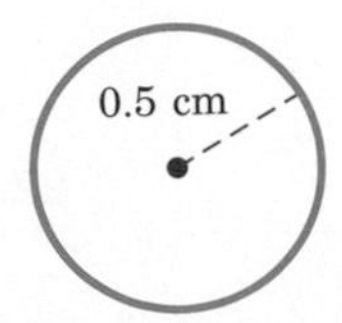

10.

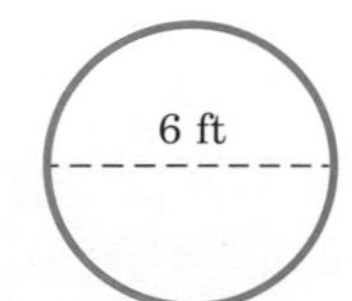

ANSWERS

11. ______

12. ______

13. ______

14. ______

11.

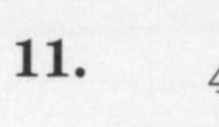

$4\frac{1}{2}''$

12.

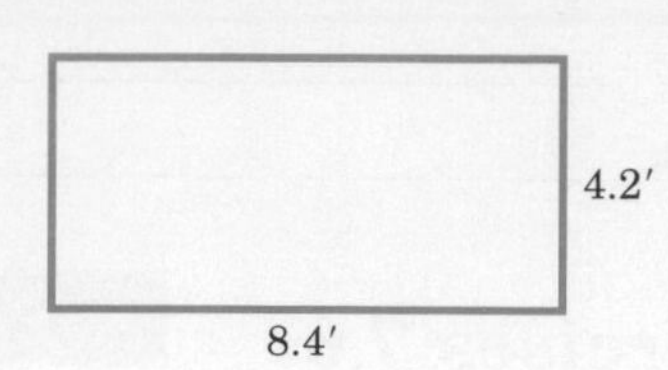

13.

10″

9.2″

24″

26″

14.

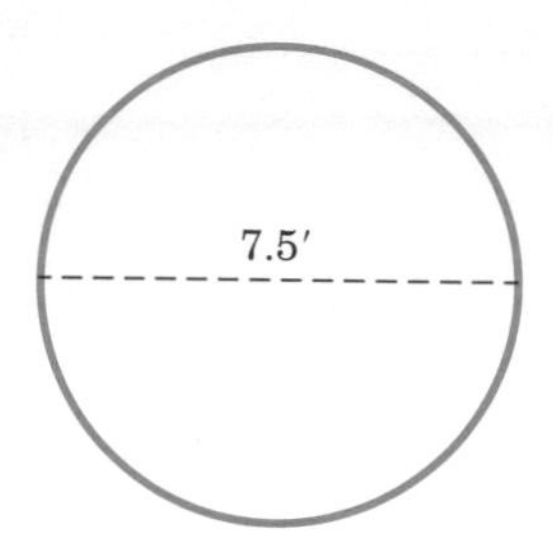

B

15. ______

16. ______

17. ______

18. ______

19. ______

20. ______

21. ______

22. ______

15.

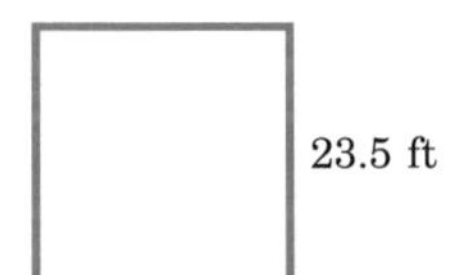

16.

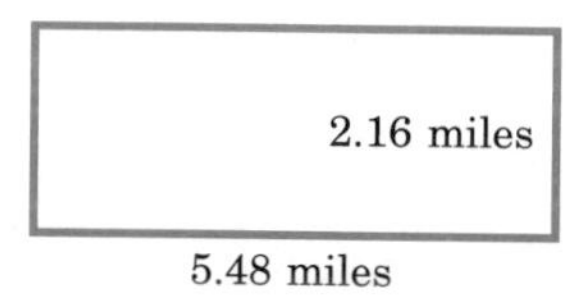

17.

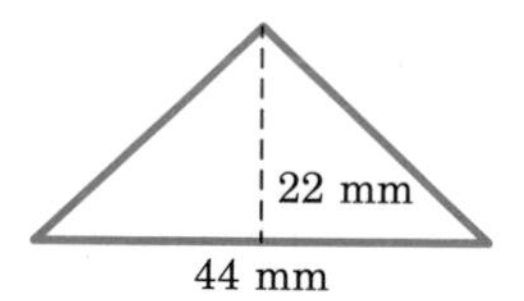

18.

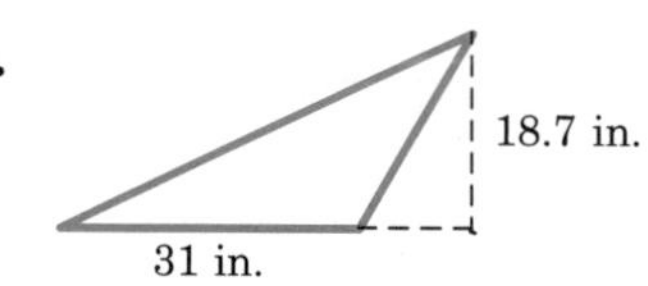

19.

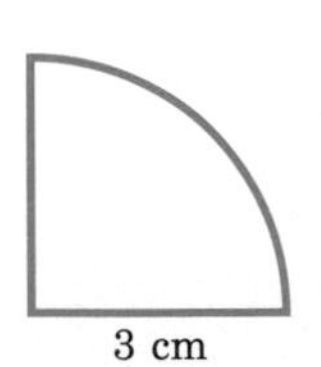

20.

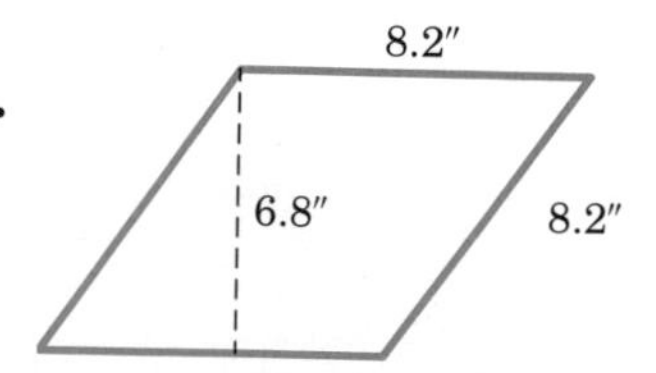

21.

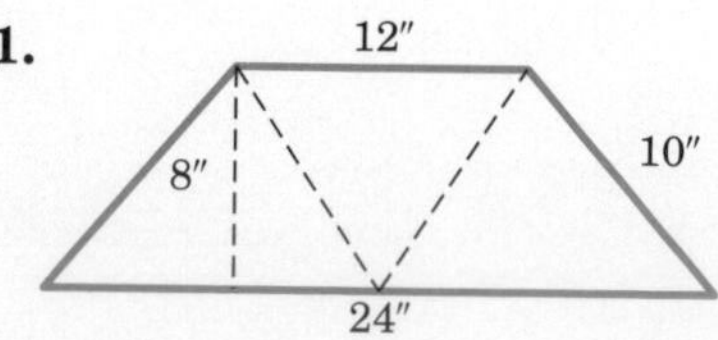

22.

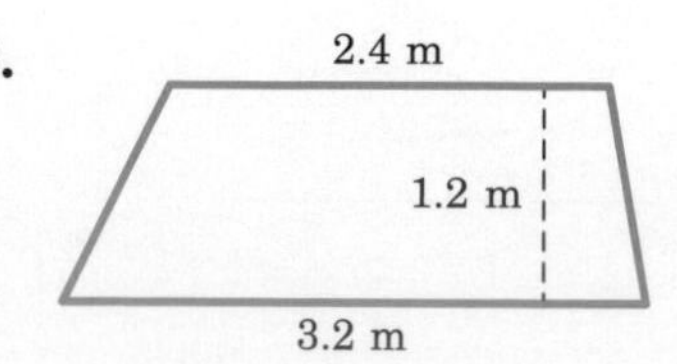

NAME ______________________

CLASS ____________ SECTION ____________ DATE ____________

ANSWERS

23. ____________

24. ____________

25. ____________

23.

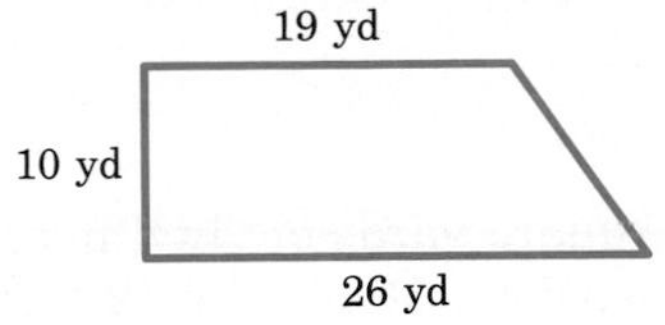

19 yd

10 yd

26 yd

24.

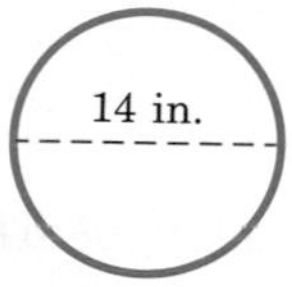

14 in.

25.

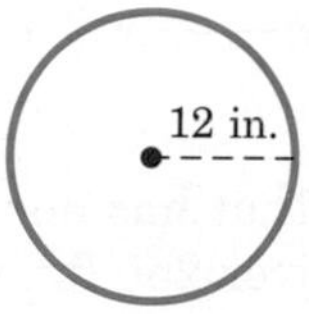

12 in.

C

26. ____________

26. Find the area of a rectangle with a length of 16 inches and a width of 12 inches.

27. ____________

27. Find the area of a circle with a diameter of 16 centimeters.

28. ____________

28. Find the area of a parallelogram that has a base of 18.9 meters and a height of 17.3 meters.

29. ____________

29. Find the area of a trapezoid with bases of 2 ft 4 in. and 3 ft 2 in. and a height of 9 in.

ANSWERS

30. ______

31. ______

30. One roll of wallpaper will cover 32 square feet. How many rolls are needed for a windowless wall that measures 15 feet by 8 feet? (You can not buy part of a roll.)

31. How many square yards of vinyl floor covering are needed for a kitchen floor that measures 10 feet by 12 feet?

D

32. ______

33. ______

34. ______

35. ______

36. ______

32. The side of Jane's house that has no windows measures 35 ft by 22 ft. If one gallon of stain will cover 250 ft^2, will 2 gallons of stain be enough to stain this side?

33. If the south side of Jane's house measures 85 ft by 22 ft and has two windows, each 4 ft by 6 ft, will 4 gallons of the stain be enough for this side? (See Exercise 32.)

34. How many square inches of glass are in a circular mirror of diameter 36 inches?

35. How many square yards of carpet will Peggy need to carpet a rectangular floor that measures 21 ft by 30 ft?

36. If one ounce of weed killer treats one square meter of lawn, how many ounces of weed killer will Debbie need to treat a rectangular lawn that measures 30 m by 8 m?

NAME ____________________

CLASS ____________ SECTION ____________ DATE ____________

ANSWERS

37. ____________

37. How many acres are contained in a rectangular plot of ground if the length is 1320 ft and the width is 528 ft? (43,560 ft^2 = 1 acre)

38. How many square tiles, each 30 cm on a side, will Kirsten need to cover a floor that is 3 m wide and 6 m long?

38. ____________

39. How many square feet of sheathing is needed for the gable end of a house that has a rise of 9 ft and a span of 36 ft? (See the drawing.)

9 ft

36 ft

39. ____________

E *Maintain your skills* (Sections 6.8, 6.9)

40. ____________

40. On a test Mildred answered 18 questions correctly and missed 7. What percent did she answer correctly?

41. Bill answered 37 questions correctly and missed 3. What percent did he miss?

41. ____________

42. The local high school basketball team won 21 games and lost 7. What percent did it win?

42. ____________

ANSWERS

43. ______________________

43. In the eight baseball games that Lew Slugger played last week he hit safely 12 times, struck out 8 times, and flied out 12 times. What percent of the time did Lew hit safely?

44. ______________________

44. In Exercise 43, what percent of the time did Lew strike out?

45. Inflation last year was 4.5%. If John earned $7000 last year, what should he earn this year, to the nearest 10 dollars, to keep up with inflation?

45. ______________________

46. Approximately 6% of the keys on Myra's harpsichord do not function. If there are 51 keys on the keyboard, how many are in working order?

46. ______________________

47. Bonita discovered that 13% of the 455 forms she had to check had been filled out incorrectly. To the nearest ten, how many of the forms were filled out correctly?

47. ______________________

48. On a particular math test you can pass if you get only 45% of the problems correct. There are fifty questions on the test. How many can you miss and still pass?

48. ______________________

49. One family saves 5% of their monthly income and uses 85% for rent, food, heat, and other monthly expenses. They have $52.10 left for entertainment. What is the dollar amount of its monthly expenses?

49. ______________________

A Second Look at Areas

OBJECTIVE

98. Find the area of geometric figures that are a combination of two or more common geometric figures.

APPLICATION

If an oval piece of sheet metal is to be made from a rectangular piece as shown, how much of the metal will be wasted? (Let $\pi \approx 3.14$.)

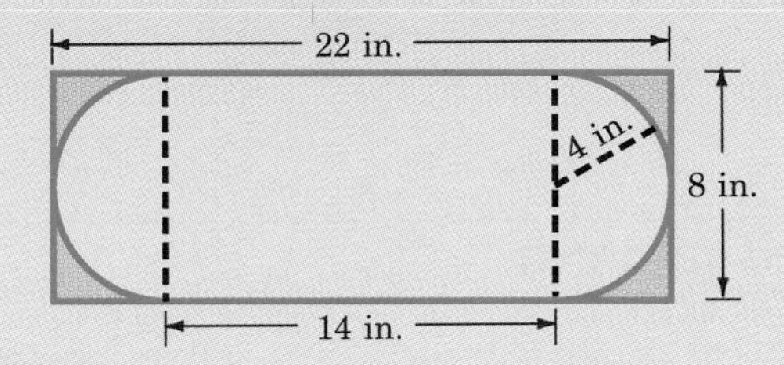

(The shaded portion is the waste.)

VOCABULARY

No new words.

HOW AND WHY

Some figures can be divided into two or more of the common shapes. The sum of the areas of each of these common figures is the area of the entire region.

Consider the following shapes:

The figure on the left can be divided into a rectangle with a triangle attached, as shown on the right. We can find the area of the rectangle and the area of the triangle. The sum of these areas is the area of the entire region.

Consider the following shapes:

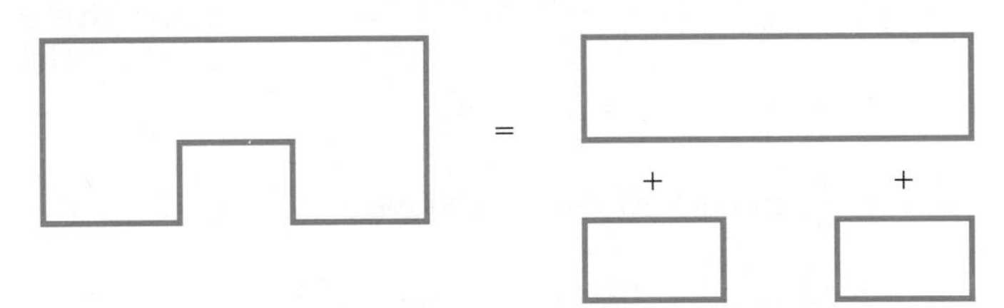

The figure on the left can be divided into three rectangles, as shown on the right. We can find the area of each rectangle. The sum of these areas is the area of the entire region.

In some figures it is helpful to attach a region to the original figure so that it can be divided into the common figures. For example:

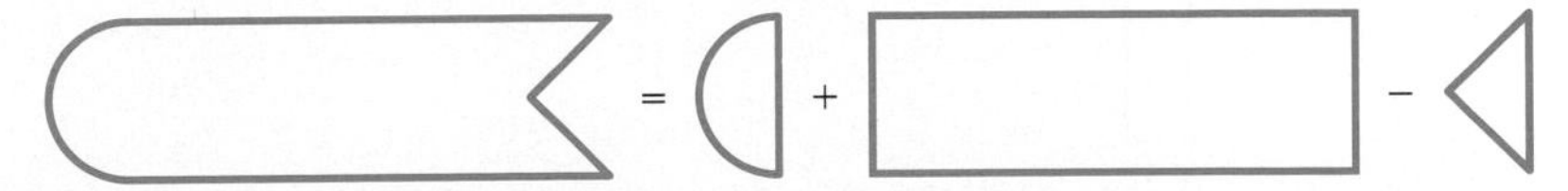

The figure on the left is a semicircle and a rectangle minus a triangle, as shown on the right. We can find the area of the semicircle, the rectangle, and the triangle. Then we find the sum of the areas of the semicircle and the rectangle, and subtract the area of the triangle to find the area of the entire region.

PROCEDURE

To find the area of a geometric figure that is a combination of two or more common geometric figures:

1. **Divide the figure into common geometric figures for which the information is given to find the area, or attach a region or regions and then divide into common geometric figures for which you can find the area.**
2. **Find the area of each of the common figures.**
3. **Find the sum or difference of those areas.**

MODEL PROBLEM-SOLVING

Examples | **Strategy**

a. Find the area of the following figure:

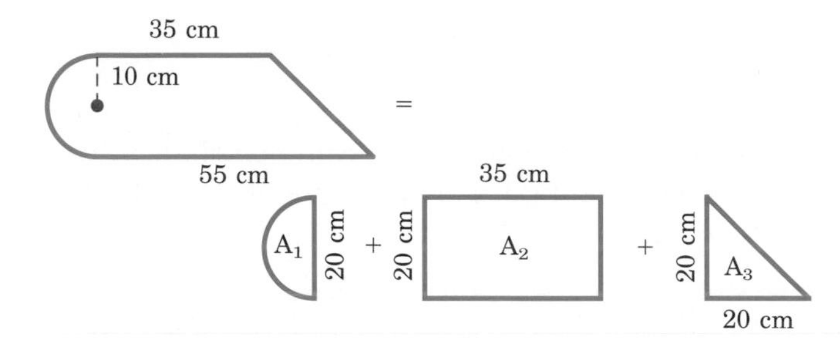

The area of the entire region is:

A = area of semicircle + area of rectangle + area of triangle

Divide the figure into a semicircle, a rectangle, and a triangle.

$A = A_1 + A_2 + A_3$

$A_1 = \frac{\pi r^2}{2} \approx \frac{(3.14)(10 \text{ cm})^2}{2} = 157 \text{ cm}^2$

Find the sum of the areas of each figure to find the total area.

$A_2 = \ell \cdot w = (35 \text{ cm})(20 \text{ cm}) = 700 \text{ cm}^2$

$A_3 = \frac{1}{2} \cdot b \cdot h = \frac{1}{2}(20 \text{ cm})(20 \text{ cm}) = 200 \text{ cm}^2$

$A \approx 157 \text{ cm}^2 + 700 \text{ cm}^2 + 200 \text{ cm}^2 = 1057 \text{ cm}^2$

Warm Up a. Find the area of the following figure:

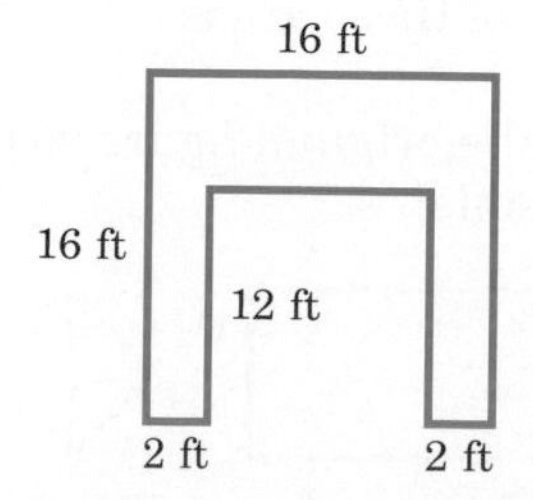

ANSWERS TO WARM UPS (7.6)

a. 112 ft^2

b. Find the area of the following figure:

20 in. 8 in. 8 in. = A_1 8 in. + 8 in. 28 in. A_2 − 8 in. A_3 8 in.

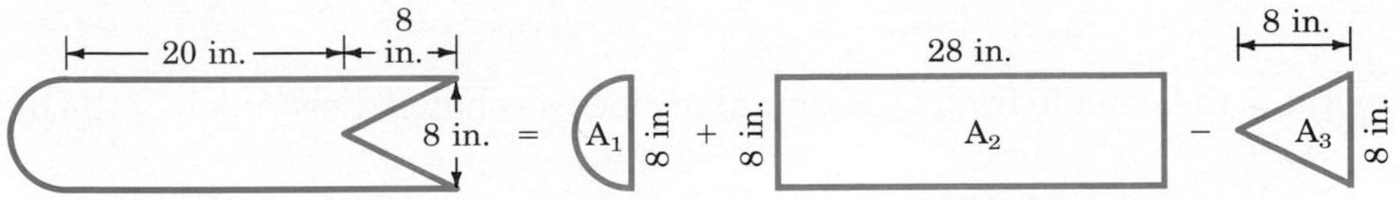

The area of the entire region is:

$A =$ area of semicircle + area of rectangle − area of triangle

$A = A_1 + A_2 - A_3$

$A \approx 25.12 \text{ in.}^2 + 224 \text{ in.}^2 - 32 \text{ in.}^2$

$A \approx 217.12 \text{ in.}^2$

Find the area formed by closing in the opening opposite the semicircle. This area is the sum of the areas of a rectangle and a semicircle.

Now subtract the area of the triangle that was formed by closing in the open area.

Warm Up b. Find the area of the following figure:

15 m 3 m

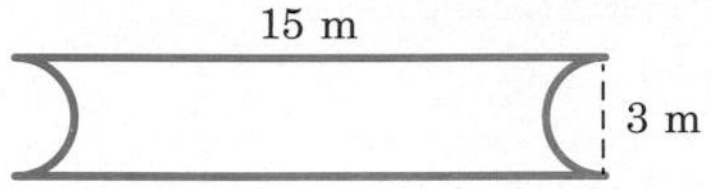

ANSWERS TO WARM UPS (7.6)

b. 37.935 m^2

APPLICATION SOLUTION

c. If an oval piece of sheet metal is to be made from a rectangular piece as shown, how much of the metal will be wasted? (Let $\pi \approx 3.14$.)

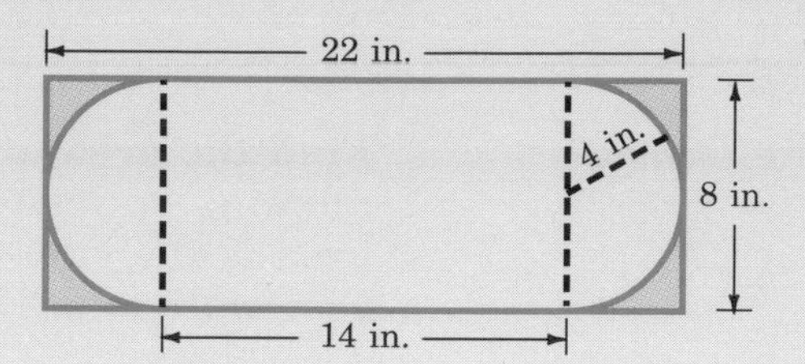

(The shaded portion is the waste.)

$$\begin{aligned} A_1 &= \ell \cdot w \\ &= (22 \text{ in.})(8 \text{ in.}) \\ &= 176 \text{ in.}^2 \end{aligned}$$

To find the amount of waste, find the area of the original piece of metal (a rectangle).

$$\begin{aligned} A_2 &= \ell \cdot w + \pi r^2 \\ &\approx (14 \text{ in.})(8 \text{ in.}) + 3.14(4 \text{ in.})^2 \\ &\approx 112 \text{ in.}^2 + 50.24 \text{ in.}^2 \\ &\approx 162.24 \text{ in.}^2 \end{aligned}$$

Now find the area of the oval piece (a rectangle and two semicircles).

$$\begin{aligned} A &= A_1 - A_2 \\ &\approx 176 \text{ in.}^2 - 162.24 \text{ in.}^2 \\ &\approx 13.76 \text{ in.}^2 \end{aligned}$$

The wasted metal is the difference between the two areas.

So, the waste is approximately 13.76 in.^{2}.

Warm Up c. A city park has dimensions of 40 ft by 65 ft. The park is all grass except for three circular flower beds. The flower beds have diameters of 8 ft, 10 ft, and 8 ft. Find the area of the park that is in grass. Let $\pi \approx 3.14$. Find to the nearest square foot.

ANSWERS TO WARM UPS (7.6)

c. 2421 ft^2

NAME ____________________

CLASS ____________ SECTION ____________ DATE ____________

Exercises 7.7

A

ANSWERS

1. ____________

2. ____________

3. ____________

4. ____________

5. ____________

6. ____________

Find the area of each of the following figures. Let $\pi \approx 3.14$.

1.

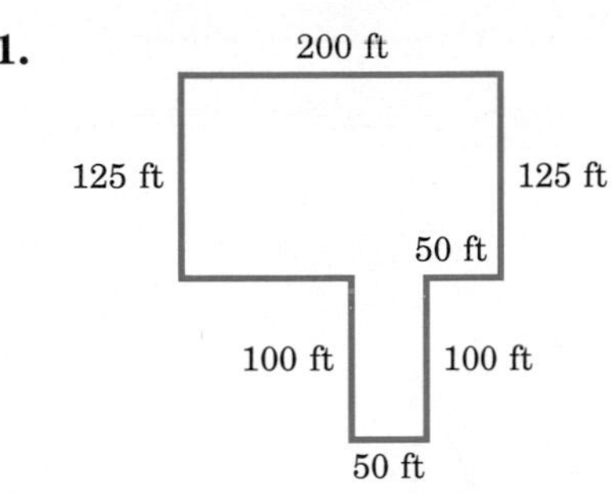

2.

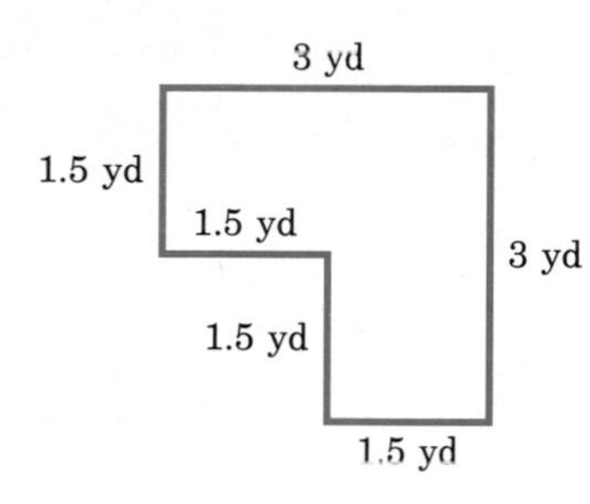

3.

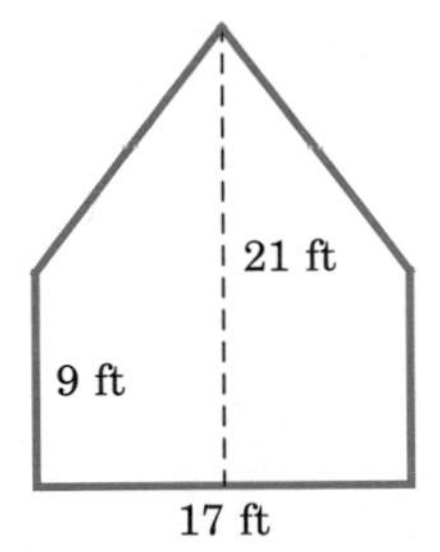

4.

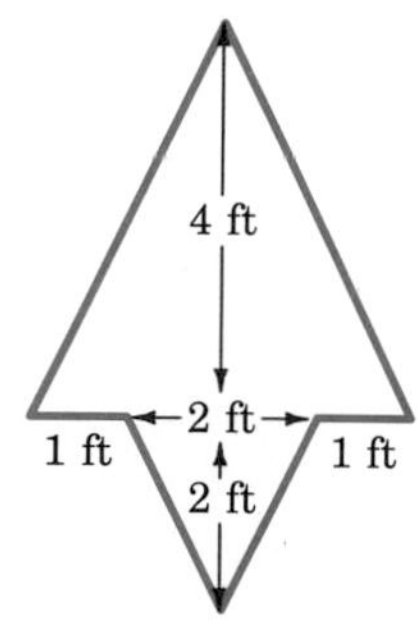

5.

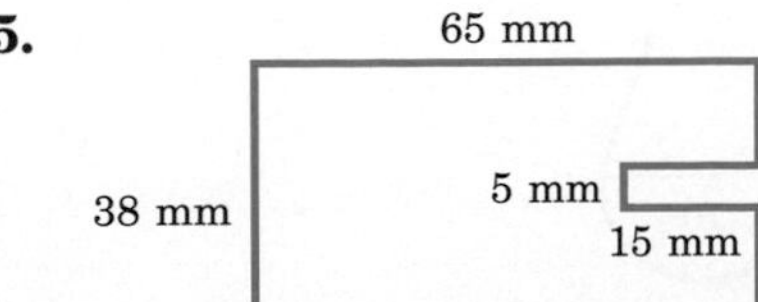

6.

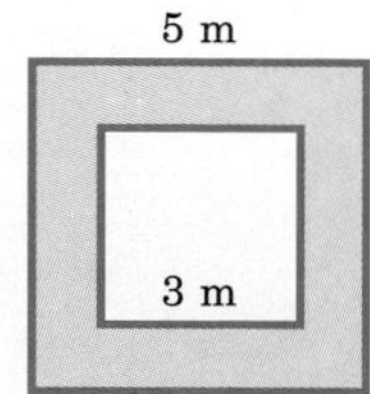

The shaded area

B

ANSWERS

7. ____________

8. ____________

9. ____________

10. ____________

11. ____________

12. ____________

13. ____________

14. ____________

7.

8 in.

24 in.

8.

10 cm

50 cm

9.

30 cm

7.5 cm

15 cm

30 cm

10.

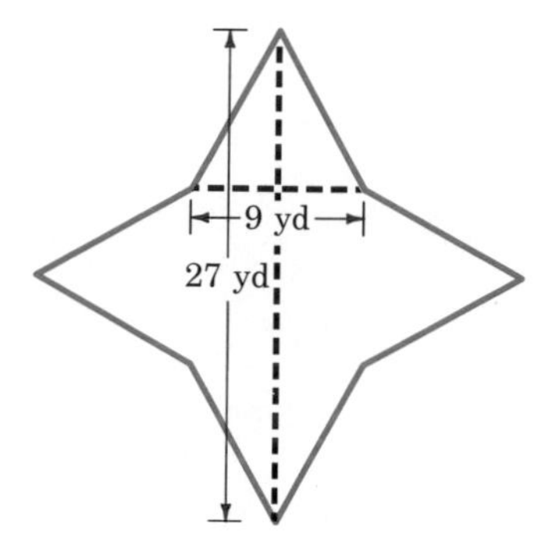

11.

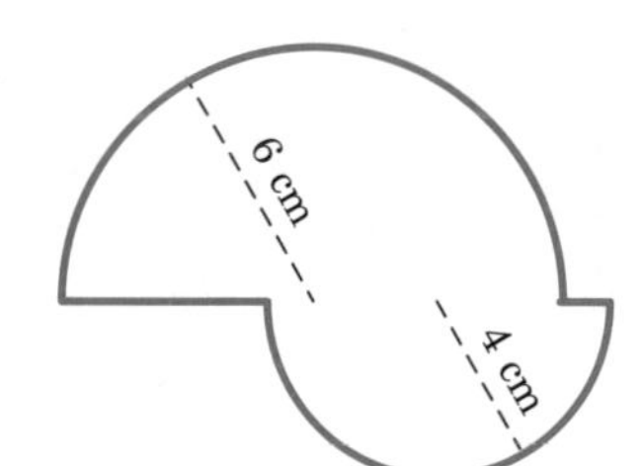

12.

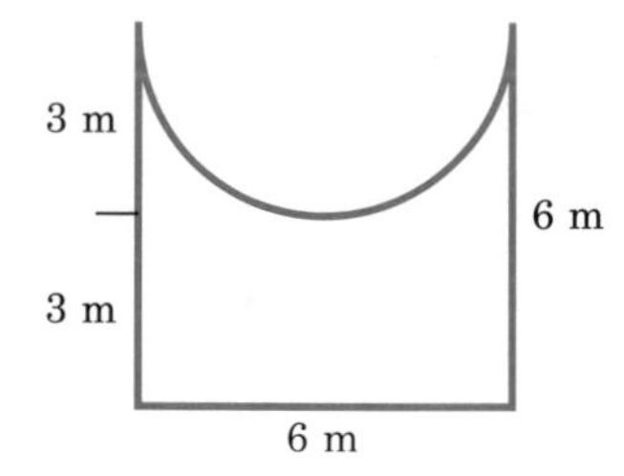

13.

10″

14″

10″

10″

14.

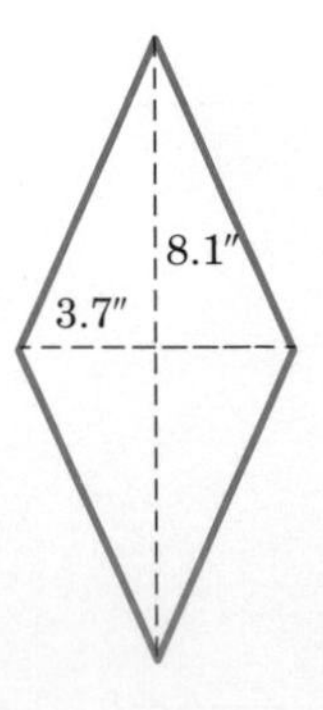

NAME ______________________________

CLASS ______________ SECTION ______________ DATE ______________

ANSWERS

15. ______________

16. ______________

15.

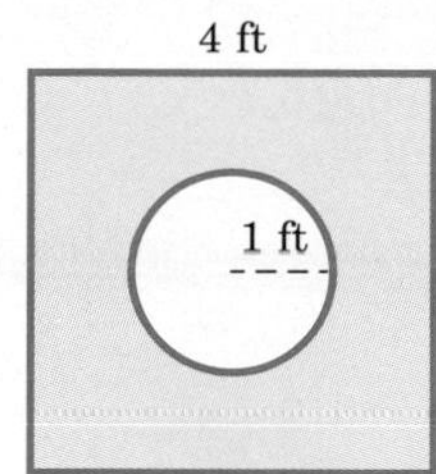

The shaded area

16.

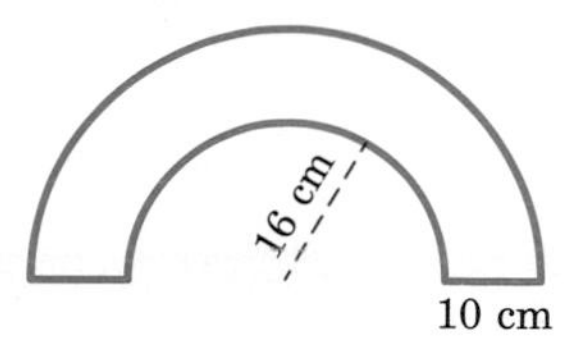

C

17. ______________

18. ______________

17.

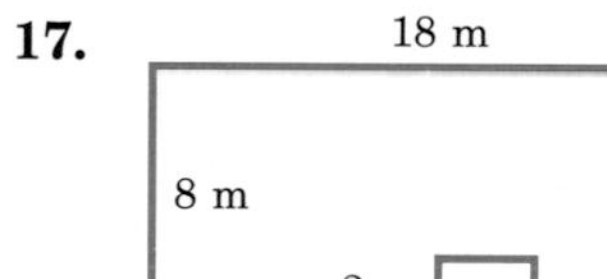

18. Find the area of the shaded portion:

6.4 in.

1.3 in.

4.75 in.

2 in.

19. ______________

20. ______________

19. Round off to the nearest hundredth:

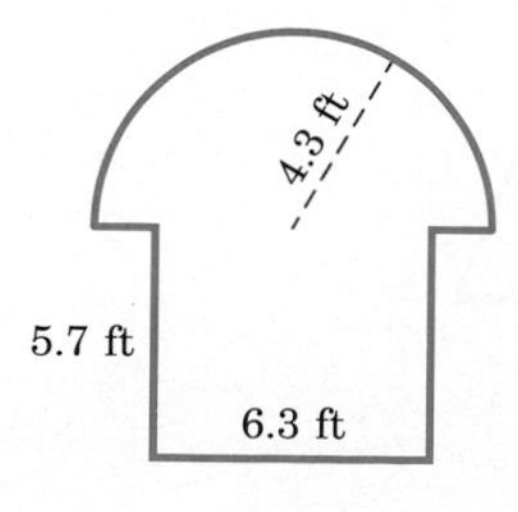

20. Round off to the nearest tenth:

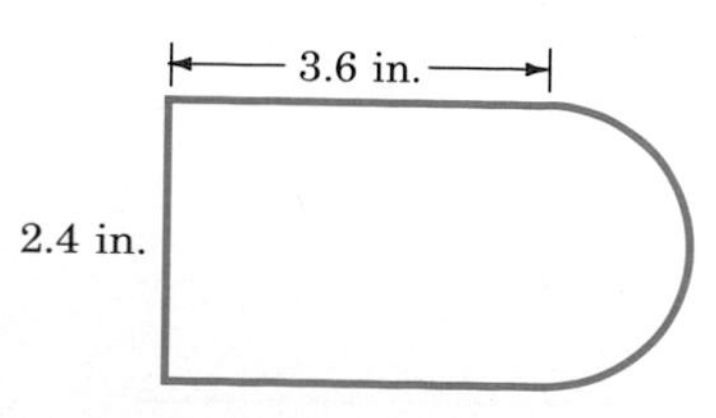

ANSWERS

21. ______________________

22. ______________________

23. ______________________

24. ______________________

25. ______________________

26. ______________________

21. Round off to the nearest hundredth:

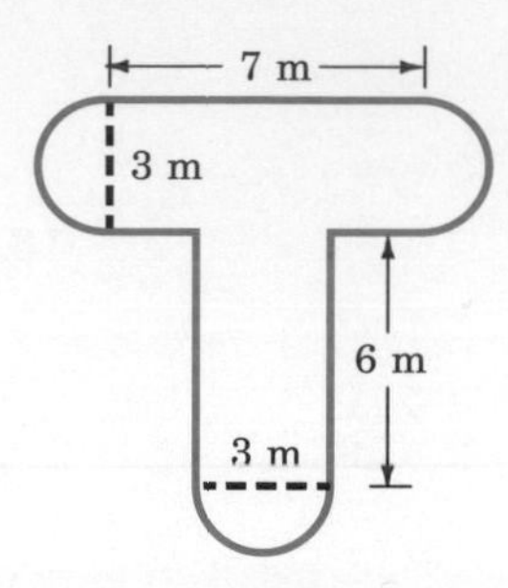

22. Round off to the nearest tenth:

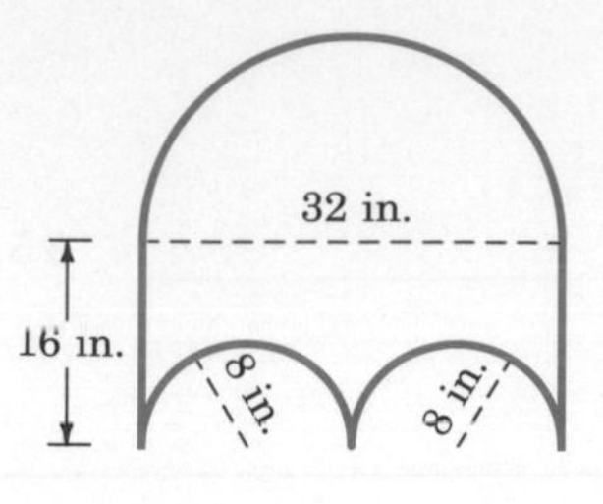

Find the area of the shaded portion:

23.

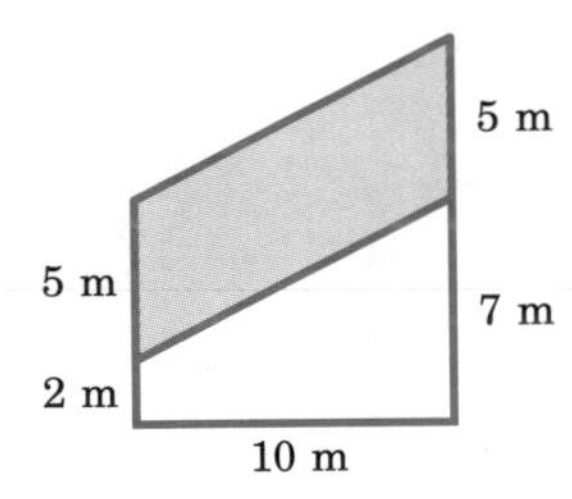

24.

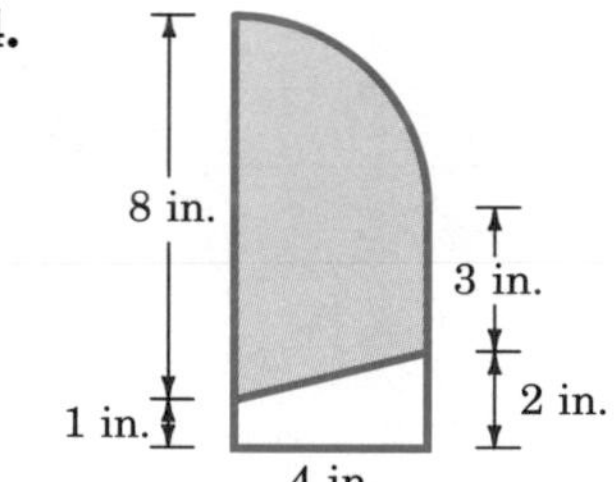

25.

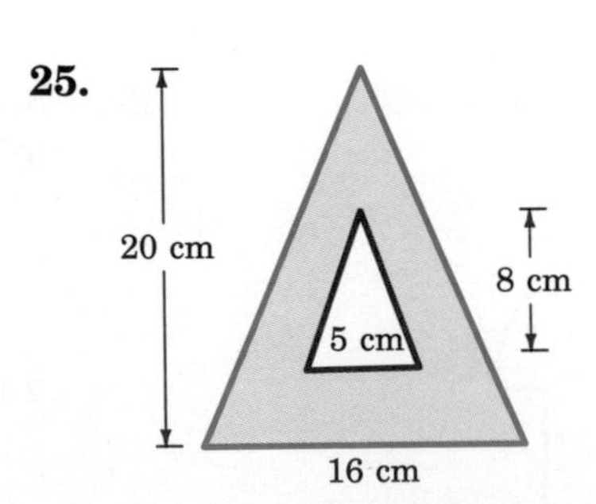

26.

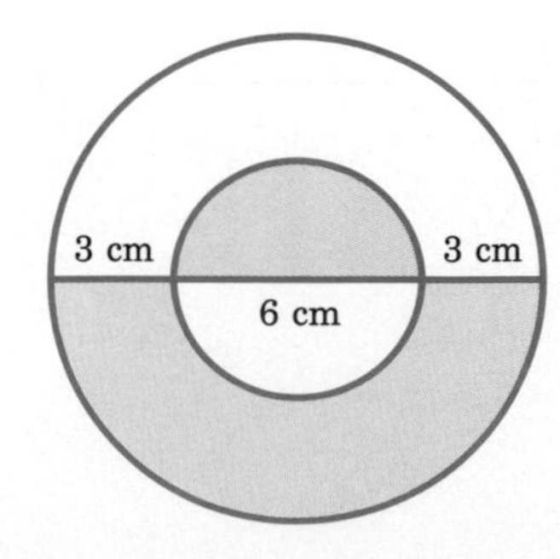

NAME ______

CLASS ______ SECTION ______ DATE ______

D

ANSWERS

27. ______

28. ______

29. ______

30. ______

27. The following diagram shows the Jones' yard with respect to their house. How much grass seed is needed to sow the lawn if one pound of seed will sow 1000 ft^2? Find the answer to the nearest pound.

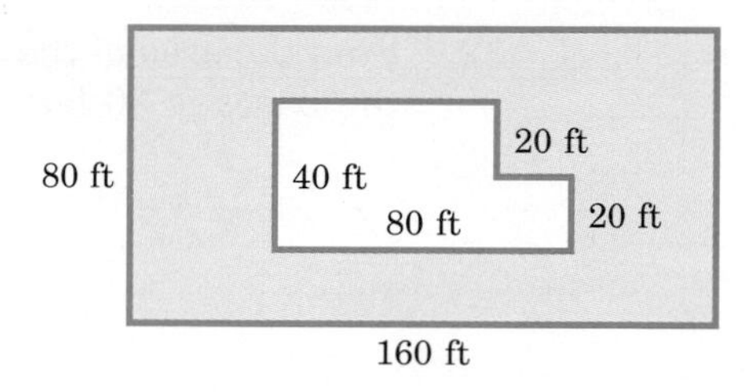

28. What is the area of the infield of a race track in which the straight-aways are 100 yards long and the ends are semicircles with 30-yard diameters?

29. The floor of a shop (floor plan shown below) is to be poured concrete. If concrete costs \$3.35 a square yard, what will the floor cost?

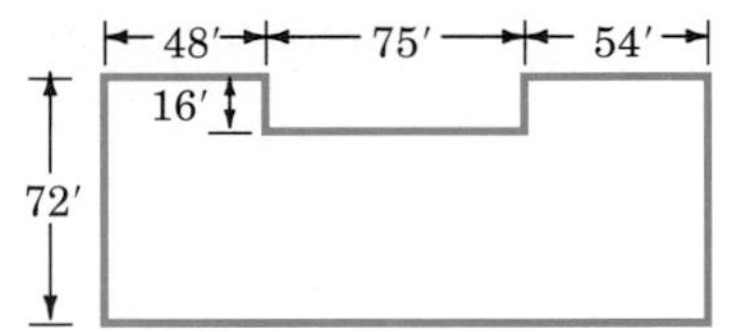

30. How many squares of aluminum siding are needed for the shed shown below? Assume that there are no windows and that the door will be made of the siding. A different material will be used for the roof. (100 ft^2 = 1 square of siding)

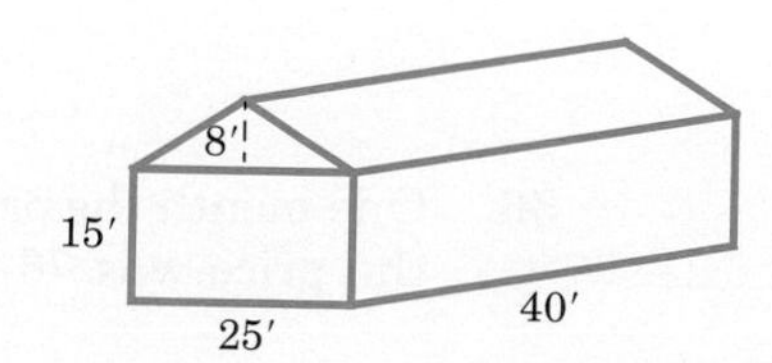

ANSWERS

31. ______________

32. ______________

31. A tin can has a radius of 5 inches and a height of 8 inches. It is painted red on top and bottom and green on the sides. (Hint: Take the can apart and look at the shapes!)

a. What area of the can is painted green?

b. What area of the can is painted red?

32. Find the area of the infield of a baseball diamond if the distance between the bases is 90 feet.

E *Maintain your skills* (Sections 6.8, 6.9)

33. ______________

33. If a saline solution is 20% salt and 80% water, how much salt is in 40 pounds of the mixture?

34. ______________

34. If a saline solution is 40% salt and 60% water, how much solution can be made using 36 pounds of salt?

35. ______________

35. An attorney collected a debt of $500 for a client. The attorney charged $75 for her services. What rate of commission did she charge?

36. ______________

36. One month the price of eggs was 95 cents per dozen. The following month the price was 76 cents. What was the percent of decrease in the price?

NAME ______________________

CLASS ______________ SECTION ______________ DATE ______________

ANSWERS

37. ______________

37. The pilot of a transcontinental airplane was instructed to increase the speed of his airplane from 480 miles per hour to 510 miles per hour. What was the percent of increase?

38. ______________

38. If the cost of gasoline is \$1.20 per gallon and the tax on a gallon of gasoline is 12 cents, what percent of the cost of gasoline is the tax?

39. The price of hamburger rose 12% last month. This was 15 cents more than it had been. What was the price before?

39. ______________

40. Wayne and Susan caught one salmon on 10% of the times they went fishing. On 2% of the trips they caught 2 salmon. On the other 88% of their fishing trips they caught nothing. On their fifty trips together, how many salmon did they catch?

40. ______________

41. The Class Rock Music Market sells cassettes for cost plus $16\frac{2}{3}\%$. What is the retail price of a cassette that cost the store \$6.18?

41. ______________

ANSWERS

42. ______________________

42. The Designer Ware Mart has jeans and tops on sale. Any combination is priced at \$33.85. There is a 25% discount on a second combination if bought at the same time. If the sales tax is 6%, how much will two sets of jeans and tops cost?

Volume of Common Geometric Solids

OBJECTIVE

99. Find the volume of common geometric solids.

APPLICATION

How many cubic yards of concrete will be needed to pour a sidewalk that is 3 ft wide, 4 in. thick, and 54 ft long?

VOCABULARY

Volume is the name given to the amount of space that is contained inside a three-dimensional figure. The volume of a box is the answer to the question, "How much does the box hold?" Volume is measured by a cubic unit, that is, a *cube* (a box) that is one unit on each edge. Volume can be thought of as the number of cubic units needed to form the figure. An example of a unit of volume measure is a *cubic inch* (in.^{3}), which is shown in Figure 7.5:

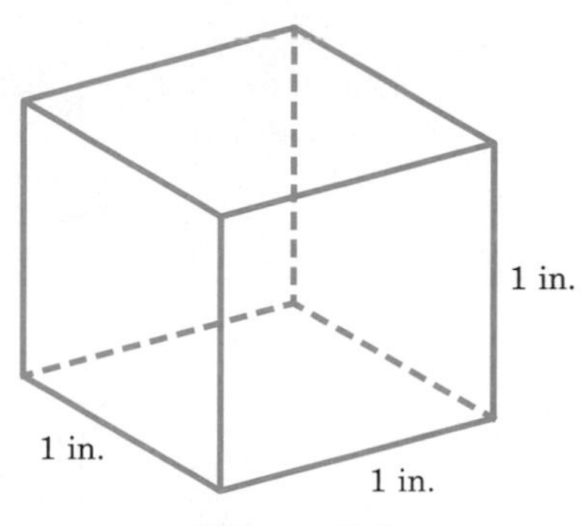

Figure 7.5

A cube that is one inch on each edge is a *cubic inch*. A cube that is 1 centimeter on each edge is a *cubic centimeter* (cm^3). A cube that is one foot on each edge is a *cubic foot* (ft^3). Any of these cubes may be used as a unit of volume.

HOW AND WHY

To find the volume of a geometric solid means to determine how many cubes or parts of cubes are contained within the solid. Each cube is the same size and is called a cubic unit.

FORMULA

The volume of a cube that is of length e on each side has the formula:

$V = e \cdot e \cdot e = e^3$

The volume of a rectangular solid (a box is a rectangular solid) is $V = \ell \cdot w \cdot h$, where V is the volume, ℓ is the length, h is the height, and w is the width. (See Figure 7.6.)

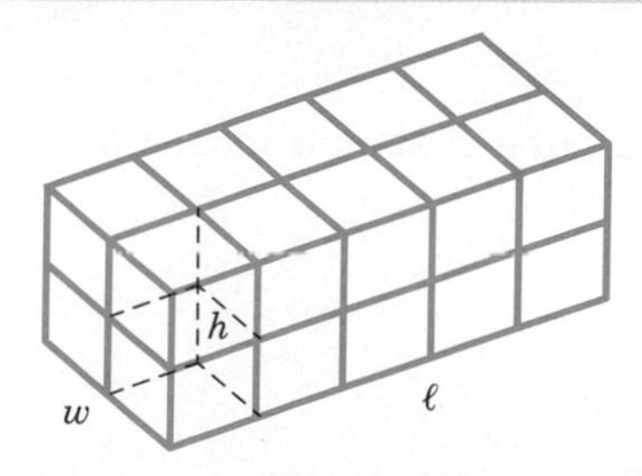

Figure 7.6

The length, width, and height must all be measured with the same unit of measure or converted to the same unit.

A cube is a rectangular solid in which the length, width, and height are all equal.

Pictures of a cylinder, a sphere, a cone, and a pyramid are shown in Figure 7.7. The formula for the volume of each is as follows:

FORMULA

Cylinder $V = \pi r^2 h$

Sphere $V = \frac{4}{3}\pi r^3$

Cone $V = \frac{1}{3}\pi r^2 h$

Pyramid $V = \frac{1}{3}Bh$ **(where B is the area of the base)**

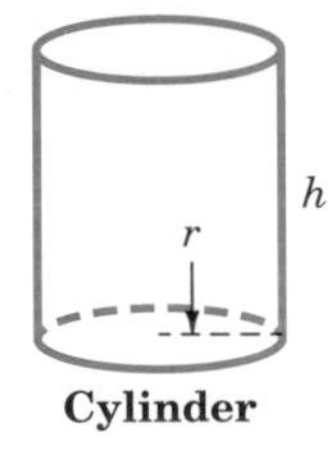

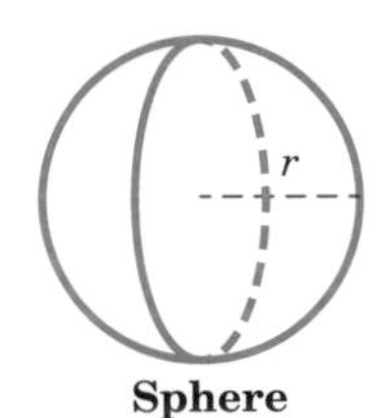

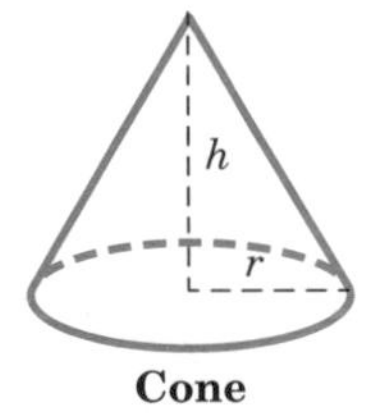

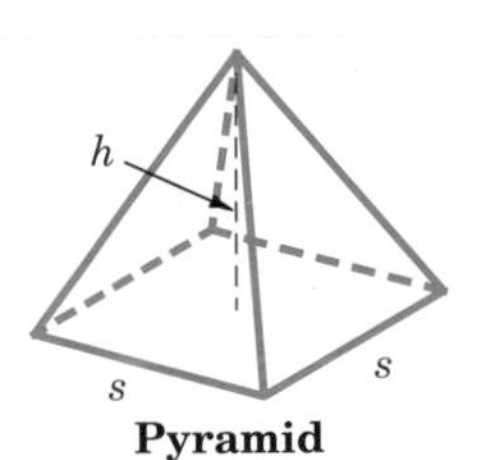

Figure 7.7

MODEL PROBLEM-SOLVING

Examples	Strategy
a. Find the volume of a cube that is 6 inches on each edge:	
$V = e^3$ or $V = e \cdot e \cdot e$	**Formula for the volume of a cube.**
$V = (6 \text{ in.})^3$	**Substitute.**
$V = (6 \text{ in.})(6 \text{ in.})(6 \text{ in.})$	**Multiply.**
$V = 216 \text{ in.}^3$	
The volume is 216 in.^3.	

Warm Up a. Find the volume of a cube that is 15 m on each edge.

b. Find the volume of the rectangular solid (box) shown here:

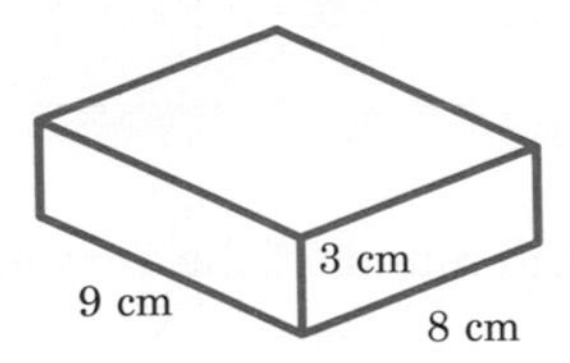

$V = \ell \cdot w \cdot h$	**Formula for the volume of a rectangular solid.**
$V = (9 \text{ cm})(8 \text{ cm})(3 \text{ cm})$	**Substitute.**
$V = 216 \text{ cm}^3$	**Multiply.**

The volume is 216 in.^{3}.

Warm Up b. Find the volume of a box with a base that measures 15 in. by 17 in. and a height of 12 in.

c. Find the volume of the following cylinder:

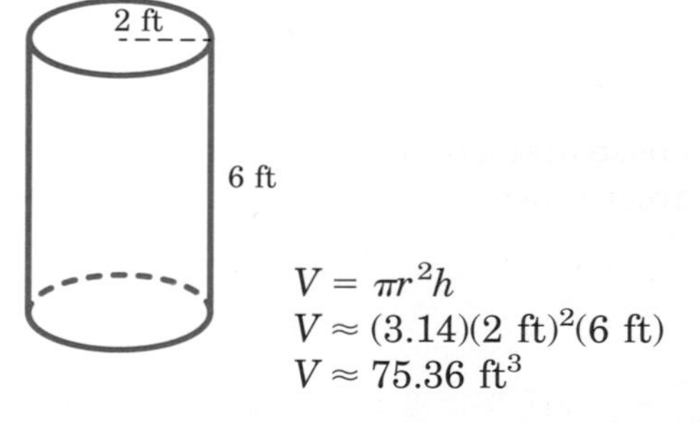

$V = \pi r^2 h$	**Formula for the volume of a cylinder.**
$V \approx (3.14)(2 \text{ ft})^2(6 \text{ ft})$	**Substitute; let $\pi \approx 3.14$.**
$V \approx 75.36 \text{ ft}^3$	**Multiply.**

The volume is approximately 75.36 ft^3.

Warm Up c. Find the volume of a cylinder that has a circular base with a radius of 12 cm and a height of 13 cm. Let $\pi \approx 3.14$, and round off the answer to the nearest tenth.

ANSWERS TO WARM UPS (7.8) **a.** 3375 m^3 **b.** 3060 in.^{3} **c.** 5878.1 cm^3

d. Find the volume of a sphere with a diameter of 18 centimeters:

$V = \frac{4}{3}\pi r^3$ **Formula for the volume of a sphere.**

$V \approx \frac{4}{3}(3.14)(9 \text{ cm})^3$ **Substitute. Let $\pi \approx 3.14$; $r = 9$ cm since $d = 18$ cm.**

$V \approx \frac{4}{3}(3.14)(729 \text{ cm}^3)$

$V \approx 3052.08 \text{ cm}^3$ **Multiply.**

The volume is approximately 3052.08 cm^3.

Warm Up d. Find the volume of a sphere with a radius of 25 feet. Let $\pi \approx 3.14$, and round off the answer to the nearest cubic foot.

e. Find the volume of the cone shown here:

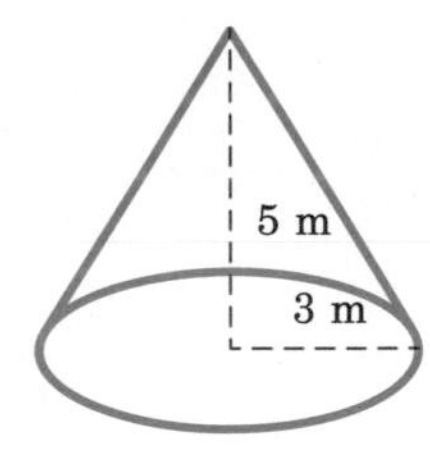

$V = \frac{1}{3}\pi r^2 h$ **Formula for the volume of a cone.**

$V \approx \frac{1}{3}(3.14)(3 \text{ m})^2(5 \text{ m})$ **Substitute; let $\pi \approx 3.14$.**

$V \approx 47.1 \text{ m}^3$ **Multiply.**

The volume is approximately 47.1 m^3.

Warm Up e. Find the volume of a right circular cone that has a base diameter of 3 ft and a height of 7 ft. Let $\pi \approx 3.14$, and round off the answer to the nearest tenth.

f. Find the volume of a pyramid that has a rectangular base with length 6 in. and width 4 in. and a height of 5 in.:

$V = \frac{1}{3}(\ell w)h$ **Since the base is a rectangle we use (ℓw) in place of B, the area in the formula.**

ANSWERS TO WARM UPS (7.8) **d.** 65,417 ft^3 **e.** 16.5 ft^3

$V = \frac{1}{3}(6 \text{ in.})(4 \text{ in.})(5 \text{ in.})$ **Substitute.**

$V = 40 \text{ in.}^3$ **Multiply.**

The volume is 40 in.^{3}.

Warm Up f. Find the volume of a pyramid whose base is a square 2.4 m on a side and has a height of 3.5 m.

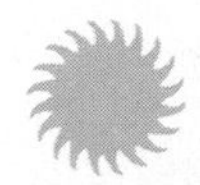

APPLICATION SOLUTION

g. How many cubic yards of concrete will be needed to pour a sidewalk that is 3 ft wide, 4 in. thick, and 54 ft long?

Width: $3 \text{ ft} = \frac{3 \text{ ft}}{1} = \frac{3 \not{\text{ft}}}{1} \cdot \frac{1 \text{ yd}}{3 \not{\text{ft}}} = 1 \text{ yd}$

To find the number of cubic yards, first change all measurements to yards.

Length: $54 \text{ ft} = \frac{54 \text{ ft}}{1} = \frac{54 \not{\text{ft}}}{1} \cdot \frac{1 \text{ yd}}{3 \not{\text{ft}}} = 18 \text{ yd}$

Thickness: $4 \text{ in.} = \frac{4 \text{ in.}}{1} = \frac{4 \not{\text{in}}.}{1} \cdot \frac{1 \text{ yd}}{36 \not{\text{in}}.} = \frac{1}{9} \text{ yd}$

$V = \ell \cdot w \cdot h$

The formula for the volume of a rectangular solid.

$V = (18 \text{ yd})(1 \text{ yd})\left(\frac{1}{9} \text{ yd}\right)$

$V = 2 \text{ yd}^3$

Therefore, 2 cubic yards of concrete are needed for the sidewalk.

Warm Up g. A roadbed 35 feet wide and 1 mile long is to be paved to a depth of 3 inches. How many cubic feet of paving are needed for the project?

ANSWERS TO WARM UPS (7.8)

f. 6.72 m^3 **g.** 46,200 ft^3

NAME ______________________

CLASS ______________ SECTION ______________ DATE ______________

Exercises 7.8

A

ANSWERS

1. ______________

2. ______________

3. ______________

4. ______________

5. ______________

6. ______________

7. ______________

Find the volume. Let $\pi \approx 3.14$.

1.

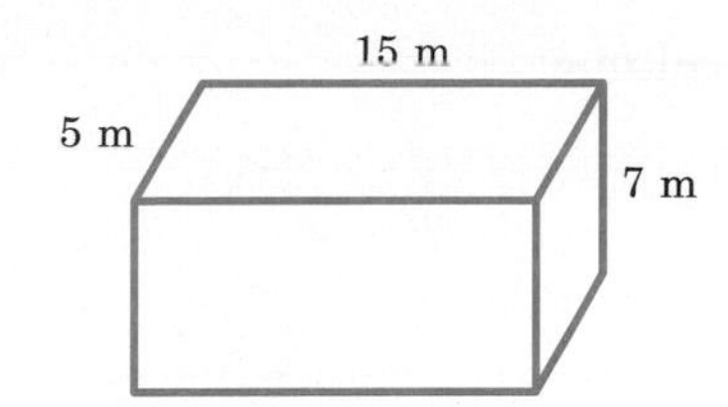

2.

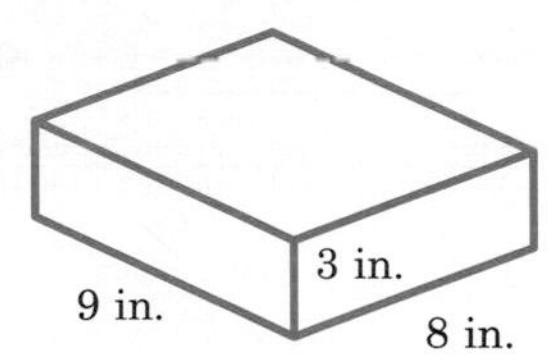

3.

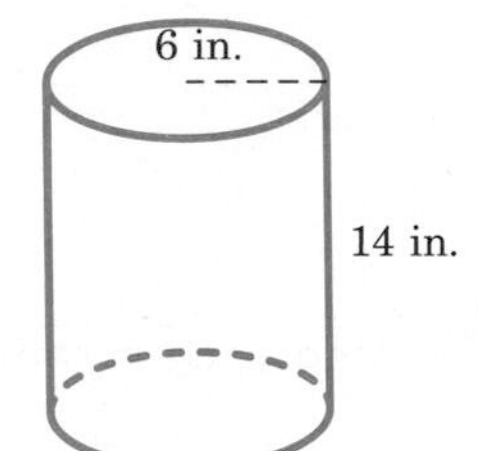

4.

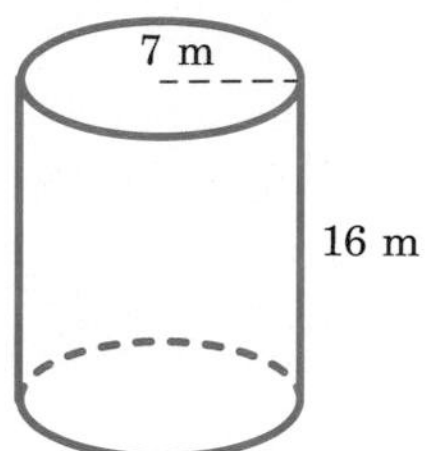

5. Find the volume of a cube that is 5 inches on each edge.

6. Find the volume of a cube that is 7.3 cm on each edge.

7. Find the volume of a golf ball (without dimples) whose diameter is 5 cm.

ANSWERS

8. ____________________

8. Find the volume of a beachball whose radius is 10 inches.

B

9. ____________________

9. Find the volume:

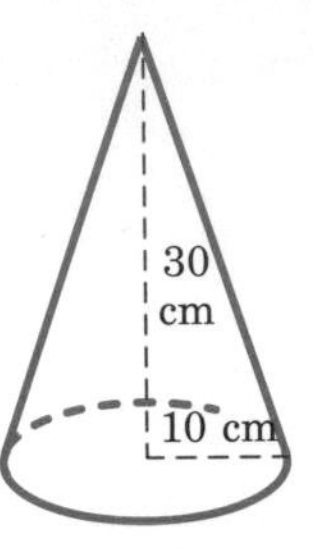

10. ____________________

10. Find the volume of a cone with a radius 6 ft and a height 8 ft.

11. Find the volume. Round off to the nearest tenth.

11. ____________________

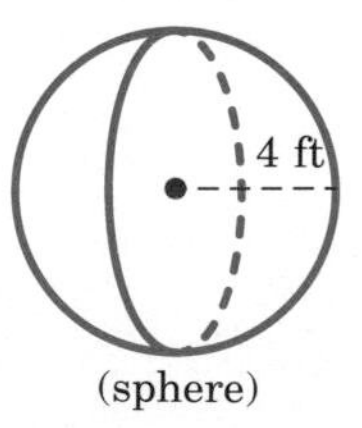

(sphere)

12. ____________________

12. Find the volume of a sphere with a diameter of 30 meters.

NAME ____________________

CLASS ____________ SECTION ____________ DATE ____________

ANSWERS

13. ____________

13. Find the volume:

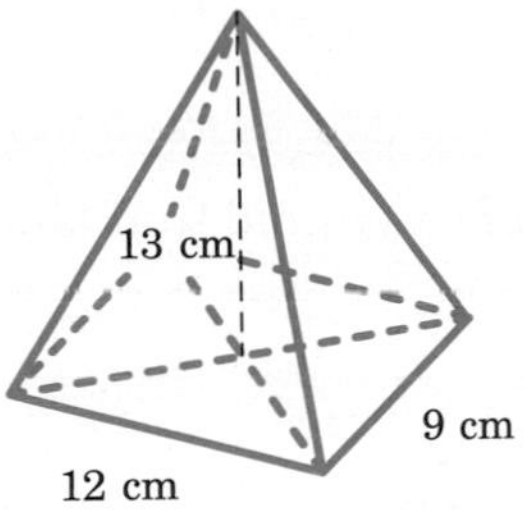

13 cm

9 cm

12 cm

14. ____________

14. Find the volume of a pyramid whose base is a square with a side 150 feet and whose height is 30 feet.

15. ____________

15. Find the volume of two identical fuzzy dice tied to the mirror of a '57 Chevy if one edge measures 6 inches.

16. ____________

16. Find the volume of a can if the radius is 7 centimeters and the height is 18 centimeters.

C

17. ____________

17. Find the volume in cubic inches of a cylinder that has a radius of 9 inches and a height of 2 feet.

ANSWERS

18. ______

19. ______

20. ______

21. ______

22. ______

18. Find the volume of a cylinder that has a 13″ diameter and is 4′ tall. (To the nearest cubic inch.)

19. Find the volume of a pyramid that has a height of 37 ft and the triangular base shown below:

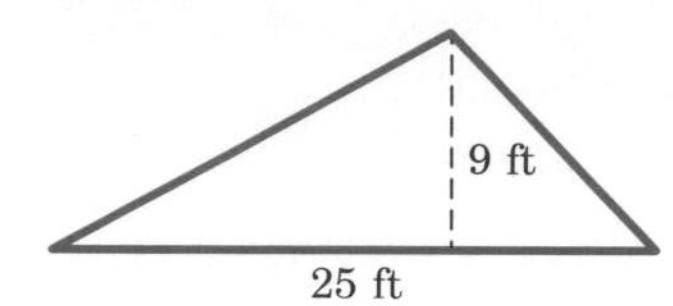

20. Find the volume of a pyramid with a height 37 meters and a square base that is 25 meters on each side.

21. Find the volume. Round off to the nearest tenth.

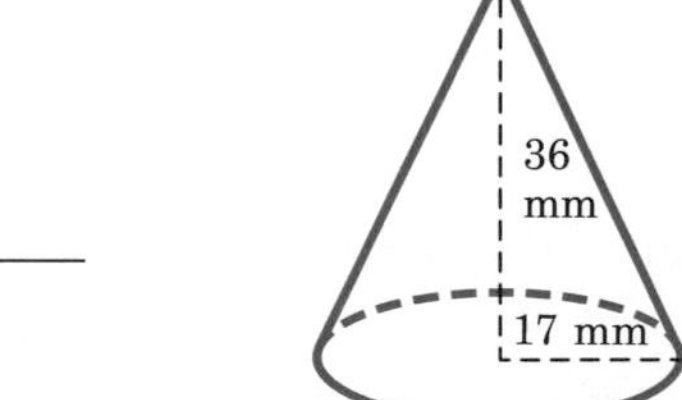

22. Find the volume of a sphere with a diameter of 2 ft 4 in. (To the nearest cubic inch.)

NAME ______________________________

CLASS ______________ SECTION ______________ DATE ______________

D

ANSWERS

23. ______________

23. A water tank is a cylinder that is 22 inches in diameter and $3\frac{1}{2}$ feet high. If there are 231 in.^{3} in a gallon, how many gallons of water will the tank hold?

24. ______________

24. How many boxes that are 12 in. wide, 8 in. high, and 18 in. long can be loaded into a truck bed that is 8 ft wide, 6 ft high, and 21 ft long?

25. ______________

25. An excavation is being made for a basement. The hole is 24 ft wide, 35 ft long, and 9 ft deep. If the bed of a truck holds 14 yd^3, how many truckloads of dirt will need to be hauled away?

26. ______________

26. A swimming pool that is 30 ft long and 10 ft wide is filled to a depth of 5 ft.

a. How many cubic ft of water are in the pool?

b. If one cubic foot of water is approximately 7.5 gallons, how many gallons of water are in the pool?

E *Maintain your skills* (Section 7.1)

ANSWERS

27. ______________________

28. ______________________

29. ______________________

30. ______________________

31. ______________________

32. ______________________

33. ______________________

34. ______________________

35. ______________________

36. ______________________

Multiply:

27. (36 gallons)4 = ? gallons

28. $\left(3\frac{7}{16}\text{ inches}\right)5 = ?$ inches

Divide:

29. (35 pounds) ÷ 7 = ? pounds

30. $\left(12\frac{1}{3}\text{ yards}\right) \div 6 = ?$ yards

Change units as shown:

31. 4.75 feet = ? inches

32. 1.8 hours = ? seconds

Add:

33.

15 hr 36 min
12 hr 52 min

34.

2 miles 2135 feet
1 mile 3015 feet

35. What is the total length of 10 sections of pipe if each section is 2 ft 6 in. long?

36. The diameter of a wire is 10.2 mils. What is the diameter in inches?

Reasoning

OBJECTIVE

100. Use of inductive and deductive reasoning.

APPLICATION

The problems of everyday life require reasoning to solve. This section is intended to give practice in the use of reasoning skills.

VOCABULARY

A *prediction* is the act of foretelling, or forecasting, a particular event, or happening. A *conclusion* is a decision or conviction that is reached as a consequence of an investigation.

Inductive reasoning is the name given to the process in which one draws conclusions based on a recognition of patterns. A study is made of the given information, and conclusions are based on the patterns and repetitions observed. Predictions are based on those patterns.

Deductive reasoning is the name given to the process through which one draws conclusions based on known facts, previously proven facts, basic assumptions, and given information.

HOW AND WHY

Inductive reasoning is used by scientists as they perform experiments over and over again. If a certain event happens each time the experiment is performed, the conclusion or observation is that the same event will occur each and every time the experiment is performed. In this case, specific examples are used to state a general conclusion. For example, we assume that each time we toss an object into the air it will return to the ground. The reason for this assumption is that our past experience has shown that the object has always returned to the ground.

Look at the sequence of numbers below to see if you can determine the number that will appear next.

3, 9, 15, 21, 27, . . .

Did you predict 33 would be the next number? If you did, then you made an observation and drew a conclusion. Perhaps you observed that

$3 = 3 \cdot 1$

$9 = 3 \cdot 3$

$15 = 3 \cdot 5$

$21 = 3 \cdot 7$

$27 = 3 \cdot 9$

so you predicted that the next number will be $3 \cdot 11$, or 33.

You could have also used a different pattern. Since there is a difference of 6 between any two consecutive numbers in the sequence, the next term of the sequence can be found by adding 27 and 6 to get 33. Both methods lead to the same conclusion. However, each of the processes makes the assumption that the sequence of numbers will continue in the same pattern.

A second type of reasoning (deductive reasoning) is used in formal proof. This involves using known facts and previously proven data to draw conclusions. For example:

It is known that if it rains, James will stay home.

It did rain. Therefore, we can conclude that James stayed home.

Mathematically this can be stated as follows:

If A then B.
A occurred; therefore B must occur.

It is also true that

If A then B.
B did not occur; therefore A did not occur.

However, one must be careful to draw a correct conclusion. This is shown in the following example.

It is known that if it rains, James will stay home.
James did stay home; therefore we conclude that it did rain.

The fact that James stayed home does not mean that it rained. Perhaps he was ill or had homework to do. He could have stayed home for any of several reasons.

Consequently one needs to use caution when using either kind of reasoning. Make sure that enough information and true data are given to draw the correct conclusion.

MODEL PROBLEM SOLVING

Examples — **Strategy**

a. Predict what numbers seem most likely to appear next in the following sequence.

2, 4, 6, 6, 4, 2, 2, 4, 6, ___, ___, ___,

Note that the digits 2, 4, and 6 are used. Their order of appearance keeps reversing each time they appear. Use this pattern to make the prediction.

The next three numbers in the sequence will probably be 6, 4, 2.

Warm Up a. Predict the next three numbers in the following sequence.

1, 3, 5, 7, 5, 3, 1, 1, 3, 5, 7, ___, ___, ___

b. Predict what figure seems most likely to appear next in the following sequence.

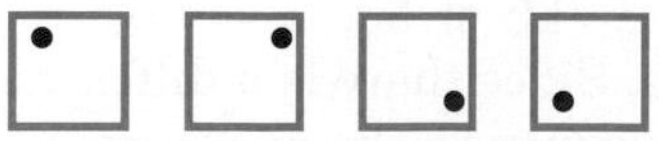

The dot appears to be moving in a clockwise direction.

Because of the pattern, we conclude that the next figure is

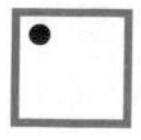

ANSWERS TO WARM UPS (7.9)

a. 5, 3, 1

Warm Up b. Predict what figure seems most likely to appear next in the following sequence.

c. Based on the first statement (assumption), use deductive reasoning to conclude whether the second statement is true or false.

(1) If Jill sleeps too late, she will be late for class.
(2) Jill slept too late; therefore Jill was late for class.

The statement is true since the given statement indicated that if she did sleep too late then she would be late for class.

Warm Up c. Based on the first statement (assumption), use deductive reasoning to conclude whether the second statement is true or false.

(1) If the sun is shining, John will not wear a jacket.
(2) The sun is shining; therefore, John does not wear a jacket.

d. Based on the first statement (assumption), use deductive reasoning to conclude whether the second statement is true or false.

(1) If Jill sleeps too late, she will be late for class.
(2) Jill is late for class; therefore, she slept too late.

False. Since there could be other reasons that she was late for class. For example, her auto may have run out of gasoline, or she may have met a friend and talked too long. There are any number of reasons that could have caused her to be late.

Warm Up d. Based on the first statement (assumption), use deductive reasoning to conclude whether the second statement is true or false.

(1) If the sun is shining, John will not wear a jacket.
(2) John did not wear a jacket; therefore, the sun is shining.

ANSWERS TO WARM UPS (7.9) **b.** **c.** True **d.** False

e. Predict what number seems most likely to appear next in the following sequence.

2, 4, 12, 24, 72, __

Observe: $2 \cdot 2 = 4$, $4 \cdot 3 = 12$, $12 \cdot 2 = 24$ and $24 \cdot 3 = 72$. We use this pattern to make the prediction. Other patterns are possible.

The pattern indicates that the next term probably will be $72 \cdot 2$, or 144; therefore, we say the next number will probably be 144.

Warm Up e. Predict what number seems most likely to appear next in the following sequence.

3, 9, 18, 54, 108, __

f. Predict what number seems most likely to appear next in the following sequence.

5, 11, 23, 47, 95, __

Observe: $5 \cdot 2 + 1 = 11$, $11 \cdot 2 + 1 = 23$, $23 \cdot 2 + 1 = 47$, and $47 \cdot 2 + 1 = 95$

We conclude that the next number in the sequence probably will be

$95 \cdot 2 + 1 = 191$

Warm Up f. Predict what number seems most likely to appear next in the following sequence.

4, 10, 22, 46, 94, __

ANSWERS TO WARM UPS (7.9)

e. 324 **f.** 190

NAME ______________________________

CLASS ______________ SECTION ______________ DATE ______________

Exercises 7.9

A

ANSWERS

1. ______________
2. ______________
3. ______________
4. ______________
5. ______________
6. ______________
7. ______________
8. ______________
9. ______________
10. ______________
11. ______________
12. ______________

Predict what numbers or figures seem most likely to appear next in the following sequences.

1. 1, 4, 9, 16, 25, __

2. 6, 10, 14, 18, 22, __

3. 5, 9, 13, 17, 21, __

4. 1, 2, 3, 6, 5, 4, 7, 8, 9, 12, 11, 10, __

5. __ __ __

6. __ __

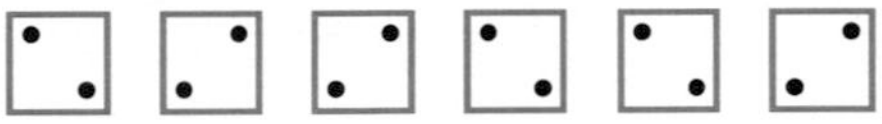

7. 0, 7, 14, 11, 18, 25, 22, 29, __, __

8. 1, 11, 20, 28, 35, __

9.

10.

11.

12.

B

ANSWERS

13. ______
14. ______
15. ______
16. ______
17. ______
18. ______
19. ______
20. ______
21. ______
22. ______
23. ______
24. ______
25. ______
26. ______
27. ______
28. ______
29. ______
30. ______

State whether or not the following arguments are valid. Answer true or false.

13. If a triangle has three equal angles, then it is equiangular. This triangle has three equal angles; therefore it is equiangular.

14. If a triangle has three equal sides, then it is equilateral. This triangle has three equal sides; therefore it is equilateral.

15. If Jill sleeps too late, she will be late for class. Jill is not late for class; therefore she did not sleep too late.

16. If the sun is shining, then John will not wear a jacket. John did wear a jacket; therefore the sun is not shining.

17. All high school students take math. Mary takes math; therefore Mary is a high school student.

18. All entering freshmen must take a placement test. Greg took a placement test; therefore Greg is a freshman.

19. All rectangles are parallelograms. This figure is not a parallelogram; therefore it is not a rectangle.

20. All regular polygons have equal sides. This polygon does not have equal-sides; therefore it is not a regular polygon.

Predict what numbers seem most likely to appear next in the following sequences.

21. 0, 5, 12, 21, 32, __

22. 4, 7, 13, 25, 49, __

23. 6, 8, 12, 18, __

24. 3, 8, 18, 33, 53, __

25. 78, 63, 48, 33, __

26. 20, 10, 5, 2.5, 1.25, __

27. 1, 3, 7, 15, 31, __

28. 1, 5, 15, 75, 225, __

29. 1, 2, 3, 5, 8, 13, __

30. 3, 4, 7, 16, 43, __

NAME ____________________

CLASS ____________ SECTION ____________ DATE ____________

C

ANSWERS

31. ____________

32. ____________

33. ____________

34. ____________

35. ____________

36. ____________

37. ____________

38. ____________

State whether the following are true or false.

31. All fishermen using worms for bait caught fish. Joe caught a fish; therefore Joe used worms for bait.

32. All mountain climbers are physically fit. Mary is a mountain climber; therefore Mary is physically fit.

33. All asparagus is grown in Washington. Marge loves asparagus; therefore Marge lives in Washington.

34. All Olympic skiers train on Mt. Hood. John trains on Mt. Hood; therefore John is an Olympic skier.

35. All corn grown in Kansas is at least six feet tall. Thelma's corn grows to a height of six feet; therefore Thelma's corn was grown in Kansas.

36. All zinnias in Peter's garden are red. Lucy's zinnias came from Peter's garden; therefore Lucy's zinnias are red.

37. All students must take writing to graduate. Jenny did graduate; therefore Jenny took writing.

38. All students must take writing to graduate. Bill did not graduate; therefore Bill did not take writing.

D

39. ____________

40. ____________

41. ____________

Decide whether the reasoning in the following is a result of inductive or deductive reasoning.

39. All three sided polygons are triangles. Polygon ABC has three sides; therefore polygon ABC is a triangle.

40. Every circle is bisected by its diameter. A clock is circular in shape; therefore the clock is bisected by its diameter.

41. A student observes that all prime numbers that she has seen end in 1, 3, 5, 7, or 9 (i.e., 1, 3, 5, 7, or 9 is the unit's digit). She suspects that the number 1003 is a prime number.

ANSWERS

42. ______________

43. ______________

44. ______________

45. ______________

46. ______________

42. A sailor observes the number of times that bad weather follows when the sky is very red at sunrise. He concludes, "Red sky in the morning, sailor take warning."

43. After measuring several swinging weights, Galileo concluded that the time of the swing of a pendulum is directly related to the square root of the length of the pendulum.

44. After washing his car on eight different occasions, Jorge observed that it rained the next day. Jorge decides that it will rain the next time he washes his car.

45. On various occasions, Kendall has burned his dinner while studying. He notices that chicken, carrots, beans, and other foods turn black if they are heated for a long time. He concludes that anything heated will eventually turn black.

46. Elan worked these exercises on her calculator.

$$\begin{aligned} 0(9) + 1 &= 1 \\ 1(9) + 2 &= 11 \\ 12(9) + 3 &= 111 \\ 123(9) + 4 &= 1111 \\ 1234(9) + 5 &= 11111 \end{aligned}$$

She concluded, without using her calculator, that

$$12345(9) + 6 = 111111$$

E *Maintain your skills* (Sections 7.4, 7.6)

47. ______________

48. ______________

49. ______________

50. ______________

51. ______________

52. ______________

47. Find the area of a parallelogram whose base is 42 inches and height is 2 feet.

48. Find the area of a triangle if the base is 9 inches and the height is 1 foot 4 inches.

49. Find the area of a circle with diameter 1 foot. Let $\pi \approx 3.14$.

50. Convert 30 miles per hour to kilometers per hour. Round to the nearest tenth.

51. $\dfrac{40 \text{ lb}}{1 \text{ in.}^2} = \dfrac{? \text{ kg}}{1 \text{ cm}^2}$ Round to the nearest tenth.

52. $\dfrac{75 \text{ km}}{1 \text{ hr}} = \dfrac{? \text{ mi}}{1 \text{ hr}}$ Round to the nearest tenth.

NAME ______________________________

CLASS ______________ SECTION ______________ DATE ______________

ANSWERS

53. ______________

54. ______________

55. ______________

56. ______________

53. $\frac{60 \text{ ft}}{1 \text{ sec}} = \frac{? \text{ m}}{1 \text{ sec}}$ Round to the nearest tenth.

54. An automobile race is to be 500 miles. How many kilometers is this, to the nearest kilometer?

55. Find the sum of 18 feet and 37 meters in feet, to the nearest tenth of a foot.

56. Which is the longer distance, 84 miles or 120 km?

NAME ______________________

CLASS ____________ SECTION ____________ DATE ____________

ANSWERS

1. ____
2. ____
3. ____
4. ____
5. ____
6. ____
7. ____
8. ____
9. ____
10. ____
11. ____
12. ____
13. ____
14. ____
15. ____
16. ____
17. ____
18. ____
19. ____
20. ____

CHAPTER 7 TRUE-FALSE CONCEPT REVIEW

Check your understanding of the language of arithmetic. Tell whether each of the following statements is True (always true) or False (not always true).

1. Equivalent measures are those taken by different people.

2. One mile is equivalent to 5280 feet.

3. The metric system utilizes the base ten place value system.

4. 1 kiloliter = 100 milliliters.

5. English to metric system conversions are exact decimals.

6. The distance around a geometric figure is called the perimeter.

7. The circumference of a circle is 2.14 times the length of the diameter.

8. The perimeter of a circle is called the circumference.

9. A liter is a measure of volume.

10. The prefix "milli" means 0.001.

11. Volume can be thought of as the amount of liquid an object can hold.

12. An inch is smaller than a centimeter.

13. The area of a parallelogram is found by multiplying the lengths of two adjacent sides.

14. The height of a triangle is never the length of one of the sides of the triangle.

15. A quart is larger than a liter.

16. Volume is measured in cubic units.

17. The area of a geometric figure can be thought of as the number of squares, or parts of squares one unit on a side, needed to cover the inside of the geometric figure.

18. The measurement 25 feet is equivalent to 8 yards 1 foot.

19. The area of a trapezoid is given by $A = \pi r^2$.

20. The radius of a circle is one-half of the circumference.

NAME ______________________________

CLASS ______________ SECTION ______________ DATE ______________

ANSWERS

1. ______________

2. ______________

3. ______________

4. ______________

5. ______________

6. ______________

7. ______________

8. ______________

CHAPTER 7 POST-TEST

1. **(Section 7.2, Obj. 92)** 7.85 m = __?__ mm

2. **(Section 7.5, Obj. 96)** Find the perimeter of a square that is 3.4 cm on a side.

3. If a certain stock dropped $\frac{3}{8}$ of a point each day for 5 consecutive days, what is the total drop for the 5 days?

4. **(Section 7.7, Obj. 98)** Find the area of the following geometric figure. (Let $\pi \approx 3.14$.)

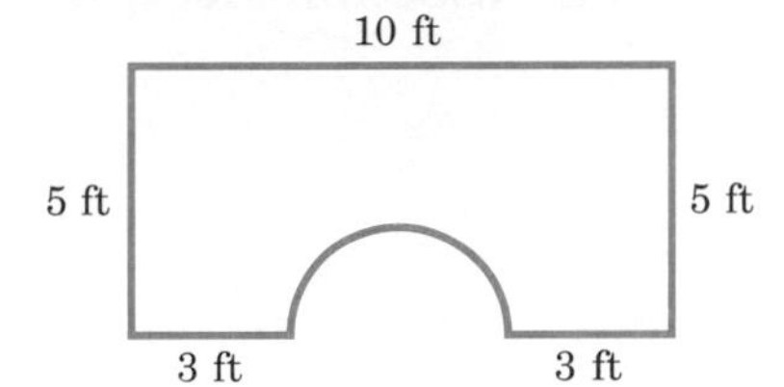

5. Mary has 124 lb of strawberries to divide equally among her 5 children. How many pounds of berries will each one receive?

6. The Golden Silver Company has a bar of silver weighing 687 g. If it is melted down to form six bars of equal weight, what will each bar weigh?

7. **(Section 7.5, Obj. 96)** Find the perimeter of the following geometric figure. (Let $\pi \approx 3.14$.)

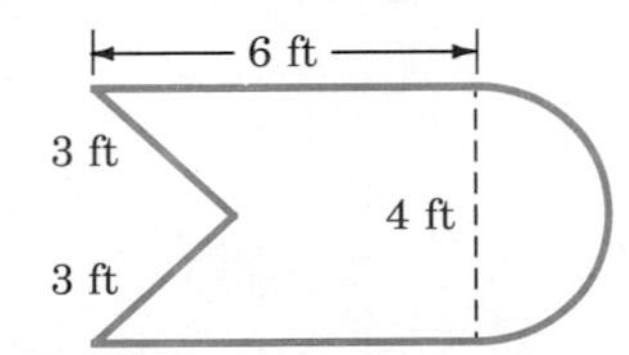

8. **(Section 7.6, Obj. 97)** Find the area of this triangle:

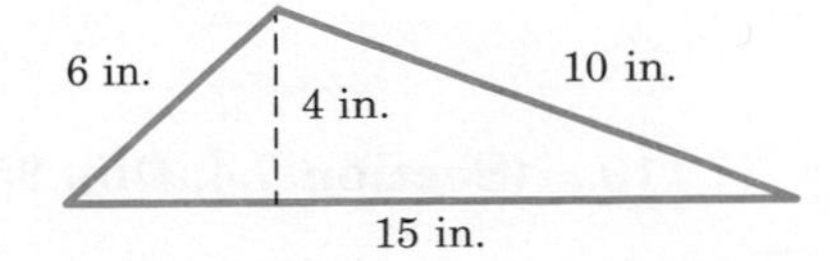

ANSWERS

9. ____________

10. ____________

11. ____________

12. ____________

13. ____________

14. ____________

15. ____________

16. ____________

17. ____________

18. ____________

19. ____________

9. (Section 7.1, Obj. 90) Subtract: 5 gal 2 qt
3 gal 2 qt 1 pt

10. (Section 7.3, Obj. 94) Convert to gallons: 20.3 pints

11. (Section 7.1, Obj. 89) Change to inches: $12\frac{2}{3}$ feet

12. If a burning oil well consumes 30 barrels of oil per hour, how many gallons per minute does it consume? (Let 1 barrel = 42 gallons.)

13. Dr. Green orders an average of 2.6 liters of blood to perform a certain type of surgery. How many liters will he order to perform 3 such operations?

14. (Section 7.6, Obj. 97) Find the area of this parallelogram:

8 cm
7 cm
6 cm

15. (Section 7.8, Obj. 99) Find the volume (ft^3) of a hot-water tank that has a circular base of radius 1 ft and a height of 54 inches.

16. (Section 7.2, Obj. 93) Add: 2 cm 3 mm
4 cm 9 mm

17. (Section 7.9, Obj. 100) Predict what number seems most likely to appear next in the following sequence.

12, 18, 24, 29, 34, 40, 46, 51, ___

18. (Section 7.9, Obj. 100) State whether or not the following argument is valid. Answer true or false.

All houses have windows. This building has no windows, therefore it is not a house.

19. (Section 7.4, Obj. 95) Convert: 7 pounds = ? kilograms (Round off to the nearest tenth.)

CHAPTER 8

Algebra Preview: Signed Numbers

NAME ____________________

CLASS ____________ SECTION ____________ DATE ____________

ANSWERS

1. a. ____________

b. ____________

2. ____________

3. ____________

4. ____________

5. ____________

6. ____________

7. ____________

8. ____________

9. ____________

10. ____________

11. ____________

12. ____________

13. ____________

14. ____________

15. ____________

16. ____________

17. ____________

18. ____________

19. ____________

20. ____________

21. ____________

22. ____________

23. ____________

24. ____________

25. ____________

26. ____________

CHAPTER 8 PRE-TEST

The problems in the following pre-test are a sample of the material in the chapter. You may already know how to work some of these. If so, this will allow you to spend less time on those parts. As a result, you will have more time to give to the sections that gave you difficulty. Please work the following problems. The answers are in the back of the text.

Perform the indicated operations:

1. **(Section 8.1, Obj. 101, 102) a.** $-(-17)$
   b. $|-17|$
2. **(Section 8.2, Obj. 103)** $(-8) + (7) + (-23) + (-19)$
3. **(Section 8.2, Obj. 103)** $(7.45) + (-3.12)$
4. **(Section 8.2, Obj. 103)** $\left(-\frac{1}{2}\right) + \left(\frac{3}{4}\right) + \left(-\frac{3}{8}\right)$
5. **(Section 8.2, Obj. 103)** $(-35) + (-16)$
6. **(Section 8.3, Obj. 104)** $(-21) - (-15)$
7. **(Section 8.3, Obj. 104)** $(36) - (-11)$
8. **(Section 8.3, Obj. 104)** $(-3.4) - (2.6)$
9. **(Section 8.3, Obj. 104)** $\left(-\frac{5}{6}\right) - \left(-\frac{5}{12}\right)$
10. **(Section 8.4, Obj. 105)** $(-3)(-8)(2)$
11. **(Section 8.4, Obj. 105)** $(-11)(-2)(-1)(-1)$
12. **(Section 8.4, Obj. 105)** $\left(-\frac{3}{8}\right)\left(\frac{4}{5}\right)$
13. **(Section 8.4, Obj. 106)** $(-18) \div (-6)$
14. **(Section 8.5, Obj. 106)** $(-48) \div (12)$
15. **(Section 8.5, Obj. 106)** $\left(\frac{7}{15}\right) \div \left(-\frac{3}{5}\right)$
16. **(Section 8.5. Obj. 106)** $(-3.18) \div (-3)$
17. **(Section 8.6, Obj. 107)** $(8 - 10)(-3 + 5)$
18. **(Section 8.6, Obj. 107)** $(6 - 12) + (-2 + 7) + (5)$
19. **(Section 8.6, Obj. 107)** $(8 - 20) \div (-4) + (-8)(3)$
20. **(Section 8.6, Obj. 107)** $(-17)(-5) - (-3)(12)$
21. **(Section 8.6, Obj. 107)** $(5)(8 - 4)(-3) - (3)(-2 + 4) + (20) \div (5)$
22. **(Section 8.6, Obj. 107)** $(-6)(5) - (12)$
23. **(Section 8.7, Obj. 108)** Solve: $-5x - 12 = 3$
24. The temperature outside the Terminal Ice building was 82°F. Greg went into the building to work where the temperature was −11°F. What change in temperature did he experience? (Express as a signed number.)
25. The fullback for the Allstar Team recorded the following yardage on five carries: −4 yd, 7 yd, −2 yd, 11 yd, and −1 yd. What is his net gain?
26. What Celsius temperature is equal to a reading of 5°F? Use the formula:

$$C = \frac{5}{9}(F - 32)$$

Signed Numbers, Opposites, and Absolute Value

OBJECTIVES

101. Find the opposite of a signed number.
102. Find the absolute value of a signed number.

APPLICATION

In business a loss can be represented by a negative number and a gain by a positive number. Lucky Buck, the gambler, ended up the day's gambling in Reno with a loss of \$50.

a. Write the loss as a signed number.
b. Write the opposite of the loss (a gain of \$50) as a signed number.

VOCABULARY

Positive numbers are numbers greater than zero. Negative numbers are numbers less than zero. Positive numbers and negative numbers are called *signed numbers.*

The *opposite* of a signed number is that number on the number line which is the same distance from zero but on the opposite side of zero. Zero is its own opposite. The opposite of 5 is written -5. This can be read "the opposite of 5" or "negative 5," since they both name the same number.

The *absolute value* of a signed number is the number of units between the number and zero. $|7|$ is read "the absolute value of 7."

HOW AND WHY

Problems such as:

$3 - 4$ $\quad 8 - 22$ $\quad 16 - 17$ $\quad$ and $\quad 100 - 561$

do not have answers in the numbers of arithmetic.

Signed numbers (which include numbers not in the numbers of arithmetic) are used to represent quantities with opposite characters. For instance:

right and left

up and down

above zero and below zero

gain and loss

can be represented as opposites.

A few signed numbers are shown on the following number line:

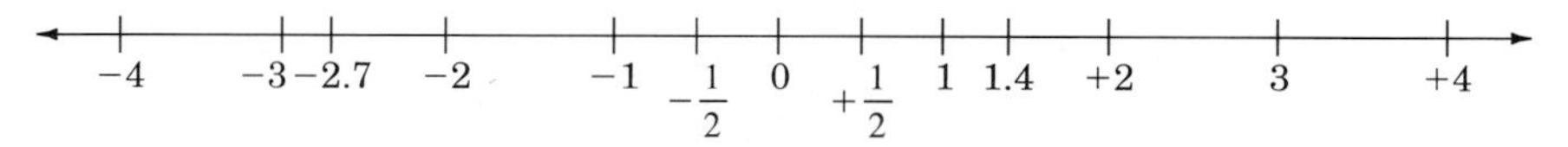

Numbers that are to the right of zero are those that are used in arithmetic. The negative numbers are to the left of zero and have a dash, or minus sign, in front of them. The numbers to the right of zero are called positive (and can be written with a plus sign). Zero is neither positive nor negative.

7	Seven or positive seven
$+\frac{1}{2}$	One-half or positive one-half
-3	Negative three
-0.12	Negative twelve hundredths
0	Zero is neither positive nor negative.

Positive and negative numbers are used many ways in the physical world, as shown below:

POSITIVE	NEGATIVE
Temperatures above zero (72°)	Temperatures below zero (−10°)
Feet above sea level (5000 ft)	Feet below sea level (−50 ft)
Profit ($75)	Loss (−$23)
Right (7)	Left (−4)

In any use where quantities can be measured in opposite directions, positive and negative numbers can be used to show direction.

The dash in front of a number is read two different ways:

-19 The opposite of 19

-19 Negative 19

RULE

The opposite of a positive number is negative.
The opposite of a negative number is positive.

For instance:

$-(5) = -5$ The opposite of positive 5 is negative 5.

$-(-3) = 3$ The opposite of negative 3 is positive 3.

$-(0) = 0$ The opposite of 0 is 0.

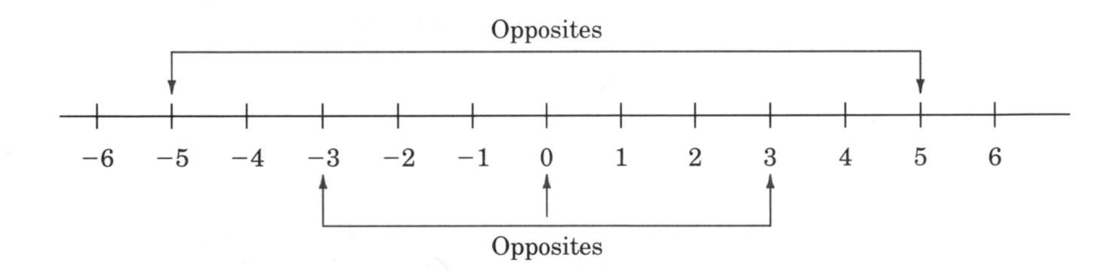

The expression -5 is sometimes referred to as the "negative of" 5, and 5 is sometimes referred to as the "negative of" -5.

The absolute value of a signed number is the distance between the number and zero (on the number line) and is never negative.

RULE

The absolute value of a positive number is the number itself.
The absolute value of zero is zero.
The absolute value of a negative number is its opposite.

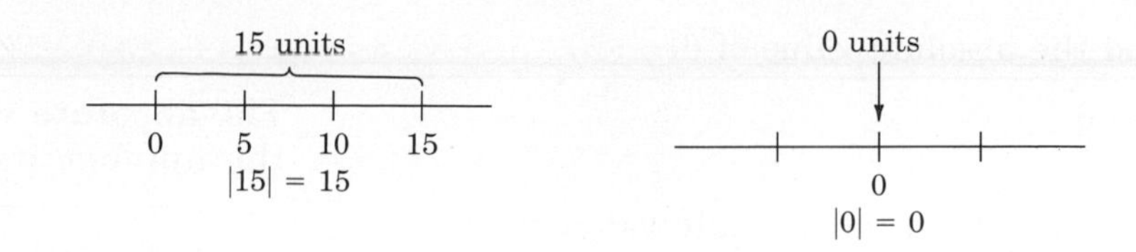

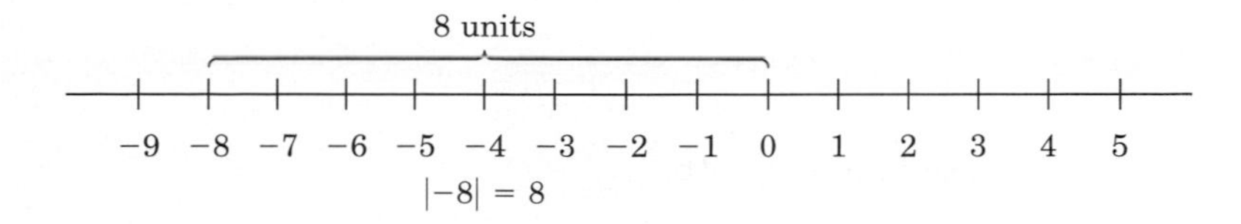

MODEL PROBLEM-SOLVING

Examples | **Strategy**

a. Find the opposite of 8:

-8

The opposite of 8 is negative 8. Since 8 is 8 units to the right of zero, the opposite of 8 is 8 units to the left of zero.

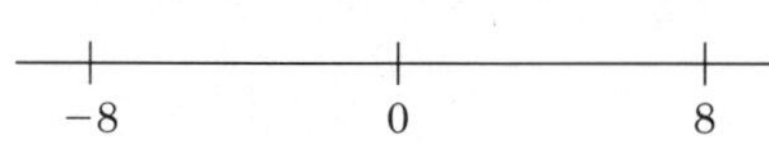

Warm Up a. Find the opposite of 47.

b. Find the opposite of -9:

$-(-9) = 9$

The opposite of negative 9 is written $-(-9)$ and is found 9 units on the opposite side of zero from -9.

Warm Up b. Find the opposite of -64.

c. Write the word name for -12:

"negative twelve"
"the opposite of twelve"

The number -12 can be thought of in two ways: as the negative number that is 12 units to the left of zero, or as the number that is 12 units on the opposite side of zero from 12.

Warm Up c. Write the word name for $-(-17)$.

ANSWERS TO WARM UPS (8.1) **a.** -47 **b.** 64 **c.** The opposite of negative seventeen

d. Find the absolute value of 6:

$|6| = 6$ **The absolute value of a positive number is the number itself.**

Warm Up d. Find the absolute value of 110.

e. Find the absolute value of -16:

$|-16| = -(-16) = 16$ **The absolute value of a negative number is its opposite.**

Warm Up e. Find the absolute value of -72.

f. Find the absolute value of 0:

$|0| = 0$ **The absolute value of zero is zero.**

Warm Up f. Find the absolute value of -1.

APPLICATION SOLUTION

g. Lucky Buck, the gambler, ended up the day's gambling in Reno with a loss of $50.

a. Write the loss as a signed number.
b. Write the opposite of the loss (a gain of $50) as a signed number.

a. $-\$50$ **Losses are usually represented by negative numbers.**

b. $-(-\$50) = \$50$ **The opposite of a loss is a gain.**

Warm Up g. The stock market dropped 13.5 points on a down day. Write the loss as a signed number.

ANSWERS TO WARM UPS (8.1) **d.** 110 **e.** 72 **f.** 1 **g.** -13.5

NAME ______________________

CLASS ____________ SECTION ____________ DATE ____________

Exercises 8.1

A

ANSWERS

1. ________
2. ________
3. ________
4. ________
5. ________
6. ________
7. ________
8. ________
9. ________
10. ________
11. ________
12. ________
13. ________
14. ________
15. ________
16. ________
17. ________
18. ________
19. ________
20. ________
21. ________
22. ________

Find the opposite of each of the following numbers:

1. -3 **2.** 5 **3.** 17 **4.** -23

5. $\frac{1}{2}$ **6.** $-\frac{3}{4}$ **7.** 4.7 **8.** -2.1

9. $-\frac{1}{3}$ **10.** -9 **11.** -8.08 **12.** 0.003

Find the indicated absolute value for each of the following:

13. $|-1|$ **14.** $|-5|$ **15.** $|7|$ **16.** $\left|-\frac{2}{3}\right|$

17. $\left|\frac{1}{6}\right|$ **18.** $|4|$ **19.** $|-1.2|$ **20.** $|-7.3|$

21. $|-3.11|$ **22.** $|0.061|$

B

ANSWERS

23. ______
24. ______
25. ______
26. ______
27. ______
28. ______
29. ______
30. ______
31. ______
32. ______
33. ______
34. ______
35. ______
36. ______
37. ______
38. ______
39. ______
40. ______
41. ______
42. ______

Find the opposite of each of the following numbers:

23. -31 **24.** 13 **25.** $-\frac{2}{3}$ **26.** -2.35

27. $3\frac{1}{8}$ **28.** $-22\frac{2}{3}$ **29.** 0.23 **30.** -103.6

31. $-4\frac{15}{16}$ **32.** -0.625 **33.** -40.7 **34.** -65.9

35. 103.7 **36.** -304.5 **37.** -14.7 **38.** -28.25

39. $13\frac{7}{8}$ **40.** $\frac{15}{4}$ **41.** -8.135 **42.** -6.234

C

43. ______
44. ______
45. ______
46. ______

Find the indicated absolute value for each of the following:

43. $|-6|$ **44.** $|0.035|$ **45.** $\left|-\frac{4}{5}\right|$ **46.** $|81|$

NAME ______________________

CLASS ______________ SECTION ______________ DATE ______________

ANSWERS

47. ______________
48. ______________
49. ______________
50. ______________
51. ______________
52. ______________
53. ______________
54. ______________
55. ______________
56. ______________
57. ______________
58. ______________
59. ______________
60. ______________
61. ______________
62. ______________

47. $|-0.71|$ **48.** $\left|-\frac{25}{8}\right|$ **49.** $|21.75|$ **50.** $|2.03|$

51. $|6 - 1.5|$ **52.** $|0|$ **53.** $|2 + 5|$ **54.** $|-0.003|$

55. $|-253|$ **56.** $|-357|$ **57.** $|0.0035|$ **58.** $|-0.0079|$

59. opposite of $\left|-\frac{2}{7}\right|$ **60.** opposite of $|7.5|$

61. opposite of $|42|$ **62.** opposite of $|-6|$

D

63. ______________

63. At the New York Stock Exchange positive and negative numbers are used to record changes in stock prices on the board. What is the opposite of a gain of three-eighths $\left(+\frac{3}{8}\right)$?

64. ______________

64. At the American Stock Exchange a stock is shown to have taken a loss of five-eighths point $\left(-\frac{5}{8}\right)$. What is the opposite of this loss?

ANSWERS

65. ______

66. ______

67. ______

68. ______

69. ______

70. ______

71. ______

65. On a thermometer temperatures above zero are listed as positive and those below zero as negative. What is the opposite of a reading of $-12°C$?

66. On a thermometer such as that in Exercise 65, what is the opposite of a reading of 23°C?

67. The modern calendar counts the years after the birth of Christ as positive numbers (A.D. 1976 or $+1976$). Years before Christ are listed using negative numbers (2045 B.C. or -2045). What is the opposite of 1875 B.C. or -1875?

68. The empty-weight center of gravity of an airplane is determined. A generator is installed at a moment of -300. At what moment could a weight be placed so that the center of gravity remains the same? (Moment is the product of a quantity, such as weight, and its distance from a fixed point. In this application the moments must be opposites to keep the same center of gravity.)

69. A cyclist travels up the mountain 1275 feet and down the mountain 1173 feet. Represent each trip as a signed number.

70. How far is the cyclist from her starting point? (See Exercise 69.) Write the distance as a signed number.

71. If 80 miles north is represented by $+80$, how would you represent 80 miles in the opposite direction?

NAME ______________________

CLASS ______________ SECTION ______________ DATE ______________

E *Maintain your skills* (Section 7.2)

ANSWERS

72. ______________

73. ______________

74. ______________

75. ______________

76. ______________

77. ______________

78. ______________

79. ______________

80. ______________

81. ______________

Multiply:

72. (48 g)25 = ? g

73. (2.65 kℓ)5.1 = ? kℓ

Divide:

74. (48 kg) ÷ 15 = ? kg

75. (78.4 cm) ÷ 5.6 = ? cm

Change units as shown:

76. 4.75 km = ? meters

77. 0.51 kg = ? cg

Add:

78. 5 m 250 cm
7 m 460 cm
= ? m

79. 7 ℓ 88 mℓ
2 ℓ 95 mℓ
= ? mℓ

80. A 10-meter coil of wire is to be divided into eight equal parts. How many meters are in each part?

81. Some drug dosages are measured in grains. For example, a common aspirin tablet contains 5 grains. If there are 15.43 grams in a grain, how many grams do two aspirin tablets contain?

8.2 Addition of Signed Numbers

OBJECTIVE **103. Add signed numbers.**

APPLICATION

John owns some stock that is traded on the American Stock Exchange. On Monday the stock gained \$3, on Tuesday it lost \$4, on Wednesday it lost \$2, on Thursday it gained \$5, and on Friday it gained \$3. What was the net change in the price of the stock for the week?

VOCABULARY No new words.

HOW AND WHY

Positive and negative numbers are used to show opposite quantities:

+482 lb may show 482 pounds loaded.

−577 lb may show 577 pounds unloaded.

+27 dollars may show 27 dollars earned.

−19 dollars may show 19 dollars spent.

Using this idea we find the sum of signed numbers:

$(27) + (-19) = ?$

Think of this as 27 dollars earned (positive) and 19 dollars spent (negative). The result is 8 dollars left in your pocket (positive). So, $(27) + (-19) = 8$.

$(-23) + (15) = ?$

Think of this as 23 dollars spent (negative) and 15 dollars earned (positive). The result is that you still owe 8 dollars (negative). So, $(-23) + (15) = -8$.

$(-5) + (-2) = ?$

Think of this as 5 dollars spent (negative) and 2 dollars spent (negative). The result is 7 dollars spent (negative). So, $(-5) + (-2) = -7$. Shortcut: $-(5 + 2) = -7$.

RULE

The sum of two positive numbers is positive.

The sum of two negative numbers is negative.

The sum of a positive and a negative number is found by subtracting and

a. is positive if the larger number (in absolute value) is positive;
b. is negative if the larger number (in absolute value) is negative;
c. is zero if the numbers are opposites.

MODEL PROBLEM-SOLVING

Examples

a. Add: $(37) + (-15)$

$|37| - |-15|$ or $(37) - (15)$
$(37) + (-15) = 22$

Strategy

Since the numbers are opposite in sign, subtract their absolute values. The result is positive, since 37 has the larger absolute value.

Warm Up a. Add: $(-57) + (38)$

b. Add: $(-48) + (29)$

$|-48| - |29|$ or $48 - 29$
$(-48) + (29) = -19$

Subtract. The result is negative because the number with the larger absolute value is negative.

Warm Up b. Add: $(77) + (-23)$

c. Add: $(-0.33) + (-1.7)$

$(-0.33) + (-1.7) = -2.03$

The signs are the same, so add the absolute values and keep the common sign.

Warm Up c. Add: $(-56) + (-43) + (-12)$

d. Add: $(-8.3) + (25)$

$(-8.3) + (25) = 16.7$

Warm Up d. Add: $(-19.4) + (13.7)$

ANSWERS TO WARM UPS (8.2) **a.** -19 **b.** 54 **c.** -111 **d.** -5.7

e. Add: $\left(\frac{4}{5}\right) + \left(-\frac{3}{2}\right)$

$$\left(\frac{4}{5}\right) + \left(-\frac{3}{2}\right) = -\frac{7}{10}$$

Warm Up e. Add: $\left(\frac{7}{8}\right) + \left(-\frac{5}{6}\right)$

f. Add:

$(-13) + (72) + (-20) + (-23)$

$(-13) + (72) + (-20) + (-23) = 16$

Hint: **It is often easier to add the numbers with the same sign first:**

−13	
−20	**−56**
−23	**+72**
−56	**+16**

Warm Up f. Add: $(-34) + (76) + (-124) + (46)$

g. Add: $(-0.1) + (0.5) + (-3.4) + (0.8)$

$(-0.1) + (0.5) + (-3.4) + (0.8) = -2.2$

Warm Up g. Add: $(0.32) + (-0.54) + (-0.73) + (1.2)$

h. Add: $\left(-\frac{3}{4}\right) + \left(\frac{7}{8}\right) + \left(-\frac{1}{2}\right) + \left(-\frac{7}{8}\right)$

$$\left(-\frac{3}{4}\right) + \left(\frac{7}{8}\right) + \left(-\frac{1}{2}\right) + \left(-\frac{7}{8}\right) = -\frac{5}{4}$$

Warm Up h. Add: $\left(-\frac{5}{7}\right) + \left(-\frac{3}{2}\right) + \left(\frac{8}{5}\right) + \left(-\frac{6}{7}\right)$

ANSWERS TO WARM UPS (8.1)

e. $\frac{1}{24}$ **f.** −36 **g.** 0.25 **h.** $-\frac{103}{70}$

i. Calculator example

Add: $(-63) + (48) + (-61) + (-14)$

ENTER	DISPLAY	
63	63.	If your calculator has a +/− key, you can use it to add positive and negative numbers. Pressing this key displays the opposite of the entry.
+/−	−63.	
+	−63.	
48	48.	
+	−15.	This is the result of the first addition.
61	61.	
+/−	−61.	
+	−76.	This is the result of the second addition.
14	14.	
+/−	−14.	
=	−90.	This is the result of the final addition.

Warm Up i. Add: $(-345) + (637) + (-871) + (-45) + (129)$

APPLICATION SOLUTION

j. John owns some stock that is traded on the American Stock Exchange. On Monday the stock gained $3, on Tuesday it lost $4, on Wednesday it lost $2, on Thursday it gained $5, and on Friday it gained $3. What was the net change in the price of the stock for the week?

Monday	gain $3	3
Tuesday	lost $4	−4
Wednesday	lost $2	−2
Thursday	gain $5	5
Friday	gain $3	3

To find the net change in the price of the stock, write the daily changes as signed numbers and then find the sum of these numbers.

$3 + (-4) + (-2) + 5 + 3 = 5$

The stock gained $5 during the week.

Warm Up j. A second stock that John owns had the following changes for the week: Monday, gained $1.25; Tuesday, gained $0.875; Wednesday, lost $2.625; Thursday, lost $1.5; Friday, gained $1.375. What was the net change in the price of the stock for the week?

ANSWERS TO WARM UPS (8.1) i. −495 j. −0.625

NAME ______________________

CLASS ______________ SECTION ______________ DATE ______________

Exercises 8.2

A

ANSWERS

1. ______
2. ______
3. ______
4. ______
5. ______
6. ______
7. ______
8. ______
9. ______
10. ______
11. ______
12. ______
13. ______
14. ______
15. ______
16. ______
17. ______
18. ______
19. ______
20. ______
21. ______
22. ______

Add:

1. $(-16) + (12)$ **2.** $(-8) + (2)$ **3.** $(6) + (-7)$

4. $(-15) + 10$ **5.** $(-2) + (-11)$ **6.** $(-10) + (-8)$

7. $(-7) + (7)$ **8.** $(-6) + (6)$ **9.** $(0) + (-13)$

10. $(-22) + (0)$ **11.** $(-10) + (-3)$ **12.** $(-5) + (10)$

13. $(25) + (-17)$ **14.** $(7) + (-14)$ **15.** $(-8) + (3)$

16. $(11) + (-3)$ **17.** $(-15) + (-12)$ **18.** $(-17) + (9)$

19. $(-24) + (-56)$ **20.** $(-31) + (45)$ **21.** $(-3) + (-6) + (5)$

22. $(-2) + (-5) + (-6) + (10)$

B

ANSWERS

23. ______

24. ______

25. ______

26. ______

27. ______

28. ______

29. ______

30. ______

31. ______

32. ______

33. ______

34. ______

35. ______

36. ______

37. ______

38. ______

39. ______

40. ______

41. ______

42. ______

43. ______

44. ______

23. $(-72) + (-72)$ **24.** $(-34) + (34)$ **25.** $(-75) + (75)$

26. $(-26) + (26)$ **27.** $(-87) + (94)$ **28.** $(48) + (-39)$

29. $(-239) + (225)$ **30.** $(-432) + (439)$ **31.** $(-28) + (-47)$

32. $(-41) + (-144)$ **33.** $(-2.3) + (-4.3)$ **34.** $(-8.2) + (-3.2)$

35. $(6.3) + (-3.7)$ **36.** $(-1.4) + (4.1)$ **37.** $(-7.5) + (9.3)$

38. $(4.6) + (-2.8)$ **39.** $\left(\frac{5}{12}\right) + \left(-\frac{7}{12}\right)$ **40.** $\left(-\frac{4}{15}\right) + \left(-\frac{6}{15}\right)$

41. $\left(-\frac{2}{3}\right) + \left(\frac{1}{6}\right)$ **42.** $\left(\frac{5}{8}\right) + \left(-\frac{3}{4}\right)$

43. $\left(-\frac{5}{6}\right) + \left(\frac{2}{3}\right) + \left(-\frac{1}{9}\right)$ **44.** $\left(-\frac{3}{5}\right) + \left(\frac{7}{10}\right) + \left(-\frac{2}{15}\right)$

NAME ______________________________

CLASS ______________ SECTION ______________ DATE ______________

C

ANSWERS

45. ______
46. ______
47. ______
48. ______
49. ______
50. ______
51. ______
52. ______
53. ______
54. ______
55. ______
56. ______
57. ______
58. ______
59. ______
60. ______
61. ______
62. ______
63. ______
64. ______

45. $(-72) + (81) + (53)$

46. $(115) + (-102) + (17)$

47. $(-81) + (-32) + (-16)$

48. $(-75) + (72) + (-18)$

49. $(-31) + (18) + (-63) + (22)$

50. $(-52) + (59) + (-26) + (-44)$

51. $(-19) + (54) + (-68) + (6)$

52. $(32) + (11) + (-68) + (21)$

53. $(-17.3) + (14.6) + (6.9)$

54. $(-11.23) + (15.36) + (-27.22)$

55. $(-65.52) + (19.51) + (-87.72)$

56. $(16.49) + (-23.45) + (8.35)$

57. $\left(-\frac{1}{2}\right) + \left(-\frac{7}{8}\right) + \left(-\frac{5}{6}\right)$

58. $\left(1\frac{2}{3}\right) + \left(-4\frac{5}{6}\right) + \left(-2\frac{1}{4}\right)$

59. $\left(13\frac{2}{9}\right) + \left(-8\frac{2}{3}\right) + \left(-2\frac{2}{3}\right)$

60. $\left(-15\frac{1}{2}\right) + \left(24\frac{3}{4}\right) + \left(-18\frac{5}{8}\right)$

61. $(0.345) + (-1.203) + (-0.211)$

62. $(-0.035) + (0.751) + (-0.111)$

63. $(-13.56) + (-18.49) + (21.07)$

64. $(-39.5) + (12.81) + (-3.7)$

ANSWERS

65. ______

66. ______

67. ______

68. ______

69. ______

70. ______

65. An airplane is being reloaded; 577 pounds of baggage and mail are removed (−577 pounds) and 482 pounds of baggage and mail are loaded on (+482 pounds). What net change in weight should the cargo master report?

66. At another stop the plane in Exercise 65 unloads 1,234 pounds of baggage and mail and takes on 1,184 pounds. What net change should the cargo master report?

67. During the current fiscal year Le Baroque Coffee House recorded the following quarterly earnings (positive numbers represent profit, negative numbers represent loss): $3,456, −$507, −$498, $4,007. What was the total profit (or loss) for the year?

68. The Pacific Northwest Book Depository handles most textbooks for local schools. On September 1 the inventory was 18,340 volumes. During the month the company made the following transactions (positive numbers represent volumes received, negative numbers represent shipments): 1800, −356, −843, −500, 250, −650. What is the inventory at the end of the month?

69. The Pacific Northwest Book Depository had 8066 volumes on November 1. During the month the depository had the following transactions: −1,044, 213, −555, −178, −840, 88, −229. What was the inventory at the end of the month?

70. The change in altitude of a plane in flight was measured every 10 minutes. The figures between 3 P.M. and 4 P.M. were as follows:

3:00 P.M. 30,000 ft initially	(+30,000)
3:10 P.M. increase of 220 ft	(+220)
3:20 P.M. decrease of 200 ft	(−200)
3:30 P.M. increase of 55 ft	(+55)
3:40 P.M. decrease of 110 ft	(−110)
3:50 P.M. decrease of 25 ft	(−25)
4:00 P.M. increase of 40 ft	(+40)

What was the altitude of the plane at 4 P.M.? (*Hint:* Find the sum of the initial altitude and the six measured changes between 3 and 4 P.M.)

NAME ______

CLASS ______ SECTION ______ DATE ______

ANSWERS

71. ______

72. ______

73. ______

74. ______

71. Nordstrom stock had the following changes in one week: up $\frac{3}{8}$, down $\frac{1}{2}$, down $1\frac{1}{4}$, up $2\frac{3}{4}$, up $\frac{7}{8}$. What is the net change for the week?

72. If the Nordstrom stock (previous problem) started at $72\frac{5}{8}$ at the beginning of that week, what is the closing price?

73. The Buffalo Bills made the following plays during a recent Monday night football game: 9-yard gain, 8-yard loss, and a 6-yard gain. A first down requires a gain of 10 yards. Did they get a first down?

74. The Seattle Seahawks had these plays one Sunday: 18-yard loss, 25-yard gain, and a 4-yard gain. Did they get a first down?

E *Maintain your skills* (Sections 7.3, 7.5)

75. ______

76. ______

77. ______

78. ______

79. ______

Convert the units as shown:

75. $\frac{450 \text{ miles}}{1 \text{ hour}} = \frac{? \text{ feet}}{1 \text{ second}}$

76. $\frac{5 \text{ pounds}}{1 \text{ foot}} = \frac{? \text{ ounces}}{1 \text{ inch}}$

77. 7824 inches = ? feet

78. $\frac{\$8.40}{1 \text{ hour}} = \frac{? \text{ cents}}{1 \text{ minute}}$

79. Find the perimeter of a rectangle that is 37 meters long and 25 meters wide.

ANSWERS

80. ______

81. ______

82. ______

83. ______

84. ______

80. Find the circumference of a circle with a diameter of 48 cm. (Let $\pi \approx 3.14$.)

81. Find the perimeter of a square that is 16 in. on each side.

82. Find the perimeter of a triangle with sides of 28 cm, 42 cm, and 16 cm.

83. Bars of soap are sold at 3 bars for $1.86. What is the cost of one bar?

84. The cost of pouring a 3-foot-wide cement sidewalk is estimated to be $16 per square foot. If a walk is to be placed around a rectangular plot of ground that is 24 feet along the width and 48 feet along the length, what is the cost of pouring the walk?

8.3 Subtraction of Signed Numbers

OBJECTIVE

104. Subtract signed numbers.

APPLICATION

Mount Everest is 29,028 ft above sea level and Death Valley is 282 ft below sea level. What is the difference in height between Mount Everest and Death Valley? (Above sea level is positive and below sea level is negative.)

VOCABULARY

No new words.

HOW AND WHY

The expression $(11) - (8) = ?$ asks $(8) + (?) = 11$; we know $(8) + (3) = 11$, so $(11) - (8) = 3$.

The expression $(-3) - (5) = ?$ asks $(5) + (?) = -3$; we know $(5) + (-8) = -3$, so $(-3) - (5) = -8$.

The expression $(-4) - (-7) = ?$ asks $(-7) + (?) = -4$; we know $(-7) + (3) = -4$, so $(-4) - (-7) = 3$.

Compare:

$(11) - (8)$	and	$(11) + (-8)$	Both equal 3.
$(-3) - (5)$	and	$(-3) + (-5)$	Both equal -8.
$(-4) - (-7)$	and	$(-4) + (7)$	Both equal 3.

Every subtraction problem can be worked as an addition problem.

SUBTRACTION	THINK	WRITE
$(11) - (8)$	11 plus the opposite of 8	$(11) + (-8) = 3$
$(-8) - (-9)$	-8 plus the opposite of (-9)	$(-8) + (9) = 1$
$(-6) - (3)$	-6 plus the opposite of 3	$(-6) + (-3) = -9$

PROCEDURE

To subtract two signed numbers, find the sum of the first number and the opposite of the second number.

MODEL PROBLEM-SOLVING

Examples / **Strategy**

a. Subtract: $(45) - (33)$

$(45) - (33) = 12$ — **Since both numbers are positive, subtract.**

Warm Up a. Subtract: $(82) - (97)$

ANSWERS TO WARM UPS (8.3)

a. -15

b. Subtract: $(-47) - (29)$

$(-47) + (-29) = ?$
$(-47) + (-29) = -76$

Rewrite as an addition problem by adding the opposite of the second number, then add.

Warm Up b. Subtract: $(103) - (-43)$

c. Subtract: $(-38) - (-56)$

$(-38) + (56) = 18$

Add the opposite of −56.

Warm Up c. Subtract: $(-145) - (261)$

d. Subtract: $(88) - (114)$

$(88) + (-114) = -26$

Add −114.

Warm Up d. Subtract: (−34[illegible] − (−321)

e. Subtract: $\left(\frac{4}{5}\right) - \left(-\frac{1}{2}\right)$

$\left(\frac{4}{5}\right) + \left(\frac{1}{2}\right) = 1\frac{3}{10}$

Add $+\frac{1}{2}$.

Warm Up e. Subtract: $\left(-\frac{6}{7}\right) - \left(\frac{3}{4}\right)$

f. Subtract: $(16) - (-4) - (-7)$

$(16) + (4) + (7) = 27$

Change all subtractions to additions, then add.

Warm Up f. Subtract: $(-57) - (23) - (-83) - (36)$

ANSWERS TO WARM UPS (8.3)

b. 146 **c.** −406 **d.** 287

e. $-\frac{45}{28}$ **f.** −33

g. Subtract: $(-0.21) - (-3.4) - (-0.15) - (0.42)$

$(-0.21) + (3.4) + (0.15) + (-0.42) = 2.92$

Change all subtractions to additions, then add.

Warm Up g. Subtract: $(-0.76) - (0.67) - (-0.54) - (-1.32)$

h. Subtract: $\left(-\frac{3}{4}\right) - \left(\frac{7}{8}\right) - \left(-\frac{1}{2}\right) - \left(-\frac{1}{8}\right)$

$\left(-\frac{3}{4}\right) + \left(-\frac{7}{8}\right) + \left(\frac{1}{2}\right) + \left(\frac{1}{8}\right)$

Change all subtractions to additions.

$\left(-\frac{4}{8}\right) + \left(-\frac{7}{8}\right) + \left(\frac{4}{8}\right) + \left(\frac{1}{8}\right) = -\frac{8}{8} = -1$

Write each fraction with a common denominator and add.

Warm Up h. Subtract: $\left(-\frac{3}{5}\right) - \left(-\frac{3}{4}\right) - \left(-\frac{7}{10}\right) - \left(\frac{7}{20}\right)$

i. Calculator example

Subtract: $(-34.8) - (-49.3)$

ENTER	DISPLAY
34.8	34.8
+/−	−34.8
−	−34.8
49.3	49.3
+/−	−49.3
=	14.5

Recall that the +/− key displays the opposite of the entry.

Warm Up i. Subtract: $(-346.87) - (-245.76) - (407.43)$

ANSWERS TO WARM UPS (8.3)

g. 0.43 **h.** $\frac{1}{2}$ **i.** −508.54

APPLICATION SOLUTION

j. Mount Everest is 29,028 ft above sea level and Death Valley is 282 ft below sea level. What is the difference in height between Mount Everest and Death Valley? (Above sea level is positive and below sea level is negative.)

Mount Everest	29,028 above	29,028
Death Valley	282 below	−282

To find the difference in height, change each to a signed number and then subtract the lower height from the higher height.

$$29{,}028 - (-282) = 29{,}310$$

The difference in height is 29,310 ft.

Warm Up j. One night last winter the temperature dropped from 23°F to −8°F. What was the difference between the high and low temperatures?

ANSWERS TO WARM UPS (8.3)

j. 31°F

NAME ______________________

CLASS ______________ SECTION ______________ DATE ______________

Exercises 8.3

A

ANSWERS

1. ________
2. ________
3. ________
4. ________
5. ________
6. ________
7. ________
8. ________
9. ________
10. ________
11. ________
12. ________
13. ________
14. ________
15. ________
16. ________
17. ________
18. ________
19. ________
20. ________
21. ________
22. ________
23. ________
24. ________
25. ________
26. ________

Subtract:

1. $(6) - (4)$ **2.** $(-6) - (4)$ **3.** $(6) - (-4)$

4. $(-6) - (-4)$ **5.** $(-8) - (5)$ **6.** $(10) - (7)$

7. $(-15) - (-3)$ **8.** $(-15) - (3)$ **9.** $(11) - (-5)$

10. $(-9) - (4)$ **11.** $(-3) - (4)$ **12.** $(-9) - (-13)$

13. $(17) - (-15)$ **14.** $(-14) - (-23)$ **15.** $(-16) - (15)$

16. $(-18) - (-11)$ **17.** $(10) - (-3)$ **18.** $(12) - (-8)$

19. $(14) - (-22)$ **20.** $(8) - (-7)$ **21.** $(-5) - (-5)$

22. $(-5) - (5)$ **23.** $(-7) - (-10)$ **24.** $(-6) - (-15)$

25. $(-9.5) - (-2.3)$ **26.** $(-7.3) - (-10.7)$

B

ANSWERS

27. ______
28. ______
29. ______
30. ______
31. ______
32. ______
33. ______
34. ______
35. ______
36. ______
37. ______
38. ______
39. ______
40. ______
41. ______
42. ______
43. ______
44. ______
45. ______
46. ______
47. ______
48. ______

27. $\left(-\frac{1}{4}\right) - \left(\frac{2}{4}\right)$

28. $\left(-\frac{2}{3}\right) - \left(-\frac{1}{3}\right)$

29. $(16) - (-25)$

30. $(56) - (-15)$

31. $(-42) - (-33)$

32. $(-49) - (-43)$

33. $(-45) - (45)$

34. $(-83) - (83)$

35. $\left(-\frac{3}{8}\right) - \left(-\frac{4}{8}\right)$

36. $\left(-\frac{4}{10}\right) - \left(\frac{4}{10}\right)$

37. $(63) - (-17)$

38. $(-36) - (28)$

39. $(-17) - (-18)$

40. $(56) - (-14)$

41. $(98) - (-32)$

42. $(-73) - (-56)$

43. $(-45) - (72)$

44. $(92) - (-41)$

45. $(-23) - (-45)$

46. $(-72) - (-115)$

47. $(112) - (120)$

48. $(182) - (215)$

C

49. ______
50. ______

49. $(3.45) - (4.22)$

50. $(7.33) - (-2.48)$

NAME ____________________

CLASS ____________ SECTION ____________ DATE ____________

ANSWERS

51. ____________

52. ____________

53. ____________

54. ____________

55. ____________

56. ____________

57. ____________

58. ____________

59. ____________

60. ____________

61. ____________

62. ____________

63. ____________

64. ____________

65. ____________

66. ____________

67. ____________

68. ____________

51. $(-8.92) - (4.68)$

52. $(-43.8) - (-62.9)$

53. $(37) - (-12.3)$

54. $(-25) - (-18.7)$

55. $(-25) - (-32.8)$

56. $(17) - (-21.7)$

57. $\left(\frac{3}{4}\right) - \left(1\frac{1}{2}\right)$

58. $\left(-\frac{2}{3}\right) - \left(\frac{1}{2}\right)$

59. $\left(-\frac{5}{8}\right) - \left(-\frac{2}{5}\right)$

60. $\left(\frac{6}{7}\right) - \left(\frac{9}{10}\right)$

61. $\frac{7}{2} - \left(-\frac{19}{4}\right)$

62. $-\frac{11}{2} - \left(-\frac{7}{4}\right)$

63. $(-18) - (-54) - (21)$

64. $(34) - (-21) - (45)$

65. $(-61) - (-43) - (-32)$

66. $(-91) - (-56) - (-12)$

67. $\left(-\frac{5}{8}\right) - \left(-\frac{3}{4}\right) - \left(\frac{5}{6}\right)$

68. $\left(-\frac{2}{15}\right) - \left(\frac{3}{5}\right) - \left(-\frac{5}{6}\right)$

ANSWERS

69. ______

69. Viking II recorded high and low temperatures of $-22°$ and $-107°$ for one day on the surface of Mars. What was the change in temperature for that day?

70. ______

70. The surface temperature of one of Jupiter's satellites was measured for one week. The highest temperature recorded was $-75°C$ and the lowest was $-139°C$. What was the difference in the extreme temperatures for that week?

71. ______

71. At the beginning of the month, Joe's bank account had a balance of \$315.65. At the end of the month, the account was overdrawn by \$63.34 ($-\$63.34$). If there were no deposits during the month, what was the total amount of the checks Joe wrote? (*Hint:* Subtract the ending balance from the original balance.)

72. ______

72. At the beginning of the month Jack's bank account had a balance of $-\$3.45$. At the end of the month the balance was $-\$8.73$. If there were no deposits, find the amount of checks Jack wrote. (Refer to the previous problem.)

73. ______

73. *Viking II* recorded a high temperature of $3°$ and a low temperature of $-56°$. What was the temperature change?

74. ______

74. Carol started school owing her Mom \$12. by school's end she borrowed \$85 more from her Mom. How does her account with her Mom stand now?

NAME

CLASS SECTION DATE

ANSWERS

75.

76.

77.

78.

75. At the beginning of the month Jana's bank account had a balance of \$237.65. At the end of the month the account was overdrawn by \$14.22. If there were no deposits during the month, what was the total amount of the checks Jana wrote? (*Hint:* Subtract the ending balance from the original balance.)

76. What is the difference in altitude between the highest point in the United States and the lowest point?

Highest point: Mt. McKinley is 20,320 ft above sea level (+20,320).

Lowest point: Death Valley is 282 ft below sea level (−282).

77. Marie's account had a balance of \$82.75. She writes a check for \$100. What is her balance now?

78. Thomas started with \$8.15 in his account. He writes a check for \$25. What is his account balance?

E *Maintain your skills* (Sections 7.3, 7.5)

79.

80.

79. Find the area of a circle that has a radius of 16 in. (Let $\pi \approx 3.14$.)

80. Find the area of a square that is 18 m on each side.

ANSWERS

81. ____________

82. ____________

83. ____________

84. ____________

85. ____________

86. ____________

87. ____________

88. ____________

81. Find the area of a rectangle that is 36 cm long and 24 cm wide.

82. Find the area of a triangle that has a base of 3 m and a height of 3 m.

83. If a secretary can type 120 words per minute, how many words can she type in 1 second?

84. Find the area of a trapezoid that has a height of 1 meter and bases of 34 cm and 42 cm.

85. Find the area of a circle that has a diameter of 13.5 km. (Let $\pi \approx 3.14$.)

86. Find the area of a triangle that has a base of 2 ft and a height of 18 inches.

87. How many square yards of wall-to-wall carpeting are needed to carpet a rectangular floor that measures 28 ft by 36 ft?

88. How many square tiles, which are each 14 inches on a side, are needed to cover a floor that is 21 feet by 28 feet?

▲▲ *Getting Ready for Algebra*

Now that we have learned about signed numbers, we can solve equations in which the solution is a negative number. Recall that when solving equations, we perform the inverse of the operation involving the variable to eliminate the operation.

▲▲ MODEL PROBLEM-SOLVING

Examples / **Strategy**

a. Solve:

$$x - 8 = -10$$

$$x - 8 + 8 = -10 + 8$$

Eliminate the subtraction by adding 8 to both sides.

$$x = -2$$

Add.

Check: $x - 8 = -10$

$$-2 - 8 = -10$$

$$-10 = -10$$

Check by substituting -2 for x in the original equation.

So, the solution is $x = -2$.

Warm Up a. Solve: $x - 13 = -27$

b. Solve:

$$y + 17 = -3$$

$$y + 17 - 17 = -3 - 17$$

Eliminate the addition by subtracting 17 from both sides.

$$y = -3 + (-17)$$

Change subtraction to addition.

$$y = -20$$

Add.

Check: $y + 17 = -3$

$$-20 + 17 = -3$$

$$-3 = -3$$

So, the solution is $y = -20$.

Warm Up b. Solve: $y + 63 = -18$

GETTING READY FOR ALGEBRA

a. $x = -14$ **b.** $y = -81$

NAME ____________________

CLASS ____________ SECTION ____________ DATE ____________

▲▲ EXERCISES

ANSWERS

1. ____________
2. ____________
3. ____________
4. ____________
5. ____________
6. ____________
7. ____________
8. ____________
9. ____________
10. ____________
11. ____________
12. ____________
13. ____________
14. ____________
15. ____________

Solve:

1. $7 = y + 10$ **2.** $-20 = x - 10$ **3.** $-4 = w + 6$

4. $x - 7 = -5$ **5.** $x - 19 = -27$ **6.** $x + 14 = 5$

7. $x + 21 = -6$ **8.** $x - 7 = -45$ **9.** $x - 15 = -8$

10. $x + 56 = -76$ **11.** $x + 23 = 16$ **12.** $x + 0.54 = -0.32$

13. $x + 1.05 = 0.85$ **14.** $x + \frac{3}{4} = \frac{1}{2}$ **15.** $x + \frac{11}{12} = \frac{2}{3}$

8.4 Multiplication of Signed Numbers

OBJECTIVE

105. Multiply signed numbers.

APPLICATION

In order to attract business the Family Grocery ran a "loss leader" sale last week. The store sold eggs at a loss of \$0.20 (−\$0.20) per dozen. If 242 dozen eggs were sold last weekend, what was the total loss? Express this loss as a signed number.

VOCABULARY

No new words.

HOW AND WHY

Consider the following multiplications:

$(3)(4) = 12$

$(3)(3) = 9$

$(3)(2) = 6$

$(3)(1) = 3$

$(3)(0) = 0$

$(3)(-1) = ?$

$(3)(-2) = ?$

Each product is 3 smaller than the one before it. Continuing this pattern:

$(3)(-1) = -3$ and $(3)(-2) = -6$

RULE

The product of a positive number and a negative number is negative.

A similar pattern shows the product of two negative numbers:

$(-3)(+4) = -12$

$(-3)(+3) = -9$

$(-3)(+2) = -6$

$(-3)(+1) = -3$

$(-3)(0) = 0$

$(-3)(-1) = ?$

$(-3)(-2) = ?$

Each product is three larger than the one before it. Continuing this pattern:

$(-3)(-1) = 3$

$(-3)(-2) = 6$

RULE

The product of two negative numbers is positive.

MODEL PROBLEM-SOLVING

Examples | **Strategy**

a. Multiply: $(-6)(5)$

$(-6)(5) = -30$

The product of a positive number and a negative number is a negative number: $(+)(-) = (-)$ or $(-)(+) = (-)$.

Warm Up a. Multiply: $(-11)(4)$

b. Multiply: $(2)(-3.3)$

$(2)(-3.3) = -6.6$

The product of two factors with unlike signs is negative.

Warm Up b. Multiply: $(25)(-6)$

c. Multiply: $\left(-\frac{2}{3}\right)\left(\frac{1}{5}\right)$

$\left(-\frac{2}{3}\right)\left(\frac{1}{5}\right) = -\frac{2}{15}$

Warm Up c. Multiply: $\left(\frac{4}{7}\right)\left(-\frac{3}{2}\right)$

d. Multiply: $(5)(-0.7)$

$(5)(-0.7) = -3.5$

Warm Up d. Multiply: $(-0.03)(0.24)$

e. Multiply: $(-6)(-7)$

$(-6)(-7) = 42$

The product of two negative numbers is positive: $(-)(-) = (+)$.

Warm Up e. Multiply: $(-14)(-12)$

ANSWERS TO WARM UPS (8.4)

a. -44 **b.** -150 **c.** $-\frac{6}{7}$

d. -0.0072 **e.** 168

f. Multiply: $(-1.5)(-0.7)$

$(-1.5)(-0.7) = 1.05$

The product of two factors with like signs is positive.

Warm Up f. Multiply: $(-3.2)(-0.8)$

g. Multiply: $\left(-\frac{2}{7}\right)\left(-\frac{3}{8}\right)$

$\left(-\frac{2}{7}\right)\left(-\frac{3}{8}\right) = \frac{6}{56} = \frac{3}{28}$

Warm Up g. Multiply: $\left(-\frac{7}{12}\right)\left(-\frac{8}{25}\right)$

h. Multiply: $(-121)(-5)$

$(-121)(-5) = 605$

Warm Up h. Multiply: $(-305)(-12)$

i. Multiply: $(3)(6)$

$(3)(6) = 18$

Warm Up i. Multiply: $(34)(6)$

j. Multiply: $(-6)(-3)(-4)$

$$\begin{aligned}(-6)(-3)(-4) &= (18)(-4) \\ &= -72\end{aligned}$$

Multiply the first two factors.
Multiply again.

Warm Up j. Multiply: $(-12)(6)(-5)$

ANSWERS TO WARM UPS (8.4)

f. 2.56 **g.** $\frac{14}{75}$ **h.** 3660

i. 204 **j.** 360

k. Calculator example

Multiply: $(-28)(-4.6)(-2.9)$

ENTER	DISPLAY	
28	28.	
+/−	−28.	
×	−28.	
4.6	4.6	
+/−	−4.6	
×	128.8	**Product of the first two factors.**
2.9	2.9	
+/−	−2.9	
=	−373.52	**Final product.**

The product is -373.52.

Warm Up k. Multiply: $(-36.3)(-1.45)(-0.35)$

APPLICATION SOLUTION

l. In order to attract business the Family Grocery ran a "loss leader" sale last week. The store sold eggs at a loss of \$0.20 (−\$0.20) per dozen. If 242 dozen eggs were sold last weekend, what was the total loss? Express this loss as a signed number.

$(242)(-0.20) = -48.40$ **To find the total loss, multiply the loss per dozen by the number sold.**

Therefore, the loss, written as a signed number, was −\$48.40.

Warm Up l. John lost \$25 on each of 17 consecutive rolls of the dice at the craps table. Express his total loss as a signed number.

ANSWERS TO WARM UPS (8.4)

k. −18.42225 **l.** −\$425

NAME ______________________

CLASS ______________ SECTION ______________ DATE ______________

Exercises 8.4

A

ANSWERS

1. ______
2. ______
3. ______
4. ______
5. ______
6. ______
7. ______
8. ______
9. ______
10. ______
11. ______
12. ______
13. ______
14. ______
15. ______
16. ______
17. ______
18. ______
19. ______
20. ______
21. ______
22. ______
23. ______
24. ______

Multiply:

1. $(-3)(2)$

2. $(-5)(-6)$

3. $(9)(-6)$

4. $(-5)(-2)$

5. $(-1)(6)$

6. $(-17)(0)$

7. $(8)(-6)$

8. $(-6)(9)$

9. $(-1)(-1)(-1)$

10. $(-12)(-3)$

11. $(-7)(-2)(-3)$

12. $(2)(5)(-6)$

13. $(-3)(4)(-5)$

14. $(2)(0)(-3)$

15. $(+2)(-3)(4)$

16. $(-3)(-2)(-4)(-5)$

17. $(-1)(-2)(-3)(-4)$

18. $(-10)(-4.5)$

19. $(8.2)(-20)$

20. $(-5)(-6)(7)$

21. $(-1)(-1)(-1)(-1)(-3)$

22. $(-1)(-1)(-1)(-1)(-1)(9)$

23. $\left(-\frac{3}{5}\right)(-10)$

24. $(-30)\left(-\frac{5}{6}\right)$

B

ANSWERS

25. ______
26. ______
27. ______
28. ______
29. ______
30. ______
31. ______
32. ______
33. ______
34. ______
35. ______
36. ______
37. ______
38. ______
39. ______
40. ______
41. ______
42. ______

25. $(-3.02)(6)$ **26.** $(-2.03)(-7)$ **27.** $\left(-\frac{3}{4}\right)\left(-\frac{1}{2}\right)$

28. $\left(-\frac{3}{8}\right)\left(-\frac{4}{5}\right)$ **29.** $(-0.6)(-0.4)(0.2)$ **30.** $(-0.1)(0.3)(-0.5)$

31. $(-0.06)(-0.4)$ **32.** $(-2.6)(3)$ **33.** $\left(-\frac{9}{16}\right)\left(\frac{8}{15}\right)$

34. $\left(-\frac{9}{17}\right)(0)$ **35.** $\left(-\frac{3}{4}\right)(-1.2)$ **36.** $\left(\frac{1}{4}\right)(-2.4)$

37. $(-0.25)(-100)$ **38.** $(0.39)(-100)$ **39.** $(15 - 8)(5 - 12)$

40. $(13 - 16)(2 - 6)$ **41.** $(25 - 36)(5 - 9)$ **42.** $(48 - 15)(16 - 21)$

C

43. ______
44. ______
45. ______
46. ______
47. ______
48. ______

43. $(13)(-2)(-10)$ **44.** $(21)(-10)(2)$

45. $(-2)(-3)(-4)(-5)$ **46.** $(-4)(-5)(6)(2)$

47. $(-3.18)(-1.6)(0.1)$ **48.** $(-1.7)(5.3)(-0.2)$

NAME ______________________________

CLASS ______________ SECTION ______________ DATE ______________

ANSWERS

49. ______

50. ______

51. ______

52. ______

53. ______

54. ______

55. ______

56. ______

57. ______

58. ______

59. ______

60. ______

61. ______

62. ______

49. $(-1)(-1)(-1)(-1)(-1)(3.07)$

50. $(-1)(25)(-1)(-1)(-1)$

51. $(-2)(3)(-1)(0)(-5)$

52. $(-4)(-5)(2)(3)(-2)(-1)$

53. $(-0.06)(0.2)(-10)(-100)$

54. $(-0.3)(0.05)(-10)(-10)$

55. $\left(-\frac{2}{3}\right)\left(-\frac{3}{4}\right)\left(-\frac{4}{5}\right)\left(-\frac{5}{6}\right)$

56. $\left(-\frac{5}{12}\right)\left(\frac{7}{8}\right)\left(-\frac{3}{14}\right)\left(-\frac{8}{15}\right)$

57. $(-4)(-0.3)(-5)(-0.6)(-1)$

58. $(-7)(-1)(1)(-6)(-5)(-1)(1)$

59. $\left(\frac{4}{5}\right)(-0.2)\left(\frac{1}{2}\right)(-0.5)$

60. $\left(-\frac{2}{3}\right)(-0.06)\left(-\frac{3}{4}\right)(-0.16)$

61. $(-35)(-46)(-1)$

62. $(-41)(55)(-1)(-2)(1)$

D

63. ______

63. The formula for converting a temperature measurement from Fahrenheit to Celsius is $C = \frac{5}{9}(F - 32)$. What Celsius measure is equal to 18°F?

64. ______

64. Use the formula in Exercise 63 to find the Celsius measure that is equal to -13°F.

ANSWERS

65. ______________

66. ______________

67. ______________

68. ______________

69. ______________

70. ______________

65. For five consecutive weeks Ms. Rilea recorded a loss of 2.5 lb. If each loss is represented by -2.5 lb, what was her total weight loss for the five weeks, expressed as a signed number?

66. For six consecutive weeks Mr. Rilea recorded a loss of 1.9 lb. If each loss is represented by -1.9 lb, what was his total weight loss for the six weeks, expressed as a signed number?

67. The Dow Jones Industrial Average sustained 12 straight days of a 2.83 point decline (-2.83 each day). What was the total decline during the 12-day period, expressed as a signed number?

68. The Dow Jones Industrial Average sustained eight straight days of a loss of $1\frac{3}{4}$. What was the total decline in this period, expressed as a signed number?

69. Safeway Inc. offered as a loss leader 10 lb of sugar at a loss of 12¢ per bag (-12¢). If 560 bags were sold during the sale, what was the total loss, expressed as a signed number?

70. Albertsons offered a loss leader of coffee at a loss of 18 cents per can. If they sold 235 cans of coffee, find the total loss, expressed as a signed number.

NAME ______________________________

CLASS ______________ SECTION ______________ DATE ______________

ANSWERS

71. ______________

72. ______________

73. ______________

71. Thriftway's loss leader was a soft drink that lost 8 cents per six pack. They sold 251 of these six packs. What is Thriftway's total loss expressed as a signed number?

72. Fred Meyer's loss leader was soap powder that lost 12 cents per carton. They sold 326 cartons. What is Fred Meyer's total loss expressed as a signed number?

73. Safeway's loss leader was 1 dozen eggs that lost 14 cents per dozen. The store sold 712 dozen eggs that week. Express Safeway's total loss as a signed number.

E *Maintain your skills* (Sections 7.3, 7.5, 7.6, 7.7)

74. ______________

75. ______________

74. What is the equivalent piecework wage (dollars per piece) if the hourly wage is \$15.84 and the average number of articles completed in 1 hour is 4.4?

75. A wheel on Ellie's automobile is 13 inches in diameter. To the nearest whole number, how many revolutions will the wheel turn if the automobile travels 2 miles?

ANSWERS

76. ______

77. ______

78. ______

79. ______

80. ______

81. ______

76. How far does the hour hand of a clock travel in 6 hours if the length of the hand is 2 inches?

77. The following diagram shows the Smith's yard with respect to their house. How much grass seed is needed to sow the lawn if 1 pound of seed will sow 1000 ft^2? Find the weight to the nearest pound:

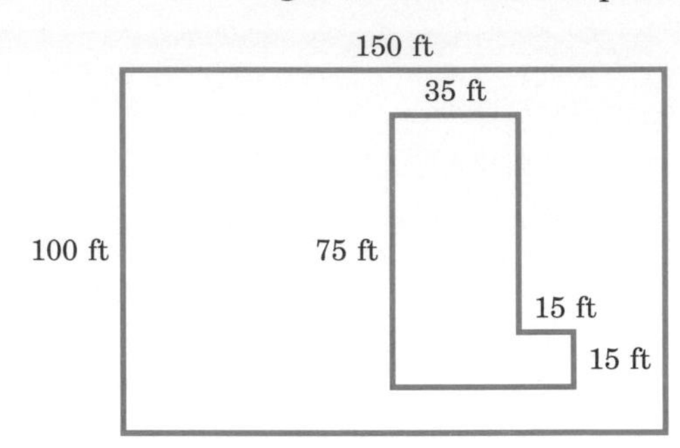

78. If carpeting costs $15.95 per square yard, what is the cost of wall-to-wall carpeting needed to cover the floor in a rectangular room that is 18 ft wide and 24 ft long?

79. How many square feet of sheet metal are needed to make a box without a top that has measurements of 5 ft 3 in. by 4 ft 8 in. by 9 in.?

80. A cylindrical tank designed to hold acidic liquids must be coated on the inside to prevent corrosion. The tank has an inside diameter of 7 ft 6 in. and a height of 4 ft. If 1 gallon of the coating will cover 3.5 square feet, how many gallons are needed to coat the tank? (Include top and bottom.)

81. A mini-storage complex has one unit that is 40 ft by 80 ft and rents for $1200 per year. What is the cost of a square foot of storage for a year?

8.5 Division of Signed Numbers

OBJECTIVE **106. Divide signed numbers.**

APPLICATION

Over a period of 16 weeks Mr. Richey lost a total of \$5824 (−\$5824) in his stock market account. What was his average loss per week, expressed as a signed number?

VOCABULARY No new words.

HOW AND WHY

To divide two signed numbers, we must find the number that when multiplied by the divisor equals the dividend.

The expression $(-8) \div (4) = ?$ asks $(4)(?) = -8$; we know $(4)(-2) = -8$, so $(-8) \div (4) = -2$.

The expression $(-21) \div (-3) = ?$ asks $(-3)(?) = -21$; we know $(-3)(7) = -21$, so $(-21) \div (-3) = 7$.

The expression $(15) \div (-5) = ?$ asks $(-5)(?) = 15$; we know $(-5)(-3) = 15$, so $(15) \div (-5) = -3$.

These examples lead to rules similar to those for multiplication:

RULE

The quotient of two negative numbers is positive.
The quotient of a positive and a negative number is negative.

MODEL PROBLEM-SOLVING

Examples | **Strategy**

a. Divide: $(-6) \div (3)$

$(-6) \div (3) = -2$

The quotient of a positive number and a negative number is negative: $(+) \div (-) = (-)$ or $(-) \div (+) = (-)$

Warm Up a. Divide: $(-32) \div (16)$

b. Divide: $(8.6) \div (-4.3)$

$(8.6) \div (-4.3) = -2$

The quotient of two numbers with unlike signs is negative.

Warm Up b. Divide: $(78) \div (-6)$

ANSWERS TO WARM UPS (8.5) **a.** −2 **b.** −13

c. Divide: $(1.44) \div (-0.3)$

$(1.44) \div (-0.3) = -4.8$

Warm Up c. Divide: $(-20.8) \div (1.3)$

d. Divide: $(68) \div (-4)$

$(68) \div (-4) = -17$

Warm Up d. Divide: $(-108) \div (4)$

e. Divide: $\left(-\frac{3}{4}\right) \div \left(-\frac{1}{2}\right)$

$\left(-\frac{3}{4}\right) \div \left(-\frac{1}{2}\right) = \frac{3}{2}$

The quotient of two negative numbers is positive: $(-) \div (-) = (+)$

Warm Up e. Divide: $\left(-\frac{8}{25}\right) \div \left(-\frac{4}{15}\right)$

f. Divide: $(-7) \div \left(-\frac{4}{3}\right)$

$(-7) \div \left(-\frac{4}{3}\right) = \frac{21}{4}$

The quotient of two numbers with like signs is positive.

Warm Up f. Divide: $\left(-\frac{13}{20}\right) \div (-4)$

APPLICATION SOLUTION

g. Over a period of 16 weeks Mr. Richey lost a total of \$5824 (−\$5824) in his stock market account. What was his average loss per week, expressed as a signed number?

$(-5824) \div (16) = -364$

To find the average loss per week, divide the total loss by the number of weeks.

Therefore, Mr. Richey had an average loss of \$364 (−\$364) per week.

Warm Up g. Ms. Richey lost \$720 in 15 consecutive hands in a game of twenty-one. What was her average loss per hand, expressed as a signed number?

ANSWERS TO WARM UPS (8.5)

c. −16 **d.** −27 **e.** $\frac{6}{5}$

f. $\frac{13}{80}$ **g.** −\$48

NAME ______________________________

CLASS ______________ SECTION ______________ DATE ______________

Exercises 8.5

A

ANSWERS

1. ______
2. ______
3. ______
4. ______
5. ______
6. ______
7. ______
8. ______
9. ______
10. ______
11. ______
12. ______
13. ______
14. ______
15. ______
16. ______
17. ______
18. ______
19. ______
20. ______
21. ______
22. ______
23. ______
24. ______

Divide:

1. $(-10) \div (5)$ **2.** $(10) \div (-2)$ **3.** $(-10) \div (-5)$

4. $(8) \div (-2)$ **5.** $(-16) \div (-4)$ **6.** $(15) \div (-3)$

7. $(18) \div (-1)$ **8.** $(-12) \div (-3)$ **9.** $(15) \div (-3)$

10. $(-14) \div (-2)$ **11.** $(-33) \div (11)$ **12.** $(6.06) \div (-6)$

13. $(-3) \div (-100)$ **14.** $(17) \div (-100)$ **15.** $(-51) \div (-17)$

16. $(48) \div (-16)$ **17.** $(-20) \div (5)$ **18.** $(-30) \div (2)$

19. $(-15) \div (+6)$ **20.** $(-210) \div (6)$ **21.** $(-7) \div 0$

22. $(-9) \div 0$ **23.** $0 \div (-3)$ **24.** $0 \div (-5)$

B

ANSWERS

25. ______
26. ______
27. ______
28. ______
29. ______
30. ______
31. ______
32. ______
33. ______
34. ______
35. ______
36. ______
37. ______
38. ______
39. ______
40. ______
41. ______
42. ______
43. ______
44. ______

25. $(-100) \div (-5)$ **26.** $(-81) \div (9)$ **27.** $(-12.12) \div (-3)$

28. $(8.08) \div (-4)$ **29.** $(-0.25) \div (1000)$ **30.** $(-2.33) \div (-10)$

31. $(24) \div (-16)$ **32.** $(-24) \div (20)$ **33.** $(-40) \div (2.4)$

34. $(40) \div (-25)$ **35.** $(-4.95) \div (-0.9)$ **36.** $(-15.5) \div (0.05)$

37. $(0.65) \div (-0.13)$ **38.** $(-0.056) \div (-0.4)$ **39.** $(-540) \div (12)$

40. $(-1071) \div (-17)$ **41.** $(3956) \div (-46)$ **42.** $(-5832) \div (-72)$

43. $(3364) \div (-29)$ **44.** $(-9996) \div (-42)$

C

45. ______
46. ______
47. ______
48. ______

45. $\left(-\frac{3}{8}\right) \div \left(-\frac{3}{4}\right)$ **46.** $\left(-\frac{1}{2}\right) \div \left(\frac{2}{7}\right)$

47. $\left(-\frac{6}{7}\right) \div \left(\frac{2}{7}\right)$ **48.** $\left(-\frac{4}{3}\right) \div \left(\frac{5}{3}\right)$

NAME ______________________

CLASS ____________ SECTION ____________ DATE ____________

ANSWERS

49. ____________

50. ____________

51. ____________

52. ____________

53. ____________

54. ____________

55. ____________

56. ____________

57. ____________

58. ____________

59. ____________

60. ____________

61. ____________

62. ____________

63. ____________

64. ____________

49. $\left(\frac{3}{4}\right) \div (-0.625)$

50. $(0.125) \div \left(-\frac{7}{8}\right)$

51. $[(-5)(-2)] \div (-10)$

52. $(36) \div [(-3)(-6)]$

53. $(7 - 10) \div (8 - 5)$

54. $(6 - 15) \div (8 - 5)$

55. $(-7 + 4 - 9) \div (4 - 2)$

56. $(-6 + 11 - 9) \div (3 - 5)$

57. $(43 - 65) \div (-7)$

58. $(-75 - 30) \div (-35)$

59. $[7 - (-7)] \div (-2)$

60. $[7 - (-7)] \div (-4)$

61. $\left(-\frac{2}{3}\right)\left(-\frac{3}{4}\right) \div \left(-\frac{5}{6}\right)$

62. $\left(-\frac{3}{5}\right)\left(-\frac{7}{15}\right) \div \left(-\frac{1}{30}\right)$

63. $(0.43 - 0.69 - 0.3) \div (-3.5)$

64. $(-0.72 - 0.34 + 0.16) \div (-1.5)$

ANSWERS

65. ____________________

65. The coldest temperatures in Eycee Northland for each of five days were $-15°$, $-5°$, $-4°$, $2°$, and $-3°$. What is the average of these temperatures?

66. ____________________

66. In northern Norway the daily low temperature was recorded for six days. The temperatures were $-22°$, $-18°$, $-14°$, $-18°$, $-19°$, and $-11°$. What is the average of these temperatures?

67. ____________________

67. Mr. Gambit lost \$836 ($-\$836$) in 12 straight hands of poker at a casino in Reno. What was his average loss per hand, expressed as a signed number?

68. ____________________

68. Ms. Gambit lost \$133 ($-\$133$) in 7 straight hands of twenty-one at a casino in Atlantic City. What was her average loss per hand, expressed as a signed number?

69. ____________________

69. The membership of the Burlap Baggers Investment Club took a loss of \$183.66 ($-\$183.66$) on the sale of stock. If there are six co-equal members in the club, what is each member's share of the loss, expressed as a signed number?

70. ____________________

70. The temperature in Fairbanks, Alaska, dropped from 10° above zero ($+10°$) to 15° below zero ($-15°$) in an eight-hour period. This is a drop of 25° ($-25°$). What was the average drop per hour, expressed as a signed number?

NAME ______________________________

CLASS ______________ SECTION ______________ DATE ______________

ANSWERS

71. ______________

71. Find the average weekly temperature if these temperatures were recorded during one week: $-2°$, $8°$, $-6°$, $-5°$, $10°$, $12°$, $4°$.

72. ______________

72. Ms. Obese lost 82 pounds on her diet in 16 weeks. Find her average weekly loss, expressed as a signed number.

73. ______________

73. Mr. Obese lost 90.5 pounds on his diet in 20 weeks. Express the average weekly loss as a signed number.

74. ______________

74. A certain stock lost $22\frac{1}{2}$ points in 9 days. Express the average daily loss as a signed number.

E *Maintain your skills* (Section 7.8)

75. ______________

75. Find the volume of a cylinder that has a radius of 12 cm and a height of 18 cm. (Let $\pi \approx 3.14$.)

76. ______________

76. Find the volume of a sphere with a diameter of 12 cm. (Let $\pi \approx 3.14$.)

77. ______________

77. Find the volume of a cone that has a radius of 22 in. and a height of 12 in. (Let $\pi \approx 3.14$.)

ANSWERS

78. ______

79. ______

80. ______

81. ______

82. ______

78. Find the volume of a pyramid with a square base of 10 cm on each side and a height of 8 cm, to the nearest cubic centimeter.

79. A hot-water tank is a cylinder that is 36 inches in diameter and 6 feet high. If there are 231 in.^{3} in a gallon, how many gallons of water will the tank hold? Round off the answer to the nearest gallon.

80. A swimming pool is to be dug and the dirt hauled away. The pool is to be 28 ft long, 12 ft wide, and 9 ft deep. How many cubic yards of dirt must be removed?

81. To remove the dirt for the swimming pool in Exercise 80, trucks that can haul 8 yd^3 per load are used. How many truckloads will there be?

82. A real estate broker sold a lot that measured 66.75 ft by 150 ft. The sale price was 22 cents per square foot. If the broker's commission is 8%, how much did she make?

▲▲ Getting Ready for Algebra

We can now solve equations in which the solution involves multiplication and division by signed numbers. Recall that the product of (-3) and x is $-3x$ and that x divided by 4 can be written as $\frac{x}{4}$.

▲▲ MODEL PROBLEM-SOLVING

Examples

a. Solve: $-3x = 12$

$\frac{-3x}{-3} = \frac{12}{-3}$ — Strategy: **Eliminate the multiplication by dividing both sides by −3.**

$x = -4$ — **Divide.**

Check: $-3x = 12$

$(-3)(-4) = 12$ — **To check, replace x by −4 in the original equation.**

$12 = 12$

The solution is $x = -4$.

Warm Up a. Solve: $-14y = -98$

b. Solve: $\frac{x}{-7} = -21$

$\frac{x}{-7}(-7) = (-21)(-7)$ — **Eliminate the division by multiplying both sides by −7.**

$x = 147$ — **Multiply.**

Check: $\frac{x}{-7} = -21$

$\frac{147}{-7} = -21$ — **To check, replace x by 147 in the original equation.**

$-21 = -21$

So, the solution is $x = 147$.

Warm Up b. Solve: $\frac{x}{-6} = 45$

GETTING READY FOR ALGEBRA

a. $y = 7$

b. $x = -270$

NAME ______________________

CLASS ______________ SECTION ______________ DATE ______________

▲▲ EXERCISES

ANSWERS

1. ______
2. ______
3. ______
4. ______
5. ______
6. ______
7. ______
8. ______
9. ______
10. ______
11. ______
12. ______
13. ______
14. ______
15. ______
16. ______
17. ______
18. ______
19. ______
20. ______
21. ______

Solve:

1. $-18 = -2c$

2. $-20 = 5w$

3. $-2 = \dfrac{d}{-4}$

4. $10 = \dfrac{g}{-0.5}$

5. $6 = -0.24x$

6. $-7x = 21$

7. $8y = -32$

8. $-9x = -99$

9. $-5y = -75$

10. $\dfrac{x}{2} = 18$

11. $\dfrac{y}{-7} = -7$

12. $\dfrac{x}{-9} = -12$

13. $\dfrac{x}{-8} = 20$

14. $-0.25x = -8$

15. $0.65y = -1.3$

16. $\dfrac{x}{0.4} = -2.1$

17. $\dfrac{x}{-0.6} = 0.07$

18. $-3.5y = -16.45$

19. $-5.6x = 78.4$

20. $\dfrac{y}{-0.5} = -46$

21. $\dfrac{x}{-33} = 0.12$

Review: Order of Operations

OBJECTIVE **107. Do any combination of operations with signed numbers.**

APPLICATION

Hilda was told that a temperature of $-40°F$ is also $-40°C$. She wanted to check this, so she found the formula for changing degrees Fahrenheit to degrees Celsius. It is;

$$C = \frac{5}{9}(F - 32)$$

Use this formula to find out whether $-40°F$ does equal $-40°C$.

VOCABULARY No new words.

HOW AND WHY The order of operations for signed numbers is the same as that for whole numbers.

MODEL PROBLEM-SOLVING

Examples	Strategy
a. Perform the indicated operations:	
$(-64) + (-22) \div (2)$	
$(-64) + (-22) \div (2) = (-64) + (-11)$	**Multiplication and division are performed before addition and subtraction, so divide first.**
$= -75$	**Add.**

Warm Up a. Perform the indicated operations: $(-75)(-2) + (-123)$

b. Perform the indicated operations:	
$(-12)(5) - (54) \div (-3)$	
$(-12)(5) - (54) \div (-3) = (-60) - (-18)$	**Multiply and divide.**
$= -42$	**Subtract.**

Warm Up b. Perform the indicated operations:

$(-34) + (-14)(-12) - (56) \div (-4)$

ANSWERS TO WARM UPS (8.6) **a.** 27 **b.** 148

c. Perform the indicated operations:

$(3) \div (-0.8) + (2)(-1.7)$

$(3) \div (-0.8) + (2)(-1.7) = (-3.75) + (-3.4)$ **Multiply and divide.**

$= -7.15$ **Add.**

Warm Up c. Perform the indicated operations:

$(-8.6) \div (0.43) - (1.4)(-5.2) - 6.04$

d. Perform the indicated operations: $(6) - \left(\frac{1}{2}\right)(-5)$

$(6) - \left(\frac{1}{2}\right)(-5) = (6) - \left(-2\frac{1}{2}\right)$ **Multiply.**

$= 6 + 2\frac{1}{2}$ **Change to addition.**

$= 8\frac{1}{2}$ **Add.**

Warm Up d. Perform the indicated operations:

$\frac{1}{2} + \left(\frac{3}{4}\right)(-12) - \left(\frac{3}{2}\right) \div (-5 + 9)$

e. Perform the indicated operations: $(-7)^2 - 4^2$

$(-7)^2 - 4^2 = 49 - 16$ **Do exponents first.**

$= 33$ **Subtract.**

Warm Up e. Perform the indicated operations: $(-5)(-3)^2 + (-32) - (-4)^2$

ANSWERS TO WARM UPS (8.6)

c. -18.76 d. $-\frac{71}{8}$ e. -93

f. Calculator example

Perform the indicated operations: $(13)(-12) - (-42) \div (-7)$

If your calculator has algebraic logic:

ENTER	DISPLAY	
13	13.	When using a calculator with algebraic logic, we ignore the rules for order of operations. The calculator is programmed to yield the correct answer as you enter the numbers and operations from left to right.
×	13.	
12	12.	
+/−	−12.	
−	−156.	
42	42.	
+/−	−42.	
÷	−42.	
7	7.	
+/−	−7.	
=	−162.	

If your calculator has parentheses:

ENTER	DISPLAY	
(	0.	The parentheses are to indicate that the result is to be stored for recall.
13	13.	
×	13.	
12	12.	
+/−	−12.	
)	−156.	Result of the multiplication.
−	−156.	
(	0.	The parentheses indicate that another operation will be performed before subtracting.
42	42.	
+/−	−42.	
÷	−42.	
7	7.	
+/−	−7.	
)	6.	Result of the division.
=	−162.	Result of the subtraction.

The answer is -162.

Warm Up f. Perform the indicated operations:

$(-112)(306) - (10{,}551) \div (-3)$

ANSWERS TO WARM UPS (8.6)

f. $-30{,}755$

APPLICATION SOLUTION

g. Hilda was told that a temperature of $-40°F$ is also $-40°C$. She wanted to check this, so she found the formula for changing degrees Fahrenheit to degrees Celsius. It is:

$$C = \frac{5}{9}(F - 32)$$

Use this formula to find out whether $-40°F$ does equal $-40°C$.

$$C = \frac{5}{9}(F - 32)$$

To find out whether $-40°F$ equals $-40°C$, substitute -40 for F in the formula.

$$C = \frac{5}{9}(-40 - 32)$$

$$C = \frac{5}{9}[-40 + (-32)]$$

$$C = \frac{5}{9}(-72)$$

$$C = -40$$

Therefore, it is true that $-40°F$ equals $-40°C$.

Warm Up g. How many degrees Celsius is $14°F$?

ANSWERS TO WARM UPS (8.6)

g. $-10°C$

NAME ____________________

CLASS ____________ SECTION ____________ DATE ____________

Exercises 8.6

A

ANSWERS

1. ________
2. ________
3. ________
4. ________
5. ________
6. ________
7. ________
8. ________
9. ________
10. ________
11. ________
12. ________
13. ________
14. ________
15. ________
16. ________
17. ________
18. ________
19. ________
20. ________

Perform the indicated operations:

1. $(-8)(6) + (-18)$

2. $(25) + (-3)(-8)$

3. $(2)(-8) + (9)$

4. $(19) - (4)(-2)$

5. $(-2)(3) + (5)(-2)$

6. $(-7) + (2)(-3)$

7. $(-5)(-3) - (2)(6)$

8. $(-3)(-7) - (10)$

9. $(2)(-6 + 5) - (3)$

10. $(-7)(6) \div (-3)$

11. $(-6)(-5) \div (-2)(-3)$

12. $(-8)(-5) \div (-4)(+2)$

13. $(-6 + 3)[-2 + (-3)]$

14. $(-3 + 6)[-2 + (-3)]$

15. $7 + (-2) - (7 - 10)$

16. $-8 + (-10) - (8 - 12)$

17. $(8 - 10) - (7 - 10)$

18. $(2 - 6)(3 - 5) - (10 - 12)$

19. $(-2)^2 + 8$

20. $(-4) + (-3)^2$

ANSWERS

21. ____

22. ____

23. ____

24. ____

25. ____

26. ____

21. $(3)(-2) + (-12) - (2)$

22. $(-4)(-3) - (2) + (5)$

23. $(7.7) - (1.3)(-2)$

24. $(-5.5) + (-2)(-3.1)$

25. $[10 + (-3)(-8)] - (-2)$

26. $(-9) - (3)[(-4) - (-5)]$

B

27. ____

28. ____

29. ____

30. ____

31. ____

32. ____

33. ____

34. ____

35. ____

36. ____

37. ____

38. ____

27. $(-13)(-3) + (15)(-2)$

28. $(14)(-2) - (8)(-5)$

29. $(6)(-10 + 4) - (33) \div (-11)$

30. $(84) - (9)(9) + (-6)(2)$

31. $(16)(-2) \div (-4) + (-2)$

32. $(-3)(-8) \div (-6) + 10$

33. $(-120) \div (-20) - (9 - 11)$

34. $-16 + (-80 \div -5) - 20$

35. $-2^3 - (-2)^4$

36. $4^4 - (-4)^3$

37. $(-35) \div (-5) + (-2)(-3)$

38. $(3)(-11) - (14)(-2) + (3)(-2)$

NAME ______________________________

CLASS ______________ SECTION ______________ DATE ______________

ANSWERS

39. ______________

39. $(10) - (3)(-6) + (-8)(-9) - (3)(-12)$

40. ______________

40. $(15) + (-2)(4) - (7)(8) + (-2)(-13)$

41. ______________

41. $(-4)(-1)(-3) - (-9)(-2) + (-4)(3)$

42. ______________

42. $(-24) - (-3)(2) + (-1)(4) - (-3)(-1)$

43. ______________

43. $(-5)^2 + (-2)^3 - (-5)(-2)$

44. ______________

44. $(-9)(-8) + (-4)^2(-5) + (-3)(-6)$

45. ______________

45. $(-1)(-6)^2(-1) - (-2)^2(-3)^2$

46. ______________

46. $(-7)^2 - (-6)^2 + (-3)(4) - (-5)^2$

C

ANSWERS

47. ______

48. ______

49. ______

50. ______

51. ______

52. ______

53. ______

54. ______

55. ______

56. ______

57. ______

58. ______

47. $[(-8)9 + 20] \cdot (-3) - 25$

48. $28 \div (-4) - 7 + (-1)(-8) - 6$

49. $(-126) - [(-6)(6) - 6]$

50. $[6 + (-8)(-5)] - (-8)$

51. $(-8)(+9) \div (-3)(-4)$

52. $2(7 - 9) - 4(5 - 6)$

53. $(4 - 6)^2 - (-2)(8 - 12)$

54. $(3 - 8)^3 - (-25)(7 - 2)$

55. $[(-5)(6) - (3)(8)] \div [6(-3) + (6)(6)]$

56. $[(-8)(-6) + (4)(11)] \div [(7)(-8) - (6)(-8)]$

57. $(-6 + 8 - 2)^2 + (-4)^3$

58. $-3[-2 - (-6)(-4) - 8]$

NAME ______________________________

CLASS ______________ SECTION ______________ DATE ______________

ANSWERS

59. ______________

59. $-5(-2 - 8 - 4 - 6)^2$

60. ______________

60. $[(-4)^2 - (-2)](4)(3 - 5)$

61. ______________

61. $(-6)^2(-2) - (-3) - 8 + 6$

62. ______________

62. $[7 - 10]^2 + [5 - 2]^3 - [8 - 11]^3$

63. ______________

63. $[(-3) + (-6)]^2 - [(-2)(-3)]^2$

64. ______________

64. $[(-7)(9) - (-5)^2]^2 - (-8)(-1)^3$

65. ______________

65. $[(-50) - (-5)^2]^2 - [(-8)(-2) - (-2)(-4)]^2$

66. ______________

66. $[56 - (3)(-4)^2]^2 - [(-7)(-1)^3 + (-5)(-8)]$

ANSWERS

67. ______

67. Does 14°F equal −10°C? To check, substitute 14 for F and −10 for C in the formula:

$$F = \frac{9}{5}C + 32$$

and tell whether the statement is true.

68. ______

68. The temperature at 5 P.M. was 18°C. If the temperature dropped 0.6° every hour until midnight, we can find the midnight temperature by calculating the value of:

$$T = 18 + (7)(-0.6)$$

What was the midnight temperature?

69. ______

69. If the temperature at 6 A.M. was −10°F and rose 3.4°F every hour until 11 A.M., what was the temperature at 11 A.M.?

70. To check whether $x = -6$ is a solution to the equation:

$$x^2 - 4x - 7 = 53$$

70. ______

we substitute −6 for x and find the value of:

$$(-6)(-6) - 4(-6) - 7$$

Check whether this expression equals 53.

71. ______

71. Check whether 10 is a solution in the previous equation.

72. Substitute $x = -\frac{3}{2}$ in the following equation and check whether it is a solution:

$$\frac{3}{4}\left(x + \frac{2}{3}\right) - \frac{3}{8} = -1$$

72. ______

NAME ______________________

CLASS ______________ SECTION ______________ DATE ______________

ANSWERS

73. ______________

73. Check whether $1\frac{1}{2}$ is a solution in the previous equation.

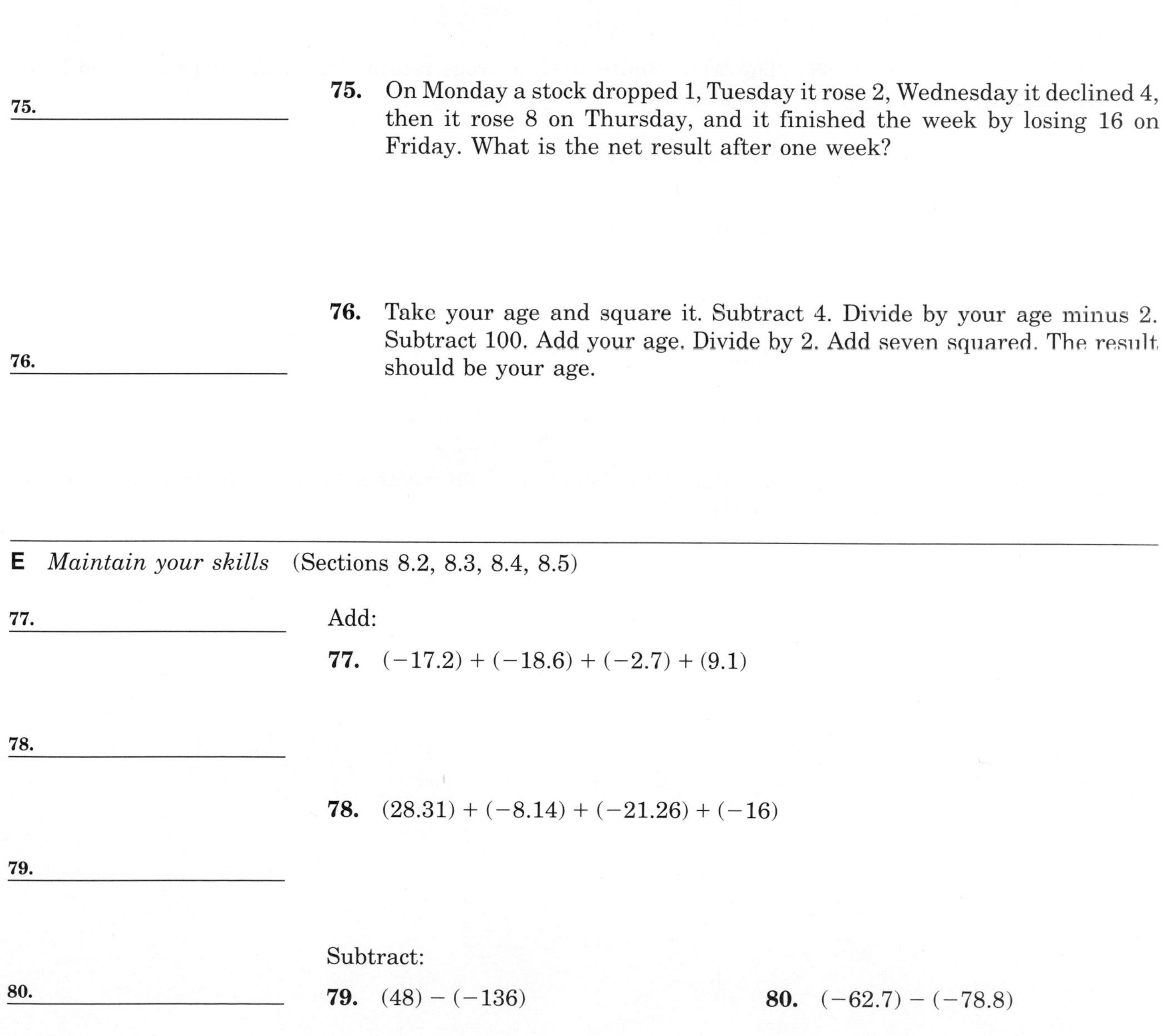

74. Does $-10°F$ equal $-10°C$? To check, substitute -10 for F and also for C in the formula:

74. ______________

$$F = \frac{9}{5}C + 32$$

Tell whether the statement is true.

75. ______________

75. On Monday a stock dropped 1, Tuesday it rose 2, Wednesday it declined 4, then it rose 8 on Thursday, and it finished the week by losing 16 on Friday. What is the net result after one week?

76. Take your age and square it. Subtract 4. Divide by your age minus 2. Subtract 100. Add your age. Divide by 2. Add seven squared. The result should be your age.

76. ______________

E *Maintain your skills* (Sections 8.2, 8.3, 8.4, 8.5)

77. ______________

Add:

77. $(-17.2) + (-18.6) + (-2.7) + (9.1)$

78. ______________

78. $(28.31) + (-8.14) + (-21.26) + (-16)$

79. ______________

Subtract:

80. ______________

79. $(48) - (-136)$

80. $(-62.7) - (-78.8)$

ANSWERS

81. ______

82. ______

83. ______

84. ______

85. ______

86. ______

Multiply:

81. $(-36)(84)(-21)$

82. $(-62)(-22)(-30)$

Divide:

83. $(-900) \div (-36)$

84. $(-18.415) \div (2.5)$

85. The four Zaple brothers formed a company. The first year the company lost \$5832 (-5832). The brothers shared equally in the loss. Represent each brother's loss as a signed number.

86. The WOW-Smith stock average recorded the following gains and losses for the week:

Monday	loss $2\frac{1}{4}$
Tuesday	loss $3\frac{1}{8}$
Wednesday	gain $4\frac{1}{2}$
Thursday	gain 2
Friday	loss $1\frac{1}{8}$

Use signed numbers to find out whether the stock average gained or lost for the week.

8.7 Equations of the Form $ax + b = c$ or $ax - b = c$

OBJECTIVE

108. Solve equations of the form $ax + b = c$ or $ax - b = c$, where a, b, and c are signed numbers.

APPLICATION

Using the formula $s = v + gt$, find the initial velocity (v) in feet per second, of a skydiver if after 3.5 seconds (t) she reaches a speed (s) of 125 feet per second and $g = 32$.

VOCABULARY

No new words.

HOW AND WHY

The solutions of equations using signed numbers involves two operations to isolate the variable.

PROCEDURE

To find the solution of an equation of the form $ax + b = c$ or $ax - b = c$

1. **Add (subtract) the constant to (from) each side of the equation to isolate the variable.**
2. **Divide both sides by the coefficient of the variable.**

To review these skills refer to *Getting Ready for Algebra* sections 1.4, 1.6, 1.8, 3.7, 3.14, 4.7, 4.10, 4.12, 8.3, and 8.5.

MODEL PROBLEM SOLVING

Examples | **Strategy**

a. Solve: $-5x + 17 = -8$

$-5x + 17 = -8$ — Original equation.

$-5x + 17 - 17 = -8 - 17$ — Eliminate the addition by subtracting 17 from both sides

$-5x = -25$ — Subtract.

$\dfrac{-5x}{-5} = \dfrac{-25}{-5}$ — Eliminate the multiplication by dividing both sides by -5.

$x = 5$

Check: $-5x + 17 = -8$

$-5(5) + 17 = -8$

$-25 + 17 = -8$

$-8 = -8$

The solution is $x = 5$.

Warm Up a. Solve: $-7x - 32 = 17$

b. Solve: $-31 = 13x + 34$

$-31 = 13x + 34$	**Original equation.**
$-31 - 34 = 13x + 34 - 34$	**Eliminate addition by subtracting 34 from both sides.**
$-65 = 13x$	**Subtract.**
$\frac{-65}{13} = \frac{13x}{13}$	**Divide both sides by 13.**
$-5 = x$	

Check: $-31 = 13x + 34$

$-31 = 13(-5) + 34$
$-31 = -65 + 34$
$-31 = -31$

The solution is $-5 = x$ or $x = -5$

Warm Up b. Solve: $9 = -3x + 81$

APPLICATION SOLUTION

Using the formula $s = v + gt$, find the initial velocity (v) in feet per second, of a skydiver if after 3.5 seconds (t) she reaches a speed (s) of 125 feet per second and $g = 32$.

Formula:

$s = v + gt$

Substitute:

$125 = v + 32(3.5)$	**Replace s with 125, g with 32, and t with 3.5**

Solve:

$125 = v + 112$	**Multiply.**
$125 - 112 = v + 128 - 112$	**Subtract 112 from both sides.**
$13 = v$	

Check:

$125 = v + 112$	**Original equation.**
$125 = 13 + 112$	**Replace v with 13.**
$125 = 125$	**Add.**

The skydiver's initial velocity was 13 feet per second.

Warm Up c. Using the formula $s = v + gt$, find the initial velocity (v) in feet per second, of a skydiver if after 5 seconds (t) she reaches a speed (s) of 180 feet per second and $g = 32$.

ANSWERS TO WARM UPS (8.7) **a.** $x = -7$ **b.** $x = 24$ **c.** 20 feet per second

NAME ______________________

CLASS ______________ SECTION ______________ DATE ______________

Exercises 8.7

A

ANSWERS

1. ______
2. ______
3. ______
4. ______
5. ______
6. ______
7. ______
8. ______
9. ______
10. ______
11. ______
12. ______

1. $-7x + 32 = 4$

2. $-9y + 14 = 59$

3. $-8 + 5x = 17$

4. $-9 + 3y = 12$

5. $4y - 67 = -35$

6. $8x - 49 = -73$

7. $2a + 12 = 12$

8. $4b + 15 = 15$

9. $-5x - 19 = -64$

10. $-11y - 45 = -12$

11. $3x + 7 = -8$

12. $9y + 11 = -25$

B

13. ______
14. ______
15. ______
16. ______
17. ______
18. ______
19. ______
20. ______

13. $-10 = 2x - 6$

14. $12 = 3x - 6$

15. $-40 = 5x - 10$

16. $-30 = -5x - 10$

17. $6 = -4x - 6$

18. $8 = -3x + 2$

19. $-10 = -4x + 2$

20. $20 = -8x + 4$

ANSWERS

21. ______

22. ______

23. ______

24. ______

21. $-14y - 1 = -99$

22. $-16x + 5 = 149$

23. $-3 = -8a - 3$

24. $-12 = 5b - 12$

C

25. ______

26. ______

27. ______

28. ______

29. ______

30. ______

31. ______

32. ______

33. ______

34. ______

35. ______

36. ______

37. ______

38. ______

25. $-0.6x - 0.15 = 0.15$

26. $-1.05y + 5.08 = 1.72$

27. $0.03x + 2.3 = 1.55$

28. $0.15x - 7.8 = -7.734$

29. $-135x - 674 = 1486$

30. $94y + 307 = -257$

31. $-125y + 6 = 506$

32. $-76c + 12 = 468$

33. $5x + 9 + (-12) = 12$

34. $8z + 6 + (-12) = 50$

35. $-3b - 12 + (-4) = 11 + (-6)$

36. $-5z - 15 + 6 = -21 - 18$

37. *Using the formula* $s = v + gt$, find the initial velocity (v) in feet per second, of a skydiver if after 4 seconds (t) she reaches a speed (s) of 150 feet per second and $g = 32$.

38. Using the formula $s = v + gt$, find the initial velocity (v) in feet per second, of a skydiver if after 6 seconds (t) he reaches a speed (s) of 200 feet per second and $g = 32$.

NAME ____________________

CLASS ______________ SECTION ______________ DATE ______________

ANSWERS

1. ______
2. ______
3. ______
4. ______
5. ______
6. ______
7. ______
8. ______
9. ______
10. ______
11. ______
12. ______
13. ______
14. ______

CHAPTER 8 TRUE–FALSE CONCEPT REVIEW

Check your understanding of the language of arithmetic. Tell whether each of the following statements is True (always true) or False (not always true).

1. Negative numbers are found to the left of zero on the number line.

2. The opposite of a signed number is always negative.

3. The absolute value of a number is always positive.

4. The opposite of a number is the same distance from zero as the number on the number line but in the opposite direction.

5. The sum of two signed numbers is always positive or negative.

6. The sum of a positive signed number and a negative signed number is always positive.

7. To find the sum of a positive signed number and a negative signed number, subtract their absolute values and use the sign of the number with the larger absolute value.

8. To subtract two signed numbers, add their absolute values.

9. If a negative number is subtracted from a positive number the difference is always positive.

10. The product of two negative numbers is never negative.

11. The sign of the product of a positive number and a negative number depends on which number has the larger absolute value.

12. The sign of the quotient, when dividing two signed numbers, is the same as the sign obtained when multiplying the two numbers.

13. The order of operations for signed numbers is the same as the order of operations for positive numbers.

14. Subtracting a number from both sides of an equation results in an equation that has the same solution as the original equation.

NAME ______

CLASS ______ SECTION ______ DATE ______

ANSWERS

1. ______

2. ______

3. ______

4. ______

5. ______

6. ______

7. ______

8. ______

9. ______

10. ______

11. ______

12. ______

13. ______

14. ______

CHAPTER 8 POST-TEST

Perform the indicated operations:

1. **(Section 8.2, Obj. 103)** $(-21) + (-17) + (42) + (-18)$

2. **(Section 8.5, Obj. 106)** $\left(-\frac{3}{8}\right) \div \left(\frac{3}{10}\right)$

3. **(Section 8.3, Obj. 104)** $\left(-\frac{7}{15}\right) - \left(-\frac{3}{5}\right)$

4. **(Section 8.6, Obj. 107)** $(36 - 42)(-18 + 6)$

5. **(Section 8.6, Obj. 107)** $(-3 + 18) - (21 - 7) + (3)$

6. **(Section 8.2, Obj. 103)** $(-3.65) + (4.72)$

7. **(Section 8.1, Obj. 101, 102)** a. $-(-21)$
   b. $|-21|$

8. **(Section 8.3, Obj. 104)** $(-37) - (-41)$

9. **(Section 8.6, Obj. 107)** $(-16 + 4) \div (3) + (2)(-6)(-3)$

10. **(Section 8.5, Obj. 106)** $(-88) \div (-22)$

11. **(Section 8.4, Obj. 105)** $(-7)(-9)(2)$

12. **(Section 8.4, Obj. 105)** $(|-6|)(-4)(-1)(-1)$

13. **(Section 8.3, Obj. 103)** $(-57.9) - (32.5)$

14. **(Section 8.6, Obj. 107)** $(-21)(3) - (-6)(-7)$

ANSWERS

15. ______

16. ______

17. ______

18. ______

19. ______

20. ______

21. ______

22. ______

23. ______

24. ______

25. ______

26. ______

15. (Section 8.5, Obj. 106) $(30.66) \div (-0.6)$

16. (Section 8.2, Obj. 103) $\left(-\frac{1}{3}\right) + \left(\frac{5}{6}\right) + \left(-\frac{1}{2}\right) + \left(-\frac{1}{6}\right)$

17. (Section 8.5, Obj. 106) $(-56) \div (-7)$

18. (Section 8.3, Obj. 104) $(18) - (-25)$

19. (Section 8.6, Obj. 107) $(-7)(3 - 11)(-2) - (4)(-3 - 5)$

20. (Section 8.4, Obj. 105) $\left(-\frac{1}{3}\right)\left(\frac{6}{7}\right)$

21. (Section 8.2, Obj. 103) $(-17) + (-36)$

22. (Section 8.6, Obj. 107) $(3)(-7) + (25)$

23. (Section 8.7, Obj. 106) Solve: $-10 = 3x + 5$

24. The temperature in Chicago ranged from a high of 12°F to a low of −9°F within a 24-hour period. What was the drop in temperature, expressed as a signed number?

25. A stock on the West Coast Exchange opened at $6\frac{5}{8}$ on Monday. It recorded the following changes during the week: Monday, $+\frac{1}{8}$; Tuesday, $-\frac{3}{8}$; Wednesday, $+1\frac{1}{4}$; Thursday, $-\frac{7}{8}$; Friday, $+\frac{1}{4}$. What was its closing price on Friday?

26. What Fahrenheit temperature is equal to a reading of −10°C? Use the formula:

$$F = \frac{9}{5}C + 32$$

APPENDIX I

The Properties of Zero

Historically, the number zero is a recent development in mathematics. There is evidence that notched sticks or bones were used for counting as long as 30 million years ago, but the use of a symbol for zero began from about 1200 to 1800 years ago (the development was gradual and not a sudden decision).

Zero is *not* a counting number.

When counting a group of objects, we usually count "one, two, three, four," and so on.

Zero is a whole number.

The classification of zero as a whole number is arbitrary although useful.

Zero is an integer.

All whole numbers are also integers.

Zero is a rational number.

Zero can be written as a fraction $\frac{a}{b}$, where a and b are integers, $b \neq 0$, since

$0 = \frac{0}{1} = \frac{0}{2}$.

Zero is a real number.

All rational (and irrational) numbers are called real numbers.

The zero power of any nonzero real number is 1: $x^0 = 1, \quad x \neq 0$

Examples: $1^0 = 1$, $2^0 = 1$, $6^0 = 1$, $10^0 = 1$, and so on. This is an arbitrary definition in most arithmetic and algebra texts. It is a useful definition because it fits mathematical patterns such as these:

$$\begin{aligned} 3^3 &= 1 \cdot 3 \cdot 3 \cdot 3 = 27 \\ 3^2 &= 1 \cdot 3 \cdot 3 \quad\; = 9 \\ 3^1 &= 1 \cdot 3 \qquad\;\; = 3 \\ 3^0 &= 1 \qquad\qquad = 1 \end{aligned}$$

and

10^3 has place value "thousand."
10^2 has place value "hundred."
10^1 has place value "ten."
10^0 has place value "one" or "units."

Any number plus zero is that number: $a + 0 = a$

Examples: $0 + 0 = 0$, $1 + 0 = 1$ (also $0 + 1 = 1$), $2 + 0 = 2$ (also $0 + 2 = 2$), and so on. This is called the Addition Property of Zero or the Identity Property of Addition.

Any number times zero is zero: $a \cdot 0 = 0$

Examples: $0 \cdot 0 = 0$, $0 \cdot 1 = 0$ (also $1 \cdot 0 = 0$), $0 \cdot 6 = 0$ (also $6 \cdot 0 = 0$), and so on. This is called the Multiplication Property of Zero.

Zero divided by any nonzero real number is zero: $0 \div a = 0, \quad a \neq 0$

Examples: $0 \div 1 = 0$, $0 \div 2 = 0$, $0 \div 5 = 0$, and so on. Division does not work in reverse (division is not commutative). Division is the inverse (a kind of opposite) of multiplication. The statement $0 \div 1 = 0$ is true because $1 \cdot 0 = 0$. The statement $0 \div 5 = 0$ is true because $5 \cdot 0 = 0$. Because of this, zero can be used as a numerator for a fraction if the denominator is a nonzero real number.

Division by zero is not defined: $a \div 0$ has no value

Examples: $1 \div 0$, $2 \div 0$, and $6 \div 0$ have no value.

$1 \div 0$ is not 0 because $0 \cdot 0$ is not 1.
$1 \div 0$ is not 1 because $0 \cdot 1$ is not 1.
$6 \div 0$ is not 0 because $0 \cdot 0$ is not 6.
$6 \div 0$ is not 6 because $0 \cdot 6$ is not 6.
$0 \div 0$ is not defined because it is ambiguous.

By the definition of division, if zero were not an exception, any answer would check:

$0 \div 0 = 17$ (?) because $0 \cdot 17 = 0$.
$0 \div 0 = 2$ (?) because $0 \cdot 2 = 0$.
$0 \div 0 = 0$ (?) because $0 \cdot 0 = 0$.

Such problems are of no use, so division by zero is not defined for *any* dividend. Because of this, zero cannot be used as the denominator of a fraction.

APPENDIX II

Perimeter and Area

Square

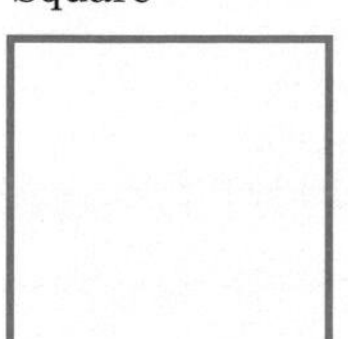

Perimeter: $P = 4s$

Area: $A = s^2$

Rectangle

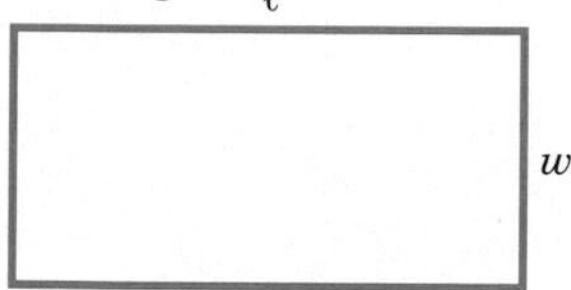

Perimeter: $P = 2\ell + 2w$

Area: $A = \ell w$

Triangle

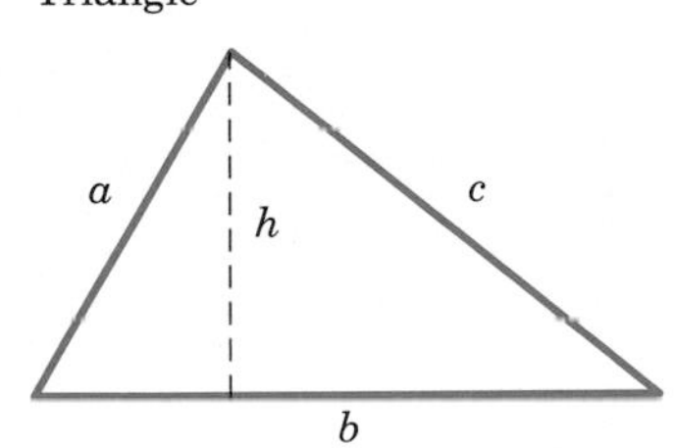

Perimeter: $P = a + b + c$

Area: $A = \dfrac{bh}{2}$

Parallelogram

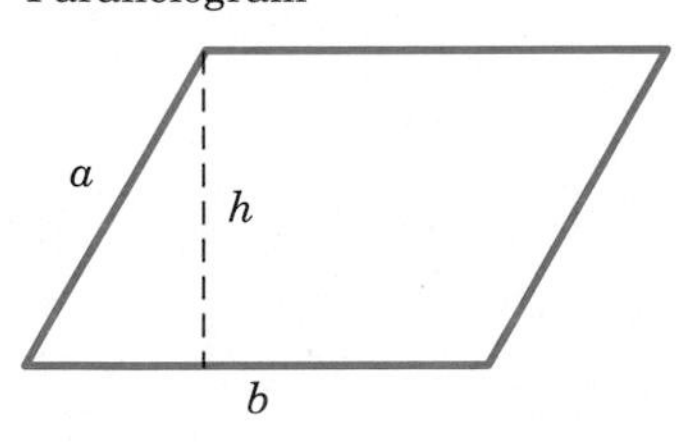

Perimeter: $P = 2a + 2b$

Area: $A = bh$

Trapezoid

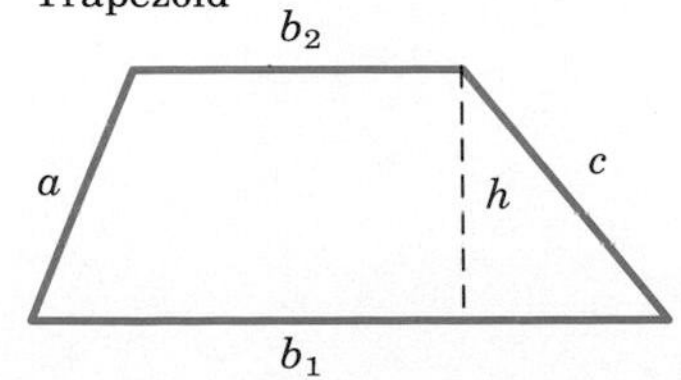

Perimeter: $P = a + b_1 + c + b_2$

Area: $A = \dfrac{1}{2}(b_1 + b_2) \cdot h$

Circle

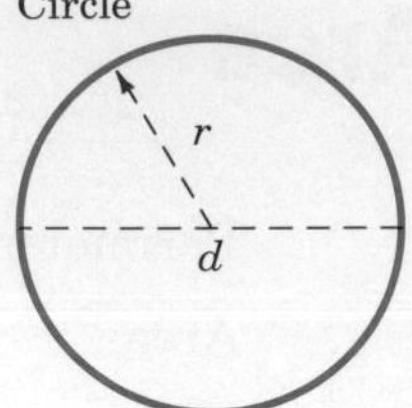

Circumference: $C = 2\pi r$

$C = \pi d$

Area: $A = \pi r^2$

$A = \dfrac{\pi d^2}{4}$

Semicircle

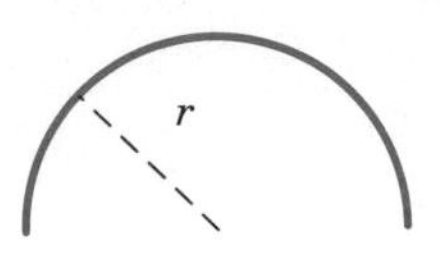

Length: $L = \pi r = \dfrac{1}{2}\pi d$

APPENDIX III

Volume

Rectangular solid

$V = \ell wh$

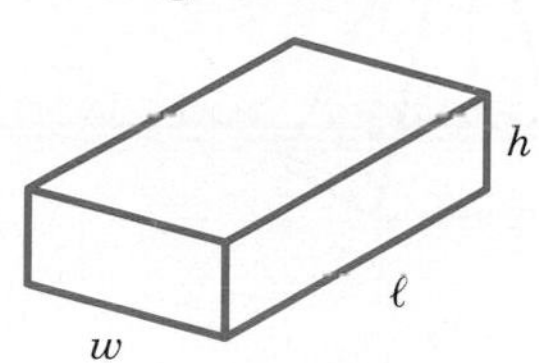

Cube

$V = e^3$

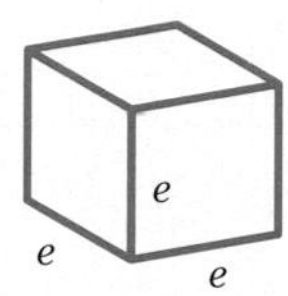

Sphere

$V = \frac{4}{3}\pi r^3$

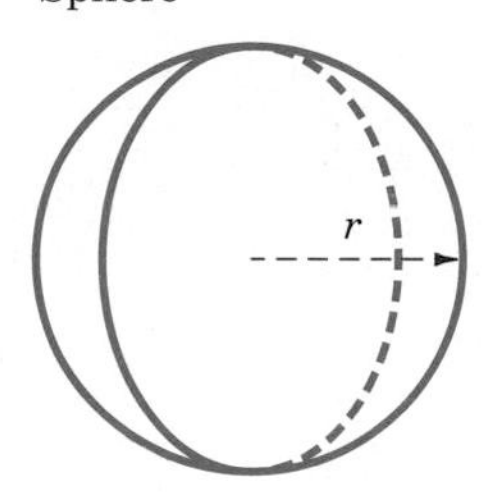

Cylinder

$V = \pi r^2 h$

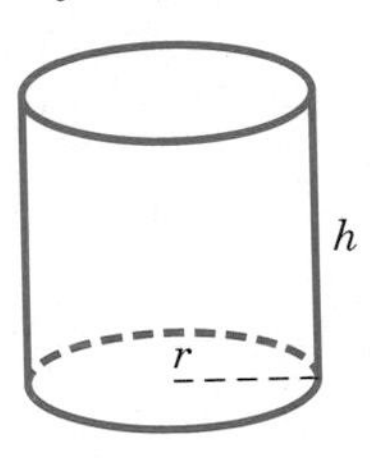

Cone

$V = \frac{1}{3}\pi r^2 h$

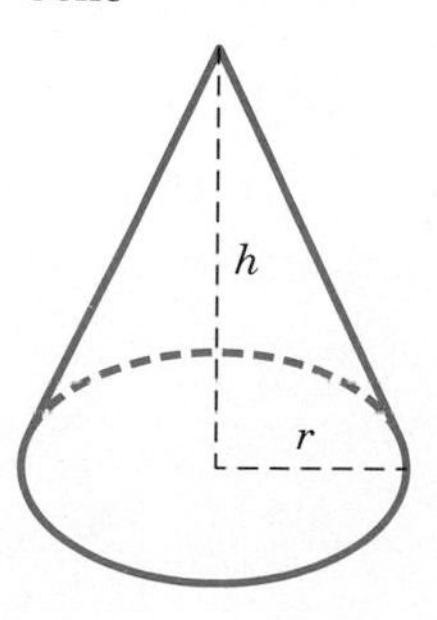

Pyramids

$$V = \frac{1}{3}\text{ (area of base)(height)}$$

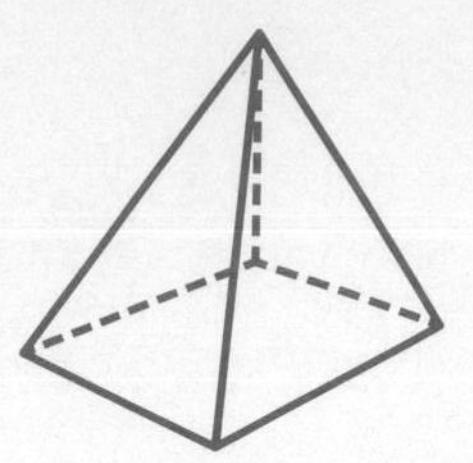

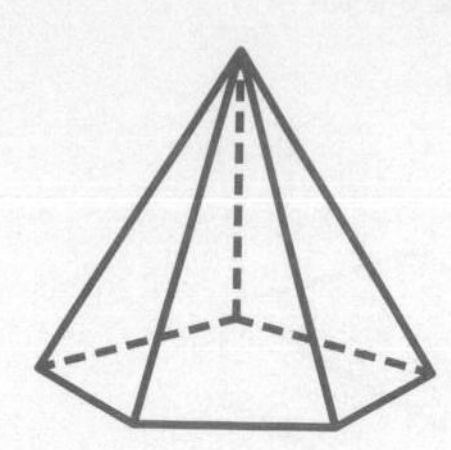

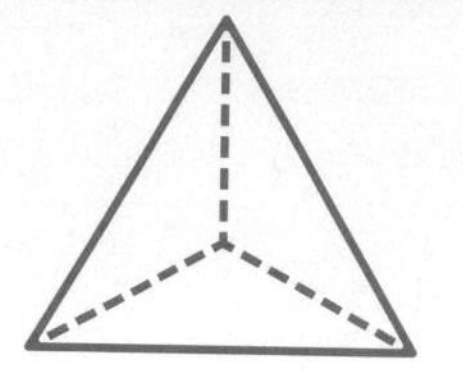

APPENDIX IV
Calculators

The wide availability and economical price of current hand-held calculators make them ideal for doing time-consuming arithmetic operations. You are encouraged to use a calculator as you work through this text. Calculator examples throughout the text show where the use of a calculator is appropriate. Your calculator will be especially useful for

1. doing the fundamental operations of arithmetic (add, subtract, multiply, and divide),
2. checking solutions to equations,
3. checking solutions to problems,
4. finding square roots of numbers, and
5. finding powers of numbers.

To practice solving the problems in this appendix with your calculator, you will need to know whether the calculator is a basic calculator or a scientific calculator. Here are examples of each.

Basic calculator

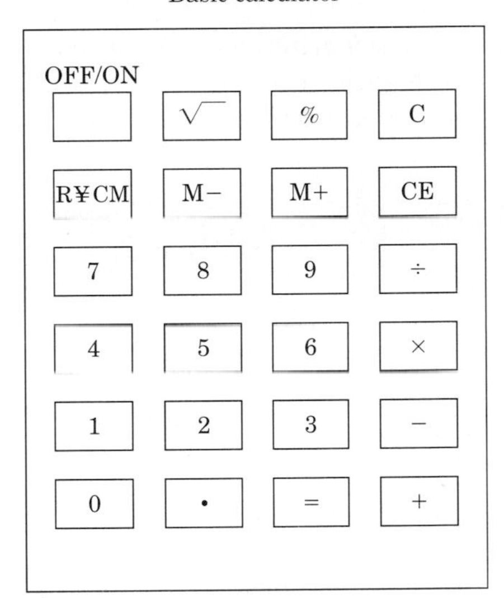

Scientific calculator

$1/x$	x^2	$\sqrt{x}$	OFF	ON/C
INV	SIN	COS	TAN	DRG
K	EE ↓	LOG	ln x	y^x
Σ+	$x \gtrless y$	(	)	÷
STO	7	8	9	×
RCL	4	5	6	−
SUM	1	2	3	+
EXC	0	•	+/−	=

Most computers, many scientific calculators, and some other calculators have the fundamental order of operations built into their circuitry. To tell if your calculator has this feature, do the following exercise:

6 + 4(9)

ENTER	DISPLAY
6	6.
+	6.
4	4.
×	4.
9	9.
=	42.

If the display reads "42," your calculator finds values of combinations according to the agreed-upon order. If the display reads "90," you can get the correct result, 42, by entering the problem in the calculator following the rules for the order of operations.

ENTER	DISPLAY
4	4.
×	4.
9	9.
+	36.
6	6.
=	42.

A calculator that has parentheses keys can also override the order of operations.

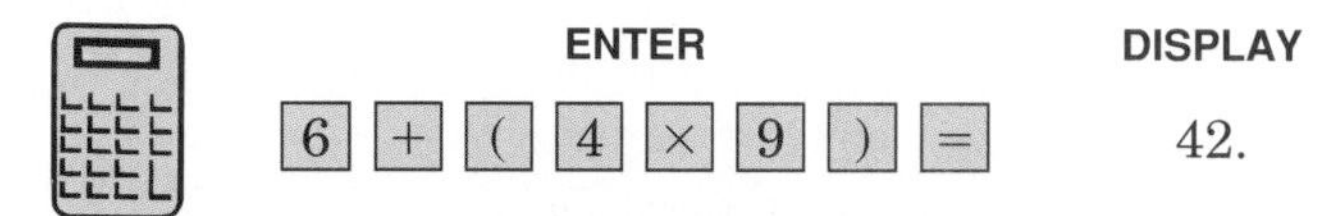

If yours is a scientific calculator it has other keys that you may find useful in later mathematics courses. Recall that parentheses and fraction bars are used to group operations to show which one is done first.

EXPRESSION	BASIC CALCULATOR Enter		SCIENTIFIC CALCULATOR Enter	DISPLAY
$144 \div 3 - 7$	144 ÷ 3 − 7 =		144 ÷ 3 − 7 =	41.
$\frac{28 + 42}{10}$	28 + 42 = ÷ 10 =		28 + 42 = ÷ 10 =	7.
$\frac{288}{6 + 12}$	6 + 12 =	STEP 1		18.
	288 ÷ 18 =	STEP 2		16.
			6 + 12 = $1/x$ × 288 =	16.
$13^2 + 4(17)$	13 × 13 =	STEP 1		169.
	4 × 17 =	STEP 2		68.
	169 + 68 =	STEP 3		237.
			13 × 13 + 4 × 17 =	237.
			or	
			13 x^2 + 4 × 17 =	237.

MODEL PROBLEM-SOLVING

Study the following and practice on your calculator until you can get the results shown in the answer column.

		ANSWERS
a.	$47 + \frac{525}{105}$	52
b.	$\frac{45 + 525}{38}$	15
c.	$\frac{648}{17 + 15}$	20.25
d.	$\frac{140 - 5(6)}{11}$	10
e.	$\frac{3870}{9(7) + 23}$	45
f.	$\frac{5(73) + 130}{33}$	15

Round answer to the nearest hundredth:

g.	$\frac{388 - 16(14)}{27}$	6.07
h.	$\frac{4567}{45(13) - 6(25)}$	10.50

APPENDIX V

Plane Geometry Supplement

For the section on Reasoning, please turn to section 7.9.

Geometry of Angles

OBJECTIVES

2. **Classify angles according to size.**
3. **Find the supplement or the complement of a given angle.**

APPLICATION

The ability to classify angles and compute complements and supplements of given angles is a skill needed to work with triangles.

VOCABULARY

In the geometry of this chapter, there are four undefined words: point, line, plane, and straight. We generally say that a *point* has no dimensions. A *line* has no width or thickness, only length. A *plane* has only length and width, no thickness. *Straight* is usually thought of as not curving or bending.

A quantity or object is said to be *finite* if it has bounds, or limits. Such quantities or objects can be counted or measured. A quantity is said to be *infinite* if it has no bounds, or limits. In that case the quantity or object cannot be counted or measured.

Points are named (usually) by capital letters.

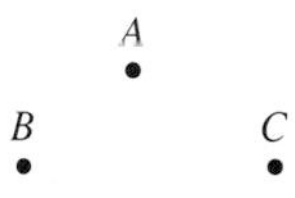

A, B, and C are points.

A line extends infinitely in both directions. A line can be named with a single lower-case letter or by two of the points on the line.

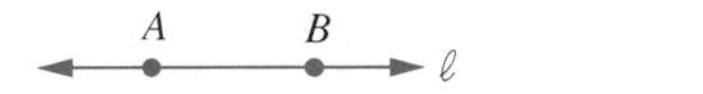

line ℓ, or line AB

A *ray* has an endpoint and extends infinitely in one direction. A ray is named by the endpoint and one other point on the ray.

ray AC

A *line segment* is a portion of a line with two distinct endpoints. A line segment is named by the two endpoints.

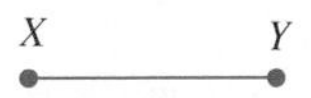

line segment XY

Equal line segments have the same length.

Two lines *intersect* when they have one point in common.

An *angle* is formed by two rays or two line segments with a common endpoint. $\angle$ is a symbol used to denote an angle. The common endpoint is called the *vertex*. An angle can be named in several ways. It can be named by the letter at the vertex (common endpoint) if it is the only angle with that vertex. It can be named by a number placed inside the angle. It can also be named by three points, one on each side and the third at the vertex. (When this name is used, the vertex is in the middle.) An angle is the amount of rotation necessary to bring one of the rays (or line segments) into alignment with the other ray (or line segment) which formed the angle.

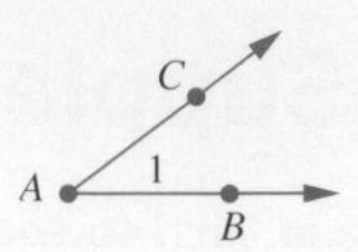

$\angle$A, $\angle$1, or $\angle$CAB, or $\angle$BAC

Adjacent angles have a common vertex and a common side between them.

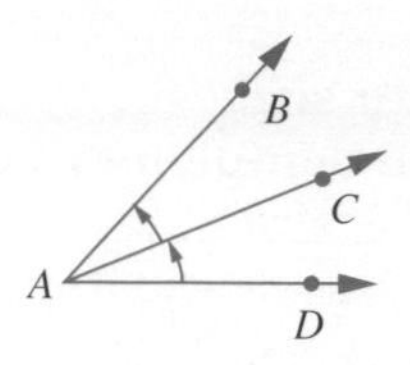

$\angle$DAC and $\angle$CAB are adjacent angles

A common unit of measure for angles is a *degree.* A degree is 1/360 of a complete revolution. The symbol for degree is °. *Equal angles* have the same degree measure.

The *bisector* of an angle is a line segment with one endpoint at the vertex of the angle which divides the angle into two equal angles.

Angles can be classified by their measure.

Acute angle

Measures between 0° and 90°

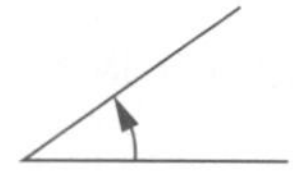

Right angle

Measures exactly 90° (one-quarter of a complete rotation)

A square at the vertex is used to symbolize a right angle.

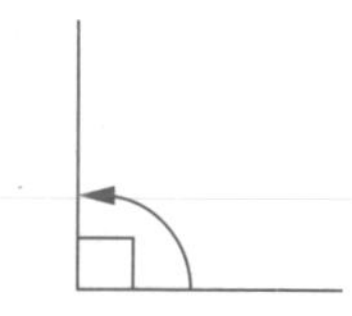

Obtuse angle

Measures between 90° and 180°

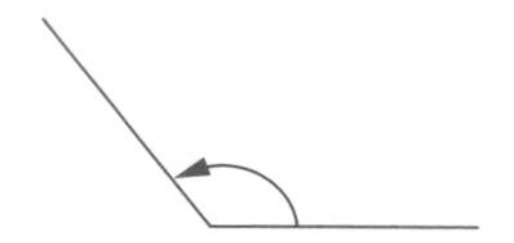

Straight angle

Measures exactly 180°; the sides form a straight line

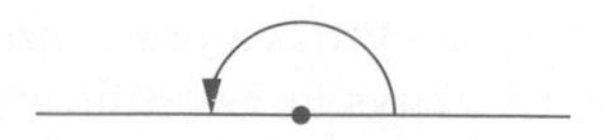

Reflex angle

Measures greater than 180°

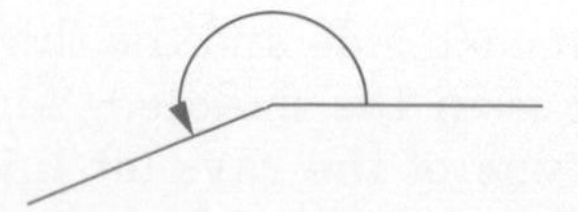

Two angles are *complementary* if the sum of their measures is 90°, or if they are adjacent angles, they form a right angle. Two angles are *supplementary* if the sum of their measures is 180°, or if they are adjacent angles, they form a straight angle.

Two lines or line segments are said to be *perpendicular* when they intersect so that the angles formed are equal (right angles). The symbol used to indicate that two lines or segments are perpendicular is ⊥.

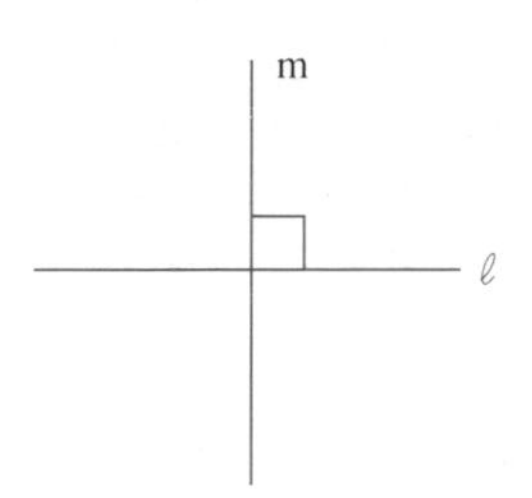

⊥ m (is perpendicular to m)

HOW AND WHY

Recall, from Chapter 8, that a polygon is a figure bounded by straight sides. Examples of polygons are triangles, rectangles, parallelograms, squares, pentagons, and hexagons. Each has an angle formed wherever two of its sides meet. Thus, a polygon has as many angles as it has sides. For example, a triangle has three sides and three angles; a rectangle has four sides and four angles; and a hexagon has six sides and six angles.

Basic properties of mathematics are needed to solve the problems of geometry. Many of the properties used here are comparable to those used in algebra. These properties are often used in steps to form "a proof." A *proof* is a series of steps, each of which is true based on assumptions, definitions, properties of algebra and geometry, or previously proven statements. (See Examples d, e, and f.)

Here is a comparison of some of the properties of algebra and geometry:

ALGEBRA	GEOMETRY
Reflexive Property $a = a$ "a number is equal to itself"	Identity ∠A = ∠A, or CE = CE "a quantity is equal to itself"
Transitive Property If $x = y$ and $y = z$, then $x = z$,	Quantities equal to the same or equal quantities are equal to each other.
Division Property of Equality If $2x = 4$, then $x = 2$ Recall that division by zero is not defined.	Equals divided by equals are equal. Halves of equals are equal, thirds of equals are equal, etc.
Addition or Subtraction Property of Equality If $x + 30 = 50$, then $x + 30 - 30 = 50 - 30$	Equals added to equals or equals subtracted from equals are equal.
Substitution Property If $x + y = 75$ and $y = 30$, then $x + 30 = 75$	Substitution: If ∠A = ∠B and ∠A + ∠B + 50° = 180°, then ∠A + ∠A + 50° = 180°

Another property of geometry is that the whole is equal to the sum of its parts. For example:

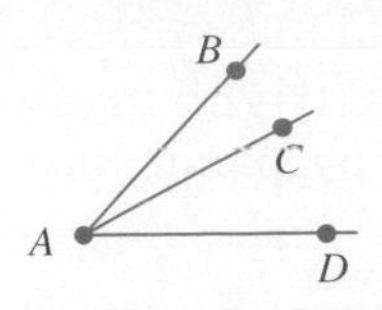

($\angle$CAB and $\angle$DAC are adjacent angles.)
$\angle$CAB + $\angle$DAC = $\angle$DAB

Since two angles are complementary when their sum is 90°, to find the complement of a given angle, subtract the given angle measure from 90°. Thus, angles whose measures are 18° and 72° are complementary since

90° − 72° = 18° and 90° − 18° = 72°

Since two angles are supplementary when their sum is 180°, to find the supplement of a given angle, we subtract the given angle measure from 180°. Thus angles whose measures are 113° and 67° are supplementary since

180° − 113° = 67° and 180° − 67° = 113°.

MODEL PROBLEM SOLVING

Examples

a. Find the complement of 16°.

90° − 16° = 74°

Strategy: **Two angles are said to be complementary when their sum is 90°, so we subtract the given angle from 90°.**

The complement of 16° is 74°.

Warm Up a. Find the complement of 39°.

b. Find the supplement of 16°.

180° − 16° = 164°

Two angles are said to be supplementary when their sum is 180°, so we subtract the given angle from 180°.

The supplement of 16° is 164°.

Warm Up b. Find the supplement of 116°.

c. What kind of an angle is one whose measure is 62°?

An acute angle

Since the measure of the angle is less than 90°, it is an acute angle.

Warm Up c. What kind of an angle is one whose measure is 143°?

ANSWERS TO WARM UPS **a.** 51° **b.** 64° **c.** Obtuse

d. Given: $\angle A = \angle B$, AC bisects $\angle A$ and BD bisects $\angle B$
Prove: $\angle 1 = \angle 2$

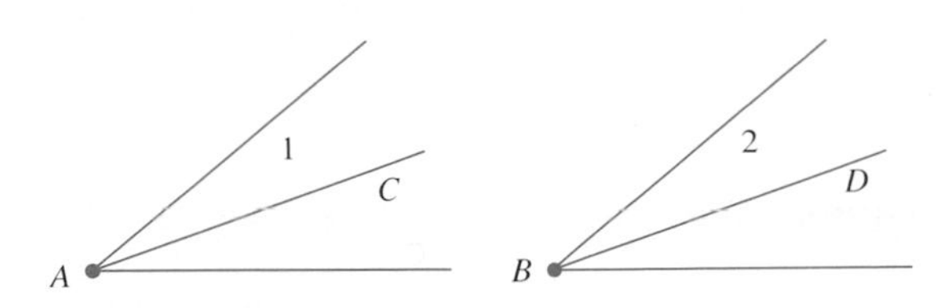

Statements	Reasons
1. $\angle A = \angle B$	**1. Given**
2. $\angle 1$ is half of $\angle A$	**2. Definition of bisect**
3. $\angle 2$ is half of $\angle B$	**3. Definition of bisect**
4. $\angle 1 = \angle 2$	**4. Halves of equals are equal**

Warm Up d. Given: $\angle 1 = \angle 2$, $\angle 2 = \angle 3$, $\angle 3 = \angle 4$
Prove: $\angle 1 = \angle 4$

e. Given: $\angle EAC = \angle DAB$
Prove: $\angle EAD = \angle CAB$

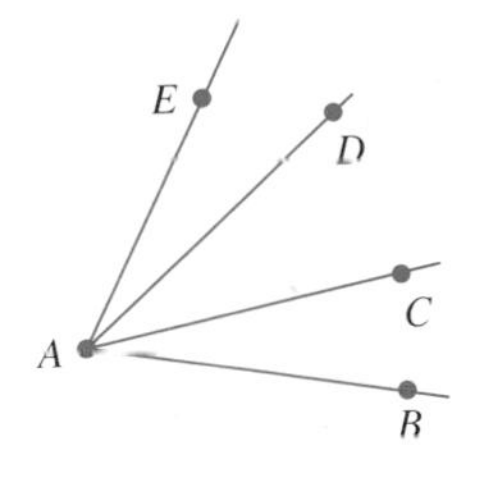

Statements	Reasons
1. $\angle EAC = \angle DAB$	**1. Given.**
2. $\angle DAC = \angle DAC$	**2. Identity.**
3. $\angle EAC - \angle DAC = \angle DAB - \angle DAC$	**3. Equals subtracted from equals are equal.**
4. $\angle EAD = \angle CAB$	**4. Substitution in Step 3. $\angle EAC - \angle DAC = \angle EAD$ and $\angle DAB - \angle DAC = \angle CAB$.**

Warm Up e. Given: AC = BD
Prove: AB = CD

ANSWERS TO WARM UPS

d.

$\angle 1 = \angle 2$	Given
$\angle 2 = \angle 3$	Given
$\angle 1 = \angle 3$	Quantities = to the same quantity are = to each other.
$\angle 3 = \angle 4$	Given
$\angle 1 = \angle 4$	Quantities = to the same quantity are = to each other.

e.

AC = BD	Given
BC = BC	Identity
AC − BC = BD − BC	= subtracted from =
AB = CD	Substitution in step 3. AC − BC = AB and BD − BC = CD.

f. Given: $\angle 1 + \angle 2 + \angle 3 = 150°$
$\angle 4 + \angle 5 + \angle 6 = 150°$
$\angle 1 = \angle 4$
$\angle 2 = \angle 5$

Prove: $\angle 3 = \angle 6$

Statements	Reasons
1. $\angle 1 + \angle 2 + \angle 3 = 150°$	1. **Given**
2. $\angle 4 + \angle 5 + \angle 6 = 150°$	2. **Given**
3. $\angle 1 + \angle 2 + \angle 3 = \angle 4 + \angle 5 + \angle 6$	3. **Quantities equal to the same quantity are equal to each other.**
4. $\angle 1 = \angle 4$ and $\angle 2 = \angle 5$	4. **Given**
5. $\angle 1 + \angle 2 + \angle 3 - \angle 1 - \angle 2 = \angle 4 + \angle 5 + \angle 6 - \angle 4 - \angle 5$	5. **Equals subtracted from equals are equal.**
6. $\angle 3 = \angle 6$	6. **Simplify**

Warm Up f. Given: $\angle 1$ and $\angle 2$ are supplementary
$\angle 3$ and $\angle 4$ are supplementary
$\angle 1 = \angle 3$

Prove: $\angle 2 = \angle 4$

ANSWER TO WARM-UP

f.

$\angle 1 + \angle 2 = 180°$	Supplementary angles
$\angle 3 + \angle 4 = 180°$	Supplementary angles
$\angle 1 + \angle 2 = \angle 3 + \angle 4$	Quantities equal to the same quantity are equal to each other
$\angle 1 = \angle 3$	Given
$\angle 1 + \angle 2 - \angle 1 = \angle 3 + \angle 4 - \angle 3$	Equals subtracted from equals are equal
$\angle 2 = \angle 4$	Simplify

ANSWERS

1. ______
2. ______
3. ______
4. ______
5. ______
6. ______
7. ______
8. ______
9. ______
10. ______
11. ______
12. ______
13. ______
14. ______
15. ______

Exercises

Classify each of the following angles as acute, right, obtuse, straight, or reflex.

1. 80° **2.** 305° **3.** 98°

4. 90° **5.** 149° **6.** 180°

7. 4° **8.** 193°

9. 135° **10.** 72°

Find the complement of each of the following angles.

11. 40° **12.** 25° **13.** 85°

14. 36° **15.** 15°

Find the supplement of each of the following angles.

16. 80° **17.** 125° **18.** 25°

19. 115° **20.** 55°

Use the following list of angle measures to answer 21–25.
95°, 18°, 210°, 45°, 180°, 82°, 121°, 175°, 315°, 90°, 282°

21. Identify the acute angles.

22. Identify the straight angles.

23. Identify the reflex angles.

24. Identify the right angles.

25. Identify the obtuse angles.

26. If $\angle A$ and $\angle B$ are supplementary and $\angle B = 134°$, find $\angle A$.

27. If $\angle A$ and $\angle B$ are supplementary and $\angle B = 37°$, find $\angle A$.

28. If $\angle C$ and $\angle D$ are complementary and $\angle D = 15°$, find $\angle C$.

29. If $\angle C$ and $\angle D$ are complementary and $\angle D = 78°$, find $\angle C$.

30. If the sum of $\angle C$ and $\angle D$ is a right angle and $\angle C = 32°$, find $\angle D$.

Identify each of the following statements as either true (always true) or false (not always true).

31. The complement of an acute angle is an acute angle.

16. ______
17. ______
18. ______
19. ______
20. ______
21. ______
22. ______
23. ______
24. ______
25. ______
26. ______
27. ______
28. ______
29. ______
30. ______
31. ______

ANSWERS

32. ______________________

33. ______________________

34. ______________________

35. ______________________

36. ______________________

37. ______________________

38. ______________________

39. ______________________

40. ______________________

41. ______________________

42. ______________________

43. ______________________

32. The supplement of an obtuse angle is an obtuse angle.

33. The supplement of an acute angle is a reflex angle.

34. The sum of a right angle and an acute angle is an obtuse angle.

35. The sum of two acute angles is an obtuse angle.

36. The sum of two obtuse angles is a reflex angle.

37. The sum of an obtuse angle and an acute angle is a reflex angle.

38. A straight angle minus an obtuse angle is an acute angle.

39. A line that bisects a straight angle forms two right angles.

40. A line that bisects an obtuse angle forms two acute angles.

Prove each of the following.

41. Given: $\angle A + \angle B = 90°$
$\angle A + \angle C = 90°$
Prove: $\angle B = \angle C$

42. Given: $\angle A + \angle B + \angle C = 140°$
$\angle B + \angle C = \angle A$
$\angle B = \angle C$
Prove: $\angle C = 35°$

43. Given: $\angle A$ and $\angle B$ are supplementary
$\angle B$ and $\angle C$ are complementary
$\angle C = 36°$
Prove: $\angle A = 126°$

44. Given: $\angle$ A is bisected to form $\angle$ 1 and $\angle$ 2
$\angle$ 1 is a right angle
Prove: $\angle$ A is a straight angle.

ANSWERS

44. ____________________

Geometry of Triangles

OBJECTIVES

4. **Identify triangles by the lengths of sides.**
5. **Identify triangles by the measures of the angles.**

APPLICATION

Rosemary is installing a triangular dog run behind her house. One side of the dog run is formed by a side of the house; a second side of the run is at right angles to the first side; and the third side of the run forms an angle of 42° with the second side. Find the measure of the third angle of the run.

VOCABULARY

A *triangle* is a closed figure in a plane bounded by three straight sides. Each side is a line segment. The symbol for triangle is Δ.

An *angle of a triangle* is formed by the intersection of two sides of the triangle. The two sides of the triangle are the sides of the angle. A triangle has three angles.

Triangles are classified by their angles or by their sides. Triangles classified by their angles include

Acute triangle	All three angles are acute
Right triangle	One angle of the triangle is a right angle
Obtuse triangle	One angle is obtuse
Equiangular triangle	All angles have equal measure

Triangles classified by their sides are called

Scalene triangle	All sides are of different length
Isosceles triangle	Two sides are equal in length
Equilateral triangle	All sides are equal in length

The *vertex angle* of an isosceles triangle is the angle formed by the two equal sides. The *base angles* of an isosceles triangle are the angles opposite the equal sides. The *legs* of an isosceles triangle are the two equal sides.

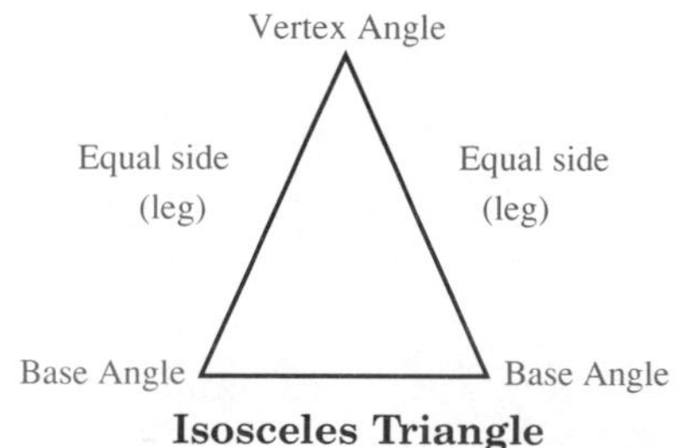

Isosceles Triangle

An angle is said to be an angle that is *included* between two sides when it is formed by those two sides.

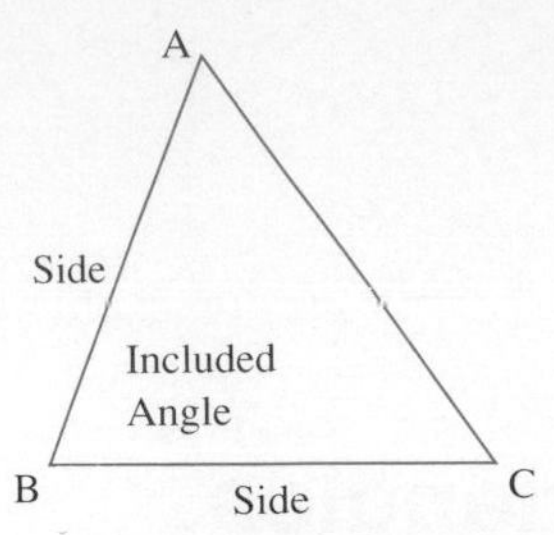

∠B is included between AB and BC

A side is said to be an *included side* when it is included between two angles that is, it is a side of each of the angles.

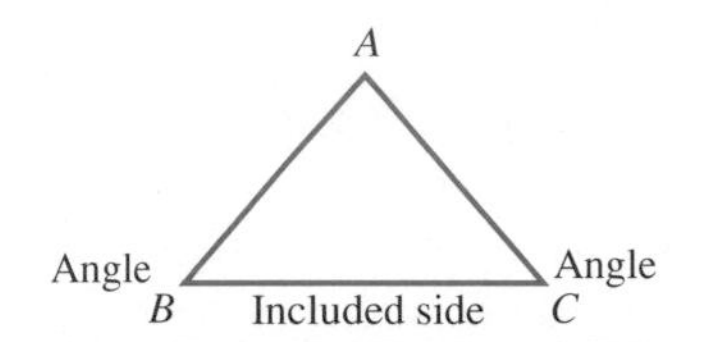

Side BC is included between ∠B and ∠C

HOW AND WHY

So that we may work with triangles in later applications, we need to become familiar with the different classifications. We must recognize that a triangle in which none of the sides are equal is called scalene, whereas a triangle with two sides equal is isosceles. If all three sides have the same length, the triangle is equilateral.

At this time, we assume that the sum of the measures of the angles of a triangle is 180°. In a later section, we will prove this statement.

MODEL PROBLEM SOLVING

Examples	Strategy
a. Classify Δ ABC if AB = 12, AC = 15, and BC = 18.	
Δ ABC is scalene.	**The length of each side is given. So we classify by the sides. None of the sides are equal.**

Warm Up a. Classify ΔABC if side AB = 16, side BC = 11, and side AC = 16.

b. Classify ΔABC if ∠A = 60°, ∠C = 80°, and ∠B = 40°.	
ΔABC is acute.	**In this case the measure of each angle is known. The measure of each angle is less than 90°.**

Warm Up b. Classify ΔABC if ∠A = 48°, ∠B = 48°, and ∠C = 84°.

c. Classify ΔABC if ∠C = 90°, ∠A = 42°, and ∠B = 48°.	
ΔABC is a right triangle.	**The triangle has a right angle.**

ANSWERS TO WARM-UPS

a. Isosceles **b.** Acute

Warm Up c. Classify ΔABC if $\angle B = 134°$, $\angle A = 14°$, and $\angle C = 32°$.

d. Given: ΔABC has $\angle A = 92°$ and $\angle C = 44°$
What is the measure of $\angle B$?

Statements	Reasons
1. $\angle A + \angle B + \angle C = 180°$	**1. The sum of the angles of a triangle is 180°**
2. $92° + \angle B + 44° = 180°$	**2. Substitution**
3. $\angle B + 136° = 180°$	**3. Simplify**
4. $\angle B + 136° - 136° = 180° - 136°$	**4. Equals subtracted from equals**
5. $\angle B = 44°$	**5. Simplify**

Warm Up d. Given ΔABC with $\angle B = 99°$ and $\angle C = 44°$. What is the measure of $\angle A$?

e. Given: ΔABC with $\angle C$ a right angle and $\angle A = \angle B$. What is the measure of $\angle A$ and $\angle B$?

Statements	Reasons
1. $\angle A + \angle B + \angle C = 180°$	**1. The sum of the angles of a triangle is 180°**
2. $\angle C = 90°$	**2. Given**
3. $\angle A = \angle B$	**3. Given**
4. $\angle A + \angle A + 90° = 180°$	**4. Substitution**
5. $2\angle A + 90° = 180°$	**5. Simplify**
6. $2\angle A + 90° - 90° = 180° - 90°$	**6. Equals subtracted from equals**
7. $2\angle A = 90°$	**7. Simplify**
8. $\angle A = 45°$	**8. Division property of equality (halves of equals are equal)**
9. $\angle B = 45°$	**9. Substitution**

Warm Up e. Given: ΔABC is isosceles with $\angle A$ and $\angle B$ the base angles. $\angle C$, the vertex angle, is 72°. What is the measure of each of the base angles?

ANSWERS TO WARM UPS **c.** Obtuse **d.** 37° **e.** 54°

Application Solution

f. Rosemary is installing a triangular dog run behind her house. One side of the dog run is formed by a side of the house; a second side of the run is at right angles to the first side; and the third side of the run forms an angle of 42° with the second side. Find the measure of the third angle of the run.

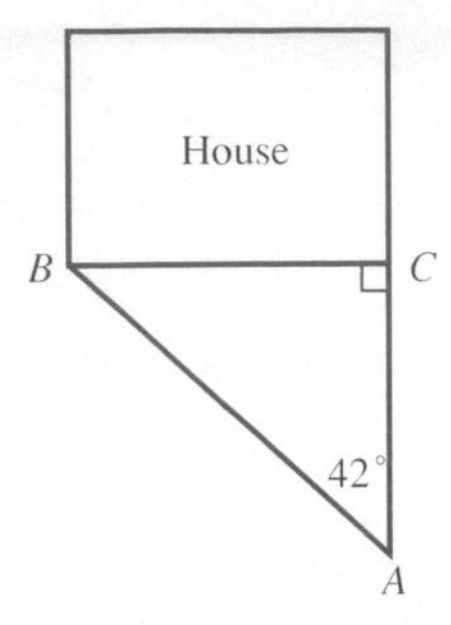

Given: $\angle A = 42°$

$\angle C$ is a right angle

Find: $\angle B = ?$

Statements	**Reasons**
1. $\angle A = 42°$	1. Given
2. $\angle C$ is a right angle	2. Given
3. $\angle C = 90°$	3. Definition of a right angle
4. $\angle A + \angle B + \angle C = 180°$	4. The sum of the angles of a triangle is 180°
5. $42° + \angle B + 90° = 180°$	5. Substitution
6. $\angle B + 132° = 180°$	6. Simplify
7. $\angle B + 132° - 132° = 180° - 132°$	7. Equals subtracted from equals
8. $\angle B = 48°$	8. Simplify

The measure of the third angle is 48°.

Warm Up f. Rosemary is installing a triangular dog run behind her house. One side of the dog run is formed by a side of the house; a second side of the run is at right angles to the first side; and the third side of the run forms an angle of 56° with the second side. Find the measure of the third angle of the run.

ANSWER TO WARM UP

f. 34°

Exercises

ANSWERS

For 1–4, classify the following triangles by sides.

1. Given ΔABC with $AB = 7$, $BC = 7$, and $AC = 15$, what kind of triangle is ΔABC?

2. Given ΔABC with $AB = 9.8$, $BC = 3.2$, and $AC = 9.7$, what kind of triangle is ΔABC?

3. Given ΔDEF with $DE = 3.21$, $DF = 3.21$, and $EF = 3.21$, what kind of triangle is ΔDEF?

4. Given ΔDEF with $DE = 83$, $DF = 73$, and $EF = 63$, what kind of triangle is ΔDEF?

For 5–8, classify the following triangles by angles.

5. Given ΔABC with $\angle A = 15°$, $\angle B = 15°$, and $\angle C = 150°$, what kind of triangle is ΔABC?

6. Given ΔABC with $\angle A = 35°$, $\angle B = 56°$, and $\angle C = 89°$, what kind of triangle is ΔABC?

7. Given ΔDEF with $\angle D = 13°$, $\angle E = 90°$, and $\angle F = 77°$, what kind of triangle is Δ DEF?

8. Given ΔDEF with $\angle D = 91°$, $\angle E = 75°$, and $\angle F = 14°$, what kind of triangle is $\angle DEF$?

9. Find the measure of $\angle R$ in ΔPQR if $\angle P = 45°$ and $\angle Q = 30°$.

10. Find the measure of $\angle R$ in ΔPQR if $\angle P = 35°$ and $\angle Q = 55°$.

11. Find the measure of $\angle T$ in ΔRST if $\angle R = 26°$ and $\angle S = 124°$.

1. ______
2. ______
3. ______
4. ______
5. ______
6. ______
7. ______
8. ______
9. ______
10. ______
11. ______

ANSWERS

12. ______

13. ______

14. ______

15. ______

16. ______

17. ______

18. ______

19. ______

20. ______

12. Find the measure of $\angle T$ in ΔRST if $\angle R = 85°$ and $\angle S = 80°$.

13. Find the measure of $\angle C$ in ΔABC if $\angle A = 20.3°$ and $\angle B = 94.7°$.

14. Find the measure of $\angle A$ in ΔABC if $\angle B = 27.83°$ and $\angle C = 28.17°$.

15. Find the measure of $\angle D$ in ΔDEF if $\angle F = 129.6°$ and $\angle E = 37.5°$.

16. Find the measure of $\angle E$ in ΔDEF if $\angle D = 10.1°$ and $\angle F = 88.8°$.

For 17–20, classify the following triangles by angles.

17. Given ΔPQR with $\angle P = 45°$ and $\angle Q = 30°$, what kind of triangle is ΔPQR?

18. Given ΔPQR with $\angle P = 35°$ and $\angle Q = 55°$, what kind of triangle is ΔPQR?

19. Given ΔRST with $\angle R = 28°$ and $\angle S = 124°$, what kind of triangle is ΔRST?

20. Given ΔRST with $\angle R = 85°$ and $\angle S = 80°$, what kind of triangle is ΔRST?

State whether the following are true (always true) or false (not always true).

ANSWERS

21. A triangle can be both a right triangle and an isosceles triangle.

21. ______

22. A triangle can be both an acute triangle and an equilateral triangle.

22. ______

23. A triangle can be both an obtuse triangle and an equilateral triangle.

23. ______

24. A triangle can be both an equiangular triangle and a scalene triangle.

24. ______

25. Every acute triangle is also a scalene triangle.

25. ______

26. Every equilateral triangle is also an isosceles triangle.

26. ______

27. Some isosceles triangles are scalene triangles.

27. ______

28. Some right triangles are obtuse triangles.

28. ______

29. Every equilateral triangle is an acute triangle.

29. ______

30. Every obtuse triangle is an isosceles triangle.

30. ______

31. Some right triangles are isosceles triangles.

31. ______

32. Some scalene triangles are right triangles.

32. ______

Congruent Triangles

OBJECTIVES

6. **Prove that two triangles are congruent.**
7. **Use congruent triangles to prove that angles and sides are equal.**

APPLICATION

James wants to construct a bridge across a pond on his farm. So he needs to find the distance across the pond. To do this, he constructs a triangle in his field as shown in the diagram. The triangle is constructed in the following way. He continues AC to the point E so that CE = AC. He continues BC to the point D so that CD = BC and ∠DCE = ∠ACB. Show that DE = AB, which is the distance across the pond.

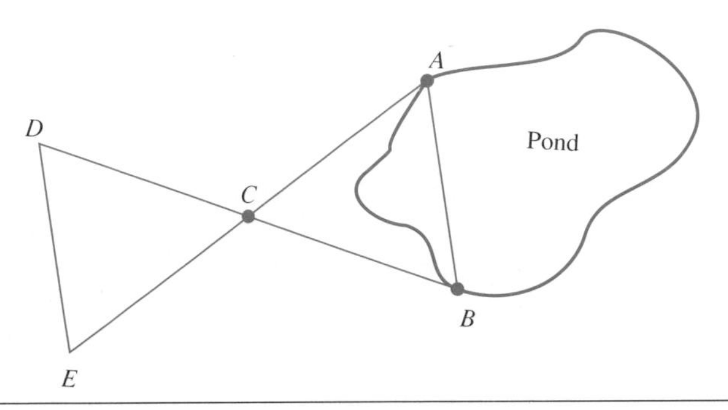

VOCABULARY

Axioms are statements (assumptions) that are accepted without proof. *Postulates* are geometric statements (assumptions) that are accepted without proof. *Theorems* are statements that are proven to be true by using deductive reasoning.

To *bisect* is to divide a quantity or figure into two parts that have equal measures and equal shapes. That which bisects a quantity or figure is referred to as a *bisector*. The bisector may be a point or a line. The *midpoint* of a segment bisects the segment. A bisector of a line segment is a line which passes through the midpoint of the segment. The *perpendicular bisector* of a line segment is the line (segment) that passes through the midpoint of the segment and is also perpendicular to the segment.

Congruent triangles are two or more triangles with exactly the same size and shape; that is, there is a correspondence between each vertex of one triangle with a vertex of the other triangle such that corresponding sides and angles are equal. The symbol used for congruent is ≅. If two triangles are congruent, corresponding parts are equal.

Corresponding parts of congruent triangles refers to pairs of angles or pairs of sides that must occupy the same relative position in their respective figures.

The symbol ∴ is an abbreviation of the word *therefore*.

HOW AND WHY

We accept the following three statements as postulates. Each is related to congruent triangles.

Congruent Triangle Postulate

If three sides of one triangle are equal to three sides of another, the two triangles are congruent. (Side, Side, Side, abbreviated SSS.)

If two sides and the included angle of one triangle are equal to two sides and the included angle of another, then the two triangles are congruent. (Side, Angle, Side, abbreviated SAS.)

If two angles and the included side of one triangle are equal to two angles and the included side of another, then the two triangles are congruent. (Angle, Side, Angle, abbreviated ASA.)

If two triangles have sides or combinations of sides and angles that satisfy any one of these three sets of conditions, then the two triangles are congruent. If they are congruent, corresponding parts are equal. Thus, if it is possible to show that two sides (line segments) or two angles are corresponding parts of congruent triangles, those line segments or angles are equal. This will be designated *corresponding parts of congruent triangles* or simply CPCT.

CAUTION: If three angles of one triangle are equal respectively to the three angles of another, the two triangles are not necessarily congruent.

A common way to show the corresponding parts (sides and angles) in a figure is to mark them in the following way.

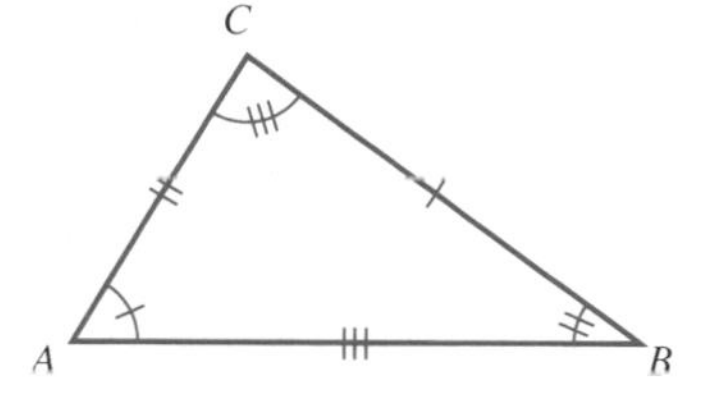

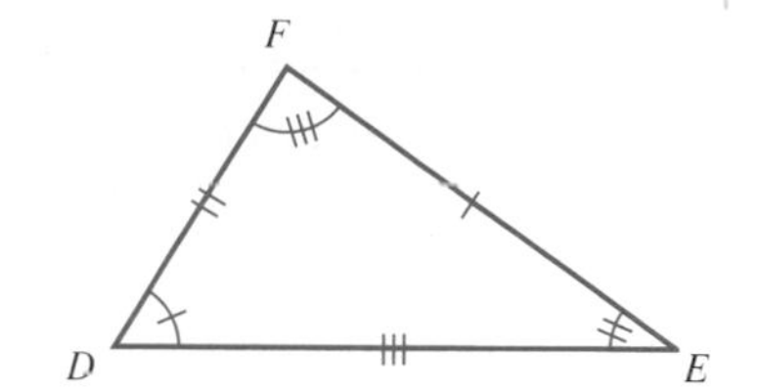

The following pairs of parts of the two triangles are equal.

ΔABC	ΔDEF
$\angle A$, arc with one mark	$\angle D$, arc with one mark
$\angle B$, arc with two marks	$\angle E$, arc with two marks
$\angle C$, arc with three marks	$\angle F$, arc with three marks
AB, side with three marks	DE, side with three marks
BC, side with one mark	EF, side with one mark
AC, side with two marks	DF, side with two marks

In figures marked in this way, we understand that angles or sides with the same marks are equal.

To say that $\Delta ABC \cong \Delta DEF$ is to say also that $\angle A$ corresponds to $\angle D$, $\angle B$ corresponds to $\angle E$, $\angle C$ corresponds to $\angle F$, side AB corresponds to side DE, side BC corresponds to side EF, and side AC corresponds to side DF.

MODEL PROBLEM SOLVING

Examples

a. True or False: If $\Delta ABC \cong \Delta DEF$, then $\angle B = \angle F$

The statement is not necessarily true. $\angle B$ corresponds to $\angle E$.

Strategy

The order in which the letters are written show that $\angle B$ and $\angle E$ are corresponding parts. It is possible that the angles are equal but not because the two triangles are congruent as stated.

Warm Up a. True or False: If $\Delta ABC \cong \Delta DEF$, then $AB = DE$.

b. Is $\Delta ABC \cong \Delta DEF$?

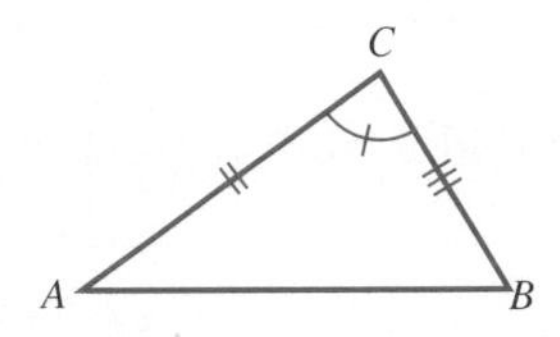

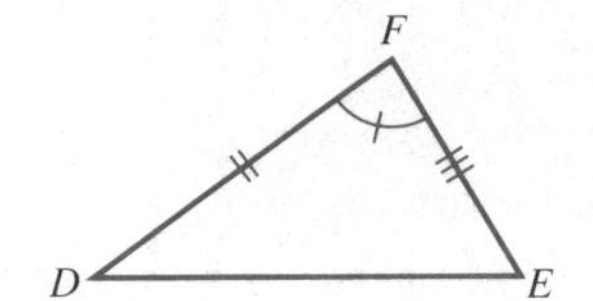

We see that $AC = DF$, $BC = EF$, and $\angle C = \angle F$. The angles are included between the pairs of equal sides. This satisfies the SAS postulate.

Yes, the two triangles are congruent.

Warm Up b. Is $\Delta ABC \cong \Delta DEF$?

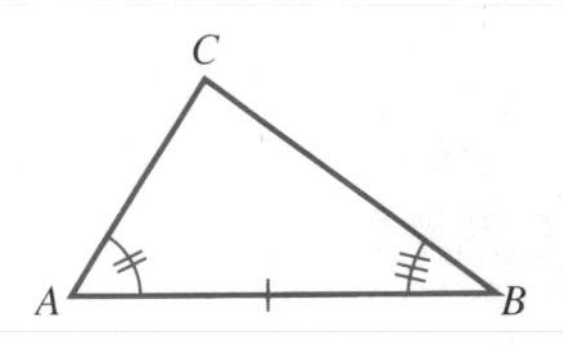

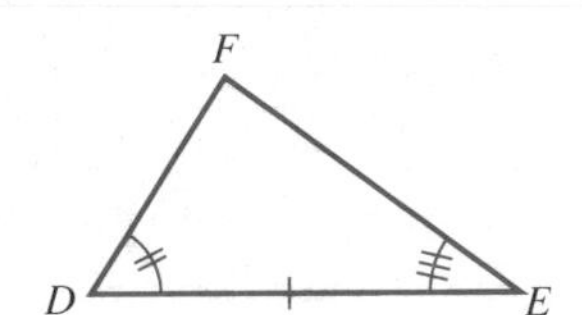

c. Select the pairs of congruent triangles from the following group.

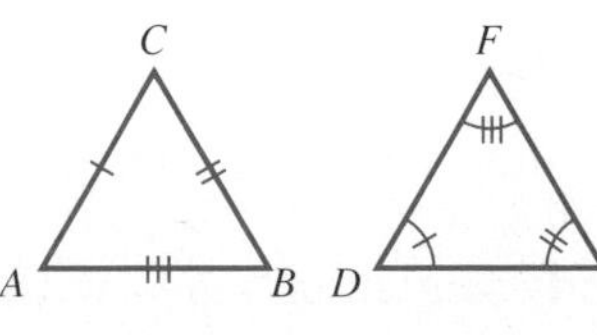
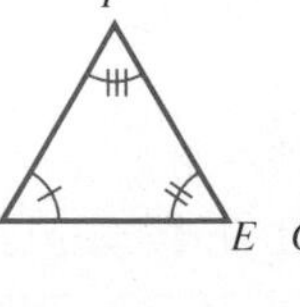
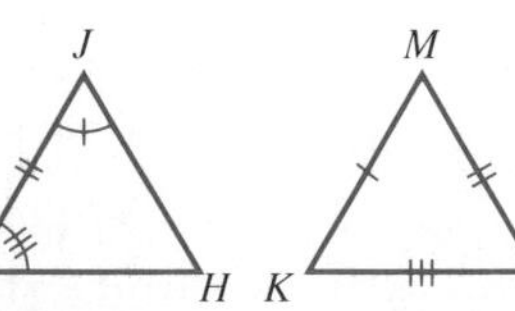
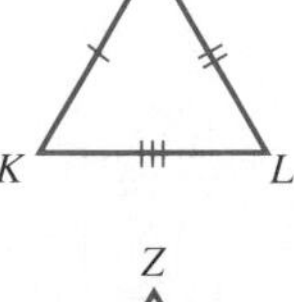

We look at the marks that indicate sides and angles are equal, keeping in mind the three postulates: SSS, SAS, and ASA.

ANSWERS TO WARM UPS

a. True

b. Yes; ASA

ΔABC ≅ ΔKLM, SSS;
ΔGHJ ≅ ΔNPQ, ASA;
ΔRST ≅ ΔUVW, SAS;
ΔDEF and ΔXYZ are not necessarily congruent.

CAUTION: **There is no congruent postulate for angle, angle, angle.**

Warm Up c. Select the pairs of congruent triangles from the following group.

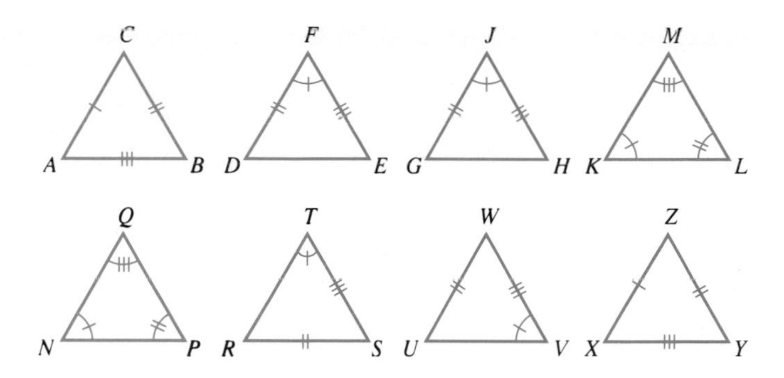

d. Use the following figure to show that the base angles of an isosceles triangle are equal.

Given: Isosceles ΔABC with AC = BC
where CD bisects side AB
Prove: ∠A = ∠B

Recall that an isosceles triangle has two equal sides. The base angles of an isosceles triangle are opposite the equal sides. We first make a drawing to show the figure and label the equal parts.

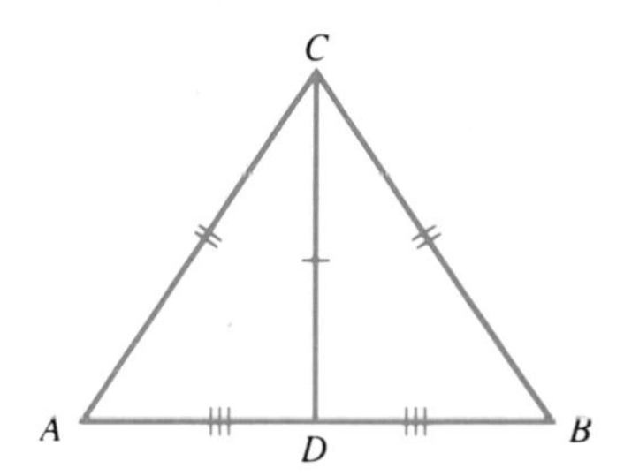

We see that there are two triangles within the isosceles triangle. Is it possible that this pair of triangles (ΔADC and ΔBDC) are congruent and that the base angles of the isosceles triangle are corresponding parts of these triangles?

Statements	Reasons
1. AC = BC	**1. Given**
2. CD bisects AB	**2. Given**
3. AD = BD	**3. Definition of bisect**
4. CD = CD	**4. Identity**
5. ∴ ΔADC ≅ ΔBDC	**5. SSS**
6. ∴ ∠A = ∠B	**6. CPCT**

Therefore the base angles of an isosceles triangle are equal.

The following statement is true; however, we will not prove it. If two angles of a triangle are equal, the sides opposite those angles are equal. So if two angles of a triangle are equal, it is an isosceles triangle.

ANSWER TO WARM UPS

c. ΔABC ≅ ΔXYZ, SSS;
ΔDEF ≅ ΔGHJ, SAS;
ΔKLM and ΔNPQ are not necessarily congruent. There is no AAA postulate for congruent triangles.
ΔRST and ΔUVW are not necessarily congruent. The angle is not included between the sides.

Warm Up d. Use the figure from Example d to prove that CD is the perpendicular to the base. (Prove that CD ⊥ AB; that is, angles formed are equal.)

Application Solution

e. James wants to construct a bridge across a pond on his farm. He needs to find the distance across the pond. To do this, he constructs a triangle in his field as shown in the following diagram. The triangle is constructed in the following way. He continues AC to the point E so that CE = AC. He continues BC to the point D so that CD = BC and ∠DCE = ∠ACB. Show that DE = AB, which is the distance across the pond.

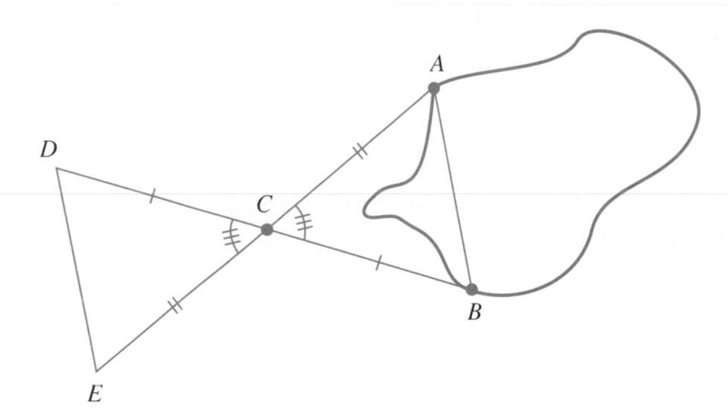

The figure is drawn and the equal parts are marked. We see that ΔDCE ≃ ΔBCA by the SAS postulate. Therefore, DE and AB are equal since they are corresponding parts of congruent triangles. Thus, Jamie will know how far it is across the pond by measuring the length of DE.

ANSWER TO WARM UP

d.

ΔADC ≅ ΔBDC	Given
∠BDC = ∠ADC	CPCT
∠BDC + ∠ADC = 180°	Supplementary angles
∠BDC = ∠ADC = 90°	Halves of equals are equal
CD ⊥ AB	Angles formed are equal

Warm Up e. If Jamie has a surveyor's instrument to measure angles in a vertical direction, show that the following method could also be used to compute the distance across his pond. He sets the instrument up on one side of the pond (perpendicular to the ground). He then sights through the instrument and adjusts it until he can see the far bank of the pond. He then turns the instrument and sights through it to a landmark on his side of the pond. He then measures the distance from the instrument to the landmark and claims that this is the distance across the pond. Prove him correct.

ANSWER TO WARM UPS

e.

angle of depression
instrument
across pond

angle of depression
instrument
across land

The corresponding equal parts of each triangle are marked in the figure. The height of the instrument forms one side of the triangle in each case. The instrument is ⊥ to the ground, so a right triangle is formed in each case. The angle of depression is the same in each triangle; therefore the two triangles are congruent by ASA. The distance across the pond and the distance across the land are the same since they are corresponding parts of congruent triangles.

Exercises

ANSWERS

In exercises 1–6, assume $\Delta ABC \cong \Delta DEF$. Answer true or not necessarily true.

1. $AC = DF$

2. $\angle A = \angle D$

3. $AB = EF$

4. $\angle F = \angle A$

5. If $AC = 12$ in, what side in ΔDEF is equal to 12 in?

1. ______

2. ______

3. ______

4. ______

5. ______

ANSWERS

6. If $\angle E = 60°$, what angle in $\Delta ABC = 60°$?

6. ______________________

7.

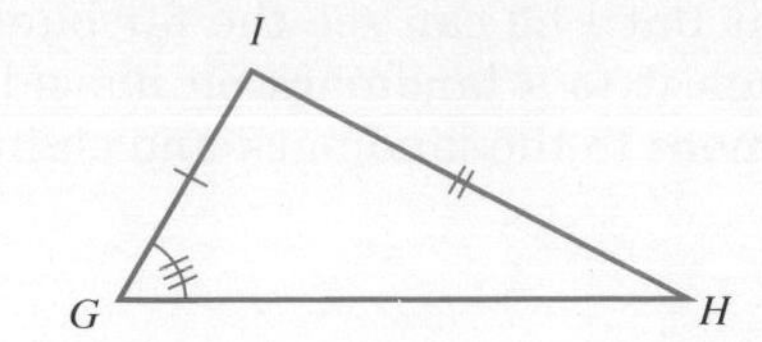

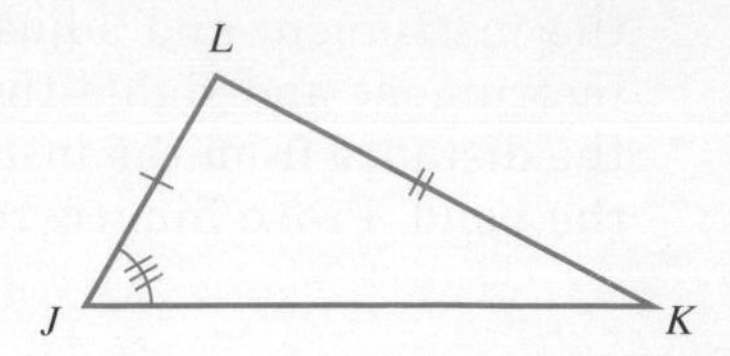

7. ______________________

Name a third pair of parts needed to prove that the two triangles shown are congruent by

a. ASA b. SAS c. SSS

In exercises 8–10, there are two triangles that can be proved congruent. Identify the pair of triangles and give the postulate that proves congruency.

8.

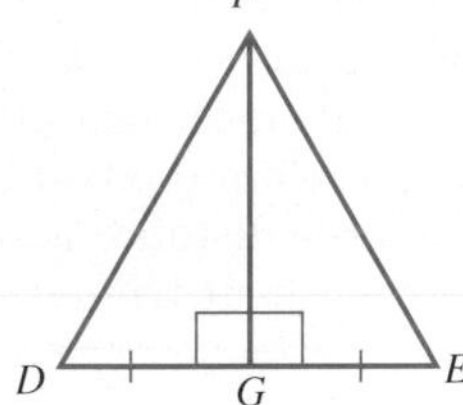

8. ______________________

9.

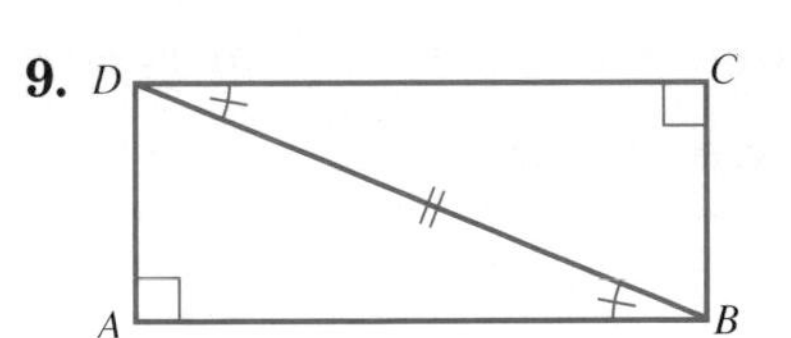

9. ______________________

10.

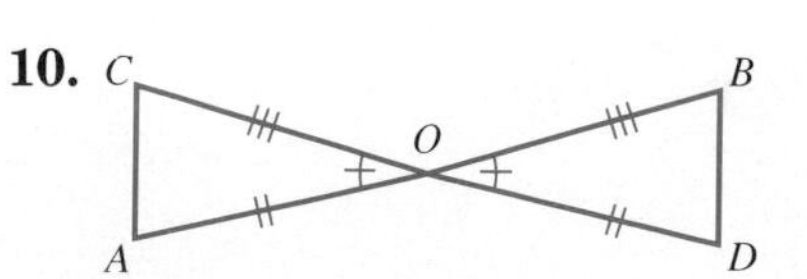

10. ______________________

Parallel Lines

OBJECTIVES

8. **Classify angles formed by parallel lines cut by a transversal.**
9. **Find the measure of angles formed by parallel lines cut by a transversal.**
10. **Prove statements about parallel lines cut by a transversal.**

APPLICATION

Louise constructs a table so that the supports bisect each other. Show that the table top is parallel to the floor.

VOCABULARY

Parallel lines are lines that are in the same plane and never intersect. The symbol || is used to indicate lines that are parallel.

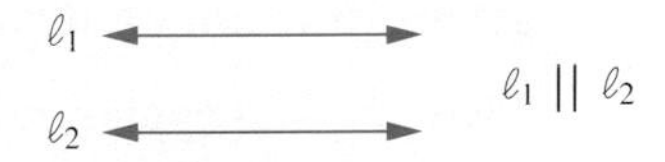

If lines are not parallel, they will intersect. When two lines intersect, they form four angles. These four angles are four pairs of adjacent angles. They also form two pairs of *vertical angles*, or *opposite angles*.

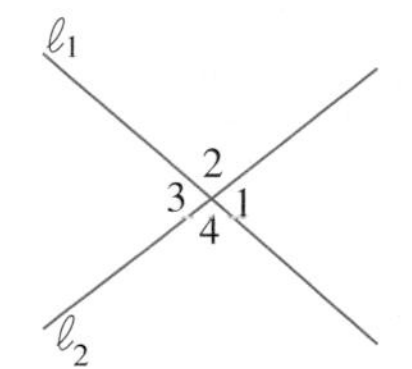

$\angle 1$ is adjacent to $\angle 2$ and $\angle 4$
$\angle 2$ is adjacent to $\angle 1$ and $\angle 3$
$\angle 2$ and $\angle 4$ are vertical angles
$\angle 1$ and $\angle 3$ are vertical angles

A *transversal* is a line that intersects two or more other lines. There are four angles formed at each intersection.

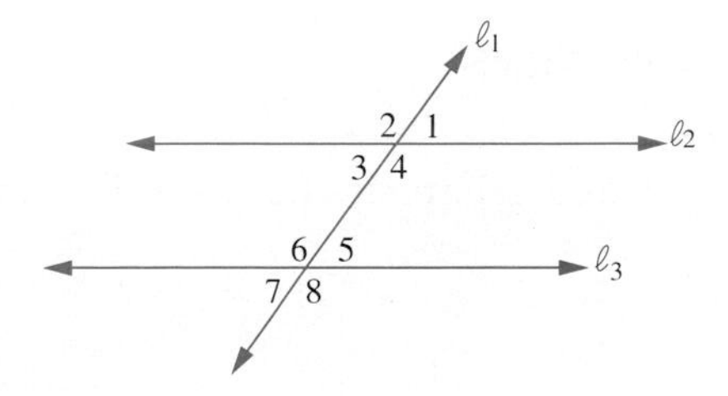

The following names are given to angles formed by a transversal intersecting two lines. In this case, transversal ℓ_1 intersects lines ℓ_2 and ℓ_3. $\angle 3$, $\angle 4$, $\angle 5$, and $\angle 6$ (those between the lines) are called *interior* angles. $\angle 1$, $\angle 2$, $\angle 7$, and $\angle 8$ (those outside the lines) are called *exterior* angles. $\angle 4$ and $\angle 6$ form a pair of *alternate interior* angles since they are on opposite sides of the transversal and are interior angles. Likewise $\angle 3$

and $\angle 5$ are a pair of alternate interior angles. $\angle 1$ and $\angle 5$, $\angle 2$ and $\angle 6$, $\angle 3$ and $\angle 7$, and $\angle 4$ and $\angle 8$ are pairs of *corresponding* angles. Note that they are in corresponding positions at each intersection. $\angle 1$ and $\angle 2$, $\angle 2$ and $\angle 3$, $\angle 3$ and $\angle 4$, and $\angle 1$ and $\angle 4$ are supplementary since together they form straight angles. $\angle 1$ and $\angle 3$, $\angle 2$ and $\angle 4$, $\angle 5$ and $\angle 7$, and $\angle 6$ and $\angle 8$ are pairs of vertical angles. $\angle 1$ and $\angle 7$ and $\angle 2$ and $\angle 8$ are pairs of *alternate exterior* angles.

HOW AND WHY

We will accept the following five statements about points and lines in the same plane as postulates.

Parallel and Perpendicular Line Postulates

Through a point not on a line, one and only one line parallel to the given line can be drawn.

Through a point not on a line, one and only one line perpendicular to the given line can be drawn.

Two lines perpendicular to the same line are parallel.

If two lines are cut by a transversal so that the alternate interior angles are equal, the lines are parallel.

If two parallel lines are cut by a transversal then the alternate interior angles are equal.

MODEL PROBLEM SOLVING

Examples

Strategy

a. Classify $\angle 3$ and $\angle 8$.

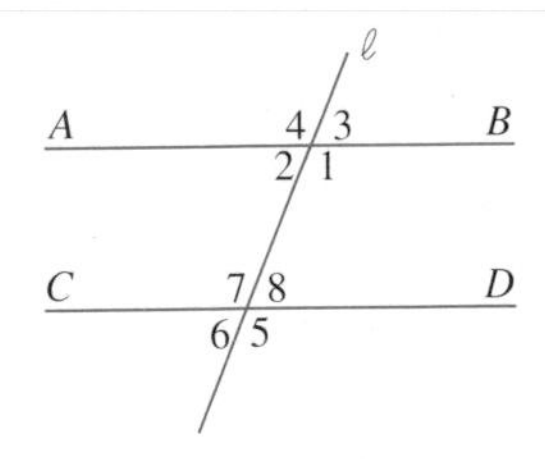

They are corresponding angles.

They are in corresponding positions at each intersection.

Warm Up a. Classify $\angle 1$ and $\angle 7$ in the figure in Example a.

ANSWER TO WARM UPS

a. Alternate interior angles

b. Find the measure of $\angle 1$ if AB || CD and $\angle 8 = 58°$.

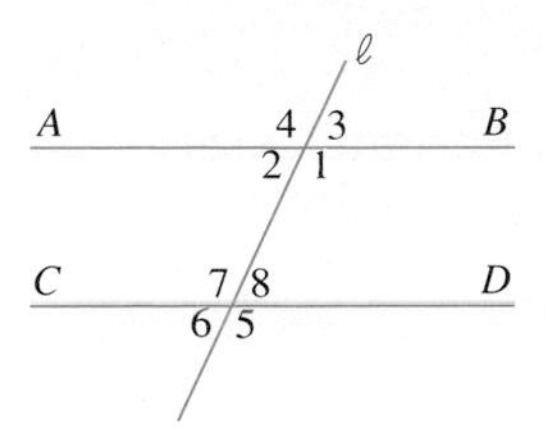

1. $\angle 8 = 58°$	1. **Given**
2. $\angle 2 = \angle 8$	2. **Alternate interior angles are equal**
3. $\therefore \angle 2 = 58°$	3. **Substitution**
4. $\angle 1 + \angle 2 = 180°$	4. **Supplementary angles**
5. $\angle 1 + 58° = 180°$	5. **Substitution**
6. $\angle 1 = 180° - 58°$	6. **Solve**
7. $\angle 1 = 122°$	7. **Simplify**

Warm Up b. Use the figure from Example b and find the measure of $\angle 3$ if $\angle 7 = 132°$

c. Prove: If two straight lines intersect, vertical angles formed are equal.
We make a drawing and call the intersecting lines l_1 and l_2. The pairs of vertical angles formed are named $\angle 1$ and $\angle 3$ and $\angle 2$ and $\angle 4$.

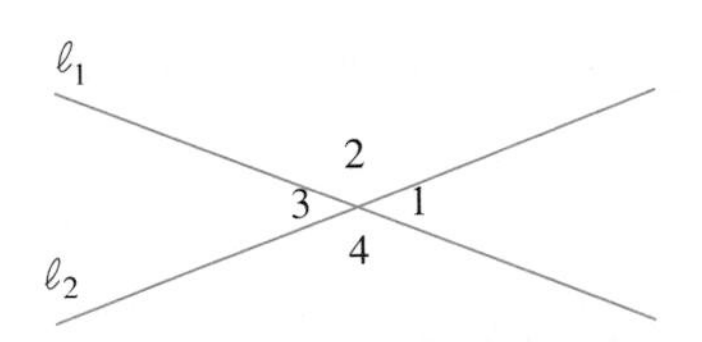

1. $\angle 2 + \angle 3 = 180°$	1. **Supplementary angles.**
2. $\angle 2 + \angle 1 = 180°$	2. **Supplementary angles.**
3. $\angle 2 + \angle 3 = \angle 2 + \angle 1$	3. **Quantities equal to the same quantity are equal.**
4. $\angle 1 = \angle 3$	4. **Equals subtracted from equals are equal.**
5. $\angle 3 + \angle 4 = 180°$	5. **Supplementary angles.**
6. $\angle 3 + \angle 2 = 180°$.	6. **Supplementary angles.**
7. $\angle 3 + \angle 4 = \angle 3 + \angle 2$	7. **Quantities equal to the same quantity are equal.**
8. $\angle 4 = \angle 2$	8. **Equals subtracted from equals are equal**

Since $\angle 1 = \angle 3$ and $\angle 2 = \angle 4$, we say that vertical angles formed by intersecting straight lines are equal.

ANSWER TO WARM UPS

b. 48°

Warm Up c. Prove: If two lines intersect so that one angle formed is a right angle, the lines are perpendicular.

d. Given that AB || CD and that they are cut by the transversal l, show that the alternate exterior angles are equal.

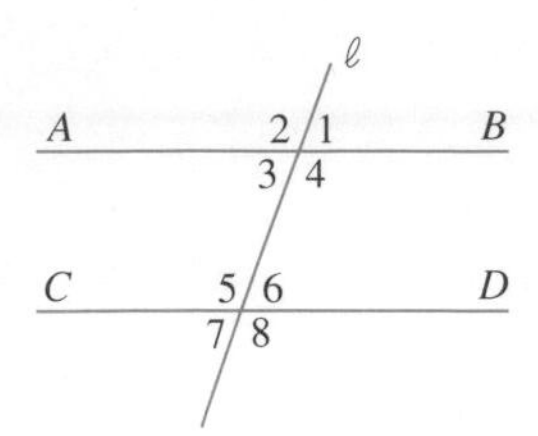

We make a drawing and label it as shown. Our job is to prove $\angle 1 = \angle 7$ and $\angle 2 = \angle 8$.

1. AB || CD	1. **Given**
2. $\angle 3 = \angle 6$	2. **Alternate interior angles are equal**
3. $\angle 1 = \angle 3$	3. **Vertical angles**
4. $\angle 6 = \angle 7$	4. **Vertical angles**
5. $\therefore \angle 1 = \angle 7$	5. **Quantities equal to equal quantities are equal**
6. $\angle 2 = \angle 4$	6. **Vertical angles**
7. $\angle 5 = \angle 8$	7. **Vertical angles**
8. $\angle 4 = \angle 5$	8. **Alternate interior angles are equal**
9. $\therefore \angle 2 = \angle 8$	9. **Quantities equal to equal quantities are equal**

Thus if two parallel lines are cut by a transversal the alternate exterior angles formed are equal.

Warm Up d. Prove: If two lines are cut by a transversal so that the corresponding angles are equal, the lines are parallel.

ANSWERS TO WARM UP

c.

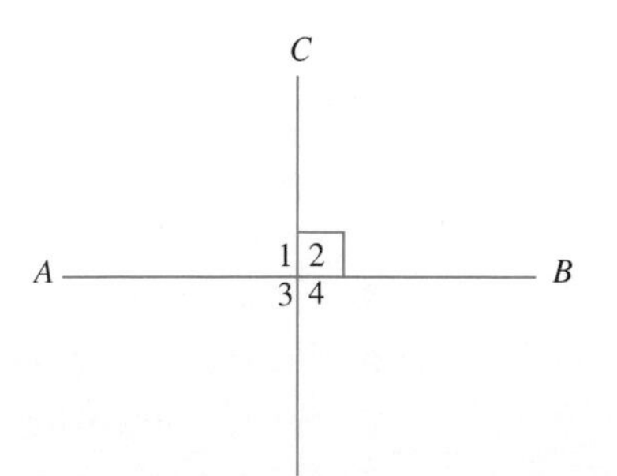

1. $\angle 2 = 90°$	1. **Given**
2. $\angle 1 + \angle 2 = 180°$	2. **Supplements**
3. $\therefore \angle 1 = 90°$	3. **180° − 90°**
4. $\angle 2 = \angle 3$	4. **Vertical angles**
5. $\therefore \angle 3 = 90°$	5. **Quantities equal to the same quantity are equal**
6. $\angle 1 = \angle 4$	6. **Vertical angles**
7. $\therefore \angle 4 = 90°$	7. **180° − 90°**
8. $\therefore$ AB $\perp$ CD	8. **All angles are right angles**

d.

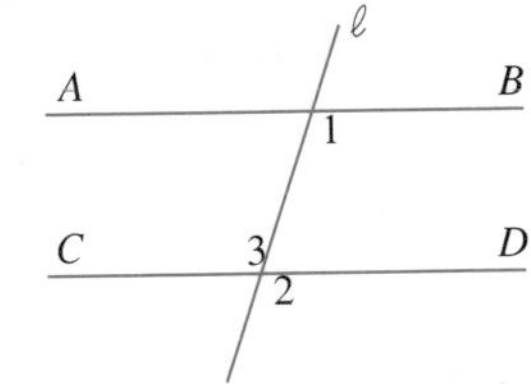

1. $\angle 1 = \angle 2$	1. **Given**
2. $\angle 2 = \angle 3$	2. **Vertical angles**
3. $\angle 1 = \angle 3$.	3. **Quantities equal to the same quantity are equal**
4. $\therefore$ AB || CD	4. **Alternate interior angles are equal so the lines are parallel**

e. If AB || CD and they are cut by a transversal, show that the interior angles on the same side of the transversal are supplementary.

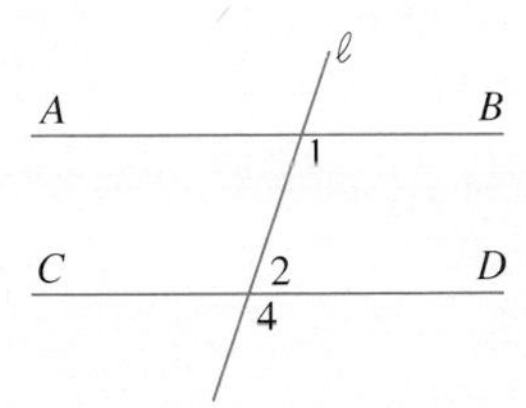

We make a drawing and label it as shown. We must show that $\angle 1 + \angle 3 = 180°$. We know that $\angle 1 + \angle 2 = 180°$ since they form a straight line. AB || CD, so we know that $\angle 2 = \angle 3$ since they are alternate interior angles. Therefore, we substitute $\angle 3$ for $\angle 2$, and we have $\angle 1 + \angle 3 = 180°$.

Warm Up e. If two parallel lines are cut by a transversal, show that a pair of exterior angles on the same side of the transversal are supplementary.

Application Solution

f. Louise constructs a table so that the supports bisect each other. Show that the table top is parallel to the floor.

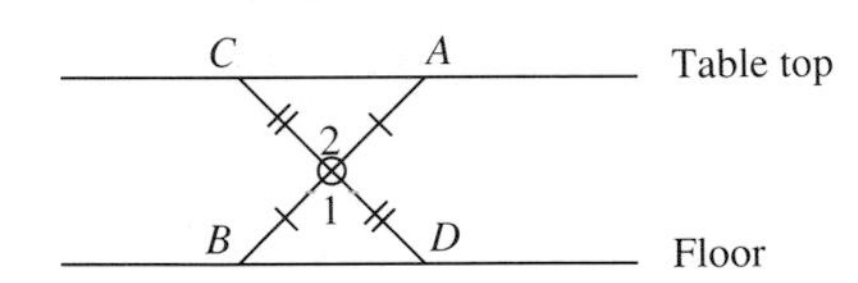

Given: AO = OB
CO = OD
Show: CA || BD

We make a drawing and label it as shown. We also mark all of the known equal line segments. We will prove that $\Delta COA \cong \Delta DOB$. We see that AO = OB and CO = OD. We note that $\angle 1 = \angle 2$ since they are vertical angles. Therefore, the two triangles are congruent by the SAS postulate. Therefore $\angle CAO = \angle DBO$ since they are corresponding parts of congruent triangles. Therefore, CA || BD since the alternate interior angles are equal.

ANSWER TO WARM UP

e.

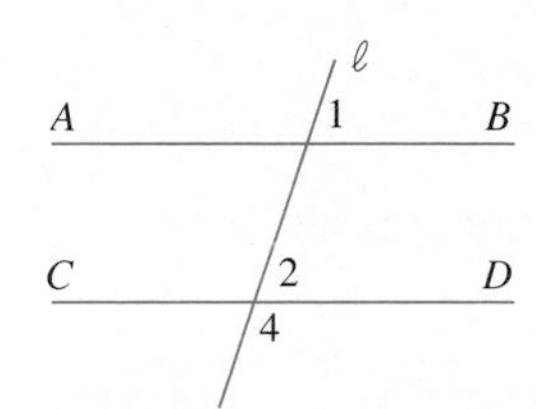

1. $\angle 1 = \angle 2$ — **1. Corresponding angles**
2. $\angle 2 + \angle 4 = 180°$ — **2. Supplementary angles**
3. $\angle 1 + \angle 4 = 180°$ — **3. Substitute**

Warm Up f. Show that two shelves that are built perpendicular to a vertical wall are parallel.

ANSWER TO WARM UP

f.

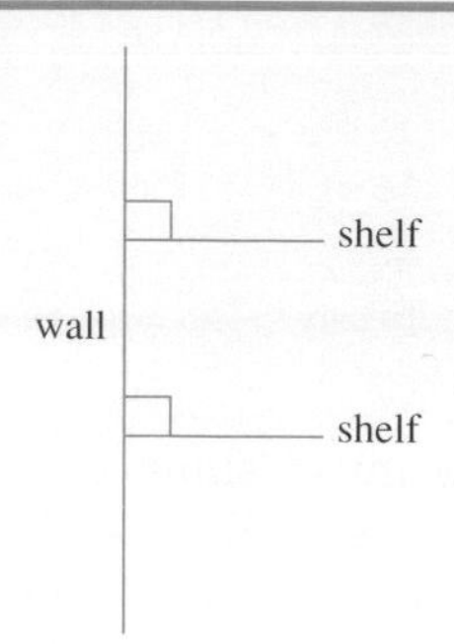

A pair of corresponding angles are equal; therefore the shelves are parallel.

ANSWERS

1. ______

2. ______

3. ______

4. ______

Exercises

In Exercises 1–8, use the following figure to answer the questions or to find the measure of the indicated angle. It is known that the lines AB and CD are parallel.

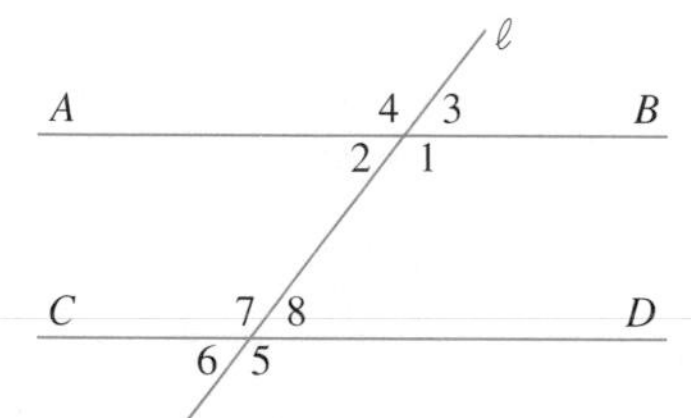

1. Specify all pairs of alternate interior angles.

2. Specify all pairs of corresponding angles.

3. Specify all interior angles.

4. Specify all exterior angles.

5. Find the measure of $\angle 1$ if $\angle 4$ is 120°.

ANSWERS

5. ________

6. Find the measure of $\angle 7$ if $\angle 3$ is 60°.

6. ________

7. Find the measure of $\angle 1$ if $\angle 6$ is 65°.

7. ________

8. Find the measure of $\angle 5$ if $\angle 4$ is 146°.

8. ________

In problems 9–12, use the following figure to answer the questions.

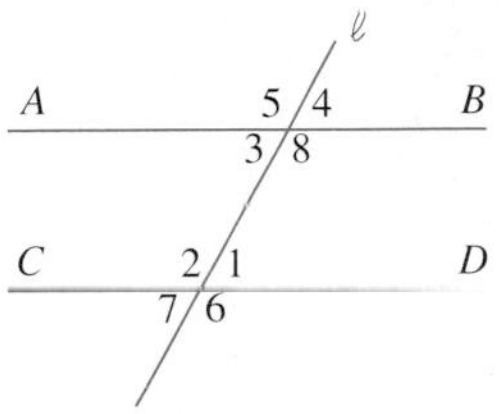

ℓ
A 5 4 B
3 8
C 2 1 D
7 6

9. What angle(s) in the diagram is (are) equal to $\angle 6$?

9. ________

10. What angle(s) in the diagram is (are) equal to $\angle 1$?

10. ________

11. What angle(s) in the diagram is (are) supplementary to $\angle 4$?

11. ________

12. What angle(s) in the diagram is (are) supplementary to $\angle 8$?

12. ________

ANSWERS

13. ______

14. ______

15. ______

16. ______

17. ______

13. Use the following figure to prove that the sum of the interior angles of a triangle is 180°.

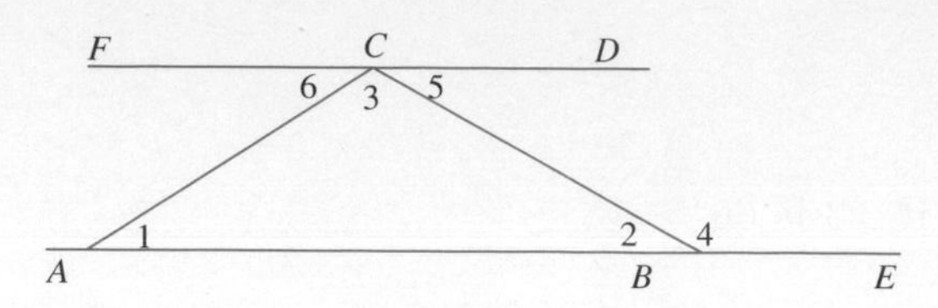

Given: ΔABC with $FCD \parallel ABE$
Prove: $\angle 1 + \angle 2 + \angle 3 = 180°$

14. Prove: If two lines are perpendicular to the same line, the two lines are parallel.

15. Given: $AB \parallel CD$ and $PQ \parallel RS$
Prove: $\angle 1 = \angle 2$

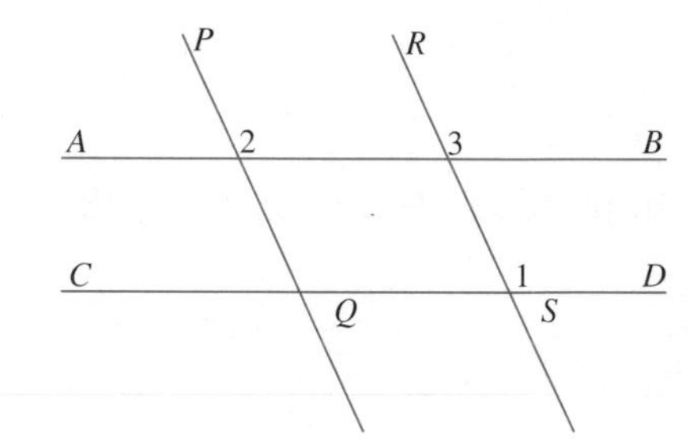

16. Use the following figure to prove that $AB \parallel CD$. $\angle 1 = \angle 2$

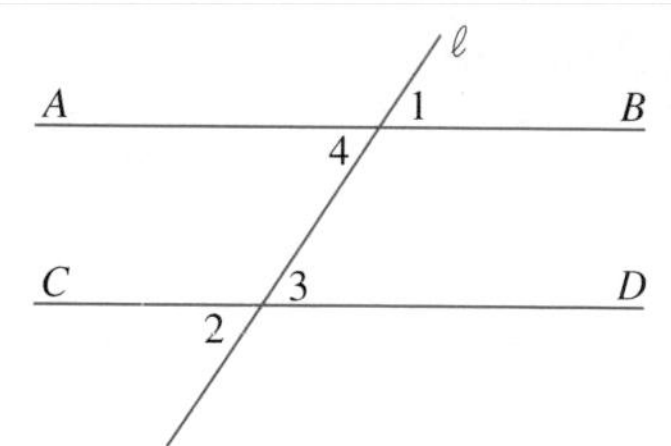

17. Use the following figure to prove $AB \parallel CD$ if $\angle 5 + \angle 7 = 180°$

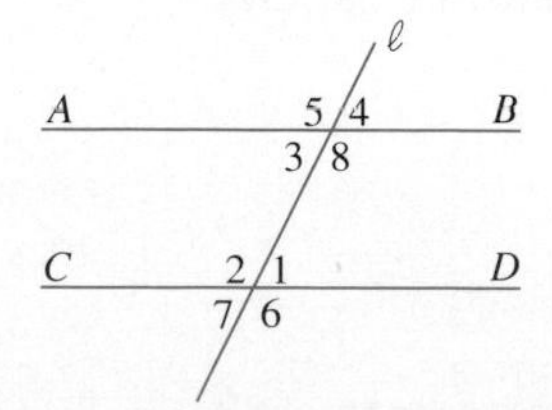

18. Given: EH ∥ AD
∠FBC = ∠GCB

ANSWERS

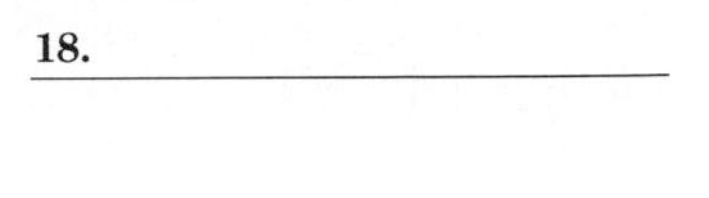

18. ________________

Prove: ∠EFB = ∠KCD

19. Given: ΔADC is isosceles with CA = CD
BE ∥ AD
Prove: ΔBEC is isosceles

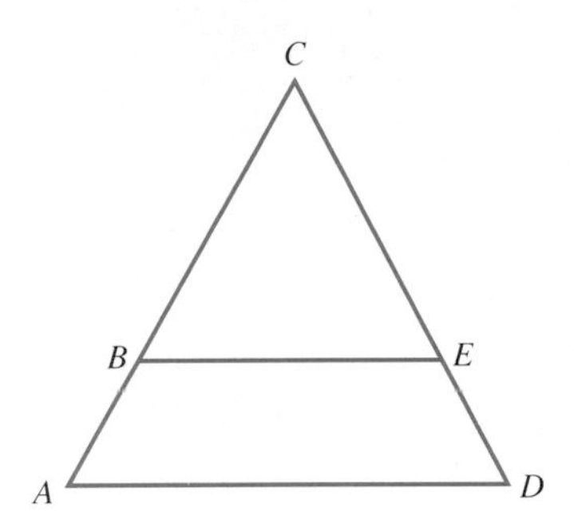

19. ________________

Similar Triangles

OBJECTIVES

11. Given two similar triangles, determine the measurements of designated sides and angles.
12. Show two triangles are similar.

APPLICATION

Florence decides to build a fence around a triangular shaped field. Last year, she built a fence around a similar triangular field with sides of 300 ft, 200 ft, and 280 ft. How much fencing will she need if she knows that the length of the smallest side of the new field is 320 ft?

VOCABULARY

When two or more polygons are *similar*, they have the same shape but are not necessarily the same size. Their corresponding sides are *proportional*. A *proportion* is a statement that says that two ratios are equal. Thus, if two figures are similar, their corresponding sides are in the same ratio. The symbol ~ is used to indicate similar.

HOW AND WHY

We accept the following three statements as postulates.

Similar Triangle Postulates

If three angles of a triangle are equal to three angles of another, the two triangles are similar.

If one angle of one triangle is equal to one angle of another and the two sides that include the angles are proportional, then the triangles are similar.

If the sides of two triangles are respectively proportional, the two triangles are similar.

To show that two or more triangles are similar, we need to show that their corresponding angles are equal and their corresponding sides are in the same ratio. Thus, the sides will be proportional.

We use these postulates together with deductive reasoning to find parts of similar triangles.

MODEL PROBLEM SOLVING

Examples | **Strategy**

a. Given: ΔABC ~ ΔDEF, AC = 10, AB = 12, BC = 14 and EF = 7
Find: The length of DF and DE

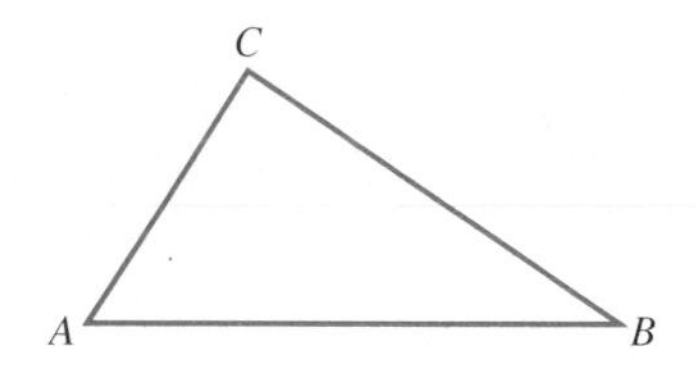

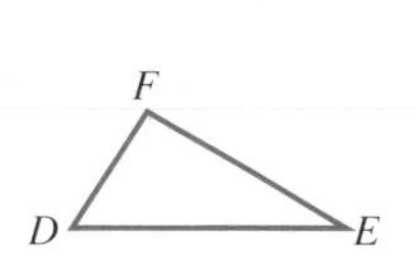

The ratio of EF to BC is

$$\frac{7}{14} = \frac{1}{2}$$

BC and EF are corresponding sides.

DF corresponds to AC so,
since AC = 10, DF = 5
DE corresponds to AB so,
since AB = 12, DE = 6

All corresponding sides must be in the same ratio. Each side in ΔDEF must be one-half its corresponding side in ΔABC.

Warm Up a. Given: ΔABC ~ ΔDEF, AC = 21, AB = 27, BC = 24 and EF = 8
Find: The length of DF and DE

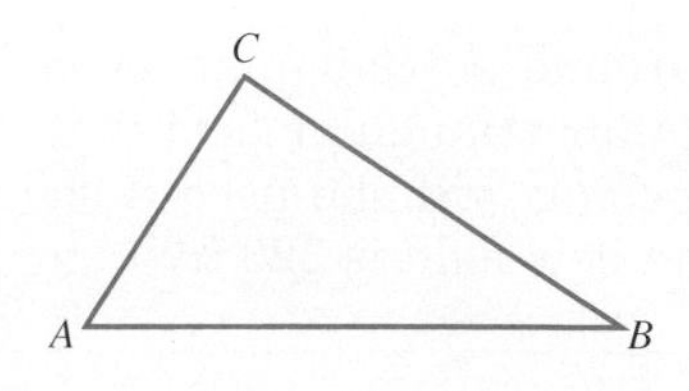

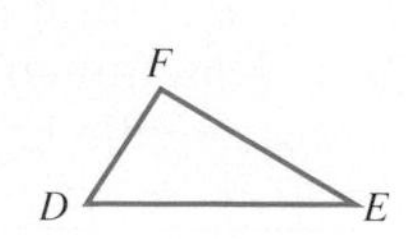

ANSWER TO WARM UPS

a. DF = 7, DE = 9.

b. Show that a line parallel to one side of a triangle and intersecting the other two sides forms a triangle similar to the original triangle.

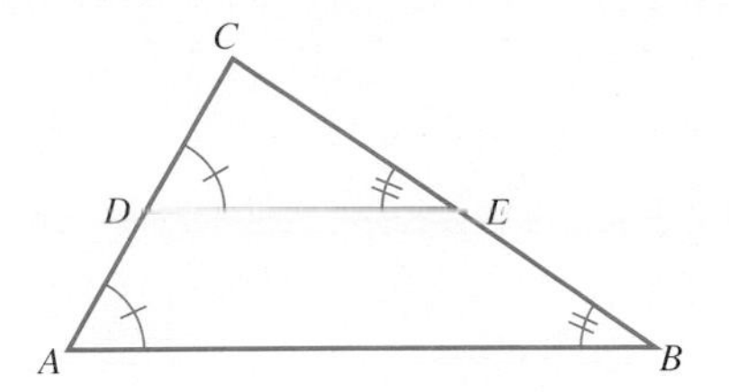

We draw a figure and label it as shown.

We know that DE || AB, so ∠CED and ∠CBA are equal corresponding angles. For the same reason, ∠CDE = ∠CAD. ∠C is an angle in each of the triangles, so the triangles are similar since the three angles of one are equal to the three angles of the other.

Warm Up b. Show that a line parallel to one side of a triangle and intersecting the other two sides forms a triangle whose sides are proportional to the original triangle.

c. Given: Triangle ACD with BE || CD
ED = 6, AE = 8, BC = 5
Find: AB = ?

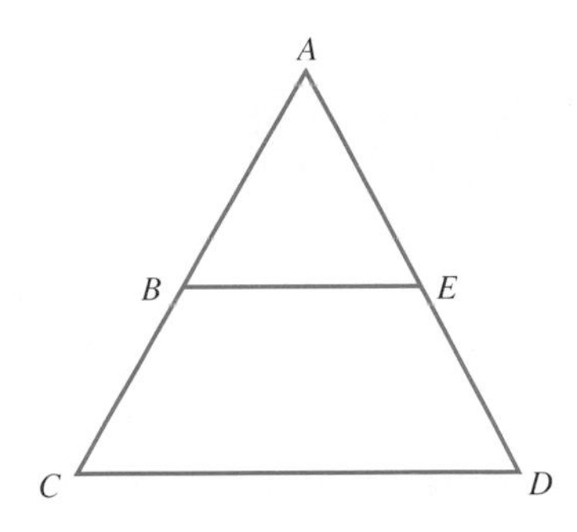

Since BE || CD, we know that ΔABE and ΔACD are similar. That means that the corresponding sides are proportional. So

$\frac{AB}{AC} = \frac{AE}{AD}$; AC = AB + BC = AB + 5; AD = AE + ED = 8 + 6 = 14.

Therefore, we solve the proportion

$\frac{AB}{AB + 5} = \frac{8}{14}$. **Substitute AB + 5 for AC, 8 for AE, and 14 for AD**

$14(AB) = 8(AB + 5)$ **Cross multiply**

$14(AB) = 8(AB) + 40$ **Simplify**

$6(AB) = 40$

$AB = 6\frac{2}{3}$

ANSWER TO WARM UPS

b. Since the triangles are similar, corresponding sides are proportional.

Warm Up c. Given: ΔACD with $BE \parallel CD$
$ED = 8$, $AE = 12$, $BC = 6$
Find: $AB = ?$

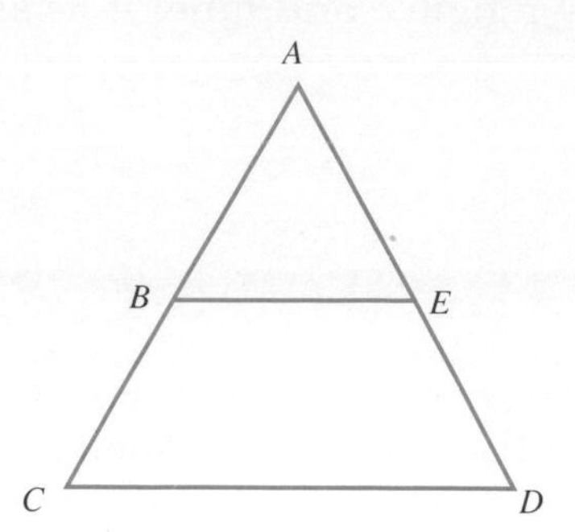

Application Solution

d. Florence decides to build a fence around a triangular shaped field. Last year, she built a fence around a similar triangular field with sides of 300 ft, 200 ft, and 280 ft. How much fencing will she need if she knows that the length of the smallest side of the new field is 320 ft?

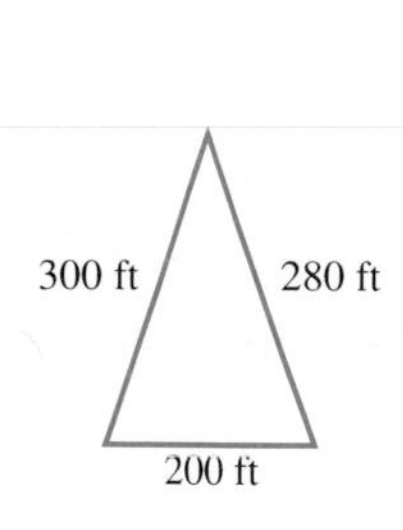

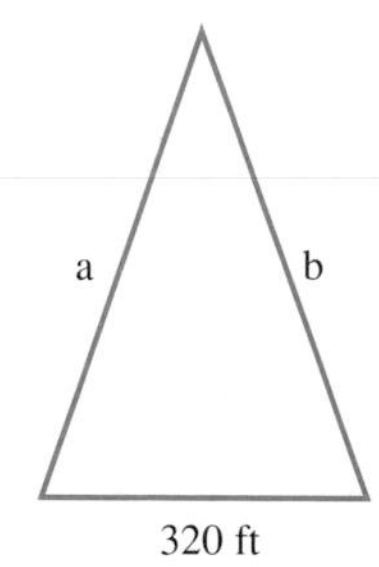

Since the two triangles are similar, the corresponding sides must be proportional. We compute the ratio of the two smallest sides. The other corresponding sides have the same ratio.

The ratio of the smallest sides in the new triangle to the old triangle is

$$\frac{320}{200} = \frac{8}{5},$$

the other corresponding sides have the same ratio. For the longest side, we have

$$\frac{a}{300} = \frac{8}{5}$$

where a is the longest side in the new triangle.

ANSWER TO WARM UPS c. $AB = 9$

$5a = 8 \cdot 300$	**Solve the proportion**
$5a = 2400$	**Simplify**
$a = 480$	**The longest side is 480 ft**

For the remaining side, we have

$$\frac{b}{280} = \frac{8}{5}$$

where b is the remaining side in the new triangle.

$5b = 8 \cdot 280$	**Solve the proportion**
$5b = 2240$	**Simplify**
$b = 448$	**The remaining side is 448 ft**

The distance around the new field is 320 ft + 448 ft + 480 ft = 1248 ft. She needs 1248 ft of fencing.

Warm Up d. How much fencing would Florence need to fence her triangular field if the longest side measured 375 ft?

ANSWER TO WARM UPS

d. 975 ft of fencing

Exercises

ANSWERS

Use the following data to answer 1–5.
Given: $\Delta ABC \sim \Delta DEF$, $\angle A = 46°$, $\angle C = 76°$, AC = 12, CB = 10, AB = 13, and DF = 15

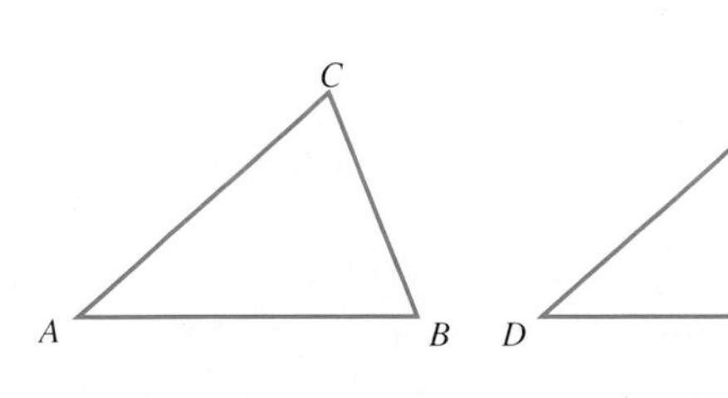

Find:

1. $\angle E$

2. $\angle D$

3. $\angle F$

4. DE

5. FE

1. ______
2. ______
3. ______
4. ______
5. ______

ANSWERS

6. ______________

7. ______________

8. ______________

9. ______________

10. ______________

11. ______________

12. ______________

13. ______________

14. ______________

15. ______________

16. ______________

Use the following to answer 6–10.

Given:

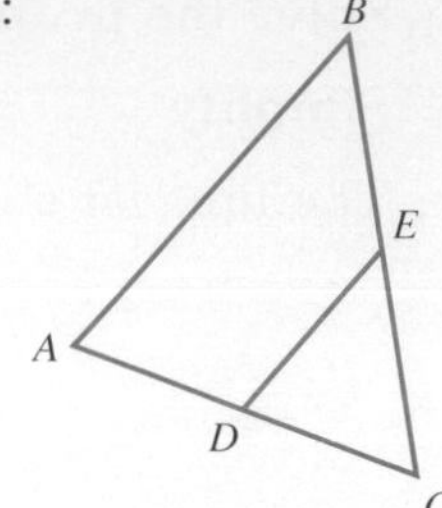

ED || BA
AB = 20, EB = 15
EC = 10, AD = 6
∠CED = 40°, ∠C = 75°

Find:

6. ED **7.** DC **8.** AC

9. ∠B **10.** ∠A

Use the following to answer 11–16.
Given: AB = 13, AC = 5, CB = 12, and ΔABD ~ ΔACB ~ ΔBCD

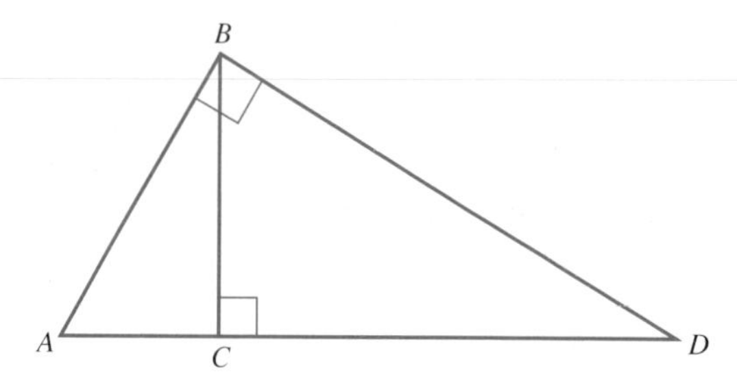

Find:

11. BD **12.** AD **13.** DC

Given: BC = 9, CD = 12, and BD = 15
Find:

14. AB **15.** AC **16.** AD

Use the following to answer 17–20.
Given: ∠ABC = ∠CDB, AB = 21, CB = 24, and CD = 18

ANSWERS

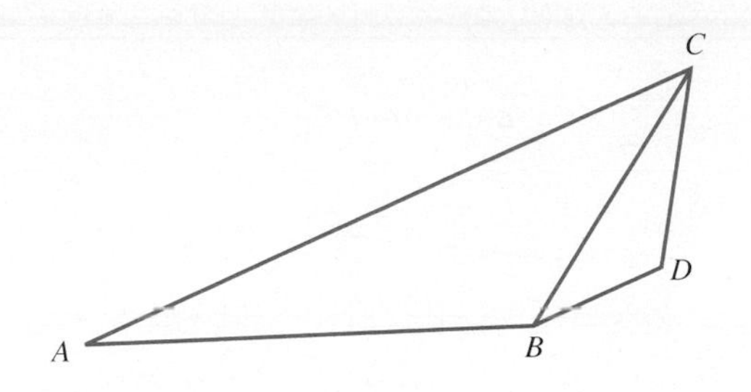

ΔABC ~ ΔCDB

Find:

17. AC

18. DB

Given: ∠ABC = ∠CDB, BD = 90, CB = 120, and AB = 100
Find:

19. AC

20. CD

21. To find the height of a tree, Cami uses similar triangles. She first forms a triangle that has one side six feet up on the tree and another side eight feet from the tree. She then backed off an additional twenty feet from the tree so that the angle to the top is the same as the angle to the six foot height. Find the height of the tree.

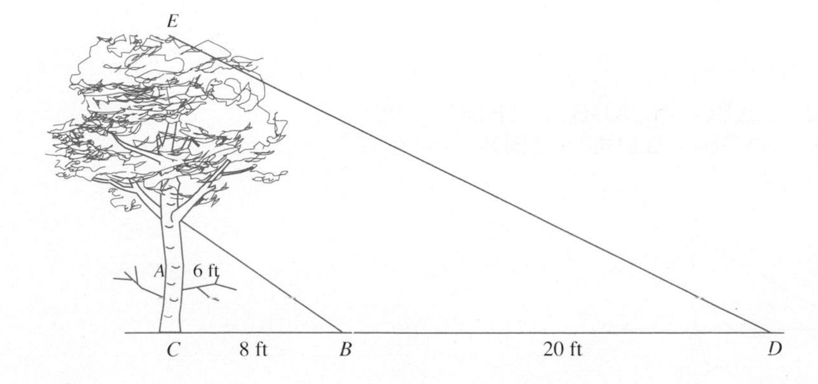

17. ______

18. ______

19. ______

20. ______

21. ______

22. ______________

23. ______________

24. ______________

22. To find the distance across a pond, Marvin forms two similar triangles.

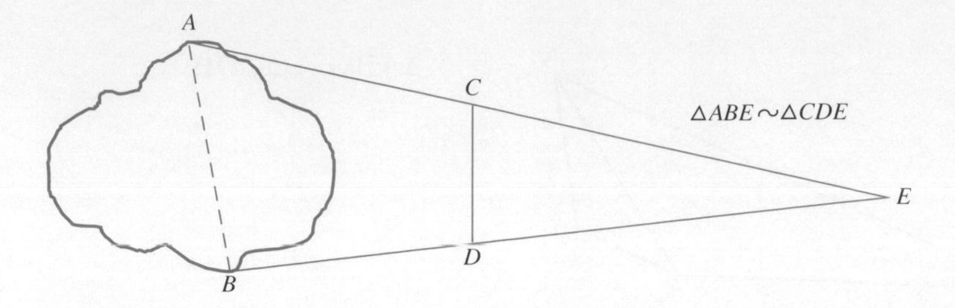

ΔABE ∼ ΔCDE

He also finds that BE = 450 yds, BD = 180 yds, and CD = 120 yds. Find the distance across the pond.

23. A triangular plot of ground is surrounded by a sidewalk, as shown in the diagram, with ΔABC ∼ ΔDEF.

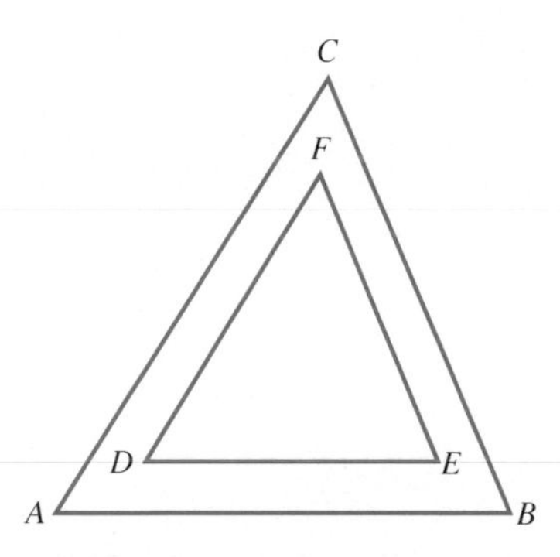

Given: CB = 40 ft, AC = 34 ft,
AB = 52 ft and
FE = 34 ft

Find: the length of fence it will take to fence the perimeter of the plot of ground (ΔDEF).

24. Given: ∠ABC = ∠ADB = ∠BDC = 90°
Prove: ΔADB ∼ ΔABC ∼ ΔBDC

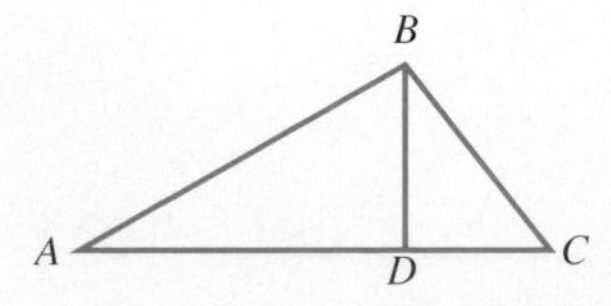

25. Given: Two squares with sides of 1 foot and 2 feet respectively with a diagonal drawn from one vertex to the opposite one in each square

Prove: The triangles formed in the squares are similar

ANSWERS

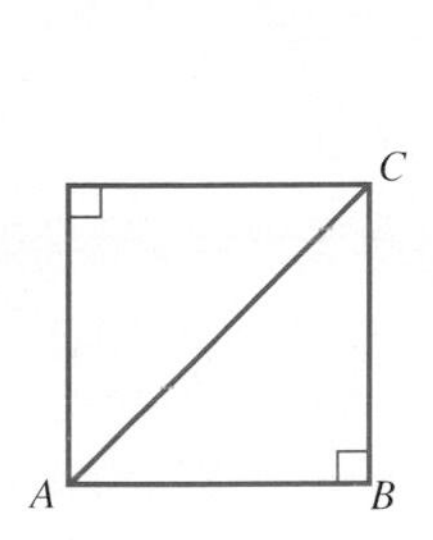

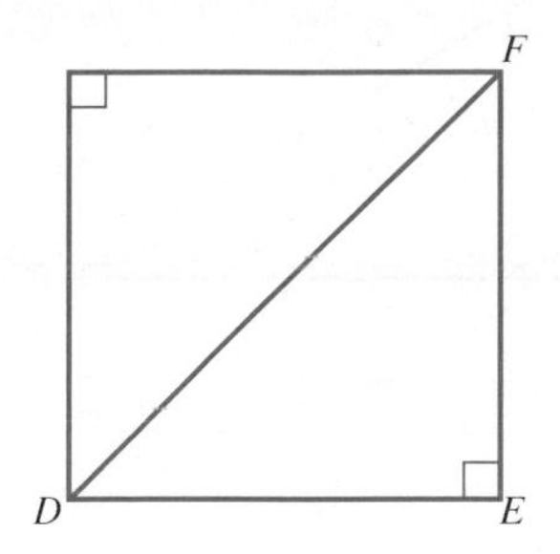

26. If two intersecting straight lines are cut by parallel lines, as shown in the figure below, prove that $\Delta CDE \sim \Delta ABC$.

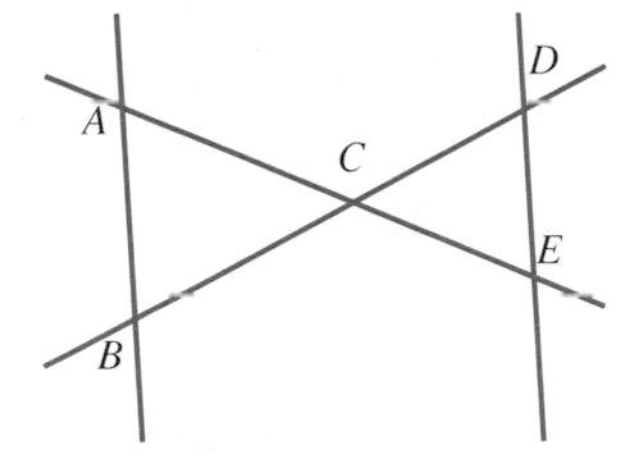

26. ______________________

27. Two triangles are inscribed in circles of different diameters so that each triangle has a vertex at the center of the circle. If $\angle A = \angle D$, prove that the triangles are similar.

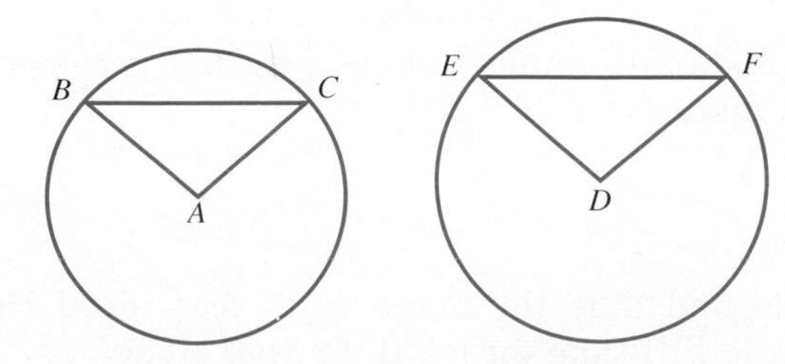

27. ______________________

ANSWERS

28. ______

29. ______

30. ______

28. Given: A trapezoid with diagonals drawn
Prove: That ΔCAB and ΔDAE are similar

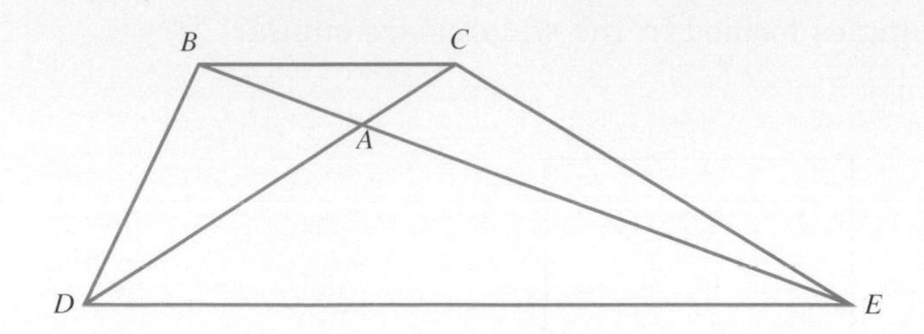

29. If two angles of one triangle are equal to two angles of another, the two triangles are similar.

Given: ΔABC and ΔDEF
$\angle A = \angle D$, $\angle B = \angle E$
Prove: ΔABC ~ ΔDEF

30. If two isoceles triangles have their vertex angles equal, the two triangles are similar.

Given: Isosceles ΔABC and isosceles ΔDEF with vertex $\angle C =$ vertex $\angle F$
Show: ΔABC ~ ΔDEF.
(Hint: $\angle A = \angle B$, $\angle D = \angle F$ and the sum of the angles of a triangle is 180°)

Square Roots

OBJECTIVE

13. Find either the square root or the approximate square root of a number.

APPLICATION

The formula for finding the approximate number of seconds that it takes a free falling body to fall a given distance is

$$t = \sqrt{\frac{d}{16}}$$

where t is the time in seconds and d is the distance in feet. Find the approximate number of seconds it will take for a ball to drop 10,000 ft.

VOCABULARY

A *perfect square* is a whole number or fraction that is the square of another whole number or fraction.

For example, $16 = 4^2$, $64 = 8^2$, and $\frac{1}{9} = \left[\frac{1}{3}\right]^2$, so 16, 64, and $\frac{1}{9}$ are perfect squares. Numbers such as 13 and $\frac{1}{5}$ are not perfect squares.

A *square root* of a positive number is one of the two equal factors of the number. The square roots of 25 are 5 and -5. The symbol for the positive square root is called a *radical sign:* $\sqrt{\ }$. The symbol for the negative square root is $-\sqrt{\ }$. Thus, $\sqrt{25} = 5$ and $-\sqrt{25} = -5$ since $5^2 = 25$ and $(-5)^2 = 25$.

In this text we will be working problems in which only the positive square root of a number will be considered. Thus we will use the phrase "the square root" to indicate the positive square root. In algebra the negative square root of a positive number will be studied.

A number such as 2 is not a perfect square and does not have an exact square root. In these cases we use an *approximate square root.* Some calculators display 1.414213562 for the square root of 2. Like π this is a never-ending decimal and we will use approximations for its value. Hence it is not uncommon to say that $\sqrt{2} \approx 1.414$ correct to the nearest thousandth.

HOW AND WHY

The whole number 81 is the square of 9, so 9 is the square root of 81. We write $9^2 = 81$ and $\sqrt{81} = 9$.

If a whole number is a perfect square, the square root can be found by trial and error, a square root table, or with a calculator.

To find $\sqrt{196}$ by trial and error we discover that

10 is too small, since $10 \times 10 = 10^2 = 100$

12 is too small, since $12 \times 12 = 12^2 = 144$

15 is too large, since $15 \times 15 = 15^2 = 225$

14 is the square root, since $14 \times 14 = 14^2 = 196$

so $\sqrt{196} = 14$.

Most whole numbers are not perfect squares. To find $\sqrt{95}$, we discover that

9 is too small, since $9 \times 9 = 81$

10 is too large, since $10 \times 10 = 100$

So $\sqrt{95}$ is between 9 and 10. It is the case that 95 is not a perfect square and thus we can not write an exact value for its square root. The number, $\sqrt{95}$, can be approximated by using a calculator.

APPROXIMATIONS			
Square Root	Tenth	Hundredth	Thousandth
$\sqrt{95}$	9.7	9.75	9.747
$\sqrt{3}$	1.7	1.73	1.732
$\sqrt{111}$	10.5	10.54	10.536

It is assumed for the following exercises that you have access to a calculator or will use the square root table in Appendix V to find the approximations.

If your calculator has a square root key, (indicated by $\sqrt{\ }$), see Example c.

If your calculator does not have a square root key, use the table in the back of the text.

MODEL PROBLEM SOLVING

Examples	Strategy
a. Find $\sqrt{64}$ and $\sqrt{25}$.	
$\sqrt{64} = 8$	$8 \times 8 = 8^2 = 64$.
$\sqrt{25} = 5$	$5 \times 5 = 5^2 = 25$.

Warm Up a. Find $\sqrt{36}$ and $\sqrt{225}$.

b. Find $\sqrt{\frac{9}{25}}$ and $\sqrt{\frac{16}{121}}$.

$\sqrt{\frac{9}{25}} = \frac{3}{5}$	Since $\frac{3}{5} \times \frac{3}{5} = \frac{9}{25}$.
$\sqrt{\frac{16}{121}} = \frac{4}{11}$	Since $\frac{4}{11} \times \frac{4}{11} = \frac{16}{121}$.

Warm Up b. Find $\sqrt{\frac{4}{49}}$ and $\sqrt{\frac{81}{64}}$.

c. Calculator example

Find the square root of 2925 using a calculator with a square root key. Round the answer to the nearest tenth.

$\sqrt{2925} = ?$

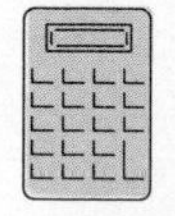

ENTER	DISPLAY
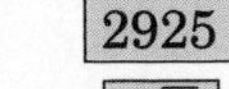	2925.
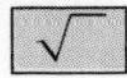	54.08326913

Therefore, after rounding, $\sqrt{2925} \approx 54.1$.

ANSWERS TO WARM UPS **a.** 6, 15 **b.** $\frac{2}{7}, \frac{9}{8}$

Warm Up c. Use a calculator with a square root key to find $\sqrt{3529}$. Round the answer to the nearest tenth.

Application Solution

d. The formula for finding the approximate number of seconds that it takes a free falling body to fall a given distance is

$$t = \sqrt{\frac{d}{16}}$$

where t is the time in seconds and d is the distance in feet. Find the approximate number of seconds it will take for a ball to drop 10,000 ft.

$t = \sqrt{\frac{d}{16}}$	**Given formula. To find the approximate time the ball will take to fall 10,000 ft, we substitute 10,000 for *d* in the formula.**
$t = \sqrt{\frac{10,000}{16}}$	**Substitute.**
$t = \sqrt{625}$	**Simplify under the radical.**
$t = 25$	**The square root of 625 is 25.**

Therefore it will take about 25 seconds for the ball to fall 10,000 ft.

Warm Up d. The formula for finding the approximate number of seconds it will take a free-falling body to fall a given distance is

$$t = \sqrt{\frac{d}{16}}$$

where t is the time in seconds and d is the distance in feet. Find the approximate number of seconds it will take for a ball to drop a distance of 28,224 feet.

ANSWERS TO WARM UPS **c.** 59.4 **d.** 42 sec

Exercises

ANSWERS

A

Find the square root of each of the following:

1. $\sqrt{16}$ 2. $\sqrt{49}$ 3. $\sqrt{81}$

4. $\sqrt{64}$ 5. $\sqrt{121}$ 6. $\sqrt{144}$

7. $\sqrt{\frac{9}{25}}$ 8. $\sqrt{\frac{4}{49}}$

9. $\sqrt{\frac{121}{144}}$ 10. $\sqrt{\frac{64}{169}}$

Use a calculator and round answer to the indicated place value:

11. $\sqrt{116}$ (tenth) 12. $\sqrt{216}$ (hundredth)

13. $\sqrt{814}$ (hundredth) 14. $\sqrt{814}$ (thousandth)

15. $\sqrt{21}$ (tenth) 16. $\sqrt{21}$ (thousandth)

B

Find the square root of each of the following:

17. $\sqrt{196}$ 18. $\sqrt{484}$ 19. $\sqrt{576}$

1. ______
2. ______
3. ______
4. ______
5. ______
6. ______
7. ______
8. ______
9. ______
10. ______
11. ______
12. ______
13. ______
14. ______
15. ______
16. ______
17. ______
18. ______
19. ______

ANSWERS

20. ______

21. ______

22. ______

23. ______

24. ______

25. ______

26. ______

27. ______

28. ______

29. ______

30. ______

31. ______

32. ______

33. ______

34. ______

35. ______

20. $\sqrt{1024}$ **21.** $\sqrt{\frac{144}{169}}$ **22.** $\sqrt{\frac{529}{225}}$

Use a calculator or a table to complete the following chart.

	NEAREST TENTH	NEAREST HUNDREDTH	NEAREST THOUSANDTH	NEAREST TEN-THOUSANDTH
23. $\sqrt{30}$				
24. $\sqrt{210}$				
25. $\sqrt{150}$				
26. $\sqrt{305}$				

C

Find the square roots of each of the following:

27. $\sqrt{4489}$ **28.** $\sqrt{7921}$

29. $\sqrt{14884}$ **30.** $\sqrt{17161}$

Use a calculator or a table, to complete the following chart.

	NEAREST TENTH	NEAREST HUNDREDTH	NEAREST THOUSANDTH	NEAREST TEN-THOUSANDTH
31. $\sqrt{365.96}$				
32. $\sqrt{0.9682}$				
33. $\sqrt{20.037}$				
34. $\sqrt{0.03985}$				

D

35. The formula for finding the approximate number of seconds that it takes a free-falling body to fall a given distance is

$$t = \sqrt{\frac{d}{16}}$$

where t is the time in seconds and d is the distance in feet. Find the approximate number of seconds, to the nearest tenth of a second, that it takes a free-falling body to fall 50 feet.

36. The formula for finding the approximate number of seconds that it takes a free-falling body to fall a given distance is

$$t = \sqrt{\frac{d}{16}}$$

where t is the time in seconds and d is the distance in feet. Find the approximate number of seconds, to the nearest tenth of a second, that it takes a free-falling body to fall 200 feet.

37. What is the length of the side of a square whose area is 64 square centimeters? (The formula for the length of the side of a square is $S = \sqrt{A}$, where S is the length of the side and A is the area of the square.)

38. What is the length of the side of a square whose area is 2.89 square yards? (The formula for the length of the side of a square is $S = \sqrt{A}$, where S is the length of the side and A is the area of the square.)

39. Evaluate the formula $t = 2\pi\sqrt{\frac{\ell}{g}}$ for t if $\ell = 49$, $g = 36$, and $\pi = \frac{22}{7}$.

40. Evaluate the formula $t = 2\pi\sqrt{\frac{\ell}{g}}$ for t if $\ell = 49$, $g = 121$, and $\pi = \frac{22}{7}$.

41. To meet the city code, the attic of a new house needs a vent with a minimum area of 706 square inches. What is the radius of a circular vent that will meet the code requirement? The radius (r) of a circle in terms of the area (A) is given by the formula $r \approx 0.564\sqrt{A}$. Compute the length of the radius to one decimal place.

42. What is the perimeter of a square field whose area is 13,240 square feet? (Find the perimeter to the nearest tenth of a foot.)

ANSWERS

36. ______________

37. ______________

38. ______________

39. ______________

40. ______________

41. ______________

42. ______________

The Pythagorean Theorem

OBJECTIVE **14. Find the missing side of a right triangle.**

APPLICATION

What is the length of a brace that is needed to go diagonally across a rectangular gate that is 6 feet high and 4 feet 6 inches wide? (See the figure.)

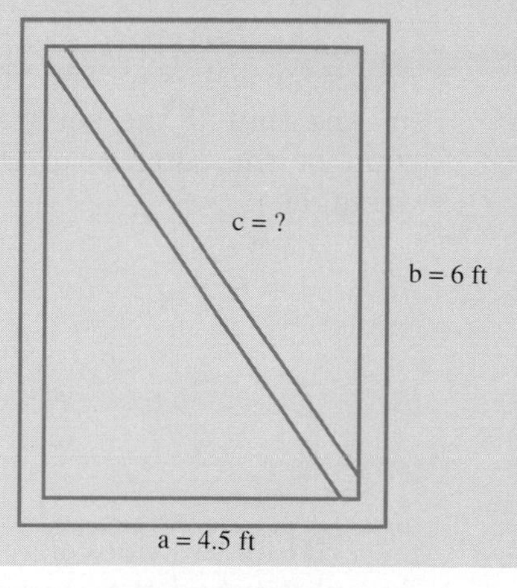

VOCABULARY

Right Triangle

The *Pythagorean theorem* (or theorem of Pythagoras) describes a relationship between the sides of a right triangle. A *right triangle* is a triangle that contains one right (90°) angle. The sides a and b, which form the right angle, are called the *legs*. The longest side (c), which is opposite the right angle, is called the *hypotenuse*. The hypotenuse is the longest side of a right triangle. Right triangle ABC with right angle at C is shown. (For all triangles in this section we assume the right angle is at C.)

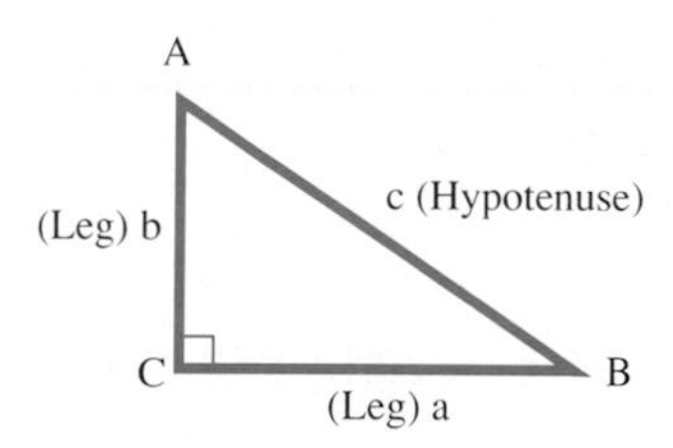

Figure 8.5

HOW AND WHY

The relationship described by the Pythagorean Theorem is:

Pythagorean Theorem

In any right triangle whose hypotenuse is c and legs a and b, $c^2 = a^2 + b^2$. It is also true that if $c^2 = a^2 + b^2$ then the triangle is a right triangle.

The first part means that in any right triangle, the square of the hypotenuse is equal to the sum of the square of the legs. The second part means that if the square of the longest side (hypotenuse) is equal to the sum of the squares of the other two sides (legs), the triangle is a right triangle.

$c^2 = a^2 + b^2$

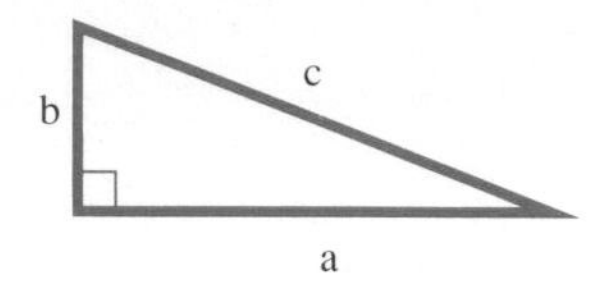

Figure 8.6

If any two sides of a right triangle are known, the third side can be found by substituting the known values into the formula.

If both legs of a right triangle are known we can find the hypotenuse in the following way.

Given: one leg of a right triangle is 6 inches and the other leg is 8 inches.

$c^2 = a^2 + b^2$
$c^2 = 6^2 + 8^2$
$c^2 = 36 + 64$
$c^2 = 100$
$c = \sqrt{100}$
$c = 10$ inches

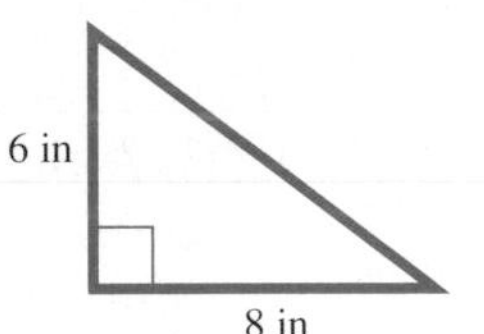

MODEL PROBLEM SOLVING

Examples **Strategy**

a. Find the length of the leg of a right triangle whose other leg is 12 cm and whose hypotenuse is 15 cm.

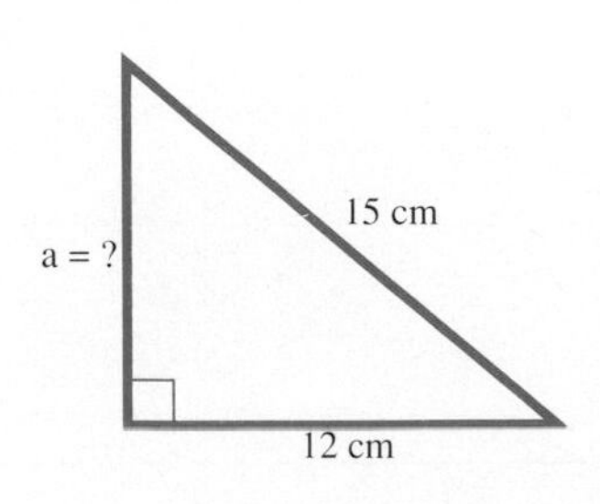

$c^2 = a^2 + b^2$	**The figure indicates that side a is unknown, so we substitute the values for b and c into the formula.**
$15^2 = a^2 + 12^2$	**Substitute 15 for c and 12 for b.**
$225 = a^2 + 144$	**Simplify.**
$225 - 144 = a^2$	**Solve for a^2, subtract 144 from both sides.**
$81 = a^2$	
$a^2 = 81$	
$a = \sqrt{81}$	
$a = 9$ cm	

Warm Up a. Find the length of the leg of a right triangle whose other leg is 12 cm and whose hypotenuse is 13 cm.

b. Find the length of the hypotenuse of the right triangle whose legs are 5 ft and 6 ft respectively.

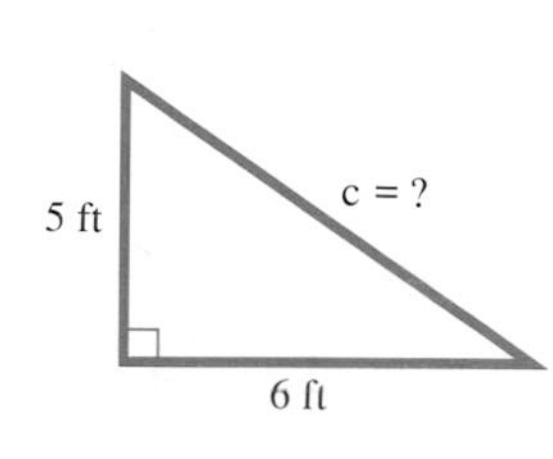

$c^2 = a^2 + b^2$

$c^2 = 5^2 + 6^2$ **Substitute 5 for *a*, and 6 for *b*.**

$c^2 = 25 + 36$ **Simplify.**

$c^2 = 61$

$c = \sqrt{61}$

$c \approx 7.81$ ft **The approximate value was found using a calculator.**

Warm Up b. Find the length of the hypotenuse of the right triangle whose legs are 5 feet and 8 feet respectively. Round the answer to the nearest hundredth of a foot.

c. Is the triangle whose sides are 12 ft, 18 ft, and 24 ft a right triangle? (*Hint*: If $c^2 = a^2 + b^2$, then they are the sides of a right triangle.)

Does $c^2 = a^2 + b^2$? We will check.

Does $24^2 = 12^2 + 18^2$? **The longest side, 24 is *c*, the other two lengths are *a* and *b*. Therefore we substitute 12 for *a*, 18 for *b*, and 24 for *c*.**

Does $576 = 144 + 324$?

Does $576 = 468$? **No.**

Since the length of the sides do not satisfy the Pythagorean relationship, the triangle is not a right triangle.

Warm Up c. Is the triangle whose sides are 6 m, 8 m, and 10 m a right triangle?

ANSWERS TO WARM UPS **a.** 5 cm **b.** 9.43 ft **c.** yes

Application Solution

d. What is the length of a brace that is needed to go diagonally across a rectangular gate that is 6 feet high and 4 feet 6 inches wide? (See the figure.)

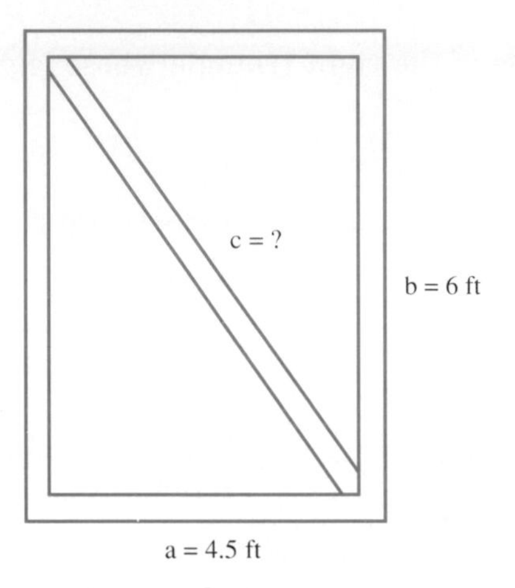

$c^2 = a^2 + b^2$	**To find the length of the diagonal, we must find side c.**
$c^2 = (4.5)^2 + 6^2$	**Substitute 4.5 for a and 6 for b.**
$c^2 = 20.25 + 36$	**Simplify.**
$c^2 = 56.25$	
$c = 7.5$	

The diagonal brace must be 7.5 feet, or 7 feet 6 inches in length.

Warm Up d. What is the length of a brace that is needed to go diagonally across a gate that is 3.75 meters high and 9 meters wide?

ANSWER TO WARM UPS

d. 9.75 m

Exercises

ANSWERS

1. ______
2. ______
3. ______
4. ______
5. ______
6. ______
7. ______
8. ______
9. ______
10. ______
11. ______
12. ______
13. ______
14. ______
15. ______
16. ______
17. ______
18. ______

A

Find the missing side of each of the following right triangles. Round decimal answers to the nearest hundredth.

1. $a = ?, \quad b = 8, \quad c = 17$

2. $a = 9, \quad b = 12, \quad c = ?$

3. $a = 12, \quad b = 5, \quad c = ?$

4. $a = 8, \quad b = ?, \quad c = 10$

5. $a = ?, \quad b = 16, \quad c = 20$

6. $a = 10, \quad b = 24, \quad c = ?$

B

7. $a = 5, \quad b = 6, \quad c = ?$

8. $a - 13, \quad b - ?, \quad c = 18$

9. $a = 40, \quad b = 30, \quad c = ?$

10. $a = 27, \quad b = ?, \quad c = 45$

11. $a = 2, \quad b = 3, \quad c = ?$

12. $a = ?, \quad b = 11, \quad c = 16$

13. $a = 5, \quad b = ?, \quad c = 7.7$

14. $a = 2.5, \quad b = 3.6, \quad c = ?$

C

15. $a = 110, \quad b = ?, \quad c = 175$

16. $a = ?, \quad b = 210, \quad c = 312$

17. $a = 5.5, \quad b = 13.2, \quad c = ?$

18. $a = 10.4, \quad b = 19.5, \quad c = ?$

ANSWERS

19. ____________

20. ____________

21. ____________

22. ____________

23. ____________

24. ____________

25. ____________

26. ____________

27. ____________

28. ____________

29. ____________

In each of the following, determine if the triangle whose sides are given is a right triangle.

19. 16, 30, and 34

20. 6, 8, and 9

21. 4, 2.4, and 3.2

22. 2.6, 1, and 2.4

23. 8.4, 9.2, and 3.5

24. 6, 7.1, and 3.2

25. 33.6, 44.8, and 56

26. 0.3, 0.72 and 0.78

D

27. What is the length of a rafter that has a rise of 6 feet and a run of 12 feet? Round the answer to the nearest hundredth of a foot. (The rise and run are the legs of a right triangle where the rafter is the hypotenuse.)

28. What is the length of a rafter that has a rise of 4 feet and a run of 13 feet? Round the answer to the nearest hundredth of a foot.

29. What is the length of a cable needed to replace a brace that goes from the top of a 50-foot power pole to a ground-level anchor that is 35 feet from the base of the pole? Round the answer to the nearest tenth of a foot.

30. What is the length of a cable needed to replace a brace that goes from the top of a 175-foot power pole to a ground-level anchor that is 42 feet from the base of the pole? Round the answer to the nearest tenth of a foot.

ANSWERS

30. ____________

31. What is the rise of a rafter that is 20 feet in length and has a run of 16 feet?

31. ____________

32. What is the rise of a rafter that is 25 feet in length and has a run of 17 feet? Round the answer to the nearest tenth of a foot.

32. ____________

33. A plane is flying south at a speed of 200 miles per hour. The wind is blowing from the west at a rate of 50 miles per hour. To the nearest tenth of a mile, how many miles does the plane actually fly in one hour, in a southeasterly direction? See the accompanying figure.

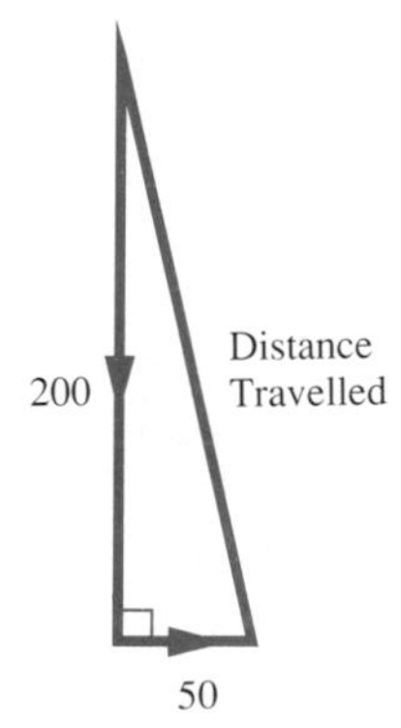

33. ____________

34. A plane is flying south at a speed of 310 miles per hour. The wind is blowing from the west at a rate of 32 miles per hour. To the nearest tenth of a mile, how many miles does the plane actually fly in one hour, in a southeasterly direction?

34. ____________

ANSWERS

35. ______

35. A baseball "diamond" is actually a square that is 90 feet on each side (between the bases). To the nearest tenth of a foot, what is the distance the catcher must throw when attempting to put out a runner who is attempting to steal second base?

36. ______

36. What is the length of the diagonal of a square that is 100 meters on each side? Round the answer to the nearest tenth of a meter.

37. ______

37. A ten-foot ladder is leaning against a building. The bottom of the ladder is 3.6 feet from the building. To the nearest tenth of a foot, how high up the building will the ladder reach?

38. ______

38. Carlotta is 8.3 miles east of Lake Begun and Jorge is 5.9 miles south of the lake. What is the shortest land distance between them, to the nearest tenth of a mile?

NAME ______________________

CLASS ____________ SECTION ____________ DATE ____________

ANSWERS

1. ______
2. ______
3. ______
4. ______
5. ______
6. ______
7. ______
8. ______
9. ______
10. ______
11. ______
12. ______
13. ______
14. ______
15. ______
16. ______
17. ______
18. ______
19. ______
20. ______
21. ______
22. ______
23. ______
24. ______
25. ______
26. ______

CHAPTERS 1 TO 4 MIDTERM EXAMINATION

1. Write the place value of the digit 6 in 59,638.

2. Write the word name for 90,053.

3. Add:
$$\begin{array}{r} 1397 \\ 42 \\ 135 \\ 89713 \\ \underline{814} \end{array}$$

4. Add: $786 + 15{,}382 + 31 + 6$

5. Subtract:
$$\begin{array}{r} 7296 \\ \underline{3759} \end{array}$$

6. Multiply:
$$\begin{array}{r} 307 \\ \underline{805} \end{array}$$

7. Multiply: (691)(53)

8. Divide: $65\overline{)8399}$

9. Divide: $42\overline{)5712}$

10. Perform the indicated operations: $7 - 3 \cdot 2 + 10 \div 5$

11. Find the sum of the quotient of 54 and 6 and the product of 11 and 3.

12. Find the average of 305, 165, 94, and 100.

13. Write the least common multiple (LCM) of 12, 10, and 18.

14. Is 107 a prime number or a composite number?

15. Is 382 a multiple of 9?

16. List the first five multiples of 18.

17. List all the factors of 204.

18. Write the prime factorization of 192.

19. Change to a mixed number: $\frac{41}{9}$

20. Change to an improper fraction: $8\frac{3}{4}$

21. Which of these fractions are improper? $\frac{3}{4}, \frac{7}{6}, \frac{8}{8}, \frac{7}{8}, \frac{9}{8}, \frac{5}{4}, \frac{4}{4}$

22. List these fractions from the smallest to largest: $\frac{5}{9}, \frac{5}{8}, \frac{7}{12}, \frac{2}{3}$

23. Reduce to the lowest terms: $\frac{48}{72}$

24. Multiply and reduce: $\frac{5}{8} \cdot \frac{2}{25} \cdot \frac{6}{5}$

25. Multiply. Write the answer as a mixed number: $\left(4\frac{2}{3}\right)\left(6\frac{1}{2}\right)$

26. Divide: $\frac{6}{7} \div \frac{14}{15}$

27. ______
28. ______
29. ______
30. ______
31. ______
32. ______
33. ______
34. ______
35. ______
36. ______
37. ______
38. ______
39. ______
40. ______
41. ______
42. ______
43. ______
44. ______
45. ______
46. ______
47. ______
48. ______

27. What is the reciprocal of $\frac{5}{9}$?

28. Add: $\frac{2}{5} + \frac{5}{12}$

29. Add:

$$\begin{array}{r} 5\frac{2}{3} \\ \underline{2\frac{5}{7}} \end{array}$$

30. Subtract: $9 - 3\frac{5}{7}$

31. Subtract:

$$\begin{array}{r} 8\frac{3}{5} \\ \underline{4\frac{7}{9}} \end{array}$$

32. Find the average of $2\frac{2}{5}$, $1\frac{7}{8}$, and $7\frac{1}{2}$.

33. Perform the indicated operations: $\frac{4}{5} - \frac{1}{2} \cdot \frac{5}{6} \div \frac{5}{6}$

34. Write the word name for 71.306.

35. Write as a decimal: $\frac{9}{40}$

36. Write as an approximate decimal to the nearest hundredth: $\frac{7}{9}$

37. What is the place value of the 2 in 39.8972?

38. Round off to the nearest hundredth: 6.8481

39. Change to a fraction: 0.2375

40. Change to a fraction: $0.166\overline{6}$

41. Is the following true or false? $0.4741 < 0.4727$

42. Add: $2.7 + 12.631 + 3.02 + 6.0032$

43. Subtract:

$$\begin{array}{r} 23.602 \\ \underline{11.671} \end{array}$$

44. Multiply:

$$\begin{array}{r} 3.603 \\ \underline{5.8} \end{array}$$

45. Divide: $28.24 \div 10{,}000$

46. Divide. Round off the answer to the nearest hundredth: $3.5\overline{)21.83}$

47. Perform the indicated operations: $0.8 - 0.7(0.002) + 0.02$

48. Letha used 5.2 gallons of gasoline on Monday, 6.2 gallons on Tuesday, 5.9 gallons on Wednesday, 6.7 gallons on Thursday, and 7.4 gallons on Friday. What was the average number of gallons she used per day? (To the nearest tenth.)

NAME ______________________

CLASS ____________ SECTION ____________ DATE ____________

ANSWERS

1. ______

2. ______

3. ______

4. ______

5. ______

6. ______

7. ______

8. ______

9. ______

10. ______

11. ______

12. ______

13. ______

14. ______

15. ______

16. ______

17. ______

18. ______

19. ______

CHAPTERS 1 TO 8 FINAL EXAMINATION

1. Add: $\frac{3}{8} + \frac{5}{16}$

2. Write the LCM (least common multiple) of 12, 14 and 21.

3. Subtract: $\begin{array}{r} 73.43 \\ \underline{65.45} \end{array}$

4. Add: $17.09 + 0.095 + 0.21$

5. Divide: $\frac{3}{4} \div \frac{1}{5}$

6. Multiply. Write the result as a mixed number: $1\frac{5}{6} \cdot 15$

7. Which of these numbers is a prime number? 99, 199, 299, 699

8. Divide. Round off the answer to the nearest hundredth: $0.67\overline{)43.45}$

9. Subtract: $10 - 4\frac{5}{8}$

10. Multiply: $(0.0098)(10{,}000)$

11. Round off to the nearest thousandth: 79.0068

12. Multiply: $(5.6)(8.09)$

13. Write as a fraction reduced to the lowest terms: 70%

14. Add: $\begin{array}{r} 5\frac{7}{8} \\ \underline{\frac{5}{12}} \end{array}$

15. Solve the proportion: $\frac{7}{9} = \frac{x}{3}$

16. What is the place value of the 3 in 23.95?

17. List these fractions from the smallest to largest: $\frac{1}{4}, \frac{2}{5}, \frac{3}{20}$

18. Change to percent: $\frac{11}{20}$

19. Divide: $18\overline{)685}$

20. ______

21. ______

22. ______

23. ______

24. ______

25. ______

26. ______

27. ______

28. ______

29. ______

30. ______

31. ______

32. ______

33. ______

34. ______

35. ______

36. ______

37. ______

38. ______

39. ______

40. ______

20. Write as an approximate decimal to the nearest thousandth: $\frac{9}{14}$

21. Write the place value name for "four thousand and four tenths."

22. Divide: $0.08\overline{)5}$

23. Write as a decimal: $8\frac{2}{5}\%$

24. Thirty-five percent of what number is 2.1?

25. Write as a percent: 2.3

26. Write as a decimal: $\frac{33}{200}$

27. Sixty-five percent of 32 is what number?

28. Multiply: (0.13)(0.2)(0.11)

29. If 4 books cost $5.20, how much would 18 books cost?

30. A hi-fi set was priced at $400. It is on sale for $335. What was the percent of discount based on the original price?

31. List the first five multiples of 13.

32. Write the word name for 505.05.

33. Is the following proportion true or false? $\frac{1.8}{30} = \frac{10}{20}$

34. Write the prime factorization of 420.

35. Reduce to the lowest terms: $\frac{72}{180}$

36. Change to a fraction and reduce to the lowest terms: 0.325

37. Change to a mixed number: $\frac{187}{6}$

38. Multiply and reduce: $\frac{4}{21} \cdot \frac{28}{64}$

39. Change to a fraction: 0.0525

40. At a service station 29 out of 50 drivers asked for a "fill-up." What percent of the drivers wanted a full tank of gas?

NAME ____________________

CLASS ____________ SECTION ____________ DATE ____________

ANSWERS

41. ____________

42. ____________

43. ____________

44. ____________

45. ____________

46. ____________

47. ____________

48. ____________

49. ____________

50. ____________

51. ____________

52. ____________

53. ____________

54. ____________

55. ____________

41. Is 553 a multiple of 3?

42. Subtract: $8\frac{1}{5} - \frac{7}{10}$

43. Change to an improper fraction: $7\frac{5}{16}$

44. List the following decimals from the smallest to largest: 1.32, 1.332, 1.299, 1.322

45. Divide: $48.73 \div 1000$

46. Divide: $3\frac{3}{8} \div 6\frac{3}{16}$

47. Write a ratio to compare 3 in. to 3 ft (using common units) and reduce.

48. A woman has calculated that she pays $1.06 for gas and oil to drive 5 miles and that, in addition, it costs her 24¢ for maintenance for each 5 miles she travels. How much will it cost her to drive 5,000 miles?

49. Twenty percent of a family's income is spent on food, 8 percent on transportation, 35 percent on housing, 10 percent on heat and utilities, 7 percent on insurance, and the rest on miscellaneous expenses. If the family's income is $1200 per month, how much is spent on miscellaneous expenses?

50. In a certain state the gasoline tax is 14 cents per gallon. What is the tax rate (to the nearest tenth of a percent) on gas that costs $1.289 per gallon (not including tax)?

51. Perform the indicated operations: $27 - 4 \cdot 3 + 12 \div 4$

52. Find the average of $\frac{7}{12}$, $1\frac{5}{8}$, and $4\frac{5}{6}$.

53. Perform the indicated operations: $3.4 - 1.8(0.4) \div 3$

54. Add:

$$\begin{array}{r} 3 \text{ yd } 2 \text{ ft } 9 \text{ in.} \\ \underline{3 \text{ yd } 2 \text{ ft } 8 \text{ in.}} \end{array}$$

55. Subtract:

$$\begin{array}{r} 4 \text{ m } 27 \text{ cm} \\ \underline{1 \text{ m } 67 \text{ cm}} \end{array}$$

ANSWERS

56. ____________

57. ____________

58. ____________

59. ____________

60. ____________

61. ____________

62. ____________

63. ____________

64. ____________

65. ____________

66. ____________

67. ____________

56. Convert 5¢ per gram to dollars per kilogram.

57. Find the perimeter of a trapezoid with bases of $10\frac{1}{2}$ ft and $12\frac{1}{4}$ ft and sides of 3 ft and $5\frac{1}{2}$ ft.

58. Find the area of a triangle with base 3.3 m and height 1.8 m.

59. Find the area of the following geometric figure (let $\pi = 3.14$):

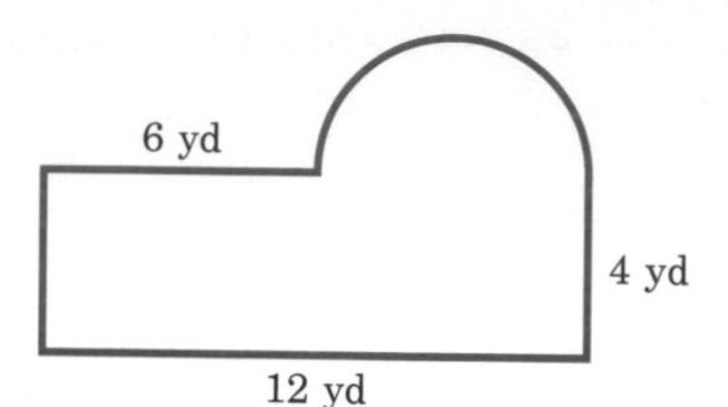

60. Find the volume of a box with length $1\frac{1}{4}$ feet, width 9 inches, and height 6 inches (in cubic inches).

61. Add: $(-26) + (-18) + 15$

62. Subtract: $(-85) - (-62)$

63. Multiply: $(3)(7)(-6)$

64. Divide: $(-88) \div (-22)$

65. Perform the indicated operations: $(-6 - 4)(-4) \div (-5) - (-10)$

66. Find the missing side:

3.3

4.4

67. Solve: $12a + 50 = 2$

ANSWERS TO SELECTED EXERCISES

CHAPTER 1

Chapter 1 Pre-Test

1. ten thousand **2.** 14,402
3. three thousand eight hundred twenty-one
4. seventy thousand six **5.** 9,000 + 800 + 70 + 4
6. 7,823 **7.** 373,000 **8.** 1,700,000 **9.** false
10. 19,196 **11.** 6,193 **12.** 2,413 **13.** 5,332
14. 2,706 **15.** 58,464 **16.** 353,028 **17.** 123
18. 402 **19.** 159 R 18 **20.** 689 R 4 **21.** 216
22. 32,600,000 **23.** 860 **24.** 27 **25.** 5
26. 241 **27.** 491 **28.** $882 **29.** 12,325 words
30. $113,700

Exercises 1.1

1. one **3.** thousand **5.** ten thousand
7. five hundred forty **9.** five thousand forty-two
11. 7,500 **13.** 243 **15.** 605 **17.** hundred
19. ten **21.** ten thousand
23. twenty-five thousand, three hundred ten
25. two thousand, five hundred thirty-one
27. two hundred five thousand, three hundred ten
29. 1 **31.** 4 **33.** 243,007 **35.** 23,407
37. 2,437 **39.** million **41.** billion
43. ten million **45.** five hundred two million, five hundred twenty thousand, fifty-two
47. fifty million, fifty thousand, five hundred
49. three million, seven hundred fifty-six thousand, four hundred eighty-nine **51.** 3 **53.** 2
55. 406,242,713 **57.** 6,000,606 **59.** 7
61. eight hundred thirty-two **63.** $34,000
65. six hundred eighty-four **67.** 4,000,000,000

Exercises 1.2

1. 400 + 0 + 1 **3.** 3,000 + 0 + 0 + 5
5. 100 + 20 + 6 **7.** 701 **9.** 33,445 **11.** true
13. true **15.** false **17.** 310 **19.** 1,660
21. 800 **23.** 60,000 + 0 + 0 + 0 + 0 **25.** 50,000 + 0 + 200 + 0 + 4 **27.** 300,000 + 0 + 2,000 + 0 + 80 + 4 **29.** 62,832 **31.** 58,873 **33.** 72,634
35. false **37.** true **39.** 130,000 **41.** 135,000
43. 135,000 **45.** 215,000 **47.** 600,000 + 50,000 + 4,000 + 300 + 70 + 7 **49.** 9,000,000 + 400,000 + 40,000 + 3,000 + 200 + 0 + 5 **51.** 507,431
53. 9,874,321 **55.** 3,069,000 **57.** true **59.** true
61. 672,457,000 **63.** 34,870,000 **65.** 75,347,200
67. $268,000 **69.** $279,000 **71.** $2,300,000

Exercises 1.3

1. 49 **3.** 99 **5.** 33 **7.** 505 **9.** 408
11. 676 **13.** 3,181 **15.** 10,387 **17.** 64
19. 1,539 **21.** 164 **23.** 183 **25.** 837
27. 168 **29.** 1,325 **31.** 2,777 **33.** 13,652
35. 17,819 **37.** 197 **39.** 919 **41.** 4,534
43. 49,922 **45.** 90,106 **47.** 66,394 **49.** 245
51. 9,074 **53.** 29,118 **55.** 22,274
57. 1,391 miles **59.** 51,908 people **61.** $721
63. 864 blocks **65.** 2,830 miles **67.** $1,200

Exercises 1.4

1. 7 **3.** 3 **5.** 11 **7.** 13 **9.** 22 **11.** 48
13. 28 **15.** 27 **17.** 19 **19.** 25 **21.** 93
23. 372 **25.** 361 **27.** 139 **29.** 549 **31.** 399
33. 313 **35.** 282 **37.** 5,063 **39.** 145
41. 1,186 **43.** 3,348 **45.** 21,786 **47.** 16,199
49. 2,285 **51.** 3,438 **53.** 691 **55.** 16,222
57. 4,889 **59.** 55,412 tickets **61.** 135 lb
63. $1,900 **65.** $4,600 **67.** 13 cubic yards

Getting Ready for Algebra

1. $x = 9$ **3.** $x = 11$ **5.** $z = 7$ **7.** $c = 29$
9. $a = 183$ **11.** $x = 34$ **13.** $y = 47$ **15.** $k = 246$
17. $37 = x$ **19.** $130 = w$

Exercises 1.5

1. 126 **3.** 236 **5.** 368 **7.** 0 **9.** 2,460
11. 2,400 **13.** 684 **15.** 9,450 **17.** 574
19. 345 **21.** 0 **23.** 882 **25.** 24,360
27. 1,833 **29.** 3,648 **31.** 7,035 **33.** 37,200
35. 6,420 **37.** 64,500 **39.** 42,588 **41.** 41,013
43. 66,151 **45.** 414,765 **47.** 182,546
49. 176,928 **51.** 2,180,304 **53.** 2,512,104
55. 2,054,536 **57.** 18,203,276 **59.** 66,150
61. 2,576 bottles **63.** 1,300 watts
65. 2,226 turns **67.** $68,000 **69.** $16,646

Exercises 1.6

1. 9 **3.** 6 **5.** 1 **7.** 51 **9.** 71 **11.** 40
13. 2 R 12 **15.** 15 R 8 **17.** 16 **19.** 35
21. 112 **23.** 261 **25.** 3,019 **27.** 161 **29.** 24
31. 74 **33.** 305 **35.** 12 R 19 **37.** 279 R 3
39. 2,093 **41.** 54 R 38 **43.** 607 **45.** 79
47. 214 **49.** 91 **51.** 683 R 20 **53.** 193 R 185
55. 218 R 160 **57.** 207 **59.** 705
61. 10,365 packages **63.** 11 houses **65.** $5,567
67. 1,955 radios, 9 res. left
69. 111 tons, 28 extra bales

Getting Ready for Algebra

1. $x = 5$ **3.** $c = 18$ **5.** $x = 4$ **7.** $b = 120$
9. $x = 12$ **11.** $y = 312$ **13.** $x = 24$
15. $b = 45,414$ **17.** $5 = x$ **19.** $1278 = w$

Exercises 1.7

1. 11 **3.** 64 **5.** 1 **7.** 1 **9.** 3,600
11. 7,000 **13.** 15 **15.** 97 **17.** 16 **19.** 125
21. 324 **23.** 1 **25.** 8,530,000 **27.** 10
29. 1,783,000 **31.** 2,001 **33.** 2,187 **35.** 512
37. 14,641 **39.** 2,401 **41.** 5,060,000,000
43. 80 **45.** 2,123 **47.** 2,001 **49.** $7 \times 10^2 + 8 \times 10^1 + 2 \times 10^0$ **51.** $6 \times 10^3 + 8 \times 10^2 + 3 \times 10^1 + 5 \times 10^0$ **53.** $7 \times 10^4 + 0 + 9 \times 10^2 + 6 \times 10^1 + 1 \times 10^0$ **55.** $1 \times 10^7 + 6 \times 10^6 + 0 + 0 + 3 \times 10^3 + 9 \times 10^2 + 3 \times 10^1 + 0$ **57.** $44,000,000
59. $160,000,000; one hundred sixty million dollars
61. $32,768

Exercises 1.8

1. 40 **3.** 8 **5.** 19 **7.** 1 **9.** 8 **11.** 18
13. 4 **15.** 45 **17.** 60 **19.** 73 **21.** 43
23. 64 **25.** 9 **27.** 1 **29.** 146 **31.** 80
33. 14 **35.** 70 **37.** 12 **39.** 36 **41.** 28
43. 634 **45.** 6 **47.** 2,821 **49.** 1,127 **51.** 775 cans **53.** $7,072 **55.** $5,225 **57.** $476

Getting Ready for Algebra

1. $x = 5$ **3.** $y = 24$ **5.** $x = 4$ **7.** $c = 88$
9. $x = 6$ **11.** $c = 12$ **13.** $a = 800$ **15.** $b = 22$

Exercises 1.9

1. 4 **3.** 6 **5.** 6 **7.** 14 **9.** 12 **11.** 10
13. 5 **15.** 3 **17.** 19 **19.** 4 **21.** 25 **23.** 24
25. 35 **27.** 13 **29.** 31 **31.** 37 **33.** 57
35. 165 **37.** 105 **39.** 494 **41.** 453 **43.** 724
45. 4,104 calories **47.** 20 points
49. 30 miles per gallon **51.** 83

Chapter 1 True–False Concept Review

1. F **2.** T **3.** T **4.** F **5.** F **6.** F
7. T, 250 to the nearest 10 is 250. **8.** T **9.** F
10. T **11.** F **12.** T, we can subtract by adding from 37: 37 + 3 + 10 + 5 = 55 and 3 + 10 + 5 = 18
13. T **14.** T **15.** F **16.** T **17.** F **18.** T
19. T **20.** F **21.** F **22.** T **23.** T **24.** F
25. T **26.** F **27.** F **28.** T **29.** T **30.** T

Chapter 1 Post-Test

1. 7,089 **2.** 693,860,000 **3.** 17,760 **4.** 1,583
5. 79 **6.** 306 **7.** four thousand, two hundred five
8. 5,600 **9.** 309,963 **10.** one hundred twenty thousand, three hundred fifty-five **11.** 1,300
12. 43,681 **13.** 3,527 **14.** 353 R 69 **15.** 5,230
16. 256 **17.** 5,748 **18.** 900 + 30 + 7
19. 486,048 **20.** 3,006,000 **21.** 209 **22.** 16,988
23. 907 **24.** 20 **25.** false **26.** thousand
27. 89 R 19 **28.** $3,373 **29.** 76,000 pounds
30. $161,440

CHAPTER 2

Chapter 2 Pre-Test

1. 6, 12, 18, 24, 30 **2.** yes **3.** no
4. 1, 2, 4, 17, 34, 68 **5.** 1 · 42, 2 · 21, 3 · 14, 6 · 7
6. no **7.** yes **8.** prime **9.** composite
10. 2 · 3 · 3 · 7 **11.** 11 · 37 **12.** 2 · 5 · 7 · 7
13. 72 **14.** 600 **15.** 210

Exercises 2.1

1. 2, 4, 6, 8, 10 **3.** 13, 26, 39, 52, 65
5. 5, 10, 15, 20, 25 **7.** yes **9.** yes **11.** no
13. yes **15.** no **17.** no **19.** 17, 34, 51, 68, 85
21. 25, 50, 75, 100, 125 **23.** 31, 62, 93, 124, 155
25. 40, 80, 120, 160, 200 **27.** yes **29.** no
31. yes **33.** yes **35.** no **37.** yes
39. 35, 70, 105, 140, 175
41. 110, 220, 330, 440, 550
43. 96, 192, 288, 384, 480 **45.** multiple of 4 and 6
47. multiple of 6 and 15
49. multiple of 4, 6, and 15 **51.** 4, 8, 12, 16, 20, 24, 28, 32, 36, 40, 44, 48, 52, 56, 60
53. 72, 80, 88, 96, 104, 112 **55.** 306, 315, 324
57. 2,568 **59.** 4,546 **61.** 2,017 **63.** 1,302
65. 4 days, 50 gallons

Exercises 2.2

1. 1, 2, 3, 4, 6, 12 **3.** 1, 31 **5.** 1, 5, 7, 35
7. 1, 2, 13, 26 **9.** 1, 2, 5, 7, 10, 14, 35, 70
11. 1 · 12, 2 · 6, 3 · 4 **13.** 1 · 16, 2 · 8, 4 · 4
15. 1 · 49, 7 · 7 **17.** 1 · 50, 2 · 25, 5 · 10
19. 1 × 45, 3 × 15, 5 × 9 **21.** 1, 2, 3, 6, 7, 14, 21, 42 **23.** 1, 3, 9, 11, 33, 99
25. 1, 2, 4, 5, 10, 20, 25, 50, 100
27. 1, 2, 4, 8, 16, 32 **29.** 1, 2, 4, 8, 11, 22, 44, 88
31. 1 · 96, 2 · 48, 3 · 32, 4 · 24, 6 · 16, 8 · 12
33. 1 · 220, 2 · 110, 4 · 55, 5 · 44, 10 · 22, 11 · 20
35. 1 · 120, 2 · 60, 3 · 40, 4 · 30, 5 · 24, 6 · 20, 8 · 15, 10 · 12 **37.** 1 · 305, 5 · 61 **39.** 1 × 210, 2 × 105, 3 × 70, 5 × 42, 6 × 35, 7 × 30, 10 × 21, 14 × 15
41. 1, 5, 7, 35, 49, 245 **43.** 1, 2, 4, 5, 10, 20, 25, 50, 100, 125, 250, 500 **45.** 1, 2, 3, 4, 5, 6, 8, 9, 10, 12, 15, 18, 20, 24, 30, 36, 40, 45, 60, 72, 90, 120, 180, 360
47. 1 · 245, 5 · 49, 7 · 35 **49.** 1 · 720, 2 · 360, 3 · 240, 4 · 180, 5 · 144, 6 · 120, 8 · 90, 9 · 80, 10 · 72, 12 · 60, 15 · 48, 16 · 45, 18 · 40, 20 · 36, 24 · 30 **51.** 1 × 510, 2 × 255, 3 × 170, 5 × 102, 6 × 85, 10 × 51, 15 × 34, 17 × 30 **53.** 1 program of 150 minutes, 150 programs of 1 minute; 2 programs of 75 minutes, 75 programs of 2 minutes; 3 programs of 50 minutes, 50 programs of 3 minutes; 5 programs of 30 minutes, 30 programs of 5 minutes; 6 programs of 25 minutes, 25 programs of 6 minutes; 10 programs of 15 minutes, 15 programs of 10 minutes **55.** 1 program of 90 minutes, 90 programs of 1 minute; 2 programs of 45 minutes, 45 programs of 2 minutes; 3 programs of 30 minutes, 30 programs of 3 minutes; 5 programs of 18 minutes, 18 programs of 5 minutes; 6 programs of 15 minutes, 15 programs of 6 minutes; 9 programs of 10 minutes, 10 programs of 9 minutes
57. 14 different ways **59.** 2,419 **61.** 18,865
63. 31 **65.** 24 R 2 **67.** 58 speakers, 10 feet

Exercises 2.3

1. yes **3.** yes **5.** no **7.** yes **9.** no **11.** no
13. no **15.** yes **17.** no **19.** yes **21.** no
23. yes **25.** no **27.** yes **29.** no **31.** yes
33. no **35.** yes **37.** no **39.** no **41.** yes
43. no **45.** no **47.** yes **49.** 2 no, 3 no, 5 yes
51. 2 and 3 no, 5 yes **53.** 2 and 5 yes, 3 no
55. 2 yes, 3 and 5 no **57.** 2 and 5 yes, 3 no

59. yes **61.** no **63.** 68,000 **65.** 6,238,500 **67.** 765,500,000 **69.** 26 **71.** 553 **73.** 9 **75.** 36 cents per liter

Exercises 2.4

1. composite **3.** prime **5.** prime **7.** prime **9.** composite **11.** composite **13.** composite **15.** prime **17.** composite **19.** prime **21.** composite **23.** composite **25.** prime **27.** prime **29.** composite **31.** prime **33.** composite **35.** prime **37.** composite **39.** prime **41.** composite **43.** composite **45.** composite **47.** composite **49.** composite **51.** composite **53.** prime **55.** composite **57.** composite **59.** numbers not crossed off: 2, 3, 5, 7, 11, 13, 17, 19, 23, 29, 31, 37, 41, 43, 47, 53, 59, 61, 67, 71, 73, 79, 83, 89, 97, 101, 103, 107, 109, 113, 127, 131, 137, 139, 149 **61.** 1979 **63.** answer varies **65.** false **67.** 16,384 **69.** 82 **71.** 105 **73.** 11

Exercises 2.5

1. $3 \cdot 5$ **3.** $5 \cdot 7$ **5.** $2 \cdot 17$ **7.** $3^2 \cdot 5$ **9.** $2 \cdot 3 \cdot 11$ **11.** $2^4 \cdot 3$ **13.** $2^4 \cdot 5$ **15.** $2 \cdot 41$ **17.** $3 \cdot 17$ **19.** $2^2 \cdot 3^2$ **21.** prime **23.** $7 \cdot 13$ **25.** $3 \cdot 5^2$ **27.** $2 \cdot 3 \cdot 5^2$ **29.** $3 \cdot 61$ **31.** $7 \cdot 19$ **33.** $2 \cdot 7 \cdot 13$ **35.** $5 \cdot 37$ **37.** $2^3 \cdot 13$ **39.** $2^2 \cdot 3 \cdot 13$ **41.** $3^2 \cdot 5^2$ **43.** $2 \cdot 3^2 \cdot 23$ **45.** $3 \cdot 5 \cdot 37$ **47.** $2^3 \cdot 3 \cdot 37$ **49.** prime **51.** $2^2 \cdot 3^2 \cdot 5^2$ **53.** $3 \cdot 7 \cdot 43$ **55.** $2^3 \cdot 3 \cdot 5 \cdot 13$ **57.** 31, 62, 93, 124, 155, 186, 217 **59.** yes **61.** yes **63.** no **65.** yes **67.** 510, 527, 544, 561, 578, 595

Exercises 2.6

1. 12 **3.** 40 **5.** 42 **7.** 90 **9.** 20 **11.** 50 **13.** 30 **15.** 10 **17.** 24 **19.** 40 **21.** 30 **23.** 180 **25.** 240 **27.** 90 **29.** 120 **31.** 360 **33.** 150 **35.** 24 **37.** 600 **39.** 360 **41.** 224 **43.** 504 **45.** 1,092 **47.** 480 **49.** 750 **51.** 630 **53.** 204 **55.** 342 **57.** 1,400 **59.** 6 **61.** 48 **63.** 108 **65.** 60 **67.** 1, 2, 4, 8, 47, 94, 188, 376 **69.** 1, 2, 4, 97, 194, 388 **71.** yes **73.** 1,350 people

Chapter 2 True–False Concept Review

1. F **2.** T **3.** T **4.** T **5.** T **6.** T **7.** F **8.** F **9.** F **10.** T **11.** F **12.** T **13.** F **14.** T **15.** F **16.** F **17.** F **18.** F **19.** T **20.** F **21.** T **22.** T **23.** F **24.** F **25.** T

Chapter 2 Post-Test

1. no **2.** 1, 2, 5, 10, 11, 22, 55, 110 **3.** yes **4.** yes **5.** 80 **6.** $1 \cdot 75$, $3 \cdot 25$, $5 \cdot 15$ **7.** $2^2 \cdot 5 \cdot 13$ **8.** 252 **9.** $2 \cdot 3^2 \cdot 47$ **10.** 11, 22, 33, 44, 55 **11.** no **12.** prime **13.** composite **14.** $7 \cdot 11^2$ **15.** 210

CHAPTER 3

Chapter 3 Pre-Test

1. $\frac{5}{8}$ **2.** $\frac{10}{11}, \frac{11}{12}, \frac{10}{12}$ **3.** $\frac{13}{13}, \frac{14}{14}$ **4.** $2\frac{4}{5}$ **5.** $\frac{17}{1}$ **6.** $\frac{14}{3}$ **7.** $\frac{3}{4}$ **8.** $\frac{3}{5}$ **9.** $\frac{8}{15}$ **10.** $\frac{1}{4}$ **11.** $\frac{2}{21}$ **12.** $\frac{5}{3}$ or $1\frac{2}{3}$ **13.** $\frac{3}{5}$ **14.** $\frac{32}{35}$ **15.** $8\frac{1}{4}$ **16.** 3 **17.** 28 **18.** $\frac{1}{4}, \frac{3}{8}, \frac{2}{5}$ **19.** true **20.** $\frac{5}{7}$ **21.** $\frac{11}{12}$ **22.** 1 **23.** $7\frac{9}{20}$ **24.** $\frac{1}{5}$ **25.** $8\frac{11}{18}$ **26.** $1\frac{5}{6}$ **27.** $6\frac{5}{6}$ **28.** $2\frac{1}{3}$ **29.** $2\frac{1}{6}$ **30.** $1\frac{1}{3}$ **31.** $11\frac{1}{6}$ inches **32.** $4\frac{11}{16}$ gallons

Exercises 3.1

1. $\frac{4}{6}, \frac{5}{6}$ **3.** $\frac{6}{12}, \frac{7}{14}, \frac{9}{12}, \frac{10}{14}, \frac{11}{22}$ **5.** $\frac{6}{6}$ **7.** $\frac{11}{11}, \frac{1}{1}$ **9.** $\frac{7}{5}, \frac{8}{3}, \frac{6}{6}$ **11.** $\frac{4}{7}$ **13.** $\frac{5}{7}$ **15.** $\frac{1}{4}$ **17.** $\frac{4}{4}$ **19.** $\frac{3}{3}, \frac{15}{15}, \frac{144}{144}$ **21.** $\frac{3}{3}, \frac{15}{15}, \frac{144}{144}$ **23.** $\frac{5}{3}$ **25.** $\frac{13}{9}$ **27.** $\frac{4}{10}$ **29.** $\frac{3}{6}$ **31.** $\frac{7}{10}$ **33.** $\frac{11}{8}$ **35.** $\frac{4}{4}$ **37.**

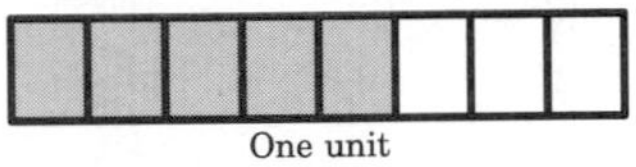

39. $\frac{8}{15}$ **41.** $\frac{23}{60}$ **43.** $\frac{25}{57}$ **45.** $\frac{7}{41}$ **47.** $\frac{5}{12}$ **49.** prime **51.** $3 \cdot 7 \cdot 11$ **53.** 6,010 **55.** no **57.** 165 miles

Exercises 3.2

1. $4\frac{1}{4}$ **3.** $4\frac{1}{2}$ **5.** $4\frac{3}{4}$ **7.** $2\frac{4}{9}$ **9.** 5 or $5\frac{0}{8}$ **11.** $\frac{31}{7}$ **13.** $\frac{9}{1}$ or $\frac{18}{2}$ or $\frac{27}{3}$ etc **15.** $\frac{23}{4}$ **17.** $\frac{43}{6}$ **19.** $\frac{42}{5}$ **21.** 65 or $65\frac{0}{3}$ **23.** $24\frac{3}{8}$ **25.** $6\frac{6}{7}$ **27.** 6 or $6\frac{0}{13}$ **29.** $21\frac{1}{7}$ **31.** $\frac{221}{8}$ **33.** $\frac{79}{5}$ **35.** $\frac{101}{6}$ **37.** $\frac{15}{1}$ or $\frac{30}{2}$ or $\frac{45}{3}$ etc **39.** $\frac{227}{7}$ **41.** 4 or $4\frac{0}{13}$ **43.** $6\frac{14}{31}$ **45.** $28\frac{9}{11}$ **47.** $3\frac{7}{25}$ **49.** $14\frac{15}{19}$ **51.** $\frac{571}{12}$ **53.** $\frac{423}{4}$ **55.** $\frac{4{,}607}{100}$ **57.** $\frac{13{,}676}{111}$ **59.** $\frac{101}{1}$ or $\frac{202}{2}$ or $\frac{303}{3}$ etc **61.** 3 pounds **63.** $3\frac{1}{10}$ dollars **65.** $19\frac{0}{100}$ dollars **67.** 50 feet **69.** \$190 **71.** \$125 **73.** 5,002 **75.** 7,318 **77.** yes **79.** 4,096 **81.** 1,240 **83.** 36,000 revolutions

Exercises 3.3

1. $\frac{1}{2}$ **3.** $\frac{1}{3}$ **5.** $\frac{2}{3}$ **7.** $\frac{1}{5}$ **9.** $\frac{6}{7}$ **11.** $\frac{9}{10}$ **13.** $\frac{7}{10}$ **15.** $\frac{4}{3}$ **17.** $\frac{2}{3}$ **19.** $\frac{4}{5}$ **21.** 4

23. $\frac{5}{3}$ **25.** $\frac{7}{12}$ **27.** $\frac{13}{18}$ **29.** $\frac{7}{9}$ **31.** $\frac{5}{6}$
33. $\frac{1}{3}$ **35.** $\frac{3}{5}$ **37.** $\frac{5}{8}$ **39.** $\frac{3}{8}$ **41.** $\frac{7}{12}$
43. 5 **45.** $\frac{45}{32}$ **47.** $\frac{5}{7}$ **49.** $\frac{11}{12}$ **51.** $\frac{2}{3}$
53. $\frac{25}{18}$ **55.** $\frac{31}{90}$ **57.** $\frac{2}{3}$ **59.** $\frac{49}{51}$ **61.** $\frac{20}{33}$
63. $\frac{3}{5}$ **65.** $\frac{2}{11}$ **67.** $\frac{2}{3}$ **69.** $\frac{1}{3}$ **71.** $\frac{9}{10}$ correct
73. $\frac{1}{4}$ **75.** $\frac{2}{3}$ **77.** 2,526 **79.** 1,123
81. 2,830 **83.** $2 \cdot 3 \cdot 5 \cdot 7$ **85.** 1,500 bricks

Exercises 3.4

1. $\frac{3}{8}$ **3.** $\frac{35}{48}$ **5.** $\frac{4}{15}$ **7.** $\frac{35}{48}$ **9.** $\frac{1}{3}$ **11.** $\frac{1}{6}$
13. $\frac{1}{2}$ **15.** $\frac{1}{6}$ **17.** $\frac{25}{8}$ or $3\frac{1}{8}$ **19.** $\frac{7}{10}$
21. $\frac{30}{11}$ or $2\frac{8}{11}$ **23.** 1 **25.** $\frac{1}{3}$ **27.** $\frac{2}{3}$ **29.** 1
31. $\frac{1}{5}$ **33.** $\frac{200}{63}$ or $3\frac{11}{63}$ **35.** $\frac{3}{10}$ **37.** $\frac{5}{7}$
39. $\frac{1}{4}$ **41.** $\frac{1}{3}$ **43.** $\frac{7}{8}$ **45.** $\frac{3}{16}$ **47.** $\frac{63}{125}$
49. $\frac{1}{49}$ **51.** $\frac{7}{9}$ **53.** $\frac{1}{15}$ **55.** $\frac{1}{2}$ cup of sugar, $\frac{1}{3}$ cup of milk **57.** $\frac{1}{2}$ inch **59.** $16 **61.** $\frac{3}{14}$
63. $2\frac{1}{3}$ feet **65.** $4\frac{1}{2}$ gallons **67.** 144
69. 2,448 **71.** 26,600 **73.** 120 **75.** yes, 10 lb

Exercises 3.5

1. $\frac{4}{3}$ **3.** $\frac{1}{5}$ **5.** $\frac{2}{3}$ **7.** $\frac{18}{35}$ **9.** $\frac{3}{4}$ **11.** $\frac{1}{3}$
13. $\frac{2}{5}$ **15.** $\frac{2}{3}$ **17.** $\frac{10}{9}$ or $1\frac{1}{9}$ **19.** 1 **21.** $\frac{8}{9}$
23. $\frac{20}{13}$ or $1\frac{7}{13}$ **25.** $\frac{11}{39}$ **27.** $\frac{24}{17}$ or $1\frac{7}{17}$
29. $\frac{16}{15}$ or $1\frac{1}{15}$ **31.** $\frac{3}{4}$ **33.** $\frac{21}{50}$ **35.** $\frac{10}{11}$
37. $\frac{1}{4}$ **39.** $\frac{200}{147}$ or $1\frac{53}{147}$ **41.** 1 **43.** $\frac{9}{8}$ or $1\frac{1}{8}$
45. $\frac{5}{6}$ **47.** $\frac{3}{4}$ **49.** $\frac{2}{33}$ **51.** $\frac{5}{21}$
53. $\frac{25}{16}$ or $1\frac{9}{16}$ **55.** 3 **57.** $\frac{3}{2}$ or $1\frac{1}{2}$
59. $\frac{28}{25}$ or $1\frac{3}{25}$ **61.** 6 rings **63.** $3\frac{1}{2}$ omelets
65. 16 pinheads **67.** 6 gerbils **69.** 120
71. $3\frac{6}{13}$ **73.** $\frac{53}{3}$ **75.** $\frac{2}{3}$ **77.** $\frac{3}{8}$

Exercises 3.6

1. $1\frac{5}{16}$ **3.** $22\frac{2}{3}$ **5.** 7 **7.** $20\frac{5}{6}$ **9.** 24
11. $13\frac{1}{3}$ **13.** 2 **15.** $\frac{3}{4}$ **17.** 4 **19.** $2\frac{1}{6}$
21. $\frac{20}{27}$ **23.** $\frac{2}{3}$ **25.** $12\frac{5}{6}$ **27.** $2\frac{11}{12}$ **29.** $1\frac{29}{48}$
31. 0 **33.** 102 **35.** $18\frac{3}{4}$ **37.** $2\frac{2}{3}$ **39.** $\frac{1}{2}$
41. $\frac{7}{30}$ **43.** $\frac{3}{20}$ **45.** $18\frac{1}{3}$ **47.** $8\frac{1}{28}$ **49.** $35\frac{1}{5}$
51. 105 **53.** $69\frac{1}{2}$ **55.** 186 **57.** $14\frac{2}{5}$ **59.** $2\frac{4}{7}$
61. $1\frac{1}{27}$ **63.** 3 **65.** $4\frac{8}{129}$ **67.** $1\frac{5}{49}$
69. 36 inches **71.** 4 boards **73.** 216 inches
75. $7\frac{7}{8}$ inches **77.** $14\frac{9}{10}$ minutes **79.** $10\frac{5}{16}$ in^2
81. 15,153 **83.** 41,580 **85.** 360 **87.** $46\frac{1}{12}$
89. no

Getting Ready for Algebra

1. $x = \frac{3}{4}$ **3.** $y = \frac{16}{15}$ or $y = 1\frac{1}{15}$ **5.** $z = \frac{15}{16}$
7. $\frac{17}{8} = x$ or $2\frac{1}{8} = x$ **9.** $a = \frac{10}{7}$ or $a = 1\frac{3}{7}$
11. $\frac{3}{2} = b$ or $1\frac{1}{2} = b$ **13.** $z = 2$
15. $a = \frac{165}{8}$ or $a = 20\frac{5}{8}$

Exercises 3.7

1. $\frac{4}{6}, \frac{6}{9}, \frac{8}{12}, \frac{10}{15}$ **3.** $\frac{2}{8}, \frac{3}{12}, \frac{4}{16}, \frac{5}{20}$
5. $\frac{10}{8}, \frac{15}{12}, \frac{20}{16}, \frac{25}{20}$ **7.** $\frac{22}{24}, \frac{33}{36}, \frac{44}{48}, \frac{55}{60}$
9. $\frac{6}{8}, \frac{9}{12}, \frac{12}{16}, \frac{15}{20}$ **11.** $\frac{10}{12}, \frac{15}{18}, \frac{20}{24}, \frac{25}{30}$ **13.** 5
15. 10 **17.** 16 **19.** 44 **21.** 21 **23.** 20
25. $\frac{6}{14}, \frac{9}{21}, \frac{12}{28}, \frac{15}{35}$ **27.** $\frac{8}{20}, \frac{12}{30}, \frac{16}{40}, \frac{20}{50}$
29. $\frac{14}{6}, \frac{21}{9}, \frac{28}{12}, \frac{35}{15}$ **31.** 14 **33.** 16 **35.** 25
37. 15 **39.** 9 **41.** 60 **43.** 75 **45.** 8
47. 46 **49.** 75 **51.** 140 **53.** 72 **55.** 57
57. 120 **59.** $\frac{24}{30}, \frac{15}{30}, \frac{9}{30}$ **61.** $\frac{12}{24}, \frac{16}{24}, \frac{4}{24}, \frac{15}{24}$
63. 32 problems **65.** 15 pieces **67.** 60 **69.** $\frac{31}{40}$
71. $\frac{159}{7}$ **73.** $13\frac{8}{9}$ **75.** $1,875

Exercises 3.8

1. $\frac{7}{9}$ **3.** $\frac{1}{3}$ **5.** $\frac{3}{4}$ **7.** $\frac{4}{6}$ **9.** $\frac{3}{7}, \frac{4}{7}, \frac{5}{7}$
11. $\frac{1}{4}, \frac{3}{8}, \frac{1}{2}$ **13.** $\frac{1}{4}, \frac{3}{8}, \frac{1}{2}$ **15.** $\frac{7}{12}, \frac{2}{3}, \frac{3}{4}$
17. true **19.** false **21.** false **23.** $\frac{4}{5}$ **25.** $\frac{13}{10}$

27. $\frac{11}{6}$ **29.** $2\frac{3}{8}$ **31.** $\frac{5}{11}$ **33.** $\frac{2}{3}$ **35.** $\frac{2}{5}, \frac{4}{7}, \frac{2}{3}$
37. $\frac{5}{8}, \frac{7}{10}, \frac{3}{4}$ **39.** $\frac{4}{5}, \frac{5}{6}, \frac{13}{15}, \frac{9}{10}$
41. $2\frac{3}{4}, 2\frac{5}{6}, 2\frac{7}{8}$ **43.** $\frac{1}{2}, \frac{5}{8}, \frac{3}{4}, \frac{11}{12}$
45. $7\frac{2}{3}, 7\frac{3}{4}, 7\frac{5}{6}$ **47.** false **49.** false
51. $\frac{11}{24}, \frac{17}{36}, \frac{35}{72}$ **53.** $\frac{6}{14}, \frac{13}{28}, \frac{17}{35}$
55. $\frac{11}{30}, \frac{17}{45}, \frac{7}{18}, \frac{2}{5}$ **57.** $\frac{11}{12}, \frac{14}{15}, \frac{19}{20}, \frac{29}{30}$
59. false **61.** false **63.** true
65. $\frac{3}{32}, \frac{1}{8}, \frac{1}{4}, \frac{5}{16}, \frac{3}{8}, \frac{1}{2}, \frac{9}{16}$
67. smallest: $\frac{1}{4}$ gallon; largest: $\frac{1}{2}$ gallon
69. Chang's **71.** 12,793 **73.** composite
75. 13, 26, 39, 52, 65 **77.** 70,000,000 **79.** $70

Exercises 3.9

1. $\frac{9}{11}$ **3.** $\frac{2}{3}$ **5.** 1 **7.** $\frac{4}{5}$ **9.** $\frac{4}{5}$ **11.** $\frac{8}{11}$
13. 1 **15.** $1\frac{1}{4}$ **17.** $\frac{3}{5}$ **19.** $1\frac{1}{7}$ **21.** $\frac{5}{16}$
23. $\frac{3}{5}$ **25.** 8 **27.** $\frac{7}{8}$ **29.** $\frac{5}{9}$ **31.** $\frac{5}{6}$
33. $\frac{5}{12}$ **35.** $\frac{3}{5}$ **37.** $\frac{7}{15}$ **39.** $\frac{4}{5}$ **41.** 1
43. $\frac{1}{3}$ **45.** $3\frac{3}{8}$ points **47.** $3\frac{1}{4}$ gallons
49. $\frac{7}{16}$ of the income **51.** $1\frac{3}{5}$ meters
53. $1\frac{1}{2}$ inches **55.** 120 **57.** 678,224,000,000
59. $\frac{55}{4}$ **61.** 327 **63.** 24 pieces

Exercises 3.10

1. $\frac{11}{18}$ **3.** $\frac{3}{4}$ **5.** $\frac{5}{12}$ **7.** $\frac{13}{16}$ **9.** 1 **11.** $\frac{7}{10}$
13. $\frac{1}{2}$ **15.** $1\frac{7}{16}$ **17.** $\frac{47}{60}$ **19.** $1\frac{7}{30}$ **21.** $\frac{63}{80}$
23. $\frac{7}{15}$ **25.** $1\frac{1}{60}$ **27.** $1\frac{3}{8}$ **29.** $1\frac{2}{5}$ **31.** $1\frac{13}{16}$
33. $1\frac{5}{16}$ **35.** $2\frac{23}{24}$ **37.** $1\frac{77}{144}$ **39.** $2\frac{19}{90}$
41. $1\frac{11}{40}$ **43.** $\frac{7}{54}$ **45.** $\frac{1,381}{2,160}$ **47.** $\frac{37}{48}$
49. $\frac{215}{432}$ **51.** $\frac{87}{100}$ **53.** $1\frac{13}{144}$ **55.** $1\frac{19}{34}$
57. $1\frac{149}{240}$ **59.** $2\frac{113}{300}$ **61.** $1\frac{13}{16}$ inches
63. $1\frac{11}{12}$ yd **65.** $\frac{3}{4}$ in. **67.** $1\frac{7}{12}$ cups
69. $3\frac{1}{6}$ yards **71.** $2^2 \cdot 3 \cdot 11$ **73.** 48 **75.** 28
77. 3,757 **79.** 34

Exercises 3.11

1. $3\frac{5}{7}$ **3.** $8\frac{1}{5}$ **5.** $5\frac{11}{12}$ **7.** $6\frac{2}{9}$ **9.** $6\frac{5}{14}$
11. $15\frac{13}{15}$ **13.** $20\frac{19}{24}$ **15.** 11 **17.** $26\frac{11}{12}$
19. $11\frac{5}{8}$ **21.** $7\frac{2}{5}$ **23.** $26\frac{3}{10}$ **25.** $49\frac{5}{6}$
27. $719\frac{31}{36}$ **29.** $86\frac{11}{30}$ **31.** $60\frac{11}{12}$ **33.** $117\frac{34}{45}$
35. $107\frac{5}{14}$ **37.** $56\frac{11}{28}$ **39.** $68\frac{61}{72}$ **41.** $19\frac{11}{36}$
43. $13\frac{1}{25}$ **45.** $48\frac{7}{60}$ **47.** $131\frac{59}{63}$ **49.** $147\frac{9}{25}$
51. $160\frac{25}{72}$ **53.** $47\frac{1}{2}$ **55.** $64\frac{3}{8}$ **57.** $118\frac{1}{75}$
59. $23\frac{3}{4}$ hours **61.** $25\frac{5}{8}$ miles **63.** $2\frac{1}{8}$ in.
65. 44 ft **67.** $10\frac{1}{2}$ **69.** $\frac{40}{81}$ **71.** $3^3 \cdot 7^2$
73. $\frac{5}{11}$ **75.** 186 pounds

Exercises 3.12

1. $\frac{1}{4}$ **3.** $\frac{1}{4}$ **5.** $\frac{1}{3}$ **7.** $\frac{1}{2}$ **9.** $\frac{7}{16}$ **11.** $\frac{7}{45}$
13. $\frac{1}{2}$ **15.** $\frac{5}{18}$ **17.** $\frac{4}{15}$ **19.** $\frac{7}{24}$ **21.** $\frac{1}{24}$
23. $\frac{3}{20}$ **25.** $\frac{4}{21}$ **27.** $\frac{19}{48}$ **29.** $\frac{1}{18}$ **31.** $\frac{13}{24}$
33. $\frac{3}{20}$ **35.** $\frac{5}{36}$ **37.** $\frac{4}{75}$ **39.** $\frac{17}{48}$ **41.** $\frac{1}{36}$
43. $\frac{11}{200}$ **45.** $\frac{7}{30}$ **47.** $\frac{1}{72}$ **49.** $\frac{23}{72}$ **51.** $\frac{19}{108}$
53. $\frac{119}{225}$ **55.** $\frac{4}{63}$ **57.** $\frac{23}{48}$ **59.** $\frac{101}{720}$
61. $1\frac{1}{8}$ inches **63.** $\frac{5}{12}$ cup **65.** $\frac{5}{24}$ lb
67. $\frac{3}{16}$ inch **69.** $\frac{9}{16}$ yard **71.** 19,964
73. 1,021 R 26 **75.** $\frac{5}{7}$ **77.** $\frac{9}{28}$
79. 9,600 bricks

Exercises 3.13

1. $5\frac{2}{7}$ **3.** $103\frac{1}{5}$ **5.** $5\frac{1}{8}$ **7.** $\frac{3}{7}$ **9.** $118\frac{1}{6}$
11. $2\frac{1}{2}$ **13.** $13\frac{3}{5}$ **15.** $155\frac{1}{9}$ **17.** $1\frac{5}{8}$
19. $4\frac{3}{8}$ **21.** $56\frac{1}{3}$ **23.** $20\frac{9}{20}$ **25.** $21\frac{3}{4}$
27. $18\frac{23}{48}$ **29.** $28\frac{1}{3}$ **31.** $5\frac{25}{32}$ **33.** $7\frac{11}{36}$
35. $9\frac{2}{5}$ **37.** $7\frac{23}{24}$ **39.** $27\frac{31}{36}$ **41.** $25\frac{5}{78}$
43. $10\frac{41}{60}$ **45.** $28\frac{19}{72}$ **47.** $85\frac{17}{48}$ **49.** $7\frac{29}{36}$

51. $8\frac{11}{14}$ **53.** $4\frac{5}{6}$ **55.** $1\frac{3}{20}$ **57.** $28\frac{24}{35}$
59. $32\frac{91}{180}$ **61.** $1\frac{1}{2}$ pounds **63.** $18\frac{9}{20}$ tons
65. $17\frac{7}{10}$ miles **67.** $7\frac{1}{4}$ gallons **69.** $\frac{13}{17}$
71. $3\frac{17}{36}$ **73.** $\frac{2}{15}$ **75.** $1\frac{3}{5}$
77. 33 miles per gallon

Getting Ready for Algebra

1. $a = \frac{1}{2}$ **3.** $c = \frac{5}{8}$ **5.** $x = \frac{11}{72}$ **7.** $y = 1\frac{38}{63}$
9. $a = 1\frac{11}{40}$ **11.** $c = 4\frac{1}{6}$ **13.** $x = 3\frac{5}{36}$
15. $3\frac{1}{6} = w$ **17.** $a = 36\frac{4}{9}$ **19.** $c = 21\frac{2}{21}$

Exercises 3.14

1. $\frac{2}{7}$ **3.** $\frac{1}{7}$ **5.** $\frac{1}{6}$ **7.** 1 **9.** 0 **11.** 1
13. $\frac{11}{12}$ **15.** $\frac{4}{9}$ **17.** $\frac{10}{21}$ **19.** $\frac{10}{21}$ **21.** $13\frac{1}{2}$
23. $\frac{3}{5}$ **25.** $\frac{1}{24}$ **27.** $1\frac{1}{12}$ **29.** $1\frac{1}{8}$ **31.** $\frac{32}{45}$
33. $1\frac{22}{75}$ **35.** $1\frac{11}{12}$ **37.** 2 **39.** 4 **41.** $\frac{95}{96}$
43. $1\frac{8}{9}$ **45.** $\frac{1}{6}$ **47.** $1\frac{2}{5}$ **49.** 2 **51.** $\frac{19}{30}$
53. $3\frac{37}{54}$ **55.** $7\frac{1}{5}$ **57.** $3\frac{125}{128}$ **59.** $11\frac{1}{2}$ yards or $5\frac{3}{4}$ yards each **61.** $8\frac{1}{8}$ lb **63.** $\frac{6}{10}$ correct
65. $\frac{1}{6}$ **67.** $1\frac{7}{8}$ **69.** $\frac{4}{9}$ **71.** $5^2 \cdot 13$
73. $1\frac{1}{4}$ in.

Chapter 3 True–False Concept Review

1. F **2.** T **3.** F **4.** T **5.** T **6.** F **7.** T
8. T **9.** T **10.** T **11.** T **12.** T **13.** F
14. F **15.** T **16.** T **17.** T **18.** T

Chapter 3 Post-Test

1. $3\frac{1}{16}$ **2.** $\frac{19}{24}$ **3.** $\frac{89}{10}$ **4.** $\frac{3}{10}, \frac{3}{8}, \frac{2}{5}$ **5.** $\frac{3}{1}$
6. 56 **7.** $9\frac{2}{15}$ **8.** $9\frac{9}{10}$ **9.** $\frac{3}{4}$ **10.** $\frac{2}{3}$
11. $5\frac{1}{8}$ **12.** $\frac{5}{8}$ **13.** $\frac{9}{16}$ **14.** $\frac{9}{20}$ **15.** $2\frac{2}{3}$
16. $9\frac{5}{6}$ **17.** $\frac{2}{3}$ **18.** $\frac{26}{35}$ **19.** $\frac{8}{29}$ **20.** $1\frac{1}{7}$
21. $\frac{7}{8}, \frac{7}{9}, \frac{8}{9}$ **22.** $\frac{4}{15}$ **23.** $9\frac{13}{40}$ **24.** $\frac{7}{12}$
25. $1\frac{4}{9}$ **26.** $\frac{5}{6}$ **27.** $1\frac{3}{8}$ **28.** $\frac{5}{5}, \frac{6}{6}, \frac{7}{7}$
29. $\frac{4}{15}$ **30.** false **31.** 22 truckloads **32.** $1\frac{1}{2}$ ft

CHAPTER 4

Chapter 4 Pre-Test

1. $\frac{1}{1000}$ **2.** $\frac{1}{10}$ **3.** 0.3006
4. two hundred three and four thousand seventy-five ten-thousandths **5.** $10 + 3 + \frac{0}{10} + \frac{1}{100} + \frac{0}{1000} + \frac{8}{10000}$ **6.** 5.3007 **7.** $\frac{27}{200}$ **8.** 2.6499, 2.6509, 2.65099, 2.651 **9.** 2.64 **10.** 200 **11.** 20.5255
12. 4.198 **13.** 5.348 **14.** 5.0875 **15.** 0.07344
16. 78,590 **17.** 0.00315 **18.** 3.4×10^5
19. 3.45×10^{-7} **20.** 1,730,000 **21.** 0.2875
22. 37.54 **23.** 0.225 **24.** 0.29 **25.** 11.778
26. 10.01 **27.** 84.7 miles **28.** $0.20

Exercises 4.1

1. tenths **3.** hundred-thousandths
5. thousandths **7.** hundredths **9.** five tenths
11. twelve hundredths **13.** sixty-seven hundredths
15. two hundred sixty-seven thousandths **17.** four thousand eight hundred sixty-five ten-thousandths
19. eight and seven thousand five hundred forty-three ten-thousandths **21.** 5 **23.** 6 **25.** 0.6
27. 0.11 **29.** 0.111 **31.** thousandths
33. hundredths **35.** tens **37.** hundredths
39. thousandths **41.** five hundred four thousandths
43. fifty and four tenths **45.** fifty and four hundredths **47.** eight and two hundred five ten-thousandths **49.** three hundred eighty-four
51. four hundred five and five hundredths **53.** 0.15
55. 0.0018 **57.** 6 **59.** tens **61.** 2 **63.** two and two hundred two ten-thousandths **65.** one hundred twenty-one and twenty-three hundred-thousandths **67.** Seventy-two and one thousand, fifty-three ten-thousandths **69.** 700.096
71. 500.005 **73.** 0.00005 **75.** 1,005.005
77. 0.231 **79.** 89,050.946 **81.** 0.200 **83.** fifty-three and ninety-eight hundredths **85.** seventy-seven and fifteen hundredths **87.** $\frac{2}{9}$ **89.** $12\frac{8}{25}$
91. $1\frac{1}{9}$ **93.** $1\frac{4}{7}$ **95.** 324 miles

Exercises 4.2

1. $\frac{2}{10}$ **3.** $\frac{2}{10} + \frac{1}{100}$ **5.** $\frac{6}{10} + \frac{1}{100}$
7. $\frac{2}{10} + \frac{5}{100} + \frac{7}{1000}$ **9.** $\frac{3}{10} + \frac{1}{100} + \frac{4}{1000}$
11. $5 + \frac{3}{10}$ **13.** 0.5 **15.** 0.16 **17.** 0.38
19. 0.121 **21.** 0.938 **23.** 0.403 **25.** $\frac{4}{10} + \frac{2}{100} + \frac{1}{1000}$ **27.** $\frac{0}{10} + \frac{2}{100} + \frac{5}{1000}$ **29.** $\frac{9}{10} + \frac{0}{100} + \frac{8}{1000}$ **31.** $\frac{0}{10} + \frac{5}{100} + \frac{0}{1000} + \frac{9}{10000}$
33. $\frac{0}{10} + \frac{0}{100} + \frac{1}{1000} + \frac{1}{10000}$ **35.** $10 + 4 + \frac{7}{10} +$

$\frac{2}{100} + \frac{3}{1000}$ **37.** 0.03 **39.** 0.011 **41.** 0.302
43. 0.7802 **45.** 0.4003 **47.** 0.45008 **49.** 73.02
51. $2 + \frac{3}{10}$ **53.** $90 + 1 + \frac{3}{10} + \frac{2}{100} + \frac{1}{1000}$
55. $600 + 20 + 5 + \frac{0}{10} + \frac{0}{100} + \frac{3}{1000} + \frac{1}{10000}$
57. 2.135 **59.** 105.508 **61.** 5,000.8096 **63.** 33
65. $\frac{15}{16}$ **67.** $\frac{25}{22}$ **69.** true **71.** $1\frac{1}{4}$ pounds

Exercises 4.3

1. $\frac{7}{10}$ **3.** $\frac{9}{10}$ **5.** $\frac{17}{100}$ **7.** $\frac{7}{100}$ **9.** $\frac{13}{100}$
11. $\frac{11}{50}$ **13.** $\frac{6}{25}$ **15.** $\frac{3}{1000}$ **17.** $\frac{17}{200}$
19. $\frac{27}{200}$ **21.** $\frac{7}{250}$ **23.** $3\frac{1}{2}$ **25.** $1\frac{17}{20}$
27. $\frac{1}{2500}$ **29.** $3\frac{1}{500}$ **31.** $\frac{7}{8}$ **33.** $3\frac{19}{20}$
35. $2\frac{1}{10}$ **37.** $\frac{61}{200}$ **39.** $\frac{111}{250}$ **41.** $8\frac{11}{20}$
43. $15\frac{13}{20}$ **45.** $68\frac{1}{2}$ **47.** $25\frac{27}{40}$ **49.** $700\frac{7}{1000}$
51. $233\frac{617}{5000}$ **53.** $2\frac{1}{100,000}$ **55.** $\frac{447}{625}$
57. $7\frac{4,321}{20,000}$ **59.** $\frac{1}{8}$ **61.** $\frac{7}{8}$ inch **63.** $\frac{5}{8}$
65. $\frac{3}{4}$ yard **67.** 0.055 **69.** $24\frac{3}{8}$ **71.** $67\frac{1}{12}$
73. $3\frac{13}{28}$ **75.** $8\frac{1}{6}$ **77.** 45 miles per gallon

Exercises 4.4

1. true **3.** true **5.** true **7.** true **9.** $<$
11. $>$ **13.** $>$ **15.** 0.2, 0.7, 0.9 **17.** 1.3, 1.4, 1.7
19. 0.05, 0.07, 0.6 **21.** 6.139, 6.14, 6.141
23. false **25.** false **27.** true **29.** false
31. $<$ **33.** $<$ **35.** $>$
37. 7.59, 7.6, 7.61, 7.62, 7.63
39. 0.555, 0.556, 0.565, 0.566
41. 0.86, 0.899, 0.9, 0.903, 0.91
43. 17.05, 17.0506, 17.057, 17.16 **45.** true
47. true **49.** true **51.** false **53.** 0.072, 0.0729, 0.073, 0.073001, 0.073015 **55.** 0.88759, 0.88799, 0.888, 0.8881 **57.** 50.004, 50.039, 50.04, 50.093
59. $>$ **61.** $<$ **63.** $>$ **65.** 98.35 cents
67. too heavy **69.** Karla **71.** less
73. $\frac{3}{4}, \frac{4}{5}, \frac{17}{20}, \frac{9}{10}$ **75.** $\frac{2}{3}$ **77.** $\frac{3}{4}$ **79.** 180
81. $38\frac{7}{10}$

Exercises 4.5

1. 321, 321.2, 321.22 **3.** 530, 529.7, 529.66
5. 1, 0.6, 0.65 **7.** 60, 55.7, 55.677
9. 50, 53.3, 53.313 **11.** \$75.62 **13.** \$123.41
15. \$18.92 **17.** 2.7, 2.65, 2.653
19. 12.3, 12.30, 12.302 **21.** 10.0, 9.99, 9.989
23. 53.3, 53.31, 53.313 **25.** 2.2, 2.18, 2.179
27. 0.8, 0.79, 0.793 **29.** 0.8, 0.79, 0.789
31. 1.0, 1.00, 1.000 **33.** \$75.60 **35.** \$123.50
37. 500, 543.54 **39.** 4,000, 3,971.24 **41.** 0, 49.51
43. 400, 396.5 **45.** 20, 19.5
47. 24,000, 23,786.223 **49.** 1,000, 965.035
51. 10,000, 9,501.994 **53.** 8,000, 7,688.267
55. 100 **57.** 5 **59.** 6,235 **61.** \$11
63. \$3,167 **65.** 3.8 **67.** 1.62 **69.** 1.6
71. \$151 **73.** 482.7 feet **75.** \$132.25
77. \$0.72 **79.** \$368.95 **81.** $4\frac{1}{13}$ **83.** $\frac{53}{3}$
85. $1\frac{1}{15}$ or $\frac{16}{15}$ **87.** $\frac{3}{4}$ **89.** 20 springs

Exercises 4.6

1. 0.7 **3.** 1.8 **5.** 0.46 **7.** 1.08 **9.** 0.548
11. 2.302 **13.** 12.2 **15.** 13.68 **17.** 10.814
19. 32.312 **21.** 9.753 **23.** 23.59 **25.** 1.23
27. 25 **29.** 2.468 **31.** 7.0575 **33.** 97.786
35. 0.13887 **37.** 44.5255 **39.** 92.356
41. 14.8556 **43.** 9.149 **45.** 872.1691
47. 37.284 **49.** 20.733 **51.** 19.604 **53.** 21.1
55. 119.38 **57.** 310 **59.** 131.4 **61.** 19.7 miles
63. yes **65.** \$84.58 **67.** \$469.96
69. 353.1323 miles **71.** $1\frac{3}{8}$ **73.** $2\frac{1}{16}$ **75.** $\frac{9}{16}$
77. $\frac{5}{36}$ **79.** 24 blanks

Exercises 4.7

1. 0.3 **3.** 0.5 **5.** 0.13 **7.** 6.2 **9.** 7.8
11. 11.1 **13.** 1.089 **15.** 8.09 **17.** 2.44
19. 5.7 **21.** 5.71 **23.** 0.457 **25.** 0.948
27. 2.89 **29.** 8.822 **31.** 2.76 **33.** 1.08
35. 0.949 **37.** 1.71 **39.** 5.32 **41.** 5.743
43. \$151.24 **45.** 105.753 **47.** 24.388
49. 4.1971 **51.** 14.0406 **53.** 2.541 **55.** 74.87
57. 0.3 **59.** \$31.08 **61.** 2.7 cc **63.** \$79.20
65. \$413.13 **67.** 3.874 gallons **69.** 0.778 second
71. $19\frac{23}{24}$ **73.** $7\frac{5}{6}$ **75.** 12 **77.** $1\frac{5}{16}$ or $\frac{21}{16}$
79. \$6 per pound

Getting Ready for Algebra

1. $11.5 = x$ **3.** $y = 14.98$ **5.** $t = 0.073$
7. $x = 12$ **9.** $2.62 = w$ **11.** $t = 7.23$
13. $a = 0.78$ **15.** $x = 6.56$ **17.** $a = 21.6$
19. $s = 6.289$ **21.** $c = 476.02$

Exercises 4.8

1. 2.4 **3.** 4.2 **5.** 1.8 **7.** 0.27 **9.** 0.35
11. 0.06 **13.** 0.84 **15.** 13.5 **17.** 0.015
19. 0.26 **21.** 2.226 **23.** 0.078 **25.** 0.1792
27. 0.02553 **29.** 5.4717 **31.** 0.8432
33. 1.52646 **35.** 6.3448 **37.** 450.375
39. 0.7742 **41.** 31.329 **43.** 42.718 **45.** 3.4524
47. 481.152 **49.** 9,591.4524 **51.** 0.096
53. 31.2708 **55.** 168.9603 **57.** 78.7626
59. 3,263.03208 **61.** 1,765.764441 **63.** \$23.36
65. \$1,889.25 **67.** 31.5 inches **69.** 60.75 yards

71. \$357.08 **73.** $14\frac{17}{42}$ **75.** 70 **77.** $15\frac{39}{40}$
79. $1\frac{3}{20}$ or $\frac{23}{20}$ **81.** $4\frac{1}{4}$ ft

Exercises 4.9

1. 4.25 **3.** 183 **5.** 821.4 **7.** 2.76 **9.** 2.14361
11. 0.0186 **13.** 1×10^3 **15.** 1×10^{-4} **17.** 1×10^0
19. 10,000 **21.** 0.01 **23.** 100 **25.** 6,274
27. 0.185 **29.** 48,700 **31.** 0.87621 **33.** 42,756
35. 0.3695 **37.** 2,600,000 **39.** 1.2231 **41.** 98.7
43. 7×10^2 **45.** 7.8×10^{-2} **47.** 1.5×10^4
49. 60,000 **51.** 0.122 **53.** 2,340 **55.** 0.35896
57. 0.0623 **59.** 0.00825 **61.** 21.4 **63.** 0.00832
65. 8.216 **67.** 7×10^4 **69.** 8.16×10^{-3}
71. 6.27×10^5 **73.** 6,000,000,000 **75.** 0.0000444
77. 785,100 **79.** \$2,229 **81.** 82.75 feet
83. \$98,500 **85.** 5.9 pounds **87.** 275 inches
89. 0.00182 ohms **91.** 5.2×10^7 square miles
93. 7.2×10^{-8} cm **95.** 11,160,000 miles per minute
97. 0.00004 centimeters **99.** $2\frac{7}{12}$ or $\frac{31}{12}$
101. $9\frac{5}{24}$ **103.** $7\frac{7}{8}$ **105.** $7\frac{19}{20}$ **107.** $1\frac{7}{8}$ hours

Exercises 4.10

1. 0.5 **3.** 26.64 **5.** 0.136 **7.** 18.025 **9.** 160
11. 1.3 **13.** 0.2 **15.** 9.1 **17.** 1.2 **19.** 43.76
21. 4.57 **23.** 410.77 **25.** 2.39 **27.** 3.2
29. 12.6 **31.** 0.52 **33.** 0.052 **35.** 0.126
37. 7.05 **39.** 5.945 **41.** 15.555 **43.** 0.221
45. 4.510 **47.** 533.333 **49.** 395.349 **51.** 97.06
53. 120.0685 **55.** 16.255 **57.** 5.09 **59.** 12.05
61. 16.5 **63.** 43.2 **65.** 90.9 **67.** 79.4
69. 18.8 **71.** 36 **73.** 29 **75.** 11
77. \$0.093 or 9.3¢ **79.** \$0.172 or 17.2¢
81. 42.6 miles per gallon **83.** 19 miles per gallon
85. 17.3 hours **87.** 1.5 volts **89.** 586 feet
91. 12 stalls **93.** $4\frac{7}{16}$ **95.** $7\frac{11}{16}$ **97.** $3\frac{1}{16}$
99. $2\frac{149}{160}$ **101.** 140 books, \$32.50

Getting Ready for Algebra

1. $x = 6$ **3.** $y = 204$ **5.** $0.06 = t$ **7.** $m = 0.04$
9. $q = 437.5$ **11.** $500 = h$ **13.** $y = 2.673$
15. $0.1032 = c$ **17.** $0.9775 = x$ **19.** $w = 0.0141$
21. $z = 30.16$

Exercises 4.11

1. 0.125 **3.** 0.375 **5.** 0.6875 **7.** 0.45
9. 0.03125 **11.** 0.4375 **13.** 0.1, 0.11
15. 0.6, 0.57 **17.** 0.5, 0.45 **19.** 0.65 **21.** 0.792
23. 0.475 **25.** 0.85 **27.** 3.2 **29.** 0.8, 0.83
31. 0.4 0.36 **33.** 0.2, 0.21 **35.** 0.5, 0.53
37. 0.6, 0.58 **39.** 0.935 **41.** 0.945 **43.** 3.625
45. 15.656 **47.** 33.7825 **49.** 0.426
51. 0.3, 0.35, 0.349 **53.** 0.4, 0.39, 0.387
55. 11.8, 11.81, 11.810 **57.** 2.8, 2.78, 2.778
59. 21.9, 21.87, 21.865 **61.** 0.75 inch
63. 0.9375 inch **65.** 0.781 inch **67.** 0.3125 yard
69. 4.875 cups **71.** 10.658 **73.** 3,890.275
75. 20.17 **77.** 1.66 **79.** 37.5 in.

Exercises 4.13

1. 0.5 **3.** 0.05 **5.** 0.17 **7.** 0.128 **9.** 0.088
11. 0.23 **13.** 2.3 **15.** 2 **17.** 2.42 **19.** 220.1
21. 102.4 **23.** 49.8 **25.** 5.62 **27.** 27.67
29. 0.06 **31.** 5.42 **33.** 5.93 **35.** 0.063
37. 21 **39.** 7.765 **41.** 1.738 **43.** 5.25083
45. 0.25 **47.** 5.6709 **49.** 7.46 **51.** 28.54
53. 13.3365 **55.** 1.92 **57.** \$17.33 **59.** \$54.38
61. 48.7 accidents **63.** \$10.15 **65.** \$10.17
67. 0.9375 **69.** 0.2125 **71.** $\frac{41}{50}$ **73.** $\frac{17}{200}$
75. \$24.79

Getting Ready for Algebra

1. $x = 9$ **3.** $x = 0.6$ **5.** $0.1 = t$ **7.** $x = 524$
9. $x = 0.32$ **11.** $m = 67$ **13.** $16.5 = y$
15. $p = 3.002$ **17.** $40 = x$ **19.** $7.28 = h$
21. $17.25 = c$

Chapter 4 True–False Concept Review

1. F **2.** T **3.** F **4.** T **5.** F **6.** T **7.** T
8. T **9.** F **10.** T **11.** F **12.** F **13.** T
14. F **15.** F **16.** T **17.** F **18.** F **19.** T
20. T

Chapter 4 Post-Test

1. 0.040 **2.** 0.7279, 0.728, 0.7299, 0.7308, 0.731
3. hundredth **4.** twenty-seven and twenty-seven thousandths **5.** 41.808 **6.** ten-thousandths
7. 0.024 **8.** 9.00 **9.** 4.211 **10.** $16\frac{37}{40}$
11. $700 + 0 + 2 + \frac{3}{10} + \frac{0}{100} + \frac{5}{1000}$ **12.** 2.4×10^{-3}
13. 0.31 **14.** 47,500 **15.** 15.328 **16.** 4.898
17. 0.266 **18.** 212.063 **19.** 21.6 **20.** 25.709
21. 3.275×10^4 **22.** 7.273 **23.** 0.0008
24. 0.01368 **25.** \$74.72 **26.** 22.989 **27.** 0.055
28. \$6.27

CHAPTER 5

Chapter 5 Pre-Test

1. $\frac{3}{8}$ **2.** $\frac{1}{8}$ **3.** true **4.** true **5.** 27.5 **6.** 42
7. 20 lb **8.** \$4.38

Exercises 5.1

1. $\frac{8 \text{ people}}{11 \text{ chairs}}$ **3.** $\frac{15}{16}$ **5.** $\frac{6}{5}$ **7.** $\frac{1 \text{ family}}{3 \text{ children}}$
9. $\frac{5 \text{ pounds}}{1 \text{ foot}}$ **11.** true **13.** false **15.** true
17. true **19.** false **21.** false **23.** $\frac{5}{2}$

25. $\frac{75 \text{ miles}}{1 \text{ hour}}$ or 75 miles per hour **27.** $\frac{55 \text{ miles}}{3 \text{ gallons}}$
29. true **31.** false **33.** true **35.** true
37. true **39.** false **41.** false **43.** true
45. $\frac{13 \text{ tickets}}{5 \text{ people}}$ **47.** $\frac{1.3 \text{ television sets}}{1 \text{ house}}$
49. $\frac{27.5 \text{ miles}}{1 \text{ gallon}}$ **51.** true **53.** false **55.** true
57. false **59.** true **61.** true **63.** false
65. (a) $\frac{1}{2}$ (b) $\frac{1}{3}$ **67.** yes **69.** $\frac{52.8 \text{ people}}{1 \text{ square mile}}$
71. $\frac{7}{12}$ **73.** $\frac{5}{12}$ **75.** $\frac{1}{3}$ **77.** $\frac{33}{10}$ **79.** $\frac{117}{500}$
81. 7.005, 7.009, 7.05, 7.059, 7.095 **83.** 1
85. 0.4184 **87.** 2.6 cubic centimeters per minute

Exercises 5.2

1. 7 **3.** 6 **5.** 6 **7.** 3 **9.** 8 **11.** 14
13. 10 **15.** 5 **17.** 3 **19.** 3 **21.** 16 **23.** 6
25. $2\frac{2}{15}$ or $\frac{32}{15}$ **27.** 20 **29.** 22.5 or $22\frac{1}{2}$
31. 150 **33.** 12 **35.** $\frac{3}{8}$ **37.** 1.5 **39.** 19
41. $2\frac{13}{16}$ or 2.8125 **43.** 82.5 **45.** 20 **47.** 75
49. 0.25 **51.** 24 **53.** 0.3 **55.** 26.3 **57.** 9.55
59. 0.57 **61.** 13.50 **63.** 86.83 **65.** 0.152304
67. 619.442 **69.** 5 **71.** 23.333 **73.** 626 rivets

Exercises 5.3

1. 3 **3.** x **5.** $\frac{3}{5} = \frac{x}{10}$ **7.** 20 feet **9.** x
11. $\frac{20}{12} = \frac{x}{18}$ **13.** 17 TVs **15.** 68 TVs
17. $\frac{17}{34} = \frac{68}{x}$ **19.** x **21.** $\frac{3}{70} = \frac{x}{980}$
23. 10 teachers **25.** y **27.** $\frac{8}{3} = \frac{120}{y}$
29.

	CASE I	CASE II
Days	$1\frac{1}{2}$	14
Garbage (lb)	30	x

31. 280 pounds **33.** 30 shirts **35.** 187.5 hours
37. $26\frac{2}{3}$ lb **39.** $8.75 **41.** $7.80 **43.** 7 hours
45. 17 **47.** $56.25 **49.** 960 **51.** 5 jobs
53. $1,560 **55.** 38 gallons **57.** 440 square feet
59. no; $1.07 **61.** 0.0043 inch **63.** 48 boys
65. 28 ounces **67.** 30 bags **69.** $2\frac{1}{7}$ quart
71. 114.3 kg **73.** $1,007 **75.** 5,000 **77.** $135
79. 51,870 drachma **81.** $48 **83.** $\frac{71}{800}$
85. 0.53 **87.** 0.10188 **89.** 1.81
91. $0.32 per quart

Chapter 5 True–False Concept Review

1. F **2.** T **3.** F **4.** T **5.** T **6.** T **7.** F
8. T **9.** T **10.** F

Chapter 5 Post-Test

1. $\frac{16}{9}$ **2.** 90 questions **3.** 1.5 **4.** true
5. true **6.** 5.25 **7.** $78.96 **8.** $\frac{1}{8}$

CHAPTER 6

Chapter 6 Pre-Test

1. 35% **2.** 170% **3.** 35.7% **4.** 0.003 **5.** 1.12
6. 162.5% **7.** 77.8% **8.** $\frac{19}{250}$ **9.** $1\frac{1}{45}$
10. 133.3% **11.** 45 **12.** 53.29 **13.** 17%
14. $577.96 **15.** $23.32 **16.** a. C b. 25 c. 20
17. a. Brand B, b. 15 g, c. 84 calories
18.

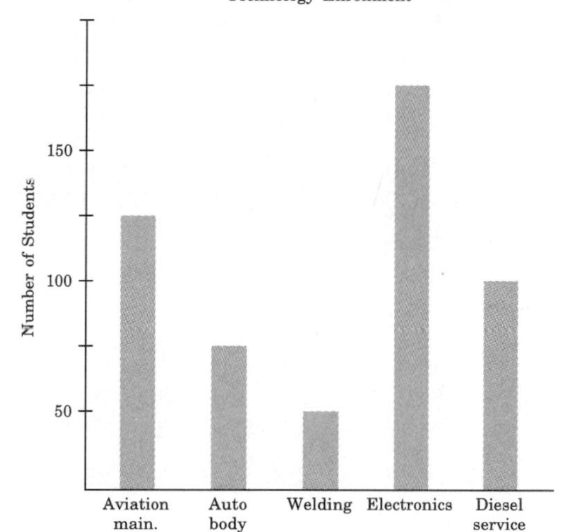

Exercises 6.1

1. 15% **3.** 63% **5.** 17% **7.** 6% **9.** 20%
11. 44% **13.** 35% **15.** 120% **17.** 156%
19. 34% **21.** 160% **23.** 48% **25.** 100%
27. 215% **29.** 300% **31.** 120% **33.** 130%
35. $33\frac{1}{3}\%$ **37.** 60% **39.** $116\frac{1}{4}\%$ **41.** $10\frac{3}{5}\%$
43. $43\frac{3}{4}\%$ **45.** $83\frac{1}{3}\%$ **47.** $87\frac{1}{2}\%$ **49.** $21\frac{7}{8}\%$
51. $66\frac{2}{3}\%$ **53.** $466\frac{2}{3}\%$ **55.** 420% **57.** $416\frac{2}{3}\%$
59. $49\frac{1}{3}\%$ **61.** 700% **63.** 65% **65.** 62%
67. 8% **69.** $5\frac{1}{4}\%$ **71.** 5% **73.** 3%
75. 0.46875 **77.** 0.94 **79.** 16.8 **81.** 0.0456003
83. yes

Exercises 6.2

1. 36% **3.** 476% **5.** 8% **7.** 160% **9.** 1,200%
11. 0.9% **13.** 53.1% **15.** 29% **17.** 100%
19. 10% **21.** 21.4% **23.** 700% **25.** 1,321%
27. 0.5% **29.** 70% **31.** 320% **33.** 3.17%
35. 284% **37.** 0.8% **39.** 0.15% **41.** 575%
43. 56.25% **45.** $74\frac{1}{6}\%$ **47.** 20.51%
49. 10.25% or $10\frac{1}{4}\%$ **51.** 5.2% or $5\frac{1}{5}\%$
53. $27\frac{2}{3}\%$ **55.** 0.09% **57.** 1,000% **59.** 123.4%
61. $20\frac{1}{3}\%$ **63.** 22% **65.** 37% **67.** 62%
69. 37.5% or $37\frac{1}{2}\%$ **71.** 187% **73.** 23.5%
75. 4,000 **77.** 0.96 **79.** 0.242 **81.** 17.9347
83. 1.6875 inches

Exercises 6.3

1. 0.16 **3.** 0.82 **5.** 0.73 **7.** 0.0215 **9.** 3.12
11. 1.106 **13.** 0.0004 **15.** 1.35 **17.** 0.0279
19. 0.179 **21.** 3.147 **23.** 0.0012 **25.** 0.005
27. 0.0025 **29.** 0.01 **31.** 2 **33.** 0.00058
35. 1.25 **37.** 0.00625 **39.** 0.00009 **41.** 0.2975
43. 4.755 **45.** 0.00875 **47.** 0.0125 **49.** 0.014
51. 0.7261 **53.** 0.293468 **55.** 0.002 **57.** 0.358
59. 0.018 **61.** 0.05 **63.** 0.1 **65.** 0.1425
67. 0.075 **69.** 0.002 **71.** 0.875 **73.** 1.1875
75. 1.8 **77.** $\frac{143}{200}$ **79.** $206.55

Exercises 6.4

1. 75% **3.** 22% **5.** 85% **7.** 50% **9.** 28%
11. 15% **13.** 105% **15.** 250% **17.** 36%
19. 37.5% **21.** $66\frac{2}{3}\%$ **23.** $116\frac{2}{3}\%$ **25.** $183\frac{1}{3}\%$
27. $5\frac{1}{4}\%$ **29.** 93.3% **31.** 44.4% **33.** 83.3%
35. 183.3% **37.** 8.3% **39.** 33.3% **41.** 53.8%
43. 142.9% **45.** 38.1% **47.** 0.3% **49.** 0.4%
51. 1.9% **53.** 1.0% **55.** 488.9% **57.** 557.1%
59. 80% **61.** 17.5% **63.** 1.9% **65.** 15%
67. 23% **69.** 6.96261 **71.** 0.00808 **73.** 15.1
74. 7.74 **77.** $5.35

Exercises 6.5

1. $\frac{1}{20}$ **3.** $\frac{7}{20}$ **5.** $1\frac{1}{4}$ **7.** 4 **9.** $\frac{7}{10}$ **11.** $\frac{14}{25}$
13. $\frac{3}{4}$ **15.** $\frac{9}{10}$ **17.** 1 **19.** $1\frac{3}{25}$ **21.** $\frac{11}{40}$
23. $\frac{63}{1000}$ **25.** $\frac{9}{20{,}000}$ **27.** $\frac{41}{200}$ **29.** $\frac{1}{300}$
31. $\frac{11}{200}$ **33.** $\frac{21}{200}$ **35.** $\frac{3}{400}$ **37.** $\frac{163}{500}$
39. $\frac{101}{200}$ **41.** $\frac{17}{2000}$ **43.** $\frac{1}{25{,}000}$ **45.** $\frac{117}{700}$
47. $\frac{23}{250}$ **49.** $\frac{1}{62{,}500}$ **51.** $\frac{201}{8000}$ **53.** $\frac{3}{800}$
55. $\frac{7}{6}$ **57.** $\frac{3}{50{,}000}$ **59.** $\frac{1}{125{,}000}$ **61.** $\frac{3}{8000}$
63. $\frac{8}{3}$ **65.** $\frac{4}{25}$ **67.** $\frac{7}{20}$ **69.** $\frac{9}{50}$ **71.** $\frac{1}{8}$
73. $\frac{22}{25}$ **75.** true **77.** false **79.** $\frac{\$13}{25 \text{ people}}$
81. $\frac{2953 \text{ gallons}}{20 \text{ people}}$ **83.** $1,875

Exercises 6.6

0.1, 10%; $\frac{3}{10}$, 0.3; $\frac{3}{4}$, 75%; 0.9, 90%; $1\frac{9}{20}$, 1.45;
0.375, 37.5% or $37\frac{1}{2}\%$; $\frac{1}{1000}$, 0.1%; $\frac{1}{1}$ or 1, 100%;
2.25, 225%; $\frac{4}{5}$, 80%; $\frac{11}{200}$, 0.055; $\frac{7}{8}$, 87.5% or $87\frac{1}{2}\%$;
$\frac{1}{200}$, 0.005; $\frac{3}{5}$, 60%; $\frac{5}{8}$, 0.625; $\frac{1}{2}$, 0.50; $\frac{43}{50}$, 86%;
$0.833\overline{3}$, $83\frac{1}{3}\%$; $\frac{2}{25}$, 8%; $0.66\overline{6}$, $66\frac{2}{3}\%$; $\frac{1}{4}$, 0.25; $\frac{1}{5}$, 20%;
$\frac{2}{5}$, 0.4; $\frac{1}{3}$, $0.33\overline{3}$; $\frac{1}{8}$, 12.5% or $12\frac{1}{2}\%$; 0.7, 70%

1. 25% **3.** $\frac{2}{3}$ **5.** client **7.** PT **9.** $x = 37.5$
11. $x = 304$ **13.** $A = 0.2$ **15.** $R = 29.6$
17. 29.5 gallons

Exercises 6.7

1. 18 **3.** 60 **5.** 300% **7.** 50% **9.** 40
11. 36 **13.** 80% **15.** 100 **17.** 1 **19.** 26.8 or $26\frac{4}{5}$ **21.** $133\frac{1}{3}\%$ **23.** 25 **25.** 5.58 or $5\frac{29}{50}$
27. 2% **29.** 125 **31.** 19.2 **33.** 200 **35.** 36
37. 60.8% **39.** 12.48 **41.** 0.0224 **43.** 128.6
45. 10.7% **47.** 205 **49.** 1,248 **51.** 156.7%
53. 18.3 **55.** $8\frac{91}{300}$ **57.** 128.7% **59.** $x = 17.5$
61. $a = 0.25$ or $\frac{1}{4}$ **63.** $w = 55.9$ **65.** 115 miles
67. $20\frac{1}{2}$ inches

Exercises 6.8

1. $4.80 **3.** $781.25 **5.** 18 boxes
7. $14,062.40 **9.** 16.3% **11.** 87% **13.** 7.5%
15. 16% **17.** $875 **19.** $1,425 **21.** $64.13
23. $23.94 **25.** $1.457 **27.** $1.08 **29.** 33%
31. $17.85 **33.** 25 questions **35.** 1,850 people
37. 142,600 **39.** 50% **41.** $55.00 **43.** 42%
45. 78% **47.** 85% **49.** 47% **51.** 334%
53. 5.7%

Exercises 6.9

1. $33\frac{1}{3}\%$ **3.** $10.92 **5.** $11.20 **7.** $24,000

9. $239.99 **11.** $118.36 **13.** $228 retirement, $23.75 United Way **15.** $493.38 **17.** $394.91 **19.** $320,600 **21.** $10.98 **23.** $8.79 **25.** $8,471 **27.** 18% **29.** $4,564; $3,651.20; $5,841.92; $4,198.88 **31.** $543.75 **33.** $326 **35.** 9.2% **37.** $33\frac{1}{3}\%$ **39.** 3.2% **41.** 2.7% **43.** $9.96 **45.** less **47.** $78; $126.36 **49.** $21.72 **51.** $439.96 **53.** $77 **55.** 24% **57.** 234% **59.** 0.006 **61.** 4 **63.** Rita: $3,280; Sally: $4,920

Exercises 6.10

1. 10–11 **3.** 250 **5.** 1,425 **7.** 30% **9.** 25% **11.** 40 vans **13.** 200 **15.** subcompacts **17.** 1984 **19.** 5,000 units **21.** tooling department **23.** $500,000 **25.** $45,000 **27.** $100,000 **29.** lumber and paint

31.

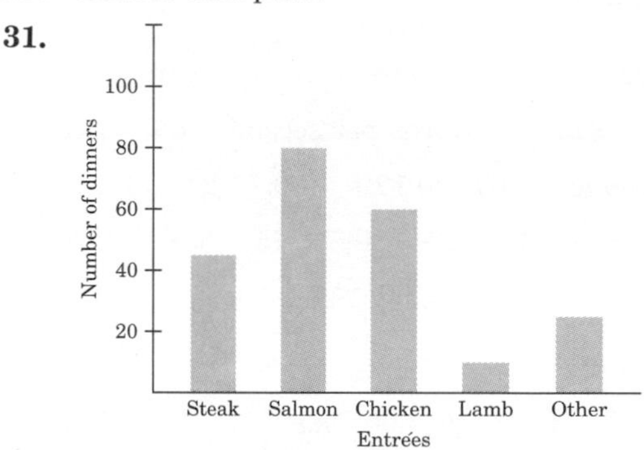

33.

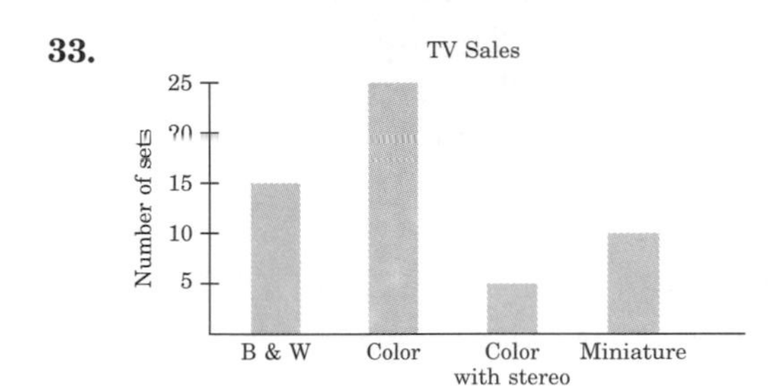

35.

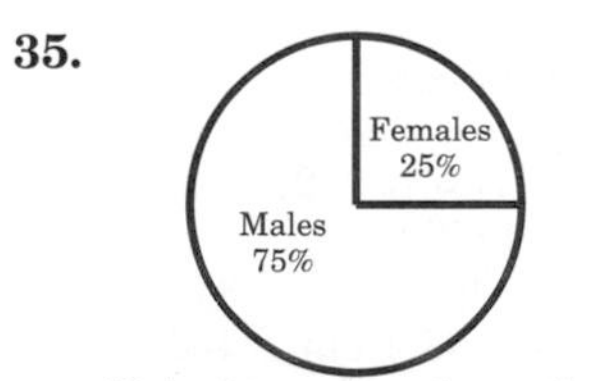

Worker by sex: Acme Corporation

37.

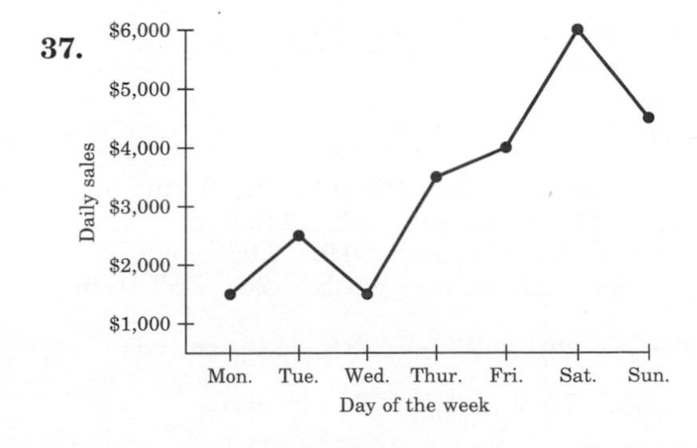

39.

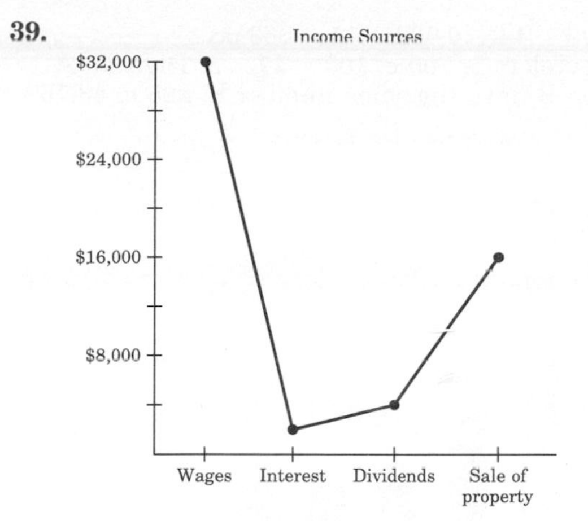

41.

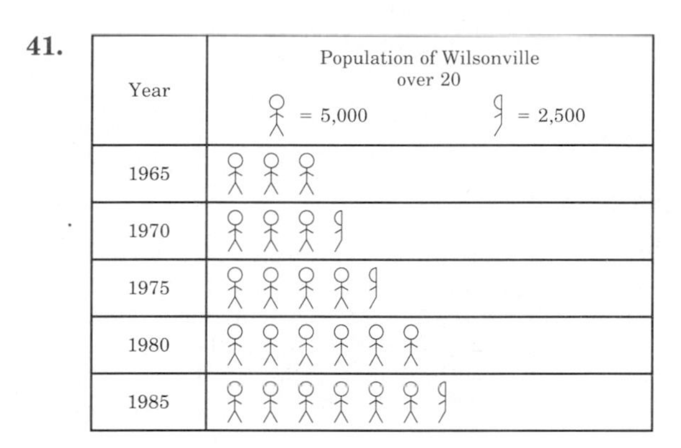

43. $x = 35$ **45.** $x = 2$ **47.** $b = 5$ **49.** $x = 12.5$ **51.** 861,437.1

Exercises 6.11

1. $18.31 **3.** Dallas **5.** $2.31 **7.** $2.31 **9.** $24.12 **11.** 7% **13.** fish cakes **15.** 133 mg **17.** 72 g **19.** 68.3% **21.** 908 calories **23.** 67.3% **25.** $15,599 **27.** $148,794 **29.** 38.7% **31.** 74.1% **33.** 63 and 65 **35.** $163,399 **37.** $71.76 **39.** $109.50 **41.** 3,793 campers **43.** 35% **45.** Fisher Zoo **47.** Utaki Park **49.** $\frac{5}{6}$ **51.** $62.72 **53.** $216 **55.** $220.08 **57.** 2.038

Chapter 6 True–False Concept Review

1. T **2.** T **3.** T **4.** F **5.** F **6.** T **7.** F **8.** T **9.** F **10.** T **11.** T **12.** T **13.** F **14.** T **15.** T **16.** F

Chapter 6 Post-Test

1. 15% **2.** 31.25% or $31\frac{1}{4}\%$ **3.** 7.125% or $7\frac{1}{8}\%$ **4.** 56 **5.** 12% **6.** $\frac{5}{7}$ **7.** 56.8% **8.** 230% **9.** 35.91 **10.** 114.3% **11.** $2\frac{3}{10}$ or $\frac{23}{10}$ **12.** 0.05

13. \$413.79 **14.** 0.002 **15.** \$63.08
16. a. Chevrolets b. 700 c. 100 **17.** a. Division A and Division B have the same number b. 225 c. 43.75%

18.

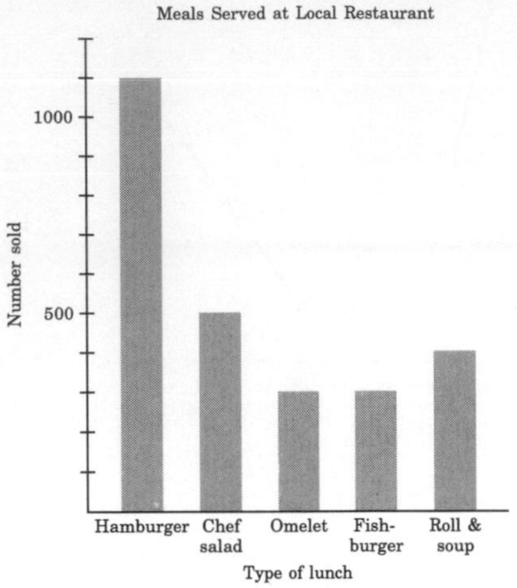

Meals Served at Local Restaurant

Number sold: 500, 1000

Hamburger, Chef salad, Omelet, Fish-burger, Roll & soup

Type of lunch

CHAPTER 7

Chapter 7 Pre-Test

1. 33.2 in **2.** 4.5 ft **3.** 400 min **4.** 11 ft 1 in.
5. 110 liters **6.** 114.5 g **7.** 2,700 m
8. 5 ℓ 9 dℓ **9.** 1.5 yd **10.** 66 feet per second
11. 99 pounds **12.** 40.8 ft **13.** 37.68 cm
14. 48 cm^2 **15.** 113.04 in^2 **16.** 38 ft^2
17. 3,052.08 in^3 **18.** 14 **19.** F

Exercises 7.1

1. 24 ft **3.** 16.1 pounds **5.** 8 feet **7.** 5 feet
9. 60 inches **11.** 48 oz **13.** 72 hours
15. 252 inches **17.** 84 inches **19.** 28 quarts
21. 8 ft 9 in. **23.** 3 yd 1 ft **25.** $17\frac{1}{2}$ ounces
27. $2\frac{2}{3}$ hours **29.** 43 in. **31.** 7 pints
33. 26 feet **35.** 176 hours **37.** 19 ft 7 in.
39. 20 hr 25 min **41.** 6 min 52 sec
43. 13 hr 52 min 42 sec **45.** 38.08 miles
47. 39.875 hr **49.** 8.2 tons **51.** 8.1 yards
53. 107 in. **55.** 4,260 pounds **57.** 5,380 yards
59. 112 hours **61.** 532.8 inches **63.** 9,300 pounds
65. 45 ft 1 in. **67.** 13 quarts **69.** \$84
71. 50 miles per hour **73.** $27\frac{1}{4}$ hours
75. 0.025 inches **77.** 28.125% **79.** 82%
81. $\frac{18}{25}$ **83.** 0.72 **85.** 1,792 kilometers

Exercises 7.2

1. 3,000 cm **3.** 0.1 kg **5.** 200 m **7.** 7,000 mℓ
9. 7 m **11.** 7.4 m **13.** 20 g **15.** 360 km
17. 6.27 ℓ **19.** 1,300 m **21.** 1,300 ℓ
23. 0.245 ℓ **25.** 360 cm **27.** 175 g **29.** 78 g
31. 3.4 km **33.** 14,035 mℓ **35.** 4,188 cg
37. 7,250,000 mm **39.** 6,800 ℓ **41.** 0.214 km
43. 17.25 ℓ **45.** 2.5 kℓ **47.** 1100 g
49. 11,330 m **51.** 422 m **53.** 7,150 m
55. 1.8 kℓ **57.** 170 ℓ **59.** 0.018 g **61.** 300 mg of calcium **63.** 2,086 grams **65.** \$2.86 per *kg*
67. 1.65 centimeters **69.** 0.75 milligram **71.** 300
73. 444 **75.** 22.56 **77.** 900 **79.** \$1,746.10

Exercises 7.3

1. 5 pounds **3.** 100 mm **5.** 5 gallons **7.** 6 kg
9. 10,000 g **11.** 2 yards **13.** 0.55 ℓ **15.** 10 lb
17. 86,400 seconds **19.** 4.6 hours **21.** 30 yards
23. 6.05 m **25.** 1,800 mm **27.** $\frac{2}{3}$ foot
29. 120 inches **31.** 1.3 tons **33.** 68 meters
35. 26,000 pounds **37.** 24 quarts **39.** 270 feet
41. 158,400 ft **43.** $\frac{66 \text{ ft}}{1 \text{ sec}}$ **45.** $\frac{1{,}500 \text{ pounds}}{1 \text{ in.}}$
47. $\frac{40 \text{ meters}}{1 \text{ second}}$ **49.** 0.345 m^2 **51.** 250,000 cm^2
53. $\frac{33\frac{1}{3} \text{ ounces}}{1 \text{ inch}}$ **55.** $\frac{12 \text{ cents}}{1 \text{ minute}}$ **57.** $\frac{1\frac{7}{9} \text{ oz}}{1 \text{ in}^2}$
59. $\frac{20\frac{5}{6} \text{ lb}}{1 \text{ min}}$ **61.** $1\frac{1}{2}$ words per second **63.** 5 cc
65. 136.5 pounds **67.** 0.12 ℓ **69.** \$246.40
71. \$3.50 per article **73.** 80% **75.** $33\frac{1}{3}\%$
77. 87% **79.** 27.1% **81.** 15%

Exercises 7.4

1. 22.5 m **3.** 4.8 ℓ **5.** 90.0 kg **7.** 367.4 ft
9. 8.7 km **11.** 10.8 cm **13.** 0.8 lb **15.** 57.0 ℓ
17. 7.2 in **19.** 10,143 m **21.** 0.2 oz
23. $\frac{45{,}000 \text{ g}}{1 \text{ m}}$ **25.** $\frac{49.6 \text{ mi}}{1 \text{ hr}}$ **27.** $\frac{1.8 \text{ g}}{1 \text{ cm}^2}$
29. $\frac{1.1 \text{ lb}}{1 \text{ qt}}$ **31.** 464.1 g **33.** 6.2 miles
35. $\frac{239 \text{ gal}}{1 \text{ hr}}$ **37.** 21.6 cm by 27.9 cm **39.** 2-ℓ bottle **41.** 130 km **43.** 125% **45.** 59.4%
47. 0.625, 62.5% **49.** $\frac{1}{125}$, 0.008

Exercises 7.5

1. 24 in. **3.** 72 in. **5.** 94.2 meters **7.** 19 in.
9. 26 ft **11.** 50.24 ft **13.** 32 in. **15.** 32 in.
17. 32.6 cm **19.** 72 in. **21.** 51.4 ft **23.** 41.12 ft
25. 82 in. **27.** 66.12 ft **29.** 11.14 ft
31. 28.56 in. **33.** 168 m **35.** 42.84 in. **37.** 70 ft
39. 67.1 in. **41.** 41.68 in. **43.** 38.4 cm **45.** 12 ft
47. \$2,880 **49.** 1,261 revolutions **51.** $\frac{249}{200}$, 1.245; 0.4375, $43\frac{3}{4}\%$; $\frac{5}{6}$, $83\frac{1}{3}\%$; $\frac{26}{5}$, 5.2 **53.** 45%

Exercises 7.6

1. 96 in^2 **3.** 243 ft^2 **5.** 169 in^2 **7.** 9 km^2
9. 0.785 cm^2 **11.** 20.25 in^2 **13.** 119.6 in^2
15. 552.25 ft^2 **17.** 484 mm^2 **19.** 7.065 cm^2
21. 144 in^2 **23.** 225 square yards **25.** 452.16 in^2
27. 200.96 cm^2 **29.** 297 in^2 **31.** $13\frac{1}{3}$ sq. yds.
33. no **35.** 70 sq. yds. **37.** 16 acres

39. 162 ft^2 **41.** 7.5% **43.** 37.5% **45.** $7320
47. 400 forms **49.** $442.85

Exercises 7.7

1. 30,000 ft^2 **3.** 255 ft^2 **5.** 2,395 mm^2
7. 418.08 in^2 **9.** 562.5 cm^2 **11.** 81.64 cm^2
13. 327.75 in^2 **15.** 12.86 ft^2 **17.** 150 m^2
19. 64.94 ft^2 **21.** 49.60 m^2 **23.** 50 m^2
25. 140 cm^2 **27.** 10 pounds **29.** $4,296.93
31. (a) 251.2 in^2, (b) 159 in^2 **33.** 8 pounds
35. 15% **37.** 6.25% **39.** $1.25 **41.** $7.21

Exercises 7.8

1. 525 m^3 **3.** 1,582.56 in^3 **5.** 125 in^3
7. $65\frac{5}{12}$ cm^3 **9.** 3,140 cm^3 **11.** 267.9 ft^3
13. 468 cm^3 **15.** 432in^3 **17.** 6,104.16 in^3
19. 1,387.5 ft^3 **21.** 10,889.5 mm^3
23. 69.08 gallons **25.** 20 truckloads
27. 144 gallons **29.** 5 pounds **31.** 57 inches
33. 28 hr 28 min **35.** 25 ft

Exercises 7.9

1. 36 **3.** 25 **5.**  **7.** 36, 33

9.  **11.**  **15.** True **17.** False

19. True **21.** 45 **23.** 26 **25.** 18 **27.** 63
29. 21 **31.** False **33.** False **35.** False
37. True **39.** Deductive **41.** Inductive
43. Inductive **45.** Inductive **47.** 7 ft^2 or 1008 in^2
49. 0.785 ft^2 or 118.04 in^2 **51.** $\frac{2.8 \text{ kg}}{1 \text{ cm}^2}$ **53.** $\frac{18 \text{ m}}{1 \text{ sec}}$
55. 139.4 ft

Chapter 7 True–False Concept Review

1. F **2.** T **3.** T **4.** F **5.** F **6.** T **7.** F
8. T **9.** T **10.** T **11.** T **12.** F **13.** F
14. F **15.** F **16.** T **17.** T **18.** T **19.** F
20. F

Chapter 7 Post-Test

1. 7,850 mm **2.** 13.6 cm **3.** $1\frac{7}{8}$ points
4. 43.72 ft^2 **5.** 24.8 pounds **6.** 114.5 g
7. 24.28 ft **8.** 30 in^2 **9.** 1 gal 3 qt 1 pt
10. 2.5375 gallons **11.** 152 inches **12.** 21 gallons per minute **13.** 7.8 liters **14.** 48 cm^2
15. 14.13 ft^3 **16.** 7 cm 2 mm **17.** 56 **18.** True
19. 3.2 kg

CHAPTER 8

Chapter 8 Pre-Test

1. a. 17 b. 17 **2.** −43 **3.** 4.33 **4.** $-\frac{1}{8}$
5. −51 **6.** −6 **7.** 47 **8.** −6.0 **9.** $\frac{-5}{12}$
10. 48 **11.** 22 **12.** $\frac{-3}{10}$ **13.** 3 **14.** −4
15. $-\frac{7}{9}$ **16.** 1.06 **17.** −4 **18.** 4 **19.** −21
20. 121 **21.** −62 **22.** −42 **23.** $x = -3$
24. −93°F **25.** 11 yd **26.** −15°C

Exercises 8.1

1. 3 **3.** −17 **5.** $-\frac{1}{2}$ **7.** −4.7 **9.** $\frac{1}{3}$
11. 8.08 **13.** 1 **15.** 7 **17.** $\frac{1}{6}$ **19.** 1.2
21. 3.11 **23.** 31 **25.** $\frac{2}{3}$ **27.** $-3\frac{1}{8}$ **29.** −0.23
31. $4\frac{15}{16}$ **33.** 40.7 **35.** −103.7 **37.** 14.7
39. $-13\frac{7}{8}$ **41.** 8.135 **43.** 6 **45.** $\frac{4}{5}$ **47.** 0.71
49. 21.75 **51.** 4.5 **53.** 7 **55.** 253
57. 0.0035 **59.** $-\frac{2}{7}$ **61.** −42 **63.** $-\frac{3}{8}$
65. 12°C **67.** AD 1875 **69.** +1,275 feet, −1,173 feet **71.** −80 miles **73.** 13.515 kℓ **75.** 14 cm
77. 51,000 cg **79.** 9.183 mℓ **81.** 154.3 grams

Exercises 8.2

1. −4 **3.** −1 **5.** −13 **7.** 0 **9.** −13
11. −13 **13.** 8 **15.** −5 **17.** −27 **19.** −80
21. −4 **23.** −144 **25.** 0 **27.** 7 **29.** −14
31. −75 **33.** −6.6 **35.** 2.6 **37.** 1.8 **39.** $-\frac{1}{6}$
41. $-\frac{1}{2}$ **43.** $-\frac{5}{18}$ **45.** 62 **47.** −129
49. −54 **51.** −27 **53.** 4.2 **55.** −133.73
57. $-2\frac{5}{24}$ **59.** $1\frac{8}{9}$ **61.** −1.069 **63.** −10.98
65. −95 pounds **67.** $6,458 **69.** 5,521 books
71. $+2\frac{1}{4}$ **73.** no **75.** $\frac{660 \text{ feet}}{1 \text{ second}}$ **77.** 652 feet
79. 124 meters **81.** 64 in. **83.** $0.62

Exercises 8.3

1. 2 **3.** 10 **5.** −13 **7.** −12 **9.** 16 **11.** −7
13. 32 **15.** −31 **17.** 13 **19.** 36 **21.** 0
23. 3 **25.** −7.2 **27.** $-\frac{3}{4}$ **29.** 41 **31.** −9
33. −90 **35.** $\frac{1}{8}$ **37.** 80 **39.** 1 **41.** 130
43. −117 **45.** 22 **47.** −8 **49.** −0.77
51. −13.6 **53.** 49.3 **55.** 7.8 **57.** $-\frac{3}{4}$
59. $-\frac{9}{40}$ **61.** $\frac{33}{4}$ or $8\frac{1}{4}$ **63.** 15 **65.** 14
67. $-\frac{17}{24}$ **69.** −85°C **71.** $378.99 **73.** −59°
75. $251.87 **77.** −$17.25 **79.** 803.84 in^2
81. 864 cm^2 **83.** 2 words per second
85. 143.06625 km^2 **87.** 112 square yards

Getting Ready for Algebra

1. $y = -3$ **3.** $w = -10$ **5.** $x = -8$
7. $x = -27$ **9.** $x = 7$ **11.** $x = -7$ **13.** $x = -0.2$
15. $x = -\frac{1}{4}$

Exercises 8.4

1. −6 **3.** −54 **5.** −6 **7.** −48 **9.** −1
11. −42 **13.** 60 **15.** −24 **17.** 24 **19.** −164
21. −3 **23.** 6 **25.** −18.12 **27.** $\frac{3}{8}$ **29.** 0.048
31. 0.024 **33.** $-\frac{3}{10}$ **35.** 0.9 **37.** 25
39. −49 **41.** 44 **43.** 260 **45.** 120
47. 0.5088 **49.** −3.07 **51.** 0 **53.** −12
55. $\frac{1}{3}$ **57.** −3.6 **59.** 0.04 or $\frac{1}{25}$ **61.** −1,610
63. $-7\frac{7}{9}$°C **65.** −12.5 lb **67.** −33.96
69. −$67.20 **71.** $20.08 **73.** −$99.68
75. 3,104 revolutions **77.** 12 pounds
79. 39.375 ft^2 **81.** $\frac{\$0.375}{1\ ft^2}$

Exercises 8.5

1. −2 **3.** 2 **5.** 4 **7.** −18 **9.** −5 **11.** −3
13. 0.03 **15.** 3 **17.** −4 **19.** −2.5
21. cannot be done **23.** 0 **25.** 20 **27.** 4.04
29. −0.00025 **31.** −1.5 **33.** $-\frac{50}{3}$ **35.** 5.5
37. −5 **39.** −45 **41.** −86 **43.** −116 **45.** $\frac{1}{2}$
47. −3 **49.** −1.2 or $-\frac{6}{5}$ **51.** −1 **53.** −1
55. −6 **57.** $\frac{22}{7}$ **59.** −7 **61.** $-\frac{3}{5}$ **63.** 0.16
65. −5° **67.** −$69.67 **69.** −$30.61 **71.** 3°
73. −4.525 **75.** 8,138.88 cm^3 **77.** 6,079.04 in^3
79. 317 gallons **81.** 14 truckloads

Getting Ready for Algebra

1. $c = 9$ **3.** $d = 8$ **5.** $x = -25$ **7.** $y = -4$
9. $y = 15$ **11.** $y = 49$ **13.** $x = -160$ **15.** $y = -2$
17. $x = -0.042$ **19.** $x = -14$ **21.** $x = -3.96$

Exercises 8.6

1. −66 **3.** −7 **5.** −16 **7.** 3 **9.** −5
11. 45 **13.** 15 **15.** 8 **17.** 1 **19.** 12
21. −20 **23.** 10.3 **25.** 36 **27.** 9 **29.** −33
31. 6 **33.** 8 **35.** −24 **37.** 13 **39.** 136
41. −42 **43.** 7 **45.** 0 **47.** 131 **49.** −84
51. −96 **53.** −4 **55.** −3 **57.** −64
59. −2,000 **61.** −71 **63.** 45 **65.** 5,561
67. true **69.** 7°F **71.** yes **73.** no
75. down 11 **77.** −29.4 **79.** 184 **81.** 63,504
83. 25 **85.** −$1,458

Exercises 8.7

1. $x = 4$ **3.** $x = 5$ **5.** $y = 8$ **7.** $a = 0$ **9.** $x = 9$
11. $x = -5$ **13.** $x = -2$ **15.** $x = -6$
17. $x = -3$ **19.** $x = 3$ **21.** $y = 7$ **23.** $a = 0$
25. $x = -0.5$ **27.** $x = -25$ **29.** $x = -16$
31. $y = -4$ **33.** $x = 3$ **35.** $b = -7$ **37.** 22 fps

Chapter 8 True–False Concept Review

1. T **2.** F **3.** F **4.** T **5.** F **6.** F **7.** T
8. F **9.** T **10.** T **11.** F **12.** T **13.** T
14. T

Chapter 8 Post-Test

1. −14 **2.** $-\frac{5}{4}$ **3.** $\frac{2}{15}$ **4.** 72 **5.** 4
6. 1.07 **7.** (a) 21, (b) 21 **8.** 4 **9.** 32 **10.** 4
11. 126 **12.** −24 **13.** −90.4 **14.** −105
15. −51.1 **16.** $-\frac{1}{6}$ **17.** 8 **18.** 43 **19.** −80
20. $-\frac{2}{7}$ **21.** −53 **22.** 4 **23.** $x = -5$
24. −21°F **25.** 7 **26.** 14°F

PLANE GEOMETRY SUPPLEMENT (APPENDIX V)

Geometry of Angles

1. Acute **3.** obtuse **5.** obtuse **7.** acute
9. obtuse **11.** 50° **13.** 5° **15.** 75° **17.** 55°
19. 65° **21.** 18°, 45°, 82° **23.** 315°, 282°, 210°
25. 95°, 121°, 175° **27.** ∠A = 143° **29.** ∠C = 12°
31. true **33.** false **35.** false **37.** false
39. true

41.

Statement	Reason
1. ∠A + ∠B = 90°	1. Given
2. ∠A + ∠C = 90°	2. Given
3. ∠A + ∠B = ∠A + ∠C	3. Quantities equal to the same quantity are equal
4. ∠A + ∠B − ∠A = ∠A + ∠C − ∠A	4. Equals subtracted from equals are equal
5. ∴∠B = ∠C	5. Addition

43.

Statement	Reason
1. ∠A + ∠B = 180°	1. Given
2. ∠B + ∠C = 90°	2. Given
3. ∠C = 36°	3. Given
4. ∠B + 36° = 90°	4. Substitution
5. ∠B = 54°	5. Equals subtracted from equals are equal
6. ∠A + 54° = 180°	6. Substitution
7. ∴∠A = 126°	7. Equals subtracted from equals are equal

Geometry of Triangles

1. isosceles **3.** equilateral **5.** obtuse **7.** right
9. ∠R = 150° **11.** ∠T = 30° **13.** ∠C = 65°
15. ∠D = 12.9° **17.** obtuse **19.** obtuse
21. true **23.** false **25.** false **27.** false
29. true **31.** true

Congruent Triangles

1. true **3.** not necessarily true **5.** DF
7. a. $\angle I = \angle L$; b. $GH = JK$ or $\angle I = \angle L$; c. $GH = JK$
9. ΔDAB ΔBCD, ASA

Parallel Lines

1. $\angle 1$ and $\angle 7$, $\angle 2$ and $\angle 8$ **3.** $\angle 1, \angle 2, \angle 7, \angle 8$
5. $\angle 1 = 120°$ **7.** $\angle 1 = 115°$ **9.** $\angle 2, \angle 5, \angle 8$
11. $\angle 5, \angle 8$

13.

Statement	Reason
1. $FCD \parallel ABE$	1. Given
2. $\angle 6 + \angle 5 + \angle 3 = 180°$	2. Exterior sides form straight line FCD
3. $\angle 2 = \angle 5$	3. If two parallel lines are cut by a transversal, the alternate interior angles are equal
4. $\angle 1 = \angle 6$	4. If two parallel lines are cut by a transversal, the alternate interior angles are equal
5. $\therefore \angle 1 + \angle 2 + \angle 3 = 180°$	5. Substitution

15.

Statement	Reason
1. $AB \parallel CD$	1. Given
2. $PQ \parallel RS$	2. Given
3. $\angle 1 + \angle 3$	3. If two parallel lines are cut by a transversal, the corresponding angles are equal
4. $\angle 3 = \angle 2$	4. If two parallel lines are cut by a transversal, the corresponding angles are equal
5. $\therefore \angle 1 = \angle 2$	5. Quantities equal to the same quantity are equal to the other

17.

Statement	Reason
1. $\angle 5 + \angle 7 = 180°$	1. Given
2. $\angle 5 + \angle 3 = 180°$	2. Exterior sides form a straight line
3. $\angle 5 + \angle 7 = \angle 5$	3. Quantities equal to the same quantity are equal to each other
4. $\angle 5 + \angle 7 - \angle 5 = \angle 5 + \angle 3 - \angle 5$	4. Equals subtracted from equals are equal
5. $\angle 7 = \angle 3$	5. Simplify
6. $\therefore AB \parallel CD$	6. If two lines are cut by a transversal such that corresponding angles are equal, the lines are parallel

19.

Statement	Reason
1. ΔADC is isosceles	1. Given
2. $CA = CD$	2. Given
3. $\angle A = \angle D$	3. Base angles of an isosceles triangle
4. $BE \parallel AD$	4. Given
5. $\Delta CEB = \angle A$	5. If two parallel lines are cut by a transversal, the corresponding angles are equal
6. $\angle CBE = \angle D$	6. If two parallel lines are cut by a transversal, the corresponding angles are equal
7. $\angle CBE - \angle CEB$	7. Quantities equal to the same quantity are equal to each other
8. $\therefore \Delta CBE$ is isosceles	8. If two angles of a triangle are equal, the triangle is isosceles

Similar Triangles

1. 58° **3.** 76° **5.** 12.5 **7.** 4 **9.** 40°
11. 31.2 **13.** 28.8 **15.** 6.75 **17.** 28 **19.** 160
21. 21 ft **23.** 107.1

25.

Statement	Reason
1. $\angle B = \angle E$	1. All angles of a square are right angles
2. ΔABC and ΔDEF are isosceles	2. All sides of a square are equal
3. $\angle CAB = \angle BCA$	3. Base angles of an isosceles triangle are equal
4. $\angle DFE = \angle EDF$	4. Base angles of an isosceles triangle are equal
5. $\angle CAB + \angle BCD = 90°$	5. The acute angles of a right triangle are complementary
6. $\angle DFE + \angle EDF = 90°$	6. The acute angles of a right triangle are complementary
7. $\angle CAB = \angle BCA = \angle DFE = \angle EDF = 45°$	7. If the sum of two angles is 90°, each angle is 45°
8. $\therefore \Delta ABC$ ΔDEF	8. Corresponding angles are equal

27.

Statement	Reason
1. $\angle A = \angle D$	1. Given
2. $BA = AC$ and $ED = DF$	2. Radii of the same circle are equal
3. ΔABC and ΔDEF are isosceles	3. Definition of isosceles triangle
4. $\angle B = \angle C$ and $\angle E = \angle F$	4. Base angles of an isosceles triangle are equal
5. $\angle B + \angle C + \angle A = 180°$ and $\angle E + \angle F + \angle D = 180°$	5. The sum of the angles of a triangle is 180°
6. $\angle B + \angle C + \angle A = \angle E + \angle F + \angle D$	6. Quantities equal to the same quantity are equal
7. $\angle B + \angle C = \angle E + \angle F$	7. Equals subtracted from equals are equal
8. $\angle B + \angle B = \angle E + \angle E$	8. Substitution
9. $2\angle B = \angle E$	9. Add
10. $\angle B = \angle E$	10. Halves of equals are equal
11. $\angle C = \angle F$	11. Substitute in step 9

12. ∴ΔABC ΔDEF	12. Corresponding angles are equal

29.

1. $\angle A = \angle D$	1. Given
2. $\angle B = \angle E$	2. Given
3. $\angle A + \angle B + \angle C = 180°$ and $\angle D + \angle E + \angle F = 180°$	3. The sum of the angles of a triangle is 180°
4. $\angle A + \angle B + \angle C = \angle D + \angle E + \angle F$	4. Quantities equal to the same quantity are equal to each other
5. $\angle A + \angle B + \angle C - \angle A = \angle D + \angle E + \angle F - \angle D$	5. Equals subtracted from equals are equal
6. $\angle B + \angle C = \angle E + \angle F$	6. Simplify
7. $\angle B + \angle C - \angle B = \angle E + \angle F - \angle E$	7. Equals subtracted from equals are equal
8. $\angle C = \angle F$	8. Simplify
9. ∴ΔABC ΔDEF	9. Corresponding angles are equal

Square Roots

1. 4 **3.** 9 **5.** 11 **7.** $\frac{3}{5}$ **9.** $\frac{11}{12}$ **11.** 10.8 **13.** 28.53 **15.** 4.6 **17.** 14 **19.** 24 **21.** $\frac{12}{13}$ **23.** 5.5; 5.48; 5.477; 5.4772 **25.** 12.2; 12.25; 12.247; 12.2474 **27.** 67 **29.** 122 **31.** 19.1; 19.13; 19.130; 19.1301 **33.** 4.5; 4.48; 4.476; 4.4763 **35.** 1.8 sec **37.** 8 cm **39.** $t = 7\frac{1}{3}$ **41.** 15.0 in

The Pythagorean Theorem

1. $a = 15$ **3.** $c = 13$ **5.** $a = 12$ **7.** $c \approx 7.81$ **9.** $c = 50$ **11.** $c \approx 3.61$ **13.** $b \approx 5.86$ **15.** $b \approx 136.11$ **17.** $c = 14.3$ **19.** yes **21.** yes **23.** no **25.** yes **27.** 13.42 ft **29.** 61.0 ft **31.** 12 ft **33.** 206.2 mi **35.** 127.3 ft **37.** 9.3 ft

Chapters 1 to 4 Midterm Examination

1. hundred **2.** ninety thousand fifty-three **3.** 92,101 **4.** 16,205 **5.** 3,537 **6.** 247,135 **7.** 36,623 **8.** 129 R 14 **9.** 136 **10.** 3 **11.** 42 **12.** 166 **13.** 180 **14.** prime **15.** no **16.** 18, 36, 54, 72, 90 **17.** 1, 2, 3, 4, 6, 12, 17, 34, 51, 68, 102, 204 **18.** $2 \cdot 2 \cdot 2 \cdot 2 \cdot 2 \cdot 2 \cdot 3$ **19.** $4\frac{5}{9}$ **20.** $\frac{35}{4}$ **21.** $\frac{7}{6}, \frac{8}{8}, \frac{9}{8}, \frac{5}{4}, \frac{4}{4}$ **22.** $\frac{5}{9}, \frac{7}{12}, \frac{5}{8}, \frac{2}{3}$ **23.** $\frac{2}{3}$ **24.** $\frac{3}{50}$ **25.** $30\frac{1}{3}$ **26.** $\frac{45}{49}$ **27.** $\frac{9}{5}$ **28.** $\frac{49}{60}$ **29.** $8\frac{8}{21}$ **30.** $5\frac{2}{7}$ **31.** $3\frac{37}{45}$ **32.** $3\frac{37}{40}$ **33.** $\frac{3}{10}$ **34.** seventy-one and three hundred six thousandths **35.** 0.225 **36.** 0.78 **37.** ten-thousandths **38.** 6.85 **39.** $\frac{19}{80}$ **40.** $\frac{1}{6}$ **41.** False **42.** 24.3542 **43.** 11.931 **44.** 20.8974 **45.** 0.002824 **46.** 6.24 **47.** 0.8186 **48.** 6.3 gallons

Chapters 1 to 8 Final Examination

1. $\frac{11}{16}$ **2.** 84 **3.** 7.98 **4.** 17.395 **5.** $\frac{15}{4}$ or $3\frac{3}{4}$ **6.** $27\frac{1}{2}$ **7.** 199 **8.** 64.85 **9.** $5\frac{3}{8}$ **10.** 98 **11.** 79.007 **12.** 45.304 **13.** $\frac{7}{10}$ **14.** $6\frac{7}{24}$ **15.** $\frac{7}{3}$ or $2\frac{1}{3}$ **16.** one **17.** $\frac{3}{20}, \frac{1}{4}, \frac{2}{5}$ **18.** 55% **19.** 38 R 1 **20.** 0.643 **21.** 4,000.4 **22.** 62.5 **23.** 0.084 **24.** 6 **25.** 230% **26.** 0.165 **27.** 20.8 **28.** 0.00286 **29.** $23.40 **30.** 16.25% **31.** 13, 26, 39, 52, 65 **32.** five hundred five and five hundredths **33.** False **34.** $2 \cdot 2 \cdot 3 \cdot 5 \cdot 7$ **35.** $\frac{2}{5}$ **36.** $\frac{13}{40}$ **37.** $31\frac{1}{6}$ **38.** $\frac{1}{12}$ **39.** $\frac{21}{400}$ **40.** 58% **41.** No **42.** $7\frac{1}{2}$ **43.** $\frac{117}{16}$ **44.** 1.299, 1.32, 1.322, 1.332 **45.** 0.04873 **46.** $\frac{6}{11}$ **47.** $\frac{1}{12}$ **48.** $1,300 **49.** $240 **50.** 10.9% **51.** 18 **52.** $2\frac{25}{72}$ **53.** 3.16 **54.** 7 yd 2 ft 5 in. **55.** 2 m 60 cm **56.** $50 per kg **57.** $31\frac{1}{4}$ ft **58.** 2.97 m^2 **59.** 62.13 yd^2 **60.** 810 in^3 **61.** −29 **62.** −23 **63.** −126 **64.** 4 **65.** 2 **66.** 5.5 **67.** $a = -4$

INDEX